Ilyoung Chong (Ed.)

 KU-318-397

Information Networking

Wireless Communications Technologies and Network Applications

International Conference, ICOIN 2002
Cheju Island, Korea, January 30 – February 1, 2002
Revised Papers, Part II

 Springer

Series Editors

Gerhard Goos, Karlsruhe University, Germany
Juris Hartmanis, Cornell University, NY, USA
Jan van Leeuwen, Utrecht University, The Netherlands

Volume Editor

Ilyoung Chong
Hankuk University of Foreign Studies
86, Wangsan, Mohyun, Yongin, Kyonggi-Do, Korea 449-791
E-mail: iychong@hufs.ac.kr

Cataloging-in-Publication Data applied for

Die Deutsche Bibliothek - CIP-Einheitsaufnahme

Information networking : international conference ; revised papers / ICOIN
2002, Cheju Island, Korea, January 30 - February 1, 2002. Ilyoung Chong
(ed.). - Berlin ; Heidelberg ; New York ; Hong Kong ; London ; Milan ; Paris ;
Tokyo : Springer
Pt. 2. Wireless communications technologies and network applications. -
(2002)
  (Lecture notes in computer science ; Vol. 2344)
  ISBN 3-540-44255-3

QM LIBRARY
(MILE END)

CR Subject Classification (1998): C.2, D.2.12, D.4, H.3, H.4, H.5

ISSN 0302-9743
ISBN 3-540-44255-3 Springer-Verlag Berlin Heidelberg New York

This work is subject to copyright. All rights are reserved, whether the whole or part of the material is
concerned, specifically the rights of translation, reprinting, re-use of illustrations, recitation, broadcasting,
reproduction on microfilms or in any other way, and storage in data banks. Duplication of this publication
or parts thereof is permitted only under the provisions of the German Copyright Law of September 9, 1965,
in its current version, and permission for use must always be obtained from Springer-Verlag. Violations are
liable for prosecution under the German Copyright Law.

Springer-Verlag Berlin Heidelberg New York
a member of BertelsmannSpringer Science+Business Media GmbH

http://www.springer.de

© Springer-Verlag Berlin Heidelberg 2002
Printed in Germany

Typesetting: Camera-ready by author, data conversion by PTP-Berlin, Stefan Sossna e.K.
Printed on acid-free paper   SPIN 10869969   06/3142   5 4 3 2 1 0

Preface

The papers comprising Vol. I and Vol. II were prepared for and presented at the International Conference on Information Networking 2002 (ICOIN 2002), which was held from January 30 to February 1, 2002 at Cheju Island, Korea. It was organized by the KISS (Korean Information Science Society) SIGIN in Korea, IPSJ SIG-DPE (Distributed Processing Systems) in Japan, the ITRI (Industrial Technology Research Institute), and National Taiwan University in Taiwan. The papers were selected through two steps, refereeing and presentation review.

We selected for the theme of the conference the motto "One World of Information Networking". We did this because we believe that networking will transform the world into one zone, in spite of different ages, countries and societies. Networking is in the main stream of everyday life and affects directly millions of people around the world. We are in an era of tremendous excitement for professionals working in many aspects of the converging networking, information retailing, entertainment, and publishing companies. Ubiquitous communication and computing technologies are changing the world. Online communities, e-commerce, e-service, and distance learning are a few of the consequences of these technologies, and advanced networking will develop new applications and technologies with global impact. The goal is the creation of a world-wide distributed computing system that connects people and appliances through wireless and high-bandwidth wired channels with a backbone of computers that serve as databases and object servers.

Thus, Vol. I includes the following subjects related to information networking based on wired communications and management:
- New Networking and Applications
- Switching and Routing
- Optical Networks
- Network Performance Issues
- Quality of Service
- Home Networking and Local Access Protocols
- Network Management

And Vol. II includes the following subjects related to wireless communications technologies and network applications:
- Wireless and Mobile Internet
- 4G Mobile Systems
- Satellite Communications Systems
- Network Security
- Multimedia Applications
- Distributed Systems

With great pleasure we take this opportunity to express our gratitude and appreciation to Prof. Chung-Ming Huang, who acted as PC vice-chair of ICOIN 2002, for reviewing and encouraging paper submission in Taiwan and other locations, to Prof.

Leonard Barolli, who acted as PC vice-chair of ICOIN 2002, for organizing paper reviewing in Japan, and to Prof. Changjin Suh, Prof. Sungchang Lee and Prof. Kyungsik Lim, who did excellent reviewing, editing and assembling of contributions for this book.

We are confident that this book series will prove rewarding for all computer scientists working in the area of information networking.

June 2002   Ilyoung Chong

Table of Contents, Part II

IV. Network Security

V. Multimedia Applications

VI. Distributed Systems

Table of Contents, Part I

II. Switching and Routing

III. Optical Networks

IV. Network Performance Issues

V.   Quality of Services

VI.   Home Networking and Local Access Protocols

VII.   Network Management

I. Wireless and Mobile Internet

Modelling QoS Based Admission Control in Mobile Multimedia Systems Using an Extension of GSPN

Tony Tsang

Department of Computing
The Hong Kong Polytechnic University
Hung Hom, Hong Kong
ttsang@ieee.org

Abstract. Modelling and analysing for multimedia applications in a mobile network, with real-time behaviour taken into consideration, is still a subject of much research. Generalized Stochastic Petri Net (GSPN) is well known formalisms used to represent and analyse communication and computer systems. However, it does not have enough expressive power to model timing behaviours of multimedia applications in a mobile network. To address this limitation, we develop quality of services based GSPN, which is capable of doing so. The extension offers the promise of providing real-time system predictability for systems characterised by substantial stochastic behaviour. With this model we model and analyse the handover and admission control schemes in the different multimedia traffic environments. Our result, thereby, provide the optimal performance achievable for these schemes under the real-time constraints in the given setting of design parameters.

Keywords: GSPN, quality of services, multimedia, mobile systems

1 Introduction

The success of the Internet and mobile terminals has created a tremendous need to access messaging and streaming services wirelessly, in a mobile fashion. The number of Internet connections has grown almost exponentially, and the popularity of mobile devices has experienced an even more aggressive growth: the amount of mobile handsets is higher then the amount of fixed Internet connections. It is generally estimated that the number of mobile devices will match the number of personal computers during next few years. Mobile Internet is currently one of the most astounding trends in computing. Towards the mobility supports, the University Mobile Telecommunications System (UMTS) [1,2] architecture provides the building blocks for the vast deployment of next generation mobile Internet. The combination of UMTS and IP-based quality enabled networks seems the right step towards the provision of a quality assured mobile Internet. Regarding mobile services, UMTS supports a wide range of applications with different quality of service profiles. There are four different QoS classes

I. Chong (Ed.): ICOIN 2002, LNCS 2344, pp. 3–10, 2002.
© Springer-Verlag Berlin Heidelberg 2002

targeting conversational [3], streaming, interactive and background applications. The promised QoS diversity is adequate to support the future mobile Internet applications.

In packetised wireless communication in mobile Internet, the transfer of packets between base stations and efficient use of channels, channel-allocation schemes and admission control schemes are essential mechanisms. In this paper, a new methodology to model and analyse these schemes of wireless communication system including quality of services based GSPN is presented. Generalized Stochastic Petri Nets (GSPNs)[4], are well-known interpreted extensions of autonomous Petri Net (PN) models allowing the consideration of performance aspects in the design of complex concurrent systems, in addition to the PNs capability of functional validation. This model adds a new dimension to GSPN, which combines real-time queueing theory [5], by allowing tasks to have real-time traffic of mobile communication system to satisfy the task timing requirements of multimedia applications. It can also model and analyse the distribution of the lead-time profile of packets waiting for services. Moreover, the underlying handover and admission control mechanisms to improve quality of service can be introduced into the formulation, and their performance can be evaluated using this model. In the following sections, we briefly present the definition of model. Then we have applied our model to analyse the performance of connection admission control schemes in a mobile network.

2   The Extension of GSPN

The formal definition of the extension can now be presented : Let the extension of GSPN is ($GSPN$, β, ϕ)
Definition 1.
$GSPN = (P, T, I, O, H, M_0, \Pi, W)$ is the underlying Petri Net, which as usual comprises in the following.
P is the set of places; and T is the set of transitions such that $P \cap T = \varnothing$.
$I, O, H : T \times P \to \mathbb{N}$, are the input, output and inhibition function respectively.
$M_0 : P \to \mathbb{N}_0$ is the initial marking: a function that assigns a nonnegative integer to each place.
$\Pi : T \to \mathbb{N}$ is the priority function Π which associates lowest priority (0) with timed transitions and higher priorities (≥ 1) with immediate transitions:

$$\Pi(t) = \begin{cases} 0, & \text{if } t \text{ is timed,} \\ \geq 1, & \text{if } t \text{ is immediate} \end{cases}$$

$W : T \to \mathbb{R}$ is a *weight* function which associates a real value with transitions. $w(t)$ is :

- the parameter of the negative exponential probability density function (pdf) of the transition firing delay, if t is a timed transition,
- a weight used for the computation of firing probabilities of immediate transitions, if t is an immediate transition.

Definition 2.

$\beta : T \rightarrow (Q^+ \cup 0)(Q^+ \cup \infty)$ is the time interval function that defines the permitted transition firing times, where Q^+ is the set of positive rational numbers and

$\beta_i = \{[\tau i_{min}, \tau i_{max}] \mid \forall t_i \in T\}$ with $0 \leq \tau i_{min} \leq \tau i_{max}$; and
$\phi : T \rightarrow \mathbb{R}$ is a real space $\mathbb{R}$, which associate with transitions.

The state space of an ordinary GI/G/1 queue (using the method of supplementary variables) is (m, l, τ, θ), where m is the number of customers, τ is the length of time since the last customer arrived, and θ is the length of time the current customer has been in service. l_i is the lead-time (the time unit the deadline is reached) of the i the customer. In particular, we must keep track of a dynamic variable for each customer, its lead-time. At time t, the lead-time of a customer is the difference between its absolute deadline and the current time, hence a customer's lead-time decreases linearly with time. During an interval $[\, t, \, t+\delta\,]$, if a customer has not departed, then its lead-time is reduced by δ. Negative lead-times are possible and indicate that a customer is late. The lead-times ($l_1, l_2, \ldots , l_m$) can be associated with a measure on a subset of a real space $\mathbb{R}$. That measure being the number of customers with lead-times in the subset. Since m is, in general, inbounded, so is the state spaces. Thus the state space has become ($\phi(s), \tau, \theta$). A discrete transform $\phi(s)$ of the lead-times is introduced:

$$\phi(s) = \begin{cases} \sum_{j=1}^{m} e^{isl_j}, & \text{if } m \geq 1, \\ 0, & \text{if } m = 0, \end{cases}$$

where l_j is the lead-time of the j the customer and $i = \sqrt{-1}$. Note that the number of customers is implicitly contained in the state space since $\phi(0) = m$.

3   Modelling Wireless Link

A generalised methodology analyses the performance characteristics of the different call handover and admission control schemes. This method allows studying the behaviour of each scheme under different system conditions, such as traffic load, number of users assigned and real-time behaviours. The extended model provides an analytic approach, which approximates the lead-time profile where is the time until the deadline is reached, as a function of queue length in the buffer. Assuming packet departs from the queue has a lead-time, which gives the mixed distribution. This model was used to study the QoS issue in wireless networks at both the connection and packet levels in [6,7,8]. Assuming arriving packets have positive lead-time, then if the lead-time profile is negative where the departing packet is late. If the lead-time profile is positive, then a departing packer is not late.

Packets that arrive at the base station form being mobile terminal to encounter mechanism causing delay and loss. The queuing delay at the base station buffer and packet loss due to buffer overflow or packet dropping due to deadline violation while waiting for service. The queuing delay can be controlled by implementing deadline based queuing at the base station. In a wireless environment, the signal power received at the mobile terminal varies with time due to fading, changes in the interference level, movement of the scattering objects around the portable terminal. This causes the errors on the link to be highly correlated. A practical and widely used approach to this problem is to break up these correlations by using interleaving, which ensures that successive symbols of the same code word are separated by a certain time interval. The link-shaping delay at the base station, which in this case consists of interleaving and coding delay and loss due to decoding failure at the mobile terminal.

We describe the model that is used to analyse the delay performance over a wireless link. In this model, the incoming traffic at the transmitter is represent by an arrival process with the rate (λ). The state of our real-time model is characterised by the number of packets in the buffer (Q) and a lead-time profile associated with each packet (β). Figure 3 shows the extended GSPN modelling a wireless link model. When the packet arrives at the base station, indicated by a token in place link. There are two situations can arise. If there are processes waiting to be served, the service process starts, thereby using the discard and forward process form place buffer to active via transition serve. If the discard process is served, which the place passive is transferred via the transition t2 to limiter, and then feed back to mobile terminal. After each service completion, if there is no error and forward process, the transition t3 is disabled and the place active is transferred via the transition forward to place upper layer.

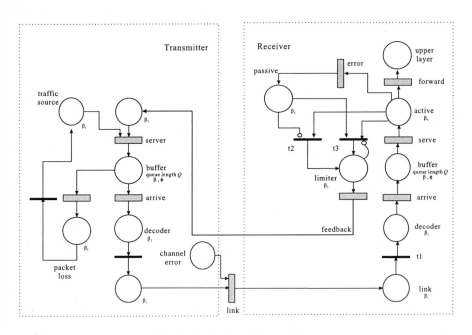

Fig. 1. Modelling Wireless Link

4   Numerical Result

There are two step methodology of determining the queue length conditions for which packets should be finishing late, then approximating the probability of this queue length condition using different offered traffic environment. One can use the closed form expressions given in this section to find a change in the design parameters that will result in some target value for the fraction of packets that are lost. For example, suppose that we allocate more processing capacity or more bandwidth to the real-time tasks, so that the service rate is increased. By how much does the service rate need to be increased in order for the fraction of packets that are late to be reduced to some goal. We next turn to First In First Out (FIFO) scheduling. FIFO is a scheduling policy, which does not consider packet deadlines. Consequently, to analyse the real-time behaviour of FIFO, for example to determine the packet lateness probabilities, we proceed in two steps. First, we ignore packet deadlines completely and assume that each packet has a zero deadline. The lead-time is then the negative of the time spent in queue. When a packet leaves the system, we can add its lead-time to its deadline to determine its true lead-time at departure. That value is negative if and only if the packet is late. Using this idea, we can determine the lateness conditions of packets for different traffic environments.

To study this approach, we consider an M/M/1 queue with traffic intensity 0.95. Packets have a relative deadline drawn from three different distributions: (1) Exponential with mean 50; (2) Uniform on [0,100]; and (3) Constant queue length of 50. All three of these distributions have a mean value of 50, and with arrival rate ($\lambda = 0.95$) , service rate ($\mu = 1.0$) (hence $\rho = 0.95$), the critical blocking value will be 50. This model would predict that packet loss would occur if and only if the queue length is on or above the critical value, 50. We simulate the threshold policy using a series of blocking values, beginning at 50 and reducing to 40. Furthermore, one can see that by implementing a blocking threshold of 50, one can reduce the probability of late packets from 0.08 down to around 0.02. Moreover, as we move the threshold down to 40, we can reduce the total probability of lost or late packets to below 0.01. The results are consistent across the three packet deadline distributions. It is important to note that the deadline distributions minimise the possibility of "intrinsic lateness", the case in which packets have an initial deadline which are shorter than their service times, hence will be late no matter policy is used.

Because of the length for the paper, we present a part of the analysis results. Figures 4 and 5 shows curves of the blocking probability, which in this case, due to the fact that no channel is reserved to handovers, coincides with the probability that a new call or a handover request finds no free channel, versus the system traffic load for different values of the system parameters. The two figures comprise six curves referring to different system parameters of CAC, as well as simulation results. The Figures show the call blocking probability against the traffic load in different call time-intervals. The blocking probability increases when the offered load increases. However, forced the blocking probability decreases when the number of reserved channels increases. Blocking begins to rise rapidly at offered loads near the maximum capacity for the CAC scheme. The blocking probability with buffer size of 50 packets is almost smaller than that the buffer size of 40 packets and the blocking probability of lower bound calls is almost not change for short and long time-interval. The result indicates that the large buffer of CAC with heavy traffic load is effective to decrease the blocking probability for long time-interval. This phenomenon is because the heavy traffic load correspond to large value of packet arrival rate, resulting in an increase in the packet queue length and hence an increase in the blocking probability. The improvement of blocking probability from the increase of buffer size is very small for short time-intervals. Therefore, the increase of buffer size adapts for long time-interval while suitable buffer size is effect for long time-interval situation.

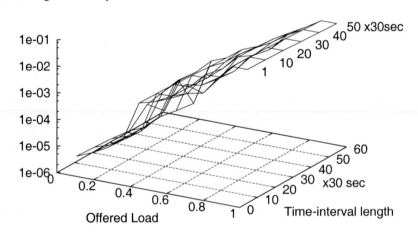

Fig. 2. Call blocking probability for buffer of 40

5   Conclusion

In this paper, we have presented the detail of the extended GSPN, which is able to describe the qualitative traffic characteristics and quantitative real-time behaviours of a wireless packetized network. This method makes use of a real-time queueing model for modelling multimedia applications of mobile network, which enabling it to be specified and analysed in our model. This model can analytically determine the distribution of the lead-time of packets waiting for service

Blocking Probability

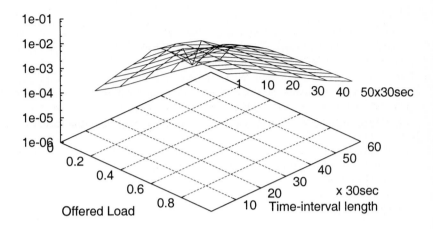

Fig. 3. Call blocking probability for buffer of 50

for the different type inter-arrival and service distribution case with stochastic lead-times. To aid the analysis further, we use the time-interval related performance indices in order to capture the blocking probability and time correlation of realistic mobile traffic for multimedia applications. We have also described a method of using the extension of GSPN for a wireless link model that implements a call admission control scheme with limited ARQ retransmissions. This scheme supports the admissibility of a prospective connection under the given QoS requirements and the availability of network resources. It includes the time varying channel characteristics and the impact of the underlying link layer error control mechanisms on the network layer QoS performance. This admission control scheme can be analysed for real-time traffic of multimedia application by our model.

We therefore conclude that our method of performance analysis using the extended GSPN is useful for predicting the performance of multimedia applications for mobile network, and our effort in developing GSPN based model is a worthwhile exercise. As such, Petri Net based model should become a more attractive formal description technique for use in development of wireless communication systems.

References

1. ETSI TS 120 101 v4.0.0, Universal Mobile Telecommunications System (UMTS); General UMTS Architecture, http://www.etsi.org/, 2001-04.

2. ETSI TS 122 105 v3.10.0, Universal Mobile Telecommunications System (UMTS); Service aspects;Services and Service Capabilities, http://www.etsi.org/, 2001-01.
3. ETSI TS 123 107 v4.0.0, Universal Mobile Telecommunications System (UMTS); QoS Concept and Architecture, http://www.etsi.org/, 2000-12.
4. Ajmone Marsan, M., Balbo, G., Conte, G., Donatelli, S. and Franceschinis, G., "Modelling with Generalized Stochastic Petri Nets", John Wiley & Sons, 1995.
5. Lehoczky, J.P., "Real-time queueing network theory", Proceedings Real-Time Systems Symposium, pps. 58–67, December 1997.
6. Das S. K., Jayaram R., Kakani N. K., and Sanjoy K. Sen, "A call admission and control scheme for quality-of-service (QoS) provisioning in next generation wireless networks", Wireless Networks, Vol. 6, pps. 17–30, 2000.
7. Ramjee R., Towsley D., and Nagarajan R., "On optimal call admission control in cellular networks", Wireless Networks, Vol. 3, pps. 29–41, 1997.
8. Marwan M. Krunz, Jeong Geun Kim, "Fluid Analysis of Delay and Packet Discard Performance for QoS Support in Wireless Networks", IEEE Journal on Selected Areas in Communications, Vol. 19, No.2, pps. 384–395, February 2001.

A Comparative Performance Evaluation of the Transport Layer Protocols for Wireless Transaction Processing in Mobile Internet Environments

Yoonsuk Choi, Howon Jung, and Kyungshik Lim

Computer Science Department
Kyungpook National University
Taegu, 702-701, Korea
{yoonsuk, won, kslim}@ccmc.knu.ac.kr

Abstract. In this paper, we design and implement Wireless Transaction Protocol(WTP) and evaluate it for wireless transaction processing in mobile computing environments. The design and implementation of WTP are based on the co-routine model that might be suitable for light-weight portable devices. For the evaluation of WTP, we use an Internet simulator that can arbitrary generate random wireless errors based on the Gilbert model. In our experiment, the performance of WTP is measured and compared to those of Transmission Control Protocol(TCP) and TCP for Transactions. The experiment shows that WTP outperforms the other two protocols for wireless transaction processing in terms of throughput and delay. Especially, WTP shows much higher performance in case of high error rate and high probability of burst errors. This comes from the fact that WTP uses a small number of packets to process a transaction compared to the other two protocols and introduces a fixed time interval for retransmission instead of the exponential backoff algorithm. The experiment also shows that the WTP performance is optimized when the retransmission counter is set to 5 or 6 in case of high burst error rate.

1 Introduction

Many conditions of wireless Internet environments should be considered for developing a transport layer protocol providing Internet service. First, compared with fixed and wired networks, the wireless networks are fragile and have high bit-error rate(BER). So, a transport layer protocol should have some error recovery schemes for offering proper contentment to a application. Second, the wireless networks have a limited bandwidth and long delay. Thus, some mechanisms are needed for solving this problems in a transport layer protocol[1].

TCP, TCP for Transaction(T/TCP) and Wireless Transaction Protocol(WTP) have been researched as a transport layer protocol for wireless transaction processing. In this paper, we describe properties of these protocols and the design and implementation of WTP. We measure the performance of WTP and compare it with those of TCP, T/TCP. We also suggest the most optimal retransmission count of WTP over wireless networks.

This paper is organized as follows. Section 2 makes a comparison among TCP,

I. Chong (Ed.): ICOIN 2002, LNCS 2344, pp. 11–21, 2002.
© Springer-Verlag Berlin Heidelberg 2002

T/TCP and WTP. Section 3 provides an implementation model and essential mechanisms of WTP. In Section 4, simulation model, experiment results and analysis of the results are given and discussed. Finally, conclusions and future works are presented in section 5.

2 Related Works

Two conditions are considered to design a transport layer protocol for wireless Internet service.
1. low bandwidth, long delay and unstable connection in wireless links and
2. wireless Internet service for browsing with a transaction(request/ response).
   Thus, we compare and contrast WTP with TCP and T/TCP based on these conditions. First, TCP is a reliable end to end transport protocol designed for the wired networks. The way that TCP works is that it keeps increasing the sending rate of packets as long as no packets are lost. However, when a packet loss occurs over wireless networks, TCP degrades a congestion window size by triggering the congestion control or fast retransmission. Unfortunately, these measures result in an unnecessary reduction in end-to-end throughput[6]. Second, T/TCP is an experimental extension for TCP protocol. It was designed to address the need for a transaction-based transport protocol such as WWW in the TCP/IP stack. Compared with TCP, T/TCP decreases the number of segment in network links, and it shortens the response time by simplifying a explicit connection set up and tear down[2,3]. However, it is not proper in wireless networks because of using congestion control and error recovery schemes of TCP[2,3]. Finally, WTP is defined to provide the services necessary for interactive browsing application. WTP has been defined as a light weight transaction oriented protocol that is suitable for implementation in thin clients and operates efficiently over wireless datagram networks. Table 1 shows the functions of TCP, T/TCP and WTP. T/TCP is almost same as TCP except for explicit connection set up and tear down.

Table 1. Comparison of the functions of TCP, T/TCP and WTP

	TCP	T/TCP	WTP
Congestion control	O	O	X
Explicit connection set up and tear down	O	X	X
Transaction ID verification	X	O	O
Segmentation & reassembly	O	O	O
Differentiated transaction service	X	X	O
User acknowledgement	X	X	O
Transaction abort	X	X	O

WTP rises the throughput by retransmitting a lost packet instead of triggering the congestion control when a packet lose occurs. Like T/TCP, WTP also decreases the number of segment in network links and the response time by simplifying a explicit connection set up and tear down[5]. As mentioned before, WTP is the most optimized protocol for wireless transaction processing. Next, we design and implement WTP and compare the performance of WTP with those of TCP, T/TCP by simulating these protocols under various wireless environments.

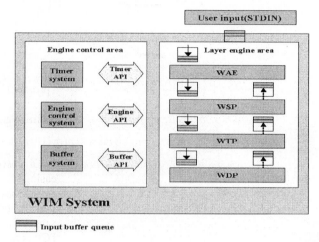

Input buffer queue

Fig. 1. WIM architecture

3   WTP Implementation

3.1   WAP Implementation Model (WIM)

WIM, which is shown in fig. 1, is a WAP implementation model based on co-routine model that is suitable for implementation in mobile devices[7]. WIM consists of engine control area and layer engine area. Engine control area has timer system for retransmission of messages, buffer system for exchanging messages between layers and engine control system for engine call and schedule. The function of each system is provided to layer engine area through systems' API. The service primitives determine the interface of each layer, and an inner function of protocols is implemented based on the state table which describes the action of protocols. Each layer makes service primitives into buffer messages in buffer system for defining the interface between layers, and it sends message to next layer using buffer I/O function. Following example code shows a chief module of engine control system for managing a protocol layer engine. Main function controls the input buffer queue of engines and calls corresponding engine by the analysis mechanism of virtual_select function. Main function is blocked while no input occurs in the buffer queue, and it is woken when an input occurs.

Chief module of engine control systme

```
begin
While(1){
  layer = virtual_select(msg_queue);
  switch(layer){
    case WAE_LAYER : wae_engine(); break;
```

```
    case WSP_LAYER : wsp_engine(); break;
    case WTP_LAYER : wtp_engine(); break;
    case WDP_LAYER : wdp_engine(); break;
  }
}
end.
```

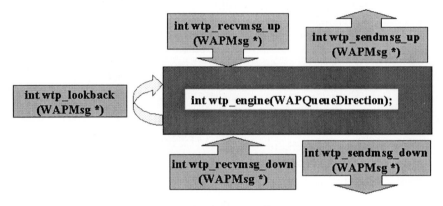

Fig. 2. WTP engine interface

3.2 WTP Engine Implementation

Fig. 2 shows the WTP engine interface for communicating between WTP and its upper or lower layer. Another layer is implemented as the same way. WTP engine has upper and lower input buffer queue for storing messages received from the other layers. We know about the input queue which has a received message and also select the input buffer queue which will send a message using wtp_engine function. The descriptions of WTP engine interface are given below:

1. wtp_recvmsg_up - It gets messages from upper input buffer queue.
2. wtp_recvmsg_down - It gets messages from lower input buffer queue.
3. wtp_sendmsg_up - It sends messages to WTP upper layer.
4. wtp_sendmsg_down - It sends messages to WTP lower layer.
5. wtp_lookback - It sends messages to own layer.

4   Simulation and Evaluation

4.1 Simulation Environment

Our experimental testbed, which is shown in Fig. 3, consists of a wireless cell and wired link. WAP client and server are implemented on LINUX using C language and GNU compiler. Internet simulator of RADCOM std. makes the section between WAP client and server into wireless link by setting a packet loss rate, link capability, packet order and jitter. In this experiment, we assume that there are 20 clients and 1 server,

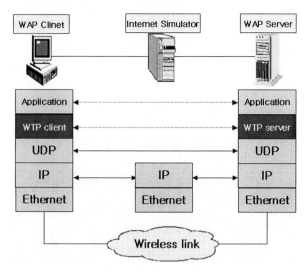

Fig. 3. Testbed for simulation

and each client is executed in a process. We generate random errors on the wireless network using the Gilbert model. We do not use packet order and jitter because our experiment environment only has one wireless hop. In these environments, we measure the throughput and system response time of TCP, T/TCP and WTP.

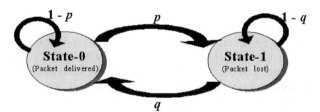

Fig. 4. Gilbert model

We use the Gilbert model which can represent a burst or non-burst error in wireless networks[8,9]. The Gilbert model, which is shown in Fig. 4, is a 2-state in which one(which we refer to as state 1) represents a packet loss, and the other state(which we refer to as state 0) represents a packet reaching the destination. Let p denote the probability of going from state 0 to state 1, and let q denote the probability of going from state 1 to state 0. Let X_n denote the probability variable which is set to 1 if packet n is lost, and 0 otherwise. Transition probability p and q are presented using X_n as follows.

$$P = \Pr[X_{n+1} = 1| X_n = 0], Q = \Pr[X_{n+1} = 0| X_n = 1] \qquad (1)$$

The steady state properties of the chain are given by $\Pr[X_n = 0]$ and $\Pr[X_n = 1]$ which symbolize the steady state probability that the channel is in the state 0 and state 1 respectively[8,10]. These are as follows.

$$\Pr[X_n = 0] = \qquad (2)$$

$$\Pr[X_n = 1] = \qquad (3)$$

We assume that packet loss rate is 50% less and the average number of consecutive packet loss is 2-10 in our experimental environment. P is calculated by equation (2) and q is also calculated using the Mean Residence Time(MRT) of state 1 in the Gilbert model. MRT given by 1/q means the average duration of state 1. Since the average number of consecutive packet loss is 2-10 by the assumption, q is 0.1•q•05. Therefore we measure the performance of the protocols when FER is 1-50%, and q is 0.1, 0.3 and 0.5.

Table 2. Traffic generation model

Process	Parameters	Distribution
Request size	min = 36bytes max = 80bytes	uniform
Result size	shape = 1.2 scale = 101bytes max = 1500bytes	truncated pareto
Transaction inter arrival time	shape = 1.5 scale = 17sec max = 46800sec	truncated pareto
Session inter arrival time	Lambda = 0.011 (mean = 90sec)	exponential
Session duration	shape = 1.5 scale = 260sec max = 900sec	truncated pareto

The traffic generation model is shown in Table 2. The traffic model consists of two parts, the client and server model. While the client model determines user behavior, the server model is responsible for the generated amount of traffic. The user behavior is modeled as an on-off model. Transaction inter arrival time and session duration are determined by the truncated pareto distribution, and session inter arrival time is determined by the exponential distribution[11]. Based on the result of WAP traffic analysis, user request packet size and the result packet size of server are determined by uniform and truncated pareto distribution respectively[11,12].

4.2 Simulation Results

In order to analyze the throughput of the protocols, let us assume these parameters as follows.
- the number of occurred transaction : N_a
- the number of succeeded transaction : N_s
- Normalized throughput : N_s / N_a

Each experiment carries out under the wireless network with a bandwidth of 144kbps and an RTT of 200ms. Q is assumed to be 0.1, 0.3 and 0.5 while FER is varied and shown on the x-axis.

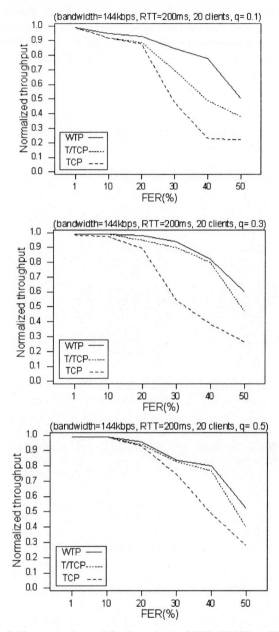

Fig. 5. The comparison of the throughput of TCP, T/TCP and WTP

Fig. 5 shows the normalized throughput of TCP, T/TCP and WTP in processing a transaction. Namely, we measure the transaction success rate over wireless network with a burst error and non-burst error. As can be seen from these figures as the FER increases, the normalized throughput of all protocols decreases, and it quickly decreases with a wide range when FER is 30%-40% except for WTP. The throughput of WTP is higher than those of the other two protocols. Especially, it is 39.5%, 166% higher than those of T/TCP, TCP respectively when q is 0.1 and FER is 30%-40%. According to the fact that the reciprocal of q gives the average number of packets lost, as the value of q increases, the throughput of all protocols improves. However, WTP shows high throughput regardless of the value of q. In other words, WTP shows high throughput even if FER and the probability of burst error is high. These are because of the number of packet in completing a transaction. As is shown in Fig. 6, WTP needs only 3 packets while TCP, T/TCP require 9 and 6 packets respectively. WTP uses a small number of packets compared to the other two protocols because it does not carry out the explicit connection setup and teardown.

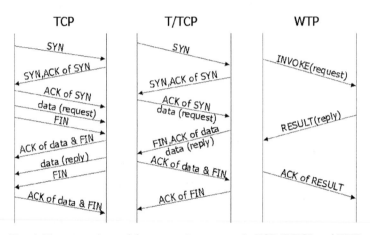

Fig. 6. The comparison of the transaction execute in TCP, T/TCP and WTP

Fig. 7 shows the system response time of TCP, T/TCP and WTP in processing a transaction. The system response time is the average duration of transaction execution including success or failure of transaction. As FER increases, the system response time rapidly increase with a wide range in TCP, T/TCP. However, it increases a little only in 50% of FER and is generally low in WTP. When q is 0.1 and FER is 30%-40%, the system response time of TCP and T/TCP is sixteen, twenty four times longer than those of WTP. This comes from the fact that WTP uses a fixed time interval for retransmission instead of the exponential backoff when a packet loss occurs.

In order to find the most optimal retransmission count of WTP over wireless network, we measure the throughput and system response time with various retransmission counts of WTP. As shown in fig. 8, as the retransmission count increases, the throughput and system response time get high. However, 7 of retransmission count is not proper due to extremely high response time. When we consider two factors, the most optimal retransmission count of WTP is 5 or 6 over wireless network.

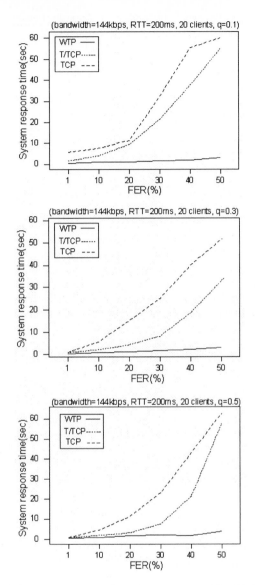

Fig. 7. The comparison of the system response time of TCP, T/TCP and WTP

5   Conclusion and Future Works

In these days, we have increasing demands on wireless Internet services. To satisfy with these demands, we need a transport layer protocol which is suitable for wireless networks that has a limited bandwidth, long delay and unstable connection. We first describe characteristics of TCP, T/TCP and WTP, and we design and implement WTP

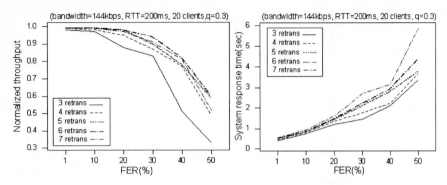

Fig. 8. The throughput and system response time with various retransmission count of WTP

based on the co-routine model. Our WAP system consists of engine control area and layer engine area. The engine control area has engine control system, timer system and buffer system, and the layer engine area has each implemented protocol. We evaluate WTP for wireless transaction processing in wireless environments. For the evaluation, we measure the throughput and system response time based on the Gilbert model which can present various packet losses in wireless network. The result of this experiment shows that WTP outperforms the other two protocols for wireless transaction processing in term of throughput and delay. Especially, WTP shows high throughput even if FER and the probability of burst error is high. From these results, we might conclude that WTP is optimized over wireless network. We also have found the WTP performance is optimized when the retransmission counter is set to 5 or 6 in case of high burst error rate.

In this paper, we only consider about IS-95C which is widely used as a wireless network. For the future works, we would like to simulate a big and long transaction which has a large size user data over next generation wireless networks such as IMT-2000.

References

1.   Wen-Tsuen Chen and Jyh-Shin Lee: Some Mechanisms to Improve TCP/IP Performance over Wireless and Mobile Computing Environment. IEEE 7th International Conference on Parallel and Distributed Systems, pp. 437–444, Japan, July 2000.
2.   R. Braden: Extending TCP for Transactions – Concepts. RFC 1397, November 1992.
3.   R. Braden: T/TCP – TCP Extensions for Transactions Functional Specification. RFC 1644, July 1994.
4.   WAP Forum: Wireless Application Protocol Architecture Specification. July 12, 2001. URL: http://www.wapforum.org/
5.   WAP Forum: Wireless Transaction Protocol Specification. February 19, 2000. URL: http://www.wapforum.org/
6.   W. Stevens: TCP Slow Start, Congestion Avoidance, Fast Retransmit, and Fast Recovery Algorithm. RFC 2001, January 1997.
7.   L. Svobodova: Implementing OSI systems. IEEE Journal on Selected Areas in Communications, vol. 7, pp. 1115–1130, September 1989.

8.   Yeunjoo Oh, Nakhoon Baek, Hyeongho Lee, Kyungshik Lim: An adaptive loss recovery algorithm based on FEC using probability models in mobile computing environments. SK Telecommunications Review, Vol. 10, No. 6, pp. 1193–1208, December 2000.
9.   Jean-Chrysostome Bolot, Hugues Crepin, Andres Vega Garcia: Analysis of Audio Packet loss in Internet. Proceedings of Network and Operating System Support for Digital Audio and Video NOSSADV'95, pp. 163–174, Durham, NH, April 1995.
10. J-C. Bolot, S. Fosse-Parisis, D. Towsley: Adaptive FEC-Based Error Control for Internet Telephony. Proceedings of IEEE INFOCOM'99, New York, USA, March 1999
11. Yingxin Zhou and Zhanhong Lu: Utilizing OPNET to Design and Implement WAP Sevice. OPNETWORK 2001, August 2001.
12.   Thomas Kunz, et al.: WAP traffic: Description and Comparison to WWW traffic. Proceedings of the 3rd ACM international workshop on Modeling, analysis and simulation of wireless and mobile systems, pp. 11–19, Boston, USA, August 2000.

Mobile Multicast Method Using the Habitude of Movement

Su-lyun Sung[1], Yongtae Shin[1], Jongwon Choe[2], and Kwangsue Chung[3]

[1] Department of Computer Science, Soongsil University,
[2] Department of Computer Science, Sookmyung Women's University,
[3] School of Electronics Engineering, Kwangwoon University
{ssl, shin}@cherry.ssu.ac.kr, choejn@sookmyung.ac.kr,
kchung@daisy.gwu.ac.kr

Abstract. This paper presents the new mobile multicast method using a habitude of movement that can reduce the total number of the graft and join by making the network including to the habitude be active. The proposed mechanism can provide a multicast service in the mobile environment with the less service delay and not too excessive overhead efficiently. Furthermore, numerical results show that our scheme compared to other mechanisms reduces a delay efficiently until the receipt of multicast service but it has a few overheads.

1 Introduction

The next generation Internet will focus on a service of the high-speed network and real-time multimedia. To make an environment like this, it is required that the support of the mobile computing technology with multicast service. Mobile IP allows a mobile host to move between different networks without disconnecting an established session. And multicast supports source conservation by transferring one copy of data instead of transmitting information from a sender to each receiver separately. But, the existing IP multicast doesn't consider a mobility of host and uses a multicast protocol within the only static hosts and routers and IETF Mobile IP doesn't consider a transfer of multicast either. As a result, to support a multicast in mobile environment, it is required that the interoperation and extension of technology like these.

IETF Mobile IP provides two approaches to provide multicast in mobile environment, which are remote subscription and bidirectional tunneling. The former always provides an optimal route but it has a delay resulted from join and graft. The latter provides a transparency to the multicast host but it has a not-optimal route. In this way, the mobile node always receives a multicast service from home agent. So wherever mobile node moves to, it can receive a multicast service without delay. It doesn't need a support of foreign agent. Like this, many plans related to the multicast in mobile environment have been studied. But, although the existing methods present a solution, they still have limitations. Consequently, this paper tries to solve the essential problems when a multicast in mobile environment is provided. In the aspect of routing efficiency, the remote subscription based on the FA is the better one. But it

I. Chong (Ed.): ICOIN 2002, LNCS 2344, pp. 22–31, 2002.
© Springer-Verlag Berlin Heidelberg 2002

has a limitation, delay for graft and join. To reduce the total number of experience of graft and join, we exploit the habitude of movement pattern of mobile node. And, by making the network including to the habitude be active, we intend to reduce the total number of the graft and join and provide an efficient multicast service in mobile environment.

The rest of the paper is organized as follows. Section 2 presents the previous works about mobile multicast methods. Section 3 analyzes the advantages and disadvantages of the existing methods and   defines a habitude of movement and introduces a mobile multicast method using a habitude. Section 4 evaluates the performance and compares proposed method with the other protocols. Finally, we conclude this paper in Section 6.

2   Related Works

Many mechanisms have proposed to provide a multicast in mobile environment. Fundamentally, IETF Mobile IP suggests two methods for handling multicast for mobile hosts as we referred [1]. One is bi-directional tunneled multicast, the other is remote subscription [2].

The bi-directional tunneled multicast is based on the HA. All the multicast service is carried by HA. In this way, HA sends a multicast packet to the mobile node by tunneling in the same way as the unicast. In other words, this way can use an advantage of multicast. The mobile node can receive a multicast service without FA's supporting of it. Also, when the mobile node moves to the other network, it can receive a multicast service without delay. But, it cannot use an advantage of the multicast, resource reservation and bandwidth savings because the mobile node receives a multicast service by the form of unicast packet. And if multiple mobile nodes on the same foreign network belong to the same multicast group, then the multicast packets will be duplicated on that foreign network. In other words, the tunnel convergence problem will happen. This problem results in overheads of home agent and network. And the mobile node always comes to receive the multicast packet by not-optimal route.

The remote subscription is based on the FA. The FA provides a multicast. In this way, the mobile node is required to re-subscribe to the multicast group on each foreign network. Therefore, the mobile node experiences a delay resulted from the join and graft in the case of not existing a consistent multicast group. But, because the mobile node receives a multicast service from the foreign agent, it can always receive it by an optimal route. And using a major advantage of multicast, this way doesn't have overheads of the home agent and network. In this way, the foreign agent has to support a multicast group.

The MoM based on bi-directional tunneling was proposed to solve the tunnel convergence problem [3]. It uses the DMSP to avoid duplicate datagram being tunneled to the common foreign agent(FA). The FA designates one home agent(HA) and only the DMSP can send a multicast packet to the consistent foreign agent. But Although the MoM solves the problem of the bidirectional tunneling, there exists still many overheads resulted from DMSP management and it experiences a not-optimal route.

3 Problem

Most of the existing mechanism can be classified into bidirectional tunneling and remote subscription. In the aspect of routing efficiency and originality, the remote subscription is superior to the bidirectional tunneling except a delay resulted from the graft and join. If the delay problem can be solved, it will be a better mechanism than any other ones.

The way to reduce a delay is classified two methods. One is to reduce the real delay time. The other is to reduce the number of experience of delay. But, the former depends on the existing multicast mechanism, for example, IGMP or DVMRP ([4][5]). Therefore, if we intend to reduce a real delay time, we have to modify many parts of existing mechanism. So, we will choose the latter case, which is to reduce the total number of experience of delay. By reducing the total number of experience of delay, we solve the problem of the remote subscription and provide an efficient multicast service in mobile environment.

4 Mobile Multicast Method Using the Habitude of Movement

4.1 Overview

The total number of the delay resulted from join and graft can be reduced by use of a habitude of movement. The habitude of movement means the repetitive movement pattern of mobile node. The HA defines a scope of habitude. The elements in the scope of habitude are the group of recently visited network. According to the specified scope, the HA maintains the habitude information. Fundamentally, this paper uses a remote subscription. If there is established multicast tree, the mobile node can receive the multicast packet without delay. To increase the above case, we intend to use a habitude of movement.

When the mobile node moves to the other network and the HA receives a registration request message by FA, the HA sends a habitude information to the mobile node with the registration response message. On the way to the mobile node, the FA manages the information of habitude. After processing the habitude information, the FA will maintain the multicast routing tree without a prune until it receives a deletion notification message, a newly defined message in this paper. Although the mobile node moves to the other network, the FA will maintain the multicast tree. While the FA maintains the multicast routing tree, the mobile node can receive the multicast service without delay at any time of returning to the foreign network. In the case of multicast tree's being pruned by the deletion notification message, the mobile node will come to join the multicast group newly. In this way, we can reduce the total number of experience of the delay.

4.2 Definition of the Habitude

This paper defines the habitude as a repetitive pattern of mobile node's movement. We intend to explain its concept by a case of one professor. He spends most of his time in home and school and company. In other words, the professor's movement does not get out of the restricted location. And his movement pattern is repetitive and restricted. So, we can say that the location where has ever been visited inclines to be re-visited by one. By using the feature like this, we will define the habitude as the refined form. The existing papers exploited this character of mobile node: [6],[7]. This paper defines the habitude as the group of network visited by the mobile node for the specified number of network. In other words, if we assume that Δn is the number of network specified in advance, we can say the definition of habitude like that, For Δn, G = { n1, n2, n3, n4 }. The maintenance of group G will be dealt with in the later part.

4.3 Method Description

In this mechanism, the scope of habitude is defined as the group of foreign agent for the specified network number. In other word, for Δn, the group can be explained as G = { FA1, FA2, FA3, FA4 }. According to the Δn value, the size of G can be changed. The HA specifies a Δn value in advance. The HA has an information of group G in the form of cache. After the HA's processing a registration response message, the HA comes to update the cache by using it. The whole procedure of the proposed mechanism is like a Figure 1.

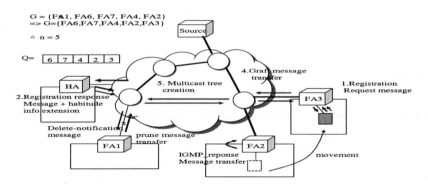

Fig. 1. The total operation of proposed mechanism

4.3.1 Operation of HA

After the HA receives a register request, the HA sends the mobile node a register reply extended by the habitude information message. By using this message, the HA notify the correspondent FA to support this mechanism. The figure 2 shows a format of habitude information message.

Type (150)	Reserved

Fig. 2. The format of habitude information message.

After sending a register reply message, the HA will update a cache of the habitude information. At this time, if the cache is full of the FAs, the HA has to discard the existing FA list in the cache. The way for cache management depends on a policy. As soon as the HA discard one FA entry, it sends a delete notification message to the correspondent FA. When the FA receives this message, it can prune the existing multicast routing tree included to the correspondent mobile node if the mobile node is the last one to want the packet of the correspondent multicast group. The figure 3 shows the deletion notification message.

Type (151)	Reserved
Mobile node address	

Fig. 3. Deletion notification message

The operation algorithm of HA

Algorithm operation_by_HA

Variables

binding$_i$ = set of binding information for mobile host i

habitude_info$_i$ = the group of network visited by the
                   mobile node for the specified number
                   of network

Initially binding = nil

    upon receiving reg_request$_j$ from MH$_i$:

            update binding$_i$;   /* create binding$_i$ */

                if habitude_info$_i$ = Full **then**

                        remove FA$_i$ from habitude_info$_i$;

                        send delete_notification to FA$_i$;

                add FA$_i$ to habitude_info$_i$;

                send reg_reply to FA$_i$

4.3.2 Operation of FA

When the FA receives the register reply message, it processes a habitude information message. The FA maintains a habitude table. In the habitude table, the FA maintains the list of mobile node using a proposed method in this paper. After receiving a register reply message, the FA adds the mobile node to the habitude table. If the mobile node has a multicast group to join, it can send an IGMP_response message to the FA. When the mobile node moves to the other network, the mobile node sends a leave message to the FA. Although the FA receives the leave message, the FA should not prune the multicast routing tree until it receives a deletion notification message, which is different from the existing IGMP mechanism. To prevent a circumstance like this, the FA should maintain the multicast routing tree by sending an IGMP response message to itself instead of the MN.

In this way, if the mobile node re-visits the FA that still maintains the multicast routing tree in the future, the mobile node can receive the multicast service without delay. Because the mobile node inclines to move into the limited region, this way can reduce the total number of experience of such a delay.

The operation algorithm of FA

Algorithm operation_by_FA

Variables

visitor_list$_i$ = set of mobile host i visited to the FA

habitude_table$_i$ = set of mobile host i using a proposed

method

members$_j$ = set of host ids for multicast group j

Initially visitor_list$_i$ = nil, members = nil

upon receiving reg_reply$_i$ from HA ;

update visitor_list$_i$;

if exist habitude_info_extension **then**

add MH$_i$ to habitude_table$_i$;

upon receiving Join$_j$ from MH$_i$

if members$_j$ = nill **then**

send Graft to MR$_j$;

add MH$_i$ to members$_j$;

upon receiving leave$_j$ from MH$_i$

```
if MHᵢ ∈ habitude_tableᵢ  then

        hold Multi_tree until receive
                        delete_notification ;

else

        send Prune to MRⱼ ;

update membersⱼ ;
```

4.4 Handoff Plan

In such a case that the HA adds a new entry to the cache of the habitude, the mobile node will experience the delay from the graft and join. For this state, we suggest that the bidirectional tunneling should be used for some times in the above circumstance. The time can be defined by a policy.

5 Performance Evaluation

To evaluate this mechanism, we compare this one to the existing mechanism such as a remote subscription and bidirectional tunneling based on the connection continuation, overhead. For this, this paper supposes that the network is composed of M subnets and the number of router to provide a multicast service is k ($0 \leq k \leq M$). And the set of mobile node using a multicast service of group G is $H_G = \{MH1, MH2, ... , MH3\}$. They are distributed uniformly and the average number of movement per unit time is λ (1/sec).

5.1 Service Delay

After the mobile node moves to the other network, it experiences a delay from the registration procedure to the receipt of multicast service.

The figure 4 shows the total delay the mobile node experiences over the movement. By using these variables, we can express the delay of bidirectional tunneling and remotes subscription like (1), (2). Basically, this paper supposes the FA provides a multicast. The P_m is a probability the FA's to support a multicast and the P_i is a probability the FA's to have already joined to the multicast group. And if the mobile node moves to the FA not supporting a multicast, the sojourn time is a delay.

$$D_{HA} = \tau_s + \tau_R + \delta_T \tag{1}$$

$$D_{FA} = P_m (\tau_s + \tau_R + (1 - P_i) \delta_M) + (1 - P_m) 1/ \lambda \tag{2}$$

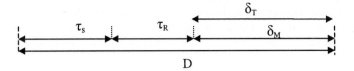

Fig. 4. Total delay D is composed to τ_s the time of searching a new agent and τ_R the time of registration and δ_M the time of graft and join to multicast routing tree and δ_T the time to tunneling

The delay of the proposed mechanism is like a (3). The P_h is the probability the FA's to be included in the scope of the habitude information. In the proposed mechanism, the mobile node will experiences the delay resulted from the join and graft when the FA doesn't include in the habitude table and it doesn't join to the multicast group at the same time. Consequently, the probability of experiencing the join and graft will be decreased.

$$D_{habitude} = P_m (\tau_s + \tau_R + (1 - P_i)(1 - P_h) \delta_M) + (1 - P_m) 1/\lambda \qquad (3)$$

The figure 5 shows an analysis on the total delay according to the each mechanism. As (1) shows us, the higher the probability of habitude is, the less the total delay value is. And our mechanism shows the superiority to other mechanisms

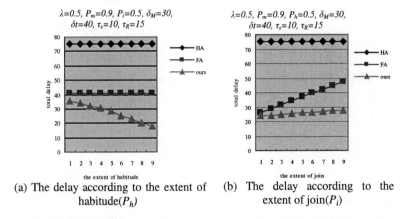

(a) The delay according to the extent of habitude(P_h)

(b) The delay according to the extent of join(P_i)

Fig. 5. Comparison result of delay according to the extent of habitude and join rate

5.2 Total Overhead

In the bi-directional tunneled multicast, the HA does a tunneling of the multicast packet. So the bidirectional tunneling has an overhead σ_T for the tunneling. The remote subscription has σ_M overheads for multicast tree construction when the mobile node's multicast group does not exist. The proposed mechanism has σ_M an overhead for multicast tree construction when the FA doesn't join a multicast group and it doesn't include in the habitude table and σ_a another overhead for a useless multicast packet when the member doesn't exist. The followings are the overhead for each mechanism.

$$O_{HA} = n\sigma_T \tag{4}$$

$$O_{FA} = n\,(1 - P_i)\,\sigma_M \tag{5}$$

$$O_{habitude} = n\,P_h\,(1 - P_i)\,\sigma_a + n\,(1 - P_h)\,(1 - P_i)\,\sigma_M \tag{6}$$

The Figure 6 shows the overheads according to the each mechanism.

Generally, the overhead of the remote subscription is the lower than one of the bi-directional tunneling. The proposed mechanism has an overhead of the middle value between the remote subscription and bi-directional tunneling. The overhead of proposed mechanism resulted from the additional packet while the correspondent FA includes to the habitude scope. Its value can be controlled by Δn. In other words, the overhead value can be beyond the consideration.

Consequently, the proposed mechanism has a superiority of the service delay to the other mechanism and the higher the extent of join is and the less the number of mobile is, the overhead of proposed mechanism has a similarity with the remote subscription.

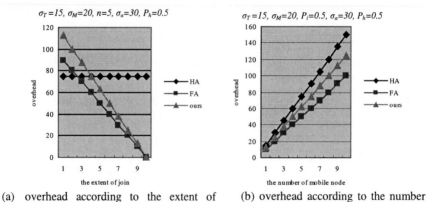

(a) overhead according to the extent of habitude(P_i)

(b) overhead according to the number of MN(n)

Fig. 6. Comparison of overheads according to the extent of habitude and number of MN

6 Conclusion

The IETF Mobile IP provides two approaches to provide multicast in mobile environment, which are remote subscription and bi-directional tunneling. The former always provides an optimal route but it has a delay resulted from join and graft. The latter provides a transparency to the multicast host but it has a not-optimal route. So in the aspect of routing efficiency, we used a remote subscription. To reduce the total number of experience of graft and join which is a disadvantage of remote subscription, we exploit the habitude of movement pattern of mobile node.

This paper presented the new mobile multicast method using a habitude of movement that can reduce the total number of the graft and join by making the network including to the habitude be active. The proposed mechanism can provide a multicast service in the mobile environment with the less service delay and not too excessive overhead. The numerical results show that our scheme compared with the other mechanism reduces a delay efficiently until the receipt of multicast service but it has a few overheads.

In the future, we plan to bring about a real outcome by embodying an algorithm of this paper through the real implementation.

References

1. C. Perkins, "IP Mobility support," *Internet RFC 2002*, Oct. 1996.
2. G. Xylomenos and G. Polyzos, "IP Multicast for Mobile Hosts," *IEEE Communications Magazine*, Jan. 1997.
3. Harrison, T.G., Williamson, C.L. mackrell, and Bunt, R. B, "Mobile Multicast (MoM) protocol: multicast support for mobile hosts," *In Proc. 3rd Annual ACM/IEEE conf. Mobile Comp and Networks*, 26-30 September, pp. 151-160.
4. Fenner, W., "Internet group management protocol, Version 2," *RFC 2236*.
5. Pusateri, T., "Distance vector multicast routing protocol," *Internet Draft*, 1999.
6. G. Cho, L.F. Marshall, An efficient location and routing scheme for mobile computing environment, IEEE J. Select. Areas Commun. 13 (5) (1995) 868-879.
7. R. Jain, et al., A caching strategy to reduce network impacts of PCS, IEEE J. Select. Areas Commun. 12 (8) (1994) 1434-1444.
8. V. Chikamane and C. L. Williamson, "Multicast Support for Mobile Host using Mobile IP: Design Issues and Proposed Architecture," *ACM/Baltzer Mobile Networks and Applications*, 1998, pp.365-379.
9. S. Deering, "Host Extensions for IP Multicasting," *Internet RFC 1112*, Aug, 1989.
10. Ajay Bakre and B.R. Badrinath, "I-TCP: Indirect TCP for mobile hosts," *Tech. Report DCS-TR-314*, Computer Science Depart-ment, Rutgers Univ., Piscataway, N.J., 1994.
11. Ramon Caceres and Venkata N. Padmanabhan, "Fast and Scalable Handoffs for wireless internetworks," ACM Mobicom '96, November 1996.
12. Cheng Lin Tan, Stephen Pink, and Kin Mun Lye, "A Fast Handoff Scheme for wireless networks," WoW-MoM '99, 1999.
13. Kuang-Hwei Chi, Chien Tseng and Ting-Lu Huang, "A framework for Mobile Multicast using Dynamic Route Reconstructions," *the computer journal*, vol. 42, No. 6, 1999.
14. C.L., Williamson, et al., "Performance evaluation of the MoM mobile multicast protocol," *ACM/Baltzer Mobile Networks and Applications, 3 (4) (1998) 189-201.*
15. Arup Acharya, Ajay Bakre, B,R,Badrinath, "IP Multicast Extensions for Mobile Internetworking," IEEE, 1966.
16. Ahmed Helmy, "A Multicast-Based Protocol for IP Mobility Support," ACM, 2000
17. Vineet Chikarmane, Rick Bunt, Carey Williamson, "Mobile IP-based Multicast as a Service for Mobile Hosts"

A Proposal of a Quasi-Solution State Evolution Algorithm for Channel Assignment Problems

Nobuo Funabiki[1], Toru Nakanishi[1], Tokumi Yokohira[1], Shigeto Tajima[2], and Teruo Higashino[2]

[1] Department of Communication Network Engineering, Okayama University, 3-1-1 Tsushimanaka, Okayama 700-8530, Japan
[2] Department of Informatics and Mathematical Science, Osaka University, 1-3 Machikaneyama, Toyonaka 560-8531, Japan

Abstract. The channel assignment problem (*CAP*) in a cellular network requires finding a channel assignment to the call requests from cells such that three types of interference constraints are not only satisfied, but also the number of channels (*channel span*) is minimized. This paper presents a three-stage iterative algorithm, called the Quasi-solution state evolution algorithm for CAP (*QCAP*). QCAP evolutes *quasi-solution states* where a subset of call requests is assigned channels and no more request can be satisfied without violating the constraint. The first stage computes the lower bound on the channel span. After the second stage greedily generates an initial quasi-solution state, the third stage evolutes them for a feasible solution by iteratively generating best neighborhoods, with help of the dynamic state jump and the gradual span expansion for global convergence. The performance is evaluated through solving benchmark instances in literature, where QCAP always finds the optimum or near-optimum solution in very short time.

1 Introduction

This paper presents a heuristic algorithm called *QCAP*, a Quasi-solution state evolution algorithm for the Channel Assignment Problem (*CAP*) in a cellular network. Traffic demands for mobile communication have been rapidly increased in voice and data services, due to portability and availability in everyplace with no requirement for hard wires. Besides, the successful introduction of packet communications using mobile networks has accelerated the explosive growth of demands. On the other hand, the electromagnetic spectrum allocated for this system has been limited, because of a variety of significant applications using radio waves. Thus, the efficient use of precious frequency band resources has been an important task in the research/development communities in mobile communication networks. The concept of the cellular network has been widely adopted as the efficiency realization [1]. This cellular network allows the reuse of the same channel in geographically separated regions simultaneously. As a result, CAP has become the critical problem to be solved for its efficient solutions. A solution for CAP is not only requested to avoid the mutual radio interference between closer channels, but also to maximize the channel utilization.

I. Chong (Ed.): ICOIN 2002, LNCS 2344, pp. 32–41, 2002.
© Springer-Verlag Berlin Heidelberg 2002

Due to the NP-hardness of CAP [2], a number of polynomial time approximation algorithms have been reported [5]-[16]. In [5], Sivarajan et al. proposed eight different greedy algorithms based on graph coloring algorithms. They referred practical benchmark instances of 21 cells. Their algorithms and these benchmarks have been widely used for performance evaluations in many literature. In [6], Kunz proposed a neural network algorithm using continuous sigmoid neurons for a limited case of CAP, which does not consider the adjacent channel constraint. He provided a practical benchmark instance of 25 cells. In [7], Funabiki et al. proposed another neural network algorithm using binary neurons for the general CAP. In [8], Wang et al. proposed a two-phase adaptive local search algorithm named *CAP3*. They showed its superiority through solving a subset of Sivarajan's benchmarks, Kunz's benchmark, and Kim's benchmarks. In [9], Sung et al. proposed a generalized sequential packing (*GSP*) algorithm with its two variations for CAP, and a lower bound on the number of channels (*channel span*). In [10], Hurley et al. proposed an integrated system called *FASoft*, which incorporates various CAP algorithms. In [11], Rouskas et al. proposed an iterative heuristic algorithm for CAP. They provided benchmark instances of 49 cells. In [12], Beckmann et al. proposed a hybrid algorithm composed of a genetic algorithm and the frequency exhaustive strategy. They evaluated the performance through solving a subset of Sivarajan's benchmarks. In [13], Funabiki et al. proposed a neural network algorithm combined with heuristic methods for CAP. In [14][15], Murakami et al. proposed a genetic algorithm using several schemes to improve the convergence property. This algorithm is only applicable to simple cases of CAP without considering the interference between different channels. They provided benchmark instances of 49, 80, and 101 cells. In [16], Matsui et al. proposed a genetic algorithm to determine the sequence of cells for assigning channels by a greedy method for CAP. Unfortunately, none of existing algorithms can find optimum solutions for small size benchmark instances whose lower bounds are known.

QCAP evolutes quasi-solution states through three stages to provide the high quality solution in short computation time. A *quasi-solution state* represents a channel assignment to a subset of call requests where no more call request in any cell can be assigned a channel without violating the constraint. When the full set of call requests is assigned channels, it becomes a solution. The *first stage* computes the lower bound on the channel span. The *second stage* greedily generates an initial quasi-solution state. The *third stage* iteratively evolves quasi-solution states to a feasible channel assignment, by iteratively generating best neighbor states, while schemes of the *dynamic state jump* and the *gradual span expansion* are used together for global convergence. The performance is evaluated through solving benchmark instances, where the comparisons with existing results confirm the superiority of QCAP.

2   Problem Formulation of CAP

CAP in this paper follows the common problem formulation defined by Gamst et al. [3] as in literature. The servicing region in a cellular network is managed as a set of disjoint hexagonal cells. Each cell occupies a unit area for providing communication services to users that are located in the cell area. When a user requests a call for communication services, a channel must be assigned to the user through which voice or data packets are communicating between the user's mobile terminal and a base station. This channel assignment must satisfy the constraints to avoid the mutual radio interference between closer channels. In CAP, the following three types of constraints have been considered:

1) *Co-Channel Constraint (CCC)*: the same or its adjacent channels cannot be reused in the cells that are located within a specified distance from each other in the network. This set of channel-reuse prohibited cells is called a *cluster*. In a cluster, any pair of channels assigned to call requests from the cells must have a specified channel distance.

2) *Adjacent Channel Constraint (ACC)*: adjacent channels cannot be assigned to adjacent cells in the network simultaneously. In other words, any pair of channels assigned to adjacent cells must have a specified distance. The distance for ACC is usually larger than that for CCC.

3) *Co-Site Constraint (CSC)*: any pair of channels in the same cell must have a specified distance. The distance for CSC is usually larger than that for ACC.

The channel distance is described by the difference on the channel indices in the channel domain. In this paper, the cell cluster size for CCC is denoted by "Nc", the channel distance to satisfy CCC is by "cij", the distance for ACC is by "acc", and the distance for CSC is by "cii" as in existing papers. The goal of CAP is to find a channel assignment to every call request with the minimum number of channels or *channel span* subject to the above three constraints. The three constraints in an N-cell network are altogether described by an $N \times N$ symmetric *compatibility matrix C*. A non-diagonal element c_{ij} $(i \neq j)$ in C represents the minimum distance to be separated between a channel in cell i and a channel in cell j. A diagonal element c_{ii} in C represents the minimum distance between any pair of channels in cell i.

A set of call requests in the N-cell network is given by an N-element *demand vector D*. The i-th element d_i in D represents the number of channels to satisfy the call requests in cell i. Let a binary variable x_{ik} represent whether channel k be assigned to cell i ($x_{ik} = 1$) or not ($x_{ik} = 0$) for $i = 1, ..., N$ and $k = 1, ..., M$. Note that M represents the channel span required for the instance. Then, CAP is defined as follows:

$$\text{minimize } M \text{ such that}$$

$$x_{ik} = 0 \text{ or } 1, \text{ for } i \in \{1, ..., N\} \text{ and } k \in \{1, ..., M\}$$

$$\sum_{k=1}^{M} x_{ik} = d_i, \text{ for } i \in \{1, ..., N\}$$

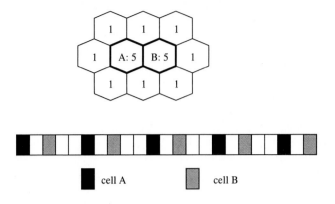

Fig. 1. A CAP example for the improved lower bound formula.

$$|k - l| \geq c_{ij}, \text{ for } k, l \in \{1, ..., M\}, i, j \in \{1, ..., N\}$$
$$\text{and } x_{ik} = x_{jl} = 1. \tag{1}$$

3 Proposal of QCAP

3.1 Lower Bound Computation

The first stage of QCAP computes the lower bound LB on the channel span for the initial value of M using the following three formulas, based on [4][9] with some improvement. The first lower bound formula gives the channel span to afford every call request in one cell while satisfying CSC by cii:

$$LB_1 \geq 1 + cii\,(d_i - 1)\,. \tag{2}$$

When two cells in the same cluster have the same number of call requests in an instance such as Rouskas's benchmarks, additional channels are required to satisfy these call requests from both cells simultaneously:

$$LB_1 \geq 1 + cii\,(d_i - 1) + cij \tag{3}$$

where cells i and j are located in the same cluster to be mutually interfered by CCC or ACC, and have the same number of call requests. Figure 1 illustrates an example channel assignment with $cii = 5$ and $cij = 2$ for this special case. Each number inside a hexagonal cell represents the number of call requests. Each of five call requests from cell A is assigned a channel with the interval of five channels, and each of four requests from cell B is assigned a channel between two neighboring channels assigned to cell A. Consequently, two additional channels are necessary to assign the last call request from cell B to satisfy $cij = 2$.

The second lower bound formula gives the channel span to afford every request call from one cell, namely the *center cell*, and from its surrounding adjacent cells while satisfying ACC:

$$LB_2 \geq 2acc + (d_i - 2)(2acc - 1) + cij\left(\sum_{j \in P_i} d_j\right) \qquad (4)$$

where P_i consists of the cells that are adjacent to cell i. The first two terms represent the number of channels that the channel assignment to the center cell forbids from the assignment to the surrounding cells to satisfy ACC. The last term represents the number of channels required to satisfy the call requests in the surrounding cells to satisfy CCC.

The third lower bound formula gives the channel span to afford every call request from the same cluster of cells while satisfying CCC:

$$LB_3 \geq 1 + cij\left(\sum_{j \in Q_i} d_j - 1\right) \qquad (5)$$

where the cell set Q_i consists of the cells that are located in the same cluster as cell i.

3.2   Greedy Initial State Generation

The second stage adopts the *requirement exhaustive strategy* in [5], to produce an initial quasi-solution state with the lower bound span, where as many call requests as possible are assigned channels greedily. The *unsatisfied cell index list* L_{cell} is initialized by sorting cells in descending order of cell degrees [5]:

$$degree_i = \left(\sum_{j=1}^{N} d_j c_{ij}\right) - c_{ii}. \qquad (6)$$

An *unsatisfied cell* represents a cell where some of its call requests are not assigned channels. A variable indicating the number of assigned channels, $assign_i$, is introduced for cell i, where $assign_i < d_i$ indicates cell i is an unsatisfied cell. In this list generation, the tiebreak is resolved randomly when two or more cells have the same degree. In addition, when LB_1 gives the lower bound LB for an instance in the first stage, each call request from the corresponding cell is assigned a channel with the interval of cii, and this assignment is fixed throughout search process. The fixed cell index is denoted by F in this paper for convenience.

3.3   Quasi-Solution State Evolution

Next State Generation. The third stage evolutes quasi-solution states by repeatedly generating best neighbor states in terms of the following cost function to evaluate the violation by the assignment of channel k to cell i:

$$cost_{ik} = \sum_{\substack{j=1 \\ c_{ij} \geq 1}}^{N} \sum_{l=k-c_{ij}+1}^{k+c_{ij}-1} w_j x_{jl}. \qquad (7)$$

A next state generation is initiated by randomly selecting one unsatisfied cell from L_{cell} to avoid biased movements. Then, one channel is selected for its assignment, such that this channel is not only assigned to this cell currently, but also minimizes $cost_{ik}$ where the tiebreak is always resolved by selecting the least index channel. The cell weight w_j is introduced to encourage the channel assignment to cells that are hard to be assigned otherwise as in [17]. Here, two auxiliary conditions are imposed in the channel selection. One is the prohibition of selecting a channel in a *tabu list* to avoid cyclic state transitions. The tabu list describes the channels that have been selected within the predefined number of iteration steps T_{tabu} since its last selection to the corresponding cell. Another is the prohibition of selecting channels conflicting with the fixed cell assignment.

However, if QCAP only repeats transitions to best neighbor states, it may cause stagnation of state changes. To provide a hill-climbing capability of escaping from local minimum, one channel is sometimes randomly selected, which is called the *random selection*. The random selection is applied when the state has not been improved during the constant number of steps. Besides, the weight w_i associated with each unsatisfied cell in L_{cell} is incremented by 1 at the same time, so as to encourage these cells to be assigned channels more progressively than others.

To retrieve a quasi-solution state after this new channel assignment, every conflicting assignment with it is sought a new feasible channel assignment. If an assignable channel is found, it is assigned to the cell. Otherwise, the channel assignment is cleared, and the cell is inserted into L_{cell} if it is not there. After every conflicting one is handled, each cell in L_{cell} is checked whether a new channel can be assigned or not. If assignable, this channel is assigned there.

Dynamic State Jump. Even several trials of the random selection may not provide enough state fluctuations to escape from local minimum. In such situations, QCAP induces the *dynamic state jump* for big changes from previous states while maintaining the achieved solution quality. Firstly, the best state in terms of the number of satisfied call requests is retrieved as the initial state for different evolutions. The best state that has been visited is memorized in QCAP. Then, channel assignments in this state are repeatedly shuffled until the half of assigned call requests in each cell may receive different channels from current ones in each dynamic state jump. In each shuffle movement, one cell is first randomly selected from cells that have movable assignments. Then, one call request with an assigned channel in this cell is randomly selected. The call request is assigned a randomly selected new channel if there is a channel that satisfies the two conditions: 1) the channel is not currently assigned to any call request in this cell, and 2) the new assignment is compatible with other channel assignments. When such a channel does not exist, the scheme is terminated. After the dynamic state jump, each cell in L_{cell} is sought a channel assignment, to retrieve a quasi-solution state, and the tabu list is cleared.

Gradual Span Expansion. In some CAP instances, the lower bound on the channel span in the first stage is too small to afford every call request. In such cases, QCAP gradually expands the channel span until it reaches a feasible solution. Actually the span expansion is carried out when the state in QCAP has not been improved after several trials of the dynamic state jump. In each span expansion, the best state is first retrieved as a current state as in the dynamic state jump. Then, the channel span M is incremented by ΔM given by the following formula:

$$\Delta M = \left\lfloor \alpha \left(\sum_{i=1}^{N} d_i - \sum_{i=1}^{N} assign_i \right) / Nc \right\rfloor,$$
$$\text{if } \Delta M < 1 \text{ then } \Delta M = 1 \tag{8}$$

where Nc is the cluster size for CCC, α is a constant parameter, and the *floor function* $\lfloor x \rfloor$ returns the maximum integer smaller than or equal to x. This equation is derived from a conjecture that a new channel can afford another call request from every cell cluster. After M is expanded, cells in L_{cell} are sequentially assigned these expanded channels by the requirement exhaustive strategy. Then, the dynamic state jump is applied for a better restarting state. At the same time, every cell weight w_i is initialized by 1 for $i = 1, ..., N$, to reset the search direction.

Table 1. Simulation Results for CAP instances by Sivarajan.

| Instance | Constraint | | | | QCAP | | | Existing results | | | | |
No.	Nc	acc	cii	LB	M	time(s)	[5]	[8]	[9]	[10]	[11]	[12]
1	12	2	5	427	427	0.238	460	-	440	427	440	-
2	7	2	5	427	427	0.094	447	433	436	427	436	427
3	12	2	7	533	533	0.002	536	-	533	-	533	-
4	7	2	7	533	533	0.036	533	533	533	-	533	533
5	12	1	5	381	381	0.003	381	-	381	-	381	-
6	7	1	5	381	381	0.000	381	381	381	-	381	381
7	12	1	7	533	533	0.003	533	-	533	-	533	-
8	7	1	7	533	533	0.002	533	533	533	-	533	533
9	12	2	5	258	258	0.087	283	-	273	258	287	-
10	7	2	5	253	253	0.329	270	263	268	253	269	253
11	12	2	7	309	309	0.026	310	309	309	-	309	-
12	7	2	7	309	309	0.019	310	309	309	-	309	309
13	12	2	12	529	529	0.007	529	529	529	-	529	-

4   Simulation Results

Benchmark instances in Tables 1- 4 are solved to evaluate the performance of QCAP. A total of 10 runs are repeated with different random numbers in each

Table 2. Simulation Results for CAP instances by Kunz.

| Instance | | QCAP | | | Existing results | |
No.	LB	M	time(s)	[8]	[10]
14	73	73	0.003	73	73

Table 3. Simulation Results for CAP instances by Rouskas.

| Instance | Constraint | | | | QCAP | | | Existing results | |
No.	Nc	acc	cii	LB	M	time(s)	[5]	[11]
15	12	2	5	468	468	0.237	486	470
16	7	2	5	468	468	0.158	481	470
17	12	2	7	484	484	0.119	520	484
18	7	2	7	484	484	0.301	523	484
19	12	1	5	413	413	0.175	422	414
20	7	1	5	346	346	0.052	349	346
21	12	1	7	484	484	0.005	484	484
22	7	1	7	484	484	0.004	484	484
23	12	2	5	273	274	3.557	307	298
24	7	2	5	253	253	0.245	275	266
25	12	2	7	273	274	5.098	330	301
26	7	2	7	262	262	0.628	297	275
27	12	2	12	447	447	1.015	447	447

Table 4. Simulation Results for CAP instances by Murakami.

| Instance | Constraint | | | | | QCAP | | Existing results | | |
No.	N	Nc	acc	cii	LB	M	time(s)	[14]	[15]	[16]
28	49	7	1	1	22	22	0.006	24	24	22
29	49	7	1	1	21	21	0.005	-	24	-
30	49	7	1	1	26	26	0.008	-	>24	-
31	80	7	1	1	22	22	0.031	>24	-	24
32	101	7	1	1	22	22	0.038	>24	-	24

instance. The tables show the instance number, the constraint parameters (Nc, acc, cii), the lower bound on the channel span (LB) in the first stage of QCAP, the average channel spans in solutions (M), and the average computation time (seconds) on Pentium-III 800 MHz by QCAP in each instance. Besides, the existing results on channel spans in literature are also summarized there. Note that only the best result among several versions of their algorithms in [5], [9], [10], and [11] is described to simplify comparisons and save the space. In Table 4, ">x" indicates that they cannot find a feasible assignment using x channels.

These tables suggest that QCAP finds the optimum solution with the lower bound on the channel span by any run for each CAP instance in less than one second, except for three instances 23, 25, and 27. In two instances 23 and 25, QCAP always finds a solution that requires one more channel than the lower bound. The more accurate lower bound formula might clarify that the obtained solutions by QCAP are their real lower bounds. On the other hand, the existing algorithms require several versions of procedures and/or many repeated runs with different random numbers to reach the optimum solution. For example, the genetic algorithm in [12] can reach the lower bound solution for the hard instance 10 by Sivarajan in only 1 run among 50 runs with different random numbers. Besides, it takes several thousand of iteration steps for convergence. The simplicity and the search efficiency reveal that the proposed QCAP is a very practical and powerful tool to solve the important task of channel assignments in the cellular network.

5   Conclusion

This paper has presented QCAP, a quasi-solution state algorithm for the channel assignment problem in the cellular network. The performance is evaluated through solving benchmark instances, where the comparisons to the existing CAP algorithms confirm the extensive search capability and the efficiency of QCAP. The study on the tighter lower bound of the channel span is essential to further improve the performance.

References

1. A. H. MacDonald, "Advanced mobile phone service: the cellular concept," Bell Syst. Tech. J., vol. 58, pp. 15–41, Jan. 1979.
2. W. K. Hale, "Frequency assignment: theory and applications," Proc. IEEE, vol. 68, no. 12, pp. 1497–1514, Dec. 1980.
3. A. Gamst and W. Rave, "On frequency assignment in mobile automatic telephone systems," in Proc. GLOBECOM, 1982, pp. 309–315.
4. A Gamst, "Some lower bounds for a class of frequency assignment problems," IEEE Trans. Veh. Technol., vol. VT-35, no. 1, pp. 8–14, Feb. 1986.
5. K. N. Sivarajan, R. J. McEliece, and J. W. Letchum, "Channel assignment in cellular radio," in Proc. 39th IEEE Veh. Technol. Conf., 1989, pp. 846–850.
6. D. Kunz, "Channel assignment for cellular radio using neural networks," IEEE Trans. Veh. Technol., vol. 40, no. 1, pp. 188–193, Feb. 1991.
7. N. Funabiki and Y. Takefuji, "A neural network parallel algorithm for channel assignment problems in cellular radio networks," IEEE Trans. Veh. Technol., vol. 41, no. 4, pp. 430–437, Nov. 1992.
8. W. Wang and C. K. Rushforth, "An adaptive local-search algorithm for the channel assignment problem (CAP)," IEEE Trans. Veh. Technol., vol. 45, no. 3, pp. 459–466, Aug. 1996.
9. C. W. Sung and W. S. Wong, "Sequential packing algorithm for channel assignment under cochannel and adjacent-channel interference constraint," IEEE Trans. Veh. Technol., vol. 46, no. 3, pp. 676–686, Aug. 1997.
10. S. Hurley, D. H. Smith, and S. U. Thiel, "FASoft: a system for discrete channel frequency assignment," Radio Science, vol. 32, no. 5, pp. 1921–1939, Sep.–Oct. 1997.

11. A. N. Rouskas, M. G. Kazantzakis, and M. E. Anagnostou, "Minimization of frequency assignment span in cellular networks," IEEE Trans. Veh. Technol., vol. 48, no. 3, pp. 873–882, May 1999.
12. D. Beckmann and U. Killat, "A new strategy for the application of genetic algorithms to the channel-assignment problem," IEEE Trans. Veh. Technol., vol. 48, no. 4, pp. 1261–1269, July 1999.
13. N. Funabiki, N. Okutani, and S. Nishikawa, "A three-stage heuristic combined neural-network algorithm for channel assignment in cellular mobile systems," IEEE Trans. Veh. Technol., vol. 49, no. 2, pp. 397–403, March 2000.
14. H. Murakami, Y. Ogawa, and T. Ohgane, "Channel allocation in mobile communications using the genetic algorithm," IEICE Tech. Report, RCS98-48, pp. 87–94, 1998.
15. H. Murakami, Y. Ogawa, and T. Ohgane, "Fixed channel allocation method in mobile radio communications using the genetic algorithm," IEICE Trans., vol. J83-B, no. 6, pp. 769–779, June 2000.
16. S. Matsui and K. Tokoro, "A fast genetic algorithm using allocation order for fixed channel assignment in mobile communications," IEICE Trans., vol. J83-B, no. 5, pp. 645–653, May 2000.
17. B. Selman and H. Kautz, "Domain-independent extensions to GSAT: solving large structured satisfiability problems," in Proc. 13th Int'l Joint Conf. Artificial Intelligence, pp. 290–295, 1993.

Software Architecture for the Adaptation of Dialogs and Contents to Different Devices

Steffen Goebel, Sven Buchholz, Thomas Ziegert, and Alexander Schill

Department of Computer Science
Dresden University of Technology, Germany.
{goebel, sb15, ziegert, schill}@rn.inf.tu-dresden.de

Abstract. Multi device service provision is a challenging problem for information service providers. To support the diverse set of existing and future devices, services have to adapt the content to the capabilities of the different devices.

In this paper we present research efforts aiming at the development of system support for automated content adaptation to different devices. We introduce a software architecture for the adaptation process covering device identification and classification, session management, data input validation, dialog fragmentation, and transcoding. The adaptation software is realized by a Java based framework.

1 Introduction

The diversity of devices for mobile and stationary information access available today and expected for the future results in new challenges for web-based services. The heterogeneity in memory size, computing power, and network connectivity as well as different content description languages must be taken into account by future information services. Unfortunately, the efforts and expenses involved in the manual adaptation of services to different devices are unreasonably high. Therefore system support for automated content adaptation is required.

When designing a system for automated content adaptation the first design choice concerns the source format of the content. Basically, there are three different approaches:

- A device-specific markup language, such as HTML or WML, is used as the source format. To support different devices, this language has to be transformed into a suitable, possibly different device-specific markup language. Due to the lack of meta information about the semantic structure of the document, however, there are restrictions on the usability of this approach.
- A device-independent markup language is used to represent content. This language contains additional semantic meta information necessary for the adaptation to different devices. Multiple devices are supported by transcoding the device-independent markup language into a device-specific format.
- Contents and presentation rules are represented separately. Thus the adaptation to different devices is realized by means of different style sheets.

I. Chong (Ed.): ICOIN 2002, LNCS 2344, pp. 42–51, 2002.
© Springer-Verlag Berlin Heidelberg 2002

For the development of our system we have chosen a device-independent markup language because of the lack of semantic information in a device-specific language. The latter, style sheet based approach is more suitable for data centric applications, such lexicons or database front-ends, not for general web-based user interactions.

We have developed an XML-based, device-independent markup language, the Dialog Description Language (DDL). DDL contains additional meta information used for navigation, multi language applications, fragmentation, and validation of user inputs. It adopts several concepts from UIML [1], XForms [2], and XML Schema [3]. The details of DDL are not subject of this paper. A description of DDL is provided by [4].

For the development of the adaptation software we assume all clients to use the HTTP protocol. This is not a restriction because even accesses from WAP phones are transformed into HTTP by the WAP gateway of the service provider.

The remainder of the paper is organized as follows. In section 2 we describe the general system architecture and discuss aspects of essential functionality. The implementation details are covered in section 3. Related work is discussed in section 4. Finally we summarize our work and discuss future directions (section 6).

2 General Architecture

This paragraph describes the general architecture and the most important components of the system are briefly introduced.

The general architecture is depicted in figure 1: The clients use a web browser to display contents retrieved from HTTP servers using the HTTP protocol. The entire application logic resides at the server side. A Java servlet engine (e.g. Tomcat [5]) runs within the HTTP server. This servlet engine is the container for the adaptation software. By using the standardized Servlet-API [6], there are no direct dependencies between the adaptation software, the servlet engine, and the HTTP server. HTTP server and servlet engine are thereby interchangeable components.

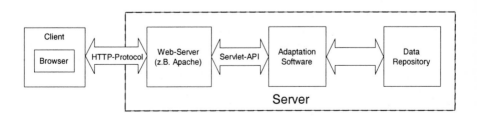

Fig. 1. System Architecture

2.1   Device Identification

A successful contents adaptation requires the properties of the device to be known. Thus mechanisms are required to transfer information about the device properties to the server.

Only limited possibilities to recognize device properties are available today: Every browser application transmits a "User-Agent" header field within the HTTP-request to the server. This string is meant for the identification of the web browser and allows conclusions about the device. Unfortunately, the syntax of the "User-Agent" header field is not standardized but is defined by the browser vendor. Thus the server needs information about all supported devices and browsers. The value of the "User-Agent" header field is used as a key to access these information. Unknown devices cannot be supported with this approach. More variable information about the device, such as the current network connectivity or the amount of memory installed on the particular device, cannot be gained as a matter of principle.

Hence, a more sophisticated mechanism of client recognition is required. A promising approach is described by the W3C's Composite Capability/Preference Profiles (CC/PP) [7]. It is a framework for client profiles describing capabilities of the devices and user preferences. CC/PP client profiles should be transmitted to the server by future browsers within the HTTP request [8]. Thus, the server does not need to store information about the supported devices anymore. New devices can be instantly handled. Furthermore, variable properties of a particular device can be transferred to the server allowing better adaptation. Unfortunately, there are no browsers available today supporting CC/PP.

In order to use CC/PP in our adaptation software prototype, we have developed an HTTP proxy inserting CC/PP profiles into an HTTP request header. By this means a CC/PP capable browser is emulated. We have defined a CC/PP vocabulary extending the UAProf vocabulary of the WAP Forum [9] by information about the input devices, the network speed, and the web browser.

2.2   Classification of Devices

To manage the complexity of the adaptation software, we allow for a device classification. Content for devices in the same class, having similar properties, is adapted in a similar manner. By this means an additional abstraction level is introduced reducing the complexity of the adaptation software.

We apply a three dimensional classification:

- the output data types supported by the browser: WML, LowHTML, HTML, Multimedia, VoiceXML,
- the interaction scheme: keypad, alphanumeric keyboard w/o pointing device, pointing device w/o keyboard (e.g. with character recognition), alphanumeric keyboard with pointing device, speech,
- the network connectivity:
  - text and low resolution monochrome images only ($\leq$ 9.6 Kbit/s)

- text and low resolution images ($\leq$ 64 Kbit/s)
- full text, images and limited multimedia ($\leq$ 1 Mbit/s)
- full text, images and multimedia ($>$ 1 Mbit/s)

All classification is subject to a trade-off between the complexity of the adaptation software and the quality of the device specific adaptation.

2.3 Transcoding

The transcoding component of the adaptation software is responsible for the conversion of the device-independent into a device-specific markup language, such as HTML or WML. The conversion is based on information gained from the client recognition and classification. Since our device independent markup language (DDL) is based on XML, XSLT style sheets are used for the transcoding process. XSLT-based transcoding has the advantage that changes in the transcoding rules do not require recompilation of the transcoding component. Therefore changes are easily possible even in the deployment phase.

XSLT transcoding lacks of performance due to the interpretation. Nevertheless, there are already approaches to address this problem by using compiled XSLT, for example the SUN XSLT-compiler [10]. The advantage of flexible changes during deployment time, however, gets lost.

2.4 User Input Validation

Our content description language DDL allows to specify validity constraints on user inputs. This information enables automatic input validation by the adaptation software.

If the validation process discovers invalid user inputs, the adaptation software reports this error and prompts the user for correction.

If the client browser supports a scripting language (e. g. JavaScript or WML-Script), parts of the validation process or even the entire one can be moved to the client. The scripting language code for the validation is generated automatically by the adaptation software. By client-side input validation unnecessary network interactions with the server are reduced. Nevertheless, client-side input validation is optional. There is server-side validation by the adaptation software if client-side validation is not supported by the particular client.

2.5 Dialog Fragmentation

A significant problem in supporting dialogs on heterogeneous devices is the variety in display and memory size. Whereas a browser on a desktop computer can easily display a complete dialog at once, the memory and display constraints of a PDA or WAP phone require splitting the dialog into smaller dialog fragments. This fragmentation must not be performed arbitrarily in order not to brake the logical structure of the dialog. Elements that make a logical atomic unit, such as an input element and its description, must not be separated. Therefore semantic information about the logical structure of the dialog must be contained within the content description language.

Automatic dialog fragmentation requires the adaptation software to provide means to navigate between the dialog fragments. Invalid data inputs of dialog fragments can be reported to the user either after completing a dialog fragment or after completing the complete dialog. Therefore, user input of different dialog fragments must be stored by the server until the completion of the whole dialog. This is done using the session management of the adaptation software (cf. section 2.6).

2.6   Session Management

A session management between the adaptation software and the clients is necessary for the adaptation process, mainly for dialog fragmentation and user input validation. Furthermore, sessions can be used to cache client profiles at the server. By this means multiple recognition within the lifetime of a session is avoided. For the personalization of applications a specific user context is assigned to a session after the user authentication. The Java servlet engine (e. g. Tomcat) comes with full session support realized by cookies or URL rewriting.

Those sessions are limited to a single device. However, it might be reasonable to start a session on one device and resume it on another one later on. This requires additional mechanisms because the session ID is not known at the second device. A possible solution is assigning the session ID to the user ID. After authentication, a user can choose to continue uncompleted sessions at the new device.

3   Implementation Details

We have implemented a prototype of our adaptation software architecture. It is based on Java servlets and makes use of the Xalan-XSLT-processor [11] as well as the Xerces-XML-parser [12] by the Apache-Group.

The main component is the so-called Transcoding Servlet (cf. fig. 2). All HTTP requests from the clients are forwarded to this servlet by the HTTP server and the servlet engine. The adaptation is performed via a chain of filters. The filters are successively invoked by the Transcoding Servlet.

A filter is implemented as a Java class with a simple interface (Transcoding-Filter) consisting of two methods: init() and start(). Additional filters can be implemented easily. We distinguish four different types of filters depending on their functionality:

- Request Modifiers – alter a HTTP request, e.g. include additional HTTP headers. Request Modifiers are processed at first.
- Generators – supply the requested content. The simplest way to do this is to read a file from the hard disk. A more sophisticated solution may retrieve the contents from a DBMS, another servlet, or an arbitrary URL. Basically

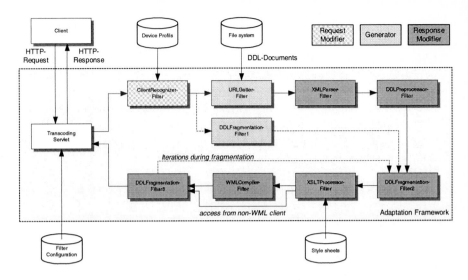

Fig. 2. Adaptation Framework

there is exactly one Generator invoked during the processing of a single
HTTP-request.

– Response Modifiers – adapt the retrieved content to the capabilities of a
particular device. They are linked into the filter chain behind a Generator.
– Monitors – are primarily used to observe or debug an application and do not
perform any modification. They may be linked at any position within the
filter chain.

Filters are distinguished semantically only. The interface syntax of all filters
is the same.

The sequence of filters in the request processing chain is determined by a
configuration file. It contains a mandatory sequence control attribute and an
optional test attribute for each filter. The sequence control attribute sets the
sequence of the filter invocation.

The test attribute, that contains an XPath expression about the client profile,
provides conditional invocation of a particular filter. For instance, only requests
from WML clients require the invocation of the WMLCompilerFilter (cf. fig. 2).

The transcoding process may include iterations, i.e. loops in the filter chain.
To allow for loops, a filter may optionally determine its successor. An example
of a loop is the iterative fragmentation of dialogs by DDLFragmentationFilter2
and DDLFragmentationFilter3 (cf. fig. 2).

Currently we have implemented a number of filters empowering the adaptation framework to support the transformation of a DDL dialog into a device
specific dialog representation. Additional functionality can be easily added by
implementing extra filters.

In the following we give a brief overview of the filter classes and their functionality. All filter classes inherit some common functionality from the abstract su-

perclass TranscodingFilterSupport and implement the common interface TranscodingFilter.

ClientRecognizerFilter (Request Modifier)
This filter is responsible for the client recognition and passes the device information as a client profile to the subsequent filters.

In the current implementation we use two different ways of device recognition: CC/PP profiles and the "User-Agent" header filed in the HTTP request. The recognition based on the "User Agent" header field is supported to be a fall-back solution in case the client does not provide a CC/PP profile.

With CC/PP-based recognition the adaptation process relies on a particular device class rather than on a specific device. Hence, new devices will be supported without any modification of the system configuration.

URLGetterFilter (Generator)
This filter retrieves a file from the local file system or from the internet based on the URL in the HTTP request.

ServletRunnerFilter (Generator)
This entity invokes an external servlet on the application server. By this means the system is able to generate dynamic contents in DDL.

XMLParserFilter (Response Modifier)
This filter converts an XML document (e.g. a DDL document) into an org.w3c.-dom.Document object (Document Object Model, DOM [13]). A DTD can be specified to validate the document. Subsequent filters work on the DOM instance of the document.

DDLPreprocessorFilter (Response Modifier)
This filter preprocesses a DDL document to resolve all external references and inheritance hierarchies. This results in a simplified DDL document. By this means the style sheet-based transformation is eased. Even the preprocessing may be style sheet-based. However, as this process is very time consuming, we decided to use a DOM-based transformation.

XSLTProcessorFilter (Response Modifier)
This filter transforms a preprocessed DDL document into a device specific format (HTML, WML). The transformation is based on XSLT style sheets. The style sheets have access to the information in the HTTP request and to the context of the processing environment. The information is presented to the style sheets as XSLT parameters.

There is a style sheet for each device (or device class). Currently we have implemented style sheets for HTML, tinyHTML and WML. The selection of the style sheet is based on the client profile generated by the ClientRecognizerFilter.

ImageTranscodingFilter (Response Modifier)
The image transcoder converts images, e.g. BMP into JPEG or WBMP, high resolution into low resolution, or full color into grayscale. Several format specific parameters can be specified, e.g. the "quality"-parameter for JPEG images or the "interlaced"-parameter for PNG images.

DDL Fragmentation Filters
These filters perform the dialog fragmentation in case the client's restrictions

forbid the particular dialog to be displayed as a whole. Besides the actual fragmentation they are responsible for the user input validation and they store input data until the final dialog part is completed. The filters must appear in the following order:

DDLFragmentationFilter1 (Generator)
The first fragmentation filter manages the caching of dialog fragments and the fragment by fragment delivery to the clients. Furthermore it stores input data until the dialog is completed. Therefore input data is transferred to the filter by the query component of the URL of the HTTP request for the next dialog fragment.

FragmentationFilter2 (Response Modifier)
This component fragments a dialog if it has exceed the resource restrictions of a device (cf. FragmentationFilter3). First, the document is fractionalized into the smallest and indivisible parts. Then these parts are combined to as few as possible fragments that still meet the resource constraints of the client. A more detailed description of the fragmentation algorithm can be found in [4].

Furthermore this component validatates user input.

FragmentationFilter3 (Response Modifier)
This component checks if the size of the rendered document exceeds the resource restrictions of a particular device. If this is true, the filter invokes FragmentationFilter2 again to trigger another fragmentation.

WMLCompilerFilter (Response Modifier)
WAP devices do not process a textual WML document but a compact binary representation (binary WML). Therefore a WAP gateway, an intermediary between the server and the WAP device, compiles the textual into the binary representation. However, this means the memory restrictions of the device do not apply to the size of the textual WML document but to the size of the compacted version.

To check if a WML document fits the resource restrictions of the client, a WMLCompilerFilter is interposed between the XSLTProcessorFilter and the FragmentationFilter3 to perform the conversion to binary WML.

4   Related Work

IBM WebSphere Transcoding Publisher [14] uses a similar chain of filters for the adaptation as in our architecture. However, the Transcoding Publisher does not support loops within the filter chain. It was developed mainly to transform existing HTML pages into other markup languages (e.g. WML, HDML, simplified HTML). The client identification is based on the User-Agent header field within a HTTP request. A network profile can be chosen based on the TCP port used by the client to access the server. CC/PP profiles are not supported. The Transcoding Publisher does not come with user input validation at the server side because HTML source documents do not contain necessary information.

Brazil Project of Sun Microsystems [15] uses a similar architecture based on filters.

5   Evaluation

In order to evaluate our architecture, we have measured the performance for processing a complex dialog. We used a Pentium III with 800 MHz and 256 MB running Windows 2000, JDK 1.3.1 and Tomcat 3.2. The total processing time is about 840 ms. Thereby the major share of the processing time is consumed by the XSLTProcessorFilter and the DDLPreprocessorFilter (cf. fig. 3).

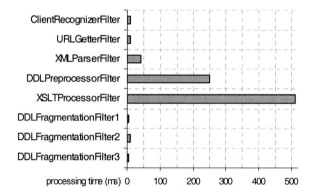

Fig. 3. Processing times of the adaptation filters

6   Summary

In this paper we have discussed aspects of a system support for multi device service provision. We have introduced a system architecture transcoding a device-independent description of dialogs (DDL) into a dialog in a device-specific markup language taking device features (e.g. display, memory) into account. A prototype implementation based on XSLT transcoding has proven the feasibility of the approach.

The evaluation of the prototype has shown that the XSLT-based transcoding performs rather poor. Therefore we are investigating possibilities for the optimization of the transcoding process. Caching of transcoded contents is one conceivable approach. Another one may be to use compiled XSLT style sheets. This may lead to a significant speed-up and preserves the advantages of XSLT transcoding, such as flexibility.

In the future we plan to integrate personalization into the adaptation framework, e.g. personal settings regarding the adaptation process. Integrating those mechanisms into the existing architecture should be easily done by introducing new filters.

References

1. User Interface Markup Language (UIML) v2.0 Draft Specification, Harmonia, Inc., 2000. (http://www.uiml.org/)
2. Dubinko, M. et al: XForms 1.0. W3C Working Draft, 2001. (http://www.w3.org/TR/xforms/)
3. Biron, P., Malhotra, A.: XML Schema Part 2: Datatypes. W3C Recommendation, 2001. (http://www.w3.org/TR/xmlschema-2/)
4. Goebel, S., Buchholz, S., Ziegert, T., Schill, A.: Device Independent Representation of Web-based Dialogs and Contents. Youth Forum in Computer Science and Engineering (YUFORIC'01), Valencia, Spain 2001.
5. The Jakarta Project, Apache Software Foundation. (http://jakarta.apache.org/tomcat/index.html)
6. Java Servlet Specification: Version 2.3, Sun Microsystems, Inc. (http://jcp.org/aboutJava/ communityprocess/first/jsr053/index.html)
7. Klyne, G. et al: Composite Capability/Preference Profiles (CC/PP): Structure and Vocabularies. W3C Working Draft, 2001. (http://www.w3.org/TR/CCPP-struct-vocab/)
8. Ohto, H., Hjelm, J.: CC/PP Exchange Protocol based on HTTP Extension Framework. W3C Note, 1999. (http://www.w3.org/1999/06/NOTE-CCPPexchange-19990624)
9. WAG UAProf, Proposed Version 30-May-2001, Wireless Application Protocol Forum Ltd. , 2001. (http://www1.wapforum.org/tech/document/SPEC-UAProf-19991110.pdf)
10. XSLT-Compiler, Sun Microsystems, Inc. (http://www.sun.com/xml/developers/xsltc/)
11. Xalan-Java version 2.2.D9, The Apache Software Foundation. (http://xml.apache.org/xalan-j/)
12. Xerces Java Parser, The Apache Software Foundation. (http://xml.apache.org(xerces-j/)
13. Le Hors, A.: Document Object Model (DOM) Level 2 Core Specification, W3C Recommendation, 2000. (http://www.w3.org/TR/DOM-Level-2-Core/)
14. WebSphere Transcoding Publisher, IBM Corp. (http://www-4.ibm.com/software/webservers/transcoding/)
15. Brazil Project: Home. Sun Microsystems, Inc. (http://www.sun.com/research/brazil/)

TCP Throughput Improvement for Semi-soft & Soft Handoff

Tae-Chul Hong, In-Ho Roh, and Young-Yong Kim[*]

Dep. of Electrical & Electronics Engineering, Yonsei University, Shinchon-Dong,
Seodaemoon-ku, Seoul 120-749, Korea
{taechori, ihroh, y2k}@win.yonsei.ac.kr
http://win.yonsei.ac.kr

Abstract. As wireless networks are evolving to IP based network, issues to support micro mobility becomes more and more important. One of the prominent solutions to support mobility management is Cellular IP proposed by Columbia University. TCP performance with hard handoff and semi-soft handoff was evaluated by authors of Cellular IP. However, it's not well known that how TCP performs with Cellular IP and soft handoff.
In this paper, we evaluate performance of TCP with various handoff schemes and find that TCP performs well with soft handoff, in general. However, it is observed that combining soft handoff and in-sequence transport of TCP may cause throughput to decrease substantially. We propose method to improve TCP performance with soft handoff by *delayed ACK* and *prediction of RTO*. Simulation results show effectiveness of our scheme.

1 Introduction

The advance in wireless transport technology accelerates data transmission rate significantly. One of the main problems in wireless network is a mobility management. Mobility management scope is divided into two areas, macro and micro mobility. Macro mobility covers campus area or more wide area, while micro mobility manages the cellular network mobility.

IETF(Internet Engineering Task Force) suggest Mobile IP[1] for mobility management. Mobile IP shows good performance for macro mobility management, but it has problems in micro mobility management, such as large latency and triangular routing. Therefore, new methods are proposed to solve micro mobility problems. HAWAII [2], Mobile IPv6 [3], and Cellular IP [4] are some of methods for micro mobility management. Micro mobility protocols have been proposed to handle local movement of mobile hosts without interaction with the Mobile-IP-enabled Internet. These protocols has the benefit of reducing delay and packet loss during handoff, and eliminating registration between mobile hosts and distant HAs when mobile hosts remain inside their local coverage areas.

---

[*] Corresponding Author

I. Chong (Ed.): ICOIN 2002, LNCS 2344, pp. 52–61, 2002.
© Springer-Verlag Berlin Heidelberg 2002

In this paper, we consider Cellular IP, which is one of prominent solutions for micro mobility management, and investigate way to adopt Cellular IP in third generation CDMA networks [5]. Especially, we compare CDMA soft handoff [6] to semi-soft handoff during TCP transmission and propose method for TCP throughput improvement during soft & semi-soft handoff. Using computer simulation, we compare soft handoff and semi-soft handoff. We also verify our proposal to improve TCP throughput over 3G cellular network.

The remaining of this paper is composed as follows. We provide an overview of Cellular IP and 3G CDMA network. After that, we provide comparison of semi-soft and soft handoff through simulation and discuss the proposed schemes to improve TCP performance. Finally, we present some concluding remarks.

2 Cellular IP

Figure 1 illustrates basic architecture of Cellular IP network. Mobile IP (MIP) in macro network and Cellular IP (CIP) in micro network are used.

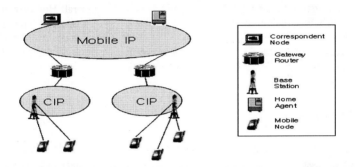

Fig. 1. Cellular IP Network Architecture

Base station and gateway router are defined by Cellular IP[4]. Base stations are built on a regular IP forwarding engine with the exception that IP routing is replaced by Cellular IP routing and location management and foreign agent of Mobile IP is replaced by gateway router. In other words, Cellular IP access network are connected to the Internet via gateway routers.

In this way, movements in side gateway router do not need to report home agent, reducing the registration latency. Therefore network management becomes very efficient.

2.1 Routing

Gateway router periodically broadcasts a beacon packet that is flooded in the access network. Base stations record the neighbor they last received this beacon from and use it to route packets toward the gateway.

In addition to that, each base station maintains a routing cache. When a data packet originated by a mobile host enters a base station, the local routing cache stores the IP address of the source mobile host and the neighbor from which the packet entered node. This soft-state mapping remains valid for a system-specific time called route-time-out. Data packets are used to maintain and refresh mappings but a mobile host may sometimes wish to maintain its routing cache mappings even though it is not regularly transmitting data packet. Therefore to keep its routing cache mappings valid, the mobile host transmits route-update-packets on the uplink at regular intervals called route-update time. These packets are special ICMP packets addressed to the gateway.

2.2 Paging

Typically, fixed hosts connected to the Internet remain online for extended periods of time, even though most of the time they do not communicate hence such hosts are excluded from routing cache. When a correspondent node transmits packets for a host, which is excluded from routing cache, gateway router broadcast in their domain to find a host. This action incurs large overhead. Therefore Cellular IP uses a paging technique. Base station does not only have a routing cache but also a paging cache. Paging cache mappings have a longer timeout period called paging-timeout hence a longer interval exists between consecutive paging-update packets. In order to remain reachable, mobile hosts transmit paging-update packets at regular interval defined by a paging update time. In addition to that, any packet sent by mobile hosts, including route-update packets, can update paging cache. However paging-update packets cannot update routing cache.

2.3 Handoff

Cellular IP supports two types of handoff scheme. Cellular IP hard handoff is based on a simple approach that trades off some packet loss for minimizing handoff signaling rather than trying to guarantee zero packet loss. Cellular IP semi-soft handoff exploits the notion that some mobile hosts can simultaneously receive packets from new and old base station during handoff. Semi-soft handoff minimizes packet loss, providing improved TCP and UDP performance over hard handoff.

Figure 2 illustrate semi-soft handoff. Semi-soft handoff is used for reducing handoff latency. Mobile host sends a semi-soft packet to the new base station and immediately returns to listening to the old base station before it hands off to a new access point. In addition to, that crossover base station has the delay device which needs to provide sufficient delay to compensate, with high probability, for the time difference between the two streams traveling on the old and new paths.

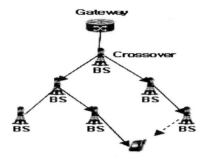

Fig. 2. Semi-Soft Handoff

3   Cellular IP in 3G CDMA Network

As shown in figure 3, it is not difficult to adopt the Cellular IP into 3G network[5]. Since IP BS(base station) in 3G network is not implemented perfectly, PDSN, PCF, BSC and BTS consist of hierarchical structure.

Table 1 shows the comparison of 3G network components and the Cellular IP components. We can deduce that the difference between each component is not large in table 1. In this paper, handoff is our focus. Since Cellular IP is designed without considering CDMA network, soft handoff is ruled out in original Cellular IP.

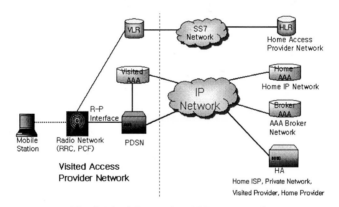

Fig. 3. Third Generation CDMA Network

Table 1. Comparison of Cellular IP and 3G CDMA Network

Cellular IP	3G CDMA Network
Gateway Router	PDSN
BS	BTS, BSC, PCF
Beacon Packet	Pilot Signal
Hard Handoff	Hard Handoff
Semi-Soft Handoff	Soft Handoff

Soft handoff [6] is more complicated, but its usage is very common in CDMA network. We compose the following simulation environment to evaluate soft handoff and semi-soft handoff.

4   Comparison of Semi-soft & Soft in CDMA Cellular IP

Figure 4 illustrates the simulation environment. In this section, we compare TCP performance with semi-soft and soft handoff by simulation.

The transmission speed of source is set at 700packets/sec and one packet size is fixed at 1000 bits. Registration Interval(RI) is defined between the moment transmission is begun from crossover node to both directions by giving previous notice at handoff and the moment physical handoff starts, and its value is set to 200ms. There is a moment of disconnection caused to in semi-soft handoff, but not in soft handoff. This time gap is defined as Connection Interval (CI) and evaluated to 150ms. Handoff rate is 10 times per minute.

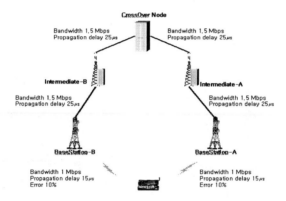

Fig. 4. Simulation Environment of comparison of semi-soft and soft handoff

Figures 5–8 show the simulation results when parameters involved in handoff changes. In figure 5, TCP throughput is measured with respect to handoff rate. Soft handoff performs better than semi-soft handoff when handoff is made frequently.

In Figure 6, the experiment is carried out by changing the RIs. Here, it is obvious that the longer the RI value becomes, the better semi-soft handoff performs. However, soft handoff performs poorer when RI increases. In semi-soft handoff, transmission from crossover is made in both directions during RI, however, packets are received from one base station over air interface. In soft-handoff, there is chance that from both directions or out of sequence packets are received because both sides can receive packets. In this situation, TCP respond by duplicate ACK, then, TCP throughput is reduced without packet loss or congestion, because source takes this as congestion. Therefore, soft-handoff needs method which can prevent Duplicate ACK to guarantee TCP performance. Delayed ACK is one of possible measures.

In figure 7, we can verify that soft handoff, which does not have connection terminated interval, shows better performance than semi-soft handoff.

In figure 8, error rate change incurs transmission delay change so the whole performance drops. So, in case of large error rate, performance of two kinds of handoff decreases. In this case, performance of soft is also better than semi-soft.

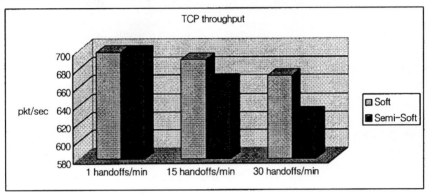

Fig. 5. TCP Throughput vs Handoff Rate

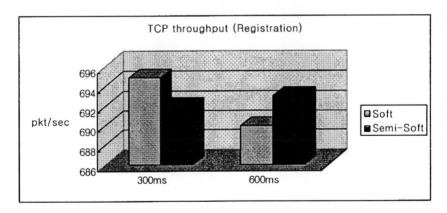

Fig. 6. TCP Throughput vs RI

5 Proposed Schemes to Improve TCP Performance

TCP throughput degradation can be caused by out of sequence or packet loss.

In [7], authors tried to improve TCP throughput with delay device. Delay device is a simple way to reduce the influence of out of sequence packet and packet loss. The delay on transmission route can be compensated with 8 buffer delay on the crossover node.

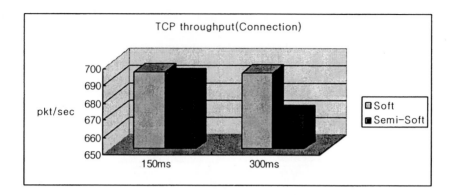

Fig. 7. TCP Throughput vs CI

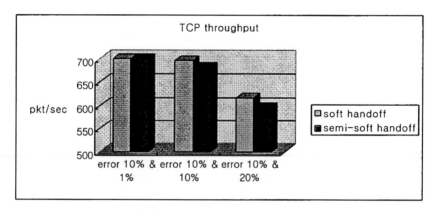

Fig. 8. TCP Throughput vs Link Error Rate

In this paper, we suggest the way to improve the throughput, using RTO (Retransmission Time Out), which can influence TCP throughput more significantly than delay device. If there exists significant RTO variance between two paths after handoff, it will be helpful to give a proper RTO value to new route.

Figure 9 shows the simulation environment to evaluate effectiveness of RTO for the TCP performance. This topology is different from previous one. We set different characteristics between intermediate node and base station for two links. Asymmetric link environment is more hostile for throughput than symmetric link environment. So we can easily validate throughput improving method.

Figure 10 to 12 show the TCP performance vs number of handoffs and CI. We test three cases. The first case is predicted delay method, which uses adaptive delay value according to delay device of cellular IP and link characteristics. Second is predicted RTO method, which uses proper RTO for the new path. The last one is hybrid method which uses predicted delay and predicted RTO together.

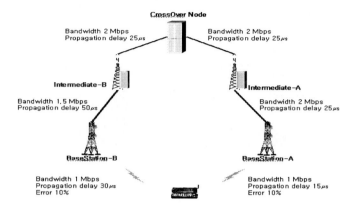

Fig. 9. Simulation Environment for Proposed Schemes

Figure 10 and 11 show the result for semi-soft handoff. As one can see, hybrid method gives the best performance. This result shows that RTO is one of the most significant factors. Therefore we can get better TCP throughput by adjusting RTO factor of new path.

Figure 12 is the result of soft handoff. As one can see, result of soft handoff is different from semi-soft handoff case. Predicted RTO is not the best throughput improving method. The essential reason of TCP throughput degradation in soft handoff is fast retransmit. It is caused by out of sequence packet and receiving two identical packets from two paths. These factors bring about fast retransmit frequently so TCP throughput drop. This problem will be solved by delayed ACK as it is described in previous section. In this simulation, predicted delay compensates difference of link delay so out of sequence packet is decreased. Therefore fast retransmit less occurs. After all, predicted delay has the best performance.

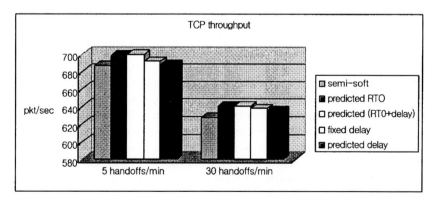

Fig. 10. TCP Throughput vs Handoff Rate

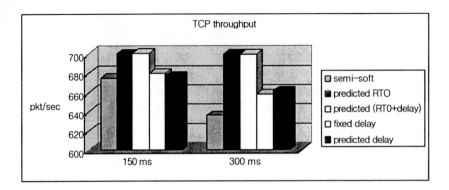

Fig. 11. TCP Throughput vs CI

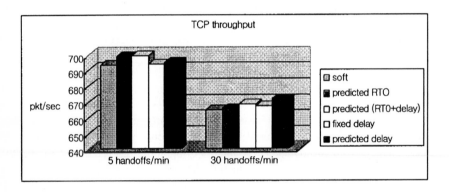

Fig. 12. TCP Throughput vs Handoff Rate

6 Conclusion

As data transmission is indispensable in 3G communication, we have investigated the handoff issues in CDMA Cellular IP. We verified that drop of TCP throughput is caused by miss prediction of RTO and out of sequence packet. We propose simple but effective methods to improve TCP throughput in Cellular IP, namely, predicted RTO and delayed ACK. Simulation shows our schemes give good performance in case of soft handoff.

The future research includes practical estimation of RTO in new path as well as detailed implementation of delayed ACK.

References

1. Charles E. Perkins, Bobby Woolf, Sherman R. Alpert: Mobile IP Design Principles and Practices, Prentice Hall (1998)
2. R. Ramjee, T. La Porta, S. Thuel, K. Varadhan, S.Y.Wang: HAWAII: A Domain-based Approachfor Supporting Mobility in Wide-area Wireless Networks,1999 IEEE
3. Claude Castelluccia: A Hierarchical Mobile IPv6 Proposal, Rapport technique n°0226-November 1998
4. Andrew T. Cambell, Javier Gomez, Sanghyo Kim, Andras G. Valko and Chieh-Yih Wan and Zoltan R. Turanyi: Design, Implementation, and Evaluation of Cellular IP, IEEE Personal Communication, August 2000
5. 3GPP2: 3G TS 23.922 version 1.0.0, Architecture for All IP network
6. Daniel Wong, Teng Joon Lim: Soft Handoffs in CDMA Mobile Systems , IEEE Personal Communications, December 1997
7. A. T. Cambell, S. Kim, J. Gomez, C-Y. Wan, Z. Turanyi, A. Valko: Cellular IP Performance, INTERNET-DRAFT <draft-gomez-celluarip-perf-00.txt>

Seamless QoS Handling Mechanism for Macro and Micro Mobility[1]

Sangheon Pack and Yanghee Choi

School of Computer Science and Engineering, Seoul National University, Seoul, Korea
shpack@mmlab.snu.ac.kr and yhchoi@snu.ac.kr

Abstract. One of the most challenging issues facing designers of wireless and mobile networks is the proper provisioning of Quality of Service (QoS), and is probably the most difficult problem in terms of meeting users' QoS requirement at various mobility levels. In this paper, we propose a seamless QoS handling mechanism for diverse mobility situations. This architecture consists of a Differentiated Service (DiffServ) model and an Integrated Service (IntServ) model. We propose the class upgrade scheme in packet forwarding for smooth QoS handling of macro mobility and propose a multiple path reservation scheme based on the Resource Reservation Protocol (RSVP) for seamless QoS handling of micro mobility. The proposed mechanism can meet constant QoS requirements after mobility. Furthermore, it is so flexible and scalable that it can be utilized as the QoS model in the next wireless and mobile networks.

1 Introduction

To guarantee Quality of Service (QoS) is the most critical problem in the current Internet. To address the QoS problem, several models have been proposed by the IETF, such as the Integrated Service (IntServ) model based on RSVP (Resource Reservation Protocol) and the Differentiated Service (DiffServ) model. However, these models are designed for wired networks and they are unsuitable in some respects for wireless and mobile networks. Unlike wired networks, wireless networks have several different characteristics, namely, a high loss rate, bandwidth fluctuations, and mobility. Therefore, the QoS mechanism in the wireless and mobile networks should consider these characteristics.

Recently, there has been much research on the provision of QoS in wireless and mobile networks. These studies have focused on several features of wireless networks such as low bandwidth, high loss rate, and the constraints of terminals [1]. However, due to the increased number of portable devices, the QoS guarantee mechanism must consider the mobility of terminals as well as wireless network characteristics.

In this paper, we propose a seamless QoS handling mechanism under various mobility situations. We assume a QoS architecture combined DiffServ model with

---

[1] This work was supported in part by the Brain Korea 21 project of the Ministry of Education, in part by the National Research Laboratory project of the Ministry of Science and Technology, 2001, Korea.

I. Chong (Ed.): ICOIN 2002, LNCS 2344, pp. 62–73, 2002.
© Springer-Verlag Berlin Heidelberg 2002

IntServ model. The former handles the QoS guarantee in a wired core network and the latter provides the QoS guarantee in a wireless access network. In case of macro mobility, it is possible to support smooth handoff by the packet-forwarding scheme using a class upgrade mechanism. In terms of micro mobility, we use the multiple path reservation scheme for seamless QoS handling.

The rest of the paper is structured as follows. Section 2 describes IntServ/RSVP and DiffServ models in brief and introduces related works. In Section 3 we propose the overall architecture based on these models. Section 4 describes the QoS handling mechanism in terms of macro and micro mobilities. Section 5 evaluates the performance of the proposed mechanism. Finally, Section 6 concludes our approaches.

2   Background and Related Work

In this section we introduce the IntServ model based on the RSVP and DiffServ models, and summarize some related work upon the QoS mechanism in wireless/mobile networks utilizing these models.

2.1   IntServ and DiffServ Model

In the IntServ model, a path between sender and receiver is reserved before establishing a session. In the path setup phase, a method of informing the application's requirements of the network elements along the path and a method of conveying QoS-related information between network elements and the application is needed. This process is achieved by using a Resource Reservation Protocol (RSVP) [2]. RSVP is a signaling protocol to carry the QoS parameters from the sender to the receiver to make resource reservations.

The protocol works as follows: (i) The sender of an application sends PATH message containing the traffic specifications to the receiver(s) of the application. (ii) The receiver on receiving this PATH message sends on RESV message to the sender specifying the flow it wants to receive. (iii) As the RSVP message flows back to the sender, reservations are made at every intermediate node along the required path. If any node along the path cannot support the request, that request is blocked. (iv) Path and reservation state are maintained at every router along the way for every session. To refresh the path and reservation states, PATH and RESV are sent periodically.

The IntServ model based on RSVP can provide three types of services to users: (i) Best effort service is characterized by the absence of a QoS specification and the network delivers at the best possible quality, (ii) Guaranteed service provides users with an assured amount of bandwidth, firm end-to-end delay bounds, and no queuing loss for flows, and (iii) Controlled load service assures that the reserved flow will reach its destination with a minimum of interference from the best-effort traffic [3].

One drawback of IntServ using RSVP is that the amount of state information increases with the number of flows. Therefore it is considered as a non-scalable solution for the Internet core network. On the other hand, DiffServ maps multiple flows into a few service classes. The 8-bit TOS (Type of Service) field in the IP header supports packet classification. The TOS byte is divided into a 6 bit Differentiated Services Code Point (DSCP) field and a 2-bit unused field [4]. DiffServ is realized by mapping the DSCP contained in the IP packet header to a particular treatment, also described as per-hop behavior (PHB). DiffServ defines various PHBs. For example, Assured Forwarding (AF) Service gives the customer the assurance of a minimum throughput, even during periods of congestion. DiffServ does not have any end-to-end signaling mechanism and works based on a service level agreement between the provider and the user. All packets from a user are marked in a border router to specify the service level agreement and are treated accordingly.

2.2 Related Works

[5] proposed a QoS supporting mechanism in mobile/wireless IP networks using a DiffServ model. [5] assumed the hierarchical FA (Foreign Agent) structures and the fast handoff [6] for mobility management. Based on these assumptions, [5] describes the Service Level Agreement (SLA) procedures in inter-FA handoff and inter-domain handoff. However, DiffServ only provide PHBs for aggregated flows so that it is impossible to meet per-flow QoS requirements, and reactive mobility managements like fast handoff are unsuitable to guarantee QoS for mobile hosts that move between small-size cells frequently.

To support the per-flow QoS requirement in a micro mobility environment, [7] proposed the extended RSVP supporting mobility. [7] proposed a path reservation scheme using a multicast tree. This scheme makes resource reservation in advance at the locations where it may visit during the lifetime of the connections. But, since [7] does not assume a specific mobility management scheme, the extended RSVP handles not only resource reservation but also mobility management. Therefore the RSVP proposed in [7] requires excessive overhead for the implementation and adaptation of the protocol.

3 Architecture

In this section we describe the overall architecture for the QoS guarantee in a wireless/mobile network, and explain the initial QoS negotiation procedure in this architecture.

3.1 Overall Architecture

Figure 1 shows the overall QoS provisioning architecture. The architecture is composed of two parts: a wired core network and wireless access network.

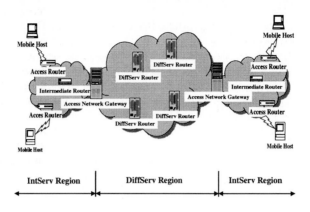

Fig. 1. Overall QoS Architecture

The core network uses the DiffServ model for QoS provisioning, which is composed of Access Network Gateways (ANGs) and DiffServ Routers (DRs). ANG plays an interface role between the IntServ region and the DiffServ region, and should classify incoming packets and mark DSCP field in the IP header. In addition, the ANG is operated as either a Foreign Agent (FA) or a Home Agent (HA) for mobility management. DR is a general DiffServ router that forwards the received packet according to corresponding PHBs.

On the other hand, the access network uses the IntServ/RSVP model. Routers in an access network are organized in a hierarchical structure, similar to Cellular IP [8] and HAWAII [9]. ANG is a root node in a wireless access network and administers resource management and admission control for all Mobile Hosts (MH). The Intermediate Router (IR) is a general router that reserves according to the QoS specification, as described in the RSVP message. Access Router (AR) is a router that acts as a Base Station (BS) for MHs. For seamless handoff, we assume the host specific routing scheme [10] is used within an access network. In host specific routing, an MH keeps its IP address while it moves between cells included in the same access network, and IP packets are routed by an entire IP address not by a network prefix. Since the host specific routing has the problem of non-scalability, it can be used only in limited regions, such as in the single access network domain.

By combining two models it might be possible to build a scalable network that would provide predictable end-to-end QoS services.

3.2 Initial QoS Negotiation Procedure

The initial signaling process to obtain end-to-end quantitative QoS starts when an MH generates an RSVP PATH message. The generated PATH message is forwarded to ANG along the AR and IRs. Then ANG sends the received PATH message towards the DiffServ region. To forward the PATH message through the DiffServ region, ANG must map the message to a DiffServ service class. The service mapping is possible by either defining a new service class [14] or utilizing existing classes. After the service mapping procedure, the PATH message is tunneled based on PHB. When the PATH message gets to the ANG in the destination access network, the message is processed according to the standard RSVP processing rules. When the PATH message reaches the destination MH, the MH generates an RSVP RESV message, and the RESV message is routed to the source MH along the reverse path. Figure 2 shows the initial QoS negotiation procedure.

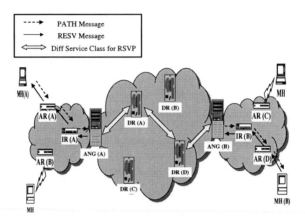

Fig. 2. Initial QoS Negotiation Procedure

Of course, the request may be rejected at any node in the IntServ region according to IntServ admission control. Also, ANG triggers the DiffServ admission control when it receives the RESV message. The ANG checks whether the resources requested in the RESV message exceed the resources defined in SLA or not. If the request fits the negotiated SLA, it is granted, if not, the RESV message is not forwarded and an appropriate RSVP error message is sent back to the receiver.

4   QoS Handling Mechanism for Mobility Management

For more efficient mobility management, [15] proposed the hierarchical mobility management. According to this proposal, the mobility management is divided into macro and micro categories. In this section, we propose the seamless and smooth QoS handling mechanism in each case.

4.1   QoS Handling Mechanism for Macro Mobility Management

Macro mobility means that an MH moves from an access network to another access network. In such a case, it is necessary to inform the home domain of the movement because of security, billing, and other considerations. Macro mobility is generally handled by the Mobile IP [12]. When an MH moves to a new access domain, it sends a registration message to a new foreign ANG in that domain.

In terms of the QoS guarantee, since an MH's IP address is changed after macro mobility, a new end-to-end QoS reservation phase is needed. The moved MH reserves a path within an access network by sending a new PATH message to the current ANG and performs an SLA procedure with the ANG for packet tunneling in the DiffServ region. In addition, the SLA between the previous ANG and the current ANG is required for packet tunneling in Mobile IP. Except that two end-point MHs move together, this QoS handling mechanism is done within the new access network. Figure 3 shows the QoS re-negotiation procedure for macro mobility management.

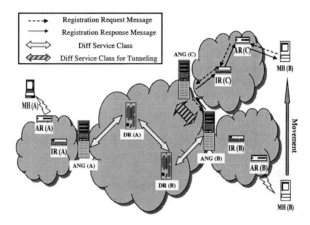

Fig. 3. QoS handling mechanism in macro mobility without route optimization

In addition to QoS re-negotiation, a packet forwarding scheme is needed for smooth handoff [11]. To minimize packet losses, the previous ANG must buffer packets during handoff and forward the buffered packets to the current ANG. Since the packet forwarding is based on the tunneling scheme in the Mobile IP, an additional IP header with same DSCP field as in the original IP header is added ahead of the existing IP packet. However, in many cases, especially in real-time multimedia applications, on time delivery of the forwarded packets is important. Therefore, forwarded packets should have a higher priority than the incoming packets with the same DSCP field value. To address this problem, we propose the class upgrade scheme. In this scheme, data packets are tunneled as packets with higher priority DSCP field values not the same DSCP field values.

Figure 4 shows the concept of the class upgrade scheme. Not to disturb packets in upper class level, the class upgrade is possible only within the same class level. And, the class update scheme is used until all buffered packets have been forwarded to the current ANG.

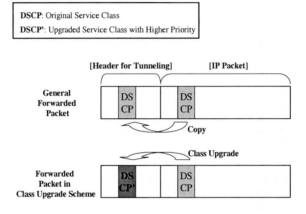

Fig. 4. Class Upgrade Scheme

If route optimization [16] by a binding procedure is supported in Mobile IP, the tunneling between the previous ANG and the current ANG is not required. In this case, the additional SLA between the DiffServ region, belonging the current ANG, and the DiffServ region belonging to the corresponding ANG is performed with a binding update procedure.

4.2   QoS Handling Mechanism for Micro Mobility Management

Unlike macro mobility, micro mobility means that a terminal moves from one cell to another adjacent cell while retaining its IP connectivity. Since a cell size will be smaller and smaller in the next mobile network, micro mobility occurs more frequently. Therefore, QoS handling for micro mobility management is a very important issue [17]. QoS negotiation in micro mobility is performed within an access network by using RSVP.

However, the standard RSVP does not consider mobility. So, we present a modified RSVP mechanism that is more suited to the mobile environment. In the modified RSVP mechanism, one PATH message reserves multiple paths. All paths are reserved only to access routers due to scarce resources in the wireless network. To minimize wasted wired network resources, we utilize two reservation styles in RSVP. Reservation in RSVP can be categorized as distinct and shared types [3]. Distinct reservation is appropriate for those applications in which multiple data sources are likely to transmit simultaneously. It requires separate admission control and queue management on the routers along its path to the receiver. On the other hand, shared reservation is appropriate for those applications in which multiple data sources are unlikely to transmit simultaneously.

In the proposed scheme, when a PATH message arrives at ANG, the ANG reserves one distinct path to the AR in the current location of the destination MH and multiple shared paths to the ARs of adjacent cells. Multiple shared paths are reserved by multicasting at the ANG.

Figure 5 presents the multiple path reservation mechanism. ANG(A) receiving a PATH message sends the PATH messages to several ARs. First, it unicasts a PATH message to the AR(B), which is an access router that attaches the destination MH. Then the MH sends an RESV message to the ANG(A) and the RESV message reserves a distinct path from the destination MH to ANG(A). In addition to unicasting, ANG(A) undertakes multicasting to adjacent ARs such as AR(A) and AR(C). AR(A) and AR(C) send the RESV message for shared path reservations.

When a MH enters into a neighbor cell, the reservation update procedure is performed as described in Figure 5. The MH sends the notification message to a new access router, AR(C), which relays the notification message to ANG(A), and ANG(A) then updates the routing path and multicasts the changed reservation styles. Using this procedure, the path styles to the AR(C) and AR(B) are changed into a distinct style and a shared style respectively. Since the AR(A) is not an adjacent access router with respect to the current access router, AR(B), the previously reserved resources are released. Also, a path to the new adjacent access router, AR(D), is reserved in shared style. In our mechanism, multiple paths are reserved in not only the specified cell but also adjacent cells in advance so that it may provide the seamless QoS handling for micro mobility management.

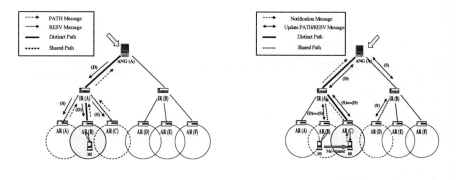

(a) Initial multiple path reservation   (b) Update procedure after movement

Fig. 5. Multiple path reservation scheme

5 Performance Evaluation

In this section, we evaluate the proposed mechanisms. First, we simulate the class upgrade scheme for macro mobility using relatively long handoff latency. For an accurate simulation, we use a DiffServ patch in NS-2 (Network Simulator). Second, we analyze the performance of the multiple path reservation scheme with respect to, latency time and resource usages.

5.1 Macro Mobility Management

To demonstrate the influence of the class upgrade scheme in smooth handoff, we measure packet loss rates for packet forwarding between the previous ANG and the current ANG.

DiffServ model defines two types of PHB; Assured Forwarding (AF) and Expedited Forwarding (EF). AF PHB guarantees low delay and has allocated bandwidth according to SLA. AF is based on the RIO (RED with In and Out) scheme [18]. In the RIO scheme, packets are classified as in-profile or out-profile according to whether the traffic exceeds the bit rate specified by the SLA. During congestion the packets tagged as out-profile will be dropped first. On the other hand, EF PHB guarantees minimum delay, low latency, and peak-limited bandwidth. The EF traffic should have rates of independency of the intensity of other traffic attempting to transit the node. Therefore, it is handled with higher priority than AF. EF uses a priority queuing scheme with a token bucket. The token bucket is used to limit the total amount of EF traffic so that other traffic will not be starved by bursts of EF traffic.

In simulation, one DiffServ node has three independent queues for each PHBs; a RIO queue for AF PHB, a simple Drop-tail queue with a token bucket for EF, and a RED queue for best effort service. We controlled the conditioner located in the incoming node to adjust the number of forwarded packets due to mobility. The packet loss was then measured by changing a drop priority in one AF service class. Since EF packets are handled independently of the other packets, we only use AF packets in simulation. Figure 6 shows the simulation results.

In Figure 6 AF11 and AF12 belong to the same service class but have different drop priorities. AF12 has a higher drop priority than AF11. In simulation results, we know that the overall packet drop ratio is decreased by giving forwarded packets a lower drop priority, which means that the class upgrade scheme can meet the requirements of time-critical multimedia applications more efficiently.

5.2 Micro Mobility Management

In the case of micro mobility, we reserve multiple paths for seamless QoS handling. In addition, we utilize two reservation types to minimize resource usages. To evaluate

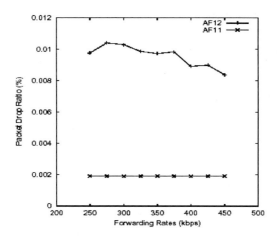

Fig. 6. Simulation Result: Macro Mobility Case

this scheme, we analyze the latency time of mobility quantitatively. For this analysis, we assume the N-dimension cell model and the K paths reservation scheme.

The total latency time of mobility is composed of the location setup time and the path setup time. Since all packets during the latency time are delivered to the previous AR or to the current AR without QoS limitations, the minimization of latency time becomes the most important problem in QoS handling in the wireless/mobile network.

Equation (1) and (2) show the latency time in the general RSVP scheme and in the proposed RSVP scheme respectively.

$$T_{LATENCY} = T_{UPDATE} + T_{PATH\_SETUP} \tag{1}$$

$$T_{LATENCY'} = T_{UPDATE} + T_{PATH\_UPDATE} \tag{2}$$

In Equation (1) and (2), T_{UPDATE} denotes the time of the location update using micro mobility management. For the general RSVP, a new path should be reserved after mobility, and therefore, a new path setup time is needed. Since the new path setup is done between two end hosts, T_{PATH_SETUP} is proportional to the end-to-end round trip time in the overall network. On the other hand, in the proposed scheme, no new path is reserved and only path types are updated after mobility. All procedures for path updates are done within an access network so that T_{PATH_UPDATE} is proportional to the round trip time in an access network. In other words, the latency time in the proposed scheme is much less than that of the general RSVP. Therefore, packet losses and miss-deliveries due to extended latency time can be efficiently eliminated.

It might be expected that the reservation of multiple paths requires more network resources. However, in our scheme, since the shared reservation is used for adjacent paths, additional resource usages can be minimized. Specifically, resources in only non-overlapped cells and not in all adjacent cells are reserved. Equation (3) shows the resource usages when all adjacent paths are reserved as distinct types, and equation (4) shows the resource usages in the proposed scheme, which utilizes distinct and shared reservation types.

$$L \cdot (Rd + K \cdot Rs) \tag{3}$$

$$L \cdot Rd + K' \cdot Rs \tag{4}$$

L: Number of flows
K': Number of non-overlapped cells to be reserved
Rd: Resources for distinct reservation
Rs: Resources for shared reservation

Although the proposed scheme requires more resource than the general RSVP, it can provide a seamless QoS handling mobility mechanism. Furthermore, since the number of reserved adjacent cells can be adjusted according to mobility patterns and packet priorities, unnecessary resource usages can be diminished.

6 Conclusion

In this paper, we propose seamless QoS handling mechanisms in diverse mobility situations. We assume a combined architecture consisting of a DiffServ model and an IntServ/RSVP model. This design can provide a scalable per-flow QoS provisioning service in a wireless/mobile network.

For macro mobility, a new service level agreement and packet tunneling between ANGs are required, and accordingly, we propose a class update scheme to meet the QoS specification of time-critical multimedia applications. For micro mobility, a multiple path reservation scheme is used for seamless mobility management, which reserves multiple paths to keep QoS specifications and connections of a moving MH. Because we use distinct and shared reservation types, resource wastage is minimized.

The most important entities in our architecture are the ANGs. For macro mobility, they play the role of the foreign agent or home agent and negotiate SLA with incoming MH for packet tunneling in the DiffServ core network. In addition, they modify the service class of forwarded packets to meet delay constraints. Besides, for micro mobility, ANGs are a root node in a hierarchical access network so that they perform packet routing for seamless handoff and multicast for multiple path reservations. Since all nodes, except ANG, keep the standard DiffServ and IntServ/RSVP mechanisms, the proposed mechanism is so flexible and scalable. The performance evaluation results show that the proposed mechanisms are better than the existing mechanisms in the wireless/mobile QoS environment, and that they could be applied as a QoS model in the next generation of wireless and mobile networks.

References

1. I. Mahadevan and K. Sivalingam, "Quality of Service Architecture for Wireless Networks: IntServ and DiffServ Models," I-SPAN, 1999.
2. R. Braden et al., "Resource ReServation Protocol (RSVP) – Version 1 Functional Specification," IETF RFC 2205, Sep. 1997.
3. R. Braden et al., "Integrated Services in the Internet Architecture: an Overview," IETF RFC 1663, June 1994.
4. S. Blake et al., "An Architecture for Differentiated Services," IETF RFC 2475, Dec. 1998.
5. S.U. Yoon et al., "QoS Support in Mobile/Wireless IP Networks using Differentiated Services and Fast Handoff Method," WCNC 2000.
6. K.E. Malki et al., "Hierarchical mobile IPv4/v6 and fast handoffs," Internet draft, draft-elmalki-soliman-hmipv4v600.txt, March. 2000
7. W.T. Chen et al., "RSVP Mobility Support: A Signaling Protocol for Integrated Services Internet with Mobile Hosts," INFOCOM 2000.
8. A. Valk, "Cellular IP: A New Approach to Internet Host Mobility," ACM Computer Communication Review, January 1999.
9. R. Ramjee et al., "IP-Based Access Network Infrastructure for Next-Generation Wireless Data Networks," IEEE Personal Communications, Aug. 2000.
10. A. Oneill et al., "Host Specific Routing," Internet draft, Nov. 2000.
11. C.E. Perkins et al.," Optimized Smooth Handoffs in Mobile IP," Computer and Communications, 1999.
12. J. Solomon, "Mobile IP: The Internet Unplugged," Prentice Hall, Sep. 1997.
13. X. Xiao et al., "Internet QoS: A Big Picture," IEEE Network, Mar. 1999.
14. G. Mamais et al., "Efficient Buffer Management and Scheduling in a Combined IntServ and DiffServ Architecture: A Performance Study," ICATM, 1999.
15. Caceres et al., "Fast and Scalable Handoffs for Wireless Internetworks," MOBICOM, 1996.
16. D. Johnson et al., "Route Optimization in Mobile IP," Internet Draft, Nov. 1998.
17. A. Cambell et al., "IP Micro-Mobility Protocols," Mobile Computing and Communication Review, Oct. 2000.
18. D. Clark et al., "An Approach to Service Allocation in the Internet," Internet draft, draft-clark-different-svc-alloc-00.txt, July 1999.

Mobility Management with Low Latency Time by Using the GSMP on the MPLS Network

Seong Gon Choi[1], Hyun Joo Kang[2], and Jun Kyun Choi[1]

[1] Information and Communications University (ICU),
P.O.Box 77. Yusong, Taejon, Korea, zip-code: 305-600
{sgchoi, jkchoi}@icu.ac.kr
http://www.icu.ac.kr
[2] Electronics and Telecommunications Research Institute (ETRI)
161 Gajong-Dong, Yusong-Gu, Daejon, Republic of Korea 306-600
hjkang@etri.re.kr

Abstract. In this paper, we aim to reduce handover latency time. For this purpose, the proposed method takes advantage of current General Switch Management Protocol (GSMP) over Multi-Protocol Label Switching network, and presents a path change procedure. For removing triangular problem, the proposed method substitutes a controller in GSMP for the role of Home Agent in Mobile IP. Consequently, we can reduce the handover latency time because a controller in using GSMP knows the overall network configuration status without rerouting and long reporting time.

1 Introduction

Due to the increased popularity of portable computer and wireless communications devices, mobile networking applications have become more and more popular. In particular mobile communications of next generation unavoidably require mobility with real-time services. In addition, the demand to provide Internet access for users is growing rapidly as they move locations.

For the efficient mobility, several requirements, such as lossless data transmission, simple signaling between the MH (Mobile Host) and its related peer, short latency time, efficiently handling handoffs, and clear location management, are required. In particular for real-time multimedia (audio, video, and text) communications, low latency time has to be reduced according to the handover.

As of today, The IETF (Internet Engineering Task Force) currently standardized MIP (Mobile IP) [1] and SIP (Session Initiation Protocol) [2] and several ideas have been proposed recently to support seamless network access to mobile users [3]–[5]. Although there have been these investigations, above requirements, especially low latency for real-time service, are not satisfied. For example, the MIP solution meets the goal of operational transparency and handover support but it is not optimized for

I. Chong (Ed.): ICOIN 2002, LNCS 2344, pp. 74–80, 2002.
© Springer-Verlag Berlin Heidelberg 2002

the latency time because it must use the HA (Home Agent) for update about handover and the process is not suitable for supporting real-time applications on the Internet.

Recently, Service providers are searching for the way they can satisfy the QoS (Quality of Service) required by customers in IP network. One of the solutions is to adopt the MPLS in IP network [6], [7]. The MPLS first uses routing protocol of layer 3 for setting up a path. And then the packet is actually transferred at the layer 2 with the switching action called label swapping.

GSMP (General Switch Management Protocol) protocol has a role of connection management of the general switching system [8]. It has also reservation mechanism for QoS based management, event mechanism for reliability and state and statistics method for the reporting function. Especially GSMP is implemented the master and slave type, which can achieved open programmable device

In this paper, we propose a method that manages the mobility. If the MH moves from a cell to another one, then the controller of GSMP can receive any event message by movement of MH from its slave which is a router including mobility function connected to the access point of MH. And then the controller can switch from the old port to a new port for connecting moving MH.

In addition, we draw the inter-domain GSMP mechanism for exchanging the information of its own controlled devices. If the handover between two domains occurs, this mechanism will be used.

The advantage of this proposed method is that the controller of GSMP in the MPLS (Multi-Protocol Label Switching) network can see overall network and have all of the configuration information. So, when link modification by handover occurs, our proposed method can have a benefit of handover latency time in MPLS network.

The paper is organized as follows. Chapter 2 describes the architecture of the hierarchical open signaling MPLS network for mobility management. Chapter 3 presents the LSP setup and the mobility management procedure, Chapter 4 presents the numerical result, and Chapter 5 has conclusion.

2   GSMP Based Open Architecture on the MPLS Network for Supporting Mobility

We approach to the mobility management in terms of open programmable interface mechanism on the MPLS network with hierarchical characteristics. For this, the mobility is based on the typical Mobile IP [1]. That is, the LSR I (Label Switched Routers I) and LSR J in Figure 1 may be the HA (Home Agent) or the FA (Foreign Agent). In this paper, we look into the only handover process based on the MIP.

As shown in Figure 1, GSMP protocol is selected for the VSC (Virtual Switch Control) interface, which is used for communicating remote control server and devices. Using the GSMP protocol's switch partitioning feature, we suggest one controller and multi-switch structure. It seems like domain based centralized network management.

In this paper, we mainly deal with handover shown in the Figure 1, where there are two LSP paths that are different FEC and different destination by movement of MH.

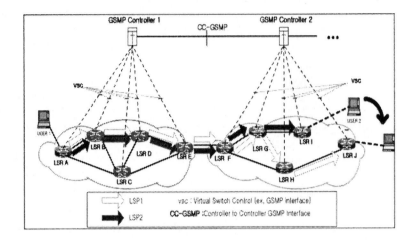

Fig. 1. An architecture using GSMP on MPLS Network

In other to have hierarchical architecture, we should define the CC-GSMP (Controller to Controller GSMP) interface. The different domain reports events through this interface. The events can be connection setup, connection release, state and statistics, handover occurrence and etc. The CC-GSMP interface can be used for communication between two GSMP controllers. For example, a MH can move between two LSRs, such as between LSR E and any LSR X, not LSR F, in other Domain X. In this case the Controller 2 can't know any information of Domain X. So, the event about the movement should be reported to the Controller 1 and Controller 2 from any Controller X managing domain X. At this time, the CC-GSMP interface can be used.

Figure 2 shows the GSMP based open signaling structure. A controller and slaves of GSMP accomplish an open signaling. These two devices are separated and the controller module is a sort of management server. So their sending and receiving information are messages type operation. For example, When link information is changed by handover, GSMP slave send the event messages to its own controller.

GSMP slave is ordered from controllers and performs the operation. This means GSMP slave directly communicates with core system management objects. Whatever network device, for example ATM (Asynchronous Transfer Mode), Frame Relay and IP, etc, can become the MPLS LSR, we don't change any interface. GSMP protocol was developed to support for label-based transmission like MPLS.

GSMP controller also refers the other modules like routing table, FIB (Forwarding Information Base), and LIB (Label Information Base). Therefore, when there is fault, controller can reroute, and order the related operation. GSMP controller has an information table for entire port configuration its own managed switch's. Simultaneously, it has backup routing path, which is easily obtained from routing table and is helpful in cases of unexpected situations.

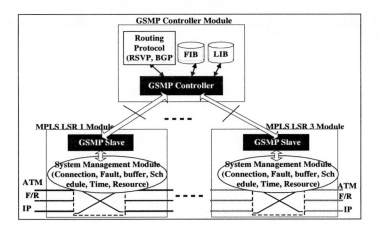

Fig. 2. LSR Components for hierarchical architecture

3 Handover Management Procedure Using GSMP on the MPLS Network

When the handover occurs from one LSR to another one, controller can easily announce the LSP release to the old LSR effectively because controller knows the resource usage in its own domain and A controller and slaves communicate with each other in terms of Port Down Event Message, Delete All Branches Message, Resource Reservation Message, and Add Branches Message.

Figure 3 shows the handover management procedure by using the mechanism of Figure 3. In this Figure 3, let's assume that LSR 1 is the router having the correspondent node. So, LSR 1 can be ingress node in the MPLS network. LSR 2 and LSR 3 are an old FA and a new FA respectively by handover. So, each FA has one or more wireless cells. Figure 3 shows that the MH moves from the LSR 2 (old FA) to the LSR 3 (new FA)

At this point, if the controller is reported from upstream LSR 3 (step ❶) corresponding to the registration of MIP by handover, it refers the resource state and statistics table and knows what labels cross over the old link in terms of output label, output port, input label, input port entities (step❷). Then it finds the backup path. It refers the routing table, which has the shortest path of its own domain (step❸). In this case, the routing table has to be made by considering traffic states since the labels mean different quality of services. Fortunately GSMP has a reservation mechanism. Therefore, after choosing the optimal path through the link and node state, GSMP controller sends the reservation message to slave for resource reservation(step❹). And then

GSMP controller orders LSR 1, LSR 2, and LSR 3 to modify the swapping table (step❺).

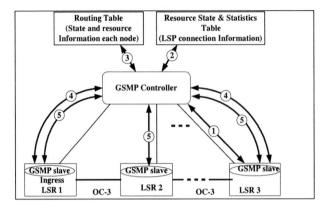

Fig. 3. Procedure for Reestablishing Path

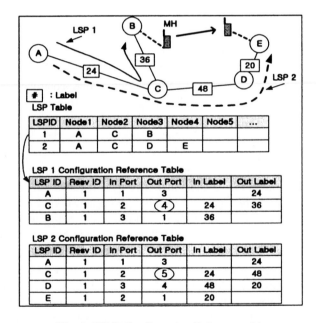

Fig. 4. GSMP Configuration Reference table

The advantage of using GSMP configuration reference table is a fast decision of the new path without referring routing table. Figure 4 shows the configuration reference table for controller's facility. Without querying the routing protocol or table, GSMP controller knows whether there is available backup path or not. In the case of Figure 4, MH moves from B to E. So, the path between C and B (LSP 1: Label Switched Path

1) can't be used. In addition, backup path (LSP2) should be searched from configuration reference table and changed from LSP1 to LSP2 that routes node A, node C, node D, and node E.

4   Effects of Proposed Method

In this chapter, we introduce several effects by using the proposed method.

– Low latency time: It is natural that the proposed method provides low latency time when compared with typical MIP. In typical MIP, Registration message must be sent from a new FA to the HA which can be located at remote site at handover time, it takes an interval for HA to update new CoA, Routing protocol may again be executed according to the new destination (new CoA), and it takes an interval for forwarding packets, which had arrived at old FA before update of new CoA at HA and after handover. But proposed method using GSMP takes only bi-directional transmission time of notification from slave (FA) to the controller and processing time at controller. Consequently, proposed method has much less time than typical MIP.

– Simplification of MIP: There is no need of HA. If proposed method is used, routing path can directly be changed at cross over router such as LSR G in Figure 1 by controller of GSMP. So, all processes about CoA related to the HA, such as registration, update, and so on, are not required any more. They should, however, be used at FA such as LSR I and LSR J in Figure 1 since the CoA is needed for supporting mobility in IP network as usual. Consequently, it is only mobility agent function that is required at both ingress router of MPLS and destination router, which are used for FA of MIP. Where, the mobility agent function can be defined with being able to manage the CoA of typical MIP.

– QoS: The Proposed method uses GSMP on the MPLS network. An important component of providing QoS is the ability to transport data reliably and efficiently. In this viewpoint, MPLS is characterized by the solution of guaranteeing the QoS and realizing VPN (Virtual Private Network). In addition, GSMP protocol Originally has a role of connection management of the general switching system [4]. It has also reservation mechanism for QoS based management, event mechanism for reliability, and state and statistics method for the reporting function. So, the proposed method improves the QoS by using the GSMP on the MPLS network.

– Network Scalability: GSMP is implemented the master and slave type, which can achieved open programmable device. Therefore we don't consider of the hardware device, for example ATM, Frame Relay, FEC based MPLS network, and etc. 5 Supplementary Material

5 Conclusion and Further Study

In this paper, we introduced hierarchical open programmable architecture for supporting mobility by using GSMP on MPLS network. Since GSMP controller can manage its own domain as a centralized server, proposed method directly changes the routing path at cross over router when the handover occurs. Therefore, proposed method can support low latency time, simple MIP function, and so on.

For using the proposed method efficiently, first, CC-GSMP should be studied more. It can be efficient for inter-domain handover. Second, we should arrange relations between the notification message, which is sent to the controller from salve (new FA), and other messages when MH executes handover. Lastly, we should investigate and define a method to register and to update CoA received from controller including correspondent node, which can nearly be used for ingress node for MPLS network.

Acknowledgment. This research was supported in part by KOSEF (Korea Science and Engineering Foundation) and OMIC (Ministry of Information and Communication) of Korean government through OIRC(Optical Internet Research Center) project.

References

1. C. Perkins, "IP Mobility Support," IETF RFC 2002, Oct. 1996.
2. E. Schooler et al., "SIP: Session Initiation Protocol," IETF RFC 2543, Mar. 1999.
3. C. Perkins and D. B. Johnson, "Route Optimization in Mobile IP," Internet draft, draft-ietf-mobileip-optim-10.txt. 2000
4. D. B. Johnson and C. Perkins, "Mobility Support in Ipv6," Internet draft, draft-ietf-mobileip-ipv6-13.txt. 2000
5. E. Gustavsson, A. Jonsson, and E.C. Perkins, "Mobile IP Regional Registration," internet draft, draft-ietf-mobileip-reg-tunnel-03.txt, July 2000.
6. Adrian Farrel, Paul Britain, Philip Matthews, "Fault Tolerance for LDP and CR-LDP", draft-ietf-mpls-ldp-ft-02.txt, May 2001.
7. Vishal Shama, Ben-Back Crane," Framework for MPLS-based Recovery", draft-ietf-mpls-recovery-frmwrk-02.txt, March 2001.
8. Avri Doria, Tom Worster, "General Switch Management Protocol" draft-ietf-gsmp-08. txt, http://www.ietf.org/html.charters/gsmp-charter.html, November, 2000.

Group-Based Location Management Scheme in Personal Communication Networks

Yungoo Huh and Cheeha Kim

Department of Computer Science and Engineering,
Pohang University of Science and Technology, Pohang, Korea
{perfect, chkim}@postech.ac.kr

Abstract. Signaling overhead incurred in location management becomes significant in PCS (Personal Communication Systems), especially with many users of high mobility. In such cases, an efficient location management scheme plays an important role in signaling. In this paper, we propose a new location management strategy which is based on the centralized database architecture with HLR/VLR concept. Our main idea is to reduce the number of location registrations by grouping MTs (Mobile Terminals) with a similar velocity and the same moving direction. Our scheme can reduce the total signaling cost of location management by reducing the cost of location registration more than the increased cost of call delivery. Simulation results show that our scheme outperforms current methods in location registration cost as long as CMR (call to mobility ratio) is not high and has a superior performance in the total signaling cost with low CMR and high population.

1 Introduction

Wireless mobile communication networks support mobile terminals (MTs) that are free to travel, and the network access point of an MT changes as it moves around the network coverage area. As a result, ID (identification) of an MT does not provide its location information. Therefore, location management is needed for effective delivery of incoming calls [1].

Current methods (IS-41, GSM MAP) [2,3] for location management employ HLR/VLR (Home Location Register / Visitor Location Register) concepts based on centralized database architecture. Currently, the centralized database network stores the location information of each MT in a location database (HLR and VLR) as MT changes its location area, a collection of cells. This location information is retrieved during call delivery.

Two standards for location management are currently available: Electronic/ Telecommunications Industry Associations (EIA/TIA) Interim Standard 41(IS-41) and the Global System for Mobile Communications (GSM) mobile application part (MAP). The IS-41 scheme is commonly used in North America, while the GSM MAP is mostly used in Europe. Both standards are based on a two-level data base hierarchy (HLR/VLR concept) [1].

I. Chong (Ed.): ICOIN 2002, LNCS 2344, pp. 81–90, 2002.
© Springer-Verlag Berlin Heidelberg 2002

Location management includes two major tasks: location registration and call delivery. Location registration procedure updates the location databases (HLR and VLRs) when an MT changes its location area. When a call for the MT is initiated, Call delivery procedure locates an MT based on the information available at the HLR and VLRs [1].

IS-41 and GSM MAP location management strategies are sufficiently similar that many commonalities exist between the two standards. For the sake of simplicity, the presentation of this paper is based on the IS-41 standard.

Two cases of database, HLR and VLR, are used to store the location information of MTs. The whole network coverage area is divided into cells. There is a base station installed in each cell and an MT within a cell communicates with the network through a base station. These cells are grouped together to form a larger area called a registration area (RA). All base stations belonging to one RA are wired to a mobile switching center (MSC) which serves as the interface between the wireless and the wired networks. In this paper, we assume that one VLR is associated with one MSC.

As the number of MTs increases, current location management methods at HLR/VLR concepts (IS-41, GSM MAP) have many problems: increasing traffic in the network, long delays at call delivery, bottlenecks by signaling traffic at HLR (since all signal messages related to location registration and call delivery always exchange between MSC/VLR and HLR).

To alleviate these problems, many proposed schemes are described in the following section. We propose a new group-based location management strategy in section 3. To show the superiority of our method, we analyze the performance with simulation under proposed strategy in Section 4. A summary and discussion are provided in the last section.

2   Location Management Scheme for PCS

Currently, many other proposed schemes are based on existing standards such as GSM MAP and IS-41 [4,5,6,7,8,9,10,11,12]. However, three different approaches, most distinctive ones, are proposed to reduce the amount of signaling traffic: a caching scheme [7,8,9,10,11,12], a local anchoring scheme [4] and a pointer forwarding scheme [5].

The basic idea of caching schemes is to reduce the volume of signaling and database access traffic used to locate an MT by maintaining a cache of location information that has been obtained at the previous call connection.

In the local anchoring scheme, when an MT moves into a new RA, the location change is reported to its local anchor instead of HLR until a new call request arrives. VLR serving an MT becomes its new local anchor at the moment when a new call directed to it is requested. The location registration may be localized so that the number of the registration messages to HLR can be reduced. However, if the communication cost between HLR and new VLR is not very high in comparison with the communication cost between the local anchor and new VLR, this method hardly reduces the registration cost.

In the pointer forwarding scheme, every time an MT switches RA, the forwarding pointer is set up from the old VLR to the new VLR without reporting the location change to HLR. The length of the pointer chain is limited to a predefined maximum value, K. When the length of the pointer chain reaches the predefined maximum value, the location change must be reported to HLR. For the pointer forwarding scheme, up to K+1 VLR queries must be performed in order to locate MT during the call delivery procedure.

However, because current proposed schemes consider individual MT as the unit of location management, all the schemes mentioned above and current standard schemes (IS-41 and GSM MAP) generate much signaling traffic in location management as the density and mobility of user is increased. That is, signaling traffic in location management increases in proportion to the number of users and the mobility of users in the system. Therefore, those schemes may not be feasible in future PCS with high density and mobility of users. A new scheme with a different point of view concering the unit of location management is needed to resolve this problem.

3   Group-Based Location Management

We assume that each MT can report its velocity and direction to its serving VLR during location registration and that each VLR can group MTs in accordance with similar velocity and moving direction. It also is assumed that there is one leader MT (which enters RA latest among involved MT when grouping MT) in each group and each MT except the leader has counter-value which records the number of crossing RA boundary before the leader MT moves into the new RA. When a new group is created, each counter-value of grouped MT initially becomes 0.

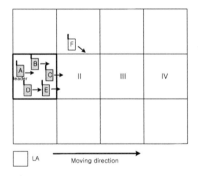

Fig. 1. Grouping MTs at VLR

As shown in Fig 1., we assume that the network is homogenous and that the shape of each RA is regular square. First of all, when an MT (A) moves into a new RA (I), serving VLR groups MTs having a similar velocity and moving direction (A, B, C, D, E) and the MT (A) becomes leader. The serving VLR notifies involved MTs of the fact by using paging channel. When grouped MTs (C,D,E) except leader MT (A) move into new RA (II) without changing the moving direction, location registrations are delayed and each counter-value of the MTs becomes 1 (see Fig. 2). Therefore, when a grouped MT is called, the paging area becomes two RAs (an RA where the leader MT currently resides and neighboring RA in the path of its moving direction).

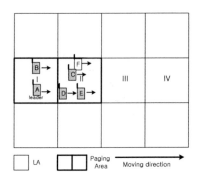

Fig. 2. Location registration of grouped MTs (example 1)

Next, when the leader MT moves into the RA (II), a new serving VLR notifies grouped MTs (C,D,E) arrived in advance of leaders arrival and old serving VLR notifies grouped MTs (B) remaining behind of leaders departure(see Fig. 3). At this time, each counter-value of grouped MTs (C, D, E) arrived in advance becomes 0. And grouped MTs (B) remaining behind are excluded from group. It means that velocity of the MT becomes slower than that of the leader MT. If the counter-value of MT (E) included in any group becomes 2, the MT is also excluded from group (see Fig. 5). It means that the velocity of the MT becomes higher than that of the leader MT.

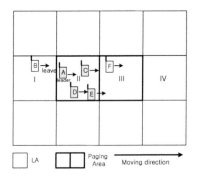

Fig. 3. Location registration of grouped MTs (example 2)

At a group-based scheme, we define two new operations, *leave* and *join*. The *leave* operation excludes specific MTs from a specific group and the *join* operation includes MTs in a specific group.

The *leave* operation can be categorized into two types, *leave I* and *leave II*. When the velocity of the MT slows down (for example, MT B at Fig. 3), *leave I* operation is executed. And if MT included in any group changes moving direction (for example, MT D at Fig. 4) or the velocity of the MT becomes higher (for example, MT E at Fig. 5. MT E moves into RA IV before leader moves into RA III.), *leave II* operation is executed.

When leader of group moves into new RA to moving direction, if MT with similar velocity and moving direction resides at the new RA, *join* operation is executed to include the MT in the group (for example, MT F at Fig. 5).

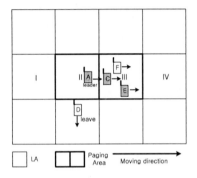

Fig. 4. Location registration of grouped MTs (example 3)

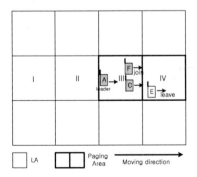

Fig. 5. Location registration of grouped MTs (example 4)

Location registration procedure in group-based scheme can be summarized by the flow chart in Fig. 6.

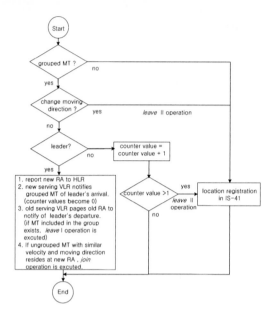

Fig. 6. Decision flow for location registration in group-based scheme

To apply group-based scheme to IS-41, first of all, HLR assigns a unique GID (Group Identification) to each group and manages the GIDs. By a *leave* or *join* operation, an MT is able to be included in or excluded from any group.

The *join* operation notifies involved MTs of being included in a specific group by using paging channel and also informs HLR of that fact during location registration of leader MT.

Therefore the *join* operation can be simply implemented by additional paging and the *leave II* operation also is able to implemented by reporting withdrawal from specified group to HLR during location registration of corresponding MT. Whenever leader MT switches RA, old serving VLR page at old RA. At this time, if grouped MT at old RA exists, *leave I* operation for reporting withdrawal from the group to HLR is needed.

If called MT is included in any group during call delivery. HLR notifies serving MSC of called MT that paging area becomes two RAs (RA where leader MT of the group currently resides and neighboring RA to its moving direction). Except for difference of paging area, the call delivery procedure in the group-based scheme is identical to the call delivery procedure in IS-41.

Location registration and call delivery procedure in MT that is not included in any group always are the same as in IS-41.

4   Performance Analysis

We analyze performance by simulation. The aim of the simulation is to analyze the effectiveness of our proposed scheme by comparing it with IS-41.

4.1   Network Model for Simulation

Network topology for simulating our scheme is a 10 x 10 grid of RAs, as shown in Fig. 7. It is assumed that the network is homogenous and that the shape of each RA is a regular square. There are only two types of velocity (high and low) and four types of moving direction (the East, the West, the North, the South). Each MT selects velocity randomly and at first move, it also selects any moving direction with an equal probability of 1/4. However, the next move is then determined with probaility of P_D for the on-going direction.

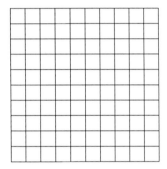

Fig. 7. Network topology for simulation

For the sake of simplicity, we assume that one location registration cost in IS-41 is equal to one call delivery cost in IS-41 and that one paging cost is half of location registration cost in IS-41. So we can define each cost as follows.

Table 1. Definition for each Notation

Description	Cost
Location registration of ungrouped	1
Location registration of leader MT	1.5
Call delivery on ungrouped MT	1
Call delivery on grouped MT	1.5
Join operation	0.5
Leave I operation	1
Leave II operation	0

That is, if one location registration cost in IS-41 is 1, one call delivery cost on grouped MT is 1.5 (one call delivery cost in IS-41 + one paging cost) and one *join* operation cost is 0.5 etc.

4.2   Simulation Results

Total location management cost is divided into the following parts: The location registration cost is the cost incurred in completing location registration procedure and the call delivery cost is the cost incurred in completing call delivery procedure. We define the total signaling cost as the sum of the location registration and the call delivery cost. In order to show the cost for easy comparison, we define the relative cost as the ratio of cost of the group-based scheme to that of IS-41

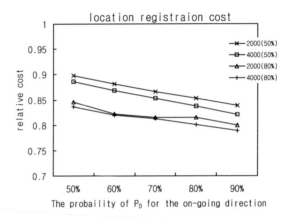

Fig. 8. Location registration cost of group-based scheme relative to that of IS-41

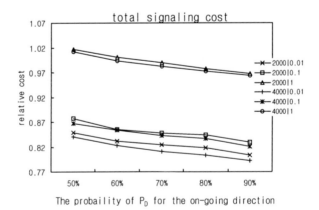

Fig. 9. Total signaling cost of group-based scheme relative to that of IS-41

First, we compare the location registration cost of the group-based scheme to that of IS-41. In Fig. 8, 2000 and 4000 refer to the number of MTs in the network and 50% and 80% represent the fraction of moving time. Fig.8 shows that location registration cost of group-based scheme is always lower than that of IS-41. As the number of MTs, the fraction of moving time and the probability of P_D for the on-going direction increases, the relative cost decreases. These results are expected. Because the group-based scheme reduces the number of location registration in proportion to the number of groups and size of each group, if the number of MTs, fraction of moving time and the probability of P_D for the on-going direction increase, the number of groups and size of each group also increase. In Fig. 9, 2000 and 4000 also signify the number of MTs in the network and 0.01, 0,1 and 1 represent CMR. Fig. 9 shows that the total signaling cost of group-based scheme outperforms the IS-41 as long as CMR stays in the low range. Because call delivery cost of group-based scheme is always higher than IS-41 and the location registration cost in total signaling cost is dominant at low CMR, a significant cost saving at low CMR is achieved. That is, it means that the signaling overhead caused at call delivery is compensated for by the cost saving obtained at location registration.

5   Conclusions

We propose a group-based scheme based on IS-41. In a group-base scheme, grouped MTs as well as individual MT are considered as the unit of location management, so a group-base scheme can reduce the number of location registrations by grouping MTs with a similar velocity and the same moving direction.

Simulation result shows that our new scheme outperforms or almost equals current methods in cost, and shows an especially better performance in the network with high density and mobility of user. That is, our scheme may be feasible in a network with a high population and show a better performance than current methods in such networks.

Under very high CMR, the group-based scheme does not gain much in comparison with IS-41. However, schemes based on caching [7,8,9,10,11,12] have much more cost reduction in comparison with IS-41 at high CMR. Therefore combining the group-based scheme with caching-based scheme may result in more efficient location registration strategy. It is possible to combine the group-based scheme with a caching-based scheme by the applying caching scheme to only ungrouped MTs in the group-based scheme.

References

1. Ian F. Akyildiz and Joseph S.M Ho, On Location Management for Personal Communications Networks, IEEE Personal Communication Magazine, Sept. 1996, pp.138–145
2. EIA/TIA, Cellular Radio-Telecommunications Intersystem Operations, Tech. Rep. IS-41 Revision B, 1991
3. S. Mohan and R. Jain, Two User Strategies For Personal Communications Services, IEEE Personal Communication, 1st Qtr. 1994, pp. 42–50

4. J.S.M. Ho and I. Akyildiz, Local Anchor Scheme for Reduce Signaling Costs in Prersonal Communications Networks, IEEE/ACM Transaction on Networking, Oct. 1996, pp. 709–725

5. R. Jain, Yi-Bing Lin, C. Lo and S. Mohan, A Forwarding Strategy to Reduce Network Impact of PCS, IEEE INFOCOM 1995, pp. 481–489

6. Yingwei Cui, Derek Lam, Jennifer Widom and Donald C. Cox, PCS mobility management using the reverse virtual call setup algorithm, IEEE/ACM Transactions on Networking, Feb. 1997, pp. 13–23

7. R. Jain, Y. B. Lin, and S.Mohan, A Caching Strategy to Reduce Network Impacts to PCS, IEEE Journal on Selected Areas in Communications, Oct. 1994, pp. 1434–1444

8. Y.B. Lin, Determining the User Locations for Personal Communications Services Networks, IEEE Transactions on Vehicular Technology, Aug. 1994, pp. 455–473

9. K. Ratnam, S. Rangarajan and A. Dahbura, An efficient fault-tolerant location management protocol for personal communication networks, IEEE Transactions on Vehicular Technology, Nov. 2000, pp. 2359–2369

10. Shivakumar and J. Widom, User Profile Replication for Faster Location Lookup in Mobile Environments, Proceedings of the 1st ACM Mobicom, Nov. 1995, pp. 161–169

11. N.Shivakumar, J. Jannink, and J. Widom, Per-user profile replication in mobile environments: Algorithms, analysis, and simulation results, ACM/Baltzer Journal of Mobile Networks and Applications, 1997, pp. 129–140

12. Subhashini Rajagopalan and B.R. Badrinath, An adaptive Location management Strategy for Mobile IP, Proceedings of the 1st ACM Mobicom, Nov. 1995, pp. 170–180

A Fast WWW Access Method for Multiple Wireless Links Shared by Mobile Computers

Yousuke Konishi[1], Susumu Ishihara[2], and Tadanori Mizuno[3]

[1] Graduate school of information, Shizuoka University,
432-8011 Hamamatsu, Japan
Phone:+81-53-478-1460, FAX:+81-53-478-1597
y-konisi@mizulab.net
http://www.mizulab.net/y-konisi
[2] Faculty of Engineering, Shizuoka University,
432-8011 Hamamatsu, Japan
ishihara@cs.inf.shizuoka.ac.jp
http://www.ishilab.net
[3] Faculty of Information, Shizuoka University,
432-8011 Hamamatsu, Japan
mizuno@cs.inf.shizuoka.ac.jp
http://www.mizulab.net/

Abstract. Wireless links used for mobile communications have some problems such as narrow bandwidth and low reliability. To offer high speed communication on wireless links, we have proposed *SHAKE*(SHAring multiple paths procedure for cluster networK Environment). In SHAKE, mobile hosts which are connected with fast local link each other use multiple wireless links owned by each host simultaneously to communicate with hosts on the internet.
In this paper, we propose a fast WWW access method with SHAKE (*Web SHAKE*). The feature of Web SHAKE is that it does not require any special software on web servers on the internet to use SHAKE. The feature is realized by an use of HTTP Proxy Server which works on each mobile host. We present the result of performance evaluation of the Web SHAKE. We have shown that data transfer speed was improved by Web SHAKE in comparison with normal WWW access from three experiments.

1 Introduction

The number of internet users has increased rapidly because of the remarkable spread of personal computers and the progress of computer network technologies in the last decade. In addition, users who access the internet from mobile / wireless computers have are common because mobile hosts (e.g., portable computers, PDAs, etc.) and cellular phones become popular.

Because of the increase in internet users, contents and services on the internet demand higher quality. For example on the WWW (World Wide Web), web pages including multimedia data (pictures, sounds, and movies, etc.) are common, and

I. Chong (Ed.): ICOIN 2002, LNCS 2344, pp. 91–103, 2002.
© Springer-Verlag Berlin Heidelberg 2002

the data size of web contents increases daily. However, current personal wireless communication technologies are not sufficient to transfer multimedia data in comparison with wired communication because of the narrow bandwidth and low reliability of wireless links.

To offer fast and high-speed communication on wireless links, we have already proposed *SHAKE* (SHAring multiple paths procedure for a cluster networK Environment). In SHAKE, mobile hosts that are connected by a fast local link (e.g., wireless LAN [IEEE802.11b], Bluetooth, etc.) each other use multiple wireless links owned by each host simultaneously to communicate with hosts on the internet. An experimental system was implemented and evaluated [1]. In this system, however, not only mobile hosts but also correspondent hosts on the internet require special software for SHAKE.

In this paper, we propose a fast WWW access method with SHAKE (we call this method the *Web SHAKE*). The feature of Web SHAKE is that it does not require any special software on web servers on the internet. The feature is achieved by the use of HTTP proxy server that works on each mobile host. The results of a performance evaluation of the Web SHAKE system are described in this paper.

This paper is organized as follows. The next section describes the outline of SHAKE and related works; Section 3 describes the detail of our fast WWW access method with SHAKE (Web SHAKE); In Section 4, we discusses the performance evaluation of this system. In Section 5, we conclude this paper and mention the unsolved subjects of the Web SHAKE.

2   SHAKE (Sharing Multiple Paths Procedure for a Cluster Network Environment)

2.1   Overview of SHAKE

When we access the internet to browse web pages outdoors or on moving, we utilize mobile communication services (e.g., cellular phone or PHS). However wireless links used by mobile hosts have some problems such as narrow bandwidth and low reliability in comparison with wired network.

To solve these problems, we have proposed a multiple paths protocol procedure. In this procedure, mobile hosts with fast wireless link (e.g., wireless LAN , Bluetooth etc.) temporarily form a network (we call this network Cluster Network), and when a mobile host access the internet, the mobile host uses not only wireless links which it uses to connect with the internet but also other wireless links that other mobile hosts have. This system can offer high quality and high speed communication on wireless links.

We call this procedure SHAKE (Sharing multiple paths procedure for a cluster network environment). A example of communication with SHAKE is shown in Fig. 1. The advantages of SHAKE are shown as follows;

− Improvement of data transfer speed on wireless links

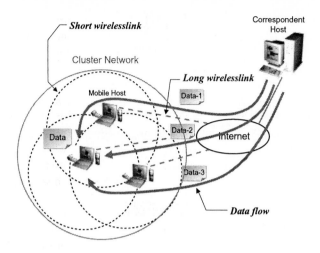

Fig. 1. Example of communication with SHAKE

- Even when a mobile host itself doesn't have external link or cannot utilize external wireless links, if it use SHAKE, it can utilize wireless link of other mobile hosts instead of its own link with SHAKE.
- Cluster networks can be formed dynamically. They do not depend on the physical location of mobile hosts.

The similar ideas are found in PPP multilink protocol [2] [3] and 384-Kbps PHS experimental system [4] that use of a special device etc. Parallel data transmission at datalink layer is realized by use of some network interfaces owed by a mobile host in those works. However SHAKE is different from these works because SHAKE offers communicate in parallel at network layer or higher.

Another technique for fast WWW access method is that using distributed Chache. For example, There are "Sharp Super Proxy Script [5]" and "Squid WWW Cache system [6]". These realize improvement in the speed by using two or more distributed cashe servers simultaneously.

An experimental system of SHAKE was already implemented and evaluated so far [1]. In this experimental system, the function of SHAKE was implemented as a library that utilizes TCP for reliable socket level communication. This library deals with dispersing of the traffic on multiple paths and sorting and combining of data blocks transmitted via multiple paths.

2.2   Problems of the Experimental Implementation of SHAKE

Cooperation between mobile hosts and the corresponding host is necessary to realize SHAKE. For example, when a mobile host receives data from the correspondent host via multiple wireless links at a time, the correspondent host has to divide the data to send and distribute the fragments to each path. Therefore the correspondent hosts need to control and manage the mobile hosts that consist

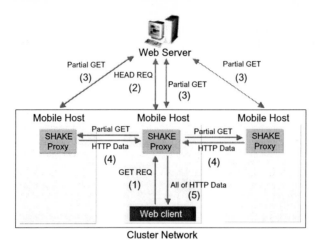

Fig. 2. Web SHAKE Architecture

of the cluster network. In addition, the mobile host receiving data must sort and combine the divided data that received from each different path.

However, in the experimental system, not only mobile hosts but also the correspondent hosts on the internet require some special software for the function of SHAKE. Therefore if SHAKE software is not implemented in the correspondent hosts, mobile hosts can't communicate with them.

3   Web SHAKE

3.1   Overview of Web SHAKE

On the experimental system of SHAKE, both mobile hosts and correspondent hosts need special software for SHAKE to communicate with hosts on the internet. However it is desirable that only mobile hosts need a SHAKE module.

In this paper, we propose a fast WWW access method with SHAKE (we call this method WebSHAKE). Furthermore the Web SHAKE solves the problems that the experimental SHAKE implementation has, although it limits the available protocol only to HTTP (Hyper Text Transfer Protocol). With Web SHAKE, mobile hosts in a cluster can communicate with any hosts on internet by an use of a HTTP proxy server which works on each mobile host. Any special software for SHAKE is not required on the correspondent hosts.

3.2   Web SHAKE Architecture

HTTP (Hyper Text Transfer Protocol) [7] [8] consists of some request methods (e.g., GET method, POST method, HEAD method etc). GET method is used to transfer data (include HTTP header and Entity-Body) from a web server to the client. HEAD method is identical to GET method except that the server doesn't return any Entity-Body in the response. The meta-information contained in the HTTP headers in the response to a HEAD request should be identical to the information sent in the response message to a GET request. In Web SHAKE, we use HEAD method and partial GET method (GET request including Range header field). We can transmit only part of the data entity with use of partial GET method. The HEAD method is used to obtain the size information of the target file, and partial GET method is used to transfer data via multiple links.

Web SHAKE architecture is shown in Fig. 2. SHAKE HTTP Proxies (HTTP Proxy which includes SHAKE modules) run on mobile hosts in a cluster. If a SHAKE HTTP Proxy (SHP) receives a HTTP request from a HTTP client (e.g., Internet Explorer, Netscape navigator, etc) running on the same mobile host, then it sends a partial GET requests to SHPs working on other mobile hosts in the same cluster network. When a SHP receives a partial GET request from SHP running on the other mobile host in the same cluster network, it behaves as a normal HTTP proxy server. That is, it sends a HTTP request instead of the mobile host that receives GET request directly from a HTTP client.

The following is the detail of the data transmission on Web SHAKE.

1. A HTTP client sends a GET request for a web server to a SHP running on the same host in stead of sending the request directly to the web server. (The HTTP client has to be configured to use the SHP as HTTP proxy)
2. The SHP that receives a GET request from a HTTP client sends a HEAD request to the web server and receives HTTP header including the file size information of the target file.
3. SHP determines the data size for each wireless link according to the whole data size and the bandwidth of each link. Then it sends n partial GET requests to the other SHPs on the same cluster network. Here n is the number of mobile hosts in the cluster and the web server. The SHPs working on each mobile host runs in parallel, forward the request and receive a part of the target data from the web server by use of the partial GET method.
4. The partial data received by each SHP is send to the SHP that receives GET request directly from the HTTP client.
5. The SHP that receives a GET request directly from the HTTP client sorts and combines the partial data and sends the combined data to the HTTP client. At that time the SHP removes the header of the response message to the partial GET request because each replying message includes useless header, and combines the extracted data in the replying message and the original header received in step (2).

We call this data transfer method F-H REQ (First-Head REQest) because the SHP receiving a GET request directly from a HTTP client sends a HEAD request first. Sending HEAD method and partial GET method, control of the cluster network, and sorting and combining parts of data are performed by SHPs. A cluster network is maintained with a "cluster table" owed by each SHP on mobile host. Each record in a cluster table includes following fields; hostname, port-number, maximum bandwidth of the external link. In Web SHAKE, web servers (i.e. correspondent hosts) need not to maintain the mobile hosts on the cluster network and disperse data to multiple paths, because SHPs working on the cluster network can disperse data by use of partial GET requests. Because SHPs behave as a normal HTTP proxy server to the local HTTP client, mobile hosts need not any special HTTP client software.

3.3   The Optimization of Data Transfer Method

If we use F-H REQ to transfer small data, we can not obtain larger throughput than normal WWW access method. This is because that the overhead of the first HEAD method is not negligible when the size of the data entity is small. As the method for preventing this degradation of the throughput, we propose a more efficient method that replaces HEAD method ((2) in Fig. 2) with partial GET method. We can receive not only partial data, but also the size of whole data from partial content header by using partial GET method. On account of this, if we use partial GET method instead of HEAD method, SHP can receive the head of data including the size information. Therefore when the size of data size is small, SHP can receive all data by a message exchange between SHP and web server at once. We call this data size that mobile host received by using of first partial GET method **Maximum Data Size at First Request (MDFR)**. When the size of data is large, SHP may receive the rest of data that can't receive at first partial GET method from web server by multiple wireless links in parallel. We call this data transfer method F-P REQ (First-Partial GET REQest), and discriminate between F-H REQ and F-P REQ.

4   Performance Evaluation

We evaluated the performance of Web SHAKE system by three experiments.

4.1   Implementation of Web SHAKE Software

We implemented the SHP on UNIX like OS (FreeBSD & Linuxs). The SHP was written in C. The SHP implementation can be utilized by a general WWW browser (e.g., Internet Explorer, Netscape Navigator etc) as a HTTP Proxy because as if SHP look like a general HTTP proxy server. When we use the Web SHAKE system, we give SHP the number of mobile hosts in the cluster, names of each mobile host, port number and the bandwidth of the external link at first. In the current implementation of SHP, dispersal rate can not be changed dynamically according to the throughput of wireless link.

4.2   Experimental Environment

Overview. We used three note PCs with PCMCIA card for PIAFS (PHS Internet Access Forum Standard) [9] communication, and wireless LAN (IEEE802.11b). The note PCs are TOSHIBA SS3010 (CPU: PentiumII 300MHz,

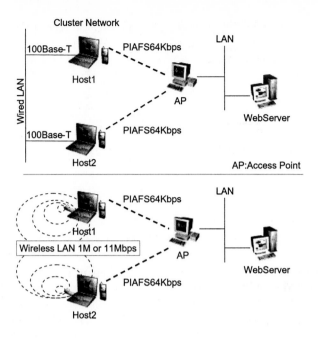

Fig. 3. Experimental Environment-1

Memory: 64Mbytes OS: Linux kernel ver. 2.2) and two Panasonic CF-M2 (CPU: Celeron 500MHz, Memory:128 Mbytes, OS: Linux Kernel ver. 2.2). PCMCIA cards for PIAFS communication are two MC-P210 (Seiko Instruments Inc.) and one P-in comp@ct (NTT DoCoMo). The cluster network consists of Ethernet (100BASE-T) or wireless LAN (IEEE802.11b).

Experiments. Followings is the details of each experment.

Experiment-1: Ideal environment (2 same wireless links). We measured the throughput in a case that throughput of all wireless links were almost equal in Web SHAKE system. Fig. 3 shows the network environment for the experiment-1. Host1 in Fig. 3 requests 1K–1M[bytes] files to the web server. Access point (AP) and the web server were connected to Shizuoka university's local area network not via the internet. The maximum bandwidth of each wireless link is 64Kbps. The cluster network consists of wired LAN (100BASE-T), or wireless LAN (IEEE802.11b). We evaluated the performance of Web SHAKE system on three cases in this environment. The first case was that the cluster network consists of the wired LAN (100BASE-T). The second case was that the cluster network consists of the wireless LAN (bit rate =11Mbps). The third case was that the cluster network consists of the wireless LAN (bit rate =1Mbps). The SHP on the same host running web client dispersed same amount of the data traffic among each path. The data transfer method was F-H REQ.

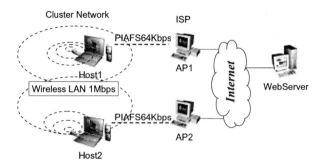

Fig. 4. Experimental Environment-2

Experiment-2: A comparison of F-H REQ with F-P REQ. Fig. 4 shows the network environment for the experiment-2. We compared the performance of F-H REQ with F-P REQ in the case that HTTP client requested the small data. In the environment of the experiment-1, both host1 and host2 were connected to the same AP (Access Point), and AP and web server were not connected to the internet. However in the environment of the experiment-2, host1, host2 were connected to different AP, and AP1 and AP2 were connected to the internet so that each wireless link might have different throughput and transmission delay. The method of measurement was the same as experiment-1. However the data size that mobile host received was not 1[KB]–1[MB] but 1[KB] – 128[KB] because we focused on evaluating the performance of Web SHAKE when the data size that host1 request to web server is small. We evaluated the performance of two different implementations of Web SHAKE system by using of this environment. In the first implementation, the data transfer method was F-H REQ. In the second implementation, the data transfer method was F-P REQ (MDFR = 10KB, 20KB, 30KB). We dispersed the same amount of data traffic among each path.

Experiment-3: Multiple links with different characteristic. We measured the throughput on Web SHAKE in the case that throughput of each wireless link was different from the experiment-2. In the environment of the experiment-2, host1, host2, and host3 were connected to different APs, and AP2 and AP3 were connected to the internet so that the wireless links might have different throughput and delay. Fig. 5 shows the network environment of the experiment-3. In this network, 1–3 wireless links (64Kbps) were used simultaneously. The data transfer method was F-H REQ and F-P REQ (MDFR = 20KB). The method of measurement was the same as experiment-1 and experiment-2. We dispersed the same amount of data traffic among each path. When the number of shared wireless link was two, used links were Host1+AP1 and Host2+AP2. When only one wireless link was shared, Host1+AP1 was used. In the all data transmission in experiment-3, the receiver was Host2 in Fig. 5 (i.e. Host2 sent HTTP requests and received HTTP data).

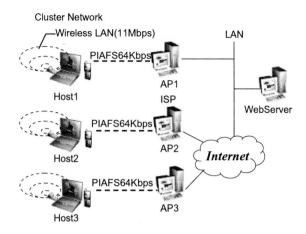

Fig. 5. Experimental Environment-3

4.3   Results and Discussion

Experiment-1. The results of experiment-1 are shown in Fig. 6. This graph shows the throughput of the data transmission on Web SHAKE and normal WWW access. We can see that as data size is larger, the throughput is larger. In particular, when data size was 1Mbytes, the throughput on Web SHAKE (2 wireless links shared) was about twice as much as normal WWW access. We tested three variations of the cluster networks (100BASE-T, wireless LAN [11Mbps], wireless LAN [1Mbps]). However the throughput was almost equal.

In SHAKE, mobile hosts which are connected with cluster link each other use multiple wireless links owned by each host simultaneously to communicate with hosts on the internet to offer high speed communication on wireless links. The cluster link needs to be the high speed network in comparison with the sum of the external links as a premise of SHAKE. If the bandwidth of the cluster link is small, this link will be a bottleneck and not be able to offer high speed communication. In this experiment, the bandwidth of the cluster link was enough large in comparison with external link (PIAFS: 64Kbps) because the cluster network consists of wireless LAN (1Mbps or 11Mbps) or wired LAN (100Mbps).

While the throughput was large when data size is large, the throughput on Web SHAKE was smaller than normal WWW access when data size was small. However in the experiment-1, when mobile hosts receive small size data less than 16Kbytes, the throughput was worse than normal WWW access (one wireless link is used). One of the reasons of this is the overhead of establishing TCP connections establishment. Because the cluster network consists of high speed LAN (100BASE-T) in this experiment, the time for TCP's 3Way Hand Shake (3WHS) between mobile hosts can be ignored. However the time for TCP's 3WHS between mobile hosts and the web server is not small enough to be ignored.

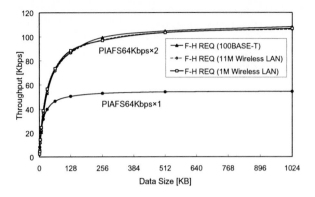

Fig. 6. Result of experiment-1

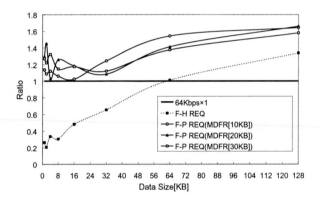

Fig. 7. Result of experiment-2(Throughput ratio)

Experiment-2. Fig.7 shows the results of experiment-2. This graph shows the ratio between the throughput on Web SHAKE (F-H REQ and F-P REQ) and the throughput on normal WWW access. We call this ratio as Performance Improvement Ratio (PIR). In this graph, the PIR of F-H REQ was substantially smaller than 1 when the data size is 0 – 64 [KB]. The reason of this can be state as follows. When F-H REQ is used to transfer data, the SHP sends a HEAD request to the web server to obtain file size information at first. When the whole data size of the request file is small, the overhead of transmitting the HEAD information can not be ignored.

On the other hand, PIR of F-P REQ was substantially greater than 1 even when the transmission data size was small. Because we can receive not only partial data but also the size of whole data from partial content header by using of F-P REQ, SHP can receive all data by a message exchange between SHP and web server at once to the small data. Therefore the overhead of F-P REQ

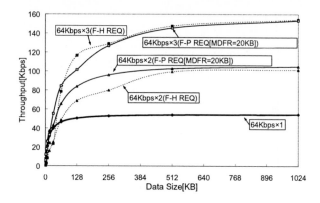

Fig. 8. Result of experiment-3 (Throughput)

was smaller than F-H REQ, and the throughput was large too when data size is small.

Experiment-3. Fig. 8 and Fig. 9 show the results of experiment-3. Fig. 8 shows the throughput on Web SHAKE (F-H REQ and F-P REQ) and normal WWW access. From these graphs, we can see that even if the throughput of each wireless link is different, the throughput becomes large with an use of Web SHAKE. When data size is 1Mbytes, the throughput on Web SHAKE (2 wireless links shared) was about 1.9 times as much as normal WWW access. And when three wireless links were shared, the throughput was about 2.8 times as much as normal WWW access by using of both F-H REQ and F-P REQ.

Fig. 9 shows the ratio between the throughput on Web SHAKE (F-H REQ and F-P REQ) and the throughput on normal WWW access. From this graph, we can see that the performance improvement ratio on F-H REQ is substantially smaller than 1 when data size is 0–64 [KB]. However the throughput of F-P REQ is greater than normal WWW access at all range (1[KB]–1[MB]). If we use F-P REQ, Web SHAKE is effective all the time.

5   Conclusion and Future Works

We have proposed the Web SHAKE as a fast WWW access method with SHAKE and implemented SHAKE HTTP Proxy. Web SHAKE enables mobile hosts in a cluster network to communicate with any hosts on the internet by an use of a HTTP proxy server which works on each mobile host. Although the available protocol is limited to HTTP, any special software for Web SHAKE is not required on the web servers.

We have shown that data transfer speed can be improved by Web SHAKE in comparison with normal WWW access from three experiments. We proposed two variations of Web SHAKE, F-H REQ and F-P REQ. When data size was

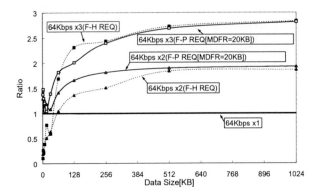

Fig. 9. Result of experiment-3 (ratio)

large, the data transfer speed was improved substantially on both F-H REQ and F-P REQ. The throughput on Web SHAKE that 2 wireless links were shared was about 1.9 times as much as normal WWW access. When three wireless links were shared, the throughput was about 2.8 times as much as normal WWW access by using of both F-H REQ and F-P REQ. In addition, even when the data size was small, the throughput on F-P REQ was larger than normal WWW access.

We assume that cluster network is Ad-Hoc. Therefore the cluster network is dynamic and mobile hosts may join to the cluster and leave at any time. In our current implementation of Web SHAKE, a cluster network is managed with a static cluster table. However because the cluster table are not configured dynamically, Web SHAKE cannot handle mobile hosts that joins to the cluster network in the halfway. In addition, if the one of the shared links is disconnected, the current this system cannot recover disconnected sharing links. As a future work, we are planning to control the members and the dispersing rate in the cluster network dynamically. The cluster table should be reconfigured according to the topology of the ad-Hoc network. When the mobile host joins or leave to the cluster network, the SHP should add or delete this host to the cluster network.

In addition to this, SHP's flow also should be controlled dynamically. The current SHP can only change dispersing rate when we run the SHP program at first. However the SHP should be able to change dispersing rate according to the network traffic by monitoring the status of the network because the traffic on the internet always changes, and the quality and the effective communication speed of wireless links always changes.

References

1. H. Mineno, et. al., "Multiple paths protocol for a cluster type network," Int. J. Commun. Syst, Vol. 12, pp. 391–403, December 1999

2. K. Sklower, B. Lioyd, G. McGregor, D. Carr "The PPP Multilink Protocol (MP)," RFC1717, November 1994
3. K. Sklower, B. Lioyd, G. McGregor, D. Carr "The PPP Multilink Protocol (MP)," RFC1990, August 1996
4. Y. Kamio, F. kojima, M. fujise, "Implementation and Performance Evaluation of 384kbps-PHS Experimental system," IEICE, Vol. E83-B, No. 8, pp. 1844–1853 July 2000
5. Sharp, "Sharp Super Proxy Script – distributed proxy servers by URL hashing ", http://naragw.sharp.co.jp/sps/
6. Squid WWW Cache system, http://www.squid-cache.org/
7. R. Fielding and H. Frystyk, "Hypertext Transfer Protocol – HTTP1.0," RFC1945, May 1996.
8. R. Fielding, et. al., "Hypertext Transfer Protocol – HTTP1.1," RFC2616, June 1999.
9. PHS Internet Access Forum meeting, "PIAFS Specifications,", March 1997.

A Study of Broadcasting Method Using Linear Clustering to Support Location Dependent Query in Mobile Computing

Young-ho Yu, Il-dong Jung, and Kyongsok Kim

Department of Computer Science, Pusan National University
Pusan, Korea 609-735
{yhyou, idjung, gimgs}@asadal.pnu.edu

Abstract. Location-dependent applications (LDA) are becoming very popular in mobile computing environments. To improve system performance and facilitate disconnection in LDA, the caching policy of the mobile host (MH) is also crucial to such applications. So, most studies of LDA are concentrated on caching policies of MH. How to broadcast data is also important in order to improve system performance and minimize energy expenditure of the MH. Broadcasting method is another issue in mobile environments. We propose a broadcasting method that minimizes energy expenditure by reducing the number of download of the MH. We divide cell region into square grids and construct broadcast schedule so that the data of the adjacent grids are linearly clustered. We propose to use space-filling curves to cluster the data linearly. And we evaluate performance of each clustering method by measuring the average setup time of MHs.

1 Introduction

Advances in wireless networking technology and portable information appliances have engendered a new paradigm of computing, called mobile computing, in which users who carry portable devices such as laptops or PDAs have access to information service through wireless networks regardless of their physical location or the movement of the user. Mobile computing is distinguished from classical computing due to the mobility of nomadic users and their computers and the mobile resource constraints such as limited wireless bandwidth and limited battery life [10]. For this reason, broadcasting is general method for the mobile support station (MSS) to deliver data to MHs in mobile computing applications.

As broadcasting is hired to deliver the data required by MHs, we take advantage of the following:

1. Broadcasting reduces energy expenditure of an MH, which is necessary for the MH to request data.
2. Although the number of MHs within a cell region has increased, broadcasting enables the MSS to service the requests of MHs without additional resources for either MHs or MSS [7].

LDA is a new application of mobile computing which is tightly associated with the geographical location of an MH. A location dependent query (LDQ), which is a

I. Chong (Ed.): ICOIN 2002, LNCS 2344, pp. 104–112, 2002.
© Springer-Verlag Berlin Heidelberg 2002

general query in LDA, is processed on location dependent data (LDD), and whose result depends on the location criteria explicitly or implicitly specified [13]. To support LDQ efficiently, the caching policy of MH is important and also the broadcasting method of MSS is important. The energy expenditure of an MH is greatly associated with the broadcasting method. The number of changes between active and doze mode during downloading depends on how well the data required by MH is clustered in broadcast disk. The number of changes between modes determines the amount of energy expenditure.

In this paper, we propose a broadcasting method using linear clustering to support LDQ efficiently and reduce the energy expenditure of an MH. Our broadcasting methods place the data into a broadcast disk according to the sequence of space-filling curve. We evaluate the performance of each proposed clustering method by measuring the setup time and active time of an MH.

2 Related Work

In mobile computing, previous researches have directed at solving the problems and overcoming the problems of mobile resource constraints such as limited wireless bandwidth and limited battery life. We focus on reducing the energy expenditure of MH in LDA system.

2.1 Location Dependent Query (LDQ)

The response to LDQ depends on the current location of an MH. The following example illustrates this point.

Example. LDQ – Hotel Information: Suppose a traveler who wants to find out information about hotels in the middle of his journey. He issues a query to obtain this information. The response to this query depends on the geographical location of MH. At one place, for example Seoul, the answer might be hotel A, and at another place, for example Cheju, the response might be hotel B.

We observe the following from the above example:
1. The semantics of a set of data and their values are tightly coupled with a particular location.
2. The answer to a query depends on the geographical location where the query originates.
3. A query may be valid only at a particular location [4].

Previous researches, related to LDQ, were concentrated on defining LDD [4] and developing caching policies [11,13].

There must be support of broadcasting method for a good quality of service because there is a difficulty in providing a good quality of service with only caching policy of MH in mobile computing environment. In this paper, we propose a broadcasting method that supports the LDQ of MH efficiently.

2.2 Broadcast

There are many researches that have proposed the broadcasting methods in order to reduce the energy expenditure of an MH. [8] proposed broadcasting methods that reduced the tuning time by using *(1,m) Indexing* and *Distributed Indexing. (1,m) Indexing* is an index allocation method in which the index is broadcast *m* times during the broadcast of one version of the file. *Distributed Indexing* is a method in which an index is partially replicated. The proposed algorithms in [8] are analytically demonstrated to lead to significant improvement of battery life.

In recent researches, [3] proposed adaptive broadcasting method that decided *priority* with *popularity* and *ignore factor* of the data. They developed a detailed cost model for their protocols and evaluated the performance through simulation. In the adaptive protocol proposed, which minimize access time and tuning time, they considered energy expenditure of an MH.

LDA is a new application of mobile computing, so the broadcasting method used in LDA must consider the location of the client. In this paper, in order to minimize energy expenditure of an MH, we propose broadcasting methods which broadcast LDD with linear clustered.

3   Broadcast to Support LDD

3.1 Preliminaries and Terms

We use the model of the mobile computing environment proposed in [1] and assume the following:
1. MSS periodically disseminates data of its database to MHs within the region covered by the MSS.
2. MHs within the region covered by MSS are able to find the current location of them.
3. Cache capacity of the MH is smaller than that of the database in the MSS, and the MHs are able to select and store data from the broadcast disk.
4. The MH stores the data contained in common region, which is overlapped between the cache region of MH and grids.
   We define the following terms:
- *Cell region:* the region covered by the MSS. This is a whole region of the cell.
- *Cache region:* the region covered by the cache of an MH. This is a set of grids that the cache wants to store in order to process LDQ. The size of cache region indicates the number of grids that can be stored in the cache [11].
- *Query region:* the region required by LDQ. This is a set of grids that is required by LDQ.
- *Setup time:* the time that MH spends tuning out of broadcast channel or tuning back in [8].
- *Active time:* the time that the MH spends in active mode in which MH is actively listening to the channel.

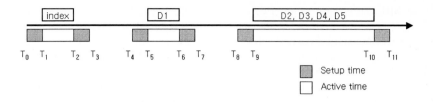

Fig. 1. Setup time and active time

Each of these terms is illustrated in Fig. 1. Consider an MH that downloads an index from T_1 to T_2 and data from T_5 to T_6 and from T_9 to T_{10}. In this scenario, setup time is { $(T_{11} - T_{10}) + (T_9 - T_8) + (T_7 - T_6) + (T_5 - T_4) + (T_3 - T_2) + (T_1 - T_0)$ }. The number of grids that the MH receives from the MSS is fixed because of the limitations of the cache region of the MH. So, download time cannot be reduced. In this paper, we propose a broadcasting method that reduces setup time such as ($T_9 - T_8$) and ($T_7 - T_6$) by broadcasting data of adjacent grids in consecutive order.

3.2 Broadcast to Support LDQ

When an MSS broadcasts data, it broadcasts an index to inform MHs of the broadcasting order in advance so as to reduce the tuning time of the MHs. We assume the data are distributed in 2-dimensional space because the data required by LDQ is related to the current geographical location of an MH. The MSS must decide their broadcasting order. If the MSS decides the broadcasting order by the geographical coordinates where the data is located, the broadcast schedule is determined without proper consideration of the geographical adjacentness of the data.

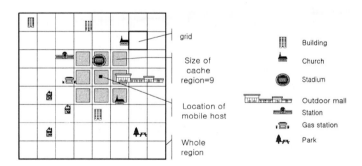

Fig. 2. Cache region when MH whose size of cache region is 9 is located in (5,4) grid

In Fig. 2, the region covered by the MSS is divided into grids. The MSS stores and broadcasts the data of each grid. As it constructs a broadcast index based on each grid, the total size of the broadcast index becomes smaller than that based on each data item. So it takes less time to broadcast the whole index, thus reducing the tuning time of an MH.

The MH stores the data of grids covered by the cache region of itself. If they are broadcast in consecutive order, the active time is reduced because of the decrease in the number of changes between active and doze mode.

For example, suppose the cache region size of the MH is 9 as illustrated in Fig. 2. The MH is located in the grid (5,4) and adjacent grids of the grid (5, 4) are broadcast from T_5 to T_6 and from T_9 to T_{10}. Before data to be cached are broadcast, the MH must change to active mode. If the data of adjacent grids are broadcast in consecutive order, $(T_9 - T_8) + (T_7 - T_6)$ are not necessary because the MH receives all data to be cached through only a setup.

As shown in the above example, if the data of adjacent grids are broadcast with clustered, the setup time of an MH can be reduced. But, when MHs are distributed in the cell region and the broadcast schedule is made suitable for a specific grid, the setup time of the MHs belonging to that grid is minimized but the setup time of the remaining MHs is much larger. This broadcasting method is not suitable to support LDQ efficiently because the average setup time of all MHs becomes larger. So, a suitable broadcasting method is needed for supporting LDQ efficiently in LDA system.

In this paper, we propose a method to decide broadcasting order using a sequence of space-filling curve, which is mapping function from a multi-dimensional space on to a one-dimensional space. This kind of mapping also makes spatially adjacent objects be stored with clustered.

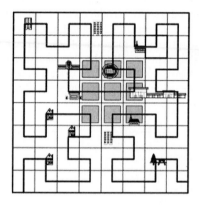

Fig. 3. Applying space-filling curve to grids dividing the cell region

Fig. 3 illustrates the broadcasting order of the data in a cell region using a sequence of space-filling curves. Because broadcast data are clustered using the space-filling curve, the data of the cell region are broadcast in adjacent order. Clustering broadcast data reduces the number of setting up MH so as to download data to be cached. So, we can use battery efficiently because the amount of energy expenditure of MH decreases in proportion to the reduction in the number of setting up.

4 Performance Study

4.1 Simulation Model

When the energy expenditure of an MH is considered, one of the most important criteria is active time. We are trying to predict and evaluate the energy expenditure of an MH by calculating active time because each kind of MH consumes a different quantity of energy in each system.

The simulator we have used is implemented in C on Linux. It simulates broadcasting methods using space-filling curves. We measure the active time of an MH in each method by using our simulator.

Table 1 shows the primary parameters we have used in our simulation study.

Table 1. Parameters used in Simulation

Parameters	Description	Value
Region	Size of Region the MSS takes charge of	3.2 (Km) * 3.2 (Km)
grid size	Size of grid	200 (m) * 200 (m)
Database	Size of Database	520
Period	Broadcast period	34.67 (sec)
Index_bucket	Size of Index bucket	16 (bytes)
Data_bucket	Size of data bucket	128 (bytes)
Clustering	Space-filling curve used in simulation	row-wise, z-order, Hilbert
Client	Number of clients in the cell	256
Bandwidth	Wireless bandwidth	28.8 (Kbps)
Cache region	Size of cache region	9, 16, 25, 36, 49, 64
Setup time	Setup time	0,04

The *clustering* parameter indicates the space-filling curve used to decide broadcasting order. Row-wise curve, Z-Order curve, Hilbert curve and Gray code curve are generally used as space-filling curves [9,6].

We divide the cell region into 256 grids. Our simulator assumes an MH starts in each grid and is randomly moving from grid to adjacent grid every few minutes. LDQ is performed on the data of grids stored in the cache of an MH. As the query region becomes larger, the MH requires a larger cache. In this paper we evaluate how clustering methods such as Row-wise curve, Hilbert curve and Z-Order curve adapt to the various sizes of the cache region.

4.2 Experiments

If the broadcast data are well clustered, the active time of an MH can be reduced. We measure only the setup time varying the size of the cache region in each method. We have excluded the download time in our experiments because it cannot be reduced as mentioned above.

Fig. 4 shows the average setup time of each method when the size of the cache region is varied from 3*3 to 9*9.

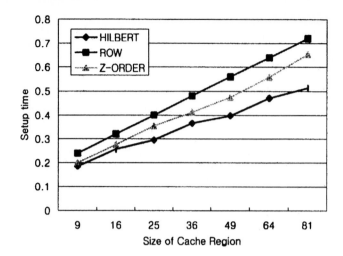

Fig. 4. Average setup time of the MH

Fig. 4 demonstrates that the performance of the broadcasting method using Hilbert curve is better than that of other methods in our experiments. As the size of cache region becomes larger vertically and horizontally with a grid, it causes the setup time to be constantly increased as much as value of *setup time* parameter in the broadcasting method using Row-wise curve. But, it does not in the broadcasting method using either Hilbert curve or Z-Order curve.

When the broadcast disk is constructed with Row-wise curve, the grids overlapped with the cache region of an MH are clustered with $\sqrt{cache\ region}$ sets of grids. For this reason, it takes as many as $\sqrt{cache\ region} * setup\ time$ for an MH to store all of the data of the grids overlapped into its own cache. When the broadcast disk is constructed with Hilbert curve or Z-Order curve, the data of the grids overlapped with the cache region are clustered with fewer sets of grids than that of Row-wise curves'. So, it takes less setup time than that of Row-wise curves'.

Fig. 5 and Fig. 6 show minimal (best) and maximal (worst) setup times in each method when the size of the cache region is varied from 3*3 to 9*9. There are few differences among the maximal setup times. But, there are significant differences among minimal setup times. Also the minimal and maximal setup time of the broadcasting method using Row-wise are same. These differences make the average setup times of them differed.

As results, we can find the consideration of the adjacentness of data makes the number of setting up of an MH.

5   Conclusions and Future Work

In this paper, we have proposed a broadcasting method using linear clustering in order to support LDQ efficiently. We have proposed constructing broadcast disk using space-filling curves. Changing the size of the cache region, we experimentally

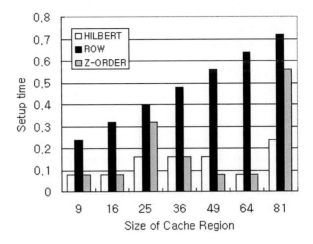

Fig. 5. Minimal setup time (best)

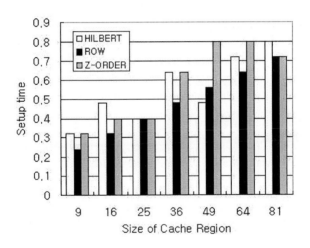

Fig. 6. Maximal setup time (worst)

evaluated the adaptability of the various methods proposed. We have estimated the setup time of the MH in each method proposed and evaluated their performances of them. The broadcasting method using Hilbert curve is used, is demonstrated to adapt well to the various size of the cache region. Also it has been shown that it leads to the significant reduction in the energy expenditure of an MH because the total setup time of the MH is reduced.

For ease of experiments, we have assumed throughout that the cell region is divided into square grids. Often different attributes may have different sizes. So, we intend to develop new broadcasting method using non-square grid in order to adapt to the various sizes of the cache region.

The initial purpose of this study was to develop an efficient caching policy for use in LDA system. To improve caching performance, a good broadcasting method is needed. We carried out this study to find a good broadcasting method. So, we also intend to make a study of caching methods and policies suitable for broadcasting method using linear clustering.

References

1.   Barbara, D., Imielinski, T.: Sleepers and Workaholics: Caching Strategies in mobile Environment. ACM SIGMOD '94. (1994) 1–12
2.   Barbara, D.: Mobile Computing and Databases – A Survey. IEEE Transactions on Knowledge and Data Engineering, Vol. 11, No. 1. (1999) 108–117
3.   Datta, A., Vandermeer, D.E., Celik, A., Kumar, V.: Broadcast Protocols to Support Efficient Retrieval from Databases by Mobile Users, ACM Transactions on Database Systems, Vol. 24, No. 1. (1999)
4.   Dunham, M.H., Kumar, V.: Location Dependent Data and its Management in Mobile Databases. Mobility in Database and Distributed System. (1998)
5.   Seydim, A.Y., Dunham, M.H.: Location Dependent Query Processing. MobiDE2001. (2001) 47–53
6.   Gaede, V., Gunther, O.: Multidimensional Access Methods. ACM Comput-ing Surveys, Vol. 30, No. 2. (1998) 170–231
7.   Imielinski, T., Viswanthan, S.: Wireless Publishing: Issues and Solutions. Mobile Computing, Kluwer Academic Publishers. (1996) 299–329
8.   Imielinski, T., Viswanthan, S., Badrinath, B.R.: Data on Air: Organization and Access. IEEE Transactions on Knowledge and Data Engineering, Vol. 9, No. 3. (1997) 353--372
9.   Jagadish, H.V.: Linear clustering of objects with multiple attributes. ACM SIGMOD '90. (1990) 332–342
10. Jing, J., Helal, A., Elmagarmid, A.: Client-Server Computing in Mobile Environments. ACM Computing Surveys, Vol. 31, No. 2. (1999) 117–157
11. Kim, S., Yong, H: Cache Replacement Strategies considering Location and Region Properties of Data in Mobile Database Systems. Journal of KISS : Databases, Vol. 27, No. 1. (2000) 53–63
12. Pitoura, E., Samaras, G.: Data Management for Mobile Computing. Kluwer Academic Publishers. (1998)
13. Ren, Q., Dunham, M.H.: Using Semantic Caching to Manage Location Dependent Data in Mobile Computing. In sixth Annual International Conference on Mobile computing and Networking, MobiCom2000. (2000) 210–221
14. Xu, J., L. D.K: Querying location-dependent data in wireless cellular environment. In W3C and WAP Workshop on Position Dependent Information Services. (2000)

Retransmission Tree Configuration for Reliable Multicast Protocols

Fumiaki Sato and Tadanori Mizuno

Faculty of Information, Shizuoka University,
3-5-1 Jo-hoku, Hamamatsu, Shizuoka 432-8011, Japan
{sato, mizuno}@cs.inf.shizuoka.ac.jp

Abstract. Some reliable multicast protocols are based on the retransmission tree that is used for the packet retransmission through the tree. In the mobile environment, the retransmission tree should be configured dynamically. The tree should be compact so that the efficient retransmission is achieved, and the link of the tree should be balanced so that the retransmission load is balanced. Although the tree is statically configurable by using the all delay information between nodes in the multicast group, the cost is high and inapplicable to the mobile environment in which the reconfiguration is frequent. In this paper, we propose a simple configuration algorithm of the retransmission tree in the mobile environment.

1   Introduction

Multicast [1,2,3] provides efficient delivery of data from a source to multiple recipients. It reduces sender transmission overhead, network bandwidth requirements, and the latency observed by receivers. This makes multicast an ideal technology for applications (e.g., teleconference, information services, distributed interactive simulation, and collaborative work) based upon a group communications model.

Especially, a reliable multicast protocol that has a retransmission mechanism is the very important protocol for the file transfer, distributed DB, and other reliable applications. SRM [4], RMTP [5,6], RMTP-II [7], MAR [8] and many other protocols have been developed. In these protocols, the mechanism based on the retransmission tree is very efficient because the retransmission is achieved in local (local recovery) and quickly. However, the configuration of the optimal retransmission tree is expensive because the calculation requires all delay information between nodes in the multicast group. In the mobile environment, reconfiguration of the retransmission tree is frequent because mobile nodes may change the geographical position and dynamically join and leave the group. Therefore, we need the efficient algorithm to configure the proper retransmission tree dynamically.

The retransmission tree should be compact so that the retransmission is achieved in local and quickly. For the balance of the retransmission load, the number of the link in the tree should be balanced. It is possible that the centralized node gather the delay

I. Chong (Ed.): ICOIN 2002, LNCS 2344, pp. 113–122, 2002.
© Springer-Verlag Berlin Heidelberg 2002

information of all nodes and calculates the optimal tree structure. But the architecture has one weak point and is not scalable. Therefore, we need the decentralized architecture and protocol to configure the proper tree based on a little information.

In this paper we propose a decentralized protocol to construct retransmission tree dynamically. The protocol does not recalculate the whole tree structure when a new node joins to the tree. The protocol serves the connection point that the new node should connect. Consequently, the cost of configuration is minimized and the effect to the retransmission can be avoided. In order to make the tree compact, the protocol must use the size information of the tree. But the whole information of the tree structure is too much to manage. Therefore, we define the value Node Depth (ND) that represents the size of the tree in the direction of the link. The largest value of the NDs in the node is called Maximum Node Depth (MND). The change of Node Depth propagates the information to the whole tree. On the other hand, for the balance of the retransmission cost, the information of the number of the retransmission nodes, i.e. the number of the links is used in the protocol. The last information is the delay between the new node and connection point. These three information, MND, number of links and delay between the new node and connection point are used to decide the connection point.

The rest of this paper is organized as follows. Section 2 describes the retransmission process in multicast protocols and related works of configuration of the retransmission tree. Section 3 presents our protocol to configure the retransmission tree. The simulation and evaluation of this method is described in Section 4. Section 5 is conclusions of this paper.

2   Retransmission in Multicast Protocols

IP multicast defines the copy of the packet and routing in the IP router. Therefore, the retransmission of the lost IP packet is managed the upper layer protocols. These are called as Reliable Multicast Protocols and many protocols have been developed. SRM is based on the NACK multicast and very scalable. RMTP, RMTP-II and MAR are based on the retransmission tree. Especially, the tree based retransmission is certain and efficient. The following is the retransmission of RMTP depicted in Figure 1.

RMTP is a tree-based reliable multicast protocol that controls the ACK implosion with aggregation of the ACK in the intermediate node. There is the designated receiver (DR) in the branch of the tree. The DR receives ACKs from all children in the branch, aggregates them and forwards the ACK to upper DR. The DR receives the multicast packet and stores the cache memory. If the child node requests the retransmission of the packet, DR retransmits the packet from this cache.

RMTP does not define the configuration of the tree. For the multicast group of the fixed nodes, the configuration can be configured by fixed parameters. However, mobile environment including the mobile terminals is not possible to assume the fixed tree structure. Therefore, the dynamic configuration protocol of retransmission is required.

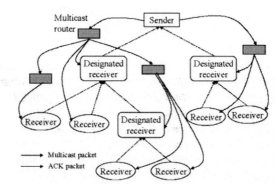

Fig. 1. An RMTP tree

As a dynamic tree configuration protocols, we have proposed a method based on the delay from a centered node (root node)[9]. Assumption in this protocol is that the sender is the root node and the retransmission is processed through the tree. Furthermore, the binary tree is assumed. As depicted in Figure 2, each node in the tree has a delay from root node. A new node can select the connection point by calculate the delay between each node of the tree and the parameter of each node. In the example of Figures 2, the smallest value of the total delay from root node is the left most node and the delay time is 8+2=10.

However, our method is not applicable to the many-to-many multicast that has multiple senders. The architecture also has a weak point that is the root node of the tree.

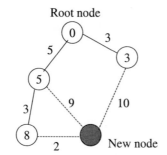

Fig. 2. Dynamic tree configuration.

3   Retransmission Tree Configuration Protocol

3.1   Basic Concept

We design the retransmission tree configuration protocol based on the following concepts.

(1) Decentralized management
(2) Simplicity of the protocol
(3) Scalability

Decentralized management of the information avoids the system failure. Simple protocol is good for the performance of the configuration and avoids the effect to the retransmission process. Scalability is required to support the great number of nodes in the group.

Configuration method we adopted does not recalculate the tree structure when a new node joins to the tree. Our method provides the connection point for the new node and constructs the tree additionally. Consequently, the cost of configuration is minimized and the effect to the retransmission can be avoided.

The retransmission tree should be compact so that the retransmission is achieved in local and quickly. For the balance of the retransmission load, the number of the link in the tree should be balanced. It is possible that the centralized node gather the delay information of all nodes and calculates the optimal tree structure. But the architecture has one weak point and is not scalable. Therefore, we propose the decentralized protocol to manage the tree information and to configure the tree.

3.2 Tree Configuration Protocol

In order to make the tree compact, the protocol must use the size information of the tree. We define the value Node Depth (ND) that represents the size of the tree in the direction of the link. The largest value of the NDs in the node is called Maximum Node Depth (MND). Figure 3 shows the ND and MND. In the Figure 3, there are four nodes, S, T, U and V in the tree. The delay between nodes is depicted as l (S-T), m (T-U) and n (U-V). Now, we consider the node U. The ND in the direction to the node T (ND_T) is defined as $l+m$ and ND in the direction to the node V (ND_V) is defined as n. The largest value of NDs, i.e. MND is $l+m$ (we assume $l+m>n$). These values of the NDs and MND are used to estimate the tree size in the direction of the link and decide the connection point.

ND and MND information are managed in each node and the changes of these values are maintained by propagation of the change. Since the data size of ND and MND is very smaller than the whole tree information, the maintenance cost is not expensive. When the size of the tree does not change even if the tree structure is changed, the ND propagation is omitted.

On the other hand, for the balance of the retransmission cost, the information of the number of the retransmission nodes, i.e. the number of the links (NOL) is used in the protocol. In the example of Figure 3, the NOL at node U is 2.

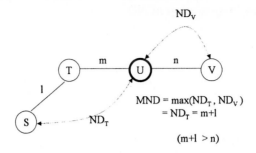

Fig. 3. Node Depth and Maximum Node Depth.

3.3 A Protocol Specification

3.3.1 Join Tree

(1) The joining node sends the join request via multicast to the all member node of the tree.

(2) The each acceptable node in the tree replies the MND and NOL.

(3) The joining node calculates the following equation and sends the link request to the member node that has the smallest value of the equation.

$$C = \alpha*MND + \beta*NOL + \gamma*DEL \qquad (1)$$

(4)The member node that receives the link request relies link response and processes the link operation.

3.3.2 Leave Tree

(1)If the leaving node is the leaf node in the tree, the node sends the leave request to the connected nodes. After receiving the leave response, the node can leave the tree.

(2)If the leaving node is the intermediate node in the tree, the node sends the pre-leave request to the connected nodes.

(3)The connected node relies the pre-leave response with NOL.

(4)The leaving node calculates the following equation and selects the node with the smallest value as the substitute node. Then, it sends the re-link request to the all connected nodes with the substitute node ID.

$$S = \mu*NOL + v*DEL \qquad (2)$$

(5)After receiving the re-link responses from all connected nodes and the all retransmission process, the node can leave the tree.

3.3.3 Setting and Propagation of ND

(A) Joining node
The ND in the direction to the connecting point is set the sum of the MND of the connecting point and delay between the connecting point and joining node. The MND is equal to the ND.

(B) The node which the new node connects Joining node
(1)The ND in the direction to the new node connecting point is set the sum of the delay to the new node.
(2)If the ND is greater than MND, the MND is set the ND and the MND propagates to the all nodes except the new node.
(3)If the ND is the second, the ND is sent to the node that is the MND.

(C) The node which the leaf node leaving
(1)If the leaving node is MND, the second ND becomes MND and third ND becomes the second.
(2)The second ND is sent to the node with MND and MND is sent to the all nodes except the node with MND.

(D) The node which the intermediate node leaving
(1)The delay between this node and the re-linked node are measured.
(2)The greatest ND among all nodes except the re-linked node is exchanged between the re-linked nodes.
(3)The ND in the direction of the re-linked node is calculated as the sum of the delay and exchanged ND.
(4)MND and the second ND are selected.
(5)If the MND is changed, the MND is sent to the all connected nodes except the new link.
(6)If the second ND is changed, the ND is sent to the node with the MND.

(E) The node which receives the notification of the ND change
(1)The ND in the direction of the changed node is set the sum of the received ND and delay between this node and the changed node.
(2)MND and the second ND are selected. If the MND and second ND are not changed, the propagation is omitted.
(3)If the MND is changed, the MND is sent to the all linked nodes except the MND node.
(4)If the second ND is changed, the ND is sent to the node with the MND.

4 Simulation and Evaluation

To evaluate the proposed method, we use the following simulation. We assume the 500msec*500msec X-Y coordinates that represent the delay between nodes in the group. We measure the mean MND and maximum NOL when 6000 nodes join and leave the tree incrementally. The mean MND means the size of the tree and maximum NOL means the balance of the tree. We assume the enough time intervals between join events or leave events for deployment of ND.

First of all, we evaluate the effect of the parameter values and chose the best combination of the parameters. We examine the mean MND and maximum NOL for different parameters of α, β and γ in the join process. Table 1 is the peak of the simulation. The smallest tree is the combination A and most balanced tree is D. For the total performance, we select the parameter combination A. Furthermore we examine the μ and ν in the leave process. Table 2 is the peak through the simulation. The tree at the start of the examination uses the result of the join examination with the parameter combination A. The smallest tree is the combination C and the most balanced tree is the combination E. As the precedent, we select the combination A.

Table 1. Parameter value evaluation (Join).

Combination	A	B	C	D	E
α	1	0.5	1.5	1	1
β	1	1	1	1	1
γ	1	1	1	0.5	1.5
Mean MND	803	868	932	1045	844
Max NOL	36	66	24	18	64

To compare with the existing method of dynamic tree configuration, we evaluate the method described in section 2. The parameters of α, β, γ, μ and ν are the selected values in the preceding examination. The parameter of the maximum branch number of the existing method is 50. It is the most suitable value that the most compact and balanced tree is generated in our simulation.

Figure 4 shows the simulation result of the mean MND when the member joins to the tree incrementally. Since the mean MND means the tree size, our algorithm can generate a more compact tree than existing method.

Figure 5 shows the result of maximum NOL when the member joins incrementally. The maximum NOL means the unbalance of the tree. The tree of our algorithm is more balanced than existing method.

Figure 6 shows the result of maximum NOL when the member leaves the tree. The maximum NOL is greater than existing method. Existing method is fixed to the 51 links because the parameter of the branch is fixed to 50. Maximum value of our simulation in Figure 6 is 62. So, it is not so different from existing method.

Table 2. Parameter value evaluation (Leave).

Combination	A	B	C	D	E
μ	1	4	8	1	1
ν	1	1	1	4	8
Mean MND	971	892	858	1106	1160
Max NOL	62	66	66	43	43

Figure 7 shows the result of the mean MND when the member leaves the tree. In this simulation, the same tree structure is used at the starting point. The mean MND of our algorithm is smaller than existing method.

At last, we evaluate the mean MND and maximum NOL in regular operation. In this situation, after the 3000 nodes join the tree, 6000 times of join event and leave event are generated randomly. In the examination, the parameter of branch number of the existing method is 10. The result is depicted in Figure 8 and 9. In our method, the number of the NOL is greater then existing method. But the NOL number is stable. On the other hand, the mean MND is smaller than the existing method. Furthermore, mean MND of our method is stable. It is good result for long-term operation.

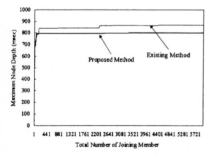

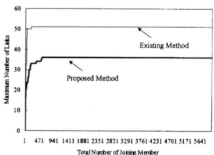

Fig. 4. The mean MNDat incremental join.   **Fig. 5.** Maximum NOL at incremental join

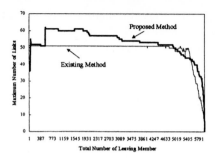

Fig. 6. Maximum ND at incremental leave.

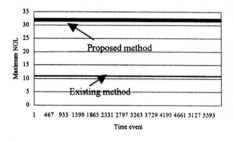

Fig. 7. Maximum ND at incremental leave.

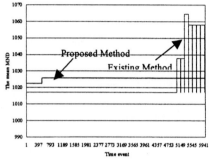

Fig. 8. The mean MND at regular operation.

Fig. 9. The maximum NOL at regular operation.

5 Conclusions

In this paper, we proposed a simple configuration algorithm of the retransmission tree in the mobile environment. Design concepts of our algorithm were the decentralized management, simplicity and scalability. Therefore, the protocol did not recalculate the whole tree structure when a new node joined to the tree. The protocol provided the connection point that the new node should connect. Consequently, the cost of configuration was minimized and the effect to the retransmission could be avoided.

In order to make the tree compact, the protocol should use the size information of the tree. But the whole information of the tree structure was too much to manage. Therefore, we defined the value Node Depth (ND) that represented the size of the tree in the direction of the link. The largest value of the NDs in the node was called Maximum Node Depth (MND). The change of Node Depth propagated the information to the whole tree. On the other hand, for the balance of the retransmission cost, the information of the number of the retransmission nodes, i.e. the number of the links (NOL) was used in the protocol. The last information was the delay between the new node and connection point (DEL). These three information, the MND, NOL and DEL were used to decide the connection point. Protocol specifications for joining tree, leaving tree and maintaining the ND were defined in this paper.

From our simulation, the tree configured by our algorithm was smaller and more balanced than the existing method. The feature is suitable for the efficient retransmission.

Future works are performance evaluation and implementation of the algorithm with the delay of the ND propagation and limited information gathering.

References

1. Deering, "Host Extensions for IP Multicasting," Indianapolis,IN:New Riders,1996
2. Eriksson, H. "MBone: The multicast backbone," Communications ACM, vol. 37 ,pp.54–60, 1994.
3. Macedonia,M.R., D.P.Brutzman, "MBone Provides Audio and Video Across the Internet," IEEE Computer, vol.27, no.4,pp.30–36,1994.
4. Sally Floyd,Van Jacobson,Steven McCanne, "A Reliable Multicast Framework for lightweight Sessions and Application Level Fraing," IEEE/ACM Transaction on Networking 1995.
5. J-C.Lin and S.Paul, "RMTP: A Reliable Multicast Transport Protocol," IEEE INFOCOM 1996, p p.1414–24, 1996.
6. S.Paul et al,. "Reliable Multicast Transport Protocol (RMTP)," IEEE JSAC, vol.15, no.3,1997.
7. B. Whetten, G. Taskale, "An Overview of Reliable Multicast Transport Protocol II," IEEE Network, Jan./Feb. 2000.
8. F.Sato, K.Minamihata, H.Fukuoka and T.Mizuno, "A Reliable Multicast Protocol with Total Ordering for Distributed Virtual Environment," In Proceedings IEEE ICDCS2000 Workshop on Group Communication Systems, 2000.
9. S. Vuong, A. Lo, A. Yung, Z.Yi and F. Sato, "Multicasting in Distributed Virtual Environment," In Proceedings the 6th International Conference on Distributed Multimedia Systems (DMS'99), pp.134–141, 1999.

An AAA Application Protocol Design and Service for Secure Wireless Internet Gateway Roaming

ByungGil Lee[1], HyunGon Kim, and KilHoum Park[2]

[1] Electronics and Telecommunications Research Institute
161 Gajeong-dong, Yuseong-gu, Daejeon, KOREA,
BGLee@etri.re.kr,
WWW ETRI, http://www.etri.re.kr
[2] Kyungpook National University, 1370 Sankyuk-dong,
Buk-gu, DaeGu, KOREA

Abstract. The Mobile IP based AAA (Diameter protocol) provides authentication, authorization, and accounting (AAA) service to a mobile network for achieving mobility in wireless IP access. This paper proposes a secured wireless Internet roaming service using wireless Internet Gateways (AAA Clients) and an AAA server. To support internetworking with wireless Internet gateways (i.e., WAP G/W), we also designed a diameter application protocol and scenarios. Traditional AAA schemes have not been used in such wireless Internet gateway networks to date. The basic strategy involves the use of the wireless Internet content of a portal that is provided independently from a mobile service provider's network and related inter-working with the outer mobile service provider's networks.

1 Introduction

Emerging IMT-2000 mobile networking standards will provide a framework for the improved handling of the mobile wireless Internet in tomorrow's wireless networks. One of the most daunting challenges facing mobile network operators is how to satisfy the pent-up demand for wireless Internet service [1].

Consumers are demanding increasing amounts of information over wireless Internet connections. To compete in this rapidly developing area, providers also must offer customers the capability of roaming through the networks of other providers, the customers to use their subscribed wireless access network.

As technologies are maturing and new wireless applications emerging, services are becoming characterized by inter-networking among mobile service providers and convergence with respect to fixed and mobile communication [2].

Today, wireless Internet Services (WAP-based, ME-based, etc.) are increasing the deployment of wireless data application devices and the development of multiple browsing function devices. With the increasing popularity of wireless Internet, users expect to be able to connect to any wireless Internet portal gateway which is convenient to their own menu. The portal infrastructure must then

I. Chong (Ed.): ICOIN 2002, LNCS 2344, pp. 123–135, 2002.
© Springer-Verlag Berlin Heidelberg 2002

be connected directly to a mobile service provider's home domain.

However, customers are interested in obtaining, not only the Internet services of their home domain, but the Internet services of any other wireless Internet gateway as well. Thus, after accessing the wireless resources of their home provider, a client often needs access to services provided by the wireless Internet services of a competitor. Mobile content service providers in a foreign domain typically require an AAA protocol to ensure a good business relationship with the client. This paper reports on the design of an the application protocol and scenarios for a Diameter protocol for a wireless Internet roaming service. The existing wireless Internet protocol structure is described in Chapter 2, and the architecture and protocol of an existing AAA is described in Chapter 3. In Chapter 4, the proposed AAA scheme of a wireless Internet gateway and an AAA server is described and, finally, in Chapter 5 some concluding remarks are offered.

2   Wireless Internet

2.1   Wireless (Mobile) Internet Service

The demand for Internet services and its applications for use in mobile devices is increasing rapidly. A wireless Internet Service structure is shown in Fig. 1. Wireless Application Protocol (WAP) is a set of protocols for optimizing standard TCP/IP/HTTP/HTML protocols in a mobile environment [3-6].

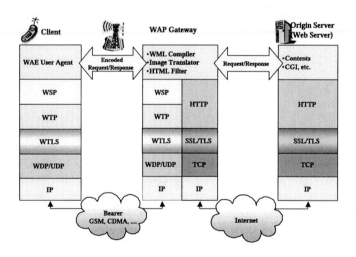

Fig. 1. WAP model

2.2   Wireless Internet Service Security

Mobile commerce (m-commerce) is a rapidly evolving area, but a lack of standards for secure transactions with the WAP protocol is clearly evident, despite the fact that wireless networks are thought to be secure [3]. Because of data exchange between TLS and WTLS protocols, the message transmitted between the handset and content web server is unencrypted for a very brief moment inside the gateway of the mobile service provider. In order to deliver these services and applications over the Internet in a secure, scalable and manageable way in a wireless environment, modifications to the wireless application protocol have been proposed [3]. However, for END-TO-END security, if we select cryptographic algorithms, user public key encryption techniques to distribute keys and authenticate each other in the upper layer, the web server of contents has no knowledge of the whether the mobile clients are properly certificated[3,4].

2.3   Wireless Internet Service in Inter-gateway Networking

Providers of wireless Internet portals will need to inter-network customer requirements and actual usage by classifying contents, so as to develop contents that are the most relevant. It is the very issue of contents and the type of services the user will require that will determine the successful implementation of service access through a wireless Internet portal.

The WAP gateway translates a user request into a hypertext transfer protocol (HTTP) or transmission control protocol (TCP) request and sends it to the appropriate origin server (or web server). The web server then sends the requested information back to the gateway via HTTP, TCP. The gateway translates and compresses the information, which can then be sent back to the micro browser via a mobile phone [5,6].

The wireless Internet gateway can be linked together so as to provide wireless subscribers with better services (Fig. 2). One possible linkage is by a wireless Internet service gateway (or proxy) which enables WAP servers and other similar servers (i.e., ME proxy) to access and negotiate information and the capabilities of the wireless network.

It also enables the wireless Internet service gateway to provide enriched services to the subscriber by interfacing with other wireless internet gateways. However issues such as Authentication, Authorization and Accounting need to be addressed.

3   AAA Protocol

3.1   The AAA Protocol and Structure

The traditional model is incapable of client-to-server authentication. Because only dial-up access service and terminal server access service in the Internet rely on the so-called RADIUS model [7]. As an AAA protocol, Diameter was heavily inspired and builds upon the tradition of the RADIUS protocol. The AAA protocol for a Mobile IP is currently under construction for standardization in IETF as RFC documents [7-14].

Within the Internet, a client belonging to one administrative domain (referred

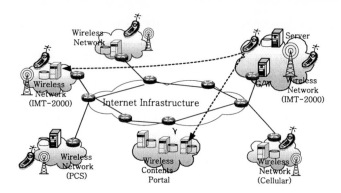

Fig. 2. Linking the wireless worlds

to as the home domain) often needs to use resources provided by another administrative domain (referred to as the foreign domain). An agent in the foreign domain that attends to the client's request (referred to as the "attendant") is likely to require the client to provide some credentials that can be authenticated before access to the resources can be permitted. When the local authority has obtained the authorization and notified of concerning the successful negotiation, the attendant can provide the requested resources to the client. There is a security trust chain model implicit in Fig. 3.

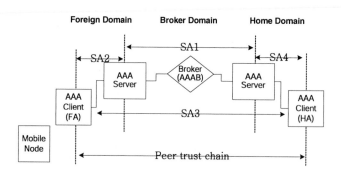

Fig. 3. AAA basic model

3.2   AAA Basic Security Function

We place the following basic security function on the AAA services in order to satisfy the wireless Internet service roaming of such clients.

- Public Key Authentication
- Security Message Support (Cryptography Message Syntax Security Application)

- Encryption (End-to-End Extension)
- Digital Signature
- Digital Signature against Authentication on Secured Message
- Authorization and Path Certification
- OCSP(Online Certificate Status Protocol)

Table 1. Security algorithm

Security Function	Algorithm
Keyed encryption	RSA
Signature	RSA with SHA-1
Message digest	SHA-1
Encryption / Decryption	Triple DES(AES)

4   AAA Protocol for Wireless Internet Roaming Service

4.1   Requirements Related to Basic Wireless Internet Roaming Service

The user may also wish to access the overlaying various outer wireless Internet data services which are provided in foreign portal gateway. We propose a wireless inter-roaming service between the home wireless Internet gateway and the outer wireless portal gateway, providing wider contents and full integration of authentication, authorization and accounting. This means providing seamless wireless Internet service mobility between different networks (cellular, PCS and IMT-2000) and outer wireless content networks as well.

The AAA system also serves as a bridge for different types of mobile operators that do not have the wherewithal to develop and maintain a rich content and need to partner with traditional Internet content providers. The basic model for the wireless Internet AAA architecture is designed to the basic AAA infrastructure shown in Fig. 4.

A security trust chain model is implicit and it is crucial to identify the specific security associations between the proposed wireless Internet gateway and the AAA infrastructures in the security model. Fig. 4 shows a Foreign Wireless Internet Gateway Agent, which is closely related to a foreign AAA server. A brokering AAA infrastructure may also exist. The AAA brokering infrastructure can be viewed as a trusted third party. It can be used and trusted to implement business agreements and act in a way that minimizes the overall number of security associations in the network. A system architecture reference model is illustrated in Fig. 5.

AAA and especially Diameter provide a wireless Internet protocol based system which includes functionalities such as

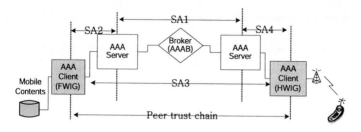

FWIG : Foreign Wireless Internet G/W
HWIG : Home Wireless Internet G/W

Fig. 4. Simplified wireless Internet supported AAA model

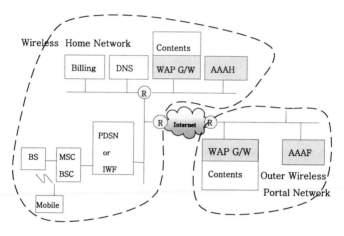

Fig. 5. System Architecture with AAA servers and Clients

- Simplified multiple service roaming management in a mobile access network
- Flexible mechanisms for new policy introduction with a wireless Internet service
- Allocation of a requested wireless Internet protocol gateway Agent
- Capabilities negotiation for a wireless Internet protocol gateway Agent
- Connection with a Mobile IP
- Reliable network with a Transport connection
- End-to-end security
- CA connection with a Broker Server
- Roaming between domains in mobile service providers
- Peer to Peer based network structure
- Support for vender specific AVPs and commands support
- Extendable (over 255 bytes) AVP, capable of wireless Internet services (numerous WAP applications)
- A Content based Accounting Mechanism
- A Real Time Accounting Mechanism

4.2   Wireless Internet Extension Command Code and AVPs

An AAA message must have AVPs as well as a command code field (24-bit address space) and must be managed from an IANA (Internet Assigned Numbers Association). At this point we have defined the new Diameter application for wireless Internet, the related protocol ID, Command Codes and AVPs as shown in Table 2 and 3.

Table 2. AAA Application Protocol for wireless Internet application

AAA Application Protocol	Protocol ID	Descriptions
NASREQ	1	
End-to-End Security	2	
Mobile IP	4	
Wireless Internet	8	New
Relay	0xffffffff	

Table 3. Command Code for wireless Internet application

	Security Function	Descriptions
	AA-Wireless-Internet-Request-Command	For Authentication
	AA-Wireless-Internet-Answer-Command	For Response
	Foreign-Gateway-WI-Request-Command	For Registration
	Foreign-Gateway-WI-Answer-Command	For Response

The AA-Wireless-Internet-Request message is used in order to request authentication and/or authorization for a given wireless Internet service user. In addition, the Foreign-Gateway-WI-Request message is used in order to request registration for a given wireless Internet service user. The User Name AVP (AVP code 1) must include a user ID (NAI style) that is a unique identifier of a foreign wireless Internet gateway. For example, in addition to an ID and network name (realm style) and soft key, a host ID must be provided.

4.3   Wireless Internet Extension Procedure

Fig. 6 depict capabilities negotiation and DSA (Diameter Security Association) set-up scenario with AAA node for the wireless Internet protocol. At this point we have defined the DSA procedure for End-to-End security of wireless internet roaming service.

Table 4. AVPs for wireless Internet application

AVPs(Major)	Descriptions
WI-Service-Binding-Update AVP	Foreign G/W-MN
WI-Service-Binding-Acknowledgement AVP	Foreign G/W-MN
WI-Service-Feature-Vector AVP	Service Feature
WI-Foreign-Gateway-ID AVP	Destination G/W Realm
WI-User-Name AVP	Mobile Node NAI
WI-Session-ID AVP	Session Verification
WI-Session-Key AVP	Security Association
WI-Accounting AVP	Accounting

1. HWIG sends a CER message to its relay agent indicating that it supports applications 1 (WI) and 2 (CMS Security).
2. The proxy server sends a CEA message to the HWIG indicating that it is a relay supporting all Diameter applications.
3. A xyz.com's foreign server sends a CER message to a proxy server indicating that it supports applications 1 (WI) and 2 (CMS Security).
4. The proxy server sends a CEA message to xyz.com's HWIG server indicating that it is a relay supporting all Diameter applications.

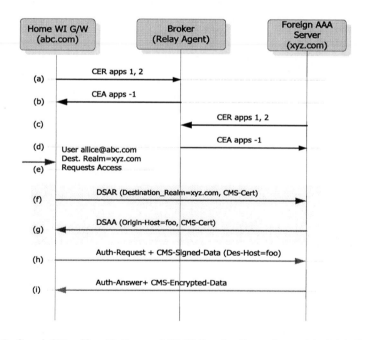

Fig. 6. Capabilities Negotiation and CMS Service Procedure with AAA Servers

5. The HWIG receives a request for access from a user (alice@abc.com). Destination Realm is xyz.com (Foreign)
6. The HWIG issues an DSAR message, with the Destination-Realm AVP set to xyz.com, and its certificate in the CMS-Cert AVP. The DSAR includes the set of AVPs that the HWIG expects to be encrypted, in the event that the home server returns messages that contain these AVPs.
7. An xyz.com's Foreign Server processes the DSAR message, and replies with the DSAA message. The DSAA also includes the set of AVPs that the HWIG server is expecting to be authenticated, as well as its certificate in the CMS-Cert AVP.
8. The HWIG issues an authentication request with the Destination-Host AVP set to the value of the Origin-Host AVP in the DSAA. The message includes the CMS-Signed-AVP, which authenticates the AVPs that were requested by the Foreign Server in the DSAA.
9. The Foreign AAA Server successfully authenticates the user, and returns a reply, which includes the CMS-Encrypted-Data AVP, whose contents include the AVPs that were specified by the HWIG in the DSAR.

Fig. 7 depicts the proposed scenario for the wireless Internet protocol gateway, inter-worked with the AAA node, taking into account the wireless Internet protocol and Diameter signaling. The WIG (Wireless Internet Gateway) is an AAA client entity that, among other wireless Internet application contents, serves as a gateway for protocol conversion in a mobile environment. The following figure and the subsequent description explain how this scenario functions.

1. Initially, a low layer connection is established and a capability negotiation procedure is performed. (SCTP or TCP)
2. The FWIG may optionally create a security association towards the AAAH using a PKI mechanism. This may involve either a PKI pre-shared key delivered by the AAA End-to-End response or by certificate exchange within the PKI.
3. The mobile node attempts to connect and to register a wireless access network using the access protocol by specific mobile communication procedures. The mobile node then initiates a packet data connection (PDSN, GPRS or IWF session) and sets up a Wireless Internet Protocol (i. e., WAP protocol) connection.
4. However, the mobile node's selections through a menu in the home access network are the wireless Internet services of the foreign network. The mobile node generates a registration request containing the NAI or MSISDN to the wireless Internet portal gateway, which is selected by the user.
5. The HWIG creates the AA-Wireless-Internet-Request (AWR) message and forwards it to the AAAH. The AAAH uses the NAI in the received AWR to forward the message to the appropriate AAAF, possibly through brokers (AAAB). The message may be delivered securely by deploying the AAA SA between the foreign (visited) and home networks.
6. Upon receipt of the AWR, the AAAF authenticates the mobile node by using the AVPs of AWR message and sends the FWR message to the FWIG.

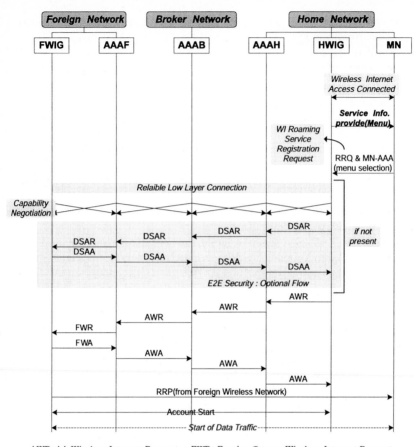

AWR :AA Wireless Internet Request   FWR: Foreign Gateway Wireless Internet Request
AWA :AA Wireless Internet Answer   FWA: Foreign Gateway Wireless Internet Answer

Fig. 7. Wireless Internet Roaming Service Procedure with AAA Servers

7. The FWIG replies to the FWA message that contains the foreign wireless
   Internet user registration.
8. The AAAF transforms the FWA to the AWA message and forwards the AWA
   to the AAAH, through the AAAB.
9. The wireless Internet roaming service may then begin. The accounting procedure also begins at this time.

4.4   Interworking with Mobile IP Network

While Mobile IP and related micro mobility protocols, as defined in the current
standards, constitute a fully operable protocol suite, there are still problems to
be solved and enhancements in wireless networks to be made [1,15].
We assume that after customers obtain Mobile IP network access, they will

want to set up a wireless Internet roaming service. In this case, the agent of the foreign network performs the AAA procedure for the Mobile IP service. In the following, the foreign network agent also performs the AAA procedure with the customer's desired network of the portal for the wireless Internet roaming service in the foreign access network. In order to provide inter-working with Mobile IP network with Diameter and wireless Internet services, the service management layer in the mobile access network has a Mobile IP supported service domain and a wireless Internet service domain [16-20]. It must construct two connections with path (1) (FA-AAAF-AAAH-AAAF♯2-FWIG) for the Mobile IP and path (2) (FA-AAAF♯1-AAAF♯2-FWIG) for wireless Internet roaming.

The location of the FWIG network may be closer to a Mobile IP foreign network than a Mobile IP home network, but it also must maintain the AAA connection with the home network in the signal. The connection with the Mobile IP is shown in Fig. 8.

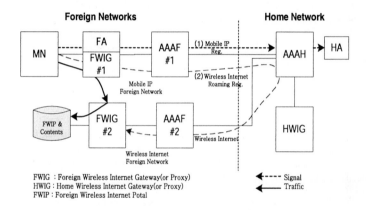

Fig. 8. Connection with Mobile IP

5   Conclusion

This paper defines a novel wireless Internet roaming service network architecture. It is based on integration of the AAA and also takes into consideration a connection with a Mobile IP service. The application of Authentication, Authorization and Accounting is a vital issue in mobile networks and service roaming between networks for wireless Internet. Of course, the challenge is delivering the appropriate foreign content or service when the subscriber wants it. Since these wireless Internet services are not designed to manage foreign gateway roaming, they need to be kept appropriate AAA mechanism. And a multi-browsing terminal can be used for operator-roaming service. Roaming service-providers must also deliver subscriber related connection information, including each user credentials (MSISDN, NAI or policy) and the infrastructure for connection must be opened. Mobile operators may hesitant to open their wireless Internet infrastructure (i.e., WAP gateway).

However, the AAA infrastructure can be securely used as a subscriber bridge to manage and maintain authentication, authorization, accounting and provisioning of services that are enabled by the mobile operator's and wireless Internet portal service provider's infrastructure.

We present AAA network connection scenarios using the SA (security Association) chain model in a mobile communication system. It is possible to search for AAA usage in many different application areas. In this paper we have limited our investigation to the use of Diameter within a wireless Internet protocol.

We propose a new application to a wireless Internet gateway that makes use of AAA (Diameter protocol) services for achieving service diversity and quality. AAA can be synergistically combined with a wireless Internet gateway to provide innovative wireless Internet roaming services. Combining the service infrastructures in wireless Internet networks and wireless Internet applications enables wireless carriers and content service providers to deliver a new opened wireless network service to wireless subscribers.

References

1. D. Ralph and C. Schephard.: Services via Mobility Portals. IEE 3G Mobile Communication Technologies Conference Publication. **477**, March (2001) 38–43
2. M. Cannataro and D. Pascuzzi.: A Component-based Architecture for the Development and Deployment of WAP-Compliant Transactional Servi. 34th Hawaii International Conference on System Sciences. IEEE (2001)
3. N. Komninos and B. Honary.: Modified WAP for Secure Voice and Video Communicati. IEE 3G Mobile Communication Technologies Conference Publication **477**, March (2001) 33–37
4. Martin Christinat, Marcus Lsler.: WTLS-The security layer in the WAP stack. keyon, Jun (2000)
5. The Wireless Application Protocol, Wireless Internet Today. Unwired Planet, Inc. Feb (1999)
6. WAP Forum.: Wireless Application Protocol Architecture Specification. Version 1.2 and 2.0, http://www.wapforum.org.
7. C. Rigney et al.: Remote Authentication Dial In User Service (RADIUS). IETF RFC **2138**, Apr. (1997)
8. IETF AAA Working Group Charter: http://www.ietf.org/html.charters/aaa-charter.html
9. IETF AAA Working Group Internet Draft.: Diameter Framework Document, http://www.ietf.org.
10. IETF AAA Working Group Internet Draft.: Diameter Base Protocol, http://www.ietf.org.
11. IETF AAA Working Group Internet Draft.:Diameter Mobile IP Extension, http://www.ietf.org.
12. IETF AAA Working Group Internet Draft: Diameter End-to-End Security Extension, http://www.ietf.org.
13. IETF AAA Working Group Internet Draft.:Proxy Nodes in AAA Configurations, http://www.ietf.org.

14. Christopher Metz.: AAA Protocols: Authentication, Authorization, and Accounting for the Internet. Cisco Systems, IEEE Internet Computing November-December (1999)
15. T. Hiller et al.: Cdma2000 wireless Data Requirements for AAA. RFC **3141**, June (2001)
16. IETF Mobile IP Working Group.: Mobile IP Authentication, Authorization, and Accounting Requirements. RFC **2977**
17. R. Caceres and L.Iftode.: Improving the Performance of Reliable Transport Protocols in Mobile Computing Environments. IEEE JSAC, vol. **13**, no. **5**, June (1995), 850–57
18. Charles E. Perkins.:Mobile IP joins Forces with AAA. IEEE Personal Communications, August (2000)
19. Charles E. Perkins and David B. Johnson.: Mobility Support in IPv6. ACM Mobicom **96**, November (1996)
20. Townsley, W. M, Valencia, A., Rubens, A., Pall, G. S., Zorn, G., Palter, B.: Layer Two Tunneling Protocol (L2TP). IETF RFC **2661**, August (1999)

Fast Move Detection over Mobile IP

Jin-Woo Jung , Jung-Hoon Cheon, Sangwoo Nam*, Youn-Joo Lee*, and
Hyun-Kook Kahng

Department of Electronics Information Engineering, Korea University
#208 Suchang-dong Chochiwon Chungnam, Korea 339-700
{grjjw, grcurly, kahng}@tiger.korea.ac.kr
Core Network Research Dept. ETRI
161 Gajeong-dong Yuseong-gu Daejeon, 305-350 Korea*
{namsw, yjlee}@etri.re.kr*

Abstract. Mobile IP has been proposed for a solution to provide IP-layer
mobility including handoffs between subnets served by different Foreign
Agents. However, the latency involved in these handoffs can be longer than the
threshold value, required to support the delay-sensitive or real-time services. To
reduce the latency for Mobile IPv4 handoffs, this paper presents a move
detection mechanism within IP-layer. Proposed method is based on strict source
routing in IP specification and includes the extension of router advertisement
message and ICMP error reports. The proposed method is verified through the
numerical results and the experimental results.

Keywords. Mobile IP, move-detection, layer 3.

1 Introduction

From a layer architecture point of view, a network layer should feature a mobility
support. Then, the users can enjoy their mobility whatever data-link devices and user
applications they use. Mobile IP has been proposed for a solution to provide IP-layer
handoffs between subnets served by different Foreign Agents. However, it still suffers
an interruption of the communications for reliable or real-time applications.

To the TCP protocol, mobile IPv4 service disruption is perceived as an indication
of congestion. Successive TCP timeouts will increase the TCP timeout interval, which
delays the mobile IP handoff. TCP is not in the position to immediately resume the
communication, even after the mobile IP handoffs has completed[3]. Due to the
increasing TCP timeout interval, most of current move detection methods using agent
advertisement for mobile IP handoffs have the fast handoff problem.

We suggest the shortest move detection method which is based on the strict source
route(SSR) option of IP version 4 for overcoming this problem. In the proposed
method, a mobile node can achieve move detection by receiving a redirection
message or ICMP error report : RSS(Redirect Subnet Switching).

The proposed techniques allow greater support for TCP services or real-time
services on a mobile IPv4 network by minimizing the period of time when a mobile
node is unable to send or receive IP packets due to the delay in the mobile IPv4 move
detection. It also requires the minimal modification of an existing mobile IP without
additional new entity.

I. Chong (Ed.): ICOIN 2002, LNCS 2344, pp. 136–142, 2002.
© Springer-Verlag Berlin Heidelberg 2002

But, this paper assumes that the underlying link layer connection is already established, and the layer 3 cannot use the function of layer 2. Therefore, paper is specified under the assumption that the technique on layer 3 is independent of the technique on underlying layer 2 for providing move detection.

To verify the performance of proposed method, we implement the RSS method on FreeBSD 4.3 and measure the performance of the implemented module between two small subnets. Also, we analyze the numerical results by comparing the RSS with the existing move detection methods.

2   Redirect Subnet Switching (RSS)

In this chapter, we will investigate the existing move detection method for mobile IP and show the RSS method in last part.

2.1   Related Work

- LCS(Lazy Cell Switching): It uses agent advertisement lifetime expiration as an indication of mobile node movement. If mobile node misses three successive agent advertisements, the mobile node must discover a new agent and register one of its care-of addresses with the Home agent (Figure 3(a)).
- ECS(Eager Cell Switching) : It is similar to LCS, but it is appropriate for nodes to handoff immediately upon encountering a new agent (Figure 3(b)).
- Prefix Matching : This method has a similar functionality to the LCS with the only difference being that it includes a prefix-length extension to agent advertisements (Figure 3 (a)).

2.2   Overview of RSS

Our approach specifies that IP-layer technique is independents of the underlying link layer techniques for providing move detection. RSS is based on the characteristic of SSR option. That is, when mobile node moves from one subnet to other subnet, the address of a default router is changed, resulting in redirect route setup message sent to mobile node. The overall RSS method is summarized in Figure 1.

Following steps provide more detail on this method :

- A mobile node sends modified datagrams with setting SSR .
- If a default foreign router acts mobility agent, it sends Redirect Route Setup message[2] to mobile node.
- The mobile node can achieve move detection upon receipt of Redirect Route Setup and, if appropriate, it sends a Registration Request[1] to Default Foreign Router requesting a simultaneous binding.

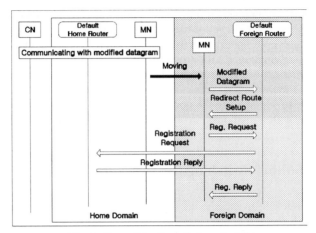

Fig. 1. General Redirect Subnet Switching

Figure 1 specified the general RSS handoff. Because mobile node transmits packets, which involve strict source route option, the default foreign router replies with a Redirect Route Setup Message. When mobile node receives this message from router that takes the place of mobile agent function, this message is similar to the agent advertisement message of Mobile IPv4. To be opposed, if a default foreign router does not act as mobility agent(or no foreign agent within foreign network) or does not exist in foreign network, follwing procedures are adapted.

When mobile node is received a redirect route setup message from router that does not take the place of agent function, this redirect route setup message does not contain the location information of agent: Generic ICMP error message.

In this case(does not exist FA), when mobile node receives an original ICMP error message, the mobile node may alternatively operate foreign agents by replacing them with a dynamic host configuration service(DHCP). Then, the mobile node sends registration request message with using allocated CCOA to its home agent.

Following steps provide more detail when does not exist foreign agent:
- A mobile node sends modified datagrams with strict source route option.
- If default foreign router does not act as mobility agent, the default foreign router sends ICMP errors message to the mobile node.
- The mobile node move detection upon receipt of ICMP error message and, if appropriate, it triggers the procedure by using DHCP.
- If the mobile node obtains CCOA in successful, it sends registration request message with using allocated CCOA to HA.

2.3 Improved Redirect Subnet Switching

General RSS has two critical problems. : RSS method has overhead per datagram(e.g. for Ethernet Link, about 0.8%), and cannot achieve move detection when mobile node does not send datagrams. For solving this problem, we suggest improved RSS combined with LCS in [1].

Combined procedures of LCS and RSS method is divide following steps:

- When mobile node receives Advertisements from any FA, mobile node should record the lifetime received in any advertisements, until that lifetime expires. Then, MN sends datagrams using the RSS method during specified time(i.e., half of advertisement period).
- If the mobile node fails to receive another advertisement from the same agent within the advertisement lifetime, it should assume that it has lost contact with that agent. Then, mobile node should send datagrams using the RSS method during specified time.
- When the mobile node receives Redirect Route Setup message or ICMP error report, it can achieve move detection.

The overall Improved RSS move detection method is summarized in Figure 2 below: In figure 2, a half of advertisement interval is notated to X.

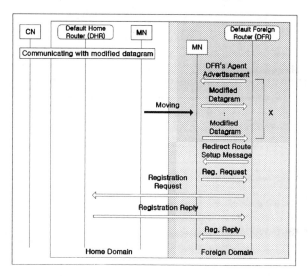

Fig. 2. Procedure of Improved RSS

3   Performance Analysis

This chapter shows the performance of each method. For the simple analysis, we assume that subnets have a single-agent and overlapped area dose not exist.

3.1   Numerical Results

Mobile IP hand-off delay is consisted of the movement detection delay and mobile node registration delay. As registration delay is the same time for three methods, we assumed that mobile node registration delay is equivalent to the RTT(Round Trip

Time between mobile node and home agent) for each method. Figure 3 illustrates an advertisement timing graph between Home Network and Foreign Network.

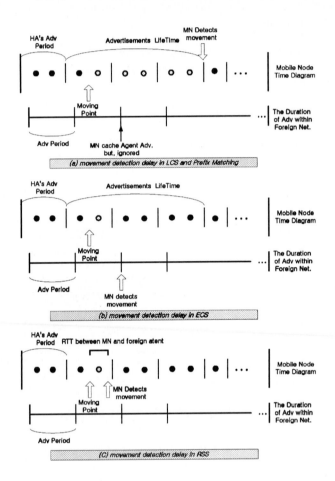

Fig. 3. Advertisement timing graph

In figure 3, both advertisements occur at the fixed period interval x. Therefore, the average delay of LCS is 5x/2 and that of ECS is x/2[3]. But, RSS is independent x and only depends on the RTT(Round Trip Time between mobile node and foreign agent). Figure 4 shows the movement detection delay for each method.

In figure 4, we assume that the advertisement period is 500ms and the transmission rate of mobile node varies form 20 kbps to 80kbps. Also, it is assumed that the movement point of mobile node is random variable. The LCS and ECS average delays are about 1.25 and 0.25 seconds respectively. It can also be seen that TCP reacts to the network disruption with successive timeouts and packet retransmissions. The functionality of the TCP retransmission scheme is to dynamically determine the

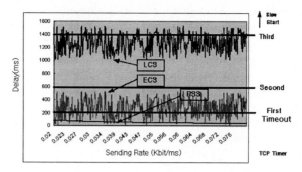

Fig. 4. Movement detection delay for each method

value of the TCP timeout interval with every packet transmission. In the case of a timeout, TCP doubles that interval. During Mobile IP handoffs, A TCP experiences several successive timeouts that increase the timeout interval beyond the duration of the handoff. As a result the TCP communication remains halted even after the completion of the mobile IP handoff. In figure 4, LCS method always experiences third timeout and may cause the slow start. Though even ECS have the average delay time of 0.25ms, it can communicate with peer node after 600ms. However, as proposed method has the average delay of 57 ms, RSS never experiences the TCP's timeout. Therefore, RSS dose not cause the degeneration of TCP's performance.

3.2 Experimental Results

From figure 5, it can be seen that the experimental results verify the theoretical results. In our test, we have used the RSS implementation on FreeBSD 4.3. This implementation is fully compliant with general RSS[2] and is implemented as a part of mobile node function. For the test, traffic has been generated using by ftp program and the performance of TCP in mobile node is measured by using traffic analyzer. As a sequence, the mobile node can detect within about 40ms. Therefore, RSS method is not affected by an agent advertisement period.

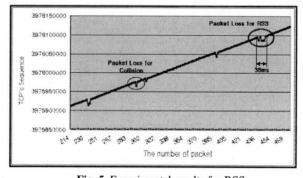

Fig. 5. Experimental results for RSS

4 Conclusion

In this paper, we suggested a new method to provide fast handoffs between subnets in IP-layer. Our approach is applicable to the support of delay-sensitive or real-time services and is useful to the subnet that has many small cells. Because a subnet is consisted of multi-cells and one router, the above statement is clear.

Also, a TCP of the existing move detection algorithms suffers severally from the exponential backoff, but proposed method has rarely suffers from TCP's exponential backoff due to only depend on transmission rate. It was verified with theoretically and experimentally results.

Because our algorithm has two problems, we suggested improved method that is combined with LCS. This method can compensate the problem of the RSS, but it has disadvantage that increases the traffic of advertisement in the subnet. Also, it is the alternative method for solving the problem of RSS that the RSS has the keepalive timer and periodically sends keepalive message when mobile node does not have any packet for transmission.

References

1. C.Perkins, "IP Mobility Support", RFC2002, Oct. 1996.
2. Hyun-Kook Kahng, "Redirect Subnet Switching in Mobile IPv4", draft-kahng-korea-rss-mipv4-00.txt, Agust 2001.
3. N. Fikouras, K. El Malki, S. Cvetkovic, C. Smythe, "Handoffs in Single-Agent Sub-networks", WCNC'99 New Orleans, USA.
4. Stevens, R., "TCP/IP Illustrated Volume I.", Addison Wesley Professional Computing Series, 1995.

A New Location Caching Scheme Employing Forwarding Pointers in PCS Networks[*]

Ki-Sik Kong[1], Joon-Min Gil[2], Youn-Hee Han[1], Ui-Sung Song[1], and Chong-Sun Hwang[1]

[1] Dept. of Computer Science and Engineering, Korea Univ.
1, 5-Ga, Anam-Dong, Sungbuk-Gu, Seoul 136-701, Korea
{kskong, yhhan, ussong, hwang}@disys.korea.ac.kr
[2] Dept. of Computer Science, Univ. of Illinois at Chicago,
851 S. Morgan, Chicago IL 60607, U.S.A.
joon@cs.uic.edu

Abstract. This paper proposes a new location caching scheme that can reduce the signaling cost for location management in PCS networks. In the existing cache scheme, the use of cache information can effectively reduce the signaling traffic for locating frequently called mobile users. However, when the cache information is obsolete, it results in much more signaling traffic than that of the IS-41. In order to solve this problem, we propose a new location caching scheme called the PB-Cache scheme, which exploits a user's movement locality as well as call locality. Even if the cached information is not up-to-date, the called user can be found by tracing forwarding pointers in the VLR's without querying the HLR. Thus, the PB-Cache scheme can effectively reduce the signaling traffic for location management, and makes it possible to establish a rapid call setup. Besides, it distributes the signaling and database access load on the HLR to the VLR's. The analytical results indicate that the PB-Cache scheme significantly outperforms the other schemes when a user's call-to-mobility ratio is high or the signaling traffic to the HLR is heavy.

1 Introduction

Personal Communications Service (PCS) networks provide wireless services to subscribers that are free to travel, and the network access point of a mobile terminal (MT's) changes as it moves around the network coverage area. A location management scheme, therefore, is necessary to effectively keep track of the MT's and locate a called MT when a call is initiated [1]. There are two commonly used standards for location management: IS-41 and GSM [2]. Both are based on a two-level database hierarchy, which consists of Home Location Register (HLR) and Visitor Location Registers (VLR's).

The whole network coverage area is divided into cells. There is a Base Station (BS) installed in each cell and these cells are grouped together to form a larger

---

[*] This work was funded by the University Research Program supported by Ministry of Information & Communication in South Korea.

I. Chong (Ed.): ICOIN 2002, LNCS 2344, pp. 143–153, 2002.
© Springer-Verlag Berlin Heidelberg 2002

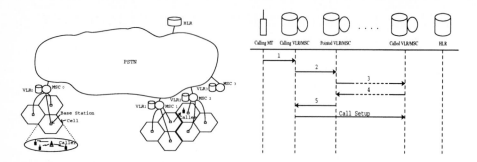

Fig. 1. An example of locating an MT   **Fig. 2.** Call delivery under the PB-Cache
under the PB-Cache scheme   scheme (cache hit)

area called a Registration Area (RA). All BS's belonging to one RA are wired to a
Mobile Switching Center (MSC). The MSC/VLR is connected to the Local Signal
Transfer Point (LSTP) through the local A-link, while the LSTP is connected
to the Regional Signal Transfer Point (RSTP) through the D-link. The RSTP
is, in turn, connected to the HLR through the remote A-link [3].

Location management is one of the most important issues in PCS networks.
As the number of MT's increases, location management scheme under the IS-41
has gone through many problems such as increasing traffic in network, bottleneck
to the HLR, and so on. To overcome these problems under the IS-41, a number
of works have been reported. A caching scheme [4] was proposed to reduce the
signaling cost for call delivery by reusing the cached information about the called
MT's location from the previous call. When a call arrives, the location of the
called MT is first identified in the cache instead of sending query messages to
the VLR. When a cache hit occurs, the cache scheme can save one query to the
HLR and traffic along some of the signaling links as compared to the IS-41. This
is especially significant when the call-to-mobility ratio (CMR) of the MT is high.
However, a penalty has to be paid when there is "location miss" since the cache
information is not always up-to-date. In this paper, we propose an enhanced
cache scheme called PB-Cache scheme. By exploiting a user's movement locality
as well as call locality at the same time, this scheme reduces the access to the
HLR and the signaling traffic for location management throughout the networks.

This paper is organized as follows. Section 2 introduces a new location caching
scheme employing forwarding pointers (PB-Cache scheme). Analytical model for
comparing location management costs is given in Section 3. Section 4 shows an
analysis of the location management costs under the PB-Cache scheme. Finally,
conclusions are given in Section 5.

2   A New Location Caching Scheme

Fig. 1 shows an illustrative example of locating the MT under the PB-Cache
scheme. The VLRo represents the calling VLR where the caller resides. We as-

sume that the cache information exists in the MSC_0 and the cache entry for the called MT currently points to the VLR_1.

Let's consider that the called MT has moved from VLR_1 to VLR_2 after the last call arrived. Then, the current location of the called MT is the RA of the VLR_2. When the next call arrives, the MSC_0 first queries the pointed VLR, that is, VLR_1. In this case, since the existing cache scheme has to perform the call delivery procedure of the IS-41 after an unsuccessful query for the cache, it results in the waste of the signaling traffic as compared to the IS-41. However, under the PB-Cache scheme, even if the cache information is obsolete, it traces the pointer chain without querying the HLR until the current location of the called MT is found within the maximum pointer chain length of K (See Table 1). So, the saving of one query to the HLR and traffic along some of the signaling links can be obtained. Note that the pointer chain length has to be limited due to the maximum pointer setup delay requirement. In Fig. 1, we assume that the maximum pointer chain length, denoted by K, is preconfigured to be one. Therefore, unless the called MT moves into the RA of the VLR_3, it can be found through both cache and forwarding pointer information without querying the HLR. This is motivated from locality pattern in a user's movement. In the following, more detailed procedures for the PB-Cache scheme are described.

2.1   Location Registration

Compared with the IS-41, most procedures for location registration under the PB-Cache scheme are exactly the same as those of the IS-41 except that the forwarding pointer is additionally set up between the two VLR's.

2.2   Call Delivery

Most procedures for call delivery under the PB-Cache scheme are almost the same as those of the cache scheme except that the called MT is traced through the pointer chain length of K. When the cache information is obsolete, the pointer chain is traced to find a called MT starting from the VLR pointed in the cache. If the called MT is located within the pointer chain length of K from that VLR, it can be found without querying the HLR.

Fig. 2 describes the procedures for a cache hit under the PB-Cache scheme. The cache hit under the PB-Cache scheme contains two situations: One is the situation that the cache information is correct. Thus, the called MT is found after the only one query to the pointed VLR. The other is the situation that the cache information is not correct. In this case, after querying the pointed VLR, the called MT is found by tracing through the pointer chain length of K (See step 3 and 4 in Fig. 2). The cache miss under the PB-Cache scheme occurs when the called MT is not found even if forwarding pointer chain has been traced until the length of K. After this, the same call delivery procedure as that of the IS-41 is performed. In this case, the current location of the called MT is transmitted from the HLR to the calling VLR together with the cache update request message.

Table 1. Costs and parameters

Parameter	Description
C_{la}	Cost for sending a signaling message through the local A-link
C_d	Cost for sending a signaling message through the D-link
C_{ra}	Cost for sending a signaling message through the remote A-link
C_v	Cost for a query or an update of the VLR
C_h	Cost for a query or an update of the HLR
t	Probability (VLR pointed in the calling MSC's cache is the VLR within the same LSTP area)
s	Probability (Any given movement or searching by the MT is within its current LSTP area)
j	Length of pointer chain traced until the MT is found ($0 < j \leq K$)
K	Maximum allowable length of pointer chain which can be traced to locate the MT
q	The MT's CMR
p	Cache hit ratio under the PB-Cache scheme
$\bar{p}$	Cache hit ratio under the cache scheme
$\hat{p}$	Probability (The MT is found by tracing forwarding pointers when the cache information is not up-to-date)

3 Performance Modeling

3.1 Basic Assumptions and Parameters

The basic assumptions for the performance modeling are as follows:

- The call arrivals to an MT follows the Poisson distribution with a mean arrival rate λ_c.
- The residual time of an MT in an RA follows the Gamma distribution with mean $1/\lambda_m$, where λ_m is the movement rate.
- We use the extended meaning of the cache hit ratio under the PB-Cache scheme. In other words, if the MT can be found without querying the HLR even if the cache information is obsolete, we include this case with a cache hit. Therefore, $p = \bar{p} + \hat{p}$ (See Table 1).

Table 1 shows the costs and parameters used for the performance modeling.

3.2 Comparison of the Cache Hit Ratios

The cache hit ratio is determined by an MT's CMR. We define a CMR, which is the average number of calls to an MT from a given originating MSC during the period that the MT visits an RA. Thus, the CMR can be denoted by $\frac{\lambda_c}{\lambda_m}$, where λ_c and λ_m represent a mean call arrival rate and a mean movement rate, respectively. With the λ_c and λ_m, the Laplace-Stieltjes Transform of a

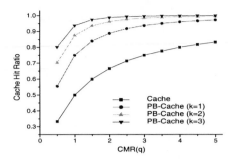

Fig. 3. Comparison of the cache hit ratios

Gamma random variable is expressed as follows: $f_m^*(\lambda_c) = \left(\frac{\lambda_m \gamma}{\lambda_c + \lambda_m \gamma}\right)^r$. In order to simplify the analysis, we set $\gamma = 1$ such that the residual time of an MT in an RA follows the exponential distribution, thus $f_m^*(\lambda_c) = \frac{\lambda_m}{\lambda_c + \lambda_m}$.

The cache hit ratio under the PB-Cache scheme can be obtained by the sum of the probabilities that the called MT resides in each RA between two consecutive phone calls. In the following, $\alpha(K)$ means the probability that the called MT moves across K RA's between two consecutive phone calls [5]. Thus, we can derive the cache hit ratio under both the PB-Cache scheme and the cache scheme from $\alpha(K)$.

$$\alpha(K) = \begin{cases} 1 - \frac{1}{q}[1 - f_m^*(\lambda_c)] & (K = 0) \\ \frac{1}{q}[1 - f_m^*(\lambda_c)]^2[f_m^*(\lambda_c)]^{K-1} & (K > 0) \end{cases} \tag{1}$$

Using equation (1), $p, \bar{p}$, and $\hat{p}$ can be derived as follows:

$$\bar{p} = \alpha(0) = \frac{q}{q+1} \tag{2}$$

$$\hat{p} = \sum_{i=1}^{K} \alpha(i) = \frac{1}{q+1}\left(1 - \frac{1}{(q+1)^K}\right) \tag{3}$$

$$p = \bar{p} + \hat{p} = \sum_{i=0}^{K} \alpha(i) = 1 - \frac{1}{(q+1)^{K+1}} \tag{4}$$

Equation (2) indicates the cache hit ratio of the cache scheme, and equation (4) indicates the cache hit ratio of the PB-Cache scheme. As mentioned above, equation (3) means the probability that the MT is found by tracing forwarding pointers when the cache information is not up-to-date. Fig. 3 shows that the cache hit ratio under the PB-Cache scheme is much higher than that of the cache scheme. Using the higher cache hit ratio, the PB-Cache scheme can reduce the access to the HLR and the delay in call delivery for location management.

3.3   Analytical Model for Location Management Costs

In PCS networks, the delays for connection establishment, location registration, and call delivery are considered to be some of the most important performance criteria. In our analytical model, the signaling and database access costs are used to measure the performance of both the PB-Cache scheme and the cache scheme. These costs are associated with the delays needed to perform the signal transmission and database update/query. For the analytical model, we also assume that there are q call arrivals for an MT between two RA crossings. Based on these assumptions, we describe location management costs for the IS-41, the cache scheme, and the PB-Cache scheme, respectively.

Location Management Costs under the IS-41:
The signaling cost for location registration (UNC_{IS}) and the signaling cost for call delivery (SNC_{IS}) are $4(C_{la} + C_d + C_{ra})$, respectively.
Then, the total signaling cost (NC_{IS}) is

$$NC_{IS} = UNC_{IS} + qSNC_{IS} = 4(1+q)(C_{la} + C_d + C_{ra}) \tag{5}$$

The database access cost for location registration (UDC_{IS}) and the database access cost for call delivery (SDC_{IS}) are $2C_v + C_h$, respectively.
Then, the total database access cost (DC_{IS}) is

$$DC_{IS} = UDC_{IS} + qSDC_{IS} = 2C_v + C_h + q(2C_v + C_h) \tag{6}$$

From (5) and (6), the total cost (TC_{IS}) is calculated by

$$TC_{IS} = NC_{IS} + DC_{IS}$$
$$= 4(1+q)(C_{la} + C_d + C_{ra}) + 2C_v + C_h + q(2C_v + C_h) \tag{7}$$

Location Management Costs under the Cache Scheme:
The signaling cost for location registration (UNC_{Ca}) is

$$UNC_{Ca} = UNC_{IS} = 4(C_{la} + C_d + C_{ra}) \tag{8}$$

The signaling cost for call delivery (SNC_{Ca}) is expressed as

$$SNC_{Ca} = \bar{p}SNC_{CaH} + (1 - \bar{p})SNC_{CaM} \tag{9}$$

where SNC_{CaH} and SNC_{CaM} are the signaling costs for call delivery when a cache hit and a cache miss occur under the cache scheme, respectively.

$$SNC_{CaH} = 4C_{la} + 4(1-t)C_d = 4[C_{la} + (1-t)C_d] \tag{10}$$
$$SNC_{CaM} = 4[C_{la} + (1-t)C_d] + 4(C_{la} + C_d + C_{ra}) \tag{11}$$

Replacing SNC_{CaH} and SNC_{CaM} by (10) and (11), (9) is re-written as

$$SNC_{Ca} = \bar{p}SNC_{CaH} + (1 - \bar{p})SNC_{CaM}$$
$$= 4[C_{la} + (1-t)C_d] + 4(C_{la} + C_d + C_{ra})(1 - \bar{p}) \tag{12}$$

Then, the total signaling cost (NC_{Ca}) is

$$
\begin{aligned}
NC_{Ca} &= UNC_{Ca} + qSNC_{Ca} \\
&= 4(C_{la} + C_d + C_{ra}) + q\{4[C_{la} + (1-t)C_d] \\
&\quad + 4(C_{la} + C_d + C_{ra})(1-\bar{p})\}
\end{aligned}
\tag{13}
$$

The database access cost for location registration (UDC_{Ca}) is

$$
UDC_{Ca} = UDC_{IS} = 2C_v + C_h
\tag{14}
$$

The database access cost for call delivery (SDC_{Ca}) is expressed as

$$
SDC_{Ca} = \bar{p}SDC_{CaH} + (1-\bar{p})SDC_{CaM}
\tag{15}
$$

where SDC_{CaH} and SDC_{CaM} are the database access costs for call delivery when a cache hit and a cache miss occur under the cache scheme, respectively.

$$
SDC_{CaH} = 2C_v
\tag{16}
$$
$$
SDC_{CaM} = 3C_v + C_h
\tag{17}
$$

Replacing SDC_{CaH} and SDC_{CaM} by (16) and (17), (15) is re-written as

$$
SDC_{Ca} = \bar{p}SDC_{CaH} + (1-\bar{p})SDC_{CaM} = 3C_v + C_h - \bar{p}(C_v + C_h)
\tag{18}
$$

Then, the total database access cost (DC_{Ca}) is

$$
DC_{Ca} = UDC_{Ca} + qSDC_{Ca} = 2C_v + C_h + q[3C_v + C_h - \bar{p}(C_v + C_h)]
\tag{19}
$$

From (13) and (19), the total cost (TC_{Ca}) is calculated by

$$
\begin{aligned}
TC_{Ca} &= NC_{Ca} + DC_{Ca} \\
&= 4(C_{la} + C_d + C_{ra}) + q\{4[C_{la} + (1-t)C_d] + 4(C_{la} + C_d + C_{ra})(1-\bar{p})\} \\
&\quad + 2C_v + C_h + q[3C_v + C_h - \bar{p}(C_v + C_h)]
\end{aligned}
\tag{20}
$$

Location Management Costs under the PB-Cache Scheme:
The signaling cost for location registration (UNC_{PB}) is

$$
UNC_{PB} = 4(C_{la} + C_d + C_{ra}) + 4[C_{la} + (1-s)C_d]
\tag{21}
$$

The signaling cost for call delivery (SNC_{PB}) is expressed as

$$
SNC_{PB} = pSNC_{PBH} + (1-p)SNC_{PBM}
\tag{22}
$$

where SNC_{PBH} and SNC_{PBM} are the signaling costs for call delivery when a cache hit and a cache miss occur under the PB-Cache scheme, respectively. To quantify j, we assume that on average $j = \frac{K}{2}$. Therefore, the signaling cost for call delivery when a cache hit occurs under the PB-Cache scheme is

$$
SNC_{PBH} = 4[C_{la} + (1-t)C_d] + 2K[C_{la} + (1-s)C_d]
\tag{23}
$$

The signaling cost for call delivery when a cache miss occurs under the PB-Cache scheme is

$$SNC_{PBM} = 4[C_{la} + (1-t)C_d] + 4K[C_{la} + (1-s)C_d]$$
$$+ 4(C_{la} + C_d + C_{ra}) \tag{24}$$

Replacing SNC_{PBH} and SNC_{PBM} by (23) and (24), (22) is re-written as

$$SNC_{PB} = pSNC_{PBH} + (1-p)SNC_{PBM}$$
$$= 4(C_{la} + C_d + C_{ra})(1-p) + 2K(2-p)[C_{la}$$
$$+ (1-s)C_d] + 4[C_{la} + (1-t)C_d] \tag{25}$$

Then, the total signaling cost (NC_{PB}) is

$$NC_{PB} = UNC_{PB} + qSNC_{PB}$$
$$= 4(C_{la} + C_d + C_{ra}) + 4[C_{la} + (1-s)C_d] + q\{4(C_{la} + C_d + C_{ra})(1-p)$$
$$+ 2K(2-p)[C_{la} + (1-s)C_d] + 4[C_{la} + (1-t)C_d]\} \tag{26}$$

The database access cost for location registration (UDC_{PB}) is

$$UDC_{PB} = 4C_v + C_h \tag{27}$$

The database access cost for call delivery (SDC_{PB}) is expressed as

$$SDC_{PB} = pSDC_{PBH} + (1-p)SDC_{PBM} \tag{28}$$

where SDC_{PBH} and SDC_{PBM} are the database access costs for call delivery when a cache hit and a cache miss occur under the PB-Cache scheme, respectively. Let $j = \frac{K}{2}$, the database access cost for call delivery when a cache hit occurs under the PB-Cache scheme is

$$SDC_{PBH} = (\frac{K}{2} + 2)C_v \tag{29}$$

The database access cost for call delivery when a cache miss occurs under the PB-Cache scheme is

$$SDC_{PBM} = (K+2)C_v + C_h + C_v = (K+3)C_v + C_h \tag{30}$$

Replacing SDC_{PBH} and SDC_{PBM} by (29) and (30), (28) is re-written as

$$SDC_{PB} = pSDC_{PBH} + (1-p)SDC_{PBM}$$
$$= (K+3)C_v + C_h - p[(\frac{K}{2} + 1)C_v + C_h] \tag{31}$$

Then, the total database access cost (DC_{PB}) is

$$DC_{PB} = UDC_{PB} + qSDC_{PB}$$
$$= 4C_v + C_h + q\{(K+3)C_v + C_h - p[(\frac{K}{2} + 1)C_v + C_h]\} \tag{32}$$

Table 2. Cost analysis

set	C_{la}	C_d	C_{ra}
1	1	3	3
2	1	3	5
3	1	5	5
4	1	5	10

(a)Signaling cost

set	C_v	C_h
5	1	2
6	1	3
7	1	5

(b)Database access cost

From (26) and (32), the total cost (TC_{PB}) is calculated by

$$TC_{PB} = NC_{PB} + DC_{PB}$$
$$= 4(C_{la} + C_d + C_{ra}) + 4[C_{la} + (1-s)C_d] + q\{4(C_{la} + C_d + C_{ra})(1-p)$$
$$+ 2K(2-p)[C_{la} + (1-s)C_d] + 4[C_{la} + (1-t)C_d]\} + 4C_v + C_h$$
$$+ q\{(K+3)C_v + C_h - p[(\frac{K}{2}+1)C_v + C_h]\} \tag{33}$$

4   Performance Analysis

In this section, we evaluate the performance of the PB-Cache scheme by comparing with the IS-41 and the cache scheme based on the analytical model described in Section 3. We define the relative cost of the PB-Cache scheme as the ratio of the total cost for the PB-Cache scheme to that of the IS-41. A relative cost of 1 means that the costs under both schemes are exactly the same. In order to estimate s, we use the model of the called MT's mobility across LSTP areas (See [6] for more details). Then, the probability of intra-LSTP movement can be estimated to be approximately 0.87, and t is assumed to be 0.2.

4.1   Signaling Costs

In the following analysis, we first evaluate the case when the signaling cost dominates by setting the database access cost parameters, C_v and C_h to 0. Parameter sets 1 and 2 show the cases when the costs for sending a message to the HLR is relatively low. Parameter sets 3 and 4 show the cases when the costs for sending a message to the HLR is relatively high. As the number of the mobile users keeps increasing rapidly, parameter set 4 is especially expected to be the common case. Fig. 4 shows the relative signaling costs for both the cache scheme and the PB-Cache scheme when the parameter sets 2 and 4, as given in Table 2(a), are used. We can see that the PB-Cache scheme for $K = 1$, on the whole, results in the lowest signaling cost as compared with other schemes.

4.2   Database Access Costs

In the following, we evaluate the case when the database access cost dominates by setting the signaling cost parameters, $C_{la}, C_d,$ and C_{ra} to 0. Fig. 5 shows

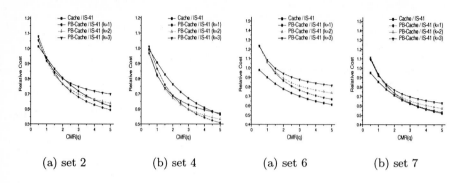

(a) set 2   (b) set 4   (a) set 6   (b) set 7

Fig. 4. Signaling costs   **Fig. 5.** Database access costs

the relative database access costs for both the cache scheme and the PB-Cache scheme when the parameter sets 6 and 7, as given in Table 2(b), are used. Fig. 5(b) shows that as the HLR access cost becomes relatively higher, the increase of K affects only a little of the database access cost. Note that the situation in Fig. 5(b) offsets the additional cost incurred by tracing more VLR queries against less HLR queries, and this will be more apparent as the cost for accessing the HLR becomes higher.

4.3   Total Costs

We compare the total cost of the PB-Cache scheme with that of the IS-41 and the cache scheme. Fig. 6 demonstrates the relative total costs for the cache scheme and the PB-Cache scheme when the parameter sets 2 and 6, and the parameter sets 4 and 7 are used. We can see that the PB-Cache scheme for $K = 1$, on the whole, results in the lowest total cost compared with other schemes. As we can see in Fig. 6(a) and (b), for high CMR, the reduction in the total cost of the PB-Cache scheme is very significant when the cost for accessing to the HLR is relatively high (See Fig. 6(b)). However, when C_{ra} and C_h are relatively low (See Fig. 6(a)), the total cost of the PB-Cache scheme for $K = 1$ is almost the same as that of the cache scheme. These results are expected because the PB-Cache scheme tries to reduce the number of messages going to the HLR by searching the MT based on the VLR pointed in the cache. However, with the PB-Cache scheme, pointer chains can be formed which involve a number of VLR's. Thus, the decision of the appropriate maximum pointer chain length is essential to improve the performance of the PB-Cache scheme than that of the IS-41 or the cache scheme. Note that the decision of the optimal K value which minimizes the location management cost is crucial factor for the performance of the PB-Cache scheme, which may be changeable according to the user's CMR and the signaling traffic condition.

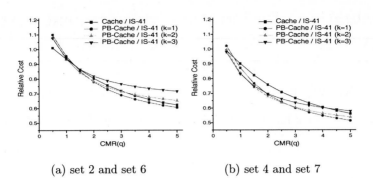

(a) set 2 and set 6   (b) set 4 and set 7

Fig. 6. Total costs

5   Conclusions

In this paper, a location management scheme called the PB-Cache scheme was proposed. The PB-Cache scheme is an enhanced variation of the alternative location algorithm proposed in [4]. In short, the primary idea is to exploit a user's movement locality as well as call locality at the same time. In the PB-Cache scheme, the information of the VLR pointed in the cache is used as a hint of the called MT's location information, and if it is not correct, the called MT can be ultimately found by tracing forwarding pointers from the pointed VLR instead of querying the HLR. By reducing the access to the HLR, the PB-Cache scheme, in most cases, results in the significant reduction in the total location management cost compared with other schemes, and distributes the signaling traffic and database access load on the HLR to the VLR's; The bottleneck to the HLR can be effectively reduced. The analytical results have indicated that the PB-Cache scheme significantly outperforms the IS-41 and the existing cache scheme when the CMR is high or the signaling traffic to the HLR is heavy.

References

1. F. Akyildiz et al., "Mobility Management in Next-Generation Wireless Systems, " *Proc. IEEE*, vol. 87, no. 8, pp. 1347–84, Aug. 1999.
2. S. Mohan and R. Jain, "Two user location strategies for personal communications services," *IEEE Personal Commun.*, pp. 42–50, First Quarter 1994.
3. D. R. Wilson, "Signaling system no.7, IS-41 and cellular telephony networking," *Proc. IEEE*, vol. 80, pp. 664–652, Apr. 1992.
4. R. Jain, Y. B. Lin, C. Lo, and S. Mohan, "A caching strategy to reduce network impacts of PCS," *IEEE J. Sel. Areas Comm.*, vol.12, no.8, pp. 1434–1444, Oct. 1994.
5. Y. B. Lin, "Reducing location update cost in a PCS network," *IEEE/ACM Trans. Network*, vol. 5, no. 1, pp. 25–33, 1997.
6. R. Jain and Y. B. Lin, "An auxiliary user location strategy employing forwarding pointers to reduce network impacts of PCS," *ACM-Baltzer J. Wireless Networks*, vol.1, no.2, pp. 197–210, July 1995.

Evaluation of a Location Management System for Wireless Communicating Mobile Computers

Hideki Shimada[1], Shigeaki Tagashira[2], Tsuneo Nakanishi[1], and Akira Fukuda[3]

[1] Graduate School of Information Science, Nara Institute of Science and Technology, JAPAN. {hideki-s, tun}@is.aist-nara.ac.jp
[2] Graduate School of Engineering, Hiroshima University, JAPAN. shigeaki@huis.hiroshima-u.ac.jp
[3] Graduate School of Information Science and Electrical Engineering Kyushu University, JAPAN. fukuda@f.csce.kyushu-u.ac.jp

Abstract. Mobile hosts can acquire location-dependent information at any place and be connected to a network from anywhere with wireless communication devices such as cellular phones, wireless LAN, *etc.* In such an environment we aim to construct a system software on which mobile hosts can share the location-dependent information with other hosts. It is, however, difficult for users who request location-dependent information to identify mobile hosts providing the information, since the system presents no mechanism to specify hosts by geographic location. In this paper we propose a location management system that manages locations of mobile hosts and enables users to specify location-dependent informations by their geographic locations. Since the location management system is employed for mobile computing, the system has the mechanism to reduce the overhead of location management on wireless communication. Moreover, we implement a part of this system and show that this system is suitable for mobile environment through the experiment.

1 Introduction

Mobile computers have become more portable and powerful in recent years. Moreover, portable input and communication devices such as digital video cameras, wireless network cards, and personal handy phones have been widely spreaded over the market. The global positioning system receiver, or the GPS receiver for short, is no longer a particular device. With these devices mobile computers cannot only obtain information from the Internet but also provide their holding or live information to the Internet with geographic locations out of door. These backgrounds are motivating researchers to develop a lot of communication systems and applications utilizing geographic locations [1], [2], [3].

Technical success of Gnutella and Napster demonstrates that peer-to-peer communication is applicable enough to share information no less than server-client communication. Peer-to-peer communication, which does not require servers prepared in advance, will befit for directly sharing of live location-dependent information better than server-client communication.

I. Chong (Ed.): ICOIN 2002, LNCS 2344, pp. 154–164, 2002.
© Springer-Verlag Berlin Heidelberg 2002

We are constructing a system sharing location-dependent information among mobile computers connected in a peer-to-peer manner. The location-dependent information sharing system provides a natural interface to specify location-dependent informations by their locations. Its clients can search for mobile hosts around them or their destinations and acquire lists of services and informations available in the mobile hosts. Mobile hosts in the location-dependent information sharing system register their locations to immobile hosts, referred to as *location management hosts*, to be found by clients with their specifed locations. Frequent communication between mobile hosts and location management hosts can consume limited bandwidth of wireless communication and battery charged power of mobile hosts. We have proposed a location management system in Reference [4] which manages exact locations of mobile hosts with low pressure on wireless networks of limited bandwidth. This paper evaluates the location management system with some experiments.

This paper is organized as follows: Section 2 presents a location management scheme assumed in this paper. Section 3 and Section 4 describe the overview and the detailed design of the location management system, respectively. Section 5 evaluates the location management system and shows some considerations. Finally, Section 6 concludes this paper.

2   The Location Management Scheme

In our location management system, only required equipments of mobile hosts are devices for acquiring geographic locations and connecting to the Internet such as GPS receivers and cellular phones. The mobile hosts can provide location-dependent informations freely without any other particular equipments.

Locations of mobile hosts must be managed to enable clients to search for mobile hosts at their specified locations. In our location management system, the mobile hosts register locations acquired from their own GPS receivers to immobile hosts, referred to as *Location Management Hosts* or *LMHs* for short, with their IP addresses. The LMHs are logically connected in a hierarchical manner to distribute overheads of location management.

To keep exact locations of mobile hosts moving frequently at reasonable overheads, it must be carefully designed how to update LMH registration without increasing packets which flow wireless links of limited bandwidth. We can consider the following three manners for updating LMH registration.

Periodical Update: LMH registration is updated periodically. To keep more accurate locations of mobile hosts, the interval of updates must be shorter at the expense of network bandwidth consumption by packets for updating LMH registration.

Update-on-Moving: LMH registration is updated when a mobile host moves.

On-Demand Update: LMH registration is updated when a client requests mobile hosts at a specified location.

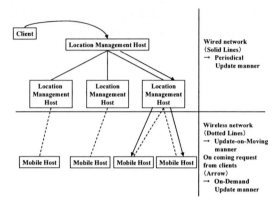

Fig. 1. Update of LMH Registration

Any mobile host updates its LMH registration on leaving its current geographical region, not its current geographical point, in the update-on-moving manner. That contributes to avoid pressure on the limited bandwidth of the wireless network by packet issues for updating LMH registration. Note that exact locations of mobile hosts are actually needed when a client searches for some mobile hosts by their locations.

If a mobile host changes its location within its current geographical region, the mobile host does not update its LMH registration. Thus LMH registration does not always keep exact location of mobile hosts. When a client searches for mobile hosts by a specified location, a responsible LMH inquires exact locations of the mobile hosts in the geographical region containing the client specified location, that is, on-demand update of LMH registration is performed. It requires longer turnaround time for on-demand update of LMH registration in general. However, since LMH registration keeps rough locations of mobile hosts in the update-on-moving manner, there are only a limited number of mobile hosts whose exact locations are inquired by the LMH. That contributes to reduce turnaround time for on-demand update of LMH registration.

On the other hand, LMHs are immobile computers connected by stable wired links of broader bandwidth enough to exchange LMH registration frequently. Therefore, LMH registration is updated among LMHs in the periodical update manner.

Fig. 1 shows how to update LMH registration mentioned above.

3   Design of the Location Management System

In this section we describe the design of our location management system. The location management system consists of two kinds of hosts, the mobile host and the LMH. The mobile host provides location-dependent information. The LMH is an immobile computer managing IP addresses and locations of mobile hosts as mentioned in Section 2.

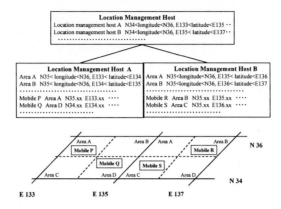

Fig. 2. Overview of location management system.

LMHs are placed in a wired network as shown in Fig. 2. LMHs provide clients with IP addresses and locations of mobile hosts, not location-dependent informations. LMHs form a logical hierarchy of two layers. Our location management system geographically partitions the world into cells. Each cell is governed by an LMH placed in the lower layer, or an L-LMH for short. Moreover, the L-LMH geographically partitions the cell into some regions. Update-on-moving of LMH registration is performed based on these regions as mentioned in Section 2. The LMH placed in the upper layer, or the U-LMH for short, manages its L-LMHs.

3.1   Acquiring Locations of Mobile Hosts

Clients of the location management system acquire locations of mobile hosts as follows.

1. A client sends a request of the mobile hosts staying a specified location.
2. The U-LMH forwards the client's request to the L-LMH governing the location specified by the request.
3. The L-LMH finds the mobile hosts in the geographical region containing the location specified by the forwarded request according to LMH registration. The L-LMH requests the current locations of the mobile hosts.
4. The mobile hosts send their current locations to the L-LMH.
5. The L-LMH updates the locations of the mobile hosts.
6. The L-LMH sends the IP addresses and the locations of the mobile hosts to the client.

3.2   Updating Locations of Mobile Hosts Registered in LMHs

The U-LMH has a database on L-LMHs and the L-LMH has a database on its governing mobile hosts. The U-LMH and the L-LMH communicate each other and update the databases periodically as follows.

U-LMH: The U-LMH manages the database, referred to as *L-LMH database*, storing the areas governed by the L-LMHs and the numbers of mobile hosts managed by the L-LMHs. The U-LMH can be duplicated to guarantee scalability of the location management system. An entry of the U-LMH database, which corresponds to an L-LMH, consists of the IP address, the boundaries of the governing area, and the number of mobile hosts of the L-LMH. The U-LMH communicates with L-LMHs and updates its L-LMH database periodically.

L-LMH: The L-LMH manages the database, referred to as *mobile database*, storing the locations of the registered mobile hosts in its governing area. An entry of the mobile database, which corresponds to a mobile host, consists of the IP address, the location, and the list of the available services of the mobile host. The L-LMH communicates with the U-LMH and updates its L-LMH database periodically.

3.3   Registering and Unregistering Mobile Hosts to LMHs

The mobile host must register or unregister itself to an appropriate L-LMH to provide its services to clients. Registration and unregistration of the mobile host correspond to starting and stoping the services provided by the mobile host, respectively.

1. The mobile host sends a request of registration with its current latitude and longitude to the U-LMH, since the mobile host has no information on the L-LMH governing the area where the mobile host stays.
2. The U-LMH searches its L-LMH database for the L-LMH governing the area where the mobile host stays by the latitude and the longitude of the mobile host. The U-LMH forwards the request of registration to the L-LMH found out of the L-LMH database.
3. The L-LMH registers the mobile host at the mobile database and sends an acknowledge to the mobile host with the IP address, the boundaries of the governing area, and the boundaries of the regions of the L-LMH.
4. Now the mobile host can communicate directly with the L-LMH and provide its services to clients.
5. The mobile host sends a request of unregistration to the L-LMH.
6. The LMH unregisters the mobile host at the mobile database and sends an acknowledge to the mobile host.

   The mobile host can leave the area where the current L-LMH governs. When the mobile host leave the area where the current L-LMH X governs for the area where an L-LMH Y governs, unregistration to L-LMH X and registration to L-LMH Y are preformed as follows.

1. The mobile host sends a request of unregistration to L-LMH X with its latitude and longitude.
2. L-LMH X forwards the latitude and the longitude of the mobile host to the U-LMH and acquires information on L-LMH Y.

3. L-LMH X sends an acknowledge to the mobile host with information on L-LMH Y.
4. L-LMH X sends information on the mobile host to L-LMH Y and unregisters the mobile host at its mobile database.
5. L-LMH Y registers the mobile host at its mobile database based on information received from L-LMH X.

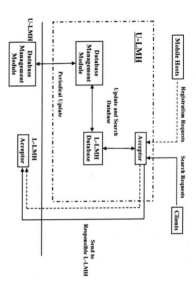

Fig. 3. Structure of U-LMH.

4   Architecture of the Location Management System

In this section we describe the architectures of the U-LMH and the L-LMH. The U-LMH is constructed with the following modules as shown in Fig. 3.

L-LMH Database: The L-LMH database stores information on the L-LMHs. For each L-LMH the L-LMH database keeps its IP address, the boundaries of its governing cells represented by the latitude and the longitude, and the number of registered mobile hosts in its governing cell.

Database Management Module: The database management module communicates with the L-LMHs periodically and updates the numbers of the mobile hosts in their governing cells. Moreover, the database management module synchronizes L-LMH database periodically among the other U-LMHs.

Acceptor: The acceptor accepts both requests to search for mobile hosts by clients and requests to register mobile hosts by the mobile hosts. The acceptor searches the L-LMH database for responsible LMHs by the locations specified by requests from clients or mobile hosts and forwards the responsible LMHs the requests.

The L-LMH is constructed with the following modules as shown in Fig. 4.

Acceptor: The acceptor accepts requests from clients and mobile hosts forwarded from the U-LMH. The acceptor forwards requests from clients and mobile hosts to the registration module and the mobile host management module, respectively.

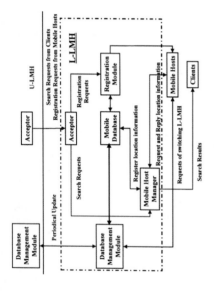

Fig. 4. The L-LMH

Mobile Database: The LMH is responsible for a cell consisting of some regions as mentioned in Section 3. Each region has a database storing information on registered mobile hosts in the region. For each mobile host the database keeps its IP address, its location represented by the latitude and longitude, and its URI. The mobile database is a union of those databases defined for the regions of the cell.

Database Management Module: The database management module communicates U-LMHs periodically and sends the number of the mobile hosts registered in the L-LMH. When a mobile host leaves the cell governed by an L-LMH, the mobile host sends a request to unregister itself from the L-LMH. The database management module receives that request, communicates with the database management module of the U-LMH, and transfers information on the mobile host to the L-LMH governing the cells where the mobile host moves.

Registration Module: The registration module registers mobile hosts to the mobile database based on the mobile host registration requests forwarded from the acceptor.

Mobile Host Management Module: The mobile host management module receives a request to search for mobile hosts at a specified location from the acceptor and requests the current locations of the mobile hosts in the region containing the specified location. Moreover, the mobile host management module updates the mobile database and sends the names and the locations of the mobile hosts to the client.

5   Evaluation

We evaluate our location management scheme with a prototype location management system. Especially, we investigate contribution of update-on-moving of

LMH registration with hierarchical location management on reducing location management overheads. In the evaluation our location management scheme is compared with another location management scheme which updates locations of mobile hosts only in the periodical manner.

5.1   Environment for Evaluation

The environment of this evaluation is as follows. Links between a client and a U-LMH and between a U-LMH and an L-LMH are wired links of 100MBits/s bandwidth. A link between an L-LMH and a mobile host is a wireless link of 11MBits/s bandwidth. In this experiment one client and one mobile host emulate multiple clients and multiple mobile hosts.

Our prototype location management system has one U-LMH and two L-LMHs connected with the U-LMH. The L-LMHs govern two different cells of the same size and each cell is partitioned into four regions. Mobile hosts set their initial locations at random and moved either north, sourth, east, or west at the same speed. Fig. 5 shows the overview of our prototype location management system employed for this evaluation.

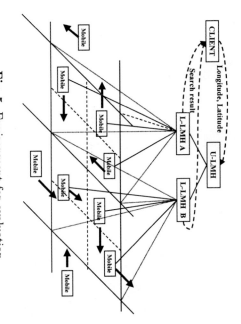

Fig. 5. Environment for evaluation

5.2   Experimental Results and Considerations

We count the number of packets which mobile hosts send to L-LMHs for updating LMH registration. The results are shown in Fig. 6 and 7 where the number of requests from clients is 50 and 100, respectively. Fig. 6 and 7 shows that our location management scheme issues about half as many number of packets to update LMH registration as the periodically updating method.

162   H. Shimada et al.

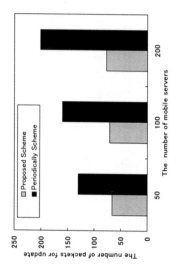

Fig. 6. The number of packets to update LMH registration (50 requests)

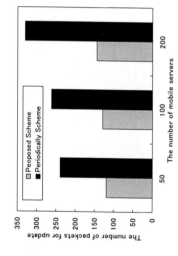

Fig. 7. The number of packets to update LMH registration (100 requests)

Moreover, we measure time required for a client to acquire the locations of the mobile hosts at a specified location. The average results are shown in Fig. 8 and 9 where the number of requests from the clients is 50 and 100, respectively. Fig. 8 and 9 shows that our location management scheme takes about 1.1 times as long average time to acquire the locations of the mobile hosts at a specified location as the periodically updating method.

6   Conclusion

In this paper we have proposed the location management system which enables clients to search for location-dependent information of mobile hosts by their locations and have described its design. The location management system reduces the overhead for updating locations of mobile hosts, which are registered in location management hosts, in a wireless network of limited bandwidth.

Exact locations of mobile hosts are actually needed when a client searches for some mobile hosts by their locations. The location management system keeps

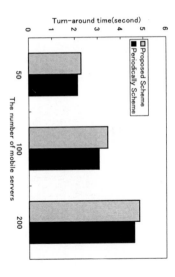

Fig. 8. Average time to acquire the locations of mobile hosts (50 requests)

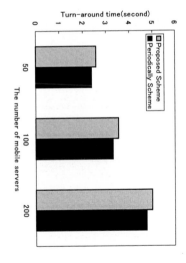

Fig. 9. Average time to acquire the locations of mobile hosts (100 requests)

rough locations of mobile hosts in a update-on-moving manner to reduce the number of packets for updating the locations of the mobile hosts registered in location management hosts. The location management system inquires and keeps the exact locations of the mobile hosts in an on-demand update manner. In this way the location management system provides clients with exact locations of mobile hosts with keeping the cost to update locations of mobile hosts registered in location management hosts as little as possible.

This paper evaluated our location management scheme with a prototype location management system. As a result our location management scheme kept the number of packets to update locations of mobile hosts registered in location management hosts half as many as a naive periodically updating scheme. Moreover, time to acquire the locations of the mobile hosts at a specified location was only 1.1 times as long as the periodically updating scheme.

It is a future work to evaluate our location management system with changing the sizes of the areas where location management hosts govern and the velocities of mobile hosts. It is also another future work extending our location management system to enable location management hosts to change their governing areas dynamically for more effective search of mobile hosts.

References

1. T. Imielinski and J. C. Navas: *"Geoglaphic Addressing, Routing, and Resource Discovery with the Global Positioning System"*, Comm. ACM, Vol.42, No.4, pp.86–92, 1999.

2. Y. Watanabe, A. Shinozaki, F. Teraoka, and J. Murai: *"The Design and Implementation of the Geographical Location Information System"*, Proc. Inet96, 1996.

3. S. Tagashira, K. Saisho, and A. Fukuda: *"Design and Implementation of an Information Announcement Toolkit for Mobile Computers"*, LNCS 1987, Springer, pp.65–76, 2001.

4. H. Shimada, S. Tagashira, T. Nakanishi, and A. Fukuda: *"Design and Implementation of Location Management System for Mobile Servers"*, Proc. the 2001 Int'l Conf. on Parallel and Distributed Processing Techniques and Applications, pp. 1000–1005 (2001).

The Enhancement of End-to-End TCP on Wireless CDMA Network to Support Interactive Data Service

Insoo Park and Yongjin Park

Division of Electrical and Computer Engineering
Hanyang University
17 Haengdang-dong, Seongdong-Gu, Seoul, 133-791, Korea
{ispark,park}@nclab.hyu.ac.kr
http://nclab.hyu.ac.kr/

Abstract. CDMA specifications provide three types of data services: Short message service (SMS), circuit switched, and packet switched data. Circuit-switched data services require a dial-up to a data service. The benefit of using a circuit-switched data service is the ability to utilize existing wireless equipment i.e. a cell phone as a modem to connect to a data network, thereby reducing overall hardware costs. Once connected with packet data services, a mobile terminal utilizes a constant data rate for wireless data services. Packet data services are cost effective, especially for highly frequent use. We have increased the cost-effectiveness of Qualcomm software by reducing or in some cases eliminating setup and transmission costs associated with wireless data calls. As no PSTN call setup or modem training time is required for implementation, the end-to-end connection setup time is drastically reduced. The software module is implemented on the Qualcomm MSM3000 chipset to support support interactive applications linked with serial interface. This module enables the mobile terminal to directly connect the server with TCP using DSSock Socket API in DMSS3000. It can also be adapted to many other PC-based applications via direct-linked RS232C control.

1 Introduction

1.1 Background

Since Qualcomm first introduced the commercial CDMA into the market place in 1994, it is expected that over 700 million people worldwide will use CDMA by 2003 and over one billion by 2005. CDMA initially supported voice communications, however, data services such as SMS, WAP (Wireless Application Protocol) and VM (Virtual Machine) have increased in importance. Although the WAP browser enables mobile users to access the Internet within a limited service area, dynamic applications such as multi-user visual games and mobile karaoke are not yet available. In addition video applications such as video telecommunication and VoD (Video on Demand) can be the highest density applications in broadband mobile networks. Due to the service charges, most of these applications require mobile networks to support low-cost

I. Chong (Ed.): ICOIN 2002, LNCS 2344, pp. 165–175, 2002.
© Springer-Verlag Berlin Heidelberg 2002

packet data services. As network and air link technologies are improved, the cost of packet data services will decrease.

Qualcomm, the worldwide leader in mobile communications, is continuously developing CDMA technologies. CDMA provides three types of data services: short message service (SMS), circuit switched, and packet switched data. Circuit-switched data services require dial-up to a data service. The benefit of using a circuit-switched data service is the ability to utilize existing wireless equipment i.e. a cell phone as a modem to connect to a data network, thereby reducing overall hardware costs. Once connected with packet data services, a mobile terminal utilizes a constant data rate for wireless data services. Packet data services are cost effective, especially with highly frequent use. The established channel can be maintained cost-free if no data transmission occurs within the session [1].

1.2 CDMA Data Service Model

Fig.1 shows the architecture of the Qualcomm mobile data service system. An air connection using TCP/IP protocol uses RLP (Radio Link Protocol) between an MS (mobile station) and its BS (Base Station). RLP Synchronization occurs between the MS and the BS. At this time, a link is made with the traffic channel in the cell. After RLP Synchronization, PPP negotiation occurs between the MS and its IWF (Inter-Working Function) and a temporary IP is assigned to the MS. The IWF attempts to connect the destination server via PSTN by using its IP address and port information sent from the mobile terminal. The IWF and the destination server are successfully connected, and PPP negotiation occurs between the MS (TE2) and the destination server. TCP synchronization then occurs between the end-to-end systems, and successfully connects.

This data service has three modes: the packet mode service, packet application and the circuit mode service. The Packet mode service, composed of TCP/IP, PPP stack in TE2, is defined in IS-707 [2]. The MS acts as a wireless modem and controls RLP synchronization between MS and BS. The performance of the upper layer protocols depends on the protocol stack of TE2. Although packet application mode operates similar to the packet mode service, the operation of TE2 in packet mode occurs in the mobile terminal protocol stack with a socket interface. That is to say that all protocols (TCP/IP, PPP and RLP) are implemented in the mobile terminal. This applies to Internet services used in an embedded micro browser. The circuit mode (FAX…) service is defined in IS-99 and summarized in IS-707. PPP and TCP in a mobile terminal connect with PPP and TCP in IWF. PPP and TCP of TE2 establish each connection with the PPP and TCP of a server.

Location-based interactive data service will be the key component in future mobile applications such as mobile commerce and user locater services. Since user locating service such as GPS is uni-directional, most applications may require mobile terminals with GPS modules to interact with application servers using the CDMA channel as a backward interface. To support reliable data connection between the mobile terminal and a location tracking server, the GPS server, which serves as the existing data service architecture of Qualcomm, must be modified and enhanced. In

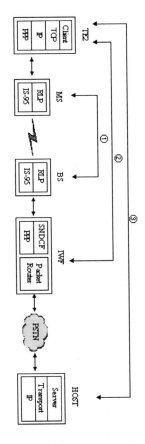

(a) Packet data service

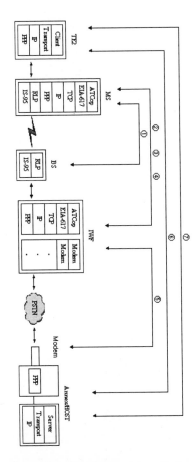

(b) Circuit-switched data service

Fig. 1. Wireless Data Service

this paper, mobile terminal software based on Qualcomm MSM3000 is developed using socket API to support interactive data service such as location tracking. Section 2 describes the Qualcomm software architecture in detail. Section 3 shows how mobile software and TE2 software are implemented to support the control packet application service of Qualcomm software. Finally, the details of the experiment are described including the testing environment and the test results.

2 Qualcomm System Architecture

2.1 DMSS3000 Software Structure

The DMSS OS is REX-real time OS. REX is composed of tasks and regulates the order of service in each task by priority. Each task communicates with signals. DMSS consists of TX, RX, MC (Main Control), UI (User Interface), HS (Handset), SND (Sound), DIAG (Diagnostics), DS (Data Call Control), and PS (Protocol Services). DS and PS use tasks for data service. The architecture software in DMSS3000 is shown in layers 2 and 3 of Fig.2.

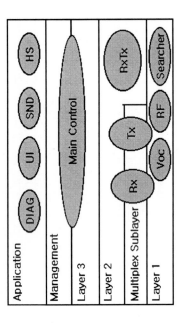

Application	DIAG	UI	SND	HS
Management				
Layer 3		Main Control		
Layer 2				
Multiplex Sublayer	Rx	Tx		RxTx
Layer 1		Voc	RF	Searcher

Fig. 2. CDMA Terminal S/W architecture

Layer 1(Physical Layer) covers physical specifications for wireless CDMA communication. Its hardware and software functions include Channel interleaving, symbol repetition and RF control. The Multiplex Sublayer separates voice data from signal data in the communication channel. Voice data is processed by vocoder which transfers signaling data to the upper layer. Layer 2 (Data Link Layer) supports end-to-end reliable communication and provides Ack/Nack signaling, duplicate message detection and CRC check. Layer 3 (Network Layer) processes two-end system messages and performs protocol request and response functions to their counterparts. It also manages call processing states and performs authentication and message encryption in the Main Control. Instead of being assigned the same IS-95 specifications, the Management and Application Layer should be designed by IS-95 system manufacturers. This layer contains a management function for monitoring and controlling terminal operations and provides a user interface function. MC performs the management function and UI, SND, HS regulate the user interface function [4].

2.2 Data Service Structure

Fig. 3 shows task interaction in data service. The solid arrow represents the data flow. PS performs all service layer protocol operations for mobile. DS performs all data service call controls. For each socket, the PS maintains a separate send queue and receive queue of internal buffers. PS receives data from SIO (Serial Input/Output Services for phone serial port) and transforms it into packets. TX indicates that RLP TX has received data from theUM interface and transfers them to PS via protocol stack. These packets are then transmitted through a serial interface.

The dotted arrow between SIO and DS operates in Auto-detect mode. Data input to Serial interface is automatically detected by DS and transferred to PS via SIO.

Interaction DS-Call Mgmt generates Inter-task Communication when DS requests Call Mgmt to open a traffic channel (Call processing and phone user interface) [3].

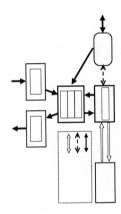

Fig. 3. Basic Servie Task

3 Implementation

3.1 Service Model

When TE2 is attempted with the ATDT command through RS232C, the MS opens a traffic channel by interpreting this command and attempts to connect BS and RLP. If the RLP connection between the MS and the BS is successful, TE2 attempts a PPP (Point-to-Point) connection with IWF and is assigned a temporary IP from IWF. After successful PPP negotiation between TE2 and IWF, the MS performs an end-to-end TCP connection with an IP address and port number in Host. [5].

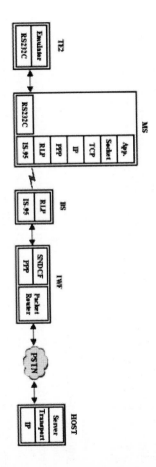

Fig. 4. Data service model

3.2 Service Scenario

To implement an end-to-end TCP connection mechanism in the present mobile communication network (IS-95), DSSocket API in DMSS3000 is utilized. Because DSSocket API for DMSS3000 was developed for phone applications such as the Embedded Browser, it cannot provide flow control of serial communication data through SIO. To solve this problem, this communication module attempts to connect a fixed command through RS232C. If the module receives a predetermined command through RS232C after connection, it attempts to connect and communicate data through RS232C. TCP/IP and PPP are used by the stack-provided communication module itself [5].

A communication module embedded in an MSM3000 based on Windows 2000, DMS3000 S/W and ARM250 Cross Complier implements the test environment. Additionally, the communication module is tested with a diagnostic monitor and SK Telecom program.

DSSocket API 2.0 in DMSS 3000 implements the service explained in this paper. Fig.5 shows the connection procedure of this communication module. The typical functions provided by DSSocket API are shown in Table 1.

Table 1. Sockets API interface functions

Function	Description
dss_open_netlib()	Opens the network library
dss_pppopen()	Brings up the network subsystem(PPP connection)
dss_socket()	Creates a socket and returns socket descriptor
dss_connect()	Initiates an active open for TCP connection. Does not support UDP sockets
dss_write()	Sends specified number of bytes across the TCP transport
dss_read()	Reads specified number of bytes from the TCP transport
dss_async_select()	Specifies which events will occur in a particular socket
dss_close()	Closes a socket, freeing it for reuse. For TCP, used as a non-blocking call
dss_pppclose()	Closes the network subsystem(PPP connection)

As shown in Fig.5 and 6, the procedure related to this paper is composed of four states. Two states establish connection and two states terminate connection. A state machine divides each state due to the blocking mode during connection and termination [6]. Fig.5 shows the connection procedure using DSSock socket API.

DSAPP_POPEN_PSTATE

This step processes initialization for end-to-end TCP connection.

(1) Call dss_open_netlib() to assign network resource, and to register network callback function and socket callback function to the PS.

(2) Converts Service Mode into Socket Mode.

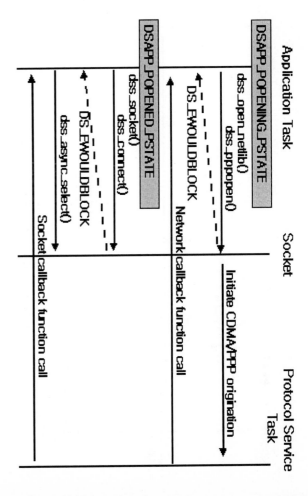

Fig. 5. End-to-end TCP establishment

(3) Call dss_open_pppopen() to attempt RLP connection between MS and BS, and attempt PPP connection between MS and IWF.

(4) When calling dss_pppopen(), the socket returns DS_EWOULDVLOCK and waits to generate network event from the PS.

(5) After PPP connection is established, this step is continuously processed through the network callback function.

DSAPP_POPENED_PSTATE

This step attempts an end-to-end TCP connection.

(1) Call dss_socket() to open a socket, and to setup connection environments (Internet family, socket type, protocol, etc.).

(2) Use dss_connect() to attempt an end-to-end connection, and to send IP address and destination server port information from application task to protocol services task.

(3) If end-to-end TCP connection is synchronized, an event (Write or Read) is generated in the socket by calling the socket callback function.

DSAPP_PCLOSING_PSTATE

This step releases the end-to-end TCP connection.

(1) Call dss_close() to close the socket, and to exit the end-to-end TCP connection.

(2) When the end-to-end TCP session terminates, the socket returns DS_EWOULDBLOCK and waits to generate DS_CLOSE_EVENT.

(3) If the socket shows DS_CLOSE_EVENT, this step is continuously processed.

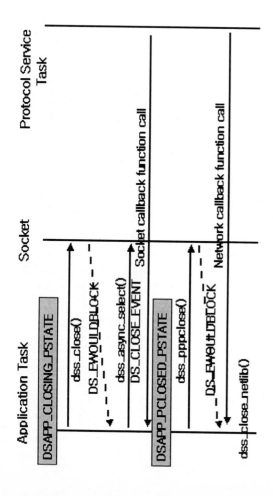

Fig. 6. End-to-end TCP termination

DSAPP_PCLOSED_PSTATE

This step converts service mode into auto-detect mode.

(1)   When the socket closes and the end-to-end TCP session terminates, the PPP session and traffic channel are closed by dss_pppclose().

(2)   all dss_close_netlib() to return used network library, and to convert service mode into auto-detect mode.

Fig.7 presents a procedure to write data into a socket or read data from a socket after successful connection. Data communication using the socket blocking mode dss_async_select() function registers events corresponding to each socket read/write to PS, and then communicates the data. After data communication occurs, the PS generates a related event, and the DS reads data from the read queue or writes data to the write queue by using the read/write functions.

Because a Read/Write event is only used once, an application task must be registered again by using dss_async_select() to process it another time. Additionally, to solve the data processing speed imbalance between RS232C and Air interface, data received from RS232C during PS-generated DS_WRITE_EVENT must be written to the socket. More specifically, the socket state waits until the socket generates DS_WRITE_EVENT, and the flow control mechanism implements the communication module to overcome the data processing speed imbalance[7][8].

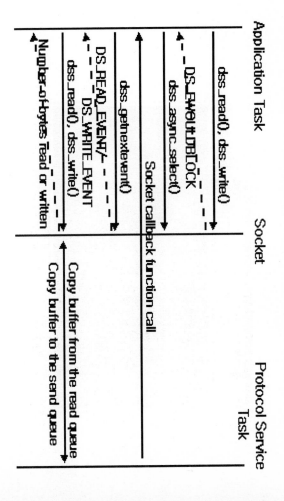

Fig. 7. Socket I/O operation

4 Conclusions

In this paper, the software module of a mobile terminal makes TCP connection with a host using the IP address and port number of the host. Since it has no UI (User Interface: Key-pad, LCD), all functions related to the establishment of connection are performed by the RS-232C interface via hyper terminal or other commercial emulators. The mobile terminal attempts to connect to a server through a TE2 command. After successful connection, the mobile terminal notifies TE2 and transfers the data from TE2 to the open socket.

The test server, or echo server, is implemented by Java language and displays and echoes data from the mobile terminal. To verify software performance, we generate a 40Kb text file and transmit it to the server which compares the echoed file and original file. A commercial PC-based emulator, PROCOMM, is used to establish TCP connection between the mobile terminal and the server. The 'AT+CRM=130' command, for example, confirms that the data service is packet data service. The 'ATDT' command attempts to make a connection. Fig. 8 and 9 show the process of connecting the echo server with the data received by the server respectively.

The results depend on the condition of the air link and network state of the server. The average connection latency is 1-2 seconds with a success rate of over 90%. With the exceptions of air link and network congestion problems, the echo data error rate is always 0%. If large packets are lost in the network, the TCP retransmission mechanism guarantees a perfect error recovery.

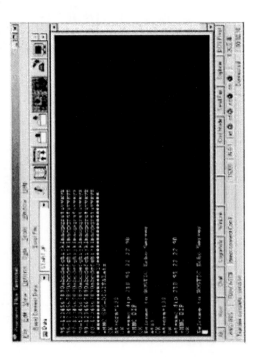

Fig. 8. Server connection emulator

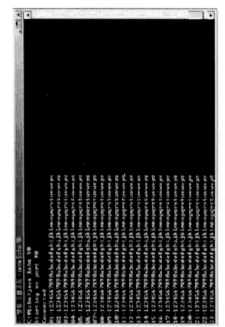

Fig. 9. Server data

Commercially, the implemented software uses the DSSock socket API of the MSM3000 chipset and DMSS3000 software. Although it is supported by DMSS3000, there are few services using the DSSock socket API. Since DSSock socket API was developed for mobile terminal applications only, it cannot be used for serial interfaces (to other devices). In this paper, software is developed to interlink RS-232C and the socket API, and can be used to support any type of mobile terminals and devices such as PDA and logistics terminals.

References

1. Douglas N. Knisely, Sarath Kumar, Subhasis Laha, and Sanjiv Nanda, ,,Evolution of Wireless Data Services:IS-95 to cdma2000,'' *IEEE Comm. Magazine*, 140–149, Oct. 1998.
2. TIA/EIA/IS-707, *Data Service Option for Wideband Spread Spectrum System*, Feb. 1988.
3. QCT Software Team, *Real Time Executive(REX) – User guide*, Qualcomm, Mar. 1999.
4. QCT Software Team, *DSSock Socket API Version 2.0*, Qualcomm, Nov. 1999.
5. QCT Software Team, *DMSS Data Service*, Qualcomm, Sept. 1999.
6. QCT Software Team, *Data Service, Browser Interface*, Qualcomm, May. 1999.
7. Geoff Huston, ,,TCP in a Wireless World,'' *IEEE Internet Computing*, 82–84, Mar./Apr. 2001.
8. George Xylomenos and George C. Polyzos, Greece Petri Mahonen, Mika Saaranen, ,,TCP Performance Issues over Wireless Links,'' *IEEE Comm. Magazine*, 52–58, Apr. 2001.

A QoS Provision Architecture for Mobile IPv6 Using RSVP

Zhaowei Qu and Choong Seon Hong

School of Electronics and Information, Kyung Hee University, 449-701 Korea
qzw@networking.kyunghee.ac.kr cshong@khu.ac.kr

Abstract. It has become very significant to enhance Quality of Service (QoS) capabilities of real-time data transmission in the Internet. The Resource Reservation Protocol (RSVP) provides a signaling mechanism for end-to-end QoS in Integrated Services Internet. Provision of QoS in wireless networks is more complex than in wired networks because of frequent mobility of mobile users. In this paper, in order to obtain more efficient use of scarce wireless bandwidth, increase data rate and reduce QoS signaling delay and data packet delay during handoff in Integrated Services Internet, we propose a novel scheme that improves the efficiency using a Hierarchical Mobile Agents Tree (HMAT) based on the definition of new option called QoS Object Option (QOO). Mobile agents are required to manage QOO, resource reservation and other mobility related tasks on behalf of mobile hosts. This scheme is based on Mobile IPv6.

1 Introduction

Mobile users want to enjoy multimedia and other real-time services in the Internet. Thus the Internet Engineering Task Force (IETF) has introduced the Mobile IPv4 [1] and Mobile IPv6 [2] to interoperate seamlessly with protocols that provide real-time services in the Internet. Resource Reservation Protocol (RSVP) [3] [4] is a resource reservation setup protocol designed for a wired network, provides resource reservation signaling support and has been facing a great challenge due to the mobile hosts. Provision of end-to-end QoS in wireless networks is more complex [5] than in wired networks because of the user mobility. Especially as recent wireless networks have been implemented based on micro-cell and handoff takes place more frequently, making QoS guarantees becomes more difficult.

In this paper, we will propose a novel scheme using the Hierarchical Mobile Agents Tree (HMAT) based on the definition of new option called QoS Object Option (QOO) [11] to improve the efficiency. Mobile agents of HMAT manage QOO, resource reservation and some tasks related with mobility on behalf of mobile hosts.

In section 2, we provide related works, and in section 3, we describe our scheme to provide a new QoS mobility support in the Internet. In section 4, we present simulation results to prove the efficiency of our scheme. Finally, we give our conclusions in section 5.

I. Chong (Ed.): ICOIN 2002, LNCS 2344, pp. 176–187, 2002.
© Springer-Verlag Berlin Heidelberg 2002

2 Related Works

Recently there have been some works [6–12] about RSVP support in mobile and wireless networks. The focus of these researches is the handoff management problem. In [6], the architecture for QoS using RSVP in the Integrated Services Packet Network has been described, and a resource reservation protocol, MRSVP, for mobile hosts has been proposed by Talukdar. The main feature of this protocol is the concept of active and passive reservations that is used to provide mobility independent service guarantees. However, the architecture requires a mobile to know all the subnets it will be visiting. The mobile obtains the identity of the proxy agents, which help with mobile RSVP in all the subnets, using a proxy discovery protocol. The mobile instructs the proxy agent in the region it is currently located to make passive reservations with all the proxy agents in all other regions. Four additional messages are used in additional to the messages already present in RSVP. The drawback of this architecture is that a mobile knows the addresses of all the subnets it is going to move into and which is not always possible. It also places a burden of finding the proxy agents in all these subnets on the mobile. In [7], the proposal proposed by Mahadevan also suggests two kinds of resource reservations that contain passive reservation and active reservation. This architecture is based on the assumption that a base station knows the addresses of the base stations in all the neighboring cells, and then solves the burden placing on the mobile host in [6].

In [8], the protocol proposed by Zhang works by combining pre-provisioned RSVP tunnels with mobile IP routing mechanism. However, the tunnels with Mobile IP may result in triangle routing problem, and the pre-provisioned RSVP tunnels are not flexible and efficient. In [9], Chen describes another signaling protocol for mobile hosts to reserve resource in Integrated Services Internet. This protocol extends the RSVP model based on IP Multicast Tree. The mobility of a host is modeled as a transition in Multicast group membership. The Multicast Tree is modified dynamically every time a mobile host is roaming to a neighboring cell. This protocol proposes that a mobile host has to make Conventional Reservation along the data flow from the sender to current location in current cell and Predictive Reservation along the Multicast Tree from the source to the neighboring cells surrounding the current cell of mobile host. Before a mobile receiver launches a reservation, the mobile host should join a Multicast group in which the sender is the root of Multicast Tree through mobile proxy also informs all of the neighbor mobile proxies surrounding the current location to join the Multicast group. Once these new branches of the Multicast Tree have formed, path messages from the sender are forwarded to mobile proxies along the Multicast Tree. Upon receiving the path messages, current mobile proxy and neighbor mobile proxies will issue reservation requests. Conventional Reservation message from the current mobile proxy is propagated toward the sender along the Multicast Tree and the Predictive reservation messages from neighbor mobile proxies are followed in the same manner. The data packets can be transmitted over the Conventional Reservation link. Figure 1 shows an example for reservations. When the mobile receiver is moving from one location to a neighbor location, the Multicast Tree will be modified. After the new Multicast Tree is formed, the Predictive Reservation from Merge Point to this mobile proxy is switched into Conventional Reservation. On the other hand, the original Conventional Reservation from the Merge Point to the original mobile proxy is switched to Predictive Reservation, and some new Predictive Reservations along the new Multicast Tree from the source to the neighboring cells surrounding the current cell of mobile host should be set up. Then the flow of data packets can be transmitted over that new Conventional Reservation link. In this protocol, there are 8 additional messages presented to complete the functions of Multicast Tree modifying and RSVP setting up.

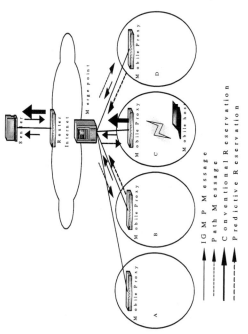

IGMP Message
Path Message
Conventional Reservation
Predictive Reservation

Fig. 1. RSVP Mobility Based on Multicast Tree

Talukdar [6], Mahadevan [7] and Chen [9] also have a challenge that is how to predict the Mobile node's movement behavior so that pre-reservations can be done only in necessary cells. If prediction is not available, resource pre-reservations can have to be performed in all neighboring cells, which wastes resource. Although this waste may be alleviated through definition of new RSVP reservation models (like Active Reservations and Passive Reservations, Conventional Reservations and Predictive Reservations), the expense would be extra protocol complexity and increasing the QoS signaling delay, the data packet delay.

In [10], Qi proposes a Flow Transparent Mobile IP and RSVP integration scheme. When the mobile host moves from subnet A to subnet B, a new reservation from Router at subnet B to a Nearest Common Router will be added, and the original reservation from Router at subnet A to the Nearest Common Router will be obsolete. However, it is difficult to choose a proper router as the Nearest Common Router, and this scheme is not feasible.

In [11], when the mobile node is receiver in Access Network, the Binding Acknowledgment has to be used so that the proposal is not efficient, and has more data packet delay.

In [12], the handoff message has to be used only for Access Network, thus the flexibility is not better.

3   Proposed Scheme

In this section we propose a framework using a Hierarchical Mobile Agents Tree (HMAT) based on the definition of new option called QoS Object Option (QOO) [11] to get more efficient use of scarce wireless bandwidth, minimize the QoS signaling delay, the data packet delay and losses and get higher data rate during handoff in mobile environment.

3.1   QoS Object Option (QOO)

This option is included in the hop-by-hop extension header of certain packets carrying Binding Update message in Mobile IPv6. The composition of a QOO [11] is shown in Table 1 by using TLV format. A QoS Object is an extension of RSVP QoS that can be used not only in the Access Network, but also in the Core Network to get a better QoS support.

Table 1. Composition of a QoS Object

		Option type 5bit	Option Data Len 8bit
Reserved	0 0 1	Object Length 8bit	QoS require-ment 8bit
Max Delay (ms) 16bit		Delay Jitter (ms) 16bit	
Average Data Rate 32bit			
Burstiness : Token Bucket Size 32bit			
Peak Data Rate 32bit			
Minimum Policed Unit 32 bit			
Maximum Packet Size 32 bit			
Values of Packet Classification Parameters			

In QOO, the QoS Requirement describes the QoS requirement of the MN's packet stream, the fields Max Delay and Delay Jitter specify the delay that packet stream can tolerate, the fields Average Data Rate, Burstiness, Peak Data Rate, Minimum Policed Unit and Maximum Packet Size describe the volume and nature of traffic that the corresponding packet stream is expected to generate, the field Packet Classification Parameters provide values for parameters in packet headers that can be used for packet classification.

3.2   Hierarchical Mobile Agents Tree (HMAT)

Recently wireless networks have been implemented based on micro-cell and the moving host may cross small cells very often. Therefore handoff takes place more frequently. Then the QoS signaling delay, the data packet delay will increase and the data rate will decrease and the packet losses, possible service degradation may occur.

Our hierarchical mobile agents tree is aimed at solving these problems. HMAT means Hierarchical Mobile Agents Tree that contains mobile agents of several levels, and can be chosen and configured in any way as the network administrator thinks appropriate.

3.3   Mobile Agent

A mobile agent is an entity that manages QOO, resource reservations and other mobility related works. Mobile agents in a HMAT can be divided into two kinds. First type is the mobile agent in a domain and the first level of the HMAT, similar to home agents in Mobile IPv6, manages QOO for QoS support, processes the mobile related RSVP messages and maintains the mobile soft state for mobile hosts, is organized into a hierarchy to handle local movements of Mobile hosts within the domain. And the second type is the mobile agent in higher levels of the HMAT can manage QOO for QoS support, merge path message and reservation message. This kind of mobile agent will be a point where merging causes no resulting state change when the Path or Resv refresh message establish path or reservation state respectively along the new route after a handoff takes place. The first type mobile agent's function includes the second type mobile agent's function.

3. 4   QoS Support in HMAT

There are two scenarios in QoS support based on our HMAT scheme.

When the Mobile Node (MN) is sender, the Correspondent Node (CN) is receiver, after a handoff, the MN sends a Binding Update with QOO to CN along HMAT, in the first level mobile agent, this agent examines QOO and immediately performs the resource reservation, sends the new path message to CN with the same source flow identity as the one before handoff, and also sends the Binding Update with QOO to the CN. Then the path message can be merged at some mobile agent that has already a path state in HMAT for that flow which is created before. This will make RSVP to have a Local Repair for sender route. Therefore the mobile agent sends a Resv message associated with the flow along the new path in HMAT to the MN upstream at once, also sends the Binding Update with QOO to the CN downstream. The flow path reserved resources previously from the mobile agent to the CN can be reused. After the CN receives the Binding Update with QOO, the CN will send the Binding Acknowledgment to the MN's current location through the HMAT. Then the data packets will be sent from the MN's new location to the CN. Figure 2 shows this scenario.

When the MN is receiver, the CN is sender, and a handoff occurs, the MN sends a Binding Update with QOO upstream along HMAT to some mobile agent that has already the path state for the flow that is created before. This mobile agent examines QOO and immediately performs the resource reservation, sends the new path message to the MN downstream and at the same time sends the Binding Update with QOO upstream to the CN. When the MN receives the new path message, it sends a Resv

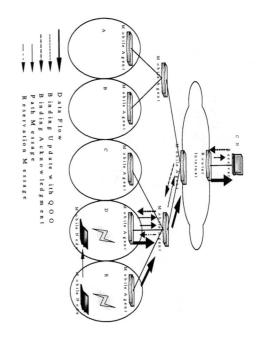

Data Flow
Binding Update with QOO
Binding Acknowledgment
Path Message
Reservation Message

Fig. 2. MN as Sender in HMAT Model

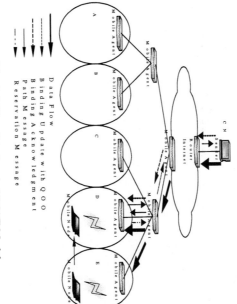

Data Flow
Binding Update with QOO
Binding Acknowledgment
Path Message
Reservation Message

Fig. 3. MN as Receiver in HMAT Model

message associated with the flow along this path in HMAT to the mobile agent. The flow path reserved resources previously from the mobile agent to the CN can be reused. And after the CN receives the Binding Update with QOO, the CN will send the Binding Acknowledgment to the MN's current location through the HMAT. Then the data packets will be sent from the CN to the MN's new location. Figure 3 shows this scenario.

Our scheme provides smooth handoff QoS provision without extra QoS delay. It is sure that the RSVP messages traverse shorter than the Binding Update and Binding Acknowledgment, because the RSVP messages traverse between some mobile agent of HMAT and the MN in a part of the route where the Binding Update and Binding Acknowledgment have to traverse between the CN and MN. Therefore the RSVP renegotiation can be finished before the CN is updated with MN's new care-of address, especially when there are congested links within the path between the CN and the mobile agent of HMAT. Thus resources have been set up before CN starts to send or receive packets with MN's new location. In other words, all packets subsequently between MN's new location and the CN will be offered QoS as desired and no any extra handoff delay may occur due to handoff. But in Multicast Tree scheme [9], when the handoff takes place, at first, the MN sends the Binding Update to the CN and receives the Binding Acknowledgment from the CN, then the Multicast tree should be modified to set up new QoS, at last, the data packets can be transmitted between the MN and CN in Mobile IPv6. We use Rational Rose 2000 to show the Sequence Diagram of the Multicast Tree scheme in Figure 4, the Sequence Diagram of our HMAT scheme in Figure 5 and Figure 6.

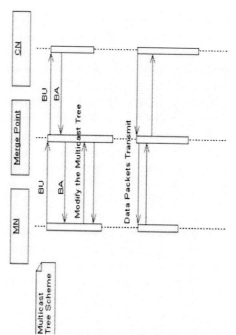

Fig. 4. Sequence Diagram of the Multicast Tree Scheme

4   Simulation Results

We use the OPNET Modeler v8.0 to simulate our scheme and compare our scheme with RSVP Mobility Based on Multicast Tree [9]. For simplicity our simulation is based on an assumption that the capacity of the links between the mobile agents is not limited. And we only considered the unicast data flows from a single mobile sender roaming freely in wireless domain to a fixed static receiver for simplicity. Figure 7 shows the network topology used for our simulation. There are two cells in this network, and each cell has a mobile agent as the first type mobile agent in HMAT that has two levels.

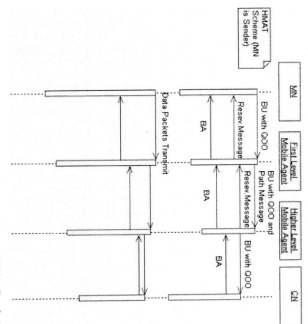

Fig. 5. Sequence Diagram of the HMAT Scheme (MN is Sender)

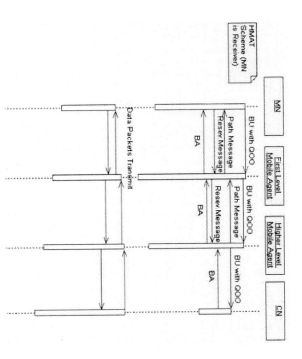

Fig. 6. Sequence Diagram of they HMAT Scheme (MN is Receiver)

184   Z. Qu and C.S. Hong

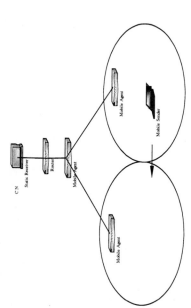

Fig. 7. Network Configuration for Simulation

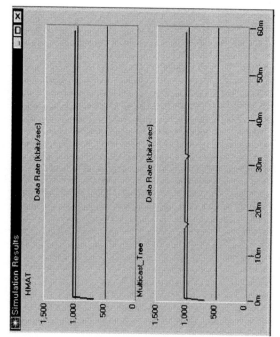

Fig. 8. Simulation Results (Data Rate)

The goal of our simulation work is for evaluating the QoS, such as data rate, packet loss ratio and packet delay, by using our scheme and comparing with RSVP Mobility Based on Multicast Tree [9] when the handoff occurs. We use a real time traffic source at a peak rate of 1Mbps to get the variations of data rate, packet loss ratio and packet delay received by the fixed static receiver CN due to the handoff.

Figure 8 shows the simulation results of the data rate using our HMAT scheme and Multicast Tree scheme over simulation time. In these figures, the X-axis represents the simulation time (minute) and the Y-axis represents the relative data rate (kbps). We can see that the data rate of Multicast Tree scheme is obviously decreased at the moment a handoff takes place, and our HMAT scheme has a real smooth handoff.

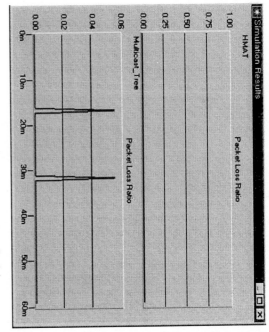

Fig. 9. Simulation Results (Packet Loss Ratio)

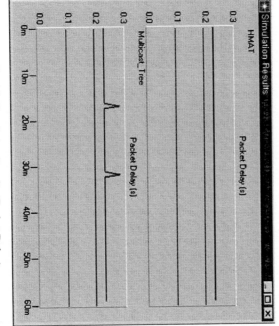

Fig. 10. Simulation Results (Packet Delay)

Figure 9 shows the simulation results of the packet loss ratio. The results show that the packet loss ratio of the Multicast Tree scheme is more than that of the HMAT scheme when handoff takes place.

Figure 10 shows the simulation results of the packet delay. The results also show that the Multicast Tree scheme has more packet delay than our HMAT scheme when handoff takes place.

These simulation results have the following reasons. First, in the Multicast Tree scheme, when the handoff occurs, the Binding Update is sent to the CN and the mobile node receives the Binding Acknowledgment, then the Multicast Tree should be modified dynamically and some QoS signaling messages should be used, the new Conventional Reservation and all of the Predictive Reservations should be made again, then the data packets is be sent continually. Second, in our HMAT scheme, there are no any extra QoS setting up delay due to handoff. Therefore the Multicast Tree scheme has more QoS signaling delay, packet loss ratio, data packet delay and lower data rate than our HMAT scheme whenever a handoff occurs.

In summary, our scheme that extends RSVP using the HMAT based on the QOO provides more efficient QoS signaling mechanism in the Integrated Service Internet with mobile hosts. We reduce the QoS signaling delay, packet loss ratio and packet delay, get higher data rate for real time multimedia applications when handoff occurs.

5 Conclusions

In this paper, a novel framework for QoS support in Mobile IPv6 using the HMAT based on QOO [11] in Integrated Services Internet has been proposed. HMAT can be configured and chosen by the network administrator in any way. Mobile agent can manage QOO, resource reservation and other mobility related works. Mobile agent can be divided into two types according to its function. When a mobile host moves to a new location, the RSVP will be made a Local Repair only between the mobile host and some mobile agent of HMAT in a part of the route where the Binding Update and Binding Acknowledgment are traversed between the MN and the CN. Therefore, our scheme can provide smooth handoff, more efficient QoS provision.

Moreover, we use OPNET Modeler to simulate our scheme and compare with RSVP Mobility Based on Multicast Tree [9]. The simulation results prove that our scheme can provide higher data rate, lower packet delay and packet loss ratio, and improve the efficiency by using HMAT based on QOO when handoff takes place in Mobile IPv6.

In the future, we will keep on researching to give more efficient QoS support in Core Network over MPLS based on the Mobile IPv6 and QOO.

Acknowledgement. This work was supported by Korea Research Foundation Grant. (KRF-2001-003-E00201).

References

1. C.Perkins, " IP Mobility Support, " RFC 2002, October 1996.
2. D.B.Johnson and C.Perkins, "Mobility Support in IPv6," IETF Internet-Draft, work in progress, November 2000.

3. R.Braden, L.Zhang, S.Berson, S.Herzog, and S.Jamin, "Resource ReSerVation Protocol (RSVP) Version 1 Functional Specification," RFC 2205, September 1997.

4. J.Wrclawski, "The Use of RSVP with IETF Integrated Services," RFC 2210, September 1997.

5. A.Terzis, J.Krawczyk, J.Wroclawski, L.Zhang, "RSVP Operation Over IP Tunnels," RFC 2746, January 2000.

6. A.K.Talukdar, B.R.Badrinath, and A.Acharya, "MRSVP: A Reservation Protocol for an Integrated Services packet Network with Mobile Hosts," Tech. Rep. Dcs-tr-337, Department of Computer Science, Rutgers University, U.S.A., 1997.

7. I.Mahadevan and K.M.Sivalingam, "An Experimental Architecture for providing QoS guarantees in Mobile Networks using RSVP," IEEE PIMRC, Boston, September 1998.

8. A.Terzis and M.Srivastava and L.Zhang, "A Simple QoS Signaling Protocol for Mobile Hosts in the Integrated Services Internet," INFOCOM 1999.

9. W.Chen and L.Huang, "RSVP Mobility Support: A Signaling Protocol for Integrated Services internet with Mobile Hosts," INFOCOM 2000.

10. Qi.Shen and W.Seah and A.Lo, "Flow Transparent Mobility and QoS Support for IPv6-based Wireless Real-time Services," IETF Internet-Draft, work in progress, February 2001.

11. H.Chaskar and R.Koodli, "A Framework for QoS Support in Mobile IPv6," IETF Internet-Draft, work in progress, March 2001.

12. ZhaoWei Qu and ChoongSeon Hong, "QoS Provision Architecture for Mobile IP using RSVP," Proceedings of KICS National Conference, Vol.23, pp.1551, Cheju, Korea, July 2001.

Performance Analysis of Wireless TCP

Yuan-Cheng Lai[1], Shuen-Chich Tsai[2], and Arthur Chang[1]

[1] Department of Information Management
National Taiwan University of Science and Technology, Taipei, Taiwan, R.O.C.
laiyc@cs.ntust.edu.tw

[2] Department of Computer Science and Information Engineering
National Cheng Kung University, Tainan, Taiwan, R.O.C.
sctsai@locust.csie.ncku.edu.tw

Abstract. TCP performs reasonably well over Internet where packet losses are mainly due to network congestion. However, TCP originally targeted on the wire networks is inapplicable for the wireless networks where packet losses are mainly caused by link errors. TCP treats a packet loss as an indication of network congestion and misleads the packet loss caused by the link error in the wireless networks. This confusion makes the over-reduction of window size in TCP, resulting in poor performance. Wireless TCP (WTCP) was proposed to promote the performance of TCP in the wireless networks [1]. WTCP splits a TCP connection between a fixed host and a mobile host into two segments with end-to-end semantics. It is installed into the base station to more correctly estimate the retransmission timeout and speed up the packet retransmission in the wireless link.

In this paper, we analyzed the WTCP based on the analysis of TCP Reno [2]. WTCP treats a timeout or a duplicate ACK as an indication of packet loss. We first analyzed the throughput under only considering the loss caused by a duplicate ACK, and then expanded to the case with timeouts. The simulation results show that WTCP outperforms TCP in the wireless environment where link error rate is high, while has the similar throughput in the wire environment with low link error rate.

1 Introduction

Many applications, such as file transfer (FTP), e-mail (SMTP), world wild web (HTTP) and remote access (Telnet), are based on the TCP/IP, one of the most popular protocols on the Internet. The congestion control algorithm of TCP is a major role and significantly influence the performance of these applications. Many different TCP versions, Reno, NewReno, SACK, and Vegas, are proposed and Reno is the most popular version in today's Internet [3]. TCP Reno is a window-based congestion control and aggressively expands its window size until the transmitted packets are lost. The sender halves its window size to avoid congestion when receiving triple-duplicate ACKs and reduces it to one when detecting a timeout. Reno performs reasonably over the Internet where packet losses are mainly due to network congestion, becoming the most popular TCP version on Internet.

However, TCP designed for the wire networks is inapplicable for the wireless networks where the losses are mainly caused by link errors. TCP treats a packet loss as

I. Chong (Ed.): ICOIN 2002, LNCS 2344, pp. 188–199, 2002.
© Springer-Verlag Berlin Heidelberg 2002

an indication of network congestion, and misleads the packet loss caused by the link error in the wireless networks. This confusion makes the over-reduction of window size in TCP, resulting in poor performance. To improve efficiency of TCP in such an environment, some modifications for the TCP congestion control mechanism have been proposed [4-11]:

(1) Modifying TCP specifications to take the link errors into account so that congestion control acts only in case of a genuine network congestion [12]. This approach is hard to carry out on Internet because millions of hosts must reinstall their TCP softwares. Also it is extremely difficult to know the actual cause of a packet loss for a TCP connection travelling wire and wireless links.

(2) Splitting a TCP connection between a fixed host and a mobile host into two separate segments without end-to-end semantics [13]. One segment is from the source to the base station and the other is from the base station to the mobile host. Separating two distinct transmission medias, wire and wireless link, is the original thought, and is very reasonable. The base station must maintain the corresponding connections for two segments: the base station receives a packet from a fixed host and responds an ACK, and transmits a packet to a mobile host and receives its ACK. Using such an approach, an ACK received by the source means the corresponding packet was successfully received by the base station, rather than received by the mobile host. Thus the important characteristic, end-to-end semantics, of TCP is not maintained. The other drawback of this approach is that a large size of buffer is needed in the base station to avoid many packet losses herein.

(3) Splitting a TCP connection between a fixed host and a mobile host into two segments with end-to-end semantics [1]. This approach, called Wireless Transmission Control Protocol (WTCP) in this paper, lightly modifies the base station and keeps the end-to-end semantics of a TCP connection. This approach different from the previous one lies that the base station passes the ACK to a fixed host when receiving an ACK from the mobile host, rather when receiving the corresponding packet from the fixed host. Due to the high link error rate in the wireless environment, WTCP treats an duplicate ACK as an indication of link error rather than network congestion. Furthermore reducing the waiting time for packet retransmission, WTCP do not take the variance of RTT caused by link error into account for the retransmission timeouts.

2   TCP Congestion Control Algorithm

Reno uses a congestion window (*cwnd*) to control the amount of transmitted data in one *RTT* and a maximum window (*mwnd*) to limit the maximum value of *cwnd*. The control scheme of Reno can be divided into five parts, which are interpreted in the following. version specified with these parts.

Slow-start: As a connection starts or a timeout happens, the slow-start state begins. The initial value of *cwnd* is set to one packet in the beginning of this state. The sender increases *cwnd* exponentially by adding one packet each time it receives an ACK. Slow-start controls the window size until *cwnd* achieves a preset threshold, slow-start threshold (*ssthresh*). When *cwnd* reaches to *ssthresh*, the "congestion avoidance" state begins.

Congestion avoidance: Since the window size in the slow start state expands exponentially, the packets sent in this increasing speed would quickly make the network congestion. To avoid the phenomenon, the "congestion avoidance" state begins when $cwnd$ exceeds $ssthresh$. In this state, $cwnd$ is added by $1/cwnd$ packet every receiving an ACK to make the window size grow linearly.

Fast retransmission: The duplicate ACK is caused by an out-of-order packet received in the receiver. The sender treats it as a signal of a packet loss or a packet delay. If three or more duplicate ACKs are received in a row, the occurrence of packet loss is very possible. The sender performs a retransmission of what appears to be the missing packet, without waiting for a coarse-grain timer to expire.

Fast recovery: At the time the fast retransmission is performed, $ssthresh$ is set to half of $cwnd$ and then $cwnd$ is set to $ssthresh$ plus three packet sizes. $Cwnd$ is added by the packet size every receiving a duplicate ACK. When the ACK of the retransmitted packet is received, $cwnd$ is set to $ssthresh$ and the sender reenters the congestion avoidance. In other words, $cwnd$ is reset to half of the old value of $cwnd$ after fast recovery.

Retransmission timeout: The sender activates a timer for every packet sent to check whether the corresponding ACK will return within a retransmission timeout (RTO). If any timeout occurs, the sender resets $cwnd$ to one and restarts slow-start phase. The initial value of RTO, T_0, for each packet is set to $RTT + 4VRTT$ where RTT is a round-trip time and $VRTT$ is its variance. RTO is doubled when the sender encounters the continuous timeouts, until $64T_0$.

3   Wireless TCP

WTCP, installed in a base station, receives a packet from the fixed host, saves it to the buffer, and then sends it to the mobile host. Also, WTCP receives an ACK from the mobile host, saves it to the buffer, and then sends it to the fixed host. WTCP can be thought as virtual destination for the TCP sender (fixed host) and virtual source for the receiver (mobile host). Figure 1 shows the architecture of WTCP, including three parts which is explained in the following:

(1) WTCP receives packets from the fixed host: When receiving a packet, which sequence number equals the expected value, from the the fixed host on the wire link, WTCP stores it in the buffer and increases the expected sequence number to the next one. When the sequence number of a received packet is larger than the expected one, this out-of-order packet is still stored in the buffer, but the expected sequence number is not updated. When the sequence number of the received packet is smaller than the expected one, this duplicate packet is dropped.

(2) WTCP sends packets to the mobile host: For each wireless segment, the base station maintains an extra variable wireless window size ($wcwnd$) which indicates that how many packets are allowed to send to the mobile host by the base station. Each time WTCP is allowed to transmit packets to the mobile host, it sends a packet stored in the buffer and starts a timer to count the round-trip time in the wireless link. If the timer is expired, the packet has been lost and WTCP enters retransmission timeout phase. To reduce the detecting time of this kind of losses, RTO is set to $1.5RTT$. If the retransmission packet encounters the continuous timeouts, RTO is set and fixed to

3RTT. When any timeout occurs, WTCP thinks the wireless link is in a bad status and sets the *wcwnd* to 1. Later, receiving an ACK from the mobile host, WTCP believes the wireless link has recovered and restores *wcwnd* to the original value.

(3) WTCP receives ACKs from mobile host and pass these ACKs to the fixed host: WTCP send the ACK to the fixed host only after it successfully receives the ACK from the mobile host, to maintain end-to-end semantics. When receiving an ACK number larger than or equal to what is expected, WTCP thinks all packets having the sequence number larger than this ACK number is successfully received by the mobile host, and then forwards this ACK to the fixed host. However, a duplicate ACK is regarded as an indication of the wireless link error and the possibly lost packet is retransmitted in time.

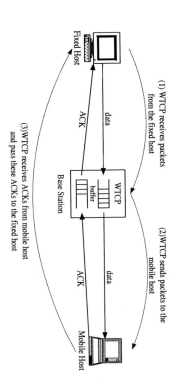

Fig. 1. The architecture of WTCP

(1) WTCP receives packets from the fixed host

(2) WTCP sends packets to the mobile host

(3) WTCP receives ACKs from mobile host and pass these ACKs to the fixed host

Fixed Host

Base Station

Mobile Host

data / ACK / WTCP buffer / data / ACK

In summary, three main differences between Reno and WTCP are observed: (1) Packet retransmission: TCP Reno retransmits the lost packets when receiving triple duplicate ACKs, while WTCP do it when receiving one duplicate ACK. (2) Initial RTO: The initial RTO, T_0, of TCP Reno and WTCP are set $RTT + 4VRTT$ and $1.5RTT$ respectively. (3) Successive timeouts: TCP Reno doubles the value of RTO until $64T_0$ when encountering successive timeouts while WTCP doubles T_0 only when two continuous timeouts occur.

4   Analysis WTCP

Some reasonable assumptions are used to simplify the analysis in our paper, listed as below:

(1) Although two segments exist in a end-to-end TCP connection from server to mobile host, the segment from the base station to mobile host is a bottleneck, since the bandwidth on the wire networks is always greater than wireless network. Thus the throughput of a end-to-end TCP connection is similar with that on the wireless link. We focus on the analysis of the throughput on wireless link below. (2) Each TCP source is backlog. The time spent in slow start is negligible in the long term compared to

the duration of congestion avoidance and fast retransmission. Thus, our analysis does not take the effect of slow start into account because of its trivial effect on the overall throughput. (3)The packet loss rate is assumed as a constant p from the base station to the mobile host and is assumed as zero in the reverse path, i.e., the ACKs are never lost. (4)If a packet is lost, the sequential packets in the same round are also lost.

WTCP treats a duplicate ACK or a timeout as an indication of packet loss. We call the former as a duplicate loss and the latter as a timeout loss. Section 4.1 analyzes the WTCP throughput in the case under considering the former, and then Section 4.2 expands to the case with both.

4.1 Throughput without Timeout

As definition, the throughput at time t, $B^D(t)$, is

$$B^D(t) = \frac{N(t)}{t},$$

where $N(t)$ is the number of packets sent in the interval $[0, t]$. Thus, the long-term steady-state throughput of a WTCP connection is got as

$$B^D = \lim_{t\to\infty} B^D(t) = \lim_{t\to\infty} \frac{N(t)}{t}.$$

Let T_j^D be the time interval, called the j-th DA period, between j-th and $(j+1)$-th duplicate loss and N_j^D be the number of packets sent during this interval. When the WTCP connection stays in the stable state, its behavior can be observed during a DA period between two duplicate losses. A new DA period starts immediately after a duplicate loss, and thus the current window size is equal to half the size of window before the duplicate loss occurred. As shown in Figure 2. If the maximum window size during the previous DA period is W_{j-1}, the first congestion window during the sequential DA period is thus $W_{j-1}/2$. For each DA period in the stable state, the duration is denoted as T^D, and the packet transmitted during this period is denoted as N^D. Thus, the throughput only with duplicate loss can be approximated as

$$B^D = \frac{E[N^D]}{E[T^D]}.$$

To derive $E[N^D]$, considering the duration T_j^D in Figure 2, we have

$$N_j^D = a_j + (W_j - 1),$$

where a_j is the first packet lost during the period T_j^D and W_j is the maximum window size during T_j^D, i.e., $(W_j - 1)$ packets sent after first packet loss. Thus, the expected value N^D is expressed as the sum of the expected values of the individual terms.

$$E[N^D] = E[a] + E[W] - 1. \tag{1}$$

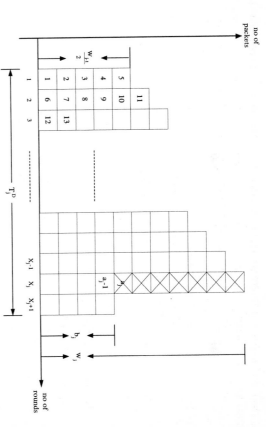

Fig. 2. Packets sent during T_j^D

Since the packet loss probability is p, the expected value of the first loss is

$$E[a] = \sum_{k=0}^{\infty} k(1-p)^k p = \frac{1}{p}.$$ (2)

Combining (1) and (2), we get

$$E[N^D] = \frac{1}{p} + E[W] - 1.$$ (3)

On the other hand, every round $wcwnd$ is increased by one and some more packets are sent in the last round. Thus, we can directly count the number of sent packets during T_j^D round by round, as

$$
\begin{aligned}
N_j^D &= \sum_{k=0}^{X_j-1}\left(\frac{W_{j-1}}{2} + k\right) + b_j \\
&= \sum_{k=0}^{X_j-1}\frac{W_{j-1}}{2} + \sum_{k=0}^{X_j-1} k + b_j \\
&= \frac{X_j}{2}(W_{j-1} + X_j - 1) + b_j
\end{aligned}
$$ (4)

where X_j is the round occurring the first loss, W_j is the maximum window size, and b_j is the number of packets sent in the last round, during the period T_j^D. Because the congestion window size is increased by one every round and halved when encountering a duplicate loss, we have

$$W_j = \frac{W_{j-1}}{2} + X_j. \tag{5}$$

The mean of N^D and W, derived from (4) and (5), are

$$E[N^D] = \frac{E[X]}{2}\left(E[W] + E[X] - 1\right) + E[b] \tag{6}$$

and

$$E[W] = 2E[X]. \tag{7}$$

From (6) and (7), it follows that

$$E[N^D] = \frac{E[W]}{4}\left(E[W] + \frac{E[W]}{2} - 1\right) + E[b]. \tag{8}$$

We consider that b_j is uniformly distributed between 1 and $W_j - 1$. Thus,

$$E[b] = \frac{1 + (E[W] - 1)}{2} = \frac{E[W]}{2}. \tag{9}$$

Replacing (9) into (8), and solving the system of Equations (3) and (8), we get

$$E[W] = 1 + \sqrt{\frac{8 - 5p}{3p}}. \tag{10}$$

Replacing (10) into (3), we get

$$E[N^D] = \frac{1}{p} + \sqrt{\frac{8 - 5p}{3p}}. \tag{11}$$

Due to each round takes a RTT, the mean length of T^D is easily calculated as,

$$E[T^D] = (E[X] + 1)RTT. \tag{12}$$

From (7), (10) and (12), it follows that

$$E[T^D] = \frac{1}{2}\left(3 + \sqrt{\frac{8 - 5p}{3p}}\right)RTT. \tag{13}$$

Finally, we can express

$$B^D = \frac{E[N^D]}{E[T^D]} \tag{14}$$

$$= \frac{\frac{1}{p} + \sqrt{\frac{8 - 5p}{3p}}}{\frac{1}{2}\left(3 + \sqrt{\frac{8 - 5p}{3p}}\right)RTT}. \tag{15}$$

4.2 Throughput with Timeouts

In steady state, every cycle can be viewed as some DA periods following a few timeout periods (TO periods). Such a cycle i, denoted by T_i, includes a duplicate loss subcycle T_i^{DA} and a timeout subcycle T_i^{TO}, as in Figure 3. The j-th DA period in T_i^{DA} is denoted by T_{ij}^D and the number of transmitted packets in T_{ij}^D is denoted by N_{ij}^D. Let n_i be the number of DA period during T_i^{DA}, and assume that n_i is independent of T_i^{DA} and T_{ij}^D. Thus we can easily calculate the length of the duplicate-loss subcycle,

$$T_i^{DA} = \sum_{j=1}^{n_i} T_{ij}^D \Rightarrow E[T^{DA}] = E[n]E[T^D], \qquad (16)$$

and the number of packets sent during T_i^D,

$$N_i^{DA} = \sum_{j=1}^{n_i} N_{ij}^D \Rightarrow E[N^{DA}] = E[n]E[N^D], \qquad (17)$$

where $E[T^D]$ and $E[N^D]$ can be obtained from (11) and (13). To derive $E[n]$, we want to get P^{TO}, the probability that a timeout follows a duplicate loss during T_i^{DA}. We have $E[n] = 1/E[P^{TO}]$.

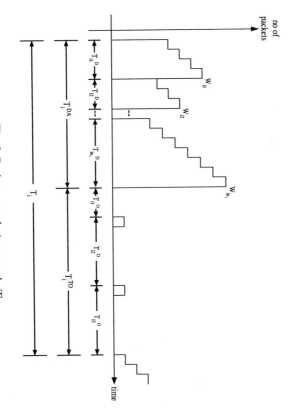

Fig. 3. Packets sent during a cycle T_i

Considering the subcycle T_i^{DA}, two cases causes a timeout. The first case is that the first packet within the last round of a DA period is lost. In such a case, because the first and all sequential packets are lost in the last round, no ACK triggers any new packet transmission, and then a timeout occurs. The probability of this case is p. The second case is that the b_i-th packet is lost in the penultimate round, and $(b_i - 1)$ new packets sent in the last round of a DA period are all lost, for $2 \leq b_i \leq w_i$. The probability of this case is the sum of the probabilities that the first lost packet is the b_i-th packet in the penultimate round and the packets sent in the last round are all lost, i.e, $\sum_{b_i=2}^{w_i} P\{$the b_i-th packet loss in penultimate round$\} \times P\{$all packets are lost in the last round$\} = \sum_{b_i=2}^{w_i}(1-p)^{b_i-1}p \times p = p(1-p)(1-(1-p)^{w_i-1})$. From above two cases, the probability of a timeout during T_i^{DA} is $P_i^{TO} = p(2 - p - (1-p)^{w_i})$. Therefore, we can approximate

$$E[P^{TO}] = p\left(2 - p - (1-p)^{E[W]}\right).$$

Let N_i^{TO} be the number of packets sent during T_i^{TO}. The throughput of WTCP in the case considering timeouts becomes

$$
\begin{aligned}
B_{wtcp} &= \frac{E[N^{DA}] + E[N^{TO}]}{E[T^{DA}] + E[T^{TO}]} \\
&= \frac{E[n]E[N^D] + E[N^{TO}]}{E[n]E[T^D] + E[T^{TO}]} \\
&= \frac{E[N^D] + E[P^{TO}] \times E[N^{TO}]}{E[T^D] + E[P^{TO}] \times E[T^{TO}]} \\
&= \frac{E[N^D] + p\left(2 - p - (1-p)^{E[W]}\right) \times E[N^{TO}]}{E[T^D] + p\left(2 - p - (1-p)^{E[W]}\right) \times E[T^{TO}]}.
\end{aligned}
$$
(18)

Observing Equation (18), we must determine $E[N^{TO}]$ and $E[T^{TO}]$. To derive $E[N^{TO}]$, observing the case that k consecutive timeouts occur, i.e., $(k-1)$ consecutive retransmitted packet losses and one successful retransmission finally. The probability of this case is

$$P[N^{TO} = k] = p^{k-1}(1-p).$$

Thus the mean of N^{TO} is

$$E[N^{TO}] = \sum_{k=1}^{\infty} kP[N^{SO} = k] = \frac{1}{1-p}.$$
(19)

To derive $E[T^{TO}]$, the j-th timeout period during T_i^{TO} is defined as

$$T_{ij}^O = \begin{cases} 1.5RTT & \text{if } k=1; \\ 3RTT & \text{if } k \geq 2. \end{cases}$$

Let C_k^{TO} be the spent time of k consecutive timeouts. Thus we get $C_1^{TO} = T_{i1}^O = 1.5RTT$ and $C_k^{TO} = \sum_{j=1}^{k} T_{ij}^O = 1.5RTT + 3(k-1)RTT$ for $k \geq 2$. Thus,

$$E[T^{TO}] = \sum_{k=1}^{\infty} C_k^{TO} P[N^{TO} = k] = \left(1.5 + \frac{3p}{1-p}\right) RTT.$$
(20)

Substituting $E[N^D]$ from (11), $E[T^D]$ from (13), $E[W]$ form (10), $E[N^{TO}]$ from (19), and $E[T^{TO}]$ (20), into (18), we have

$$B_{wtcp} = \frac{E[N^D] + p\left(2 - p - (1-p)^{E[W]}\right) \times E[N^{TO}]}{E[T^D] + p\left(2 - p - (1-p)^{E[W]}\right) \times E[T^{TO}]}.$$

$$= \frac{\left(\frac{1}{p} + \sqrt{\frac{8-5p}{3p}}\right) + p\left(2 - p - (1-p)^{1+\sqrt{\frac{8-5p}{3p}}}\right)\left(1.5 + \frac{3p}{1-p}\right)RTT}{\frac{RTT}{2}\left(3 + \sqrt{\frac{8-5p}{3p}}\right) + p\left(2 - p - (1-p)^{1+\sqrt{\frac{8-5p}{3p}}}\right)\left(\frac{1}{1-p}\right)}. \quad (21)$$

In [2], the TCP Reno throughput is calculated as

$$B_{tcp} = \frac{\frac{1-p}{p} + E[W] + \hat{Q}(E[W])\frac{1}{1-p}}{RTT(E[X]+1) + \hat{Q}(E[W])T_0\frac{f(p)}{1-p}}$$

$$= \frac{\frac{1-p}{p} + \left(1 + \sqrt{\frac{8-5p}{3p}}\right) + \frac{3}{1+\sqrt{\frac{8-5p}{3p}}}T_0\frac{f(p)}{1-p}}{\frac{RTT}{2}\left(3 + \sqrt{\frac{8-5p}{3p}}\right) + \frac{3}{1+\sqrt{\frac{8-5p}{3p}}}T_0\frac{f(p)}{1-p}}.$$

where $\hat{Q}(w) = \min(1, \frac{(1-(1-p)^3(1+(1-p)^3(1-(1-p)^{w-3}))}{1-(1-p)^w})$ and $f(p) = 1 + p + 2p^2 + 4p^3 + 8p^4 + 16p^5 + 32p^6$.

If p approaches to zero, we get the same throughput for WTCP and TCP, as

$$\lim_{p \to 0} B_{tcp} = \lim_{p \to 0} B_{wtcp}$$

$$= \frac{\frac{1}{p} + \sqrt{\frac{8}{3p}}}{\frac{RTT}{2}\left(3 + \sqrt{\frac{8}{3p}}\right)}. \quad (23)$$

5   Numerical and Simulation Results

In our simulation environment, one fixed host attached to one base station are existed in the wire networks and one mobile host is existed in the wireless environment. The wireless link has 2 Mbps transmission rate and maximum packet size (MSS) is 512 bytes, and thus the transmission time of a 512 bytes packet is 2 ms. Adding the propagation delay, RTT is assumed as 50ms. T_0 of WTCP and TCP is $1.5RTT = 75$ ms and $RTT = 50$ ms respectively when $VRTT$ is assumed to zero.

Figure 4 shows that the analytical throughputs of WTCP and TCP are almost the same in the environment where loss rate is smaller than 0.01. This observation is also obtained from the simulation results. In such a case, WTCP and TCP spend most time sending new data, rather than retransmitting the lost packets or waiting timeouts, resulting in very similar throughput. When the packet loss probability exceeds 0.01, TCP has the

This analytical result is also observed from the simulation result, which shows that WTCP and TCP have the same throughput in the low loss rate environment [1].

In a real network, the throughput of TCP or WTCP will be bounded to the link bandwidth. Let the bandwidth be Z Mbps and a packet size be s bytes. Thus, the maximum achieved throughput is $\frac{Z \times 2^{20}}{8s}$. Hence, the throughputs of WTCP and TCP become $\min(B_{wtcp}, \frac{Z \times 2^{17}}{s})$ and $\min(B_{tcp}, \frac{Z \times 2^{17}}{s})$, respectively.

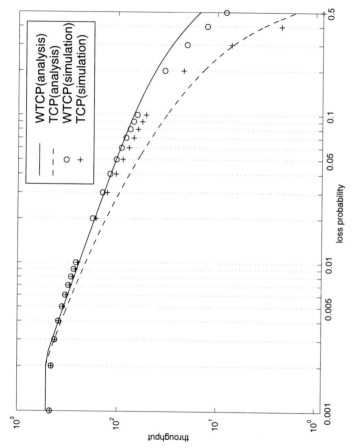

Fig. 4. Throughput comparison of WTCP and TCP Reno.

poorer throughput than WTCP because the latter reduces the detecting time of timeouts and rapidly retransmits lost packets.

NS2 is used as our simulation tool [14]. Each simulation result is collected from 5000 sec experiment. For WTCP, the throughputs from simulation results are very close to our analytical results. Generally, the analytical throughputs will be lower than simulation results since we assume that the remaining packets in that round will be lost if a packet loss occurs. This assumption cause that the analytical throughput slightly underestimates the actual throughput in the case where loss rate varying between 0.004 to 0.05. However, the analytical throughput is slightly overestimated when loss rate exceeds 0.05 because the sender spends more time in fast recovery phase. Due to many packet losses, the sender needs several rounds to recover, and causes the duration increase of the DA period, but our analysis do not take this into account. When loss rate is below than 0.004, the simulation throughput of WTCP is lower than the analytical throughput, which is equal to the link bandwidth in such cases.

6   Conclusion

In this paper, we briefly introduced the congestion control of TCP Reno and the features of WTCP, and analyzed the throughput of WTCP. From these analyses, the throughput of WTCP is proved to be better than TCP, also is observed from Figure 4, in the wireless environment where link error rate is high. On the other hand, TCP reno can achieve the similar throughput with WTCP in the wire environment where packet loss rate is small.

For deriving the throughput, some assumptions are made, and ought to relaxed for more reality in the future.

References

1. K. Ratnam and I. Matta. WTCP: an efficient mechanism for improving TCP performance over wireless links. *Computers and Communications, 1998. ISCC '98. Proceedings. Third IEEE Symposium on,* (1998) 74–78

2. J. Padhye, V. Firoiu, D. F. Towsley, and J. F. Kurose. Modeling TCP Reno performance: a simple model and its empirical validation. *Networking, IEEE/ACM Transactions on,* (2000) 133–145

3. W. Stevens. TCP Slow Start, Congestion Avoidance, Fast Retransmit, and Fast Recovery Algorithms. *RFC 2001,* (1997)

4. F. Anjum and R. Jain. Performance of TCP over Lossy Upstream and Downstream Links with Link-level Retransmissions. *Proc. IEEE Int. Conf. on Networks (ICON) 2000,* (2000)

5. D.S. Bakin, M. Joa-Ng, and A.J. McAuley. Quantifying TCP Performance Improvements in Noisy Environments Using Protocol Boosters. *Proceedings of the Fifth IEEE Symposium on Computers and Communications (ISCC 2000),* (2000)

6. N. K. G. Samaraweera. Non-congestion packet loss detection for TCP error recovery using wireless links. *IEE Proc. Comm,* **146** (1999) 222–230

7. B. S. Bakshi, P Krishna, N. H. Vaidya, and D. K. Pradhan. Improving performance of TCP over wireless networks. In *International Conference on Distributed Computing Systems,* (1997)

8. W. K. Wong and C. M. Leung. Improving End-to-end Performance of TCP using Link-layer Retransmissions over Mobile Internetworks. *IEEE International Conference on Communications '99 (ICC'99 Vancouver – Multimedia & Wireless),* (1999)

9. T. Goff, J. Moronski, D. S. Phatak, and V. Gupta. Freeze-TCP: A true end-to-end enhancement mechanism for mobile environments. (2000) 1537–1545

10. F. Peng, S. Cheng, and J. MA. An Effective way to improve TCP performance in Wireless/mobile networks. (2000)

11. J. Pan, J. W. Mark, and X. Shen. TCP Performance and Its Improvement over Wireless Links. *IEEE GLOBECOM 2000,* (2000)

12. P. Sinha, N. Venkitaraman, R. Sivakumar, and V. Bharghavan. WTCP: A reliable transport protocol for wireless wide-area networks. (1999) 231–241

13. A. Bakre and B. R. Badrinath. I-TCP: Indirect TCP for mobile hosts. *15th International Conference on Distributed Computing Systems,* (1995)

14. S. McCanne and S. Floyd. The UCB/LBNL/VINT Network Simulator (NS). http://www-mash.cs.berkeley.edu/ns/. (1997)

EM-TCP: An Algorithm for Mobile TCP Performance Improvement in a DSRC-Based ITS

KyeongJin Lee[1], SeungYun Lee[1], YongJin Kim[1], ByongChul Kim[2], and JaeYong Lee[2]

[1] Electronics and Telecommunications Research Institute,
161, Gajeong-Dong, Yuseong-Gu, Daejeon
{leekj, syl, kimyj}@etri.re.kr
[2] Chungnam National University,
220, Gung-Dong, Yuseong-Gu, Daejeon
{byckim, jyl}@cnu.ac.kr

Abstract. Since the traditional TCP is tuned to perform well in wired networks and stationary nodes environments, its performance is deteriorated in wireless networks that cause many errors and disconnections. There have been some researches such as Snoop [9][10] and M-TCP [11] to improve the TCP performance in wireless network environment. In this paper, we propose an "enhanced mobile TCP (EM-TCP)" by combining Snoop and SH-TCP in M-TCP to improve the performance of TCP connections when a mobile node uses Internet service via a roadside base-station. We implement the proposed algorithm EM-TCP using ns-2 (Network Simulator) [4] and perform simulation of a DSRC-based ITS system adopting EM-TCP, Snoop and ordinary TCP. Simulation results show that the EM-TCP improves the TCP performance in wireless network environments better than the Snoop and ordinary TCP.

1 Introduction

The traditional TCP does not guarantee reliable and efficient data transmission in wireless environment, because it does not consider properties of wireless network featuring by limited bandwidth, long latency, high bit error rate, and frequent disconnection. TCP supposes that most packet losses in wired network is caused by congestion. Thus, it invokes congestion control to provide reliable service.

However, in wireless network, there are some other reasons to cause packet losses such as high BER (Bit Error Rate) and frequent disconnection by handoff. If we use TCP in this environment, it drastically reduces the congestion window size and doubles the retransmission timeout value. This doubling is called an exponential backoff [7, 13]. This unnecessary congestion control reduces the utilization rate of the bandwidth, and degrades network performance severely.

In addition, we have found that serial timeouts at TCP sender degrade overall throughput more than losses due to bit errors or small congestion windows [11] do. Serial timeout is the phenomenon that TCP sender fails to receive acknowledgement continuously. It happens because of frequent disconnection, which increases a

I. Chong (Ed.): ICOIN 2002, LNCS 2344, pp. 200–211, 2002.
© Springer-Verlag Berlin Heidelberg 2002

probability that TCP sender sends segments for disconnected period continuously. Fig. 1 illustrates this phenomenon.

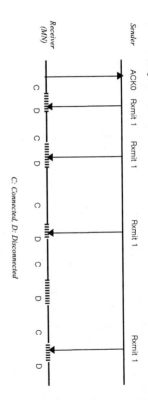

Fig. 1. Serial timeout phenomenon at the sender

ITS (Intelligent Transportation System) is an example of application using TCP in wireless environment. ITS architecture provides various services such as fare levy service, traffic control service and travel information service to the vehicles to minimize human and physical damage by heavy traffic [3]. There are various kinds of communication methods that can be used to provide intelligent transportation service in ITS. DSRC-based ITS environment is one of the promising solution that use DSRC (Dedicated Short-Range Communication) standard for communication between base-station and MNs. DSRC is a method to be applied between RSEs (Rode Side Equipments) and vehicles which are moving fast. It provides packet mode communication in short time, with high speed and within short distance not more than100m.

According to the significant increase of MNs (Mobile Nodes) which want to use Internet service, a RSE (Road Side Equipment) in ITS (Intelligent Transportation System) architecture is also required to provide Internet services. In order to meet such requirements, Mobile IP [1] is studied to support mobile routing and addressing functions, and Mobile TCP [7] is studied to guarantee reliable end-to-end connection in wireless network.

Therefore, it is required to propose a new protocol for wireless network that reflects the high bit error ratio, frequent disconnection, special property like handoff and so on to serve Internet service effectively in DSRC-based ITS system [5,6]. In order to provide Internet service to ITS system, IP and TCP should be modified to suit in wireless network. However, the previous researches cannot guarantee reasonable performance on DSRC-based ITS system, because some of them are only partially tuned to improve the TCP performance for wireless side, while others to improve wired side. Although some schemes provide improved performance of both sides, it is difficult to implement or violate TCP end-to-end semantics [7, 13, 14].

In this paper we propose a suitable algorithm called EM-TCP which enhances mobile TCP for DSRC-based ITS in which frequent handoff and error can occur. We combine Snoop [9] for wireless region and SH-TCP in M-TCP [11] for wired region, in order to improve the performance of TCP connections when a mobile node uses Internet service via a RSE. We simulate the proposed algorithm using ns-2 simulator [4] and verify that EM-TCP provides better performance than traditional TCP when it is applied to DSRC-based ITS.

In section 2, we explain the related works on mobile TCP, and describe the proposed EM-TCP in section 3. Section 4 shows the simulation results and analysis of the proposed mechanism, and we conclude in Section 5 with further works.

2 Related Work

Several schemes have been proposed to solve wireless TCP problems. Those schemes consider the different properties between wired network and wireless network and choose various mechanisms such as local retransmission, split-TCP connections, and forward error correction [7].

There are two fundamentally different approaches to improve TCP performance in wireless network: end-to-end proposals and split connection proposals, represented by Snoop and M-TCP, respectively [9, 11, 7, 8].

2.1 Snoop

Snoop [9] introduces snooping agent at the base-station to monitor every packets sent across the TCP connection that have not yet been acknowledged by the receiver and maintain a buffer of TCP segments. A packet loss is detected by the arrival of a small number of duplicate acknowledgements from the receiver or by a local timeout. The snooping agent, which is in the middle of a TCP connection, retransmits the lost packet from its buffer and discards the duplicated acknowledgements. So it prohibits unnecessary congestion control, avoids serial timeout and improves TCP performance using local retransmission [9, 10, 12]. However the disadvantage of snoop is that it cannot detect disconnection due to handoff. If there is no received packet for the handoff and the retransmission timer of snoop is expired then finally the sender invoke congestion control.

2.2 M-TCP

M-TCP [11] splits TCP into M-TCP (Mobile TCP) and SH-TCP (Supervisor Host TCP). The SH is connected to the wired network and it handles most of the routing and other protocol details for the mobile users. When SH-TCP receives a segment from the TCP sender it passes the segment on to the M-TCP client and M-TCP send acknowledgement that is from the MN. M-TCP client maintains one last byte unacknowledged. After its determination of TCP disconnection it transmits an indication of this fact to SH-TCP. SH-TCP sends ACK for the last byte to the sender. This ACK will also contain a TCP window size update that sets the sender's window size to zero. When the TCP sender receives the window updates, it is forced into *persist mode* [13]. While in this state, it will not suffer from retransmit timeouts and will not exponentially backoff its retransmission timer, nor will it close its congestion window. When the MN regains its connection, it sends a greeting packet to the SH. M-TCP is notified of this event and it passes on this information to SH-TCP which, in turn, sends an ACK to the sender and reopens its receive window. The window update allows the sender to leave persist mode and begin sending data again [11]. However M-TCP has 3 pitfalls. Firstly, it chooses split connection approach even though it intends end-to-end semantics. Secondly, if the sender transmits data to the MN occasionally, as in persistent HTTP connections for example, the base-station holds on to the acknowledgement of the last byte. This causes the sender to timeout and to repeatedly retransmit the last byte. Eventually, the sender gives up on the connection and terminates it. Finally, since M-TCP relies on a single ACK to put the sender into persist mode, this goal may not be achieved if the ACK is lost in the wired network [14].

3 An Algorithm for Improving Mobile TCP Performance

In this section, we propose a solution to improve the performance of TCP throughput over the wireless Internet based on the end-to-end approach. We name this algorithm EM-TCP (enhanced mobile TCP). It combines advantages of M-TCP [11] for wired region and Snoop [9] for wireless region. The advantage of Snoop is that it can detect packet losses caused by high BER of wireless network and retransmit the lost packets from its buffer. Thus, it prohibits the expiration of retransmission timer at the sender. Advantage of M-TCP is that it can detect frequent disconnection caused by frequent handoff. When the disconnection is detected, an ACK packet is transmitted, so that the sender goes into persist mode. In this case, the last sequence number of ACK is reused for the new ACK. Eventually, it prohibits the sender to reduce its window size to 1.

We design the EM-TCP as taking the benefits of the both solutions. The major functionalities of the EM-TCP are as follows:

1. Local retransmission of lost packet at wireless section,
2. Manipulation of duplicate ACKs,
3. Enforcement of the TCP sender into persist mode when it determines that the TCP connection is disconnected at wireless section.

The detailed operations for the functionalities are shown in Fig. 2 and Fig. 3. We introduce some notations to simplify the description. Let 'lastSeen' denote the largest sequence number received by the base-station from the TCP sender and 'lastAck' denote the largest sequence number acknowledged by MN.

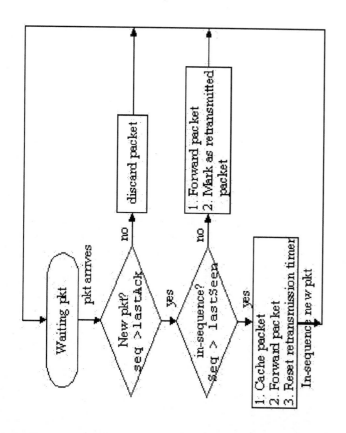

Fig. 2. Flowchart for data packet processing in an EM-TCP agent

Fig. 2 explains data packet processing in an EM-TCP agent. In Fig. 2 'seq' is a sequence number of an arrival packet. At first, the EM-TCP agent is in waiting mode for arriving packets. Once a packet is arrived, it checks whether it is new one or not. If it has been acknowledged before, the agent drops the packet without any processing. If it has not been acknowledged yet but it is in buffer, or if it is out of ordered packet, the agent considers that it is a retransmitted packet by TCP sender or a delayed packet due to congestion. The EM-TCP agent saves this information and transmits this packet to the receiver. If the new packet has correct ordered sequence number, the agent takes it to the buffer and forwards this packet to the receiver and updates retransmission timer.

Fig. 3 shows ACK processing in an EM-TCP agent. In Fig. 3 'ack' denotes a sequence number of arrived ACK. Upon arriving an ACK, the agent checks whether this connection is regained or newly established. When the EM-TCP agent receives an ACK in disconnected status, it sends all packets holding its buffer during disconnected time, and forwards the ACK to the sender in order to have it leave persist mode. When the EM-TCP agent receives an ACK in connected status, it checks whether it is new one or not. If the ACK has been already received before, the agent drops the packet without any processing. When the agent received a duplicated ACK, it sends it to the receiver only once and drops the ACK. When the agent received the duplicated ACK again, it drops the ACK without any processing. When it receives a new ACK, the agent forwards that to the receiver and resets the disconnect timer to detect disconnection.

With these processes, the EM-TCP guarantees both local retransmission in Snoop and efficient handling of disconnection in the SH-TCP of M-TCP. It makes the EM-TCP enhance the performance of TCP efficiently without breaking of the TCP end-to-end semantics.

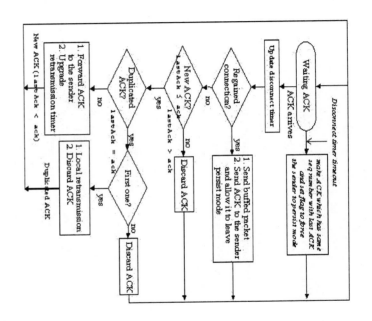

Fig. 3. Flowchart for ACK processing in an EM-TCP agent

4 Simulation Results and Analysis

In this session, we implement and verify the EM-TCP, which is proposed for improving TCP performance. The simulations are performed by ns-2 simulator [4]. We ran experimental simulations with two different environments: in a simple wireless network, and DSRC-based ITS. In simple network, we considered the situation that serial timeout occurred as well as normal situation.

4.1 Simulation 1 – In Simple Wireless Network Environment (Normal Situation)

In the simulation 1, a MN wants to communicate with the sender, S who provides Internet service. They communicate through base-station in wireless network having error ratio 2%. When it moves to the other region, handoff is occurred.

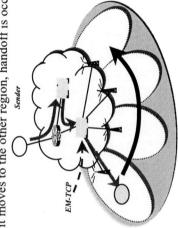

Fig. 4. Simple wireless network environment

Fig. 4 shows a network environment for this simulation. We generate TCP/FTP traffic from sender S to receiver MN. We set the bandwidth of each link as 10Mb, link delay as 30ms and error rate of the link between FA and MN as 2%. EM-TCP is run on FA to manage packets on the TCP connection. We compared performance of EM-TCP, ordinary TCP and Snoop while the MN moves from region 1 to 4 fast. It makes frequent handoff. We have scenarios that the MN is connected to each base-station for a period at 0~5s, 10~12s, 15~28s and 33~60s respectively, and disconnected at 5~10s, 12~15s and 28~33s. We generate traffic flow from 0s and the MN also starts to move from 0s.

An EM-TCP agent can observe all packets between TCP and TCPSink and takes suitable action. That is, if an ACK from the MN due to error of wireless section is lost, it retransmits packets locally to improve TCP performance before the retransmission timer is expired. And if any ACK does not come until disconnect timer is expired, the EM-TCP agent creates and transmits a new ACK to force sender S to go into persist mode. The S who receives this ACK does not transmit any packet without changing its windows size in persist mode. The sender is allowed to leave persist mode if the EM-TCP agent forwards the ACK receiving from MN whose connection is resumed. With this manner, which the EM-TCP considers properties of wireless region such as high bit error and frequent handoff, TCP performance is improved as shown in Fig. 5.

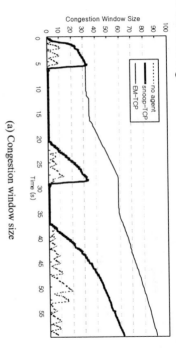

(a) Congestion window size

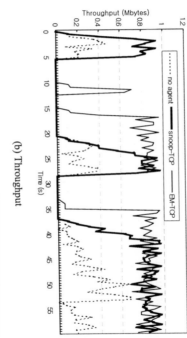

(b) Throughput

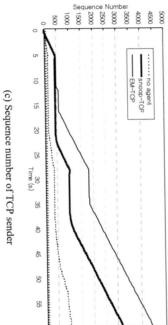

(c) Sequence number of TCP sender

Fig. 5. TCP performance improvement by using an EM-TCP agent

Fig. 5 (a) shows the congestion windows size of TCP, Snoop and the EM-TCP. We can see that TCP invokes congestion control more often than Snoop which does not perform congestion control occurring by link error. The figure shows EM-TCP has the best performance, because it prevents invocation of congestion control when fast handoff makes link disconnected as well.

208   K. Lee et al.

Fig. 5 (b) shows the experimental results of TCP throughput. In case of the EM-TCP, as soon as the connection is regained, throughput is increased straightly than the others. In TCP or Snoop, it begins data transmission by slow-start when the sender receives ACK corresponding to a segment which is used for proving the connection. However, in case of the EM-TCP, the sender can transmit data as soon as the connection is resumed with the same size of the window, because the window size does not changed in disconnection status. We can see that only the EM-TCP appropriately transmit data during 10~12s, when the connection is resumed in a short term after a disconnection period. The others cannot transmit data during the period because they spend the time to prove if the connection is resumed or not.

Fig. 5 (c) shows the sequence numbers of transmitted segment from the sender. When the connection is cut off, the sequence number is not increased any more and when the connection is regained, then the sequence number is increased again. We can see that more data is transmitted successfully in case of using an EM-TCP agent.

From the above result, we verify that EM-TCP provides more improved performance than Snoop, while Snoop provides more efficient throughput than TCP in wireless network in which high BER and frequent handoff occurs.

4.2   Simulation 2 – In Simple Wireless Network Environment (Serial Timeout Situation)

Network environment for the simulation 2 is the same with the simulation 1. Only difference is the disconnected periods. MN is connected to a base-station of each region at 0~5s, 10~12s, 15~19s and 33~60s, and disconnected at 5~10s, 12~15s and 19~33s. In Fig. 5 (b) TCP or Snoop is connected to a base-station again around at 20s and starts to transmit data. Serial timeout phenomenon is occurred due to disconnecting at 19s. Fig. 6 shows TCP throughput in case of serial timeout. It causes significant deteriorating of TCP performance.

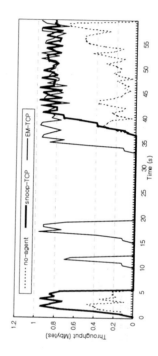

Fig. 6. Performance of an EM-TCP agent in case of serial timeout

The EM-TCP starts to transmit data as soon as the connection is regained so it communicates with MN from at 15s to 19s in this scenario. However, traditional TCP or Snoop cannot recognize the connection is opened again until the sender receives

ACK corresponding to its segment for probing. This property makes a great difference among their TCP performance.

4.3 Simulation 3 – In a DSRC-Based ITS System

In this simulation, we evaluated the performance of the three mechanisms in DSRC-based ITS. Network environment for the simulation 3 is shown in Fig. 7. We generate TCP/FTP traffic from sender S to receiver MN. We set the bandwidth of each link as 5Mb, link delay as 3ms and packet error rate of the link between agent and MN as 1%. An EM-TCP agent is on the divaricating branch of HA and FA. Traffic is transmitted since 0s, and whole simulation time is 100s and MN moves to next RSE every 9s from 0 seconds. We located each RSE 700m distances to clearly examine disconnection due to handoff, and MN moves very fast with 100m/s. Performance elevation of applying EM-TCP agent to DSRC-based ITS is shown in Fig. 8.

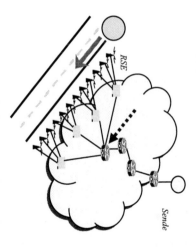

Fig. 7. Simulation environment for a DSRC-based ITS system

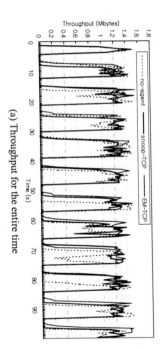

(a) Throughput for the entire time

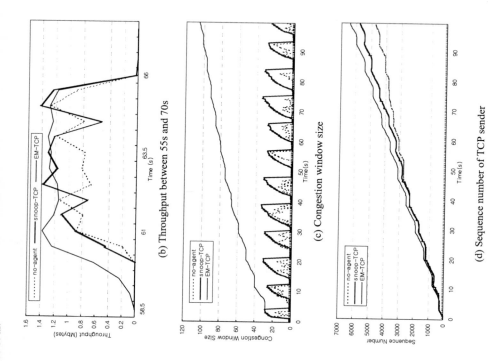

(b) Throughput between 55s and 70s

(c) Congestion window size

(d) Sequence number of TCP sender

Fig. 8. Performance of an EM-TCP agent in a DSRC-based ITS

Fig. 8 (a) and (b) shows TCP throughput. It shows that the throughput of EM-TCP is better than Snoop or TCP when the connection is cut off due to handoff. In addition, the EM-TCP and Snoop show higher throughput than TCP when there is high BER because the former 2 schemes take local retransmission. Fig. 8 (c) shows change of windows size. The EM-TCP forced the sender to persist mode when connection is cut off so window size is not decreased. Fig. 8 (d) shows the sequence number of the sender. We can see that EM-TCP guarantee more successful data transmission than Snoop, while Snoop guarantees more successful data transmission than TCP.

5 Conclusions and Future Work

In this paper, we suggested EM-TCP (Enhanced Mobile TCP) as an algorithm that can improve TCP performance when the Internet service is provided in the DSRC-based ITS (Intelligent Transportation System).

The EM-TCP can cope with wireless link section with high error ratio and frequent disconnection. For wireless link error, the EM-TCP provides us with more improved performance by local retransmission than traditional TCP. In case of handoff or disconnection, the EM-TCP let source host go to persist-mode that prevents the needless congestion control. As a result, the proposed algorithm had much bigger improvement than traditional TCP or Snoop.

However, the current version of EM-TCP is designed for data flow of downlink from a host in wired Internet to mobile node (i.e., moving car). In further study, we will consider the data flow of the reverse direction (up-link) to prevent needless congestion control and serial timeout that may happen in the mobile node.

References

1. James D. Solomon, "Mobile IP", Prentice Hall, 1996
2. ISO TC 204 WG15 Subgroup SG.L7, "Dedicated Short-Range Communication (DSRC) - DSRC Application Layer", Draft prTR2/204/15/#xx, June 1999
3. C.S Lim, etc "A Study about ITS High speed wireless packet communication system", Last Report of Research, ETRI
4. UCB/LBNL/VINT Network Simulator ns (version 2). URL: http:// www-mash.cs.berkeley.edu/ns/
5. W.G Tae, J.H. Kim, E.K. Kim, T.J. Jeong, C.S. Lim, "A Study of ATIS ETCS service using Internet, Paging and DSRC network", Telecommunications reviews, vol. 4, Aug. 1999
6. H.S. Oh, C.S. Lim, "Implementation of Dedicated Short Range Communication based Intelligent Transport System", Telecommunications review, vol. 4, Aug. 1999
7. H. Balakrishnan, V. N. Padmanabhan, S. Seshan and R. Katz, "A Comparison of Mechanisms for Improving TCP Performance over Wireless Links", IEEE Transactions on Networking, vol. 5, pp.756-769, Dec. 1997.
8. Jochen Schiller, "Mobile Communications", Addison Wesley, 2000
9. E. Amir, H. Balakrishnan, S. Seshan and R. H. Katz, "Efficient TCP over Networks with Wireless Links", ICC 2000, pp. 1604-1608, June 2000
10. H. Balakrishnan, S. Seshan, E. Amir and R. H. Katz, "Improving TCP/IP Performance over Wireless Networks", Proc. 1st ACM Conf. on Mobile Computing and Networking, Berkeley, CA, Nov. 1995
11. K. Brown and S. Singh, "M-TCP: TCP for mobile cellular networks", Computer Communication Review, vol.27, pp.19-43, Oct. 1997
12. K. Y. Wang and S. K. Tripathi, "Mobile-end transport protocol: an alternative to TCP/IP over wireless links", IEEE INFOCOM, pp.1046-1053, Apr. 1998.
13. G. Montenegro, S. Dawkins, M. Kojo, V. Magret and N. Vaidya, "Long Thin Networks", Internet Requests for Comments, no. 2757, Jan. 2000
14. M. Elaoud and P. Ramanathan, "TCP SMART: A Technique for Improving TCP Performance in a Spotty Wide Band Environment", ICC 2000, pp.1783-1787, June 2000
15. W. Stevens and NOAO, "TCP Slow Start, Congestion Avoidance, Fast Retransmit, and Fast Recovery Algorithms", Internet Requests for Comments, no. 2001, Jan. 1997

A Multi-path Support for Mobile IP with the Hierarchical Architecture

Youngsik Ma, Donghyun Chae, Wonjong Noh, Jiyoung Lee, Yeonjoong Kim, and Sunshin An

Department of Electronics Engineering, Korea University
1, 5-Ka, Anam-dong Sungbuk-ku, Seoul,136-701, Korea
{mys, hsunhwa, nwj, ljy, kyj, sunshin}@dsys.korea.ac.kr

Abstract. The explosive growth of the Internet combined with the widespread availability of highly mobile small hosts in the form of laptop and palmtop computers, and personal digital assistants, has created a big demand for the concept of Mobile IP. In the base Mobile IP, if the distance between the visited network and the home network of the mobile node is large, the signaling delay for these registrations may be long. Also the two main problems faced are the loss of registration packets and crashes of entities that maintain location caches. Therefore to solve this problem, the solution for performing registrations locally in the visited domain in order to reduce registration delay and to support micro mobility is proposed. In this paper, we propose that a hierarchical structure may be helpful to fast handoff in mobile environments. In company with the fast handoff, reliability is a great concern to the designers of mobile computing systems.

1 Introduction

In the last few years we have seen an explosion in the number of notebook computers and in the growth of the Internet. While notebook computers continue to improve with respect to size, weight, and capabilities, the Internet continues to grow at a mind-boggling pace. The mobile users expect to access the Internet's information resources and to communicate with other Internet users. The concept of Mobile IP is proposed to satisfy these demands.

Due to the characteristics of the IP routing mechanism, the Internet cannot be accessed from a point of attachment outside an accessing terminal's local subnet without changing the terminal's Internet-specific network address (IP address). In order to develop a standard that allows complete mobility throughout the Internet space, the Internet Engineering Task Force (IETF) has generated the Mobile IP Protocol specification documentation. The task force completed the standardization task by issuing IP Mobility Support RFC2002 [1].

However, several performance problems in Mobile IP need to be addressed. First, Mobile IP's tunneling scheme creates a triangle routing problem, causing packets to travel through the HA. Second, packets in flight during a handoff are often lost because they are tunneled based on out-of-date location in information.

I. Chong (Ed.): ICOIN 2002, LNCS 2344, pp. 212–220, 2002.
© Springer-Verlag Berlin Heidelberg 2002

Third, base stations with small cells result in frequent handoffs, and requiring a registration with a distant home agent (HA) for each such local handoff causes higher overhead and further aggravates packet loss.

To solve first problem, Mobile IP route optimization alleviates triangle routing and data loss during a handoff by informing the correspondent node (CN) and the previous foreign agent (FA) of the mobile host's care-of-address. Networks of small wireless cells offer high aggregate bandwidth, support low-powered mobile transceivers, and provide accurate location information. In these networks, users will often carry devices across cell boundaries in the midst of data transfers. A handoff mechanism goal is as following. It is needed to maintain connectivity as information devices move, while minimizing disruption to ongoing transfers. This mechanism should exhibit low latency, incur little or no data loss, and scale to a large internetwork. In the environment that the mobile node frequently moves into a new network, the base mobile IP standard specifies a general handoff protocol for the Internet, but does not meet these goals. The Mobile IP can be handled both local-area and wide-area movement in both wired and wireless networks. To solve third problem, the Mobile IP systems supporting local registration were introduced to reduce the number of times when home registration with the remotely located HA is needed. The local registration systems consider requirements and assumptions that may affect other aspects of the Mobile IP systems as the reliability. In an environment where the probability of failure of mobility support stations is relatively high, as in the battlefield or where those stations are located in hazardous conditions, the issue of reliability becomes of a particular interest.

In this paper, we propose that a hierarchical structure may be helpful to fast handoff in mobile environments. In company with the fast handoff, reliability is a great concern to the designers of mobile computing systems. The rest of the paper is organized as follows. We first give related work is surveyed in section 2 and describe a hierarchical foreign agent architecture for fast handoff in section 3. In the environment where the FA is failure, the method for recovery in order to support reliability will be described in section 4. Finally, we present our conclusions and future work in section 5.

2 Related Work

2.1 Mobile IP

Mobile IP [1] is a modification to IP that allows hosts to continue to receive data no matter where they happen to attach to the Internet. In Mobile IP, the HA in the home network of the mobile node intercepts packets destined for the mobile node using proxy ARP, and then delivers them to the mobile node's current attachment point to the Internet using tunneling. The current attachment point is defined by an IP address called care-of address (CoA).

There are two different types of care-of address: a foreign agent care-of address is an address used by a mobile node as tunnel exit-point when the mobile node is connected to a foreign link, and a co-located care-of address is an temporary assigned address to one of mobile node's interfaces. When a mobile node moves into a new foreign network, it receives the agent advertisement message including care-of address from new FA. At this time, the mobile node can realize that it moves into the new foreign network, and then sends the registration to the new FA. The new FA relays it to the HA of the mobile host so that the HA delivers the packets the new FA instead of the previous FA.

Mobile IP suffers from a problem known as triangle routing, which refers to the path followed by a packet from a correspondent node to a mobile node which must first be routed via the mobile node's the home agent. Triangle routing incurs potentially significant overheads in the delay and network resources consumed for communication with mobile nodes. Mobile IP with route optimization extension avoids triangle routing as follows [2].

When a mobile node's HA intercepts an IP packet, it informs the correspondent node (CN) of the mobile node's current CoA, this is called a binding update message. The CN can cache this information and send subsequent packets by tunneling them directly to the mobile node's CoA. On the other hand, data in flight that had already been intercepted by the home agent and tunneled to the old care of address are forwarded to mobile node's new care of address by previous foreign agent which is reliably notified of mobile node's new mobility binding.

2.2   Regional Registration

AS specified in Mobile-IP specifications [1], each time the MN moves to a new network, a registration request is sent to the HA to be approved and a Registration reply is transmitted back to the MN that originated the registration request. Transmitting and processing all the registration requests through the remotely located HA may become inefficient due to the fact that the registration delay contributes to the time needed to complete the handoff process. To deal with such situation, the local registration approach was proposed to decrease this overhead. In [3] and [4], the foreign agents are arranged hierarchically in the regional topology, and the mobile node is allowed to move from one local area of the regional topology to another area of the same topology without the need to send a registration request to the HA. This arrangement can be accomplished by allowing the MN to inform only the corresponding local FA each time it moves to a new area.

In Mobile IP, as specified in RFC 2002[1], a mobile node registers with its home agent each time it changes care-of address. If the distance between the visited network and the home network of the mobile node is large, the signaling delay for these registrations may be long. A solution for performing registrations locally in the visited domain is proposed. Regional registrations reduce the number of signaling messages to the home network, and reduce the signaling delay when a mobile node moves from one foreign agent to another, within the same visited domain.

When a mobile node first arrives at a visited domain, it performs a home registration, that is, a registration with its home agent. During home registration, the home agent registers the care of address of the mobile node. When the visited domain supports regional tunnel management, the care of address that is registered at the home agent is the publicly routable address of a Gateway Foreign Agent (GFA). This is care-of-address will not change when the mobile node changes foreign agent under the same GFA. When changing GFA, a mobile node must perform a home registration.

When changing foreign agent under the same GFA, the mobile node should be performed a regional registration within the visited domain. Regional registration protocol supports one level of foreign agent hierarchy beneath the GFA, but the protocol may be extended to support several levels of hierarchy. Foreign agents the support regional registrations are also required to support registrations according to RFC 2002[1]. If the mobile node chooses not to employ regional registrations, it may register a co-located care of address directly with its home agent, according to [1]. Similarly, if there is a foreign agent address announced in the Agent Advertisement, the mobile node may register that foreign agent of care of address with its home agent.

3 System Design

A handoff mechanism goal is as following. It is needed to maintain connectivity as information devices move, while minimizing disruption to ongoing transfers. This mechanism should exhibit low latency, incur little or no data loss, and scale to a large internetwork. In the environment that the mobile node frequently moves into a new network, the base mobile IP standard specifies a general handoff protocol for the Internet, but does not meet these goals [6]. If the distance between the visited network and the home network of the mobile node is large, the signaling delay for these registrations may be long. Therefore the packets destined for the mobile node will be lost.

To solve this problem, we propose that a hierarchical structure may be helpful to fast handoff in mobile environments. This may be improved the performance of handoff. In company with the fast handoff, reliability is a great concern to the designers of mobile computing systems. The two main problems faced are the loss of registration packets and crashes of entities that maintain location caches [5]. Therefore, the hierarchical registration systems consider requirements and assumptions that may affect other aspects of the Mobile IP systems as the reliability. In a situation where the probability of failure of mobility support stations is relatively high, as in the battlefield or where those stations are located in hazardous conditions, the issue of reliability becomes of a particular interest. To provide fast handoff and reliability, the hierarchical architecture is combined with a method for reliability [10]. The goal of proposed architecture is to design a mobility management framework as follows. It is fully compatible with Mobile IP and allows the deployment of different micro mobility schemes in different parts of the network transparently to the correspondent and mobile hosts. It is also as efficient as Mobile IP in terms of routing performance.

3.1 Architecture

In this paper, we propose that a hierarchy of foreign agents for registration time reduction and fast handoff. It also will be provided fault-tolerance for reliability. Figure 1 shows hierarchical foreign agents architecture, which can be divided into 3-level foreign agent in order to provide each a level mobility. We propose to separate the mobility management support into three agents as illustrated in Figure 1. Each level's mobility can be provided by the local FA for 1-level mobility, LFA (Local Foreign Agent) for 2-level and GFA (Global Foreign Agent) for 3-level. The GFA can be administrated a common network. The LFA can be administrated a local network. The remaining entity is defined in [1]. When the visited domain supports hierarchical foreign agent, mobility management should be supported regional registration. When a CN sends packets to the MN, each level foreign agent must be supported tunnel management and mobility management. If each level foreign agent is not capable of tunneling, the registration procedure is operated based on the base Mobile IP [1]. The MN is trying to minimize the amount of tracking required to maintain its connectivity by identifying the smallest region for which the mobile node has not traversed any region boundary. A hierarchical mobility management scheme will be classified into three levels for fast handoff. Each level's mobility is classified as following.

— 1-layer: Intra LFA Mobility
— 2-layer: Inter LFA Mobility
— 3-layer: Inter GFA Mobility

3.2 Operation Procedure

First, in case that when a mobile node first arrives at a visited domain, it will be received agent advertisement message from the FA in visited network. And then it performs a registration with its home network through the FA. At this registration, the home agent registers the care-of address of the mobile node. The care-of address that is registered at the home agent is the address of a GFA. The GFA should be kept a visitor list of all the mobile nodes currently registered with it. Since the care-of address registered at the home agent is the GFA address, it will not be changed in the case the mobile node move into a new foreign agent under the same GFA. For example, When the MN moves into the FA1, the Registration message sequence is FA1/LFA1/GFA1. Second, in the case of when a mobile node move from one LFA to another LFA under the same GFA, and then it perform a registration with its GFA through the FA. For example, when the mobile node moves from FA1 to FA4, the mobility management must use 2-level mobility. That is, it is called inter LFA mobility. Third, in the case of when a mobile node from one GFA from another GFA, it perform a registration with its home network through the FA. The registration message sequence is FA6/LFA4/GFA2. At this registration, the HA must be change the care of address. When the mobile node moves into the foreign network, figure 2 shows the flow chart for the registration procedure and packet delivery.

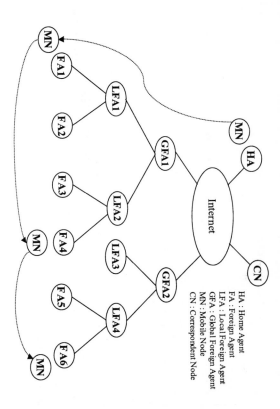

Fig. 1. Hierarchical Foreign Agent Architecture

HA : Home Agent
FA : Foreign Agent
LFA : Local Foreign Agent
GFA : Global Foreign Agent
MN : Mobile Node
CN : Correspondent Node

4   A Multi-path for Reliability

The hierarchical registration systems have to be considered requirements and assumptions that may affect other aspects of the Mobile IP systems as the reliability. In an environment where the probability of failure of mobility support stations is relatively high, as in the battlefield or where those stations are located in hazardous conditions, the issue of reliability becomes of a particular interest. A characteristic that distinguishes hierarchical registration systems from the base Mobile IP is that if any of the FAs along the path between the root FA and the leaf FA fails, this will cause the MNs located at the leaf to loose its network connectivity. The failure of any of those FAs will break the path between the root and the leaf FAs. In this paper, we will introduce two possible solutions to overcome the problem of loss of service resulting from a FA failure in a hierarchical architecture.

In this paper, we propose the method for recovery in order to support reliability. Depending on which is foreign agent failed, the possible approach will consider about using a backup agent for packet's reliability delivery and bypass routing path. The latter method is that 3-level mobility management is converted to 2-level mobility management. We assume that the anchor point will

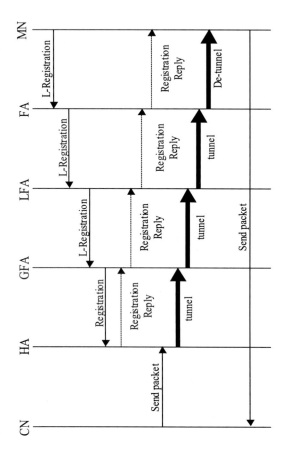

Fig. 2. Flow Chart of local registration and packet delivery

be selected according to any number of foreign agents at lower level. The anchor point manage hierarchical domain and have a backup agent. It mainly becomes the GFA. In the case of the GFA failure, all the MN serviced directly by a FA located at a lower level in the hierarchy under the GFA and on a path on which the faulty FA is an intermediate point will suffer from loss of service. In this situation the backup agent will be act as a substitute for the GFA. The anchor point has a multiple path to support reliability.

Backup agent : First, we will consider that the GFA happen to fail. We assume that the message is periodically exchanged between primary GFA and backup GFA for all the MN serviced directly by a FA located in a lower level in the hierarchy under the GFA support reliability. If the primary GFA will happen to fail, the backup agent send message to the HA and all the LFA in order to act as a substitute for the primary GFA. In the case of when the mobile node moves into the foreign network, it will be received agent advertisement message from the FA in foreign network and then it performs a registration with its home network through the LFA and GFA. At this time, the LFA will be detected the GFA's failure. And then the FA regional will be register to the backup agent. All the packets will be deliver through the backup agent. In the case of the GFA failure are shown in figure 3.

Convert : Second, in the case of when the LFA is failed. This method is based on polling principles. Also, bypass routing path. The GFA periodically sends polling message to all the LFA is located in below the GFA. If one of all the LFA is failed, the GFA will be used multi-path instead of using the LFA in order to support best service. The packet delivery can be achieved without the LFA. When the mobile node moves into the leaf node, if the leaf nodes detect

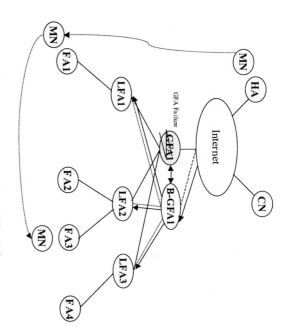

Fig. 3. In the case of the GFA failure

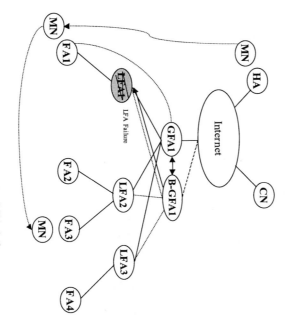

Fig. 4. In the case of the LFA failure

that the high-level agent is failed, then it register to the GFA. This method is that 3-level mobility management is converted into 2-level mobility management. Each level agent broadcast polling message to the lower level agent and higher level agent in order to detect agent's status. In the case of the LFA failure are shown in figure 4.

5   Conclusions

The base mobile IP scheme has been generated several problems because that the distance between the visited network and the home network of the mobile node is large. In this paper, to solve these problems, we propose that a hierarchical foreign agent for fast handoff and a reliability. A hierarchy of foreign agents can be divided into 3-level foreign agent in order to provide level mobility. Each level's mobility can be provided by the local FA for 1-level mobility, LFA (Local Foreign Agent) for 2-level and GFA (Global Foreign Agent) for 3-level. Reliability in mobile environments is of great importance for packet delivery. In this paper, we propose that the agent have a multiple path in hierarchical foreign agent architecture for reliability. Depending on which is foreign agent failed, the possible approach will consider about using a backup agent for packet's reliability delivery and bypass routing path. In the future work, it is also essential to provide proper QoS (Quality of Service) forwarding treatment to the packets sent by or destined to MN as they propagate along different routes in the network due to node mobility. It is necessary to study on a QoS guarantee in mobile environment and multicast in hierarchical routing and architecture.

References

1. Charles.E.Perkins.:IP Mobility Support in RFC 2002 (1996)
2. David B. Johnson, Charles.E.Perkins.:"Route Optimization in Mobile IP", Mobile IP Working Group Internet Draft (2000)
3. Charles.E.Perkins.:Mobile-IP Local Registration with Hierarchical Foreign Agents Approach. IETF Draft (1996)
4. Eva Gustafwwon et al.:Mobile IP Regional Registration. Draft-IETF-MobileIP-Reg-Tunnel-04.txt (2001)
5. Gihwan Cho, Lindsay F. Marshall.:An Efficient Location and Routing Scheme for Mobile Computing Enviroments. IEEE Journal on selected areas in communications, Vol.13. No.5 (1995)
6. Antoine Stephane, Andrej Mjhailovic, A.Hamid Aghvami.:Mechanisams and Hierarchical Topology for Fast Handoff on Wireless IP Networks
7. R.Ramjee, T.Laporta, S.Thuel, K.Varadhan.:IP Micro mobility support using HWAII. Internet Draft (1999)
8. Yingchun Xu et al.:Mobile IP based Micro Mobility Management Protocol in the Third Generation Wireless Network. Draft-IETF-MobileIP-3gwireless-ext-05.txt (2000)
9. IETF MIPv4 Handoffs Design Team.:Low Latency Handoffs in Mobile IPv4. Draft-IETF-MobileIP-Lowlatency-handoffs-v4-00.txt (2001)
10. H.Omar,T.Saadawi, M.Lee.:Support for Fault Tolerance in Local Registration Mobile-IP Systems.MILCOM 1999. IEEE, Vol 1,IEEE (1999)
11. Hesham Soliman et al.:Hierarchical MIPv6 mobility management. Draft-IETF-MobileIP-HMIPv6-02.txt (2001)

Performance Evaluation of TCP over WCDMA RLC[1]

Ji-Eun Kang, Dong-Min Kim, Yong-Chul Shin,
Hee-Yoon Park, and Jai-Yong Lee

Dept. of Electrical & Electronical Engineering, Yonsei University.
134 Shinchon-dong Seodaemun-gu, Seoul 120-749, Korea
gnee@nasla.yonsei.ac.kr

Abstract. TCP congestion control has been developed on the assumption that congestion in the network is the only cause of packet loss. Under this assumption, it performs reasonably well over wired environment. However, in the presence of high error rates and intermittent connective characteristic of wireless link, TCP suffers from significant throughput degradation over hybrid wireless networks. Therefore TCP needs to be modified or supported by other protocols to run properly over networks that contain wireless links. One of these methods is link ARQ and this is necessary to lossy links. Some considerations exists for ARQ protocols to support TCP traffics. This paper examines the TCP performance over WCDMA RLC, which provides ARQ function over wireless links.

1 Introduction

The numbers of Internet users and the demand for wireless Internet have been increasing rapidly. The integration of wired and wireless Internet is expected to provide seamless multimedia services to both mobile and fixed users anywhere at anytime in the near future.

Most of all the applications depend on the performance of TCP (Transmission Control Protocol), which is a reliable transport protocol for the Internet. However, TCP is assumed that it is mainly used in wired networks when designed but wireless link has many different characteristics affected by multi-path fading, shadowing and interference of other users'. Furthermore, these characteristics change over time and location. These result in high bit error rate and low bandwidth and frequent handoff. In these circumstances, TCP suffers from significant degradation of performance in the form of poor throughput.

There are two well-known strategies for improving the performance of TCP over wireless link. One is hiding corrupted errors from the sender and recovering lost pack-

---

[1] This job was supported by Samsung Electronics and by grant No. 1999-2-303-005-3 from the Basic Research Program of the Korea Science & Engineering Foundation.

I. Chong (Ed.): ICOIN 2002, LNCS 2344, pp. 221–228, 2002.
© Springer-Verlag Berlin Heidelberg 2002

ets by local retransmission. The other is informing the sender all the lost packets that need to be recovered.

In this paper, we focus on local retransmission technology on link level and, in particular, we consider the WCDMA (Wideband Code Division Multiple Access) RLC (Radio Link Control) protocol. WCDMA is the one of the leading proposals for the 3rd generation wireless and mobile access networks and RLC is a link layer protocols of the WCDMA air interface protocols, which provides multiple-reject ARQ (Automatic Repeat reQuest) mechanism over WCDMA wireless links.

2 Enhancement of TCP over Wireless Links

In the presence of wireless link with high bit error rate, there could exist an excessive reduction of congestion window caused by packet losses that are not mainly due to congestion. Various approaches in different layers to mitigate the deficiency have been proposed. These are classified into three categories: end-to-end protocols, link-layer protocols, and split-connection protocols. These have some merits and also have deficiencies over others.

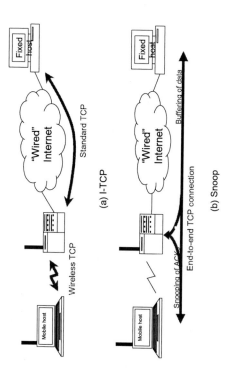

Fig. 1. I-TCP and Snoop

ARQ and FEC (Forward Error Correction) are link layer solutions. FEC is efficient when errors occur randomly. It consists of sending some additional information to correct the corrupted part of the packet. The advantages of this technique are that corrections are dealt directly without retransmission of corrupted packet and no interference with TCP mechanisms. The disadvantages of this technique are that it wastes bandwidth when link state is good and extra CPU processing time and memory. ARQ is efficient when losses are not frequent and propagation delay is not important. However, there is possibility to interfere with TCP retransmit timeout mechanisms.

TCP level solutions try to recover lost packet by retransmitting at the TCP layer instead of the link layer.

I-TCP involves splitting the TCP connection into two separate connection parts at the base station. One is between mobile host and the base station, and the other is between the base station and fixed host. Wired portion of networks uses standard TCP, and wireless portion ca use some transport protocols well tuned for wireless. This helps improvement of TCP over wireless. However, this scheme breaks end-to-end semantics, needs application relinking, and requires software overhead.

Snoop is another solution that doesn't break end-to-end semantics. This protocol just keeps the copies of data packets. It suppresses duplicate ACKs (Acknowledgements) and sends lost packet instead of the source. However, as in link level, Case, interference may happen between source and Snoop agent mechanisms.

Since all non-congested losses will not be recovered by link and TCP level retransmission and non-congestion losses are omnipresent in the network, End to end solutions are needed. The main ideas of end-to-end solutions are that notifying all lost packets to the sender to recover at the receiver. These are Explicit Loss Notification (ELN) and Explicit Congestion Notification (ECN). [1]

3   Link ARQ over Wireless Link

3.1 Link ARQ Issues

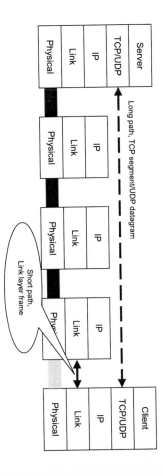

Fig. 2. End-to-end path and link layer ARQ

Performance Implications of Link Characteristic (PILC) Working Group of IETF has discussed about link ARQ issues for IP traffic. There are several benefits when we use link ARQ over lossy links. Link ARQ introduces a shorter control loop than loss recovery over the end-to-end path. It can operate on individual link frames, which are usually smaller than upper layer PDU (Protocol Data Unit). In addition, it may be able to use knowledge that is not available to end hosts.

A side effect of link ARQ is that link transit delay would be increased when frames are retransmitted. On shared high-delay links, the impact of ARQ on other UDP or TCP flows may lead to unwanted jitter. For links where error rates are highly variable,

224   J.-E. Kang et al.

high ARQ persistence may provide good performance for a single TCP flow. Lower ARQ persistence may also have merit for multiple TCP flows.

During a link outage, interactions between ARQ and multiple flows are less significant; the ARQ protocol is likely to be equally unsuccessful in retransmiting frames for all flows. High persistence may benefit TCP flows, by enabling prompt recovery once the channel is restored. [2]

Many previous studies show the interaction of TCP and link ARQ [3-5]. These considered IS-99 or IS-707 RLP (Radio Link Protocol) that provide semi-reliable NAK-based SR (Selective Repeat) ARQ mechanisms over CDMA (Code Division Multiple Access) link. They noted and discussed that even if TCP and link layer retransmission operate independently of on another, there could be retransmission redundancies and timer interactions.

3.2 WCDMA RLC

A primary motivation for third generation systems is data communication. WCMDA is one of the air interface protocols suggested by 3gpp and its protocol architecture is shown in Fig. 3.

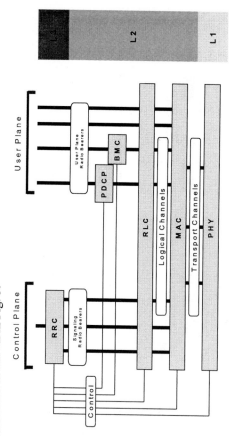

Fig. 3. Radio Interface Protocol Architecture for WCDMA

The Radio Link Control layer provides transparent, unacknowledged, or acknowledged mode data transfer to the upper layers. The main functions of the RLC are ciphering of data in unacknowledged and acknowledged mode and the support of an automatic repeat request scheme (ARQ) in acknowledged mode.

The transparent mode is mainly important for speech services and does not require much RLC functionality. In Acknowledge mode (AM), the receiver reports receiving status by sending STATUS PDU (Protocol Data Unit) using Super-Fields (SUFI). There are several cases in which receiver sends STATUS PDU. First, it is the case that sender sends polling request. Second, if the receiver detects one or several missing PU (Protocol Unit) s, it sends a STATUS PDU to the sender. Third, the receiver transmits a STATUS PDU periodically to the sender, Fourth, if not all PUs request for retransmission have been received before the EPC (Estimated PDU Counter) has expired, a new STATUS PDU is transmitted to the peer entity. [6]

Fig. 4 shows an example of AM operation.

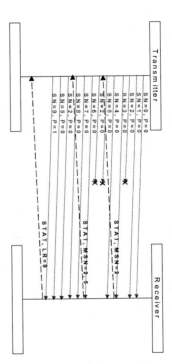

Fig. 4. An example of AM RLC Operation

4 Performance Evaluation of TCP over WCDMA

4.1 Simulation Models

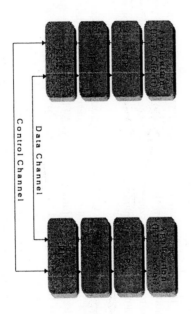

Fig. 5. Simulation Model

The goal of this simulation is to investigate TCP performance over WCDMA RLC when the parameters of RLC vary. We developed simulation models to evaluate the performance of TCP over WCDMA RLC using OPNET, which is discrete event-driven simulation tool capable of modeling large-scale communication networks [7].

Table. 1. TCP parameters

Maximum segment size	536 Bytes
Receive buffer	65536 Bytes
Maximum ack delay	200ms
Fast retransmit	enabled
Fast recovery	enabled
Karn's algorithm	enabled
Initial RTO	1.0 sec
Minimum RTO	0.5 sec
Maximum RTO	64 sec
RTT Gain	0.125
Deviation Gain	0.25
RTT Deviation Coefficient	4.0
Timer Granularity	0.5 sec

The network model we built has two nodes and two links. One of the two nodes is the client and the other is the server. The client requests file transfer to the server at 100-second after the simulation starts and the client and server make a TCP connection to receive and transfer a file if there isn't any problem. The TCP version user is Reno that performs slow-start, congestion avoidance, fast retransmission and fast recovery algorithms. Upper layer parameters that we used for simulation are shown in Table. 1.

These two nodes are connected via two point-to-point links, which could have link errors. One of the links is the data channel, which carries AM RLC data frames and the other is the logical channel, which conveys STATUS frames. Link speed of both links is 384Kbps that is expected throughput for physical layer in normal environments of WCDMA. To take advantage of shorter control loop, we added extra 10ms delay between server RLC module and server IP module. We used OPNET standard protocol models for the application module and TCP/IP module. RLC module was developed using OPNET Modeler. This module has a process model, which performs AM RLC operation. This model triggers the poll bit of a data frame, every Poll_PDU PDUs.

4.2 Results

At first we examined TCP performances for several BER settings (10^{-9}, 10^{-7}, and 10^{-5}). The PPP model allocates error to individual packets independently. The results are shown in Table. 2.

Table. 2. Download Response Time of FTP (sec)

BER	TCP with RLC	TCP without ARQ
10^{-9}	4.086550	3.8876
10^{-7}	4.086550	3.8976
10^{-5}	7.765618	N/A

Download response time of FTP means time elapsed between sending a request and receiving the response packet for the FTP application in this node. It is measured from the time a client application sends a request to the server to the time it receives a response packet. Table. 2 show that shorter control loop of RLC provides almost error-free links quality when the link has very low bite error rate. In low BER environment, RLC could introduce more delay to the TCP connection, but we can see that RLC gives much more link reliability when the link experiences relatively high BER.

The second simulation is for investigating the effect of the number of retransmission attempts on TCP performance, we set the parameter of server RLC, MaxDAT from 10 to 40 and then investigated the download response time of FTP. MaxDAT defines allowed maximum number of retransmission of RLC protocol. We ran the four simulation sets and each simulation used different seed values for randomization. Average values of the simulation sets are shown following pictures. Fig. 6 shows that too many retransmission attempts are not required. When RLC only attempts retransmission of the lost data frame for 10 times, FTP download response time is relatively high. This means that RLC gives up retransmission too early so the TCP has to recover the lost segment that contains the corrupted frame.

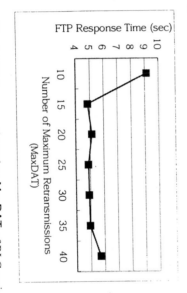

Fig. 6. FTP Response time when MaxDAT of RLC varies

5 Conclusions

Using OPNET, we examined the performance of TCP over WCDMA RLC running on lossy links with various BERs. The results showed that the link layer ARQ mechanisms causes additional delay but it provides more reliability to the relatively higher BER environment, so it can prevent expiration of retransmission timer at TCP levels. Results for FTP download delay with WCDMA RLC were obtained for the various number of retransmission attempts at the link layer. The FTP session transfer a 1,600,000-bit file using one TCP connection. The results show that too many retransmission attempts at RLC level are not needed to support TCP traffic. It is recommended to use relatively small value for the maximum number of retransmissions. Too high value could give more loads to the link and cause more delay. This means that we expect performance enhancement of the whole system when we control MaxDAT according to transmission environment.

References

1. Dong-Min Kim, Bum-Joon Kim, Won-Seo Park, Jae-Sung Park, and Jai-Yong Lee, "TCP Enhancement for Wireless Internet," Proceedings of ITC-CSCC '01, Tokusima, Japan, Page(s): 164-167 vol.1

2. Gorry Fairhurst and Lloyd Wood, "Advice to link designers on link Automatic Repeat request (ARQ)," Internet Draft

3. Gang Bao, "Performance evaluation of TCP/RLP protocol stack over CDMA wireless link," ACM Networks, 1996.

4. Yong Bai, Andy T. Ogielski, and Gang Wu, "Interaction of TCP and Radio Link ARQ Protocol," VTC 99

5. Farooq Khan, Sarath Kumar, Kameshi Medepalli, and Sanjiv Nanda, "TCP Performance over cdma2000 RLP, VTC 2000

6. Uooyeol Yoon, Seongsoo Park, Paul S. Min, Performance Analysis of Multiple Rejects ARQ at RLC (Radio Link Control) for Packet Data Service in W-CSMA System, GLOBECOM, IEEE, Volume: 1, 2000 Page(s): 48-52 vol. 1

7. OPNET Modeler Documentation, OPNET Technologies, Inc.

User Defined Location Area
for Registration and Paging

Kwang-Youn Park, Sang-Jun Park, In-Sook Cho, and Byung-Gi Kim

Dept. of Computing, Soongsil University, Seoul, Korea
uyoung@archi.soongsil.ac.kr

Abstract. In this paper, we propose a scheme that a user can define LAs along the moving path. And we also analyze the costs for registration and paging of the proposed scheme. The wired networks need to manage MS(Mobile Station) location for continuous service. For the location managements of MS, a number of cells are grouped to form a location area (LA). A MS's location is identified by the LA number it resides in. Since traffic loads on the wired networks are largely affected by LA sizes, how to assign the LA is important. User defined LA (UDLA) is a dynamic LA scheme that it defines variably LA according to routing of MS.

1 Introduction

Mobile communications should support not only voice communication but also various kinds of multimedia services. Some of them require real time services. If a packet for a real time service cannot meet the timing requirement, it must be discarded and the quality of service is degraded. To avoid such situations location of each MS is maintained by wired network.

When a MS goes out of the current LA or generates a new call, location registration is performed. The cost of this operation is called registration cost. When a request for a location registration is received, the wired network tries to identify the cell where the MS resides. This operation is called paging and its cost is paging cost. With larger LAs paging cost increases. But the registration rate is lowered and the registration cost decreases. Since there is a tradeoff between registration cost and paging cost, many researches have been performed to find out optimal LA size [3-9]. We propose a solution for the special environments where user's routing path can be predicted. With this information LAs can be dynamically assigned along the path to reduce registration and paging costs.

I. Chong (Ed.): ICOIN 2002, LNCS 2344, pp. 229–239, 2002.
© Springer-Verlag Berlin Heidelberg 2002

2   Location Registration and Paging

2.1 Location Registration

Location registration is a process that stores MS's location in VLR(Visitor Location Register) and HLR(Home Location Register). When a MS moves into a new LA, it performs location registration and updates a current LA. There are three types of updates:

① Updates involving MSC and VLR
② Updates involving MSC, VLR and HLR
③ Updates involving VLR and HLR

LA update processes are different depending on the user's movement patterns and LA sizes, and accordingly the costs are also varied.

2.2 Classification of Registration Scheme

2.2.1 Static LA

Basic conception of static LA is that LA sizes are fixed and same for all LAs.

Zone Based LA (ZBLA). Size of every LA is all the same in ZBLA scheme. MSs perform registration whenever they come into new LA. This scheme has an advantage of lower system traffic because it is very simple. However it is not responsive to the MS's moving pattern since it cannot take the MS mobility or call characteristics into consideration. Moreover, when a MS crosses an LA boundary frequently, 'ping-ponging' may occur to generate too much registration traffic [2-3].

Fixed Reporting Center (RC). RC scheme predefines a specific cell group as reporting center and it performs registration only when a MS comes into a reporting center [2].

2.2.2 Dynamically Assigned LA

In these schemes LA size is not fixed, but they don't adjust LA sizes according to the mobility or call arrival rate of a MS.

Movement based LA. In movement based LA scheme registration is performed after a MS goes through *M* cells. Each MS unit has to maintain the number of passed cells after the last registration [3].

Distance-based LA(DBLA). In DBLA, registration is performed when a MS goes far by a certain distance *D* from the last registered cell [3], [5], [6].

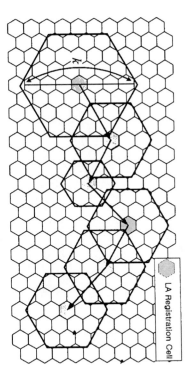

Fig. 1. 2D Distance Based Location Area

Figure 1 shows DBLA operation.

Timer based LA. In a timer based LA scheme each MS performs registration periodically with its local clock. So registrations are performed at regular interval regardless of MS's mobility or call arrival rate. When a call to a MS is requested, paging is started at the last cell where the MS registered. If not found, neighbor cells are paged until the MS is found [4].

Adaptive LA scheme. Static and dynamic LA schemes do not consider each MS's mobility and call arrival characteristics. But adaptive LA scheme dynamically assigns PLA(Personal Location Area) according to the MS's mobility and call arrival rate [8], [9], [10]. That is, when a MS enters a new LA, LA size is calculated using the MS's mobility and call arrival rate in the previous LA.

2.3 Paging

Paging search a base station for most adjacent a MS when there are need. There are two kinds of blanket polling and sequential paging. Blanket polling polls all cells in a LA at once regardless of MS's location. Sequential paging polls one cell after another. Paging starts from the center cell and is continued until the MS is found. In addition, intelligent paging has been proposed under the assumption that most MSs have repetitive mobility patterns [9]. Paging is done sequentially based on the MS's mobility profile [10].

232   K.-Y. Park et al.

3   User Defined Location Area Scheme

3.1 UDLA Environment

UDLA scheme belongs to dynamic LA scheme, but it separates DBLA and adaptively assigned LA for UDLA's performance analysis and so adaptively assigned LA called DYLA scheme. And a shape of cell and LA is a hexagon.

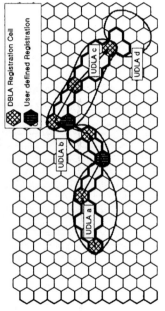

Fig. 2. 2D User defined Location Area

Figure 2 shows a LA assignment scheme for UDLA and a MS has a initial cell where a cell located time in a LA. In a DYLA scheme, MS stores each direction conversion number between cells in a previous LA so it determines a new LA. In this case, let us assume a new MS's movement pattern equals to MS's movement pattern in a previous LA.

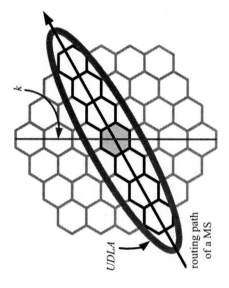

Fig. 3. Assignment UDLA

Figure 3 presents UDLA's assignment. UDLA is a LA determining a MS with user's destination input and it determines a path using a database of wireless for geographic information service network [11], [12].

In a UDLA scheme, MS stores not only each direction conversion number between cells in a current LA but also it predicts direction conversion number between the numbers of cell's direction conversion in due consideration of velocity and path where user's routing can be predicted.

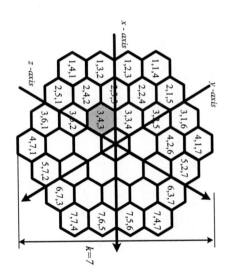

Fig. 4. Coordinates of cell in a LA

Figure 4 shows cell's coordinates in a LA. Coordinates has a format (h,i,j), where h is a value for the x-axis, i is a value for the y-axis, j is a value for the z-axis. A LA's size defines a diameter of LA(k).

3.2 Movement Model

UDLA is a LA assignment scheme in due consideration of personal MS's movement specificity for only one MS. Time is discrete with step Δt size, a MS moves along the x-axis, the y-axis or the z-axis in each step [7]. Each direction's movement is independent and movement probability can be represented as equations (1).

$$p_x = \frac{\text{number of step trasitions from left to right}}{\text{total number of step trasitions of } x}$$

$$q_x = \frac{\text{number of step trasitions from right to left}}{\text{total number of step trasitions of } x}$$

$$p_y = \frac{\text{number of step trasitions from left - top to right - bottom}}{\text{total number of step trasitions of } y}$$

$$q_y = \frac{\text{number of step trasitions from right - bottom to left - top}}{\text{total number of step trasitions of } y}$$

$$p_z = \frac{\text{number of step trasitions from left - bottom to right - top}}{\text{total number of step trasitions of } z}$$

$$q_z = \frac{\text{number of step trasitions from right - top to left - bottom}}{\text{total number of step trasitions of } z}$$

$$(1)$$

At time Δt, if a MS doesn't move any direction, at this time, the probability is $1 - p_x - q_x$, $1 - p_y - q_y$ or $1 - p_z - q_z$.

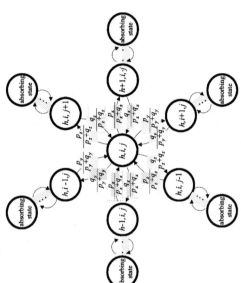

Fig. 5. Markov chain model in computation of expected time leaving a LA

In figure 5, if a MS moves to each direction during constant time $T_{h,i,j}$ for optimal location (h, i, j), a probability escaped from LA can be presented as Markov chain model. And movement probability for each direction p_x, p_y, p_z calls p and q_x, q_y, q_z calls q then discrete Markov chain transition probability(P) can be presented as equation (2).

$$P_{i,j-1} = \frac{p}{p+q}$$
$$P_{i,j+1} = \frac{q}{p+q} \tag{2}$$

In this case, MT defines a step before escapes from a LA. MT can be represented as equations (3) with gambler's ruin problem [1], [7].

$$MT = \begin{cases} \dfrac{1}{p+q} - i(k+1-i) & p=q \\[2ex] \dfrac{i}{q-p} - \dfrac{k+1}{q-p} \cdot \dfrac{1-\left(\dfrac{q}{p}\right)^i}{1-\left(\dfrac{q}{p}\right)^{k+1}} & p \neq q, \, p \neq 0, \, q \neq 0 \\[2ex] \dfrac{i}{q} & p \neq q, \, p=0 \\[2ex] \dfrac{k+1-i}{p} & p \neq q, \, q=0 \end{cases} \tag{3}$$

3.3 Costs Computation

Costs add location registration costs to paging costs and can be number of signaling bits for unit time between nodes in the mobile communication service networks [8]. In this case, each system in networks is treated nodes and costs doesn't include the amount of signals processed at nodes.

The costs of location registration multiplies location weight($\lambda > 1$) for a LA assignment by location registration rate. It is the probability that location registration occurs by only one step movement because step movement of a MS during time Δt occurs when it escapes from a LA. In this time, location registration rate is probability occurring location registration of one step movement. Location registration occurs first where the number of steps is smaller, hence the location registration probability is less than or equal to $1/\min(M_x T_h, M_y T_i, M_z T_j)$, where $M_x T_h$, $M_y T_i$, $M_z T_j$ are the number of steps in each direction. Paging cost multiplies the number of bits for MS's identification by the number of cell in a LA and call arrival rate. Therefore, total cost(C) can be presented as equation (4).

$$C = \lambda \frac{B_{MS}}{\min(M_x T_h, M_y T_i, M_z T_j)} + c \times B_{MS} \times (number\ of\ cell\ in\ LA) \tag{4}$$

The number of cells in a LA with ZBLA, DBLA, DYLA are computed as equation (5).

$$1 + \sum_{n=1}^{\left[\frac{k}{2}\right]} 6(n-1) = 3\left[\frac{k}{2}\right]^2 - 3\left[\frac{k}{2}\right] + 1, \quad k = 1,2,3,\dots \tag{5}$$

In UDLA case, as equation (6),

$$\left(3\left[\frac{k}{2}\right]^2 - 3\left[\frac{k}{2}\right] + 1\right) \times r, \quad k = 1,2,3,\dots \quad (r \le 1) \tag{6}$$

r is the paging selection rate of UDLA(the cells number of UDLA/the total number of cell in a LA). Hence the costs of UDLA can be represented as equation (7).

$$C = \lambda \frac{B_{MS}}{\min(M_x T_h, M_y T_i, M_z T_j)} + c \times B_{MS} \times r \times (number\ of\ cell\ in\ LA) \tag{7}$$

3.4 Optimization of k

The initial value of k, the size of a LA, is 2, and is increased by 1 until the total cost does not decrease by having it applied to equation (4), (7). $\Delta(k)$ can be represented as equation (8), and the optimal a LA size (k_{opt}) can be represented as equation (9).

$$\Delta(k) = C(k) - C(k-1) \tag{8}$$

$$k_{opt} = \begin{cases} 1 & if \Delta(2) > 0 \\ \max\{k : \Delta(k) \le 0\} & o.w \end{cases} \tag{9}$$

In equation (9), the condition, $\Delta(2) > 0$, is applied when the total cost with a LA size of 2 is bigger than the one with a LA size of 1. Hence, a LA size is 1 at this condition. With other conditions, the maximum k_{opt} minimizing the total cost becomes optimal LA size.

4   Performance Evaluation

Location registration of ZBLA is performed in boundary cells, so the location of MS is $(1, \langle (k+1)/2 \rangle, \langle (k+1)/2 \rangle)$, the location of DBLA is the center ($\langle (k+1)/2 \rangle$, $\langle (k+1)/2 \rangle$, $\langle (k+1)/2 \rangle$), so the optimal UDLA location (h_{opt}, i_{opt}, j_{opt}) is calculated by equation (10). Denote $\langle \overline{h_{opt}} \rangle$ as $\{h : \arg \max(MT \overline{[\frac{1}{h_{opt}}]}, MT \overline{[\frac{1}{h_{opt}}]}) \}$ and p, q are the movement probability in each direction.

The total cost can be calculated as follows when bit number of a MS's identification B_{MS} is 16, location weight λ is 32 and call arrival rate c in Δt is 0.001.

$$
h_{opt,\,i_{opt},\,j_{opt}} = \left\langle \begin{array}{ll} \dfrac{k+1}{2} & p = q \\[2ex] \ln\!\left[\dfrac{\left(\dfrac{q}{p}\right)^{k+1}}{(k+1)\ln\!\left(\dfrac{q}{p}\right)-1} \right] \Big/ \ln\!\left(\dfrac{q}{p}\right) & p \neq q,\, p \neq 0,\, q \neq 0 \\[3ex] k & p \neq q,\, p = 0 \\[1ex] 1 & p \neq q,\, q = 0 \end{array} \right. \tag{10}
$$

Example 1. A pedestrian's movement

$p_x = 0.005,\ q_x = 0.005,\ p_y = 0.005,\ q_y = 0.005,\ p_z = 0.005,\ q_z = 0.005,\ r = 0.45.$

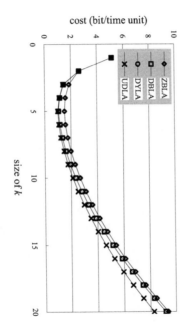

Fig. 6. The results of a pedestrian's movement model

A pedestrian's movement pattern is assumed that the movement path is repeated to the both directions with center of the initial cell. In figure 6, the results of DBLA and DYLA are identical, and ZBLA cost is the biggest because a boundary cell becomes the initial cell. A MS randomly moves, so UDLA cost varies depending on the paging selection rate, and in worst case, the cost is identical with the one of DYLA.

Example 2. A vehicle's movement

$p_x = 0.015,\ q_x = 0.0,\ p_y = 0.001,\ q_y = 0.0,\ p_z = 0.001,\ q_z = 0.002,\ r = 0.12.$

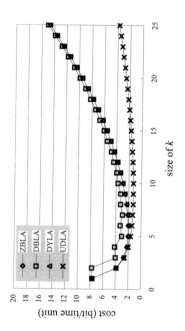

Fig. 7. The results of a vehicle's movement model

Assumes unidirectional. In ZBLA, the ping-ponging in boundary cells rarely happens, so the results of ZBLA and DYLA are identical. In DBLA, the center cell becomes the initial cell, the new and original LA's cell are overlapped, so the cost is the biggest.

Example 3. Diagonal movement

$p_x = 0.04$, $q_x = 0.003$, $p_y = 0.035$, $q_y = 0.004$, $p_z = 0.037$, $q_z = 0.004$, $r = 0.25$.

In figure 8, the case of diagonal movement is assumed unidirectional as the vehicle movement environment. The x, y and z movements are independent each other, so the 3 movements sequentially occur.

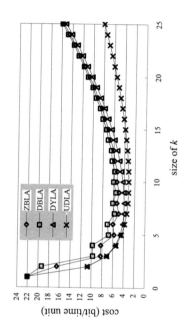

Fig. 8. The results of diagonal movement model

In figure 8, the cost for the ZBLA is the biggest because there are movements of passing through only a part of a LA. The paging selection rate for the UDLA is greater than in the vehicle's environment, but eventually the cost reduces with low paging selection rate.

5 Conclusion

A MS determines a LA with location information in UDLA. And it determines a LA with cells on the path where user's routing can be predicted in a LA then transfers to wired networks it. UDLA scheme has a value that user inputs its final destination while dynamically assigned LA has a value that it store movement probability for each direction based on current LA. Therefore, UDLA scheme supports a user's future movement patterns and prevents a waste of location registration costs for periodic LA update.

We compared UDLA scheme with ZBLA, DBLA, DYLA scheme for movement patterns in performance evaluation. And this paper is included a pedestrian and vehicle's users. In case of pedestrians, the cost's reduction is low because UDLA's paging choice rate increases. But in case of vehicle's users, the cost's reduction is high.

References

1. Sheldon M. Ross, Introduction to Probability Model, Sixth Edition, pp.183-189, Berkeley, California, 1997.

2. TIA/EIA Interim Standard, "Mobile Station-Base Station Compatibility Standard for Dual-Mode Wideband Spread Spectrum Cellular System," TIA/EIA/IS-95.

3. Bar-Noy A., I. Kessler and M. Sidi, "Mobile Users: To Update or not to Update?" ACM/Baltzer Wireless Networks, Vol. 1, No. 2, pp. 175-185, 1995.

4. C. Rose, "Minimizing the Average Cost of Paging and Registration: A Timer-Based Method," ACM/Baltzer Wireless Networks, Vol. 2, pp. 109-116, Jun. 1996.

5. Z. Naor and H. Levy, "Minimizing the Wireless Cost of Tracking Mobile Users: An Adaptive Threshold Scheme," Proc. IEEE Infocom'98, pp. 720-727, San Francisco, CA, Mar. 1998.

6. J.S.M. Ho and I.F. Akyildiz, "A Mobile User Location Update and Paging Mechanism Under Delay Constraints," ACM-Baltzer Journal of Wireless Networks, Vol. 1, No. 4, pp. 413-425, Dec. 1995.

7. Wing Ho A. Yuen, Wing Shing Wong "A dynamic location area assignment algorithm for mobile cellular systems," Proc. of IEEE International Conference on Communications, Vol. 3, pp. 1385-1389, 1998.

8. Zhuyu Lei, Christopher Rose, "Wireless Subscriber Mobility Management using Adaptive Individual Location Areas for PCS Systems," Proc. of IEEE International Conference on Communications, pp. 1390-1394, Jun. 1998.

9. S. Tabbane, "Evaluation of an alternative location strategy for future high density wireless communications systems," Technical Report WINLAB-TR-51, Rutgers University, 1993.

10. Vincent W.-S. Wong, Victor C.M. Leung, "Location Management for Next-Generation Personal Communications Networks," IEEE Network, Vol. 14, pp. 18-24, Sep. 2000.

11. Cang Ye, Danwei Wang, "A novel behavior fusion method for the navigation of mobile robots," Proc. of IEEE International Conference on Systems, Man, and Cybernetics, Vol. 5, pp.3526-353, 2000.

12. Raman B., Bhagwat P., Seshan S., "Arguments for cross-layer optimizations in Bluetooth scatternets," Proc. of Symposium on Applications and the Internet, pp. 176-184, 2001.

II. 4G Mobile Systems

A Fast Authentication Method for Secure and Seamless Handoff

Satoshi Ohzahata[1], Shigetomo Kimura[2], and Yoshihiko Ebihara[2]

[1] Doctoral Program in Engineering, University of Tsukuba
ohzahata@netlab.is.tsukuba.ac.jp
[2] Institute of Information Sciences and Electronics, University of Tsukuba
{kimura, ebihara}@netlab.is.tsukuba.ac.jp

Abstract. In mobile networks such as Mobile IP, a symmetric-key-based authentication method is a typical solution for identifying genuine users of the networks, but restricts scalability of the communications. From this reason, several asymmetric-key-based authentication methods are proposed to release the restriction. However, these methods still suffer from the heavily procedures, especially in the handoff, and public keys can not be distributed efficiently in advance. This paper introduces a key distribution and authentication mechanism for the handoff method with a certificate-based authentication. The computer simulations show that the proposed method provides low key distribution overhead of the authentication with no communications disruption.

1 Introduction

Mobile IP[1] gives one of the solutions for mobile environments instead of IP, however the mobility requires authentications to identify genuine users of the network whenever a mobile node (MN) changes its connection point to the Internet. To satisfy the requirements of security considerations, a symmetric-key-based authentication protocol is widely used because these protocols are simple and require low computation power to the MN. However, in such methods, keys for authentication must be previously distributed to each communication entity and a large number of keys are required since these keys must be prepared for all the communications. This fact restricts scalability of mobile communications.

To improve the scalable problems, a public key infrastructure (PKI) is proposed. Supporting the infrastructure, several protocols are also proposed to provide secure key distribution mechanism through the Internet[2],[3]. These protocols are also extended to support scalable and secure communications for the mobility of the MNs[4],[5],[6]. However, in the extended protocols, their authentication procedures for the MN need a quite long period. This causes disrupted communications.

In this paper, we propose a fast authentication method as a key distribution protocol with a simple movement prediction control. In the prediction, we assume that the MN can control two network interfaces for the primary and secondary foreign agents (FAs). By the control, the MN can send its registration request through the current FA to the FA predicted that the MN will connected. This registration procedure can be completed before the MN is disconnected from the current FA. The control can also be extended

I. Chong (Ed.): ICOIN 2002, LNCS 2344, pp. 243–252, 2002.
© Springer-Verlag Berlin Heidelberg 2002

to the public-key-based authentication and give scalable communications for MNs with no disrupted communications in the handoff procedures. Simulation experiments conclude that our method can switch the MN's communication route with no disrupted the communications and low key distribution overhead for cellular Internet environments.

The rest of this paper is structured as follows. Section 2 describes an authentication method in Mobile IP. Section 3 presents a fast authentication method. In section 4, simulation experiments evaluate our proposed method and the original Mobile IP. Section 5 concludes this paper.

2   Authentication Method in Mobile IP

For secure communications, Mobile IP prepares MN-HA, MN-FA, MN-FA and FA-HA authentication extensions in the registration procedures[1]. By the extensions , each node can authenticate between MN-HA, MN-FA and FA-HA, respectively. In Mobile IP, the default authentication algorithm is keyed-MD5[7] by "prefix+suffix" mode. The secret shared keys are previously distributed to the entities because Mobile IP does not have any key distribution mechanism by itself.

This authentication protocol (Protocol 1) is written as below.

[Protocol 1]
(1) FA $\rightarrow$ MN: AgentAd
(2) MN $\rightarrow$ FA: RegReq, $h(\text{RegReq})_{k_{FM}}$, $h(\text{RegReq})_{k_{HM}}$
(3) FA $\rightarrow$ HA: RegReq, $h(\text{RegReq})_{k_{FH}}$, $h(\text{RegReq})_{k_{HM}}$

k_{FM}, k_{HM}, k_{FH}: are shared secret keys between FA-MN, HA-MN and FA-HA, respectively.
$h(m)_k$: is a hashed value of m by an one-way function h, and encrypted by a key k.
AgentAd: is an agent advertisement.
RegReq: is a registration request.
In the protocol, time stamp or nonce is omitted.

□

In the Protocol 1, (1) an FA periodically broadcasts an agent advertisement which has information for MNs to connect to the FA. (2) When the MN receives the agent advertisement, it sends a registration request with authentication extensions ($h(\text{RegReq})_{k_{FM}}$, $h(\text{RegReq})_{k_{HM}}$) to the FA. The FA authenticates the MN by comparing the first part of the extension. (3) The FA forwards the registration request with $h(\text{RegReq})_{k_{FH}}$ and $h(\text{RegReq})_{k_{HM}}$ to the HA. The HA authenticates the FA and the MN with the same way in (2), and registers a care of address for the MN. The registration reply from the FA is delivered to the MN through the FA and then the authentication of the MN is completed.

Protocol 1 can provide one of the solutions for the secure communications. But it still suffers from scalability of the communications, because the secret shared keys should be previously distributed to the nodes and Mobile IP does not have any key distribution mechanism by itself. Moreover, when the number of nodes is m in the Internet, $m \times m/2$ keys are required to communicate each other. This also causes difficulty for the key management and scalability.

3   Fast Authentication Method

3.1   Introduction of Certificate-Based Authentication Method for Mobile IP

To be free from the scalable problem in the previous section, several key distribution protocols are proposed for mobile networks[4],[5],[6]. Especially, a certificate-based security protocol[6] is effective to exchange a session key through networks, and authenticate each communication entity such as HAs, FAs and MNs. In the protocol, a certificate authority (CA) plays an important role. The CA is a widely trusted agency and distributes its public key to all communication entities. It also signs a certification including a public key of each communication entity by its secret key. Each entity can verify the certification by decrypting itself with the CA's public key.

We extend the concept of the above protocol for Mobile IP and show one of the examples as Protocol 2 with the following basic principles.

- Security considerations for the attack are same with Protocol 1.
- No additional message for the original Mobile IP should be required.
- The Diffie-Hellman (DH) key exchange algorithm is used for creating a session key which is used for the registration updates.
- The MN and the HA previously obtain each other's certification.
- The MN, the HA and the FA have a PKI infrastructure like in [4] to be able to obtain certificated public keys from the CA and manage them. They also previously obtain the certification of the CA.

Protocol 2 in below ignores time stamp or nonce.

[Protocol 2]

(1) FA → MN: AgentAd, $h(\text{AgentAd})_{K^{-1}_{FA}}$, $(y_{FA}, \alpha, N)_{K^{-1}_{FA}}$, Cert$_{FA}$

(2) MN → FA: RegReq, $h(\text{RegReq})_{K^{-1}_{MN}}$, $(y_{MN})_{K^{-1}_{MN}}$, Cert$_{MN}$

(3) FA → HA: RegReq, $h(\text{RegReq})_{K^{-1}_{FA}}$, $(y_{MN})_{K^{-1}_{MN}}$,

$$(y_{FA}, \alpha, N)_{K^{-1}_{FA}}, \text{Cert}_{FA}$$

(4) HA → FA: RegRep, $h(\text{RegRep})_{K^{-1}_{HA}}$, $(y_{HA})_{K^{-1}_{HA}}$, Cert$_{HA}$

(5) FA → MN: RegRep, $h(\text{RegRep})_{K^{-1}_{HA}}$, $(y_{HA})_{K^{-1}_{HA}}$, Cert$_{HA}$

Cert$_A$: is a certification of A which includes K_A, ID$_A$, Date$_A$ and $h(K_A$, ID$_A$, Date$_A)_{CA}$, where K_A is a public key of A, and ID$_A$ and Date$_A$ are identification and expiration date of the certification, respectively.

K^{-1}_A: is a secret key of A.

N: is a prime number.

α: is a primitive element modulo of N, i.e., $\gcd(N-1, \alpha) = 1$.

$y_A = \alpha^{x_A} \bmod N$: is an A's public value for DH key exchange, where x_A is an A's secret value for the DH key exchange.

$(m, n, \cdots)_k$: is an encrypted value of the sequence $m, n, \cdots$ by a key k.

RegRep: is a registration reply.

In Protocol 2, (1) the FA creates α and N, and a public value $y_{FA} = \alpha^{x_{FA}} \bmod N$. The MN can authenticate the FA by decrypting the message authentication code $h(\text{AgentAd})_{K^{-1}_{FA}}$ with the public key K_{FA} in the certification Cert$_{FA}$ from the agent advertisement and makes a session key $k_{FM} = y_{FA}{}^{x_{MN}} \bmod N = \alpha^{x_{FA} x_{MN}} \bmod N$ between the FA and the MN. (2) The MN sends a registration request with DH public value $y_{MN} = \alpha^{x_{MN}} \bmod N$ and its certification Cert$_{MN}$. The FA authenticates the MN with the same way in (1), and also creates a session key $k_{FM} = y_{MN}^{x_{FA}} \bmod N = \alpha^{x_{FA} x_{FA}} \bmod N$. (3) The FA forwards the registration request with $y_{MN}, y_{FA}, \alpha, N$ and its certification Cert$_{FA}$. Note that α and N are shared among the MN, the FA and the HA. The HA creates session keys k_{HM} and k_{FH} between HA-MN and FA-HA with the same way in (1), and authenticates the FA. (4) The HA sends a registration reply with public value y_{HA} of DH for the FA and the MN and its certification Cert$_{HA}$. The FA can authenticate the HA and create a session key k_{FH} with the same way in above. (5) The FA forwards the registration reply to the MN. Then, the MN can create a session key k_{HM}. From these procedures, the session keys are securely distributed.

The above method succeeds in providing scalable communications for the MN. But, in general, public-key-based authentication takes longer time than the one with a symmetric key. Moreover, in handoff, the MN must release its current FA to register itself to the new FA and the new care of address to the HA, and then the MN can not communicate with any FA. This fact arises large packet disruption time. Next subsection introduces a fast authentication method which conceals the large disruption time.

3.2   Key Distribution and Authentication Mechanism

This subsection introduces the fast authentication method to conceal handoff latency by authentications of the MN. In the method, the MN, the FAs and the HA are assumed to deal with two care of addresses of the MN. The MN has two network interfaces, and then, it can set up a new communication route before it completely release the old care of address. The basic principles of the fast authentication method are same as Protocol 2, and in addition, the following principles are included.

- The key distribution and the authentication are completed before the MN changes the current FA (FA1).
- The MN can register its new FA (FA2), even if the MN stays in an area where a registration request from the MN is not reached to the FA2 by its weak signal strength, but the MN can receive an agent advertisement from the FA2.
- Protocol 2 is used only for the first time registration request to create session keys to be used again at later of registration updates by Protocol 1, and the first time authentication between the entities.
- Periodical agent advertisements contain no certification not to waste bandwidth.

Fig. 1 shows procedures of the fast authentication method. The method is added two new messages before Protocol 2(1). Two message routes of (2) and (5) differ from the orginal one. In Fig. 1, an MN has a connection to the FA1 and moves into the cell of the FA2. {1} The MN receives an agent advertisement from the FA2 in the overlap region between the cells. However, the agent advertisement has no certification of the FA2 and the MN can not authenticate the FA2. If the MN already has the certification, go to {4}. Otherwise, {2} it sends a message to require a certification of the FA2 via the FA1, not directly to the FA2. {3} When the FA2 receives the request, the FA2 broadcasts an agent advertisement with its certification (Protocol 2(1)). Then, the MN authenticates the FA2. {4} The MN sends a registration request to the FA2 through the FA1 (Protocol 2(2)). {5} When the FA2 accepts the request, it forwards the registration request from the MN to the HA (Protocol 2(3)). Then, {6} the HA registers the care of address of the FA2, and {7} sends a registration reply to the FA2 (Protocol 2(4)). {8} The FA2 receives the registration reply, and {9} forwards it to the MN through the FA1 (Protocol 2(5)).

From these procedures, session keys are securely distributed for the registration updates. In this method, the HA, the FAs and the MN must register the MN's two care of addresses, i.e., the FA1 and the FA2. Depending on the position of the MN, the signal strength from the two FA usually changes. The FA with stronger signal is called a primary FA, and the other one is a secondary FA. However, the MN register the FA2 as its secondary FA at the first time, even if its signal strength is stronger than the FA1.

In our proposal, the MN sends all registration requests and communication data to the primary FA rather than to the secondary one except for the broadcast packets. We adopt such way by the two reasons. At first, the request may not reach to the latter FA by the weak signal strength. Second, the construction of the logical link control in the MN is simplefied. The information of the primary and secondary FAs are also registered to the HA. When the HA interrupts data from the sender, the data are sent to the MN via the primary FA. Suppose the primary and secondary FAs are swapped since their signal strength are changed. Then, the MN sends a registration update message to notify this fact. Although the HA does not recognize this modification until the message is received, the MN can receive the data from the HA via the secondary FA which is the previously primary FA.

The fast authentication method is expected to provide an effective authentication and registration mechanism to Mobile IP. However, if no overlap region exists, the MN can not send registration requests to the FA1. In this case, it must send the request to the secondary FA as in Protocol 2.

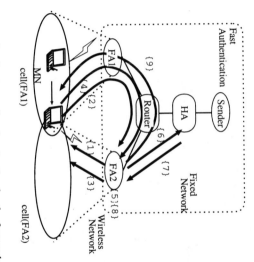

Fig. 1. An example of key distribution and authentication by the fast authentication method.

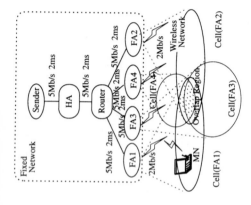

Fig. 2. Simulation model.

4   Simulation Experiments for Fast Authentication Method

To evaluate the proposed authentication method in section 3.2, we experiment with computer simulations on ns-2.1b7[8]. In our method, the MN can distribute its public keys and register for the new FA which may not become the primary FA of the MN. This overhead is also estimated in some mobility model of the MN.

4.1   Simulation Conditions

We simulate two handoff methods. The first is the proposed method in the previous section. The second one is so called the "fast handoff" which is the improved handoff method of the original Mobile IP. In the original one, the MN can not change its FA until the MN can not receive any packet from the FA. But this control causes quite large handoff latency. In the fast handoff, the MN observes the MAC layer handoff and every packet from the old FA. When it detects some packet loss or change of the BS in its MAC layer, it broadcasts an agent solicitation message to require an agent advertisement. This method is based on the concept of "access to lower-layer information" in [9]. Note that the second method uses Protocol 1 for every registration. Indeed, the second method with Protocol 2 produces much larger handoff latency than the one with Protocol 1. We observed that the average latency of the one with Protocol 2 is observed 130ms under the same conditions in this subsection. These simulation results are omitted in this paper.

The network model for the simulation experiments consists of a fixed network and a wireless network. The topology and each link speed for the experimentation is illustrated in Figure 2. The simulation conditions are below.

- The number of MNs in each model is one.
- The number of senders is one. All communication packets through the HA to the MN are transmitted only by unicast.

- The sender sends packets to the MN with a same interval (5.33ms) over UDP and the IP packet size is 256 octets. Thus, the transfer rate is 384kbps.

- An agent advertisement from each FA is periodically broadcasted every 0.2s within a whole cell area of the FA. The MN's registration life time[1] in FAs and HA are 1.0s.

- A packet size of an agent advertisement, a registration request and a registration reply are 128, 128 and 32 octets, respectively. These packet sizes includes an authentication extension, such as the hashed values in Protocol 1.

- In our proposed method, each certification size is 512 octets. When it has parameters to share secret values, such as (y_a, K_A^{-1}) or $(y_a, \alpha, N)_{K_A^{-1}}$, its size is added to 256 or 512 octets for the extension, respectively.

- All registration requests are accepted by each entity in the authentication procedures.

- In the proposed method, the MN decides its primary and secondary FAs by the signal strength of the agent advertisement, and changes its primary FA only when it receives a stronger signal of an agent advertisement than the current one. But 0.2s interval is required for changing the primary FA from the last change.

- The MN can broadcast an agent solicitation, if the MN detects a packet loss or a packet arrival interval becomes larger than 10ms.

- In the proposed method, the MN does not keep any certifications. Thus, it must require a certification for every registration procedure for the new secondary FA.

- Medium access control at the wireless network in Figure 2 is IEEE 802.11 which is implemented in [8].

- The MN's MAC medium does not deregister an FA for 0.2s after it registers the FA.

- For a wireless communication in each cell, the transfer error is ignored and propagation delay is (distance between the FA and the MN) / (speed of the light).

- In the MN, the HA, the FAs and the router, each processing capacity for routing is 100000 packets per second, and copying, encapsuling and decapsuling a packet take 10μs, respectively. A delay of the registration update by Protocol 1 at each HA and FA is 100μs.

- A delay of the registration procedure with the certificate-based authentication is 50ms at the MN and 25ms at the FA or the HA.

- There are one HA and four FAs. Each FA has a circular cell whose radius is 200m.

- The MN moves randomly within a square of 129600 (= 360×360) m² at the multiple overlap regions in Fig. 2.

- At each midpoint of all sides of the square, an FA (FA1–FA4) is placed.

- The initial location of the MN is randomly chosen within the square.

- A movement of the MN is selected combination of four speed models and three direction selection models. As the speed, a constant speed is selected uniform random value among 0–1.0m/s, 0–5.0m/s, 0–10.0m/s and 0–20.0m/s. For the direction of the MN, the MN straightly moves within the square during 1s to a random direction whose angle is restricted within ±180°, ±90° and ±45° from current direction of the MN. But the MN can select from all angle, when it stays less than 1m apart from any side of the square. The movements are repeated for 20000s.

- Every result satisfies 5% confidence interval with 90% confidence level.

Table 1. Average handoff latency (ms).

max. speed	1m/s	5m/s	10m/s	20m/s
proposed-all	5.7	5.5	5.9	5.7
proposed-half	5.5	5.4	5.5	5.4
proposed-quarter	5.4	5.4	5.4	5.5
fast-all	22.0	22.2	22.0	22.8
fast-half	21.5	22.1	22.0	21.7
fast-quarter	20.5	21.8	22.0	21.5

4.2 Simulation Results

At first, we show handoff latencies. In general, a handoff latency is used two means. The first one is a latency occurred during registration procedures, i.e., the period is from the time the MN receives an agent advertisement of the new FA to the time the MN receives the registration reply from the new FA. The authentication delay at each node is included in the period. The second latency is the packet arrival interval when no data packets arrive at the MN in the case of switching the communication route. In the following, we adopt the second latency, and just call it handoff latency.

Table 1 shows the results of handoff latencies, where our proposed method and fast handoff method are described as "proposed" and "fast," respectively. "all," "half" and "quarter" mean the available mobility model of the MN. In the table, the average handoff latencies of the two handoff methods are 5.4–5.9ms and 20.5–22.8ms. "proposed" results almost no handoff latency and no packet loss is observed in spite that it needs 149.0–153.8ms to complete the authentication procedures (and "fast" needs 16.1–16.7ms). If the MN stays for 150ms + 13ms + 200ms (handoff procedures, round trip time between the MN and the HA and the interval of an agent advertisement) at overlap region, no disrupted communications is served. In the next, we evaluate the overhead of proposed method to achieve these results.

Fig. 3 shows the average utilization rate of an authenticated secondary FA, i.e., a ratio that the secondary FA becomes as the primary within the life time. In this graph, "all" is 0.055–0.196, "half" is 0.130–0.461 and "quarter" is 0.210–0.678. For all mobility model, the rate gradually increases against the speed of the MN and also decreases against the direction range at every speed of the MN. Since higher speed or narrower direction range make the MN completely move into a new cell at higher probability. From the figure, at least 5.5%–19.6% of the secondary FA is used as the primary FA for all maximum speed. The rate may seem to be low even if the MN distributes the keys for only one FA. However, the prediction efficiency of proposed method should be rather higher than the multicast based handoff method such as [10] and [11], which are typical expansion methods for Mobile IP, because, in these method, the previous preparation is done for several FA.

Fig. 4 shows average period that the secondary FA needs to be the primary. In this graph, "all" is 10.4–116.7s, "half" is 7.1–99.1s and "quarter" is 6.5–86.8s. In lower speed, the MN dose not need to connect to the secondary FA for very longer period than that of the higher speed. In this case, many update messages for the secondary FA wastes

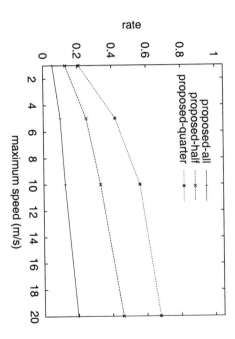

Fig. 3. Average utilization rate of previous authenticated secondary FA.

Table 2. Average number of registration requests for the secondary FA.

max. speed	1m/s	5m/s	10m/s	20m/s
proposed-all	134837.5 (129.0)	84417.0 (378.9)	84118.5 (576.6)	82864.0 (1136.7)
proposed-half	92107.0 (113.0)	83310.5 (310.5)	85218.3 (573.4)	83588.5 (1146.9)
proposed-quarter	94122.0 (187.7)	84417.0 (312.5)	81941.1 (572.4)	82617.2 (1194.3)

resource of the networks. To improve the problem, as one of the solutions, we select the registration life time of 1.0s for each FA in this simulation. But the life time should be longer to reduce the registration update.

Table 2 shows the number of registration requests from the MN to the secondary FA. The number in parentheses at the table shows that the number of certification-base authentications to the secondary FA. In the proposed methods, the MN registers to two FAs in the overlap region (covering 79.4% in the square) and must update until their registrations are expired.

5   Conclusions

In mobile networks, a handoff latency occurs when an MN changes its connection point. To reduce it, this paper proposed the improved authentication methods for cellular Internet environments in Mobile IP. Although the protocol has simple movement prediction control of the MN, the simulation experiments show our authentication method is effective for secure and real-time communications during handoff procedures. However,

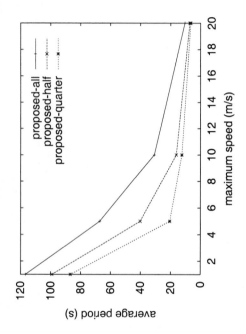

Fig. 4. Average period from authentication for the secondary FAs to be the primary.

the simulation results depends on the simulation model, especially geometrical cells arrangement. We should evaluate the other simulation models and more complicated models with many FAs, Mobile IP with the route optimization and the hierarchical BS models with expanded our control.

References

1. C. Perkins, "IP Mobility Support," RFC2002, 1996.
2. H. Krawczyk, "SKEME: A Versatile Secure Key Exchange Mechanism for Internet," Proc. Internet Soc. Symp. Network and Dist. Sys. Security, 1996.
3. A. Aziz and M. Patterson, "Design and Implementation of SKIP," http://skip.incog.com/inet-95.ps, 1995.
4. J. Zao, et al., "A Public-Key Based Secure Mobile IP," MOBICOM '97, pp.173–184 1997.
5. V. Gupta and G. Montenegro, "Secure and mobile networking," Mobile Networks and Applications 3, pp.381–390, 1998.
6. C. Park, "On Certificate-Based Security Protons for Wireless Mobile Communication Systems," IEEE Network Sep./Oct., pp.50–55, 1997.
7. R. Rivest, "The MD5 Message-Digest Algorithm," RFC 1321, 1992.
8. S. McCanne and S. Floyd, "Network Simulator ns (version2)," http://www.isi.edu/nsnam/ns/.
9. T. Camp, J. Lusth and J. Matocha. "Reduced Cell Switching in a Mobile Computing Environment," MOBICOM '00, pp.143–153, 2000.
10. J. Hoon, et al., "An Adaptive Resource Allocation Mechanism Including Fast and Reliable Handoff in IP-Based 3Gwireless Networks," IEEE Personal Communications, Dec., pp.42–47, 2000.
11. M. Stemm and R. H. Katz, "Vertical Handoffs in Wireless Overlay Networks," ACM Mobile Networking and Applications, Vol. 3, No.4, pp.335–350, 1998.

Orthogonal Multi-code CDMA Systems with Constant Amplitude Transmission Based on Level Clipping and Pulse Width Modulation

Jong Kap Oh[1], Sung Pil Kim[1], Myoung Jin Kim[1], Ho Seong Ahn[2], and Seung Moon Ryu[2]

[1] Hankuk University of Foreign Studies, San 89 Wangsan, Mohyun, Yongin, Kyonggi-do, Korea
mjkim@hufs.ac.kr

[2] CASUH Corp., C-528, Sigma II Officetel, Gumi-dong, Pundang-gu, Songnam, Kyonggi-do, Korea
{hsahn, smryu}@casuh.com

Abstract. Multi-code CDMA system provides high speed and variable rate services with different quality requirements. Since envelope fluctuation of the multi-code signal is large, the system requires highly linear power amplifier. In this paper, we propose binary level multi-code pulse width (PW-) CDMA system. In the system, linear sum of orthogonal codes is truncated and pulse width modulated. The transmitting signal has constant envelope, and the receiver structure is very simple. Hence, the proposed system is appropriate for low cost high data rate applications. We investigate the bit error performance of the binary level multi-code PW-CDMA system.

1 Introduction

As a demand for wireless multimedia communication services is growing, the transmission schemes providing high speed and variable rate with different quality of service become necessary. Possible candidates of such systems are the variable spreading gain CDMA(VSG-CDMA) system[1], [2], [3] and the multi-code CDMA system[4], [5]. The VSG-CDMA can adjust the spreading gain for variable rate users depending on the input traffics and control the signal power to meet the quality of services(QoS). When a user wants to transmit data at a very high bit rate in VSG-CDMA, the spreading gain may become too small to reject the interference among users.

In multi-code CDMA the transmitting signal is a linear sum of multiple orthogonal codes. The multi-code CDMA system changes the number of code channels based on the data rate of the source, hence a constant processing gain is maintained without the problem of very small processing gain of high rate users. However, the transmitting signal of the multi-code CDMA system has a large amplitude fluctuation as a result of linear sum of multi-code signals, and a highly linear power amplifier may be required

I. Chong (Ed.): ICOIN 2002, LNCS 2344, pp. 253–263, 2002.
© Springer-Verlag Berlin Heidelberg 2002

for transmission. In mobile communications high power efficiency of the amplifier may be important especially on the up-link. In order to obtain high power efficiency the power amplifier with nonlinear characteristic is used [6], [7]. If the multi-code signal is amplified with power efficient amplifier, the nonlinear distortion becomes very large due to large dynamic range of the signal envelope, hence the bit error rate(BER) performance significantly degrades.

To overcome envelope fluctuation problem of multi-code CDMA a number of techniques have been proposed. Precoding scheme[8] realizes the constant amplitude by inserting precoded bits into the information bit sequence before multi-code signaling. However, the precoding scheme causes a loss in the information rate. In multidimensional multi-code sheme[9], the signaling is performed both in code and time spaces to reduce the information rate loss. However, as the number of channels increases the BER performance degrades because the signal energy carried on precoding channels with no information is wasted.

Pulse width CDMA(PW-CDMA)[10] is a multi-code transmission technique that performs pulse width modulation on the multi-level signal synthesized from multiple orthogonal codes. With polar signaling the signal amplitude of PW-CDMA is maintained to be constant, and the modulation and demodulation circuits become simple. However, as the number of channels increases the bandwidth increases due to the reduction of unit pulse width, and the BER performance degrades because nonoptimal filter is used at the receiver. To compensate for an increase in bandwidth, pulse width modulation with polarity alternation or level clipping on the multi-code signal has been proposed[10].

In this paper, we propose an orthogonal multi-code CDMA system with binary level clipping. It is assumed that odd number of channels is used. Then the multi-code signal which is the sum of orthogonal codes has $M+1$ levels, $\{-M, -M+2, \cdots, -1, 1, \cdots, M\}$, where M is the number of channels. With binary level clipping, the truncated signal has constant amplitude. Due to level clipping, the BER performance degrades compared with conventional multi-code CDMA. The amount of performance degradation depends on the selection of orthogonal codes and information data as well as the number of channels[11]. We investigate the BER performance of the proposed system. It is shown that the proposed system with carefully selected code performs better than the PW-CDMA system without level clipping.

The paper is organized as follows. Section 2 describes the proposed binary multi-code CDMA system. In section 3 we analyze the BER performance of the system in AWGN. Section 4 presents simulation results. A conclusion is given in section 5.

2 The Multi-code CDMA System with Level Clipping and PW Modulation

Fig. 1 shows the baseband system model of the orthogonal multi-code CDMA system with level clipping and pulse width modulation. Bit sequences $b_1, \cdots, b_M$ with $b_m \in \{+1, -1\}$ are serial to parallel converted information bits. Information bit on each channel is spread by an orthogonal code with length of N chips. The number of channels, M, is assumed to be odd, which results in $M+1$ levels of multi-code signal, $d(n)$, with values $\{-M, -M+2, \cdots, -1, 1, \cdots, M-2, M\}$. Pulse width modulation is performed chip by chip on the multi-level signal. Pulse width of the signal during each chip time is proportional to the value of $d(n)$. If the number of channels is large, the minimum pulse width is very small, and the bandwidth of the system becomes very large. By inserting the level clipping block and the amplitude of $d(n)$ is limited to a reduced number of levels, hence the bandwidth is reduced. We will demonstrate that the level clipping gives another benefit in addition to bandwidth reduction.

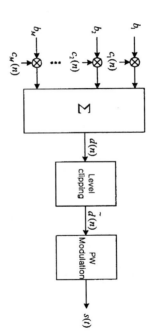

Fig. 1. Transmitter model

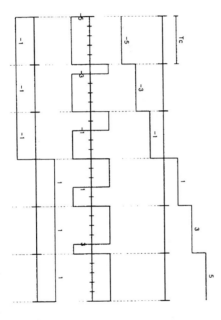

Fig. 2. Signal waveforms of multi-code CDMA, PW-CDMA, Binary level clipped PW-CDMA

Fig. 2 shows an example of waveforms of the conventional multi-code CDMA, PW-CDMA, and binary level clipped multi-code CDMA when the number of channels is 5. Fig. 3 shows the baseband receiver model of the system. The received signal is despread by the orthogonal signal and integrated to generate the decision variables as in the conventional multi-code CDMA system.

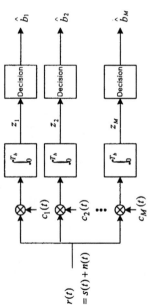

Fig. 3. Receiver model

3   Performance of the Orthogonal Multi-code CDMA System with Binary Level Clipping

3.1   System without Level Clipping

We first analyze the performance of the orthogonal multi-code PW-CDMA system without including level clipper. As shown in Figure 1 the information bit sequence is serial to parallel converted and input to the system. We assume that the system has M number of channels which is an odd integer. During bit interval T_b data bit on each channel is multiplied by the orthogonal code with N chip length. We assume that $T_b = NT_c$ where T_c is the chip time. In the analysis given below, we consider one bit interval, i.e. $0 \le t \le T_b$. The sum of spread sequences produces the multi-level signal $d(n)$, which can be represented by

$$d(n) = \sum_{m=1}^{M} b_m\, c_m(n) \quad 0 \le n \le N - 1 \tag{1}$$

where $b_m \in \{+1, -1\}$ is the information bit of the mth channel and $c_m(n)$ is the nth chip data of the spreading code for the mth channel. The spreading codes are assumed to be orthogonal:

$$\sum_{n=0}^{N-1} c_m(n)\, c_k(n) = N\delta_{m,k} \tag{2}$$

Then the multi-code signal has $M+1$ levels, $\{-M, -M+2, \cdots, -1, 1, \cdots, M-2, M\}$. Note that there is no zero level in $d(n)$ since we use an odd number of channels.

Pulse width modulation is performed on $d(n)$. Let $s_n(t)$ denotes PW modulated signal for the nth chip duration. The signal $s_n(t)$ starts with the value $+1$ and stays during the time interval determined by $d(n)$, then changes its value to -1. Fig. 4 shows an example of the nth chip PW modulated signal $s_n(t)$.

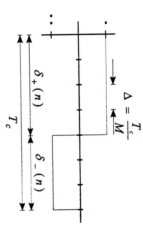

Fig. 4. An example of PW modulated signal $s_n(t)$

Let $\delta_+(n)$ and $\delta_-(n)$ be the time spans for $s_n(t)$ to be positive and negative, respectively. Then $\delta_+(n)$ and $\delta_-(n)$ can be expressed as:

$$\delta_+(n) = \frac{d(n) + M}{2} \Delta \tag{3}$$

and

$$\delta_-(n) = \frac{M - d(n)}{2} \Delta. \tag{4}$$

In the above equations Δ denotes the unit pulse width defined by the chip duration T_c divided by the number of channels M, i.e. $\Delta = T_c / M$. Then the signal $s_n(t)$ may be expressed as

$$s_n(t) = p_{\delta_+(n)}(t) - p_{\delta_-(n)}(t - \delta_+(n)) \tag{5}$$

where $p_\tau(t)$ denotes the rectangular pulse with length τ, i.e

$$p_\tau(t) = \begin{cases} 1 & \text{for } 0 \le t \le \tau \\ 0 & \text{otherwise} \end{cases}. \tag{6}$$

Note that

$$\int_0^{T_c} s_n(t)\, dt = \delta_+(n) - \delta_-(n) = d(n)\Delta. \tag{7}$$

The PW modulated signal $s(t)$ may be represented by

$$s(t) = \sum_{n=0}^{N-1} s_n(t - nT_c) \quad 0 \le t \le T_c. \tag{8}$$

We will investigate the BER performance of the system in AWGN environment. The received signal is correlated with the orthogonal spreading signal $c_m(t)$ to decode the mth channel data. The spreading signal for the mth channel $c_m(t)$ can be written as

$$c_m(t) = \sum_{n=0}^{N-1} c_m(n) p_{T_C}(t - nT_c)$$

(9)

where $c_m(n)$ is the corresponding orthogonal code. In order to derive the probability of bit error we consider the first correlator in Figure 3. Received signal $r(t)$ is the sum of the transmitting signal $s(t)$ and white Gaussian noise $n(t)$ with power spectral density $N_0/2$. The output of the correlator, i.e. the decision variable z_1 is given by

$$z_1 = \int_0^{T_b} r(t) c_1(t)\, dt = s_o + n_o$$

(10)

where, s_o and n_o are the signal component and the noise component, respectively. The signal component s_o can be determined by

$$s_o = \int_0^{T_b} s(t) c_1(t)\, dt = \sum_{n=0}^{N-1} \left(\int_0^{T_c} s_n(t)\, dt \right) c_1(n)$$

$$= \sum_{n=0}^{N-1} \Delta d(n) c_1(n) = \Delta \sum_{n=0}^{N-1} \sum_{M=1}^{M} b_m c_m(n) c_1(n) \,.$$

$$= \pm \frac{T_b}{M}$$

(11)

Since the bit energy is $E_b = T_b / M$, the probability of bit error of the system is given by

$$P_b = Q\left(\sqrt{\frac{s_o^2}{\sigma_{n_o}^2}} \right) = Q\left(\sqrt{\frac{1}{M} \cdot \frac{2E_b}{N_0}} \right) \,.$$

(12)

3.2 System with Binary Level Clipping

We now consider the orthogonal multi-code PW/CDMA system with binary level clipping as shown in Fig. 5. By binary level clipping we imply that the output of the level clipper is $\tilde{d}(n) = \pm 1$. Hence, the PW modulated signal of the nth chip is the antipodal rectangular pulse with pulse width T_c, i.e.

$$s_n(t) = \begin{cases} p_{T_c}(t) & \text{if } \tilde{d}(n) = 1 \\ -p_{T_c}(t) & \text{if } \tilde{d}(n) = -1 \end{cases} \,.$$

(13)

The PW modulated signal $s(t)$ may be represented by

$$s(t) = \sum_{n=0}^{N-1} s_n(t - nT_c) \,.$$

(14)

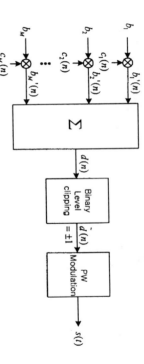

Fig. 5. Multi-code PW-CDMA with binary level clipping

As in section 3.1 we investigate the behavior of the decision variable of the first channel at the receiver. The signal component s_o of the decision variable z_1 may be written as

$$s_o = \int_0^{T_b} s(t)c_1(t)\,dt = \sum_{n=0}^{N-1} c_1(n)\int_0^{T_c} s_n(t)\,dt$$

$$= \sum_{n=0}^{N-1} c_1(n)\tilde{c}_1(n)\tilde{d}(n)T_c . \tag{15}$$

The level clipping introduces nonlinear truncation noise. The amount of the noise is dependent on the spreading codes and the information bits. As shown in Fig. 5 we denote $b_m{}'(n)$ the information bit of the mth channel multiplied by the orthogonal code assigned to that channel. To analyze the effect of nonlinear truncation noise we assume that $b_m{}'(\cdot)$ is an independent binary random sequence. We consider the case when $b_1 = 1$. Table 1 shows the correlator output at T_c for all possible combinations of $b_m{}'$ when the number of channels is 5.

The case when $b_1 = -1$ can be analyzed in the same way. By inspecting the table we can derive the mean of the signal component in decision variable z_1 as follows.

$$\bar{s}_o = E[s_o \mid b_1=1] = E\left[\sum_{n=0}^{N-1} c_1(n)\tilde{d}(n)T_c \mid b_1=1\right] \tag{16}$$

$$= \frac{{}_{M-1}C_{\frac{M-1}{2}}}{2^{M-1}}T_b$$

where

$${}_nC_r = \frac{n!}{r!(n-r)!} . \tag{17}$$

Then the probability of bit error of the system is obtained by

$$P_b = Q\left(\sqrt{\frac{\overline{s_o^2}}{\sigma_{n_o}^2}}\right) = Q\left(\sqrt{\left(\frac{M-1C_{M-1}}{2^{M-1}}\right)^2 M\,\frac{2E_b}{N_0}}\right). \tag{18}$$

Table 1. Correlator output at T_c of the binary PW-CDMA when the number of channels is 5

$b'_2\ b'_3\ b'_4\ b'_5$				$d(n)$		$\tilde{d}(n)$		$\int_0^{T_c} s(t)c_1(t)\,dt$	
				if $c_1(n)=1$	if $c_1(n)=-1$	if $c_1(n)=1$	if $c_1(n)=-1$	if $c_1(n)=1$	if $c_1(n)=-1$
1	1	1	1	5	3	1	1	T_c	$-T_c$
1	1	1	-1	3	1	1	1	T_c	$-T_c$
1	1	-1	1	3	1	1	1	T_c	$-T_c$
1	1	-1	-1	1	-1	1	-1	T_c	T_c
1	-1	1	1	3	1	1	1	T_c	$-T_c$
1	-1	1	-1	1	-1	1	-1	T_c	T_c
1	-1	-1	1	1	-1	1	-1	$-T_c$	T_c
1	-1	-1	-1	-1	-3	1	1	T_c	$-T_c$
-1	1	1	1	3	1	1	1	T_c	T_c
-1	1	1	-1	1	-1	1	-1	T_c	T_c
-1	1	-1	1	1	-1	1	-1	$-T_c$	T_c
-1	1	-1	-1	-1	-3	1	-1	T_c	T_c
-1	-1	1	1	1	-1	1	-1	T_c	T_c
-1	-1	1	-1	-1	-3	1	-1	$-T_c$	T_c
-1	-1	-1	1	-1	-3	1	-1	$-T_c$	T_c
-1	-1	-1	-1	3	-5	-1	-1	T_c	T_c

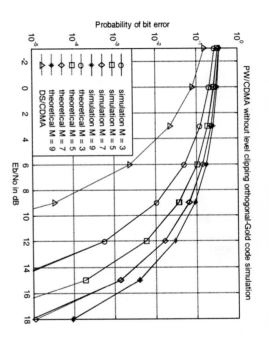

PW/CDMA without level clipping orthogonal-Gold code simulation

Fig. 6. Bit error probability of multi-code PW-CDMA system

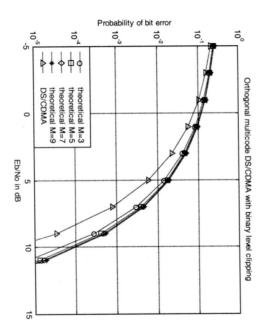

Orthogonal multicode DS/CDMA with binary level clipping

Fig. 7. Theoretical bit error probability of multi-code PW-CDMA system with binary level clipping

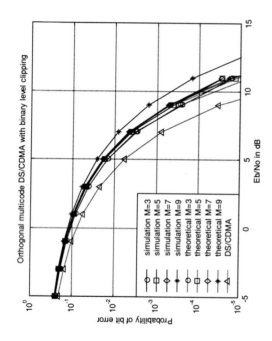

Fig. 8. Simulation result of BER performance of binary multi-code PW-CDMA system

4 Numerical Results and Discussion

We have analyzed the BER performance of multi-code PW- CDMA systems in section 3. Without level clipping probability of bit error is given by equation (12). If binary level clipping block is inserted the performance of the system is characterized by equation (18). To compare the performance of these system configurations we present numerical results. We also carried out simulation. The simulation results are given together with theoretical results.

Fig. 6 shows the bit error probability of the multi-code PW-CDMA system without level clipping. It is observed that with increase in number of channels the system exhibits severe performance degradation.

Fig. 7 shows the theoretical bit error probability of the multi-code PW-CDMA system with binary level clipping. Although the signal is truncated the binary level clipped PW-CDMA is shown to perform better than the system without level clipping. This result can be interpreted as follows. The receiver of the multi-code PW-CDMA is not optimal in the sense that the impulse response of the receiver filter is not matched to the transmitted signal. If the number of channels is large Δ in equation (11) becomes small, which results in small signal to noise ratio in decision variable.

Simulation results of PW-CDMA with binary level clipping are presented in Fig. 8. When the number of channels is large, discrepancy between theoretical result and simulation result is observed. The reason is that we assume independent binary random sequences in the derivation of probability of bit error. In reality the sequences are dependent on the information bits and orthogonal codes selected for spreading. In [14], the way of selecting code set is proposed.

5 Conclusion

Multi-code PW-CDMA with binary level clipping has been proposed. In the system, linear sum of orthogonal codes is truncated and pulse width modulated. The system has the capability of multi-code CDMA system of supporting multimedia traffics without difficulty of handling high rate traffic in VSG-CDMA. In addition, the system has the feature of constant envelope transmission. Hence, the proposed system is appropriate for low cost high data rate applications. We have investigated the bit error performance of the proposed system. Through analysis and simulations, it was shown that BER performance degrades as the number of channels increases. It was also observed that performance of multi-code PW-CDMA with binary level clipping is better than that of multi-code PW-CDMA without level clipping.

If you have supplementary material, e.g., executable files, video clips, or audio recordings, on your server, simply send the volume editors a short description of the supplementary material and inform them of the URL at which it can be found. We will add the description of the supplementary material to the online version of LNCS and create a link to your server. Alternatively, if this supplementary material is not to be updated at any stage, then it can be sent directly to the volume editors, together with all the other files.

References

1. I, C-L., Sabnani, K. K.: Variable spreading gain CDMA with adaptive control for true packet switching wireless network, Proc. ICC'95. Seattle (1995)
2. I, C-L., Sabnani, K. K.: Variable spreading gain CDMA with adaptive control for integrated traffic in wireless network, Proc. VTC'95. Chicago (1995)
3. Adachi, F., Sawahashi, M., Suda, H.: Wideband DS-CDMA for next generation mobile communications system, IEEE Commun. Mag., vol. 36, (1998) 56–69
4. I, C-L., Gitlin, R. D.: Multi-code CDMA wireless personal communications networks, Proc. ICC'95. Seattle (1995)
5. I, C-L., Pollini, G. P., Ozarow, L., Gitlin, R. D.: Performance of multi-code CDMA wireless personal communications networks, Proc. VTC'95. Chicago (1995)
6. Sawada, M., Katayama, M., Ogawa, A.: Effect of nonlinear amplifiers of transmitters in the CDMA system using offset-QPSK, IEICE Trans. Commun., vol. E76-B, no. 7, (1993) 741–744
7. Kasyap, K., Wada, T., Katayama, M., Yamazato, T., Ogawa, A.: The performance of CDMA system using π/2-shift BPSK with nonlinearity of HPA, Proc. PIMRC'96 (1996)
8. Wada, T., Yamazato, T., Katayama, M., Ogawa, A.: A constant amplitude coding for orthogonal multi-code CDMA systems, IEICE Trans. Fund., vol. E80-A, (1997) 2477–2484
9. Kim, D. I., Bhargava, V. K.: Performance of multidimensional multicode DS-CDMA using code diversity and error detection, IEEE Trans. Commun., vol. 49, no. 5, (2001) 875–887
10. Ryu, S. M., Kim, J. W., Moon, J. S., Kim, H. S.: Performance comparison of PW/CDMA and DS/CDMA, Proc. of JCCI, in Korean (2001)
11. Kim, S. P., Oh, J. K., Kim, M. J., Ahn, H. S., Ryu, S. M.: On the selection of code set for binary multi-code CDMA system, Proc. ICOIN-16 (2002)

On the Selection of Code Set for Binary Multi-code CDMA System

Sung Pil Kim [1], Jong Kap Oh [1], Myoung Jin Kim [1], Ho Seong Ahn [2], and Seung Moon Ryu [2]

[1] Hankuk University of Foreign Studies, San 89 Wangsan, Mohyun, Yongin, Kyonggi-do, Korea
mjkim@hufs.ac.kr

[2] CASUH Corp., C-528, Sigma II Officetel, Gumi-dong, Pundang-gu, Songnam, Kyonggi-do, Korea
{hsahn, smryu}@casuh.com

Abstract. Multi-code CDMA system provides high speed and variable rate services with different quality requirements. Since envelope fluctuation of the multi-code signal is large, the system requires highly linear power amplifier. In this paper, we consider a binary level multi-code pulse width (PW-) CDMA system. In the system, linear sum of orthogonal codes is truncated and pulse width modulated. Since the transmitting signal has constant envelope, and the receiver structure is very simple, the system is appropriate for low cost high data rate applications. However, the performance of the system is heavily affected by the selection of code set. In this paper, we investigate the bit error performance of the system according to the combination of orthogonal codes, and we propose algorithms for selecting code set.

1 Introduction

As a demand for wireless multimedia communication services is growing, the transmission schemes providing high speed and variable rate with different quality of service become necessary. Possible candidates of such systems are the variable spreading gain CDMA(VSG-CDMA) system[1], [2], [3] and the multi-code CDMA system[4], [5]. The VSG-CDMA can adjust the spreading gain for variable rate users depending on the input traffics and control the signal power to meet the quality of services(QoS). When a user wants to transmit data at a very high bit rate in VSG-CDMA, the spreading gain may become too small to reject the interference among users.

In multi-code CDMA the transmitting signal is a linear sum of multiple orthogonal codes. The multi-code CDMA system changes the number of code channels based on the data rate of the source, hence a constant processing gain is maintained without the problem of very small processing gain of high rate users. However, the transmitting signal of the multi-code CDMA system has a large amplitude fluctuation as a result of

I. Chong (Ed.): ICOIN 2002, LNCS 2344, pp. 264–275, 2002.
© Springer-Verlag Berlin Heidelberg 2002

linear sum of multi-code signals, and a highly linear power amplifier may be required for transmission. In mobile communications high power efficiency of the amplifier may be important especially on the up-link. In order to obtain high power efficiency the power amplifier with nonlinear characteristic is used [6], [7]. If the multi-code signal is amplified with power efficient amplifier, the nonlinear distortion becomes very large due to large dynamic range of the signal envelope, hence the bit error rate (BER) performance significantly degrades.

To overcome envelope fluctuation problem of multi-code CDMA a number of techniques have been proposed. Precoding scheme [8] realizes the constant amplitude by inserting precoded bits into the information bit sequence before multi-code signaling. However, the precoding scheme causes a loss in the information rate. In multi-dimensional multi-code shceme [9], the signaling is performed both in code and time spaces to reduce the information rate loss. However, as the number of channels increases the BER performance degrades because the signal energy carried on precoding channels with no information is wasted.

Pulse width CDMA(PW-CDMA)[10] is a multi-code transmission technique that performs pulse width modulation on the multi-level signal synthesized from multiple orthogonal codes. With polar signaling the signal amplitude of PW-CDMA is maintained to be constant, and the modulation and demodulation circuits become simple. However, as the number of channels increases the bandwidth increases due to the reduction of unit pulse width, and the BER performance degrades because non-optimal filter is used at the receiver. To compensate for an increase in bandwidth, pulse width modulation with polarity alternation or level clipping on the multi-code signal has been proposed[10].

In this paper, we consider an orthogonal multi-code CDMA system with binary level clipping. It is assumed that odd number of channels is used. Then the multi-code signal which is the sum of orthogonal codes has $M+1$ levels, $\{-M, -M+2, \cdots, -1, 1, \cdots, M\}$, where M is the number of channels. With binary level clipping, the truncated signal has constant amplitude. Due to level clipping, the BER performance degrades compared with conventional multi-code CDMA. In [11] BER performance of the system is investigated. The level clipping block in the system introduces nonlinear truncation noise. The amount of the truncation noise is dependent on the spreading codes and the information bits. In conventional orthogonal multi-code system, Walsh code is used. We may want to use other classes of orthogonal code. Possible codes include orthogonal m-sequence and orthogonal Gold code. Given a class of orthogonal code, there are number of choices of selecting M codes from the universal code set. As observed from the simulation results, choice of code set heavily affects the system performance. In this paper, we compare three classes of orthogonal code: Walsh code, orthogonal m-sequence, and orthogonal Gold code. We then propose an algorithm for finding 'good' combination of codes in the given code family.

The paper is organized as follows. Section 2 describes the binary multi-code CDMA system. In section 3 we formulate the problem and present the code selection algorithms. Section 4 presents the numerical examples and simulation results. A conclusion is given in section 5.

2   The Multi-code CDMA System with Level Clipping and PW Modulation

Fig. 1 shows the baseband system model of the orthogonal multi-code CDMA system with level clipping and pulse width modulation. Bit sequences $b_1, \cdots, b_M$ with $b_m \in \{+1, -1\}$ are serial to parallel converted information bits. Information bit on each channel is spread by an orthogonal code with length of N chips.

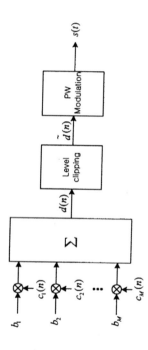

Fig. 1. Transmitter model

During bit interval T_b data bit on each channel is multiplied by the orthogonal code with N chip length. We assume that $T_b = N T_c$ where T_c is the chip time. In the analysis given below, we consider one bit interval, i.e. $0 \leq t \leq T_b$. The sum of spread sequences produces the multi-level signal $d(n)$, which can be represented by

$$d(n) = \sum_{m=1}^{M} b_m \, c_m(n) \quad 0 \leq n \leq N - 1 \tag{1}$$

where $b_m \in \{+1, -1\}$ is the information bit of the mth channel and $c_m(n)$ is the nth chip data of the spreading code for the mth channel. The spreading codes are assumed to be orthogonal:

$$\sum_{n=0}^{N-1} c_m(n) \, c_k(n) = N \delta_{m,k} \tag{2}$$

The number of channels, M, is assumed to be odd, which results in $M+1$ levels of multi-code signal, $d(n)$, with values $\{-M, -M+2, \cdots, -1, 1, \cdots, M-2, M\}$. Pulse width modulation is performed chip by chip on the multi-level signal. Pulse width of the signal during each chip time is proportional to the value of $d(n)$. If the number of channels is large, the minimum pulse width is very small, and the bandwidth of the system becomes very large. By inserting the level clipping block the amplitude of $d(n)$ is limited to a reduced number of levels, hence the bandwidth is reduced. It has been shown in [11] that level clipping in multi-code PW-CDMA system gives BER performance improvement in addition to bandwidth reduction. Fig. 2 shows an example of waveforms of the conventional multi-code CDMA, PW-CDMA, and binary level clipped multi-code CDMA when the number of channels is 5.

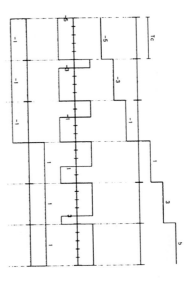

Fig. 2. Signal waveforms of multi-code CDMA, PW-CDMA, Binary level clipped PW-CDMA

In this paper, we consider binary level clipping, which implies that the output of the level clipper is $\tilde{d}(n) = \pm 1$. So, the PW modulated signal of the nth chip is the antipodal rectangular pulse with pulse width T_c, i.e.

$$s_n(t) = \begin{cases} p_{T_c}(t) & \text{if } \tilde{d}(n) = 1 \\ -p_{T_c}(t) & \text{if } \tilde{d}(n) = -1 \end{cases} \tag{3}$$

and the PW modulated signal $s(t)$ may be represented by

$$s(t) = \sum_{n=0}^{N-1} s_n(t - nT_c). \tag{4}$$

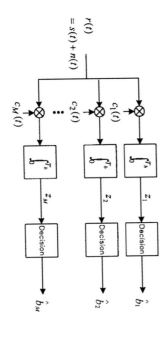

Fig. 3. Receiver model

Fig. 3 shows the baseband receiver model of the system. The received signal is despread by the orthogonal signal and integrated to generate the decision variables as in the conventional multi-code CDMA system. The decision variable for the first channel can be expressed by

$$z_1 = \int_0^{T_b} r(t)c_1(t)\,dt = s_o + n_o .$$ (5)

where, s_o and n_o are the signal component and the noise component, respectively.

3 Selection of Code Set

3.1 Problem Formulation

The system under consideration includes level clipping block, which introduces nonlinear truncation noise. The amount of the truncation noise is dependent on the spreading codes and the information bits. In conventional orthogonal multi-code system, Walsh code is used. We may want to use other classes of orthogonal code. Possible codes include orthogonal m-sequence and orthogonal Gold code which are obtained by padding zero to the usual m-sequence and Gold codes with cross-correlation -1, respectively. Given a class of orthogonal code, there are number of choices of selecting M codes from the universal code set. As observed from the simulation results, choice of code set heavily affects the system performance. In this paper, we compare three classes of orthogonal code: Walsh code, orthogonal m-sequence, and orthogonal Gold code. We then propose an algorithm for finding 'good' combination of codes in the given code family.

We will describe the transmitter and receiver operations in terms of vectors and matrices. Assume the processing gain is $N = 2^k$ for some integer k. Let

$$\underline{c}_m = [c_{m1} \ c_{m2} \ \cdots \ c_{mN}]$$ (6)

be the mth code vector composed of the orthogonal code of length N. And let a universal set $\mathbf{CS} = \{\underline{c}_1 \ \underline{c}_2 \ \cdots \ \underline{c}_L\}$ be a collection of all code vectors associated with a given type of spreading code. L is the number of code vectors in the universal code set, and M is assumed to be less than L. For Walsh code $L=N$, and for orthogonal m-sequence $L=N-1$. Now the problem can be stated as follows. For given number of channels, M, select a subset composed of M code vectors from the universal set to provide the best BER performance.

The transmitter and the receiver models as shown in Fig. 1 and Fig. 3 can be represented in discrete form as shown in Fig. 4 and Fig. 5, respectively. Note that with binary level clipping PW signaling module can be excluded in Fig. 4. Linear combination of code vectors produces multi-level signal $\underline{d}$ which can be written as

$$\underline{d} = [d_1 \ d_2 \cdots d_N]$$
$$= \underline{b}\,\mathbf{A}$$ (7)

where $\underline{b}$ is an $1 \times M$ vector composed of information bits on M channels. $\mathbf{A}$ is an $M \times N$ matrix whose row is an orthogonal code vector:

$$A = \begin{bmatrix} \underline{c}_1 \\ \underline{c}_2 \\ \vdots \\ \underline{c}_M \end{bmatrix} \tag{8}$$

with

$$\underline{c}_m^T \underline{c}_n = \begin{cases} N & \text{if } m = n \\ 0 & \text{if } m \neq n \end{cases}. \tag{9}$$

In the block diagram of transmitter model, $F[\cdot]$ indicates binary level clipping operation. Then the transmitted signal for $0 \leq t \leq T_b$ can be expressed as

$$\underline{s} = [s_1 \ s_2 \cdots s_N] = F[\underline{d}]. \tag{10}$$

Level clipping introduces nonlinear truncation noise. In order to analyze the truncation effect with relation code combination, we neglect the noise from the communication channel. The vector of decision variables cam be written as

$$\begin{aligned} \underline{z} = [z_1 \ z_2 \cdots z_M] \\ = \underline{s} A^T \\ = F[\underline{b} A] A^T \end{aligned} \tag{11}$$

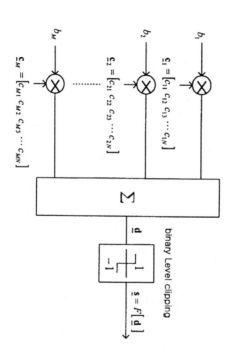

b_1

$\underline{c}_1 = [c_{11} \ c_{12} \ c_{13} \cdots c_{1N}]$

b_2

$\underline{c}_2 = [c_{21} \ c_{22} \ c_{23} \cdots c_{2N}]$

b_M

$\underline{c}_M = [c_{M1} \ c_{M2} \ c_{M3} \cdots c_{MN}]$

$\sum$

$\underline{d}$

binary Level clipping

$\underline{s} = F[\underline{d}]$

Fig. 4. Transmitter model

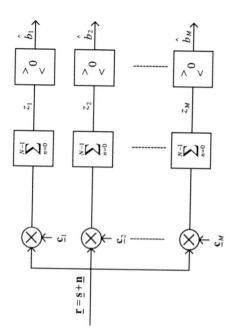

Fig. 5. Receiver model

3.2 Code Selection Algorithm

As mentioned earlier, level clipping introduces nonlinear truncation. The clipping effects appear as magnitude reduction or sign conversion of decision variables. The clipping magnitude of decision variable is reduced the output SNR decreases and the system becomes vulnerable to channel noise. If the sign of decision variable is changed to be opposite to the sign of information bit, there may be irreducible error, i.e. the BER performance curve exhibits irreducible floor.

For a given type of spreading code, suppose there are L codes in the universal code set. If there are M channels with $M < L$, then there ${}_L C_M$ are possible code combinations, where

$$_n C_r = \frac{n!}{r!(n-r)!}.$$ (12)

The truncation noise is a function of code set and information bits. If the system has M channels and information bits are binary, i.e. $b_m \in \{+1, -1\}$, $m = 1, \cdots, M$, there are 2^M possible information bit patterns input to the system. Then there are $2^M \times M$ corresponding decision variables. Our goal is to find the best code combination which leads to the best BER performance. The code selection algorithm traces the ${}_L C_M$ combinations of codes and investigates the statistical behavior of the decision variables to the input patterns. We suggest four algorithms. Algorithm A and B are based on the average distance of the decision variables. Algorithm C and D are based on the histogram of decision variables.

Before presenting the algorithms we define variables necessary to describe the algorithms. For the system with M channel input, there are 2^M possible M-tuple input vectors. Let k be the index for those input vectors, i.e. $k \in \{1, \cdots, 2^M\}$. For example,

$$\underline{b}(1) = [1, \cdots, 1], \quad \underline{b}(2) = [1, \cdots, 1, -1], \quad \underline{b}(2^M) = [-1, \cdots, -1]. \text{ Let}$$

$$\underline{z}(k) = [z_1(k), z_2(k), \cdots, z_M(k)] \tag{13}$$

be the vector of decision vector for the kth input pattern. Define

$$\underline{z}^{inv}(k) = \left[\frac{1}{z_1(k)}, \frac{1}{z_2(k)}, \cdots, \frac{1}{z_M(k)}\right]. \tag{14}$$

as a vector obtained by inversing each element in $\underline{z}(k)$. We then define variables regarding the distribution of decision variables $\{z_m(k)\}$ for all m and k, which are required to describe algorithm C and D. Let $\{v(n), n = 1, 2, \cdots\}$ be the values of $\{z_m(k)\}$ and $v(n)$'s are sorted such that $v(1) < v(2) < \cdots$. Define the number of decision variables whose value is $v(n)$, i.e.

$$N_n = \text{histogram}(v(n)) \qquad n = 1, 2, \cdots \tag{15}$$

The algorithms of code selection are stated as follows.

Common part

Traces the $_L C_M$ combinations of codes.

Discard the code set if $\text{sign}(z_m(k)) \neq \text{sign}(b_m(k))$ for any m, k.

Algorithm A

Choose the code set that minimizes

$$\eta = E_k \left\{ \max_m \frac{1}{z_m(k)} \right\} = E_k \left\{ \left\| \underline{z}^{inv}(k) \right\|_\infty \right\}. \tag{16}$$

Algorithm B

Choose the code set that minimizes

$$\eta = E_k \left\{ \left\| \underline{z}^{inv}(k) \right\|_q \right\} \qquad (q = 2). \tag{17}$$

Algorithm C

Choose the code set that minimizes

$$\eta = \frac{1}{v(1)}. \tag{18}$$

Algorithm D

Choose the code set that minimizes

$$\eta = \frac{N_1}{\{v(1)\}^q} + \frac{N_2}{\{v(2)\}^q} + \cdots \qquad (q = 3). \tag{19}$$

4 Numerical Examples and Discussion

We have analyzed the statistical behavior of decision variables to observe the effects of level clipping as a function of code combination. We have evaluated the code set selection algorithms proposed in section 3 by comparing the results with those from simulations. As spreading codes, we consider three types of codes: Walsh code, an orthogonal m-sequence, and an orthogonal Gold code.

<u>Example 1:</u> $N=32$, $M=7$

Though there are $M \times 2^M$ decision variables for all possible input vectors, the distribution characteristics are one of three types as shown in Fig. 6. For each distribution pattern there is a corresponding code set.

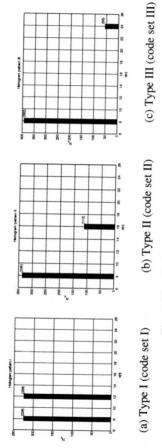

(a) Type I (code set I)   (b) Type II (code set II)   (c) Type III (code set III)

Fig. 6. Distribution patterns of decision variables

Table 1 shows the figure of merit given by each algorithm which indicates how well the code set will perform. The last column in the table indicates BER performance from simulation for given code set. The results indicate that algorithm B and D work well. Fig. 7 shows the simulation results for three code sets. The dotted line is the BER performance when a bad code set is chosen. The code selection algorithms exclude this code set. It is observed that there is irreducible BER floor for bad selection of codes.

Table 1. Comparison of code selection algorithms ($N=32$, $M=7$)

	A		B		C		D		simulation
	η	rank	η	rank	η	rank	η	rank	rank
I	.125	1	.281	1	.125	1	1.134	1	1
II	.125	1	.298	2	.125	1	1.367	2	2
III	.125	1	.311	3	.125	1	1.529	3	3

For three classes of spreading codes, Walsh code, orthogonal m-sequence, and orthogonal Gold code, we have investigated the statistics of the decision variables. We
have observed the same distribution as given in Fig. 6. This result indicates that these
three orthogonal codes provide the same BER performance if code combination is
properly selected. Fig. 8 shows the BER performance for three classes of codes with
the best code combination suggested by the algorithms. As expected, the same simulation results are obtained.

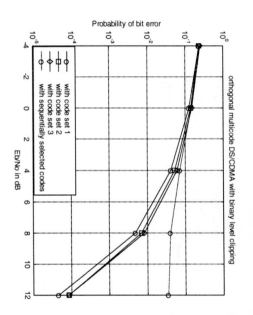

Fig. 7. BER performance for given code sets (N=32, M=7)

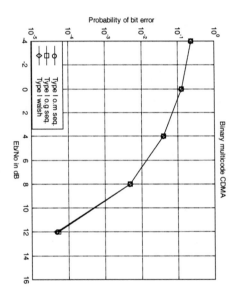

Fig. 8. Simulation results for three classes of codes

Example 2: *N=64, M=9*

We have carried out the same procedure as in example 1. In this example, we have observed six major distribution types in histogram of decision variables. Table 2 shows the figure of merit given by each algorithm which indicates how well the code set will perform. The results indicate that algorithm D works well in the sense that the algorithm expects almost the same result with that from the simulations. Fig. 9 shows the simulation results for three code sets.

Table 2. Comparison of code selection algorithms (*N=64, M=9*)

type	A		B		C		D		Simulation
	η	rank	η	rank	η	rank	η	rank	rank
I	.0647	1	.1791	1	.0714	1	1.0151	1	1
II	.0647	1	.1796	2	.0714	1	1.0278	2	2
III	.0651	3	.1847	4	.0833	3	1.1333	3	3
IV	.0764	4	.1836	4	.0833	3	1.1522	4	4
VI	.0875	5	.1855	5	.1	5	1.2002	5	5
VII	.1198	7	.2070	7	.125	6	1.9150	7	6
VIII	.1120	6	.2059	6	.125	6	1.8970	6	7

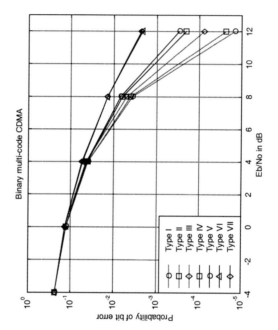

Fig. 9. BER performance for given code sets (*N=64, M=9*)

5 Conclusion

Multi-code PW-CDMA is a transmission technique that linear sum of orthogonal codes is level clipped and pulse width modulated. The system has the capability of multi-code CDMA system of supporting multimedia traffics without difficulty of handling high rate traffic in VSG-CDMA. In addition, the system has the feature of constant envelope transmission. Hence, the system is appropriate for low cost high data rate applications. We consider the multi-code PW-CDMA system with binary level clipping. The performance of the system is heavily affected by the selection of code set. In this paper, we investigate the bit error performance of the system according to the combination of orthogonal codes, and we propose algorithms for selecting code set. We also compare three classes of spreading code: Walsh code, orthogonal m-sequence, and orthogonal Gold code. From the analysis it was observed that these three orthogonal codes provide the same BER performance if code combination is properly selected.

References

1. I, C-L., Sabnani, K. K.: Variable spreading gain CDMA with adaptive control for true packet switching wireless network, Proc. ICC'95. Seattle (1995)
2. I, C-L., Sabnani, K. K.: Variable spreading gain CDMA with adaptive control for integrated traffic in wireless network, Proc. VTC'95. Chicago (1995)
3. Adachi, F., Sawahashi, M., Suda, H.: Wideband DS-CDMA for next generation mobile communications system, IEEE Commun. Mag., vol. 36. (1998) 56–69
4. I, C-L., Gitlin, R. D.: Multi-code CDMA wireless personal communications networks, Proc. ICC'95. Seattle (1995)
5. I, C-L., Pollini, G. P., Ozarow, L., Gitlin, R. D.: Performance of multi-code CDMA wireless personal communications networks, Proc. VTC'95. Chicago (1995)
6. Sawada, M., Katayama, M., Ogawa, A.: Effect of nonlinear amplifiers of transmitters in the CDMA system using offset-QPSK, IEICE Trans. Commun., vol. E76-B, no. 7. (1993) 741–744
7. Kasyap, K., Wada, T., Katayama, M., Yamazato, T., Ogawa, A.: The performance of CDMA system using π/2-shift BPSK with nonlinearity of HPA, Proc. PIMRC'96 (1996)
8. Wada, T., Yamazato, T., Katayama, M., Ogawa, A.: A constant amplitude coding for orthogonal multi-code CDMA systems, IEICE Trans. Fund., vol. E80-A. (1997) 2477–2484
9. Kim, D. I., Bhargava, V. K.: Performance of multidimensional multicode DS-CDMA using code diversity and error detection, IEEE Trans. Commun., vol. 49, no. 5. (2001) 875–887
10. Ryu, S. M., Kim, J. W., Moon, J. S., Kim, H. S.: Performance comparison of PW/CDMA and DS/CDMA, Proc. of JCCI, in Korean (2001)
11. Oh, J. K., Kim, S. P., Kim, M. J., Ahn, H. S., Ryu, S. M.: Orthogonal Multi-code CDMA Systems with Constant Amplitude Transmission Based on Level Clipping and Pulse Width Modulation, Proc. ICOIN-16 (2002)

Performance Evaluation and Improvement of an Ad Hoc Wireless Network

Takayuki Yamamoto[1], Masashi Sugano[2], Masayuki Murata[3], Takaaki Hatauchi[4], and Yohei Hosooka[4]

[1] Graduate School of Engineering Science, Osaka University, 1-3, Machikaneyama, Toyonaka, Osaka, 560-8531, Japan
tak-ymmt@ics.es.osaka-u.ac.jp

[2] Osaka Prefecture College of Health Sciences, 3-7-30, Habikino, Osaka, 583-8555, Japan
sugano@osaka-hsu.ac.jp

[3] Cybermedia Center, Osaka University, Japan 1-30, Machikaneyama, Toyonaka, Osaka 560-0043, Japan
murata@cmc.osaka-u.ac.jp

[4] Fuji Electric Co. Ltd., Japan

Abstract. An ad hoc wireless network is a self-organized network system. In this system, wireless terminals autonomously construct a multi-hop network which maintains its structural information by exchanging network information with some neighbor terminals. In such a network, the network topology changes frequently due to an unsettled wireless environment. Flexible Radio Network (FRN), one of the products applying ad hoc wireless network system, adopts an original protocol that provide a multiple routes management and a packet retransmission ability against packet transmission errors. This system has been in use in recent years. In this paper, we first evaluate its performance through simulations of data-link protocol and routing protocol to clarify its basic properties, and then we discover some problems on these protocols that degrade the performance. Furthermore, we propose some performance improvement techniques, simulate them, and then show how they improve the system performance.

1 Introduction

An ad hoc wireless network is a self-organized network built with wireless terminals that communicate with each other and exchange the network information. Those terminals have a capability to relay packets destined for another terminal, so they can construct a wide area multi-hop wireless network. The ad hoc network needs neither a wired backbone network nor base stations, and therefore network installation, expansion and removal can be performed easily and quickly. Such a wireless infrastructure covers a wide range of applications, e.g., distributed computing, disaster recovery, and military operation. Accordingly, many studies have been dedicated to analyze its characteristics and/or propose new routing methods (see, e.g.,[1,2,3,4,5,6]).

I. Chong (Ed.): ICOIN 2002, LNCS 2344, pp. 276–287, 2002.
© Springer-Verlag Berlin Heidelberg 2002

Flexible Radio Network (FRN) is one of commercially available products based on ad hoc wireless network system [7]. A large-scale network with stationary terminals can be installed in existing facilities easily by FRN. In addition, the network can be extended only by adding the radio terminal if needed. It is now used for collecting usage information of ski lifts, a sales account and monitoring information of vending machines, and for electric energy control in a factory. FRN adopts a proprietary protocol that can efficiently adapt to terminal failures or a change of network configuration. system parameters affect the performance such as throughput and packet loss rate. In the current system, these are decided by trial and error. In order to clarify the scope of this system, it needs to be evaluated in more detail.

In this paper, we propose three techniques for improving the performance of FRN. These techniques are based on some problems found through a process of simulated performance evaluation and operational experiences in the real environment. One of them is an adaptive method of system parameter setting. All data packets in FRN have a routing parameter that is called *maximum lifetime*. Nodes use this parameter to detect a long-living packet in a network and drop it efficiently. In an original system, the same value that was large enough for the network was selected for all data packets. We investigate an effect of calculating an adaptive maximum lifetime for each packet in Section 4.1. The other two of our suggestions are to decrease packet collision and to solve a problem called *packet duplication*. Preventing packet collision simply leads to better performance in any wireless network. In FRN, moreover, packet collision is one of causes that make packet duplication that leads to a higher traffic load than the actual one. Therefore, the network performance gains much more in FRN by preventing packet collision. We propose two techniques to prevent packet collisions, and at the same time, to prevent packet duplications. We will describe a packet duplication process in Section 4.2.

In this paper, we first describe the Flexible Radio Network system and evaluate its basic property. We also show the result of adaptive maximum lifetime setting and compare it with the basic performance. Next, we examine a packet duplication process in detail. We show that the network performance degrades rapidly as the number of duplicated packets increases. We suggest performance improvement techniques with consideration to packet duplication. Through simulations, we show those techniques can improve the network performance.

2 System Description of FRN

2.1 Network Configuration

In FRN, a wireless terminal is called a *node*. A node with which a certain node can communicate directly is called its *neighbor node*. Every node has the ability to select a route of packet and to relay it to one of the neighbor nodes. There are two kinds of nodes; a *host node* and a *relay node*. The host node generates and receives data packets, and the relay node simply relays the data packets, by which a multi-hop network is constructed. Every node maintains the network

Table 1. Configuration Table

	Dest. Node 0	Dest. Node 1	$\cdots$
Route 1	Neighbor Info.	Neighbor Info.	$\cdots$
Route 2	Neighbor Info.	Neighbor Info.	$\cdots$
$\cdots$		$\cdots$	$\cdots$

information in a *configuration table* that contains the routing information from the node itself to each destination node. The routing information consists of the list of neighbor nodes on the route to the destination node, and a hop count of the route. Every node exchanges a *configuration control packet* periodically, and updates its configuration table by the packets received from neighbor nodes.

2.2   Data-Link Protocol

A radio channel is divided into fixed-length time slots. In a wireless network, every neighbor node of a certain node can receive packets from the node even when it is not the source/destination of the packet. In FRN, this property is utilized for acknowledging the packet. See Fig. 1 as an example. Fig. 1(a) shows a case where packet transmission and acknowledgement at node A is successful. Node B receives the packet from node A and relays it to another node successfully. At the same time, this relayed packet is received by node A because node A is a neighbor node of node B. This acknowledgement is called *relay echo* (or simply *echo*). If the relay echo is successfully received by node A, it can recognize that the transmission to node B succeeded. The case where node A fails the first transmission is shown in Fig. 1(b). In this case, node A detects a failure because no echo from node B is received. Then, node A retransmits the packet after a pre-specified time. When a packet reaches its destination node, the destination node no longer transmit it and the previous node cannot get an echo. To delete the buffered packet in the previous node, the destination node creates an ACK packet exceptionally (Fig. 1(c)).

A maximum lifetime is defined in every data packet when it is generated in the source node. For every time slot, it is decreased by one. When the value reaches zero, the packet is discarded due to expiration of the lifetime. It is an important configuration parameter since it gives a chance to effectively remove the long-living packets from the network or a chance to try another route to the destination. It is necessary for a reliable network to set a large enough value for the longest route while too large value causes network congestion. It is necessary to decide a proper value of this parameter for the network.

2.3   Routing Protocol

In FRN, all nodes transmit network structural information from their neighbor nodes periodically. When a node receives the information packet, it updates their configuration table. Since the radio environment changes frequently, a routing protocol must select an appropriate route adaptively. Furthermore, if a node fails to transmit a packet, the same route is again chosen, and after several failures another route is selected. Every node maintains multiple routes for each destination node in the configuration table if available.

In the routing protocol, a shorter route has higher priority. If the shortest route is unavailable due to some reason such as obstacles on the route, radio noise or packet collision, the node looks up the configuration table again and selects the second shortest route. Each node sets a next hop node ID, a temporal destination, to a packet header after it decides the route. When neighbor nodes receive the packet, each of them checks the next hop field of the packet header and recognizes it is a temporal destination node or not. When it is, it looks up a new next hop of the packet and transmits. When it is not, on the other hand, they check whether it is a relayed echo (see Section 2.2).

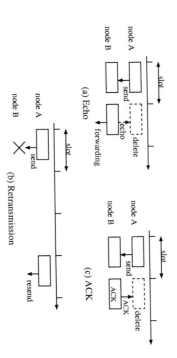

Fig. 1. Packet Transmission Timing

3 Basic Properties of FRN

3.1 Simulation Environment

In this section, we first investigate basic properties of FRN. We attend to the maximum lifetime because it greatly affects the performance. When the maximum lifetime is too long for the network, some packets stay unnecessarily long in the network and cause the degradation of performance. When it is too short, on the other hand, it may expire before relay nodes on the route try another route for the transmission and the system cannot fulfill its potential. We make some simulations where maximum lifetimes are different, and evaluate a relationship between the lifetime setting and system performance.

For simulations, ns-2 [8] is used to utilize its radio propagation model. Then, we implement the data-link and routing protocols of FRN on ns-2. We use a network model shown in Fig. 2. A circle represents a node. A line connecting two nodes means that they can communicate directly. In this model, packet losses

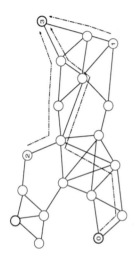

Fig. 2. Simulated Network Model

are assumed to occur only by collision of radio wave. The numbered nodes (nodes 0, 1, 2, 3) are host nodes that generate and receive data packets. Arrows from source nodes to the destination are the shortest routes that packets are expected to be passed through. In all simulations, nodes 0, 1 and 2 go on sending constant bit rate UDP packets to node 3. This network model is based on the application of FRN system that is collecting information from decentralized host nodes. The packet generation rate per one slot at each host node is assumed identical. The traffic load is defined as the number of packets generated per one slot in the whole network (i.e., the sum of the packet generation rate at three sender nodes). We simulate in the range of load from 0.01 to 0.1 where it is practical in an actual environment.

We use throughput and packet loss rate (PLR) to measure the performance. Throughput is an average number of successfully transmitted packets per one time slot. PLR is a ratio of the packet not reaching the destination.

3.2   Network Performance with Various Maximum Lifetime

The network performance is largely affected by the maximum lifetime. As described before, when it is large, the packets can be relayed to the destination node even after some retransmission trials. As the traffic load becomes higher, on the contrary, setting a smaller value is an efficient way to drop long-living packets and prevent congestion. In what follows, we will provide three results of simulations where we set the maximum lifetime value to 12, 64 and 128.

Figure 3(a) shows throughput values dependent on the traffic load. We can see that the throughput peaks at the smaller load as the maximum lifetime becomes longer. This is because the number of long-living packets increases and then congestion occurs in the network. We can achieve high throughput in the high load environment with a small lifetime value. It is difficult to clarify a throughput difference when the load is low. The performance can be compared in the next Fig. 3(b), the graph of PLR. When the load is low, the system with long lifetime shows better performance than one with short lifetime. Every node in FRN maintains multiple routes in its configuration table (see Section 2). In the long lifetime system, nodes can try these variable routes to relay packets to their destination if once a transmission fails. However, as the load of the system

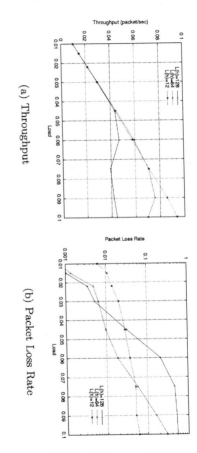

(a) Throughput

(b) Packet Loss Rate

Fig. 3. Performance Evaluation of the Original System

increases, the network performance degrades rapidly because of the network congestion caused by many long-living packets.

According to these results, it becomes clear to be important to select an adaptive value for the maximum lifetime. The best value is difficult to be determined because it depends on a scale, structure and current load of the network. We propose a method for calculating the adaptive lifetime with the route length.

4 Performance Improvement Techniques

4.1 Adaptive Maximum Lifetime

A maximum lifetime value of a packet has much effect on the network performance as described in Section 3. In an original FRN system, all packets had the same maximum lifetime that is large enough for the network scale. However, a necessary lifetime for each packet is different because a hop count against each route is various. Then it is not a good idea to set the same lifetime to all packets. Now, we suggest one performance improving method of setting an adaptive maximum lifetime to each packet.

In FRN, every node maintains a network configuration table that includes the shortest hop path to the destination. This has a close relation to a necessary packet lifetime. A host node, that initiates data packets, can calculate an adequate lifetime with this hop count.

According to the simulation results in Section 3.2, the best maximum lifetime may be between 12 and 64 in this network. We use, for example, $L(h) = 6h + 12$ (the maximum lifetime L is a function of the shortest hop length h) for an adaptive maximum lifetime calculation in this paper. This function is derived experimentally through simulations. In a real environment, many factors will affect the necessary lifetime, therefore the function may be derived for each environment.

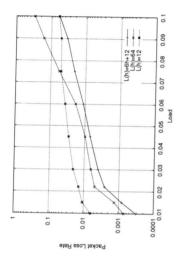

Fig. 4. PLR for the Adaptive Maximum Lifetime System

We show the performance example of adaptive maximum lifetime in Fig. 4. It is difficult to compare the results by their throughput because it shows little difference when the load is low, so we show only the graph of packet loss rate (PLR). The results of static lifetime values of 12 and 64 are shown for purpose of comparison. The simulation environment is the same as in Section 3.1. The function $L(h) = 6h + 12$ is applied to all packets at their source node to calculate their lifetime. Looking at Fig. 4, when the load is low, the modified system shows better performance of static lifetime 12. This is because the packets on longer routes can get longer maximum lifetime in the modified system. When the load is high, it does not show a rapid degradation that is shown in long lifetime system. This result means, in the modified system, a mechanism to drop long-living packets by maximum lifetime works well while it does not work in the static long maximum lifetime system. Therefore, setting an adaptive lifetime for each packet has a good ability to control the trade off between network congestion and reliability that are caused by long lifetime.

4.2   Improvement for Routing and Data-Link Protocol

Packet Duplication Problem. As mentioned in Section 2, FRN has a packet retransmission mechanism against transmission errors. While this mechanism is likely to contribute to packet reachability, it sometimes causes packet duplications. The duplicated packets are unnecessary because their original packet is not lost. The duplicated packets increase the network load more than an actually given one, and then, the network performance degrades greatly. A packet duplication process is illustrated in Fig. 5. Node A transmits a packet to node B successfully at slot 0. Node A holds the copied packet in its buffer for the purpose of retransmission in the near future. Node B relays the packet to node C at slot 1. At the same time, it is expected that this relayed packet be received by node A as a relayed echo. However, it sometimes happens that the echo is lost because of an obstacle (e.g., a person in the case of the skiing ground) or a collision with another packet. Then, node A cannot receive the echo successfully.

In such a case, the copied packet in the buffer of node A should not be removed at the end of slot 1. In other words, the same two packets exist in node A and node C. Node A retransmits the packet later. It may be retransmitted through different routes to the destination because of multiple routes information in the configuration table and affects over a large area.

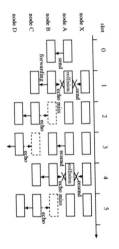

Fig. 5. Packet Duplicating Process

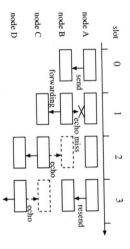

Fig. 6. Repeated Echo Loss

Collision Preventing Methods. Network performance degrades rapidly as the number of packets in the network increases. This is clear when we see Fig. 3(b) again that shows a rapid performance degradation of long maximum lifetime system. This degradation is caused by a wireless characteristic that the radio wave has the influence on all nodes within the reachable range. As the number of long-living packets increases, a probability of packet collision increases too. Moreover, the duplicated packet is unnecessary if its original packet reaches the destination properly. Therefore, decreasing a probability of packet duplication leads to much improvement on the network performance. In this subsection, we suggest two techniques that can appropriately control the number of duplicated packets. As described in previous subsection, packet duplication is caused by a loss of relay echo, which can be reduced by the following two methods: (1) to introduce the random delay before packet retransmissions (RANDOM_DELAY), and (2) to drop the packets previously which cannot reach the destination host within rest of their lifetime (EARLY_DROP).

The change RANDOM_DELAY above is an improvement of the packet retransmission method described in Section 2.2. As shown in Fig. 1(b), when a node cannot receive an echo within a specified time, it selects another route for retransmission. This waiting duration was a constant length in the original system, and therefore once the packet experiences a collision, it tends to repeat collisions until transmission through another route succeeds. By introducing the random waiting time before retransmission, the possibility of collisions would be reduced. We explain why it can decrease the number of duplicated packets. See Fig. 6. We suppose that the retransmission delay is fixed, as an original system, at three slots. Node A expects to receive an echo from node B at slot 1. At the same slot node X sends a packet and it collides with the echo. Therefore, node A cannot receive the echo and retransmits the packet at slot 3. At slot 4, the echo from node B collides again with the packet from node X because of the fixed waiting time. Collisions repeat until the transmission through another route succeeds. The change RANDOM_DELAY can inhibit the repeated losses of relay echoes, and it can be expected to reduce the number of packet duplications. In the simulations, nodes select a random time from 3 to 5, which was fixed at 3 previously. To make it 2 is bad because it is too short for an echo-based system.

The change EARLY_DROP is a technique for decreasing the number of unnecessary packets within the network. In FRN, the lifetime of the packet is reduced by one for every time slot. When no collisions occur, a packet can be passed through one hop by one time slot. That is, it should be ideally the maximum allowable hop count of the packet. With the configuration table, the hop counts to the destinations are maintained at each node. Thus, if the residual lifetime of a packet becomes less than the hop count of the shortest path to the destination, that packet of no use should be discarded. This technique prevents unnecessary packet transmissions and decrease the number of packet collisions that lead to packet duplications.

Simulation Results of Collision Preventing Methods. We investigate the effect of these proposed improvement techniques through simulations. The simulation environment and network model is the same as that used in Sections 3 and 4.1.

The results of RANDOM_DELAY, determining a packet retransmission delay randomly, are shown in Figs. 7(a) and 7(b). The label "normal" is a result of an original system with no modification, "random" is one of RANDOM_DELAY modified system, and "L(h)" means the value of maximum lifetime. See Fig. 7(a), the packet loss rate. When the maximum lifetime of packet is long, such as 64, this method is efficient because long lifetime packets can try many routes after a quick recovery from packet collisions. It shows the same or a little worse performance when the maximum lifetime is short. In this simulation, nodes select the waiting time before retransmission occurs between 3 and 5, whose average is longer than fixed 3 in an original system. This sometimes causes more expiration of packets when the lifetime is short, then this technique has less ability in short lifetime system. Fig. 7(b) shows the number of duplication occurrences per second. When the maximum lifetime is long and the load becomes higher, this technique can decrease the number of duplications and improve the performance.

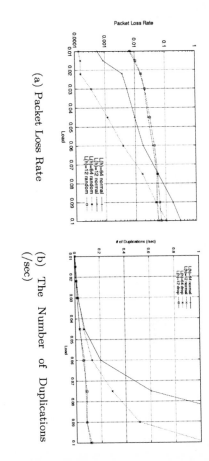

(a) Packet Loss Rate

(b) The Number of Duplications (/sec)

Fig. 7. The RANDOM_DELAY Modified System

Figures 8(a) and 8(b) are the results of EARLY_DROP, to remove packets that lack the lifetime to reach their destinations. The result of modified system is labeled "drop". On the contrary as shown the result of RANDOM_DELAY, it shows better performance improvement when the maximum lifetime is short. This technique has an effect on such packets that will be dropped forcibly because of expiration of lifetime. In a long lifetime system, there is less probability of packet time-out, and this modification shows little improvement. This technique has a little effect on packet duplications. However, in the short lifetime system, the number of duplication is little from the beginning. Therefore, this technique cannot show a great improvement on packet duplication.

Last, we show a result with all modifications. Three techniques we have proposed are independent of one another. It is expected that the network performance improves more by applying all suggestions at the same time than by applying each of them separately. Figure 9 shows the PLR transitions. We can see the network achieves the best performance with all modifications. Through these simulations, our suggestions are proved to have an ability to improve the performance of the FRN network system and show much better performance by applying them at the same time.

5   Conclusion

In this paper, we have investigated performance characteristics of Flexible Radio Network (FRN), a product of commercial use. Because of its application, collecting information from many distributed terminals, FRN has an original protocol to construct a reliable network. However, it causes high packet loss rate because long-living packets disturb other packets and sometimes the retransmission mechanism makes duplicated packets. We have introduced a dynamic adaptive maximum lifetime setting technique to control the lifetime of a packet. Next, we have described a packet duplication process in FRN and suggested

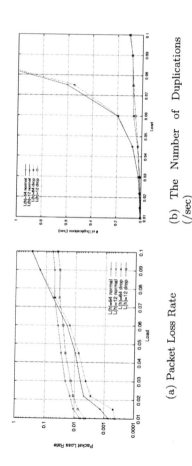

(a) Packet Loss Rate

(b) The Number of Duplications (/sec)

Fig. 8. The EARLY_DROP Modified System

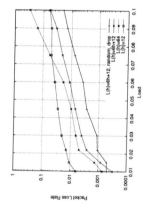

Fig. 9. PLR for the ALL Modified System

two techniques with consideration to packet duplication. We have shown those three techniques can improve the network performance through simulation experiments. In addition, we have also shown that they can be applied to the system at the same time to achieve the best performance.

As a future topic, we should develop a method to determine the best expression for adaptive maximum lifetime that considers the network load or current packet loss rate. We should also investigate end-to-end performance by using, e.g., TCP, as an upper layer protocol.

Acknowledgement. This work was supported in part by Research for the Future Program of Japan Society for the Promotion of Science under the Project "Integrated Network Architecture for Advanced Multimedia Application Systems" (JSPS-RFTF97R16301). And it was also partly supported by Special Coordination Funds for promoting Science and Technology of the Ministry of Education, Culture, Sports, Science and Technology of Japan.

References

1. P. Johansson, T. Larsson, N. Hedman, B. Mielczarek, and M. Degermark, "Scenario-based Performance Analysis of Routing Protocols for Mobile Ad-hoc Networks," in *Proc. ACM/IEEE MOBICOM'99*, Seattle, Aug. 1999, pp. 195–206.

2. Koichi Takasugi, Yasuo Suzuki, and Shuji Kubota, "Multicast Routing Protocol for Avoiding Congestion in Ad Hoc Wireless Netowork," *IEICE Trans. B (in Japanese)*, vol. J83–B, no. 7, pp. 991–998, July 2000.

3. A. Iwata, C.-C. Chiang, G. Pei, M. Gerla, and T. Chen, "Scalable Routing Strategies for Ad Hoc Wireless Networks," *IEEE J. Select. Areas Commun.*, pp. 1369–1379, Aug. 1999.

4. Z. J. Haas and M. R. Pearlman, "Determining the Optimal Conguration for the Zone Routing Protocol," *IEEE J. Select. Areas Commun.*, pp. 219–230, Aug. 1999.

5. N. Nikaein, H. Labiod, and C. Bonnet, "DDR – Distributed Dynamic Routing Algorithm for Mobile Ad Hoc Networks," in *Proc. of the MobiHOC 2000*, 2000.

6. G. Holland and N. H. Vaidya, "Analysis of TCP Performance over Mobile Ad Hoc Networks," in *Proc. of ACM/IEEE MOBICOM'99*, Seattle, 1999, pp. 219–230.

7. "Flexible Radio Network, Fuji Electric Co. Ltd," available at http://www.fujielectric.co.jp/denki/p26/ecop_contents2.html.

8. "The Network Simulator - ns-2," available at http://www.isi.edu/nsnam/ns/.

Reassignment of Mediation Function between Hosts on Mobile Ad Hoc Network

Yoshinari Suzuki[1], Susumu Ishihara[2], and Tadanori Mizuno[3]

[1] Graduate School of Information, Shizuoka University,
Hamamatsu,432-8011 Japan
Phone:+81-53-478-1460, FAX:+81-53-478-1597
suzu@mizulab.net
http://www.mizulab.net/

[2] Faculty of Engineering, Shizuoka University,
Hamamatsu,432-8011 Japan
ishihara@ishilab.net
http://www.ishilab.net/

[3] Faculty of Information, Shizuoka University,
Hamamatsu,432-8011 Japan
mizuno@mizulab.net
http://www.mizulab.net/

Abstract. In cooperative work like online games, a mediation host mediates differences of delay and the consistency of the data shared among the participating hosts (total orderings). When such applications are used on mobile ad hoc networks, one of the network hosts must take this role. If the mediation host is near the center of the network, the communication delay and traffic load are smaller, and the efficiency of application programs is better. However, the hosts move around, so the network topology changes, so meaning that the mediation host may be located far from the center. We previously proposed a method for reassigning the mediation function as the topology changes so that a host near the center plays the mediation role.
We have now evaluated the performance of our method from the view points of communication delay and the number of host reassignments on ad hoc networks by simulations, using DSR dynamic source routing as the routing protocol.

1 Introduction

Network applications for cooperative work like online games in which two or more users cooperate through the Internet are becoming popular. Moreover, because of the grpwing use of personal digital assistants, the opportunity to use cooperative work applications on personal digital assistants has also been increasing. Furthermore, the development of narrow range radio-communications technologies, such as Bluetooth [8], HomeRF [9], and wireless LAN, has enabled the realization of wireless ad hoc network environments. It is likely that narrow range radio functions will be provided in personal digital assistants, which will

I. Chong (Ed.): ICOIN 2002, LNCS 2344, pp. 288–299, 2002.
© Springer-Verlag Berlin Heidelberg 2002

enable network applications for cooperative works to be used on mobile ad hoc networks in the future.

In cooperative applications, a mediation host mediates differences of delay and the consistency of the data shared among the participating hosts (total orderings).

For example, in the impartial communication environments for game members (ICEGEM) [2], a server host arbitrates the differences in delay between the client hosts and server. In some multicast communication systems, a sequence server host guarantees the order consistency of messages [3]. In cooperative work on wired networks, a fixed server can be used as the mediation host in many cases. On mobile ad hoc networks, however, there may not be a fixed server. Therefore, one of the network hosts must act as the mediation host. If the mediation host is near the center of the network, the communication delay and traffic load will be smaller, and the efficiency of the application programs will be better.

However, since hosts on an ad hoc network move around, the mediation host may be far from the center of the network. As the mediation host moves away from the center, the communication delay and traffic load will increase, and the efficiency of the application programs will degrade. We previously proposed REMARK a method for reassigning the mediation function when the topology changes to a host close to the center of the network [1].

We have now evaluated, the performance of REMARK by simulations using DSR (dynamic source routing) as the routing protocol. The evaluation was performed from the viewpoints of communication delay and the number of host reassignments. It showed that REMARK decreases delays in communication for cooperative work applications.

2 REMARK-Reassignment of Mediation Function on Ad hoc Network

2.1 Overview

In this section, we give on overview of our method for reassign the mediation function on mobile and hoc networks. In general, a mediation host is required for applications that support cooperative work. This host arbitrates the differences in delays, causal relationships (total ordering), etc. in the group constituting the network. Mediation of differences in delay and total ordering are needed to implement cooperative applications, such as online games, and virtual space-sharing systems (shared electronic white boards, etc.) [4].

Mediating the differences in delay reduces the unfairness caused when some members of the network can reach the server host in a few seconds while other take longer. For example, in online games or auctions, delay differences impose a handicap on the members who suffer a long delay [2].

Mediating causal relationships (total ordering) ensures that objects are displayed on the members consistently by guaranteeing that the packet arrive in

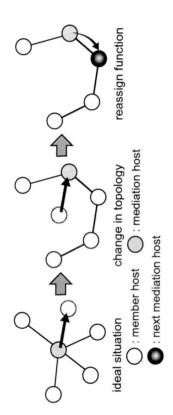

ideal situation   change in topology   reassign function

○ : member host   ◉ : mediation host

◉ : next mediation host

Fig. 1. Overview of REMARK

the same order at each screen [3]. For example, in a shared white board, mediation of the causal relationships is needed to ensure that the same objects are displayed at all hosts.

To enable arbitration of differences in delay and total ordering, the data from all hosts in the group (member hosts) must be delivered to the mediation host. Moreover, the mediation host must send the results of mediation to the member hosts.

In ad hoc network environments, one of the hosts on the network serves as the mediation host. The closer the mediation host to the center of the network, the smaller the maximum delay between the mediation host and member hosts and the smaller the traffic volume, resulting in quicker transmission of messages. The shorter the interval of message transmission between the mediation host and member hosts, the more efficient the application programs,e.g. the system will react quicker to user input. Thus, if a host near the center of the network serves as the mediation host, the applications will work more efficiently.

However, in a mobile ad hoc network environment, the topology changes frequently due to the movement of the hosts, so the host serving as the mediation host may not always be at the center of the network. If the mediation host moves far from the center, the communication delay and traffic load will increase greatly, and the efficiency of the applications will degrade.

Using our method, we can assign the mediation function to a host near the center of a network (figure 1). A decrease in application efficiency is thus prevented. Furthermore, our method needs no special packets for the reassignment, so there is no increase of in the number of the control messages.

2.2   Basic Concepts

REMARK is based on the following concepts.

- The mediation function is performed on a suitable mobile host given the topology of the network.
- The mediation host selects the next mediation host based on the topology of the network.

– Messages in the application are used to perform the reassignment.
– Information carried by the routing protocol is used by the mediation host to select the next mediation host.

The features of presumed applications are as follows.

– Messages are periodically exchanged between the mediation host and members.
– Message transmissions occurs frequently.

We assumed that the flow of messages in presumed applications as follows.

1. The mediation host sends a message to the members.
2. Each member returns a message to the mediation host.
3. After receiving all the messages from the members, the mediation host sends a new message to the members.

2.3   Control Messages

Two kinds of control messages are used in REMARK: *downlink*, and *uplink*. Downlink messages are sent from the mediation host (the host executing the mediation function) to the member hosts. Each one includes the address of the mediation host. Uplink messages are sent from the member hosts to the mediation host in reply to downlink messages. Each one includes topology-related information about the member host.

These messages are actually the messages periodically transmitted to send application data. Thus, no additional traffic is generated by the proposed method.

2.4   Reassignment of Mediation Function

Reassignment of the mediation function to a different host is triggered by a change in the topology of the network. The new mediation host is selected by the current one.

1. The initial mediation host in a group is the one with the minimum ID (IP address, etc.).
2. The mediation host constructs a complete graph of the member hosts by using the route information in their uplink messages.
3. The mediation host converts the complete graph into a spanning tree by deleting the loops.
4. The mediation host uses the tree to select a suitable member host to be the next mediation host, described in detail below.
5. The mediation host notifies the member hosts of the selection in a downlink message.
6. After receiving the notification message, the member hosts recognize the next mediation host.

7. After receiving the notification message, the next mediation host starts mediating.

8. The member hosts send an uplink message to the next mediation host (the completion of reassignment).

9. If an uplink message reaches the next mediation host before it receives he notification message, it starts mediating when the uplink message arrives.

The next mediation host is selected as follows. As described above, the mediation host constructs a spanning tree using topology-related information contained in the uplink messages. We assume that the uplink messages include route information. For example, if DSR is used as the routing protocol, the packets include the route from the sender to the destination as Source Route. The spanning tree represents the topology of the network at that moment. The leaves (nodes with only one link) of the spanning tree are nodes at the edge of the network. Therefore, if the leaves are deleted one at a time, the last node remaining is close to the center of the network. This node is selected as the next mediation host.

As described above, member hosts are notified of the next mediation host in a downlink message. The member hosts send an uplink message to the address of the next mediation host. Consequently, reassignment occures only when the next mediation host is different from the current one.

3   Evaluation

Use of our method should reduce communication delays between the mediation host and member hosts. We tested this assumption by measuring the maximum round-trip delay between the mediation host and the member hosts and by counting the number of reassignments of the mediation function. We used the ns-2 network simulator [10].

3.1   Selection of a Routing Protocol

Typical routing protocols used in mobile ad hoc networks are DSDV[5], DSR[6], and AODV[7]. With DSR, when the sender node sends a data packet to the destination node, information about the entire route, from the source to the destination, is included in the packet header (hence the name "source routing"). The mediation host can thus understand topology of the ad hoc network based on the routes that the messages (packets) have taken. The mediation function is reassigned based on this topology. Thus, by using DSR as the routing protocol, the proposed method can be executed at low cost. Accordingly, we used it in our simulations.

Using DSR, however, does entail problems. In DSR, the sender node, S, initiates a Route Discovery to find a suitable route to the destination node, D. Then, node S floods Route Request message (RREQ), and node D replies with a Route Reply message (RREP). Accordingly,

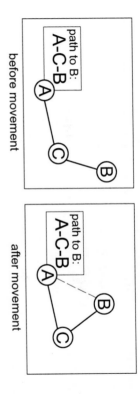

before movement   after movement

Fig. 2. Case a): Identified route is not the shortest

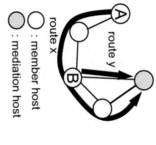

route y

route x

○ : member host

● : mediation host

Fig. 3. Case b): Different routes are selected

a) The discovered route is not necessarily the shortest one.

b) The route detected by one host may be different from that detected by another.

An example of case a) is as follows. After node A finishes Route Discovery for sending packets to node B, which is out of transmission range of node A, if node B moves into the transmission range of node A, the route discovered by node A will not be the shortest (figure 2).

An example of case b) is as following. If there are some available routes from node A to the mediation host, node A selects route X. Node B, which is on route X, selects a different route, route Y. That is, nodes A and B understand the network topology differently figure 3).

As a result, two problems may occur.

1. The spanning tree generated by the mediation host may not be the minimum spanning tree.
2. The next mediation host may not be most suitable one.

As a result, the mediation host may not always be most suitable one, and the communication delay between the mediation host and the member hosts may increase.

3.2   Simulation Model

As mentioned, we used ns-2 [10] for our simulations. We defined 20 mobile hosts in an L [m] by L [m] square domain (L was increased by 50 from 100 to 1000). Each host moved according to a random waypoint model. We simulated two cases. In one, the maximum speed of the mobile hosts (V_{max}) was 2.0 [m/s] (i.e. walking). In the other, V_{max} was 16.0 [m/s] (i.e. driving). In both cases, pause time p was 1.0 [sec]. For each value of L, we ran 10 movement scenarios.

The transport layer protocol was UDP, the size of a packet was 64 bytes, the MAC layer protocol was IEEE 802.11 (transmitting range of 250 [m]), and the execution time was 500 [sec] in simulation time.

Random Waypoint Model. In the random waypoint model used in the simulation, each mobile host moved as follows.

1. A host chooses a target point inside the domain at random.
2. It moves in a straight line at speed v to a target point; v is chosen from $[0, V_{max}]$ and remains constant.
3. After arriving at the target point, the host stops for p [sec].
4. Return to 1.

3.3   Simulation Condition and Evaluation Criteria

In the model used, connectivity between hosts is not necessarily guaranteed. Thus, to continue transmission of a message if a connectivity was refused, a timeout parameter was used. The mediation host waited 0.3 [sec] in simulation time for uplink messages. When a timeout occurred, the mediation host resent a downlink message to all member hosts from whom an uplink massage was not arrived. If a timeout occurred again after re-sending the downlink message, the uplink messages from all member hosts were cancelled, and the mediation host sent a new downlink message If the uplink messages were not cancelled, the mediation host sent a new downlink message when recieved uplink messages from all members.

Under these conditions, we tested three cases.

a) The mediation function was reassigend by REMARK.
b) The next mediation function was reassigned based on the locations of the nodes (the location-based selection).
c) The mediation function was not reassigend.

For each case, we measured the maximum round-trip delay (the maximum of the sum of the delay of the downlink message and the delay of the corresponding uplink messages) and counted the number of host reassignments.

In location-based selection, a host near the center of the minimum rectangle covering all host when all uplink messages arrive at the meditation host is selected as the next mediation host. If the mobile hosts were located uniformly in the movement domain, the next mediation host selected by this method would be located at a ideal position (near the center of the network). On the other hand, if the mobile shots were not located uniformly, the next mediation host would be located at a unideal position (far from the center of the network). However, this may not occur if there are enough mobile hosts in the group.

3.4 Simulation Results

The number of reassignment of mediation function ($V_{max} = 2.0[m/s]$). As shown in figure 4, L was increased, the number of reassignments increased. This is because as the movement domain became larger, multiple-hop routes became more common, and the topology of the network changed more often. However, when L was larger than 900, the number of reassignments decreased. This is because the number of dropped messages increased because of the time-out, which was set at 0.3 [sec] (3.3). When L was smaller than 250, there were few reassignments because the movement domain was small enough for all the nodes to be within transmission range of each other. There was thus little need to reassign the mediation function.

In contrast, when location-based selection was used, the number of assignments was large when L was small, and the number decreased as L was increased. This is because the actual topology and the connectivity between mobile hosts is not considered in location-based selection, the next mediation host is selected based only on the location of each mobile host. Thus, the smaller the L and the denser the mobile hosts, the more reassignments there are.

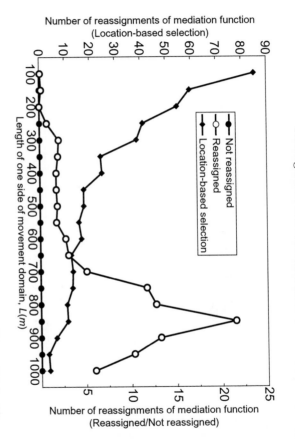

Fig. 4. Number of reassignments of mediation function (V_{max}=2.0 [m/s])

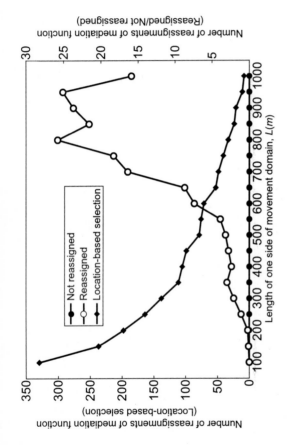

Fig. 5. Number of reassignment of mediation function (V_{max}=16.0 [m/s])

Number of reassignment of mediation function ($V_{max} = 16.0[m/s]$). As shown in Figure 5, the number of reassignments of the mediation function when V_{max} was 16.0 [m/s] was almost the same as when V_{max} was 2.0 [m/s]. The increase in the number of the reassignments was larger than when V_{max} was 2.0 [m/s]. This was because the topology of the network changed more frequently due to increased maximum speed of the mobile hosts. Moreover, the number of reassignments decreased earlier than when V_{max} was 2.0 [m/s]. This is because the connectivity among the mobile hosts weaken due to the increases maximum speed of the hosts.

In the location-based selection, almost the same as when V_{max} was 2.0 [m/s]. Furthermore, the frequency of reassignment was higher due to the increased maximum speed of the mobile hosts.

Maximum value of round-trip delay ($V_{max} = 2.0[m/s]$). Because of reassignments of mediation function, round-trip delays decreased as shown in figure 6.

When V_{max} was 2.0 [m/s], there was no difference in the maximum round-trip delay among the these cases. However, as L was increased, the delay when the mediation function was not reassigned became larger. That is, reassigning the mediation function to a more suitable host reduces the amount of the increase in the delay. When L was small, all the hosts were within transmission range of each other, so the difference in the delay between the reassigned and not reassigned cases was small.

A difference of about 0.02-0.04 [sec] was seen between the case of location-based selection and no reassignment. A difference of about 0.01-0.04 [sec] was

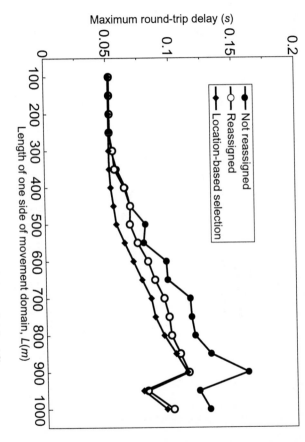

Fig. 6. Maximum round-trip delay ($V_{max} = 2.0$ [m/s])

seen between the case of reassignment and no reassignment. A difference of delay became small as L became larger, meaning that reassignment was more effective when the movement domain became large. However, when L was larger than 850, there was no difference of the delay between reassignment and the the location-based selection. These results show that when the movement domain becomes large, a mediation host selected by the location-based selection is more likely to be at an unideal position.

Maximum round-trip delay ($V_{max} = 16.0[m/s]$). As shown in Figure 7, the maximum round-trip delay when V_{max} was 16.0 [m/s] was almost same as when V_{max} was 2.0 [m/s] except that the inclinations of the round-trip delays were more sudden, and the maximum round-trip delays were larger. A difference of about 0.02-0.04 [sec] was seen between the case of location-based selection and no reassignment. A difference of about 0.01-0.04 [sec] was seen between the case of reassignment and no reassignment. The delay became smaller as L became larger, meaning that reassignment was more effective when the movement domain was large. Moreover, the inclination of the delay clearly made a sharp turn when reassignment was performed, while it did not change much when reassignment was not performed. Thus, the faster the mobile hosts move, the more reassignment reduces the round-trip delay.

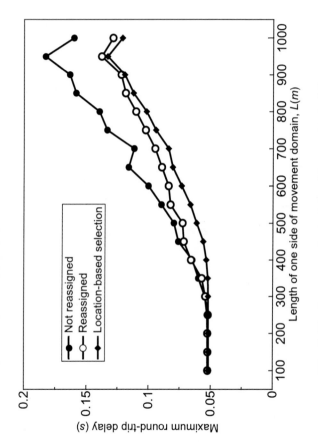

Fig. 7. Maximum round-trip delay (V_{max}=16.0 [m/s])

3.5 Summary

In summary,

– When the movement domain was sufficiently large compared to the communication range of the mobile hosts, reassignments of the mediation function by using our method effectively reduced communication delays.
– When the maximum speed of the nodes remained constant, our method worked better when the movement domain was large.

Consequently, REMARK decreases the communication delays of messages used in cooperative applications that periodically and frequently transmit messages between the mediation host and members.

4 Conclusion and Future Work

REMARK is a method for reassigning the mediation function dynamically according to the topology of ad hoc networks. Simulation of REMARK assuming an actual routing protocol showed that it decreases delays in communication for cooperative applications that periodically and frequently exchange messages between a mediation host and members. It also showed that REMARK is especially effective when the movement domain is large.

Future work includes

– Evaluation of REMARK using a larger scale simulation (more nodes, a wider movement domain, and a longer execution time).
– Examination and evaluation of reassignment methods independent of routing protocols.
– Estimation of costs for reassignment from viewpoint of time to select a next mediation host and make an expression of it.
– Examination of methods for shifting the status of a mediation from the present mediation host to the next mediation host.

References

1. Yoshinari Suzuki, Susumu Ishihara, and Tadanori Mizuno: Rearrangement of Topology Dependent Functions on Mobile Ad Hoc Networks, IPSJ SIG Notes, Vol. 2001, No. 108, pp. 87–94 2001.11) (in Japanese).

2. Takashi Ishikawa, Susumu Ishihara, Tetsuo Mizuguti, and Tadanori Mizuno: Performance Evaluation of a Method for Offering Impartiality to Game Members in Heterogeneous Network Environments, IPSJ Journal, Vol. 42, No. 7, pp. 1817–1827 (2001.7) (in Japanese).

3. Fumiaki Satoh, Kunihiko Minamihata, and Tadanori Mizuno: A Totally Ordered and Secure Multicast Protocol for Distributed Virtual Environment, Proceedings of the Seventh International Conference on Parallel and Distributed Systems: Workshop, pp. 55–60 (2000.7).

4. Y. Ishibashi and S. Tasaka: A group synchronisation mechanism for live media in multicast communications, *IEEE GLOBECOM'97*, pp. 746–752 (1997.11).

5. C.E. Perkins and P. Bhagwat: Highly Dynamic Destination-Sequenced Distance-Vector Routing (DSDV) for Mobile Computers, *Computer Communications Review*, pp. 234–244 (1994.10).

6. David B. Johnson, David A. Maltz, Yih-Chun Hu, and Jorjeta G. Jetcheva: The Dynamic Source Routing Protocol for Mobile Ad Hoc Networks, *Internet-Draft*, draft-ietf-manet-dsr-05.txt (2001.3.2).

7. Charles E. Perkins, Elizabeth M. Royer, and Samir R. Das: Ad Hoc On-Demand Distance Vector (AODV) Routing, *Internet-Draft*, draft-ietf-manet-aodv-08.txt (2001.3.2).

8. Bluetooth: http://www.bluetooth.com

9. HomeRF: http://www.home-rf.com

10. ns-2: http://www.isi.edu/nsnam/ns/

An Asymmetric Routing Protocol Using Detour in Mobile Ad Hoc Networks

Jong-Hoon Lee[1], Joo-Hwan Seo[2], Jung-Seok Lee[2], Won Ryu[1], and Ki-Jun Han[2]

[1] Network Technology Laboratory,
Electronics and Telecommunications Research Institute (ETRI),
Gajeong-Dong, Yuseoung-Gu, Daejeon, Korea
mine@etri.re.kr, wlyu@etri.re.kr

[2] Department of Computer Engineering, Kyungpook National University, Daegu, Korea
jhseo@netopia.knu.ac.kr, longlong@dreamwiz.com,
kjhan@bh.knu.ac.kr

Abstract. In mobile ad hoc networks, reliability and connectivity between all nodes is maintained by routing protocols. Most on-demand routing protocols in mobile ad hoc networks support only symmetric links that can receive messages from neighboring node on each other. Unfortunately, this assumption of the existing protocols is inconsistent with reality. In fact, the asymmetric links are considered as broken or failed links by the existing routing protocols, so connectivity between nodes cannot be maintained and finally this cause performance degradation. In this paper, we propose an asymmetric routing protocol using a detour to cope with this problem in mobile ad hoc networks. Experiments with the Global Mobile Simulation tool shows that our protocol can improve routing performance by providing better connectivity and throughput of the network.

1 Introduction

In an ad hoc network, mobile nodes communicate with each other using multihop wireless links. There is no stationary infrastructure and no base station. Each node in the network also acts as a router, forwarding data packets for other nodes [2,4,11].

Existing routing protocol for mobile ad hoc networks may generally be categorized into two classes: table-driven and source-initiated (demand-driven) [2,6]. First table driven routing protocols attempt to maintain consistent up-to-date routing information from each node to every other node in the network. These protocols require each node to maintain one or more tables to store routing information. They then respond to changes in network topology by propagating updates throughout the network in order to maintain a consistent network view. The Destination-Sequences Distance-Vector(DSDV) and the Wireless Routing Protocol(WRP) are most widely used as table-driven protocols. DSDV is a table-driven algorithm based on the classic Bellman-Ford routing mechanism [2,5,6]. Every mobile node in the network maintains a routing table in which all of the possible destinations within the network and the number of hops to each destination are recorded. Each entry is marked with a sequence number assigned by the destination node. Frequent system-wide broadcasts

I. Chong (Ed.): ICOIN 2002, LNCS 2344, pp. 300–308, 2002.
© Springer-Verlag Berlin Heidelberg 2002

limit the size of ad-hoc networks that can effectively use DSDV because the control message overhead grows as $O(n^2)$ [4].

Second, source-initiated on-demand routing creates routes only when desired by the source node. When a node requires a route to the destination, it initiates a route discovery process in the network. The representative protocols of this routing class is Ad hoc On-Demand Distance Vector routing protocol (AODV), Dynamic Source Routing (DSR), Temporally Order Routing Algorithm (TORA), and Associativity-Based Routing(ABR)[2,4,5,6]. In particular, AODV routing protocol builds on the DSDV algorithm. It provides quick and efficient route establishment between nodes desiring communication. AODV was designed specifically for ad hoc wireless networks since it provides communication between mobile nodes with minimal control overhead and minimal route acquisition latency [2].

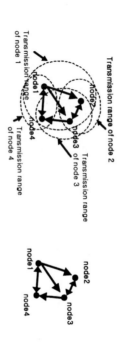

Fig. 1. Asymmetric links caused by different radio coverage.

Generally, existing routing protocols assume symmetric links that has the same radio coverage between neighboring nodes [6]. But, wireless links with reality do not necessarily conform to this assumption [6,7]. As time passes, the radio coverage of each node may decrease due to power consumption [6]. In this case, asymmetric links come out because of different radio coverage between neighbors as shown in Fig.1. If the existing routing protocols are applied to ad hoc networks which contains asymmetric links, the asymmetric links are considered broken or failed links, so connectivity and capability of network can fall off. For example, we assume that the network is partitioned into two domains that are joined by two oppositely oriented asymmetric links and that all other links are symmetric. In this case, no route can be discovered between the domains, even though asymmetric routes enable perfect network connectivity [8]. Specially, in military ad hoc networks where the topology of interconnections may be quite dynamic and there is much power consumption, the problem may be fatal.

In this paper, we propose an asymmetric routing protocol using a detour which provides better connectivity for ad hoc mobile networks. The rest of this paper is organized as follows. In Section 2, our proposed protocol is described in detail. The simulation results are discussed in Section 3, and finally, the concluding remarks are given in Section 4 .

2 Proposed Protocol

2.1 Related Works

In AODV, the Route Discovery Process is to set up a path between the source node and destination node through other intermediate nodes. It is initiated on-demand whenever a source node desires to communicate with another node for which it has no routing information in its table [2].

To begin such a process, the source node creates a Route Request packet (RREQ) and broadcasts it, which then forwards the request to its neighbors. This continues until either the destination node or an intermediate node with a "fresh enough" route to the destination is located. Once the RREQ reaches the destination node or an intermediate node with a fresh enough route, the destination/intermediate node responds by unicasting the Route Reply packet (RREP) back to the neighbor from which it first received the Route Request [2]. However, when there are asymmetric links along the route, the RREP can not be directly transmitted back to the neighbor.

2.2 Proposed Routing Protocol

Our protocol solves this problem by establishing a detour through other adjacent links when there is an asymmetric link to the neighbor. In other words, our protocol tries to transmit the RREP through a detour as shown in Fig. 2.

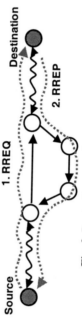

Fig. 2. Route discovery for asymmetric links

For this, our protocol first identifies the link type using hello messages. Each node periodically sends hello messages to its neighbors. The hello message in our protocol contains IP addresses of neighbors that it can hear. Unless the hello message received from the neighbors includes its own IP address at a node, the link is identified as an asymmetric link. Otherwise, it is identified as a symmetric link. In our protocol, the Neighbor Information Table (NIT) shown in Table 1 is used to indicate if the link is symmetric or not.

Table 1. Neighbor Information Table (NIT) (A: Asymmetric, S: Symmetric)

Neighbors	Link status
IP address of Node 2	A
IP address of Node 3	A
IP address of Node 4	S

When the destination/intermediate node unicasts the RREP, it first refers to NIT, and if the link to the next neighbor is asymmetric, it unicasts the RREP using the detour establishment process. Otherwise, it unicasts Route Reply packet directly.

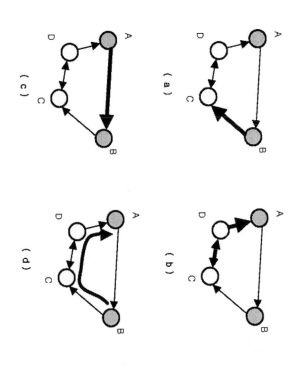

Fig. 3. An example of detour establishment: (a) node B broadcasts DEP including its own IP address and MAC address (TTL = 3) (b) nodes C and D forward it including its own IP address and MAC address (c) node A transmits a DEP-reply to node B, including information on the detour path from B to A and its own MAC address for unicasting (TTL = 1) (d) node B transmits the original packet to node A through the detour path included in the DEP-reply

The detour establishment process sets up a forwarding connection on a unidirectional link by way of other adjacent nodes. It should be noted that Time to Live (TTL) of the Detour Establishment Packet (DEP) is limited to three hops in our protocol because most of asymmetric links do not require more than four hops in the real world. Also, the DEP-reply should be broadcasted within one hop for detour establishment. If the DEP-reply packet is not received over the link, the link is assumed to be broken and unable to send any further packets.

After the detour establishment process is completed, the node can now transmit its packets to the neighbor through the detour path whose information is contained in the DEP-reply. At this time, the packet is delivered with being encapsulated in a special packet called Detour Packet (DP). The DP is used only for transmitting packets to the neighbor nodes on asymmetric links. The node adds the path information contained in the DEP-reply packet to the DP. Thus, the nodes on the detour path relay the DP using IP addresses and MAC addresses contained in the DP and finally the neighbor can receive the packets.

In addition, an Acknowledgment Request bit of the RREP is set to 1 for reliable transmission of the Route Reply packet. This enables the sender of the RREP to confirm successful transmission when it receives a RREP-ACK.

For the detour establishment process, our protocol employs another table called the Detour Establishment Table (DET) to store information on the established detour and its MAC address for local connectivity. If an asymmetric link is found, the detour is determined by looking up the DET and the RREP is unicasted round it. Our protocol described so far is expressed in Fig.4.

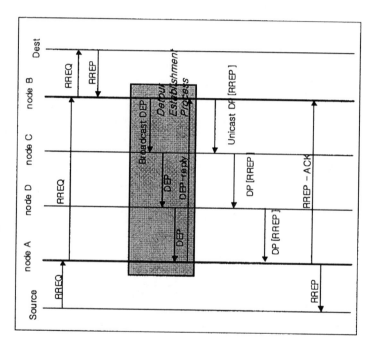

Fig. 4. Proposed protocol procedure

The Address Resolution Protocol (ARP) is an important issue when asymmetric routing protocol is based on IP because ARP assumes bidirectional connectivity. If the link between two nodes is asymmetric, the node which sends the ARP request doesn't receive an ARP reply. This causes a problem in that the node which sends the request cannot know the MAC address of the corresponding node. In this case, it is impossible to unicast a packet to the other. In order to resolve this ARP problem, the MAC address of each node must be included in the packet which is used in the detour establishment process. And each node must update ARP cache.

3   Simulation

The Global Mobile Simulation (GloMoSim) is used to evaluate the performance. The GloMoSim library is a scalable simulation environment for wireless network systems using the parallel discrete-event simulation capability provides by PARSEC[9].

3.1   Assumptions and Modeling

In our simulation, 60 nodes are initially scattered uniformly onto a 1000 meter by 1000 meter area. The maximum radio propagation range for each node is 250 meters and the channel capacity was 2 Mbps. Each run executes for 500 seconds of simulation time.

The Distributed Coordination Function (DCF) of IEEE 802.11 for wireless LANs is used as the MAC layer protocol. The 802.11 DCF uses Request-To-Send (RTS) and Clear-To-Send (CTS) control packets for "unicast" data transmission to a neighboring node. Data packet transmission is followed by an ACK. "Broadcast" data packets and the RTS control packets are sent using physical carrier sensing. An unslotted carrier sense multiple access (CSMA) technique with collision avoidance (CSMA/CA) is used to transmit these packets [12].

The application traffic used is FTP. There are two data session, each with the traffic rate of four packets per second. The size of the data payload is 512 bytes. The mobility model uses the random waypoint model [12] in a rectangular field. Each node randomly selects a position. And then moves toward that location with a speed between the minimum and maximum speeds. Once it reaches that position, it becomes stationary for a defined pause time. After that pause time, it selects another position and repeats the process. In our simulations, various pause times are used to simulate different mobilities. Longer pause times imply less mobility. The minimum and the maximum speeds are zero and 20 m/s, respectively.

In order to express the radio coverage for the asymmetric links, we define Radio Transmission Variance (RTV) as the ratio of the maximum range to the minimum range. The value of RTV is between 0 and 1. For example, an RTV value of 0.7 means that the minimum radio transmission range is 70% of the maximum range. As a result, a relatively low RTV value indicates that there are more asymmetric links in the network.

3.2   Results

In Fig. 5, we can see that our protocol offers a higher throughput than the AODV. The AODV is not capable of constructing a route if there are asymmetric links along the route because packets can no longer be unicast on the asymmetric links. This may cause failure of route discovery. On the other hand, our protocol is able to accomplish route discovery even if there are asymmetric links along the route.

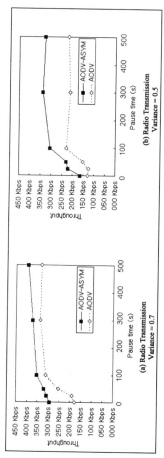

Fig. 5. Throughput

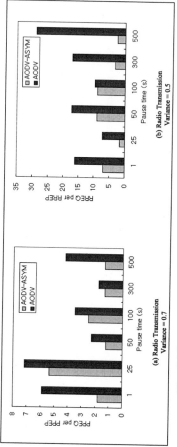

Fig. 6. The number of Route Request(RREQ) per Route Reply(RREP)

Fig.6 shows the ratio of the number of the RREQ to that of the RREP which indicates the number of query messages necessary to construct/reconstruct path towards the destination node. As shown in this figure, our protocol provides a lower ratio because if there are asymmetric links along the route in the AODV, the RREP cannot be unicasted to a source node. As a result, the source node will send the RREQ again, until it times out. As this process repeats, the number of the RREQ that the source node broadcasts for route discovery is increased. However, in our protocol, the number of Route Request packets retransmitted can be reduced.

Fig.7 depicts the routing load, which is defined as the ratio of the number of routing packets transmitted to that of data packets delivered to the destination. As shown in this figure, our protocol gives a much lower routing load than the AODV because our protocol produces less RREQ packets than the AODV. Our protocol shows a more apparent improvement as the pause time becomes smaller (that is, as mobility goes higher).

Finally, Fig. 8 depicts the data delivery ratio which represents the ratio of the data packets delivered to destinations to those generated by the FTP sources. This figure indicates that our protocol offers a similar data delivery ratio to the AODV. This is because our protocol is mainly concerned with route discovery, so the data delivery ratio is not affected once a route is set up over asymmetric or symmetric links.

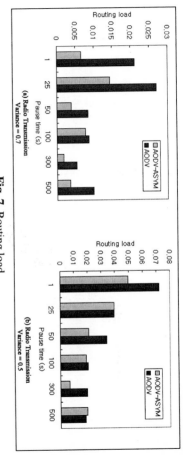

Fig. 7. Routing load

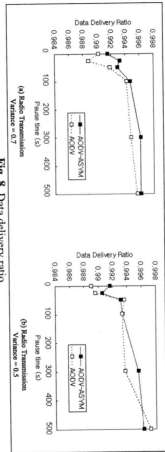

Fig. 8. Data delivery ratio

4 Conclusions

Most on-demand routing protocols in mobile ad hoc networks assume only symmetric links that have the same radio coverage between neighboring nodes [6]. This protocols are not efficient when there are asymmetric links in mobile ad hoc networks. The asymmetric links are considered as broken or failed links by the existing routing protocols, so connectivity can not be maintained and finally this causes performance degradation. Our protocol can successfully support asymmetric links without breaking ARP. Simulation result shows that our protocol can improve routing performance by providing better connectivity, higher reachability and suviviability of interconnections in mobile ad hoc network whose topology is quite dynamic.

References

1. J. P. Macker, M. S. Corson.: Mobile Ad Hoc Networking and the IETF. ACM Mobile Computing and Communications Review, Vol. 3, Number 2, Apr. (1999)
2. E. M. Royer and C-K. Toh.: A review of current Routing Protocols for Ad Hoc Mobile Wireless Networks. IEEE Personal Communications, Apr. (1999) 46–55

3. C. E. Perkins and P. Bhagwat.: Highly Dynamic Destination-Sequenced Distance-Vector Routing (DSDV) for Mobile Computers. IEEE Computer Communication Review, Oct. (1994) 234–244

4. C. E. Perkins and E. M. Royer.: Ad-hoc On-Demand Distance Vector Routing. Mobile Computing Systems and Applications, Proceedings. WMCSA '99. Second IEEE Workshop on, Feb. (1999) 90–100

5. J. Broch, D. B. Johnson, and D. A. Maltz.: The Dynamic Source Routing Protocol for mobile Ad Hoc Networks. IETF Internet draft, draft-ietf-manet-dsr-01.txt, Dec. (1998)

6. Dongkyun Kim, C-K. Toh and Yanghee Choi.: RODA: a new dynamic routing protocol using dual paths to support asymmetric links in mobile ad hoc networks. IEEE ICCCN, LasVegas, USA, (2000).

7. R. Prakash.: Undirectional Links Prove Costly in Wireless Ad Hoc Networks. Proceedings of third International Workshop on DIALM, Aug. (1999) 15–22

8. Charles E. Perkins, Ad Hoc Networkings, AddisonWesley, (2001)

9. X. Zeng, R. Bagrodia and M. Gerla.: GlomoSim : A liblary for the parallel simulation of large-scale wireless networks. IEEE Parallel and Distributed Simulation, PADS 98. Proceedings. Twelfth Workshop on, (1998) 154–161

10. R. Bagrodia, R. Meyer, M. Takai, Y. Chen, X. Zeng, J. Martin and H. Y. Song.: PARSEC: A Parallel Simulation Environment for Complex Systems. IEEE Computer , vol. 31, no. 10, Oct. (1998) 77–85

11. Lee S-J. and Gerla M.: AODV-BR: backup routing in ad hoc networks. Wireless Communications and Networking Confernce, WCNC2000, vol. 3, (2000) 1311–1316

12. C. E. Perkins, E. M. Royer, S. R. Das and M. K. Marina.: Performance comparison of two on-demand routing protocols for ad hoc networks. IEEE Personal Communications, vol. 8, Feb. (2001) 16–28

A Multi-source Multicast Routing Protocol for Ad-Hoc Networks*

Hyun-jeong Kang and Mee-jeong Lee

Ewha Womans University
{semper, lmj}@ewha.ac.kr

Abstract. We propose a multicast routing protocol for ad-hoc networks, which is particularly efficient for multi-source multicast transmissions. The proposed protocol designates one or more sources as the core sources. A tree that is rooted at a core source and reaches all the destinations of the multicast group is formed for each core source. The union of these trees constitutes the data delivery mesh, and each of the non-core sources finds the nearest core source in order to delegate its data delivery. The data delivery mesh is periodically reconfigured through the control message flooding initiated from the core sources. A local mesh recovery mechanism is also deployed to complement the periodic mesh reconfiguration. Our simulation results show that the proposed protocol achieves high multicast efficiency with low communication overhead compared with the other existing multicast routing protocols especially when there are multiple sources in the multicast group.

1 Introduction

Ad-hoc networks are multi-hop wireless mobile networks that are composed of mobile hosts without any fixed network infrastructure[1]. In an ad-hoc network, the mobile hosts take charge of data forwarding which is typically a function of the routers in a fixed network. The network topology changes over time as the mobile hosts move around, and the mobile hosts have rather a limited power and capacity. Typical applications of ad-hoc networks include outdoor special events such as conferences and festivals, as well as communications in regions with no infrastructure, in emergencies and military maneuvers. Since most of these applications have multipoint-to-multipoint communication requirements, it is important to have an efficient multicast protocol for ad-hoc networks. The unique features of ad-hoc networks pose new challenges in the design of multicast routing protocols. For example, the multicast protocols for mobile networks should be able to deal with the node mobility as well as to save the limited power and capacity of the mobile hosts.

In recent years, a number of new multicast routing protocols for ad-hoc networks, targeting at the above problems, have been proposed. The scalability of these

* This work was supported by the BK 21 project from the Korea Research Foundation in the program year of 2001.

I. Chong (Ed.): ICOIN 2002, LNCS 2344, pp. 309–320, 2002.
© Springer-Verlag Berlin Heidelberg 2002

protocols, though, has much to be studied yet. In this paper, we especially propose a multicast protocol that is scalable with respect to the size of the network as well as the number of sources in the multicast groups.

According to the constitution of data delivery structure, the protocols that have been proposed so far can mainly be classified into two categories: tree based protocols and the mesh based protocols. The tree based protocols deploy a single tree as the data delivery structure of each multicast group, with one particular node acting as the root of the tree. Each node in the delivery tree forwards the multicast data packets that come from its parent. On the other hand, the mesh based protocols usually deploy a set of trees in the data delivery structure, and each node included in the delivery structure forwards every incoming packet that belong to the multicast group of its interest.

AMRIS(Ad hoc Multicast Routing protocol utilizing Increasing id-numberS)[2], AMRoute(Adhoc Multicast Routing)[3], and MAODV(Multicast Operation of the Ad-hoc On-Demand Distance Vector Routing Protocol)[4],[5] are the examples of the tree based protocols. These tree based protocols are vulnerable to the mobility since there is only one path that leads to each group member. Maintaining the data delivery tree is critical since loss of one link on the tree means failure of data delivery. When the size of the network becomes larger, the formation of tree becomes sparser, resulting in the performance degradation and/or the increment of overhead to maintain the delivery tree. If the number of sources in the multicast group is large, it may also incur congestion on the delivery tree since the traffic from all of the sources has to be transmitted through a single tree.

ODMRP(On-Demand Multicast Routing Protocol)[6],[7],[8] and CAMP(Core-Assisted Mesh Protocol)[9] are the examples of the mesh based protocols. In ODMRP, the nodes that are on the shortest paths from each source to all the destinations constitute the data delivery mesh. That is, each source of the multicast group becomes a root of a tree that leads to all of the destinations of the multicast group. The union of these per source trees forms the data delivery mesh in ODMRP. The data delivery mesh, therefore, becomes thicker as the number of sources in the multicast group increases and as a result the amount of duplicate packets increases. In order to adapt to the node mobility, each source of the multicast group floods a control packet periodically. If the network size is large the control packet flooding overhead becomes large. Especially when the number of sources in the multicast group is large, the duplicate data packet overhead as well as the control packet flooding overhead becomes significant.

The data delivery mesh of CAMP also includes the union of the shortest paths between the source and the destinations of the multicast group. CAMP, therefore, also forms a thicker mesh as the number of sources increases. In addition to the per source trees, CAMP also includes the paths, in its data delivery mesh, that connect multicast group members to the core mesh when they first join the group. In CAMP, one or more nodes are designated as core nodes, and they are connected to one another completely. Due to this constitution, large amount of packet duplication may occur around the core mesh.

Recently, Ozaki et. al. proposed a bandwidth efficient multicast routing protocol, which builds a tree like data delivery structure with the nodes on the tree acts like a mesh node, that is, the nodes deliver any data packet that belongs to the multicast group that they take charge of [10]. For their data delivery tree, neither the concept of a root of a tree nor the effort to optimize the tree exists. Their protocol, therefore, requires less protocol overhead to maintain the delivery tree, and the performance is higher than the other tree-based schemes. When there are multiple sources, however, this protocol may also suffer overloading of the delivery tree since it is based on a single tree delivery structure.

For an efficient multi-source multicast transmission, we propose a multicast routing protocol that forms a data delivery mesh with a subset of the per source tress instead of a single tree rooted at a single node or the entire set of per source trees. The proposed protocol, named as SMMRP (Scalable Multi-source Multicast Routing Protocol), builds the data delivery mesh with the trees that are rooted at a subset of the sources called core sources. We assume that there is a server node that receives the registration from all of the sources of every multicast group, and designates an appropriate number of the sources as the core sources for each multicast group. Each of the non-core sources finds the nearest core source from itself, and delegates its packet delivery by transferring every data packet to that core node. The data delivery mesh is periodically reconfigured through a control message flooding initiated from the core sources. To complement the periodic mesh reconfiguration, we also deploy a local mesh recovery mechanism.

The number of core sources is one of the important parameters that affect the efficiency of the protocol. Enough number of core sources has to be chosen to prevent overloading the data delivery mesh and to provide robustness against node mobility. Whereas too many core sources cause excessive control message flooding overhead and unnecessary data packet duplications.

The rest of this paper is organized as follows. In section 2, the proposed protocol is explained. Simulation results comparing the proposed protocol with the other existing protocols are presented in section 3. Finally, section 4 concludes this paper.

2 SMMRP (Scalable Multi-source Multicast Routing Protocol)

This section explains the operation of the SMMRP in detail. We assume that there is a server, to which each source of a multicast group registers itself or sends quit-notification whenever it joins or leaves the multicast group. The server has to be known throughout the entire ad-hoc network domain, and it does not need to be a member of the multicast group. The purpose of the server is to collect information about the sources of each multicast group in order to select the core sources and notify this information to the multicast group members. For each multicast group, the server has to pick an appropriate proportion of the sources as the core sources. It is likely that the routers serving as access points to the rest of the Internet would serve as the server, because they are static and they must be known by the rest of the ad-hoc network for other purposes. Each of the sources of a multicast group determines the nearest core source, and transmits its data packets toward that core source. The core sources, then, relay all the incoming data packets toward the destinations.

Figure 1 shows the delivery structure of SMMRP. The delivery structure of SMMRP consists of two types of meshes: SC (Source and Core source)-mesh and CD

312   H.-j. Kang and M.-j. Lee

(Core source and Destination)-mesh. A path that starts from a source and reaches to its nearest core source constitutes the SC-mesh. A tree that is rooted at one of the core sources and has the leaves at all of the destinations of the multicast group constitutes the CD-mesh. The intermediate nodes on these two types of meshes are called as mesh nodes. The mesh nodes forward any incoming data packet that belongs to the multicast group for which they are selected to serve as the mesh node. These meshes are periodically reconfigured to connect the corresponding end nodes with the shortest paths. Between the periodic mesh refreshment intervals, each mesh node watches over the link to its upstream mesh node. When the link to the upstream mesh node is lost, it uses local flooding to find another upstream mesh node. SMMRP assumes no underlying unicast protocol for its mesh set-up.

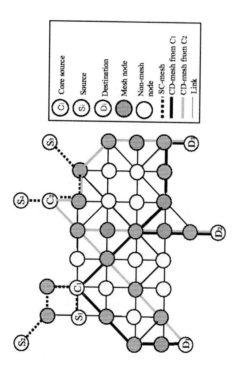

Fig. 1. SMMRP mesh structure

2.1 Periodic Mesh Reconfiguration

The server periodically floods an SV message to inform its identity and the routing information to itself. For an incoming SV message, each node marks the previous hop that relayed SV in its routing table. When a source wants to join(leave) a multicast group, it sends SN(LN) message to the server. SN(LN) messages are forwarded to the server using the routing information collected during the SV message flooding. While the SN(LN) messages are forwarded to the server, each intermediate node marks the routing information to the source of the SN(LN) message. Receiving SN(LN), the server updates its source list entry for the multicast group specified in SN(LN), and replies to the source with an SA(LA) message. SA(LA) messages are forwarded back to the source using the routing information that is collected while SN(LN) is forwarded. Whereas the LA message simply acknowledges LN, the SA message not only acknowledges SN but also informs the list of core sources. Including SN and

LN, all of the messages that require acknowledgement are issued repeatedly at regular time intervals until they are acknowledged.

Note that since a certain proportion of the sources are selected as the core sources, the change in the number of sources may require a change in the number of core sources. If adding or deleting a source requires an addition or deletion of a core source, the server floods a CoreN message. Receiving CoreN, the source that is designated as the new core source or asked to stop acting as the core source replies with CoreNA to the server. From CoreN, sources of the multicast group may find out that they need to change the core source to delegate their data delivery to, and start setting up SC-mesh to a new core source.

Each of the core sources floods a JR message periodically. While JR is flooded, the routing information to the core source is updated at each node. The sources of the multicast group collect the identities of core sources from the incoming JR messages, and choose the nearest core source among them. To facilitate this, JR messages contain the hop count from the core source that has generated the JR message. The sources periodically generate a JSA message to their nearest core source. The JSA messages are forwarded using the routing information collected while the JR messages are flooded. When an intermediate node receives JSA, it checks whether it is marked as the next hop. If it is, the node becomes a mesh node.

Similar to the source nodes, destinations of the multicast group also collect the information of core sources from the incoming JR messages. Differing from the sources, though, the destination nodes set up paths to every core source. The way that they set up paths to the core sources is similar to the way that a destination in ODMRP sets up paths from itself to every source [6], [7], [8]. A destination node periodically generates a JDA message, where the addresses of all the core sources and corresponding next hops are marked. Receiving JDA, an intermediate node checks whether it is marked as one of the next hops in JDA. If it is, it becomes a mesh node for the multicast group. It also forwards the JDA message toward the core sources for which it is marked as the next hop in the incoming JDA (called as the primary core sources of the mesh node hereafter). The outgoing JDA is modified so that it only includes the list of the primary core sources and their corresponding next hops.

Once selected as a mesh node, the node takes charge of forwarding the data packets pertaining to the multicast group for MN-LIFETIME. If a mesh node is not refreshed in MN-LIFETIME, it stops acting as a mesh node. For the local recovery procedure explained in section 2.2, the mesh nodes remember the next hop mesh node for each of its primary core sources as well as the hop count.

After issuing LN, a core source no longer floods JR, and sets a timer (Core-LEAVE) whose life time is long enough for the sources that delegate the packet delivery to itself to locate and set up a path to a new core source. The core source may not leave the multicast group, and has to continue taking charge of the core source role until it receives LA from the server as well as the Core-LEAVE timer expires.

2.2 Local Mesh Recovery Procedure

In addition to the periodic mesh reconfiguration described in section 2.1, SMMRP deploys a local recovery mechanism at each mesh node. This local recovery mechanism is a slight modification of the local recovery mechanism proposed by the authors in [11]. For the local recovery of the mesh, each of the mesh nodes keeps checking whether its upstream mesh node is within its transmission range using the beacon signals provided by the MAC layer. If a mesh node finds that it cannot reach the upstream mesh node, it starts looking for a new route to reach the primary core source that used to be connected by the lost upstream mesh node. We call this as the local mesh recovery procedure, and it consists of three phases: the request phase, the response phase, and the connection phase.

In the request phase, the mesh node, which finds the loss of upstream mesh node, advertises itself to the nearby neighbors. For the advertisement, the mesh node floods an ADVT message (Figure 2 (1)). In the ADVT message, the address of primary core source, for which the lost upstream mesh node used to provide a path, and the hop count to that primary core source are recorded. When the lost upstream mesh node has been set for more than one primary core sources, a list of this information is indicated in ADVT. The lifetime of ADVT message is set to a small number such as 2 or 3 in order to constrain the area of flooding to the nearby neighbors. Receiving an ADVT message, a node checks whether it is a member of the multicast group specified in the ADVT as well as whether the core source marked in the ADVT is one of its primary core sources. It also checks whether its hop count to the core source is smaller than the hop count marked in the ADVT message. This is to allow for only mesh nodes that are nearer to the core source than the advertising mesh node to reply. If all of these three conditions hold, the mesh node then stops forwarding the ADVT message and sends back a reply message, a PATCH message, back to the source of the ADVT (Figure 2 (2)). If any of the three conditions does not hold, it decreases the value of TTL(Time To Live) in ADVT by 1 and floods the ADVT message. If the value of TTL becomes 0, it simply discards the ADVT message. While the ADVT messages are transmitted, the routing information for the source of the ADVT message is recorded at each intermediate nodes, and this information is utilized later on to forward a PATCH message.

In a PATCH message, the address of the core source for which the ADVT message queried as well as the hop count to that core source are recorded. While the PATCH message is transmitted back to the source of the ADVT message, the routing information for the source of the PATCH is recorded at the intermediate nodes, and this information is utilized to forward the PATCHA message, which is the acknowledgement for the PATCH message. Upon receiving a PATCH message, each intermediate node increment the hop count value marked in the PATCH message. When the source of the ADVT finally receives the PATCH message, the hop count value in the PATCH message indicates the distance from the source of the ADVT to the primary core source of interest. The source of the ADVT collects the incoming PATCH messages for a certain period of time, and chooses the one that provides the smallest hop count value. It, then, sends a PATCHA message back to the source of that selected PATCH message (Figure 2 (3)), and set the previous hop that forwarded the selected PATCH message as its new upstream node for the primary core source.

While the PATCHA message is transmitted, the last phase of the local recovery procedure, the connection phase proceeds. As the PATCHA message is transmitted back to the source of PATCH message, the intermediate nodes are selected as the temporary mesh nodes. The temporary mesh nodes perform the exactly same jobs that the regular mesh nodes do, except for that they last shorter than the regular ones. Based on the experiment in [11], the MN-LIFETIME at the temporary mesh nodes are set to one third of the MN-LIFETIME value for the regular mesh nodes. This is to avoid having excessive mesh nodes be generated by the local recovery procedure. Compared to the local recovery mechanism proposed in our previous work [11], the local recovery mechanism in SMMRP may generate less mesh nodes, which, in turn, results in less packet duplication and control overhead. This is due to the additional phase, the connection phase, deployed in the local recovery procedure of SMMRP. Instead of having all of the PATCH messages set the temporary mesh nodes while they proceed toward the source of ADVT as in [11], SMMRP let the source of the ADVT gather all the PATCH messages for a while and choose the one that offers the shortest path to the core source.

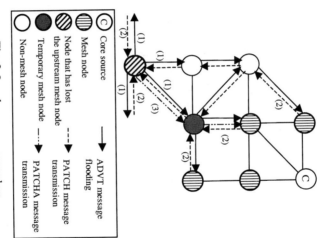

Fig. 2. Local recovery procedure

3   Performance Evaluation

In this section, we explain the simulation model and the numerical results obtained from the simulation. Among the existing schemes, the performance of the simple flooding and ODMRP are compared together. The simulation is run in the Windows NT 4.0 / Microsoft Visual C++6.0 environments. The simulation program has been implemented using the Global Mobile Simulation (GloMoSim) libraries developed in UCLA [12].

3.1 Simulation Model

The GloMoSim enables the implementation of a simulation module in a layered architecture. Table 1 shows the models that we assumed for each layer.

Table 1. Assumed models for each layer

Layer	Models
Physical (Radio propagation)	Free space with a threshold cutoff
Data link (MAC)	802.11 (CSMA/CA)
Network (Routing)	IP with SMMRP/ODMRP
Application	CBR

In our simulation network model, 150 mobile nodes are randomly located over a 2300m x 2300m area. Among the 150 nodes, 10 nodes are randomly selected as the sources, and another 10 nodes are randomly selected as the destinations. We assume that the members of the multicast group join the multicast group at the beginning of the simulation, and retain the memberships throughout the simulation. Each of the sources sends data at the rate of one packet per second, and the size of packet is 512 bytes. The radio propagation range of each host is 250meters, and the channel capacity is 2 Mbps. For the mobility model, we assume random waypoint. A node randomly selects a destination and moves towards that destination at a predefined speed. Once the node arrives at the destination, it stays there for a pause time between 0 and 10 seconds. Series of these actions are repeated over the simulation duration, and the mobility speed is varied from 15km/h to 75km/h.

Each simulation lasts for 1000 seconds of simulation time. Multiple runs with different seed numbers are conducted for each scenario and collected results are averaged over those runs.

3.2 Numerical Results

In our simulation, we vary two parameters on which the performance of SMMRP depends: the number of core sources, and the length of the periodic data delivery mesh reconfiguration interval. To measure the performance of SMMRP as well as the compared existing multicast routing protocols, we use three kinds of metric. As for the users' perspective, we measure the average packet delivery ratio. For the protocol overhead measurement, we deploy two different types of metric: one for the duplicate data packet overhead, and another for control message overhead. The duplicate data packet overhead is represented by the average number of total data packet copies generated over the entire network for each non-duplicate data packet delivered. The control message overhead is also measured in a similar way, that is, the average number of control messages generated over the entire network for each non-duplicate data packet delivered.

Figure 3–5 present the effect of the number of core sources to the performance of SMMRP. In this experiment, the periodic mesh reconfiguration interval is set to 5 seconds for SMMRP. When there is only a single core, the average packet delivery ratio of SMMRP drops rapidly as the speed of mobility increases. This is because that the mesh constitution of SMMRP for a single core source case is similar to the data delivery tree in the tree based schemes. This kind of data delivery structure is not only vulnerable to the mobility, but it is also highly loaded with all the data packets that come from all of the sources. Even though the vulnerability to the mobility can be compensated by the frequent local recovery, the overloading of the delivery mesh cannot be avoided. For the number of core sources larger than or equal to 3, SMMRP outperforms ODMRP for all the mobility speed.

The effect of increasing the number of core sources is two folded. As the number of core sources increases, the data delivery mesh becomes thicker and more robust to the mobility, whereas the duplicate data packet and control message overheads become larger and the waste of bandwidth increases. The first aspect causes the average packet delivery ratio to increase as the number of core sources increases. On the other hand, the latter aspect causes the average packet delivery ratio to go down for increasing the number of core sources. In our experiment, the data packet delivery ratio is best when we have 3 core sources among the 10 sources. Figure 4 and 5 show the effect of the number of core sources to the amount of overheads. For all the cases, SMMRP show lower degree of overhead than ODMRP, and the amount of overhead increases for larger number of core sources.

Figure 6–8 show the effect of the length of periodic mesh reconfiguration interval. The periodic mesh reconfiguration interval values are varied from 5 seconds up to 90 seconds. The number of core sources in SMMRP is set to 3 in this experiment. Among the three compared protocols, simple flooding shows the highest packet delivery ratio. However, the simple flooding protocol has high packet duplication overhead, approximately 2 times of the other two schemes, and is expected to have more serious performance degradation than the other two protocols when the traffic load in the ad-hoc network increases. For SMMRP, the packet delivery ratio slightly increases as the periodic mesh reconfiguration interval decreases. There is no sharp performance drop, though, as in the case of not having enough core sources. This indicates that the sparse periodic mesh reconfiguration can be compensated quite well by the local recovery procedure. SMMRP show higher packet delivery ratio than ODMRP in all cases.

For the duplicate data packet overhead, SMMRP show slightly lower overhead than ODMRP, except for the case when the mobility speed is $75km/h$ and the periodic mesh reconfiguration interval is 90 seconds. In this case, the mesh disruption occurs frequently due to the high mobility, and a lot of the times it has to be fixed by the local recovery procedure since the periodic mesh reconfiguration interval is large. These situations tend to generate a highly redundant non-optimal data delivery mesh most of the time, resulting in high duplicate data packet overhead.

For the control message overhead, SMMRP incurs less than one fourth of the overhead that ODMRP incurs. Note that the amount of the control message overhead of SMMRP is not directly proportional to frequency of the periodic mesh reconfiguration interval. This is because the control message overhead of SMMRP consists of two kinds: the control messages for the periodic mesh reconfiguration, and the control messages for the local recovery of the mesh. The latter kind of control message overhead decreases as the frequency of the periodic mesh reconfiguration increases, whereas the first kind of control message overhead is proportional to the frequency of the periodic mesh reconfiguration.

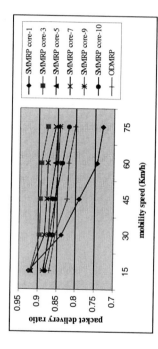

Fig. 3. Average packet delivery ratio with different number of core sources

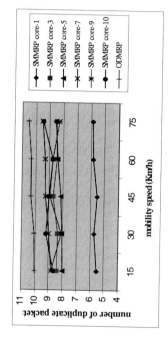

Fig. 4. Duplicate data packet overhead with different number of core sources

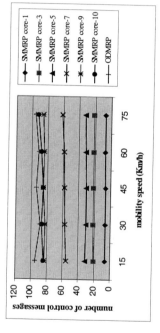

Fig. 5. Control message overhead with different number of core sources

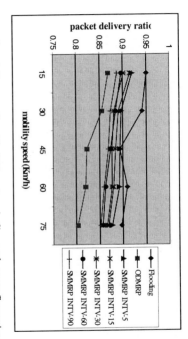

Fig. 6. Average packet delivery ratio with different periodic mesh reconfiguration intervals

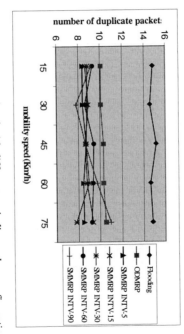

Fig. 7. Duplicate data packet overhead with different periodic mesh reconfiguration intervals

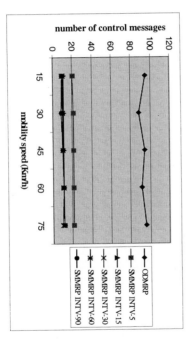

Fig. 8. Control message overhead with different periodic mesh reconfiguration intervals

4 Conclusions

The proposed multicast routing protocol for ad-hoc network, SMMRP, is scalable with respect to the number of the sources as well as the size of the network. SMMRP forms a data delivery mesh with a subset of per source trees. Among the sources of a multicast group, one or more sources are selected as the core sources, and a tree that is rooted at a core source and reaches all the destinations of the multicast group is formed for each core source. Each of the non-core sources finds the nearest core source, and delegates its data delivery to that core source. The data delivery mesh is periodically reconfigured, and local recovery mechanism is also deployed to complement the periodic reconfiguration. Simulation results show that the local recovery can be a good complement for a less frequent periodic mesh reconfiguration. It is also shown that the number of the core sources is an important parameter that affects the performance of SMMRP. Including too many per source trees in the data delivery mesh is not efficient whereas having too few per source tree in the data delivery mesh is not scalable with respect to the number of sources in the multicast group due to the delivery mesh overloading.

References

1. K. Obraczka, G. Tsudik, "Multicast routing issues in ad hoc networks," Proceedings of the IEEE ICUPC'98, Oct. 1998
2. C.W. Wu, Y.C. Tay, "AMRIS: A Multicast Protocol for Ad hoc Wireless Networks," Proceedings of MILCOM'99, Nov. 1999
3. E. Bommaiah, M. Liu, A. McAuley, and R. Talpade, "AMRoute: Ad-hoc Multicast Routing Protocol," Internet-Draft, draft-talpade-manet-amroute-00.txt, Aug. 1998
4. E.M. Royer, C.E. Perkins, "Multicast Operation of the Ad-hoc On-Demand Distance Vector Routing Protocol," Proceedings of Mobicom'99, Aug. 1999
5. E.M. Rolyer, C.E. Perkins, "Multicast Ad hoc On Demand Distance Vector (MAODV) Routing," Internet-Draft, draft-ietf-manet-maodv=00.txt, Jul. 2000
6. S.H. Bae, S.J. Lee, W. Su, M. Gerla, "The Design, Implementation, and Performance Evaluation of the On-Demand Multicast Routing Protocol in Multihop Wireless Networks, " IEEE Network, vol. 14, no. 1, Jan. 2000
7. S.J. Lee, W. Su, M. Gerla, "On-Demand Multicast Routing Protocol (ODMRP) for Ad Hoc Networks," Internet-Draft, draft-ietf-manet-odmrp-02.txt, Jul. 2000
8. S.J. Lee, M. Gerla, C.C. Chiang, "On-Demand Multicast Routing Protocol," Proceedings of IEEE WCNC'99, Sep. 1999
9. J.J. Garcia-Luna-Aceves, E.L. Madruga, "The Core-Assisted Mesh Protocol," IEEE Journal on Selected Areas in Communications, vol. 17, no. 8, Aug. 1999
10. T. Ozaki, J.B. Kim, T. Suda, "Bandwidth-Efficient Multicast Routing for Multihop, Ad-Hoc Wireless Networks," Proceedings of IEEE INFOCOM'2001, Apr. 2001
11. M.J. Lee, Y.K. Kim, "PatchODMRP: An Ad-hoc Network Multicast Routing Protocol," Proceedings of ICOIN-15, Feb. 2001
12. Wireless Adaptive Mobility Lab. Dept. of Comp. Sci., UCLA "GloMoSim: A Scalable Simulation Environment for Wireless and Wired Network Systems," http://pcl.cs.ucla.edu/ projects/glomosim

Infrastructure-Less Multicast Protocol in Mobile Ad-Hoc Networks

Sanghyun Ahn[1]*, Yujin Lim[2]**, Wonjun Lee[3], and Eunyong Ha[4]

[1] Department of Computer Science and Statistics
University of Seoul, Seoul, Korea
ahn@venus.uos.ac.kr

[2] Department of Mechanical and Information Engineering
University of Seoul, Seoul, Korea
yujin@uos.ac.kr

[3] Department of Computer Science and Engineering
Korea University, Seoul, Korea
wlee@korea.ac.kr

[4] Department of Computer Engineering
Anyang University, Anyang, Korea
eyha@aycc.anyang.ac.kr

Abstract. Ad-hoc networks have the characteristics of dynamic topology and many applications on ad-hoc networks tend to require group communication capabilities. Previously proposed multicast mechanisms on mobile ad-hoc networks build multicast-infrastructures like the tree or mesh for the group communication and these infrastructures may cause significant overhead especially in a highly dynamic mobile ad-hoc network environment. Therefore, in this paper, we propose a new multicast mechanism supporting multicast services based on only the underlying unicast routing tables, hence any multicast-related infrastructures like trees and meshes are not required to be constructed. As a result, we could achieve better performance in terms of the reliability and the control overhead.

1 Introduction

The ad-hoc network is a wireless communication network composed of mobile nodes with neither the base infrastructure network nor the base stations. In early days, mobile ad-hoc networks have been used for rescue operations and as the communication mechanism among soldiers in a battle field. However, in recent days, the proliferation of mobile communication devices like PDAs and laptop computers has made the ad-hoc network one of the needed technologies especially for the infrastructure-less communication environment. In an ad-hoc network,

---

* Corresponding author. This work was supported by grant No. R04-2001-00054 from the Korea Science & Engineering Foundation.
** The second author was supported from the Basic Research Program (for the woman scientists) of the Korea Science & Engineering Foundation

nodes can move freely, so the network topology changes continuously. Due to the characteristics of wireless channels, the mobile ad-hoc network suffers from a limited data delivery coverage, low bandwidth, high error rate, and limited battery power. These restrictions make the routing and the multicasting in the ad-hoc network difficult.

Since applications in the ad-hoc network are mostly group applications like the information sharing in a conference room and the multi-player game, efficient multicasting technology is needed in the ad-hoc network. Multicast routing protocols like the DVMRP (Distance Vector Multicast Routing Protocol) [1], the MOSPF (Multicast extensions for the Open Shortest Path First) [2], the CBT (Core-based Tree) [3]and the PIM (Protocol Independent Multicast) [4] proposed for the traditional wired network reconstruct multicast trees whenever the network topology or the group membership changes. Since the mobile ad-hoc network is highly dynamic in terms of the network topology, those multicast routing protocols are not well suited for the ad-hoc network due to the high control overhead.

To overcome these problems, new multicast routing protocols like the ODMRP (On Demand Multicast Routing Protocol) [5], the AMRIS (Ad-hoc Multicast Routing protocol utilizing Increasing id-numberS) [6], the AMRoute (Ad-hoc Multicast Routing) [7] and the CAMP (Core Assisted Mesh Protocol) [8] have been proposed for the mobile ad-hoc network. These newly proposed protocols have introduced the concept of the mesh structure instead of the tree to improve the reliability, or adopted the mechanism of periodic flooding or beaconing to reflect the frequently changing network topology. However these approaches also cause much overhead due to the extensive exchange of control messages required to adapt to the dynamically changing network topology. Therefore, in this paper, we propose a new multicast routing protocol which can provide an efficient multicast service without maintaining the multicast infrastructure like a tree or a mesh, and whose target applications are small-size multicast group communications which are typical in the ad-hoc networking environment.

This paper is organized as follows; in session 2, previously proposed multicast routing protocols for the ad-hoc network are reviewed. In session 3, we describe our newly proposed multicast routing protocol , and, in session 4, simulation results are shown and the performance evaluation is provided. Session 5 concludes this paper.

2   Related Work

The ODMRP [5] is an on-demand multicast routing protocol using the concept of the forwarding group. As shown in figure 1, a multicast source floods data packets, i.e., JOIN_DATA packets, to the entire network. JOIN_DATA packets are periodically flooded to reflect the dynamically changing group membership and the network topology. A group member receiving a JOIN_DATA packet sends a JOIN_TABLE packet to its neighbors in order to establish the path from the source to itself. The node receiving a JOIN_TABLE packet checks whether itself is on the tree, and, if it is on the tree, it sends a JOIN_TABLE packet towards the source. From this procedure, a mesh from the source to group members is established. By adopting the mesh structure, a higher connectivity can be achieved compared to the tree structure which is usually used in the wired network. Since more than one paths exist among the source and group members, a more reliable data delivery becomes possible in the case of node movements. However, in this protocol, data are periodically flooded to maintain the mesh structure, which may result in an overwhelming message overhead to the entire network.

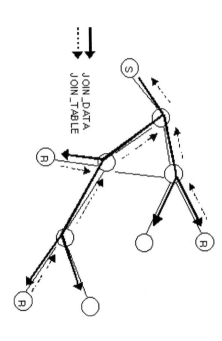

Fig. 1. The procedure of the ODMRP

The AMRIS [6] is a shared-tree-based protocol and assigns a multicast session ID number to each node on the network. A node which wants to join a multicast group sends JOIN_REQ messages to nodes with smaller ID numbers than its own ID number. Nodes maintain the information on link states and the group membership by periodically broadcasting beacon messages to neighbors. Since the AMRIS uses the tree structure, a changing topology may disturb a data delivery. To overcome this problem, the AMRIS reconstructs a tree whenever the topology changes, which becomes an overhead.

The AMRoute [7] is a protocol using both the tree and the mesh structures. The core periodically floods JOIN_REQ packets to establish a mesh structure among multicast group members. After that, the core sends TREE_CREATE packets along the mesh structure, then a virtual shared tree is established as shown in figure 2. In the case of a topology change, data packets can be delivered through a tree as long as there exist paths among group members in a mesh. However, in this scheme, loops can be formed and trees may not be optimal.

The Simple protocol [9] which is the most recently proposed one uses the Route Discovery mechanism of the unicast routing protocol DSR [10] to reduce the overhead of establishing and maintaining the tree or mesh structure. In other words, a source sends data packets to group members by simply broadcasting

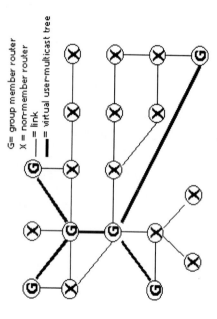

G= group member router
x = non-member router
= link
= virtual user-multicast tree

Fig. 2. The architecture of the AMRoute

them. Since this method doesn't use any multicast-related infrastructure, the control message overhead is reduced, but broadcasting data packets yields an overwhelming overhead.

3 Infrastructure-Less Multicast Protocol (ILMP)

In this paper, we propose a new protocol, called the Infrastructure-less Multicast Protocol (ILMP), which provides multicast services based on the underlying unicast routing protocols like the DSR [10] and the AODV [11], without relying on any multicast-supporting infrastructures like a tree or mesh, for dynamically changing ad-hoc networks. Our protocol doesn't have the broadcast overhead (on the other hand, the Simple protocol suffers from the significant broadcasting overhead), and can reduce the multicast-related control overhead since it doesn't use a mesh or tree structure (on the other hand, the ODMRP, the AMRIS, and the AMRoute have to setup and maintain multicast-supporting infrastructures). The operation of the ILMP is as follows. A node wishing to join a multicast group sends a JOIN message to the source. If the source receives the JOIN message, it stores the newly joining member's address to its group membership table and sends back an ACK message to the new member. When a member wants to leave a group, it sends a LEAVE message to the source.

At the source, members are classified based on their next hops which can be obtained from the unicast routing table (URT) (i.e., those members with the same next hop are classified into the same class). The ILMP doesn't rely on any specific unicast routing protocols. The source sends one data packet for each member class. In a data packet, the user data and the list of members' addresses in the same member class are included. The list of members' addresses is specified in the IP Option field [12] as shown in figure 3. The Code field specifies the copy flag, class, and option number. The copy field indicates whether or not this

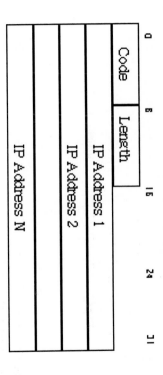

Fig. 3. The IP Option Field

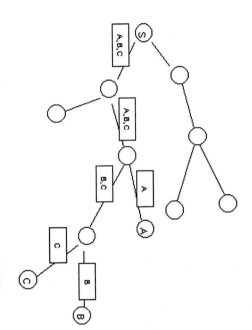

Fig. 4. An operation example of the ILMP

option field is to be copied into each fragment if the IP datagram is fragmented. The class field and option number field are set to 0 and 25, respectively. The LENGTH field indicates how many octets are in the option field, including the first two octets. The multicast group address is specified in the destination IP address field. The forwarding of a data packet is done based on the URT, and a data packet may be copied and forwarded separately if the next hops of the members in the IP Option field are different. Figure 4 shows an example of a source sending a data packet to members A, B, and C. A source can send data to group members based on only the URT without building any multicast-supporting infrastructures. Since it doesn't rely on the broadcasting, the message overhead can be reduced. And, in the case of the topology change, it can send data to members based on the up-to-date URT information. Since the ILMP is based on the URT, data packets can be delivered on the shortest paths to members, and any control overhead for establishing/maintaining the multicast-supporting infrastructure is not needed.

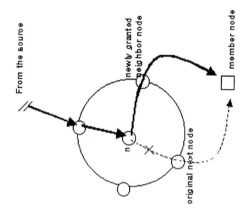

Fig. 5. Reaction to a link disconnection in the ILMP

Even though the URT gives the most recently obtained up-to-date routing information, since nodes can move around continuously, sometimes data packets may not be delivered due to link disconnection. In this case (i.e., no routing information in the URT for some members specified in the IP Option field), the ILMP requests the unicast routing protocol for an update on the routing information. To speed up this process, a node requiring a routing information update checks its neighbors whether they have the corresponding routing information, and asks a neighbor with the information to forward the data packet. As shown in figure 5, when a data packet d can not be forwarded at a node n to a member m, n floods NEXT_HOP_REQUEST messages to its neighbors with TTL set to 1 to ask whether any of them have the routing information for the member m. Neighbors with the routing information for m reply with NEXT_HOP_REPLY messages. The node n selects one of the neighbors replied with the shortest path information for m and sends a GRANT message to the selected neighbor to ask to forward d to m. If all the neighbors don't have the routing formation, the node n waits for the URT to be updated and forwards d based on the updated URT.

Our ILMP tries to reduce the multicast-related control overhead and the overhead to maintaining the multicast-related routing information by not using the multicast-supporting infrastructure. Since the IP addresses of the group members are included in the IP Option field of a data packet, the number of member addresses in a data packet is limited, so the number of packet transmissions may be increased for a large-size multicast group. However, in ad-hoc networks, most of the group communications are small-size ones, hence we can say that our ILMP is suitable for the ad-hoc network environment.

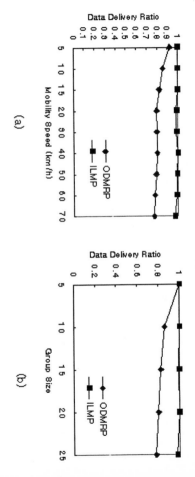

(a)

(b)

Fig. 6. Data delivery ratio

4 Performance Evaluation

To evaluate the performance of the ILMP, we used the GloMoSim simulator [13]. The GloMoSim is a simulation package for mobile network systems written with the distributed simulation language from PARSEC [14]. The simulation environment consists of 50 mobile nodes in the range of 1000m × 1000m, and the transmission range of each node is set to 250m, and the channel capacity is set to 2Mbps. The moving direction of each mobile node is chosen randomly. And we applied the free space propagation model in which the power of the signal decreases $1/d^2$ for the distance d, and assumed the IEEE 802.11 as the medium access control protocol. The AODV is assumed as the underlying unicast routing protocol, and the ODMRP, previously described in session 2, is selected as the multicast protocol to be compared with our ILMP. The ODMRP uses the mesh structure instead the tree to increase the reliability of data transmission, and periodically floods data to efficiently adjust to the dynamically changing network topology. The multicast source was randomly chosen from the group members, and the source generates two 512-byte packets per second in constant bit rates (CBR). The total simulation time was set to 100 seconds.

The performance evaluation factors that we have considered are the data delivery ratio and the total number of data or control packet transmissions for sending one data packet from a source with varying the multicast group size and the node mobility speed. These factors are from the routing protocol performance evaluation factors proposed by the IETF MANET working group [15].

Figure 6 shows the performance comparisons in terms of the data delivery ratio with changing the mobility speed (figure 6(a), with the group size 10) and the multicast group size (figure 6(b), with the node mobility 10km/hr). As the figure shows, our ILMP performs much better in terms of the reliability than the periodic-flooding-based ODMRP. Since, in the ODMRP, only the first packet of each flooding interval is flooded, those packets delivered based on the mesh may be affected by node movements. Since the ILMP is based on the URT, even when node moves, data packets can be more reliably delivered by asking the update of

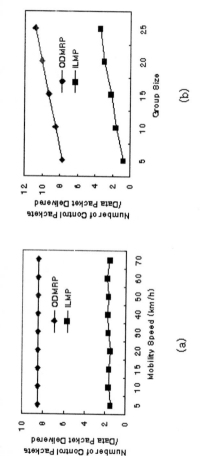

Fig. 7. Control packet overhead

the URT and/or by asking the neighbor nodes of whether they have the needed routing information.

Figure 7 shows the total number of control packets transmitted on the entire network to deliver one source data packet to group members. Both the ILMP and the ODMRP require more control packets for a larger group size. However the ILMP requires much less control overhead than the ODMRP. This is because, in the ODMRP, each node has to periodically send out JOIN_TABLE packets to its neighbors to maintain the mesh structure. On the other hand, our ILMP doesn't have to maintain any multicast-related infrastructure, so the ILMP requires only a small amount of the control overhead.

Figure 8 shows the total number of data packet transmissions required for the delivery of one source data packet to group members. Since the ODMRP is based on the flooding, this performance factor is not much affected by the node mobility and the group size. However, in the ILMP, with a higher mobility speed, the probability of a data packet transmission being failed becomes higher since the ILMP is based on the URT which may not be up-to-date at the time of a packet delivery. With the group size getting larger, the ILMP requires a larger number of data packet transmissions due to the increased number of next hops.

Figure 9 shows the summation of the results from figure 7 and 8, i.e., the total number of control and data packet transmissions per source data packet delivery. For the group sizes with less than or equal to 20 nodes, the ILMP performs better in terms of the reliability than the ODMRP with the less overall message overhead. In addition, since the ILMP doesn't require each mobile node to maintain the multicast-related state information (i.e. multicast routing tables) and is not based on the flooding mechanism, the ILMP is appropriate for the ad-hoc network in which mobile nodes have limited memory and power capacities. And most of the multicast sessions are small-size ones (actually, 20 out of 50 nodes is not a small-size group), so we can argue that the ILMP is good for multicasting in the mobile ad-hoc network.

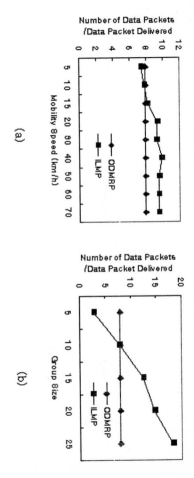

(a)

(b)

Fig. 8. Data packet overhead

5 Conclusions

Most of the applications on the ad-hoc network, which is comprised of mobile nodes without any fixed wired-networks and base stations, are based on the multicast capability. The multicast routing technology on the wired network requires nodes on a multicast tree, even if it is not a group member, to maintain the multicast infrastructure information. If we adopt this multicast technology to the frequently changing ad-hoc network, the maintenance overhead for the multicast-supporting infrastructure may increase overwhelmingly.

Therefore, in our paper, we proposed a new multicast routing protocol which is based on the underlying unicast routing table information, which doesn't burden non-member nodes with the multicast tree or mesh maintenance overhead. To show the performance of our multicast protocol, the ILMP, we compared the ILMP with the ODMRP and showed that the ILMP performs much better in terms of the reliability (i.e., the data packet delivery ratio) and the control overhead.

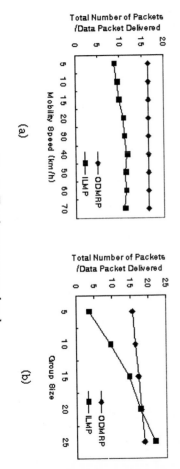

(a)

(b)

Fig. 9. Total message overhead

References

1. S.E.Deering and D.R.Cheriton, "Multicast Routing in Datagram Internetworks and Extended LANs", ACM Transactions on Computer Systems, 8(2):85-110, May 1990.

2. J.Moy, "Multicast Routing Extensions for OSPF", Communications of the ACM, 37(8):61-66, Aug. 1994.

3. T.Ballardie, P.Francis, and J.Crowcroft, "Core Based Trees(CBT) - An Architecture for Scalable Inter-Domain Multicast Routing", In Proceedings of ACM SIGCOMM'93, 85-95, Oct. 1993.

4. S.Deering, D.L.Estrin, D.Farinacci, V.Jacobson, C.G.Liu, and L.Wei, "The PIM Architecture for Wide-Area Multicast Routing", IEEE/ACM Transactions on Networking, 4(2):153-162, April 1996.

5. S.J.Lee, M.Gerla, and C.C.Chiang, "On-Demand Multicast Routing Protocol", Proceedings of IEEE WCNC'99, 1298-1304, Sep. 1999.

6. C.W.Wu, Y.C.Tay, and C.K.Toh, "Ad hoc Multicast Routing protocol utilizing Increasing id-numberS(AMRIS) Functional Specification", Internet-Draft, draft-ietf-manet-amris-spec-00.txt, Nov. 1998.

7. E.Bommaiah, M.Liu, A.McAuley, and R.Talpade, "AMRoute: Ad-hoc Multicast Routing Protocol", Internet-Draft, draft-talpade-manet-amroute-00.txt, Aug. 1998.

8. J.J.Garcia-Luna-Aceves, and E.L.Madruga, "The Core-Assisted Mesh Protocol", IEEE Journal on Selected Areas in communications, 17(8):1380-1394, Aug. 1999.

9. J.G.Jetcheva, Y.C.Hu, D.A.Maltz, and A.B.Johnson, "A Simple Protocol for Multicast and Broadcast in Mobile Ad Hoc Networks", Internet-Draft, draft-ietf-manet-simple-mbcast-01.txt, July 2001.

10. D.B.Johnson and D.A.Maltz, "Dynamic Source Routing in Ad Hoc Wireless Networks", Mobile Computing, 153-181, 1996.

11. C.E.Perkins and E.M.Royer, "Ad hoc On Demand Distance Vector Routing", In Proceedings IEEE WMCSA'99, 90-100, Feb. 1999.

12. Defense Advanced Research Projects Agency, "Internet Protocol", Internet RFC-791, IETF, Sep. 1981.

13. UCLA Computer Science Department Parallel Computing Laboratory and Wireless Adaptive Mobility Laboratory, "GloMoSim- A Scalable Simulation Environment for Wireless and Wired Network Systems",
http://pcl.cs.ucla.edu/projects/domains/glo-mosim.html.

14. R.Bagrodia, R.Meyer, M.Takai, Y.Chen, X.Zeng, J.Martin, and H.Y.Song, "PARSEC: A Parallel Simulation Environment for Complex Systems", IEEE Computer, 31(10):77-85, Oct. 1998.

15. S.Corson and J.Macker, "Mobile Ad hoc Networking(MANET): Routing Protocol Performance Issues and Evaluation Considerations", Internet RFC 2501, IETF, Jan. 1999.

III. Satellite Communications

Buffer Dimensioning of ATM Satellite Link Using Probabilistic Burstiness Curve

Yeong M. Jang

School of Computer Science
Duksung Women's University
Seoul, Korea 132-714
yjang@duksung.ac.kr

Abstract. The author has derived the maximum of *predictive* mean buffer size for a fluid model for an on-board packet switching satellite system. We analyze the traffic behavior using a *fluid bufferless model* of a statistical multiplexing system with multiple types of traffic sources, each modeled as an "On-Off" source. We propose a predictive buffer dimensioning for the onboard shared buffer of the satellite using a probabilistic burstiness curve instead of a deterministic burstiness bound, peak allocation. The result presented is a promising approach for a buffer dimensioning for the broadband OBP (on-board processing) satellite systems.

1 Introduction

ATM satellite networks with on-board ATM switch are expected to support a wide variety of multimedia services. To achieve both efficient utilization of the wireless network resources and acceptable user QoS, real-time traffic management protocols that incorporate connection control must be provided. However, large propagation delays (typically 125 ms), small on-board buffers, and low computational capabilities make some buffer dimensioning schemes [1,2] of the terrestrial broadband ATM networks inappropriate for satellite networks. Typically fluid flow approximations are used to model burst-level quality of services (QoSs) fluctuations.

Because storage capacity on the satellites is limited, buffering cells for thousands of users is impractical due to excess delay time. A typical buffer management approach to congestion control works as follows. An appropriate buffer space is allocated for every connection at every node from source to destination. When the allocated buffer at some node fills up, newly arrived cells are discarded and the upstream node has to retransmit the cell. This approach has a number of serious drawbacks, most significantly, it relies on the congestion from the congested node upstream to delay further transmission (the so-called "back-pressure" effect), thus reacting to congestion rather than avoiding it. The most promising congestion control techniques are those that avoid congestion. Buffering may be used to prevent congestion or to prevent cell loss. Most authors assume that the envelope of traffic is given [1,2]. They assume that if the

I. Chong (Ed.): ICOIN 2002, LNCS 2344, pp. 333–341, 2002.
© Springer-Verlag Berlin Heidelberg 2002

traffic is policed by a set of leaky buckets, then a deterministic constraint on the traffic is defined. The resource allocation problem for guaranteed (not probabilistic) lossless service was suggested, using the deterministic burstiness curve. The advantage of this approach is that they depend only on the deterministic traffic descriptors that the network can enforce, rather than the traffic statistics. Since the shared onboard buffer has limited capacity and some applications (e.g. simple receiver decoder design) want to avoid cell loss under the probabilistic (not deterministic) guarantee at the downlink, we need to carefully allocate the limited shared-buffer space to the downlink dynamically. Initially, allocate the maximum size of mean burst buffer space under the number of active sources for the statistical multiplexing scheme.

A more easily enforceable and verifiable specification of a bit rate function is to bound its envelop; the simplest form is a single value that limits the maximum. It is sufficient to allow lossless transfers through deterministic multiplexing. Leakey bucket is the most commonly suggested form of deterministic burstiness bounding with more than a single limit. It is characterized by the three parameters: the sustainable and the peak rates and the maximum burst duration [4].

2   Buffer Dimensioning Architecture

2.1   Network Model

The efficiency of the downlink heavily depends on the onboard switching approaches adopted. Earth stations are interconnected via a satellite switch with output buffering [13] and shared memory [14]. The shared memory approach provides more flexibility and better memory utilization. Recently, memory technology advanced rapidly, and the memory access speed is no longer a critical part of the whole structure, especially in small satellite switches. For example, as a result, the shared-buffering ATM switches seem to be the most promising architecture [15]. The satellite has a switching fabric capable of routing cells that arrive on the up-links to their destination down-link. The introduction of down-link queue improves the utility of down-link capacity by allowing for statistical multiplexing. Cells arriving from different sources are placed in a common buffer and transmitted over a downlink on a first in first out (FIFO) basis. The resource and mobility management function, residing in the satellite, is responsible to allocate the requested satellite network resources and accept all the resources allocation/deallocation requests from the communication parties. See Fig. 1 for detailed description of the system model.

Suppose that $N(= N_1 + \cdots + N_m + \cdots + N_M)$ independent heterogeneous On-Off sources (connections) are connected to a satellite downlink, where N_m denotes the number of connections of class-m. A VP therefore can be modeled as a single server system with downlink capacity of C bits/second. The buffer size at time t is denoted by K_t. Cells arriving when the buffer is full are assumed to be lost. Onboard implementation of the estimation and prediction procedure can result in a higher accuracy because of a shorter prediction time interval. We use

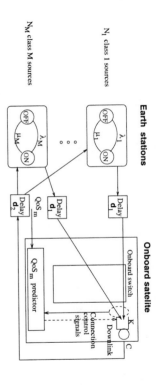

Fig. 1. Buffer dimensioning architecture for broadband satellite communication.

the statistical bufferless fluid model to predict the buffer size at time $t(=d_1+d_2)$, (i.e., the round-trip propagation delay) based on the traffic statistical behavior and the number of active sources at time 0.

2.2 Traffic Model

Although the traffic characteristics of future onboard satellite networks are hard to predict, there are a number of voice and video models reported as On-Off source models [6,7,12], and these models have been commonly used to model a voice source with speech activity detection. An On-Off source alternates between active and idle periods. A voice source is active when the talker is actually speaking, and it is idle when the speaker is silent. We assume that simple Markovian models i.e., models based on the binary On-Off model, are sufficient to capture the short range correlation for telephone speech and video telephone. Heeke [3] described a model-based characterization that can be enforced without notice-ably affecting video quality. The method forces the bit stream to obey a Markov chain model. The admissible rates will be restricted to a few level with geo-metrically distributed holding times. Heyman and Lakshman [10] suggest that long-range dependence is not an important property for most applications of a VBR-video. They show that Markov chain models can be used to describe VBR video-conference traces.

Each source is modeled as an On-Off source. We assume that a series of cells arrive in the form of a continuous stream of bits and use a fluid model. We also assume that the "OFF" and "ON" periods for sources of class-m are both expo-nentially distributed with parameters λ_m and μ_m, respectively. The transition rate from ON to OFF is μ_m, and from OFF to ON is λ_m. Hence the average lengths of the "ON" and "OFF" periods is $\frac{1}{\mu_m}$ and $\frac{1}{\lambda_m}$, respectively. In this traf-fic model, when a source is "ON", it generates cells with a constant interarrival time, $\frac{1}{R_m}$ seconds/bit. When the source is "OFF", it does not generate any cells.

3 Buffer Dimensioning Using Probabilistic Burstiness Curve

We develop the buffer dimensioning scheme bases on the fluid-flow model. We propose a buffer dimensioning scheme for non-fading channels. We allocate the maximum size of mean burst buffer space under the number of active sources for the statistical multiplexing scheme. For cost reasons, keeping the voice or video decoder design very simple is preferable in wireless network environments. In this case the decoding algorithms should reduce cell loss, but the application can still negotiate the multiplexing delay. When bursty traffic is introduced into a network that contains satellite links, we observe that when the burst size is large enough to be a significant fraction of the buffer size, it severely reduces the number of connections that can be supported in probabilistic/statistical multiplexing. To avoid burst-level cell loss, we have to allocate the buffer dynamically and use satellite buffer resources efficiently. The worst deterministic model can guarantee the deterministic (hard) guarantee [1], but the worst *stochastic* model can guarantee the probabilistic (soft) guarantee for buffer dimensioning. Each connected source is defined by (K, C) where K and C represent, respectively, the buffer space and transmission bandwidth allocated to the source at the downlink. Obviously, the allocation of (K, C) will strongly depend on the traffic characteristics and QoS_m requirements of the source. To do the allocation, we use a probabilistic burstiness curve, which allows us to take into account the buffer-sharing effect when allocating network resources. Thus, we are focusing on stochastic QoS guarantees in satellite networks.

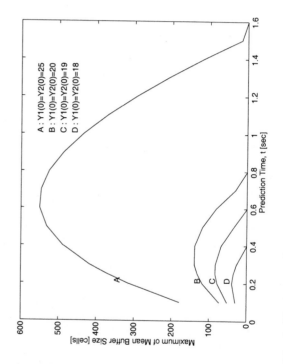

Fig. 2. Maximum of mean burst buffer size versus prediction time.

3.1   Calculating Mean Arrival Rate Using a Fluid Model

We use a statistical bufferless fluid model to dimension the buffer size at a future time $t(= d_1 + d_2)$. We use statistical bufferless fluid flow model [5] to dimension the buffer size that a cell loss occurs at time $t(= 2d)$, (i.e., the round-trip propagation delay) based on the traffic statistical behavior and the number of active sources at time 0. Let $A(t)$ denote the aggregate arrival rate from $Y_m(t)$ active sources. In a bufferless system, cell losses occur when $A(t)$ exceeds the link capacity C. Taking into consideration that each of N independent connection classes, existing connections belongs to one of the M independent connection classes, given by an arbitrary initial condition $Y(0) = Y = [Y_1(0) = i_1, Y_2(0) = i_2, \cdots, Y_M(0) = i_M]$, we obtain the conditional moment generating function of $A(t), s \geq 0$:

$$G_{A(t)|Y(0)}(s) = E[e^{sA(t)}|Y(0) = Y] = \prod_{m=1}^{M} E[e^{sA(t,m)}|Y_m(0) = i_m]$$

$$= \prod_{m=1}^{M} E[e^{sR_m Y_m(t)}|Y_m(0) = i_m] = \prod_{m=1}^{M} G_{A(t,m)|Y_m(0) = i_m}(s)$$

$$= \prod_{m=1}^{M} [p_m(t)(e^{sR_m} - 1) + 1]^{N_m - i_m}[q_m(t)(e^{sR_m} - 1) + 1]^{i_m}. \quad (1)$$

Thus,

$$E[A(t)] = G'_{A(t)|Y(0)}(0) = \sum_{m=1}^{M} A_m(t) = \sum_{m=1}^{M} R_m[i_m q_m(t) + (N_m - i_m)p_m(t)](2)$$

To derive $p_m(t)$ and $q_m(t)$, we can use the forward Chapman-Kolmogorov matrix differential equation:

$$\begin{bmatrix} \pi'_{00}(t) & \pi'_{01}(t) \\ \pi'_{10}(t) & \pi'_{11}(t) \end{bmatrix} = \begin{bmatrix} \pi_{00}(t) & \pi_{01}(t) \\ \pi_{10}(t) & \pi_{11}(t) \end{bmatrix} \begin{bmatrix} -\lambda_m & \lambda_m \\ \mu_m & -\mu_m \end{bmatrix}. \quad (3)$$

$p_m(t)$ and $q_m(t)$ are defined as:

$$\pi_{01}(t) = p_m(t) = \frac{\lambda_m}{\lambda_m + \mu_m}[1 - e^{-(\lambda_m + \mu_m)t}], \quad (4)$$

and by symmetry

$$\pi_{11}(t) = q_m(t) = \frac{\lambda_m}{\lambda_m + \mu_m} + \frac{\mu_m}{\lambda_m + \mu_m}e^{-(\lambda_m + \mu_m)t}, \quad (5)$$

where $p_m(t)$ is the transition probability that a class-m source is active at future time t, given the source is idle at the current time 0. $q_m(t)$ is the transition probability that a class-m source is active at future time t, given the source is active at time 0.

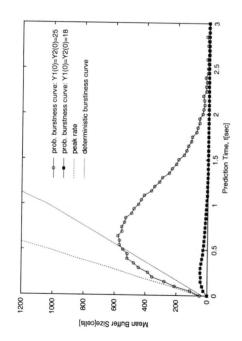

Fig. 3. Simulation result of mean burst buffer size versus prediction time.

3.2 Probabilistic Burstiness Curve Approach

Under a *fluid model*, given a connection characterized by its *instantaneous* ag-gregated peak bit rate from $Y_m(\tau)$ active sources, $\Lambda(\tau)(=\sum_{m=1}^{M} R_m Y_m(\tau))$, a function K_t is called a *burstiness function* [1,2] of an underlying connection if it satisfies

$$E[\max_{0 \leq \tau \leq t} \int_0^t (\Lambda(\tau) - C)d\tau]$$

$$= \max_{0 \leq \tau \leq t} \int_0^t E(\Lambda(\tau) - C)d\tau$$

$$= \max_{0 \leq \tau \leq t} \int_0^t E(\sum_{m=1}^{M} R_m Y_m(\tau) - C)d\tau = K_t. \qquad (6)$$

Note that this is true if $\Lambda(\tau)$ is the instantaneous *stochastic* (not deterministic) arrival rate. We assume that the busy period, t_b, of downlink server is longer than t during the temporary overload case. At $t = 0$, we know that the system (and its queue) is empty and it is never empty from $t = 0$ on, until $t = t_b$. Typically one wishes to invert the order of integration in a double integral [8]. The Fubini and Tonelli [9] theorems are frequently used in tandem. We assume that $(\Lambda(\tau) - C)$ is non-negative and $\int_0^t (\Lambda(\tau) - C)d\tau$ is finite because t is finite. If the multiplexer output is greater than C, then the mean buffer length (in the worst case) is bounded by K_t. If we use a buffer of mean length K_t and a downlink server of fixed service rate C, the mean delay under worst case condition is bounded by K_t/C and cell loss is probabilistically avoided if the buffer length is greater than or equal to K_t. The idea of this strategy is to allocate a large enough buffer to accommodate the maximum burst under worst case conditions so that cells are not lost due to buffer overflow.

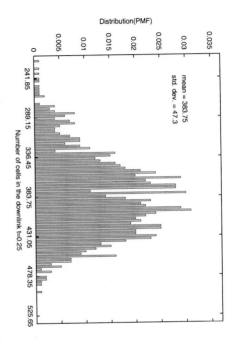

Fig. 4. Probability mass function of number of cells in the output link at t=0.25 (simulation result, 1000 samples).

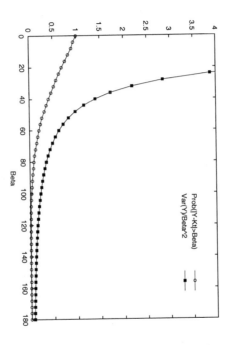

Fig. 5. Chebyshev inequality in the 1000 samples at t=0.25 (simulation result).

4   Numerical and Simulation Results

For the numerical calculation of the maximum of mean buffer size, we consider the following heterogeneous system (M=2): $N_1 = 25$ and $N_2 = 25$ PCM coded sources with $R_1 = 64kbps$, $R_2 = 32kbps$, $\lambda_1 = \lambda_2 = 0.5$ and $\mu_1 = \mu_2 = 0.833$ multiplexed onto a output link of capacity $C = 1.544Mbps$.

For probabilistic buffer dimensioning, we use the probabilistic burstiness function in Fig. 2. Applying this random input process to a single-queue, single-server work-conserving system with transmission bandwidth C and finite buffer capacity, $K_{0.25}$, one can get the following mean buffer size as a function of the number of active sources at $t = 0$. We can find the mean buffer size under the worst case (i.e. all connected sources are all active state at $t = 0$).

For steady-state solutions, we do not need buffering under the above network parameters. Whenever the initial number of active sources is less than 18 for both classes, we do not need any buffer at all during the prediction time, $0.25sec.$ But when the number of active sources is 18 for both classes, the maximum mean buffer size is 35 whereas when the number of active sources is 25, the maximum mean buffer size is 550. Buffer size depends strongly on the initial number of active sources, estimated parameters and the prediction time. Again we see big differences between the predictive and steady-state results.

For the verification of the numerical analysis, we simulated the same system that was considered the heterogeneous system (M=2) : $N_1 = 25$ and $N_2 = 25$ PCM coded sources with $R_1 = 64kbps$, $R_2 = 32kbps$, $\lambda_1 = \lambda_2 = 0.5$ and $\mu_1 = \mu_2 = 0.833$ multiplexed onto a output link of capacity $C = 1.544Mbps.$ We used the simulation tool of BONeS from Candence Design Systems, Inc. We simulated the system 1000 times

In Fig.3, when the number of active sources of both classes is 25 at t=0, the shapes of curves are very similar to that of the numerical analysis. But the maximum mean buffer size is 580 and this value is higher than that of the analysis by 30. When the number of active sources of both classes is 18, the maximum mean buffer size is 45, which is higher than that of the analysis by 10.

When the sources of both classes are always active, we can get the peak rate curve. The slope is the same to that of probabilistic burstiness curve when all of sources of both classes are active. After we found the maximum rate from active sources of both classes at the steady state, we generated traffic at the maximum rate that was found and the deterministic burstiness curve was obtained. The dimensioning from deterministic burstiness is less than that from probabilistic burstiness (the initial number of active sources of both classes is 25) at $0 < t < 0.5$.

Next we found the statistics at $t = 0.25$. The Fig. 4 shows the probability mass function with its mean and standard deviation. The statistics was obtained by using 1,000 samples.

Fig. 5 shows the Chebyshev Inequality for the random number at $t = 0.25$.

5   Conclusions

We studied a simple algorithm for buffer allocation (partition) and control in a satellite downlink with On-Off sources. We proposed a cost-effective and simple buffer allocation scheme for real-time application. We easily found the maximum of mean buffer space under worst case conditions using a fluid model and probabilistic burstiness curve. We use a predictive conditional state probability for buffer allocation. We see big differences between the predictive and steady-state results. An advantage of the predictive buffer dimensioning scheme is that the shared buffer memory may be more fully utilized because network states are almost always in the transient-state, instead of the steady-state. So this is an excellent approach for a buffer dimensioning in satellite systems.

References

1. R. L. Cruz, CA Calculus for Network Delay, Part I: Network Elements in Isolation," *IEEE Tran. on Information Theory*, 37 (1):114–141, Jan. 1991.

2. C.-F. Su and G. D. Veciana, "On the Overflow Probability of Deterministically Constrained Traffic," *IEEE ICC*, 1704–1708, June 1997.

3. H. Heeke, "A traffic-control algorithm for ATM networks," *IEEE Trans. Circuits and Systems for Video Tech.*, 3(3): 182–189, June 1993.

4. G. Karlsson, "Asynchronous transfer of video," *IEEE Comm. Magazine*, 118–126, Aug. 1996.

5. D. Anick, D. Mitra, and M. M. Sondh, "Stochastic Theory of a Data-handling System with Multiple Sources," *Bell System Technical Journals*, 61(8):1871–1894, Oct. 1982.

6. P. T. Brady, "A Statistical Analysis of On-Off Patterns in 16 Conversations," *Bell System Technical Journals*, 47(1):73–91, Jan. 1968.

7. Brady, P. T., "A Model for Generating On-Off Speech Patterns in Two-Way Conversation," *Bell System Technical Journals*, 48:2445–2472, Sept. 1969.

8. G. Klambauer, *Real Analysis*, American Elsevier Pub. Co, New York, 1973.

9. G. B. Folland, *Real Analysis: Modern Techniques and their Applications*, New York: Wiley, 1984.

10. D. P. Heyman and T. V. Lakshman, "What are the Implication of Long-range Dependent for VBR-Video Traffic Engineering," *IEEE/ACM Tran. on Networking*, 4: 301–317, June 1996.

11. A. Elwalid, D. Mitra, and R. H. Wentworth, "A New Approach for Allocating Buffers and Bandwidth to Heterogeneous, Regulated Traffic in an ATM Node," *IEEE JSACs*, 13(6): 1115–1127, August 1995.

12. B. Maglaris and et al., "Performance Models of Statistical Multiplexing in Packet Video Communications," *IEEE Tran. on Communications*, 36(7):834–843, July 1988.

13. W. D. Ivancic, M. J. Shalkhauser, and J. A. Quintana, "A network architecture for a Geostationary communication satellite," *IEEE Comm. Magazine*, July 1994.

14. H. Yamanaka and et al., "Scalable shared-buffering ATM switch with a versatile searchable queue," *IEEE JSACs*, June, 1997.

15. A. Jajszczyk, M. Roszkiewicz, and J. garcia-Haro, "Comparison of ATM shared-memory switches," *XV Int. Switching Symp.* (ISS'95), pp. 409-413, April. 1995.

Construction of Sequences for Multi-group FH-FDMA in DRT System

Seong-Bok Park[1], Kwang-Eog Lee[1], Young-Kyun Choi[1], and Soon-Ja Kim[2]

[1]Agency for Defense Development Dept.4-1-2,
Yoosung P.O. Box 35-4, Taejon, Korea(South)
{psb,kelee,ykchoi}@add.re.kr

[2]Kyungpook National University, Secure Communication Network Lab.
Taegu Korea(South)
snjkim@ee.knu.ac.kr

Abstract. This paper presents a method of constructing frequency hopping sequences for military satellite communication system which employs frequency dehop-rehop transponder(DRT). In DRT system, the dehopping and rehopping is to be performed upon a group based approach where each group has its own hopping sequence and is characterized by synchronous connection-oriented(SCO) link. The hopping patterns generated through proposed method have optimal Hamming auto-correlation property according to Lempel-Greenberger bound and assure perfect orthogonality between hopping groups i.e. no chip collisions during the period. A practical implementation of the proposed method using shift registers is also described.

1 Introduction

Present day in military communication systems, frequency hopping(FH) is generally favored spread spectrum technique to combat intentional jamming and interception. For FH satellite communication systems DRT is well known to have reasonable level of onboard processing complexity and to provide good anti-jamming capability[1][2]. In DRT system the Up-link/Down-link is generally of dehop-rehop FH-FDMA type, in which frequency hopping is performed in group i.e. all channels belonging to the same group will follow the same hopping law. And the link is characterized by SCO link that means DRT is a master and ground terminals slaves.

In this system the important requirement of the frequency hopping sequences allocated to each group is that they should be not only optimal by themselves in the sense of Hamming auto-correlation property but also orthogonal i.e. no chip collisions. While there are several methods of constructing frequency hopping sequences which assure optimal auto-correlation and cross-correlation property [3] [4] [5][6] they do not guarantee no chip collisions between groups during the period.

In this paper, a new construction method which assures optimality of each sequence and being perfect orthogonal to each other is presented. We also describe a

I. Chong (Ed.): ICOIN 2002, LNCS 2344, pp. 342–350, 2002.
© Springer-Verlag Berlin Heidelberg 2002

practical implementation of the proposed onboard frequency hopping generator using shift registers.

2 Methodology

The proposed scheme consists of two steps, coarse mapping and fine mapping. A FH sequence family with p^k symbols can be constructed from an maximal-sequence over GF(p). Given a prime p and positive integer k, let

$$P = \{0, 1, \cdots, p-1\}$$
$$P_k = \{0, 1, \cdots, p^k - 1\}$$

and let P^k denote the set of all words of length k (k-tuples) over P. The one-to-one mapping σ_k from P^k onto P_k under which $w = (w_0, w_1, \cdots, w_{k-1}) \in P^k$ corresponds to

$$w\sigma_k = \sum_{i=0}^{k-1} w_i p^{k-1-i} \in P_k. \qquad (1)$$

Let $X(j, k)$ denote the jth k-tuple $(x(j), x(j+1), \cdots, x(j+k-1))$ of successive elements in the sequence $X = \{x(j)\}$ of length q over P. If A is denoted by the set of available hopping slots within a hopping bandwidth and $\|A\| = p^k$, then the FH sequence $Y = \{y(i)\}$ of length q over P_k can be obtained from the sequence $X = \{x(j)\}$ as

$$y(j) = X(j, k)\sigma_k = \sum_{i=0}^{k-1} x(j+i) p^{k-1-i}, 0 \le j < q. \qquad (2)$$

To describe coarse mapping in the construction of an orthogonal set of FH sequences for multi-group FH-FDMA in DRT system let $N = p^r$ denote the number of groups then A can be decomposed into N subsets $A_u (0 \le u < N)$ of size $p^m (m = k - r)$.

To define one-to-one mapping from N groups onto N subsets in bijection for coarse mapping, let's construct a set of sequences $C_i = \{c(u)\}, (0 \le u, j < N)$ of length N over $\{0, 1, 2, \ldots, N-1\}$ such that none of the same element in any sequence appears by the following way. Let $C_r(i, u)$ denote the termwise mod-p sum of the r-tuple $i\sigma_r^{-1} \in P^r$ and the r-tuple $u\sigma_r^{-1} \in P^r$ which is mapped under σ_r into u. That is,

$$C_r(i,u) = i\sigma_r^{-1} \oplus u\sigma_r^{-1}, (0 \le i, u < N-1). \tag{3}$$

where $\oplus$ represents termwise mod-p sum. For each $u \in P_r$ let $C_i = \{c_i(u)\}$ be the ith sequence whose uth term is defined by

$$c_i(j) = C_r(i,u)\sigma_r. \tag{4}$$

As an example, let $p = 2$, $N = 4(r = 2)$ then we obtain the following set of sequences.

$$C_0 = 0,1,2,3.$$
$$C_1 = 1,0,3,2.$$
$$C_2 = 2,3,0,1.$$
$$C_3 = 3,2,1,0.$$

It can be easily checked that in any sequence none of the same element appears and occupies the same position. Since the value of uth element in each C_i can be regarded as a group identifier mapped onto a subset A_u, the selection of C_i determines the coarse mapping of N groups. We can determine the index of C_i pseudo-randomly by

$$i = X(j,r)\sigma_r. \tag{5}$$

Once the index i is determined by the expression above, let v, $(0 \le v < N-1)$ denote a group identifier, then the index of subset A_u for each group is identified by.

$$u = (i\sigma_r^{-1} \oplus v\sigma_r^{-1})\sigma_r = (X(j,r) \oplus v\sigma_r^{-1})\sigma_r. \tag{6}$$

Note that although C_i sequences are introduced to define one-to-one mapping from N group identifiers onto N subsets they are no longer used in the above expression for the coarse mapping, which means that the coarse mapping can be implemented with lower H/W complexity.

The next step, the fine mapping to a frequency slot in its corresponding subset $A_i (0 \le i < N)$ can be carried out by the same way in (2). But it has a weak point in a way that the relative distance between the hopping slots of each group is constant. As a result it does not provide randomness between the fine mappings of each group. To overcome this drawback N different k-tuples are taken for the fine hopping as following way.

$$f_{g_v}(j) = X(j \ominus d_v, m)\sigma_m, (0 \le v < N, 0 \le j < q) \tag{7}$$

where $\ominus$ represents modulo-q addition. The constant d_v is arbitrary but a fixed value greater than r to get m-tuple words for fine mapping and the ordering condition on d_v is $d_v < d_{v+1}$. The complete expression of the mapping algorithm including the coarse and fine mapping is

$$y_{g_v}(j) = (X(j,r) \oplus v\sigma_r^{-1})\sigma_r p^m + X(j\sigma d_v,m)\sigma_m, (0 \le v < N, 0 \le j < q). \tag{8}$$

From the realization point of view, the proposed method (8) can be easily implemented with an feedback shift registers(FSR) generating an maximal-sequence by mapping the two vectors(r-tuple, m-tuple) selected from the n stages of FSR to a hopping symbol as follows.

For simplicity let $r < d_v \le n$, $d_v < d_{v+1}$, and the first r stages of the FSR are assumed to be used for the coarse mapping, then the hopping sequence $Y_v = \{y_{g_v}(j)\}$ for a group identifier v is expressed as

$$y_{g_v}(j) = (X^j(0,r) \oplus v\sigma_r^{-1})\sigma_r p^m + X^j(d_v,m)\sigma_m, (0 \le v < N, 0 \le j < q). \tag{9}$$

Where $X^j(a,b)$ represents the a th b-tuple of successive contents of n-stage shift registers after j times forward shifting of all the contents of the FSR. The first term in the expression above indicates the coarse mapping the second term the fine mapping.

Theorem 1.

Let $X = \{x(j)\}$ be a maximal-sequence of length $q = p^n - 1$ over $GF(p)$ and $N = p^r$ be the number of groups. Then for each k, $(1 \le k \le n)$, a set of N $y_{g_v}(j)$ sequences is an orthogonal (i.e. no chip collision) set and each member is optimal by itself in terms of Hamming correlation.

Proof.

− *Optimality:*

To show optimal Hamming auto-correlation property according to Lempel-Greenberger bound we follow the similar process in [4]. Since the lower bound on the out of phase Hamming autocorrelation of any sequence of length $q = p^n$ over P_k is

$$H(Y_v) = p^{n-k} - 1$$

we should prove $H(Y_v) = p^{n-k} - 1$ for any $(0 \le v < N)$ Let's define two k-tuples $\hat{w} = (\hat{w}_1, \hat{w}_2 \cdots, \hat{w}_r, \cdots \hat{w}_{k-1}), w = (w_1, w_2 \cdots, w_r, \cdots w_{k-1}) \in P^k$ specifically such that

$$(\hat{w}_1, \hat{w}_2 \cdots, \hat{w}_r) = X^j(0,r) \oplus v\sigma_r^{-1}, (\hat{w}_{r+1}, \hat{w}_{r+2} \cdots, \hat{w}_{k-1}) = X^j(d_v,m)$$

$$(w_1, w_2 \cdots, w_r) = X^j(0,r), (w_{r+1}, w_{r+2} \cdots, w_{k-1}) = X^j(d_v,m) \,.$$

Then any element $e \in P_k$ of the sequence $Y_v = \{y_{g_v}(j)\}$ can be expressed as

$$H(Y) \ge p^{n-k} - 1 \tag{10}$$

$$\tag{11}$$

$$e = \hat{w}\sigma_k = \sum_{i=0}^{k-1} w_i p^{k-1-i} \in P_k. \tag{12}$$

Since the function $f(X^j(0,r)) = X^j(0,r) \oplus v\sigma_r^{-1} : P^r \mapsto P^r$ in (11) is a bijection for any fixed group identifier v, the function f performs just a permutation. So the distribution of $\hat{w}$ and w is exactly the same. In other words, let $\eta_X(w)$ denote multiplicity of w in X then

$$\eta_X(w) = \eta_{Y_v}(\hat{w}\sigma_k) = \eta_{Y_v}(e). \tag{13}$$

From the property of maximal-sequence mentioned above

$$\eta_X(w) = \begin{cases} p^{n-k} - 1 & if \ w = 0 \\ p^{n-k} & if \ w \neq 0 \end{cases}. \tag{14}$$

From the relationship in (12) between e and $\hat{w}$ it is obvious that for any $Y_v = \{y_{g_v}(j)\}, (0 \leq v < N)$

$$\eta_{Y_v}(e) = \begin{cases} p^{n-k} - 1 & if \ e = vp^m \\ p^{n-k} & if \ e \neq vp^m \end{cases}. \tag{15}$$

Let γ^τ denote cyclic shift operator defined by

$$Y_v \gamma^\tau = \{y_{g_v}(j + \tau)\}. \tag{16}$$

And let $\hat{w}^\tau, w^\tau$ denote k-tuple expression of $\hat{w}, w$ respectively to which the cyclic shift operator γ^τ is applied such that $j \rightarrow (j + \tau)$ in (11).

Then the out of phase Hamming autocorrelation of any sequence $Y_v = \{y_{g_v}(j)\}, (0 \leq v < N)$ is expressed as

$$H_{Y_v \ Y_v}(\tau) = \sum_{j=0}^{q-1} h[y_{g_v}(j), y_{g_v}(j + \tau)], 0 \leq \tau < q, \tag{17}$$

where

$$h[a,b] = \begin{cases} 0, if \ a \neq b \\ 1, if \ a = b \end{cases}.$$

From (9)

$$y_{g_v}(j) = y_{g_v}(j + \tau), \quad iff \quad \hat{w} = \hat{w}^\tau. \tag{18}$$

This equality between $y_{g_v}(j)$ and $y_{g_v}(j+\tau)$ also holds on the condition i.e. *iff* $w = w^\tau$ from the property of the function $f(X^j(0,r)) = X'(0,r) \oplus v\sigma_r^{-1} : P^r \mapsto P^r$, then

$$y_{g_v}(j) = y_{g_v}(j+\tau), \quad iff \quad w = w^\tau . \tag{19}$$

(19) means that the termwise mod-p difference between w and w^τ equals to all zero k-tuple, $0^k \in P^k$. Let $Z = X - Xy^\tau$ be the termwise mod-p difference between X and Xy^τ and w' denote k-tuple in Z such that

$$(w_1', w_2' \cdots, w_r') = Z^j(0,r), (w_{r+1}', w_{r+2}' \cdots, w_{k-1}') = Z^j(d_v, m). \tag{20}$$

Then it is clear that we can define the following condition on which the equality between $y_{g_v}(j)$ and $y_{g_v}(j+\tau)$ holds with respect to Z.

$$y_{g_v}(j) = y_{g_v}(j+\tau), \quad iff \quad w' = 0^k. \tag{21}$$

So (21) becomes

$$H_{Y_v \ Y_v}(\tau) = \eta_Z(0^k). \tag{22}$$

Since $Z = X - Xy^\tau$ is just another cyclic shift of X from the "shift and add " property of maximal-sequence, the multiplicity of $0^k \in P^k$ is the same for all $\tau \neq 0$. So (22) can be expressed as

$$H_{Y_v \ Y_v}(\tau) = \eta_X(0^k), \quad \tau \neq 0 . \tag{23}$$

And from (13)

$$H_{Y_v \ Y_v}(\tau) = \eta_{X_v}(vP^m), \quad \tau \neq 0 . \tag{24}$$

Finally, from (15) and (24)

$$H_{Y_v \ Y_v}(\tau) = p^{n-k} - 1, \quad \tau \neq 0 . \tag{25}$$

In case of $\tau \neq 0$, $H_{Y_v \ Y_v}(0) = q$ is self-evident. Therefore the out of phase Hamming autocorrelation of any sequence $Y_v = \{y_{g_v}(j)\}, (0 \leq v < N)$ is

$$H_{Y_v \, Y_v}(\tau) = \begin{cases} q, & if \ \tau = 0 \\ p^{n-1} - 1, & if \ \tau \neq 0 \end{cases}. \tag{26}$$

Since (26) satisfies (10) with equality for every $Y_v = \{y_{g_v}(j)\}, (0 \leq v < N)$, the optimality that each sequence in the set is optimal by itself is proved.   Q.E.D.

– *Orthogonality* :

Orthogonality between each group can be verified from the cross correlation point of view. In this context since SCO(synchronous connection-oriented) link is assumed the cross Hamming correlation between any pair of sequences is modified as

$$H_{Y_a \, Y_b}(\tau) = \sum_{j=0}^{q-1} h[y_{g_a}(j), y_{g_b}(j+\tau)], \tau = 0, a \neq b \tag{27}$$

where

$$h[a,b] = \begin{cases} 0, & if \ a \neq b \\ 1, & if \ a = b \end{cases}.$$

Let's define two k-tuples

$w'' = (w_1'', w_2'' \cdots, w_r'', \cdots w_{k-1}''), w' = (w_1', w_2' \cdots, w_r', \cdots w_{k-1}') \in P^k$ such that

$$(w_1'', w_2'' \cdots, w_r'') = X^j(0,r) \oplus a\sigma_r^{-1}, (w_{r+1}'', w_{r+2}'' \cdots, w_{k-1}'') = X^j(d_a, m)$$

$$(w_1', w_2' \cdots, w_r') = X^j(0,r) \oplus b\sigma_r^{-1}, (w_{r+1}', w_{r+2}' \cdots, w_{k-1}') = X^j(d_b, m).$$ (28)

Let w_a'', w_b', w_a, w_b, w denote $(w_1'', w_2'' \cdots, w_r''), (w_1', w_2' \cdots, w_r'), (w_1', w_2' \cdots, w_r'), a\sigma_r^{-1}, b\sigma_r^{-1}, X^j(0,r)$ which means a respectively, then the necessary condition for $H_{Y_a \, Y_b}(0) \neq 0, a \neq b$ which means a collision between Y_a and Y_b is

$$w_a'' \oplus w_b' = 0^k. \tag{29}$$

For simplicity, suppose p = 2, then the termwise mod-2 addition $\oplus$ is simply adding the corresponding components as follows

$$(w_1'' \oplus w_1', w_2'' \oplus w_2' \oplus w' \cdots, w_r'' \oplus w_r') =$$

$$(w_1' \oplus w_{a1} \oplus w_1 \oplus w_{b2}, w_2' \oplus w_{a2} \oplus w_2 \oplus w_{b2} \cdots, w_r' \oplus w_{ar} \oplus w_r \oplus w_{br}) = 0^k.$$ (30)

Since mod-2 addition $\oplus$ is commutative and the inverse of an element is itself and w is an fixed r-tuple for a given j with respect to all the group identifiers, (30) reduces to

$$(w_{a1} \oplus w_{b2}, w_{a2} \oplus w_{b2} \cdots, w_{ar} \oplus w_{br}) = 0^k . \tag{31}$$

To meet the necessary condition (22) it is obvious that all the components should be zero i.e. $a = b$. But this contradicts the basic assumption $a \neq b$. Therefore if only if the group identifiers are different each other, the cross Hamming correlation between any pair of sequences in F is 0, i.e.

$$H_{Y_a Y_b}(0) = 0, \quad \textit{iff } a \neq b . \tag{32}$$

$$\text{Q.E.D.}$$

An example of the implementation of *Theorem* 1 in case of $P(x) = 1 + x^3 + x^4 + x^6 + x^9$, $N = 4$, *and* $k = 5$ is shown in figure 1. In this example, the first 2 stages of the FSR are used for the coarse mapping and to assure the randomness between groups, $d_0 = 3, d_1 = 4, d_2 = 5, \text{and } d_3 = 6$ are assumed for the fine mapping but the values can be arbitrary but fixed values greater than 2 on condition $d_v < d_{v+1}$. As shown in figure1 the generator provides 4 optimal and independent sequences for each group with only one maximal-sequence generator, which implies lower hardware complexity and simple key management.

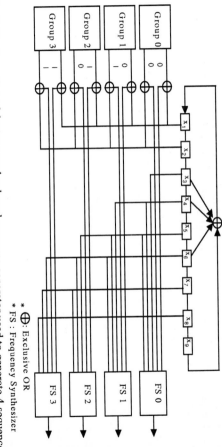

Fig. 1. Configuration of the proposed onboard sequence generator used to generate 4 sequences for **4 group-FH-FDMA** in DRT system.

* $\oplus$: Exclusive OR
* FS : Frequency Synthesizer

3   Conclusions

In DRT system, the dehopping and rehopping is to be performed upon a group based approach where each group has its own hopping sequence and is characterized by SCO link.

In this paper we proposed a method of constructing a set of sequences for multi-group FH-FDMA in DRT system such that each member has to be an optimal sequence by itself and orthogonal (i.e. no chip collisions) to any other member. Furthermore we showed that the proposed method could be easily implemented on board using shift registers.

References

1. Hovanessian, S.A., Jocic, L.B.,Kreng,J.K.: Multi-user transparent-dehop MILSATCOM system, Proceedings of MILCOM 1995, October 1995.
2. S.M. Sussman and P.Kotiveeriah: Partial Processing Satellite Relay for Frequency-Hop Antijam Communications, IEEE Transactions on Communications, Vol. Com-30, No. 8,1929–1937, August 1982.
3. S.B.Wicker and V.K.Bhargava: Read-Solomon Codes and Their Applications, IEE Press, 1994.
4. A.Lempel and H.Greenberger: Families of Sequences with Optimal Hamming Correlation Properties, IEEE Transactions on Inform. theory, Vol. IT-20, No. 1, 90–94, Jan. 1974.
5. Kwon,W.C. and Guu-Chang Yang : Frequency-hopping codes for multimedia services in mobile telecommunications, IEEE Transactions on Vehicular Technology, Vol.48, 1906–1915, Nov. 1999.
6. P.Udaya and M.U.Siddiqi : Optimal Large Linear Complexity Frequency Hopping Patterns Derived from Polynomial Residue Class Rings, IEEE Transactions on Inform. theory, Vol. IT-44, No. 4,1492–1503, Jul.1998.

Development of a High-Accurate Time Server for Measurements of the Internet

Yoshiaki Kitaguchi[1][2][3], Haruo Okazawa[3], Syunsuke Shinomiya[3],
Yutaka Kidawara[3], Katsuya Hakozaki[2], and Shin-ichi Nakagawa[3]

[1] GENESIS Project, Koganei Reserch Group,
Telecommunications Advancement Organization of Japan
4-2-1 Nukui-Kitamachi, Koganei, Tokyo 184-8795, Japan
kita@genesis.tao.go.jp
[2] Information Systems Management Laboratory,
University of Electro-Communications
1-5-1 Chofugaoka, Chofu, Tokyo 182-8585, Japan
hako@is.uec.ac.jp
[3] Next Generation Internet Group, Communications Reserch Laboratory
4-2-1 Nukui-Kitamachi, Koganei, Tokyo 184-8795, Japan
{okazawa, shinomiya, kidawara, snakagaw}@crl.go.jp

Abstract. The network delay is one of the essential elements for the
Quality of Service (QoS). The time delay is measured by placing devices
at both ends of the network and using the machine's internal time clock.
The accuracy of the measurement is dependent on the precision of the
clock, therefore the precision of the clock is very important. Network
Time Protocol (NTP) is applied for the time synchronization among
network nodes; hence, the time accuracy at each NTP server is considered
very important. Moreover, the improvement of the accuracy is required
for the recent broadband network.
Our objective in this research is to improve the time stability of the top
NTP server (stratum1) by developing a server with higher stability. The
server, which is called the high-accurate time server, was designed based
on external high accurate frequency signal, instead of the crystal oscillator on the motherboard of a Personal Computer (PC). The high-accurate
time server is capable of the time stability of several hundred nanoseconds. Also evaluated in this paper is the high precision measurement of
network delay, which was proceeded with the server.

1 Introduction

The Quality of Service (QoS) is one of the next-generation Internet technologies, and the network delay is a very important part of it. When the delay is
measured, all equipment must be synchronized for a regular number of days,
one month for example, to obtain accurate results. The widespread method to
accomplish network synchronization is called Network Time Protocol (NTP) [1].
NTP has a hierarchical structure to implement load balancing, and functions as

I. Chong (Ed.): ICOIN 2002, LNCS 2344, pp. 351–358, 2002.
© Springer-Verlag Berlin Heidelberg 2002

a synchronizer on the top NTP server (Stratum 1) via the network. Therefore, the time precision of Stratum 1 greatly influences the entire NTP structure.

In addition, as electronic commerce is widely available these days, packet time has become more and more important to avoid any time discrepancy between seller and buyer; which means, highly precise time synchronization technology is indispensable in contemporary society. The time accuracy of Stratum 1, which is synchronized by a Global Positioning System (GPS), is currently more than 10 microseconds (See the Table 1) [2]. However, on a broadband network such as OC-12 of 622Mbps, the transmission time for an average size packet (approximately 500 bytes) is about 6 microseconds; whereas the exact values of delay cannot be measured beyond the precision of the time synchronization. Therefore we believe that the time resolution of the NTP server is too low to measure accurate transmission time of the broadband network.

In this paper, we present the overview of NTP and its current situations, emphasizing the necessity of time precision in chapter 2. As for precision improvement of network time synchronization, we report our experiments with the high-accurate time server, which we designed to reduce the time jitter of the NTP server. The report and evaluation of the experiments will follow in later chapters.

Table 1. Trimble Palisade receiver operating system compatibility platform

Platform	OS	NTP Source	Accuracy
i386(PC)	Linux	NTP Distribution	$10\mu s$
i386(PC)	Windows 2000/NT	Trimble NTP	$1ms$
SUN	Solaris, SunOS 4	NTP Distribution	$50\mu s$
HP	HPUNIX 9,10,11	HP NTP	$50\mu s$
Various	FreeBSD	NTP Distribution	$20\mu s$
Cisco Router	Model 7200	Cisco NTP	$20\mu s$

2 Time Synchronization Technology in a Network

2.1 NTP

NTP which is widely used as a network time synchronization technology, constitutes its time synchronization system with a hierarchical structure. NTP acquires the Coordinated Universal Time (UTC) at the top NTP server (Stratum 1) from the external time sources (See the Figure 1). Each server obtains information from the server on the same level or one up. The information required for time synchronization is transformed into the timestamp format, and is displayed in seconds. The seconds are a continuous time starting from January 1, 1900 and is indicated by a "64-bit unsigned fixed-point number". The maximum number of seconds that can be represented is as large as 4,294,967 with the precision of approximately 200 picoseconds. As it can be seen from this fact, the NTP system itself is already capable if further precision is required.

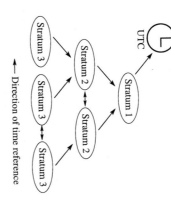

Fig. 1. Structure of NTP

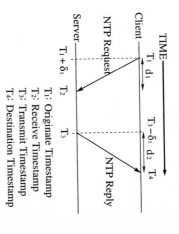

Fig. 2. Time synchronization flow at NTP

2.2 Precision of Time Synchronization

The offset value, which is acquired by measuring the difference between the reference time and internal time, is influenced by the network delay when it implements synchronization via the network. Unlike conventional time synchronizing protocols, NTP can eliminate the influence and improve the precision of synchronization by committing bilateral transmission. The concept of NTP transmission is shown in Figure 2, where each symbol has a specific meaning: T_1 is the time point when a request packet is sent from the client (Originate Timestamp); T_2 is the time point when the packet is received by the server (Receive Timestamp); T_3 is the time point when a reply packet is sent from the server (Transmit Timestamp), T_4 is the time point when the packet is received by the client (Destination Timestamp). Then the communication delay of request packet (d_1) and the communication delay of a reply packet (d_2) were figured out from the time difference between the server and the client (the offset value: δ_t):.

$$d_1 = T_2 - (T_1 + \delta_t)$$
(1)

$$d_2 = T_4 - (T_3 - \delta_t)$$
(2)

The offset value δ_t can be gained with the Formula 1 and 2.

$$\delta_t = \frac{(T_2 - T_1) + (T_3 - T_4) + (d_2 - d_1)}{2}$$
(3)

NTP normally calculates the offset value disregarding any possible time difference caused by request or reply (communication) delay ($d_1 - d_2 = 0$). Thus, when jitter causes time difference, the measurement of accurate value becomes difficult.

2.3   Precisions and Problems

Stratum 1 mainly employs a GPS as the source of external time because of its low-cost. The precision of Stratum 1 with GPS is more than 10 microseconds as mentioned before. On the other hand, Stratum 2, which refers to the time of Stratum 1, can only achieve the precision of several 10 microseconds, according to sources such as operation data [3] or research data [4]; therefore Stratum 2 and other clients are well behind Stratum 1 in precision.

Thus it was understood in this experiment that time difference is due to delay-jitter $(d_2 - d_1 \neq 0)$, which is caused by the network delay. The unresolvability of this condition is a problem with the NTP networks.

3   A Design of a High-Accurate Time Server

The concept of Stratum 1 was then applied in our designing of the high-accurate time server, which is expected to reduce the jitter that controls the precision of the internal clock.

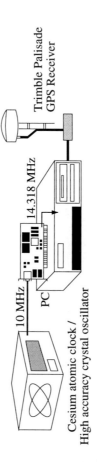

Trimble Palisade
GPS Receiver

14.318 MHz

10 MHz

PC

Cesium atomic clock /
High accuracy crystal oscillator

Fig. 3. Overview of high-accurate time server

3.1   Realization of a High-Accurate Time Sever with a PC

The high precision of the Stratum 1 server relies on the precision of its internal clock. In the case that general PC is employed as a NTP server, much less is expected with the internal clock. A $\pm 20ppm$ is the average precision of PC crystal oscillators when used as a 14.318 MHz signal source. This means that there is the error of approximately 1.8 seconds per day ($\cong 1.8s/day$). We replaced the normal crystal oscillator with a one with better precision and also employed the external signal-inputting device, which is easy to install in a PC. The signal-inputting device can convert external 10 MHz signals into those of 14.318 MHz and communicate with any time-sources capable of 10 MHz signals. We measured the stability of time accuracy of the time server that employs signals of a cesium atomic clock ($2.0 \times 10^{-12}/day$) and a high accuracy crystal oscillator ($5.0 \times 10^{-10}/day$) via the device. The composition of the NTP server used in the experiment is shown in Figure 3, and GPS via Trimble Palisade NTP was applied as the time source to the high-accurate time server.

As the result of our measurement, only small differences were found between the types of signals [5]; however, it was confirmed that the time precision of a PC server could be upgraded to the level of a microsecond order (See the Figure 2). The value acquired with the one-day measurement indicated the possibility of 0.5 microsecond precision, although it could not be confirmed. The Linux OS did not have the time resolution to measure precision of less than 1 microsecond.

Table 2. Time precision at normal Linux OS

Time source	Precision of Time source	Data (Standard deviation)
PC inside crystal oscillator	$1.8/day$	—
High accuracy crystal oscillator	$5.0x10^{-10}/day$	$0.57\mu s$
Cesium atomic clock	$2.0x10^{-12}/day$	$0.55\mu s$

3.2   The Nanokernel

As a method to improve the time resolution of the OS, a nanokernel was used in this experiment. A nanokernel, which was developed by D. Mills et. al., can be adjusted to Linux in the patch form, named PPSKit. It is implemented with the time resolution of nanosecond and PPS API that is described on RFC2783 [6]. By adding a nanokernel to our original time server, we measured its time stability of which details are shown in Figure 4. On the graphs in Figure 4, the vertical axis indicates the offset values of the time server and GPS, which are scaled by microsecond, and the horizontal axis elapsed time. These data are acquired by using loopstats that is a NTP distribution's tool.

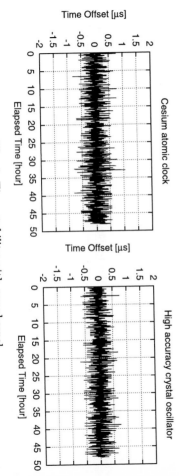

Cesium atomic clock

High accuracy crystal oscillator

Fig. 4. Time stability with nanokernel

What we found through the two-day measurement is that an average PC can achieve the precision of approximately 200 nanoseconds (standard deviation) with high accuracy crystal oscillator, and become efficient enough to be called the high-accurate time server. Although a cesium atomic clock was also applied in the measurement, it also resulted in approximately 200 nanoseconds (standard deviation). Therefore, any superiority of atomic clock to crystal oscillator was not observed in this experiment.

3.3   The Use of a High-Accurate Time Server

It is possible to improve the time precision of Stratum 1 by developing a high-accurate time server. However, as mentioned in chapter 2, time synchronization by NTP is largely influenced by network jitter, which is caused by communication delay. Thus, it is difficult to achieve time synchronization with nanosecond precision throughout networks, and is important to examine the precision of network jitter. This led us to the next stage, where we measured the time accuracy of a network using the high-accurate time server.

4   A Network Measurement Experiment with a High-Accurate Time Server

4.1   A Measurement of Influence of a Delay

The accuracy of network measurement depends on the precision of the devices at the both ends of the network. The current technology only provides time synchronization of maximum of several milliseconds. We measured the jitter on a real network using the high-accurate time server. Figure 5 shows the configuration of the measurement system, which places two high-accurate time servers, which are connected to each other via a switching HUB. One server was used

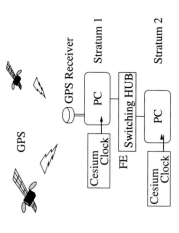

Fig. 5. System topology of network measurement

as Stratum 1, and the other as Stratum 2, and they are both applied for the measurement with NTP protocol. Signals from the cesium atomic clock control the internal clock of those servers, while GPS is also used to adjust the one in Stratum 1.

We obtained the statistic data from NTP distribution's tool. The offset value of Stratum 2 is shown in Figure 6. The delay time between Stratum 1 and Stratum 2 are shown in Figure 7. These data are acquired by using peerstats that is the NTP distribution's tool. The data indicates that when the network delay is stable, the time precision can be synchronized at ± about 10 microseconds on the Fast Ethernet, whereas, when the time stability became unstable due to the change of the network delay after approximately 250 hours of measurement, the time offset on the Fast Ethernet was swung by ± about 15 microseconds.

4.2   Consideration of the Experiment

This result indicates that the time synchronization between two NTP servers can only obtain the precision of tens of microseconds, even when they are on the same network segment of the Fast Ethernet. The implementation of NTP was designed to provide the time synchronization, by being flexible with the change of network routing. Therefore, the time precision was greatly influenced by "the change of the network delay" (delay-jitter) NTP is not yet sufficient enough to carry out synchronization over one microsecond level on a wide area network. To improve the precision of synchronization, we observed the delay-jitter and figured out that the efficient solution should be designed a system that has a NTP server in each network segment with small network delay.

The further measurement will be performed with a gigabit wide area network on a long-term basis.

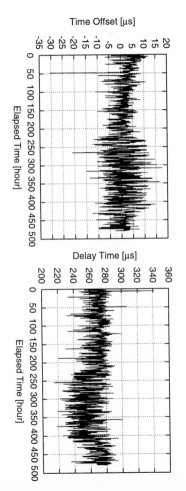

Fig. 6. Offset value between Stratum 1 and Stratum 2

Fig. 7. Delay time between Stratum 1 and Stratum 2

5   Conclusion

This research was conducted to develop the high-accurate time server and measure the time stability, aiming to achieve the future measurement precision of the network jitter on a wide area network. The time server, which we developed, employed external precision frequency signal and a nanokernel, which improve the time resolution of Linux OS. It was confirmed in the research that the ultra high accurate time source, such as cesium atomic clock couldn't act its best when used with a PC server. Therefore, it is one of major issues that we figure out the suitable time source for the current PC server, which merely keeps the time resolution of several hundred nanoseconds.

We also performed a measurement on a segment of the Fast Ethernet as the first step of our research with a wide area network. We learned that the change of the time synchronization of tens of microseconds is induced by the jitter of the network delay in this measurement. In fact, NTP is not yet sufficient enough to carry out synchronization over one microsecond level on a wide area network. Therefore it is required to place a NTP server in each network segment.

For the future, we intend to measure the jitter on a wide area network with the high-accurate time server as well as to achieve the better understanding of real time network.

References

1. D. L. Mills, Network Time Protocol (Version 3) Specification, Implementation, RFC 1305, March. 1992.
2. Trimble Navigation Ltd.: Acutime 2000 and Palisade Smart Antenna GPS receiver, "ftp://ftp.trimble.com/pub/ntp/palisadedrv/driver29.html"
3. Internet Multifeed Co.: Experimental NTP Servers (Stratum2), "http://www.jst.mfeed.ad.jp/"
4. T. Nakashima and S. Ihara, An Experimental Evaluation of the Total Cost of NTP Topology, ICOIN15, January. 2001.
5. H. Okazawa, Y. Kitaguchi, A. Machizawa, S. Nakagawa, T. Asami and A. Ito, Advanced NTP Synchronization Device for Internet Monitoring Tools, INET2001 Posters, June. 2001.
6. J. Mogul, D. Mills, J. Brittenson, J. Stone and U. Windl, Pulse-Per-Second API for UNIX-like Operating Systems, Version 1.0, RFC2783, March. 2000.

Modelling and Simulation of Phased Array Antenna for LEO Satellite Tracking

Ng Chee Kyun[1], Ashraf Gasim Elsid Abdalla[2], Nor Kamariah Noordin[1], Sabira Khatun[1], Borhanuddin Mohd Ali[1], and Ratna Kalos Zakiah Sahbuddin[1]

[1]Department of Computer and Communication Systems Engineering,
Faculty of Engineering, Universiti Putra Malaysia, UPM, 43400 Serdang,
Selangor Darul Ehsan, Malaysia.
mpnck@yahoo.com, {nknordin, sabira, borhan,
ratna}@eng.upm.edu.my
http://cc.eng.upm.edu.my

[2]Faculty of Engineering, Multimedia University (Cyberjaya Campus),
Jalan Multimedia, 63100 Cyberjaya,
Selangor Darul Ehsan, Malaysia.
asraf.abdalla@mmu.edu.my
http://foe.mmu.edu.my

Abstract. Nowadays achievements on mobile satellite systems (MSS) communication give a great concern for research in global wireless communication fields. The developments of several Low Earth Orbit (LEO) satellite systems have promised worldwide connectivity with low delay real-time voice communications. Since the LEO satellite systems revolving around the Earth overlay mobile terminals (MT) or Earth stations over several minutes only, a sophisticated LEO satellite tracking must be introduced. Although phased array antenna is seen to be the most promising solution. Although phased array antenna has been widely use for mobile cellular communication systems with advantages of its electronically beam steering, it is still not truly used in integrate with LEO MSS. In this paper, the mathematical model of phased array antenna is introduced for LEO satellite tracking purpose. Furthermore, the simulation of beam forming using phased array antenna is explored to obtain multiple and also steerable beams for tracking the satellite smoothly. Finally, a integration of phased array antenna in MT is proposed for future global wireless communication systems.

1 Introduction

An antenna is a structure for radiating or receiving electromagnetic (EM) waves that carry information. Antennas can be classified into different types. The most common one is the wire antenna used in radio, television and cellular telephones, and the reflector and horn antenna that can be found in direct broadcast satellite terminals. Nowadays, the phased array antenna technology has been developed to meet the emerging communications need. It remains the most promising type of sharply direc

I. Chong (Ed.): ICOIN 2002, LNCS 2344, pp. 359–371, 2002.
© Springer-Verlag Berlin Heidelberg 2002

tional antennas [1]. By assembling a number of antenna elements to form a phased array, the direction of the main beam can be controlled. This is accomplished through the adjustment of the signal amplitude and phase of each antenna element in the array. Accurate pointing of the beam in the desired direction minimizes radiation in the unwanted direction, and improves the signal-to-noise ratio (SNR) and the overall efficiency of the system [2].

The current development and deployment of LEO communication satellites have focused on the need to deliver increasingly larger volumes of data from the satellite to the ground and vice versa. This desire has resulted in the development of Ka frequency band systems [3]. Moreover, commercial systems have been proposed for operation in this frequency band, such as SPACEWAY, ASTROLINK, CYBERSTAR, TELEDESIC, N-STAR, WEST and EUROSKYWAY [4].

The LEO satellites are usually defined for altitudes between 500 and 2000 km above the Earth's surface. At these altitudes the velocity of the satellite relative to a fixed observer is very fast [5]. Due to this mobility, a proper beam steering system for LEO satellites tracking is necessary. The ultimate solution of this problem can be obtained by using a phased array antenna with electronic beam scanning [6].

Antenna design is an optimisation problem with a large number of variables and constraints [7]. This paper particularly deals with issues related to the design of the phased array antenna mathematical model. A number of antennas' parameters have been studied. The satellite constellation system's characteristics are also overviewed.

This paper is organized as follows. Overview of phased array antenna system is described in the next section. Then a section deal with satellite constellation system has been discussed. A large part of this paper discusses the design of phased array antenna mathematical model. Finally, the simulation results of these mathematical models with some analysis is presented. The paper concludes with a summary of the major contributions from the effective of simple mathematical model in phased array antenna design research area.

2   Overview of Phased Array Antenna System

It is often desirable to electronically scan the beam of an antenna. This can be accomplished by changing the phases of the signals at the antenna elements. If only the phases are changed, with the amplitude weights remained fixed as the beam is steered, the array is commonly known as a phased array. As shown in figure 1, a phased array consists simply of antenna elements, each of which is connected to a phase shifter, and a power combiner for adding the signals together from the antenna elements. The phase shifters control either the phase of the excitation current or the phase of the received signals. When all the signals are combined, a beam is formed in the desired direction. That is, on transmit side, a beam is formed in space, and on receiver side, the signals from the antenna elements add coherently if the signals are received from the correct region of space. A beam forming network is used to either distribute the signal from the transmitter to the elements or combine the signals from the elements to form a single signal path to the receiver. The network may also be used to provide the required aperture distribution for beam shaping and side lobe control [8].

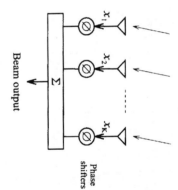

Fig. 1. Linear phased array antenna

3   Satellite Constellation System

In this section, LEO satellite constellation system here is based on currently existing Iridium like satellite system. The Iridium like system at an altitude of 780 km above the earth requires only 66 satellites to cover the entire global area. These 66 satellites are arranged in six planes, each plane containing 11 satellites. The minimum elevation angle for an earth station is 8.2 degrees, which maximizes the coverage area of the satellite, which provides average satellite in-view time of approximately 10 minutes. The satellites utilize ISL to route network traffic.

These constellation parameters resulted an orbital period of about 100.13 minutes. The minimum elevation angle for a user to see a given satellite is 8.2 degrees. Satellites with on-board switching technology are able to use ISL to route calls. A mobile user's transmitted signal is routed through several satellites and downlinked to either a regional gateway or another mobile user. This creates a network in the sky and allows the use of large regional gateways instead of gateways in each satellite footprint [9]. The parameters of Iridium satellite constellation systems are summarized in table 1.

4   Phased Array Antenna Mathematical Model

As the elements are added to the array, each with a different current, it is necessary to consider their relative field strengths as determined by their element currents. A phased linear array of n antennas with spacing d, with r_o, the distant of uniform plane waves propagating in far field, and ψ, is the angle made by the line from the origin to

Table 1. Iridium satellite constellation systems parameters

Orbits and Geometry	Iridium
Orbit class	LEO
Altitude (km)	780
Number of satellites	66
Number of planes	6
Inclination (°)	86.4
Period (minutes)	100.1
Satellite visibility time (minutes)	11.1
Minimum elevation angle (°)	8.2

the target point with the axis of the array shown in figure 2. Then assuming currents of equal amplitude I_0, progressive phase shift α and phase frequency ω in the manner $I_0 \cos \alpha t, I_0 \cos(\alpha t + \alpha), I_0 \cos(\alpha t + 2\alpha)....$ with time t for antennas 1, 2, 3,....., respectively.

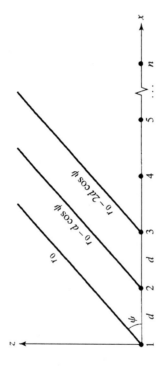

Fig. 2. Phased array of n antenna elements

We can obtain the far field ($r >> nd$) as in equation (1) below.

$$\overline{E}(\psi) = \frac{1 - e^{jn(\beta d \cos\psi + \alpha)}}{1 - e^{j(\beta d \cos\psi + \alpha)}} e^{-j\beta_0} \qquad (1)$$

The magnitude of $\overline{E}$ is given by

$$|\overline{E}(\psi)| = \left| \frac{\sin n[(\beta d \cos\psi + \alpha)/2]}{\sin[(\beta d \cos\psi + \alpha)/2]} \right| \qquad (2)$$

which has a maximum value of n for $\beta d \cos\psi + \alpha = 0, 2\pi, 4\pi, ...$

Thus, the radiation pattern becomes

$$F(\psi) = \frac{1}{n} \left| \frac{\sin n[(\beta d \cos \psi + \alpha)/2]}{\sin[(\beta d \cos \psi + \alpha)/2]} \right| \tag{3}$$

Note that for $n = 2$, equation 3 reduces to $\cos[(\beta d \cos \psi + \alpha)/2]$, which is the radiation pattern obtained for the two elements array. The nulls of the pattern occur for $n(\beta d \cos \psi + \alpha) = 2m\pi$, where m is any integer not equal to 0, n, $2m$..... For $d = k\lambda$, the equation 3 for n element array antenna reduces to

$$F(\psi) = \frac{1}{n} \left| \frac{\sin n(\pi k \cos \psi + \alpha/2)}{\sin(\pi k \cos \psi + \alpha/2)} \right| \tag{4}$$

The constant k is defined as spacing distance over wavelength ratio (SLR). Thus,

$$SLR = \frac{f\, d}{c} \tag{5}$$

Hence, for the phased array antenna design purposes, spacing distant element between element antennas is dependent on the system frequency use and SLR constant.

5 Simulation of System's Mathematical Model

Since the LEO satellite revolving around the Earth in its orbit with high velocity, the satellite visibility duration is about 10 to 20 minutes. Due to its mobility, the phased array antenna need to generate a steerable electronically controlled beam, high directivity with narrow beam, and multiple beams for call path handover management. In this section, the simulation of phase array antenna mathematical models are depicted based on equation 4 using Matlab. These simulations consist of three parts; beam steering, beam width and multiple beams of phased array antenna.

5.1 Beam Steering

Figures 3a–3e show the beam generated sequence of plots of power gain, F versus elevation angle, $\psi(0^\circ \leq \psi \leq 180^\circ)$ for values of phase shift, α ranging from -180° to 180° in steps of 90°, which will cause the beam steer to desire directivity, for $n = 6$ elements and with SLR = 0.5. It can be seen that as the value of phase shift α, is varied, the directivity of the beam varies in a continuous manner.

Figure 3a illustrate that when the phase on phase shifter is turned to -180°, the directivity of the main lobe beam will steer to 0° from phased array antenna with high gain while other sidelobes level are low. This 0° directivity beam phenomenon is known as end fire steering where the beam is steered parallel to Earth ground level and it may not be able to track the satellite in which the satellite minimum elevation angle is more than 8.2°.

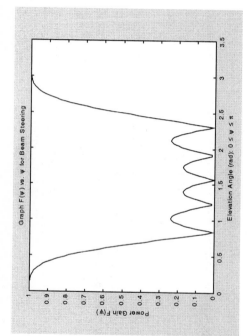

Fig. 3a. Plot of beam directivity, ψ of $0°$ for phase shift, $\alpha = -180°$

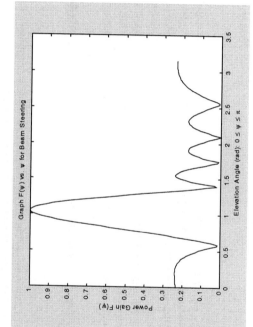

Fig. 3b. Plot of beam directivity, ψ of $60°$ for phase shift, $\alpha = -90°$

The beam steers to $60°$ by controlling the phase shift to $-90°$ as shown in figure 3b. Again the main lobe beam gain at $60°$ is high while other sidelobes level remains low. When phase shifter turn to $0°$, the directivity of the beam will steer to $90°$. This phenomenon is shown in figure 3c. Thus, this beam is generated perpendicularly from array of antenna elements. This beam is considered broadside steered where the radiation from each antenna elements is steered directly with maximum gain. Figures 3d and 3e respectively depict the beams steer to $120°$ and $180°$ on the opposite end when the phase shifter turns to $90°$ and $180°$, respectively. Thus, these electronic beams steering with constant speed has given a rational phenomenon for LEO satellite tracking from one end to another end.

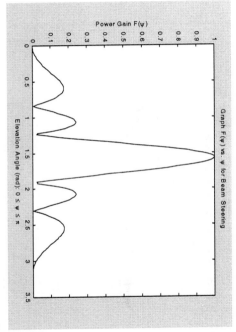

Fig. 3c. Plot of beam directivity, ψ of 90° for phase shift, $\alpha = 0^\circ$

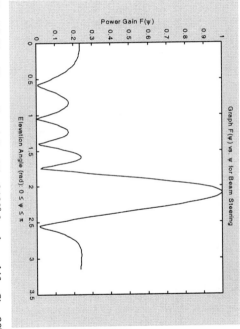

Fig. 3d. Plot of beam directivity, ψ of 120° for phase shift, $\alpha = 90^\circ$

5.2   Beam Width

Here we show how the number of antenna elements affects the performance of phased array antenna system.

If the phased array antenna consists of one element, the beam is omnidirectional or isotropic, which broadcasts to every direction as shown in figure 4a. It will not produce high gain and directional beams. Thus, these characteristics will not be considered as phased array antenna. Therefore, phased array antenna must consist at least two or more element antennas in combination array.

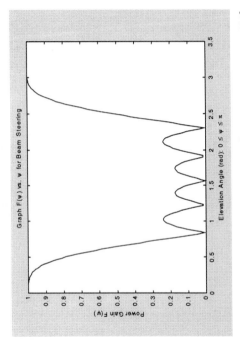

Fig. 3e. Plot of beam directivity, ψ of 180° for phase shift, $\alpha = 180°$

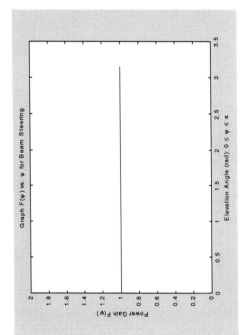

Fig. 4a. Plot of array antenna with one antenna element

Figures 4b–4e show the number of antenna elements sequence in an array plots of power gain, F versus elevation angle, ψ for phase shift, $\alpha = 0°$, and with SLR = 0.5. It can be seen that as the number of antenna elements are increased, the width of directivity beam may became narrower.

When two antenna elements are applied, the beam becomes a directional beam which steers to a particular direction. However, the beam width is wide because the interferences are only between two beams.

As depicted in figures 4c, 4d and 4e, which consist of 8, 50, and 5000 antenna elements, respectively, it is clear that when more antenna elements are used in a combination of array antenna the beam width will become narrower. The number of sidelobes is increased but their level is lower. Thus, the interference phenomenon will cause the beams field being cancelled and doubled simultaneously where it will reduce the sidelobes level besides narrowing the main lobe.

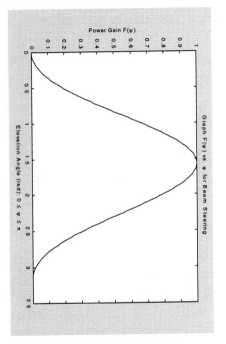

Fig. 4b. Plot of array antenna with two antenna elements

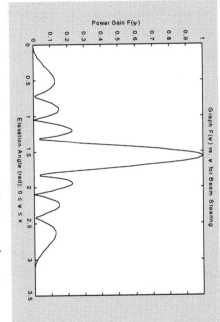

Fig. 4c. Plot of array antenna with 8 antenna elements

5.3 Multiple Beams

This is useful to generate multiple beams from a phased array antenna. The multiple beams will provide a solution especially to the call handover management. Multiple beams occur as several main lobes appear in a phased array antenna. From the definition of SLR, it is clear that when the spacing between antenna elements is larger than radiation wavelength, the small value of wavelength will enable the resulting radiation pattern to become multiple beams. In contrast, when the SLR is small ie. the spacing

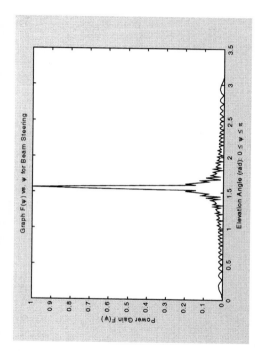

Fig. 4d. Plot of array antenna with 50 antenna elements

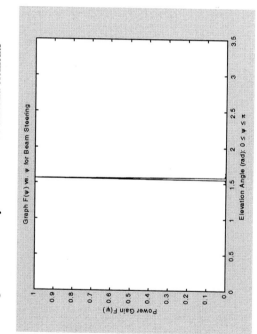

Fig. 4e. Plot of array antenna with 5000 antenna elements

between antenna elements is small compared with the radiation wavelength, it will not provide a sufficient space to generate a multiple beams. Figures 5a, 5b and 5c illustrated phased array antenna with phase shift, α of -90°, and using six antenna elements where the multiple beams is generated with SLRs of 1, 1.5, and 2 respectively.

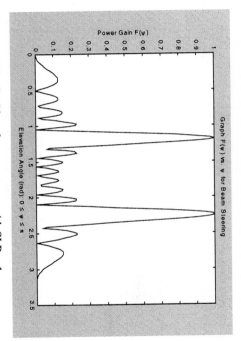

Fig. 5a. Plot of array antenna with SLR = 1

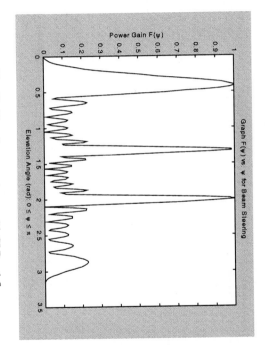

Fig. 5b. Plot of array antenna with SLR = 1.5

6 Conclusions

This paper has illustrated the mathematical model of phased array antenna which could be used for LEO satellite tracking. With only a simple mathematical function, the phased array antenna characteristics could be depicted. The simulation results provided the necessary information before the implement of real phased array antenna

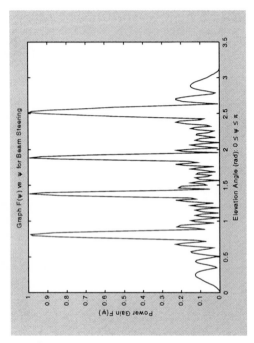

Fig. 5c. Plot of array antenna with SLR = 2

could take place. In future it is necessary to integrate a phased array antenna into mobile phone for global wireless communication, which route to the satellite communication network. This integration is possible since recently all the newly proposed satellite systems uses Ka frequency band. With this frequency band, a small phased array antenna can be implemented with the spacing between antenna elements only a few millimetres.

References

1. S. P. Skobelev, "Methods of Constructing Optimum Phased Array Antennas for Limited Field of View," *IEEE Antennas and Propagation Magazine*, vol. 40, no. 2, pp. 39–49, April 1998.

2. Per H. Lehne and Magne Pettersen, "An Overview of Smart Antenna Tehnology for Mobile Communications Systems," *IEEE Communications Surveys*, vol. 2, no. 4, Fourth Quarter, 1999.

3. J. D. Warshowsky, J. J. Whelehan and R. L. Clouse, "High Rate User Phased Array Antenna For Small LEO Satellites," *Proceedings of the Fourth Ka band Utilization Conference*, Venice, Italy, 1998.

4. G. Sfikas, "Traffic Modelling and Network Management: A Summary of The COST-252 Activities," *Information on COST 252 / 253 / 255 Joint Workshop*, 19–20 May 1999.

5. P. Carter and M. A. Beach, "Diversity in Low and Medium Earth Orbit Satellite Systems," *IEE Conference Proceedings, Ninth International Conference on Antennas and Propagation*, pp. 225–230, 4–7 April 1995.

6. S. I. Jeon, Y. W. Kim and D. G. OH, "A New Active Phased Array Antenna For Mobile Direct Broadcasting Satellite Reception," *IEEE Transactions on Broadcasting*, vol. 46, no.1, pp. 34–40, March 2000.

7. R. F. E. Guy, "General Radiation Pattern Synthesis Technique For Array Antennas of Arbitrary Configuration and Element Type," *IEE Proceedings*, vol. 135, no. 4, pp. 241–248, August 1988.

8. John Litva and Titus Kwok-Yeung Lo, "Digital Beam Forming in Wireless Communications", Artech House, Inc. 1996.

9. Stephen R. Pratt, Richard A. Raines, Carl E. Fossa Jr., and Michael A. Temple, "An Operational and Performance Overview of the IRIDIUM Low Earth Orbit Satellite System", *IEEE Communications Surveys*, Second Quarter, 1999.

IV. Network Security

An Improved ID-Based Authentication and Key Distribution Protocol

Wei-Chi Ku

Department of Computer Science and Information Engineering
Fu Jen Catholic University
Taipei, 242 Taiwan, R.O.C.
wcku@csie.fju.edu.tw

Abstract. It has been pointed out that the Shieh-Yang-Sun ID-based authentication and key distribution protocol is vulnerable to the replay attack and the unknown key share attack. We further demonstrate that the protocol is vulnerable to the forgery attack, too. In addition, we describe an improved protocol that has better resistance ability to these attacks.

1 Introduction

The ID-based authentication and key distribution protocol [1] to be improved is referred to as the Shieh-Yang-Sun protocol and can be briefly described as in the following. To initialize the system, the key information center constructs an RSA cryptosystem [2], in which p and q are two secret large prime numbers, $n = p \cdot q$ is the modulus, $e = 3$ is the public exponent, and d is the private exponent, i.e., $3 \cdot d \equiv 1 \ mod \ \phi(n)$, where $\phi(n) = (p\text{-}1)(q\text{-}1)$. The key information center then publishes n, $e = 3$, and g, which is a primitive element in both $GF(p)$ and $GF(q)$. For each user, say A, the key information center delivers $S_A = ID_A{}^d \ mod \ n$ (herein, we use ID_A instead of the EID_A in the original protocol for simplicity [3]) to him through a secure channel. It is asserted in [1] that mutual authentication and key distribution can be achieved by the protocol with two messages between the originator and the responder. Herein, we use A and B to denote the originator and the responder, respectively. To originate a session with user B, user A selects r_A (an integer $\in [1, n\text{-}1]$) and calculates x_A and y_A as follows:

$$x_A = g^{3 \cdot r_A} \ mod \ n \tag{1}$$

$$y_A = S_A \cdot t_A \cdot g^{2 \cdot r_A} \ mod \ n \tag{2}$$

I. Chong (Ed.): ICOIN 2002, LNCS 2344, pp. 375–380, 2002.
© Springer-Verlag Berlin Heidelberg 2002

where t_A is the time user A performs the calculation, and then sends $\{x_A, y_A, ID_A, t_A\}$ to user B. Upon receipt of the message from user A, user B first checks the timeliness of t_A. If the verification succeeds, user B further checks if the following equation holds:

$$ID_A \cdot t_A{}^3 \equiv \frac{y_A{}^3}{x_A{}^2} \; mod \; n. \tag{3}$$

If it is true, user B believes that the received message is freshly sent by user A. Then, user B selects r_B (an integer $\in [1, n\text{-}1]$) and calculates x_B and y_B as follows:

$$x_B = g^{3 \cdot r_B} \; mod \; n \tag{4}$$

$$y_B = S_B \cdot t_A \cdot g^{2 \cdot r_B} \; mod \; n, \tag{5}$$

and sends $\{x_B, y_B, ID_B, t_A\}$ to user A. Upon receipt of the message from user B, user A first checks if the received t_A equals the sent one. If it is true, user A further checks if the following equation holds:

$$ID_B \cdot t_A{}^3 \equiv \frac{y_B{}^3}{x_B{}^2} \; mod \; n. \tag{6}$$

If it is also true, user A believes that the received message is freshly sent by user B. In addition, the session key can be established as follows:

B side:

$$K_{BA} = x_A{}^{r_B} \; mod \; n = g^{3 \cdot r_A \cdot r_B} \; mod \; n \tag{7}$$

A side:

$$K_{AB} = x_B{}^{r_A} \; mod \; n = g^{3 \cdot r_B \cdot r_A} \; mod \; n. \tag{8}$$

Consequently, users A and B can use the session key $K_{AB} = K_{BA}$ to encrypt and decrypt their subsequent communications.

Note that there are several typos in [1]. First, Eq. (7) of [1] should read '$EID_i \cdot time_i{}^3 \equiv \frac{Y_i{}^3}{X_i{}^2} \; (mod \; n)$'. Second, '$time_j$' in Eq. (9) of [1] should be '$time_i$'. Next, Eq. (10) of [1] should read '$EID_j \cdot time_i{}^3 \equiv \frac{Y_j{}^3}{X_j{}^2} \; (mod \; n)$'. In addition, '$X_j{}^{r_i}$' and '$g^{3 \cdot r_i \cdot r_j}$' in Eq. (11) of [1] should be appended with '$(mod \; n)$'. Similarly, '$X_i{}^{r_j}$' and '$g^{3 \cdot r_i \cdot r_j}$' in Eq. (12) of [1] should also be appended with '$(mod \; n)$'. Moreover, the statement 'key $(X_i)^{r_i}$' in the eighteenth line of Section III of [1] should read 'key $(X_i)^{r_i} \; mod \; n$'.

The Shieh-Yang-Sun protocol is found to be vulnerable to two kinds of attack [3], the replay attack and the unknown key share attack. In the first attack, the opponent can easily calculate $(t_A)^{-1} \ mod \ n$ and modify the captured y_A into $y_{A\prime}$ as in the following:

$$y_{A\prime} = y_A \cdot (t_A)^{-1} \cdot t_{A\prime} \ mod \ n = S_A \cdot t_{A\prime} \cdot g^{2 \cdot r_A} \ mod \ n \qquad (9)$$

where $t_{A\prime}$ is the attacking time, and then send $\{x_A, y_{A\prime}, ID_A, t_{A\prime}\}$ to user B. User B cannot detect this attack and will falsely regard the opponent as user A. The second attack can be performed whenever user A originates the protocol. Upon seeing $\{x_A, y_A, ID_A, t_A\}$ sent by user A, the opponent chooses any integer k, calculates $R_1 = x_A{}^k \ mod \ n$, and replaces $\{x_A, y_A\}$ with $\{x_A \cdot (R_1)^3 \ mod \ n, \ y_A \cdot (R_1)^2 \ mod \ n\}$. Since the received message satisfies (3), user B will falsely take $K_{BA\prime} = K_{BA}{}^{(1+3k)} \ mod \ n$ as the session key. Next, upon seeing $\{x_B, y_B, ID_B, t_A\}$ sent by user B, the opponent calculates $R_2 = x_B{}^k \ mod \ n$, and replaces $\{x_B, y_B\}$ with $\{x_B \cdot (R_2)^3 \ mod \ n, \ y_B \cdot (R_2)^2 \ mod \ n\}$. Similarly, since the received message satisfies (6), user A will falsely take $K_{AB\prime} = K_{AB}{}^{(1+3k)} \ mod \ n$ as the session key. Hence, users A and B share a session key which is not the one supposed by them. Although this session key is unknown to the opponent, it reveals a weakness that may be employed to carry out other lethal attacks to the protocol.

In this letter, we will demonstrate that the Shieh-Yang-Sun protocol is vulnerable to the forgery attack, too. Then, we will describe an improved protocol that doesn't suffer from these attacks. Finally, the analysis of security, calculation overhead, and transmission overhead of the improved protocol will be given.

2   Forgery Attack

It is claimed in [1] that the forgery resistance of the Shieh-Yang-Sun protocol is equivalent to the difficulty of computing the discrete logarithm problem. Herein, we will show that it is incorrect. The opponent can forge $\{x_A, y_A\}$ with $\{ID_A{}^{3 \cdot u - 2} \cdot R^3 \ mod \ n, \ ID_A{}^{2 \cdot u - 1} \cdot t_A \cdot R^2 \ mod \ n\}$, where t_A is the attacking time, R is an integer $\in [1, n\text{-}1]$, and u is an integer ≥ 1. Then, the opponent sends $\{x_A, y_A, ID_A, t_A\}$ to user B. As the received message satisfies (3), user B will falsely regard the opponent as user A. Thus, the forgery attack to the Shieh-Yang-Sun protocol can be successfully performed without solving the discrete logarithm problem. Since the forgery attack does not require capturing and modifying the protocol messages previously sent by user A, it is easier than the replay attack described in [3].

3   An Improved Protocol

In this section, we will propose an improved protocol that can resist the replay attack, the unknown key share attack, and the forgery attack without incurring much overhead. The initiation of system is the same as in the original protocol except that the key information center additionally determines and publishes a keyed-hashing message authentication function, e.g., [4], $HMAC_K(m)$, which denotes the keyed-hashing message authentication code of m under the integrity key K. To originate the protocol, user A calculates x_A according to (1) and y_A according to the following equation:

$$y_A = S_A \cdot g^{2 \cdot r_A} \ mod \ n \tag{10}$$

As x_A ($= g^{3 \cdot r_A} \ mod \ n$) is a long random number generated by user A, user A can use it as a nonce. Next, user A sends $\{x_A, y_A, ID_A\}$ to user B. Upon receipt of the message from user A, user B checks if the following equation holds:

$$ID_A \equiv \frac{y_A^3}{x_A^2} \ mod \ n \tag{11}$$

If it is true, he calculates x_B according to (4) and y_B according to the following equation:

$$y_B = S_B \cdot g^{2 \cdot r_B} \ mod \ n \tag{12}$$

As x_B ($= g^{3 \cdot r_B} \ mod \ n$) is a long random number generated by user B, user B can use it as a nonce. Next, user B calculates the would-be session key K_{BA} according to (7), and calculates h_{BA} according to the following equation:

$$h_{BA} = HMAC_{K_{BA}}(x_A) \tag{13}$$

Then, user B sends $\{x_B, y_B, ID_B, h_{BA}\}$ to user A. Upon receipt of the message from user B, user A checks if the following equation holds:

$$ID_B \equiv \frac{y_B^3}{x_B^2} \ mod \ n \tag{14}$$

If it is true, he calculates the would-be session key K_{AB} according to (8) and further checks if the following equation holds:

$$HMAC_{K_{AB}}(x_A) = h_{BA} \tag{15}$$

If it is true, user A can believe that K_{AB} is the supposed session key. Simultaneously, user A can believe that the received $\{x_B, y_B\}$ has not been interfered with and user B has obtained the authentic $\{x_A, y_A\}$. At this stage, the authentication of user B has been achieved by user A. However, user B has not been confirmed that the would-be session key K_{BA} is actually the supposed session key. To achieve this, user A calculates h_{AB} as follows:

$$h_{AB} = HMAC_{K_{AB}}(x_B), \tag{16}$$

and then sends h_{AB} to user B. Upon receipt of the message from user A, user B checks if the following equation holds:

$$HMAC_{K_{BA}}(x_B) = h_{AB} \tag{17}$$

If it is true, he can believe that the received $\{x_A, y_A\}$ has not been interfered with, user A has obtained the authentic $\{x_B, y_B\}$, and K_{BA} is the supposed session key. At this stage, the authentication of user A has been achieved by user B. Thereafter, users A and B can use $K_{AB} = K_{BA}$ as the session key for subsequent communications.

4   Analysis of the Improved Protocol

If the opponent replays the previously captured $\{x_{A'}, y_{A'}, ID_A\}$ to user B, he will receive $\{x_B, y_B, ID_B, h_{BA}\}$ from user B. However, K_{AB} can be generated only from calculating $(x_{A'})^{r_B} \ mod \ n$ or $(x_B)^{r_{A'}} \ mod \ n$. Since the opponent has neither r_B nor $r_{A'}$, he cannot generate K_{AB}, which is required for calculating the h_{AB} that can be accepted by user B. Consequently, user B will terminate the protocol. Therefore, the improved protocol can resist the replay attack.

To perform the unknown key share attack, the opponent first calculates $R_1 = x_A{}^k \ mod \ n$ and $R_2 = x_B{}^k \ mod \ n$, where k is any integer. Next, the opponent replaces $\{x_A, y_A\}$ sent from user A to user B with $\{x_A \cdot (R_1)^3 \ mod \ n, y_A \cdot (R_1)^2 \ mod \ n\}$. Similarly, the opponent replaces $\{x_B, y_B\}$ sent from user B to user A with $\{x_B \cdot (R_2)^3 \ mod \ n, y_B \cdot (R_2)^2 \ mod \ n\}$. Upon receipt of the modified message, user A will detect this attack after checking (15). Therefore, the improved protocol can resist the unknown key share attack.

Next, we consider the situation that the opponent forges $\{x_A, y_A\}$ with $\{ID_A{}^{3 \cdot u - 2 \cdot R_3} \ mod \ n, ID_A{}^{2 \cdot u - 1 \cdot R_2} \ mod \ n\}$, where R is an integer $\in [1, n\text{-}1]$ and u is an integer ≥ 1. Upon receipt of the message from user B, the opponent has to either obtain r_B for calculating K_{BA} or find $r_A \ni ID_A{}^{3 \cdot u - 2} \cdot R^3 \equiv g^{3 \cdot r_A} \ mod \ n$ for calculating K_{AB}. The former approach is impractical, and the latter one is equivalent to solving the discrete logarithm problem, which is infeasible. Without K_{AB}, the opponent cannot generate the h_{AB}

that can be accepted by user B. Consequently, user B will terminate the protocol. Hence, the improved protocol can resist the forgery attack. The improved protocol provides better security at the cost of one extra transmission. The number of the transmitted items is unchanged and the total message length is increased only by $log_2 h_{BA} + log_2 h_{AB} - 2 \cdot log_2 t_A$ bits. For example, if HMAC is implemented according to [4], we have $log_2 h_{BA} = log_2 h_{AB} \leq 160$. In the improved protocol, each party further needs to perform two HMAC calculations. However, four multiplications can be saved for each party (as can be seen by examining: (2) $\rightarrow$ (10) and (6) $\rightarrow$ (14) for user A; (5) $\rightarrow$ (12) and (3) $\rightarrow$ (11) for user B). As the improved protocol doesn't use the timestamp mechanism to provide message freshness, it can be applied to less restricted environments.

Acknowledgement

This research was supported by the National Science Council, ROC, under Grant NSC-90-2213-E-030-016.

References

1. Shieh, S.-P., Yang, W.-H., Sun, H.-M.: An Authentication Protocol without Trusted Third Party. IEEE Commun. Lett., Vol. 1 (1997) 87–89
2. Rivest, R.L., Shamir, A., Adleman, L.: A Method for Obtaining Digital Signatures and Public Key Cryptosystems. Commun. ACM, Vol. 21 (1978) 120–126
3. Yen, S.-M.: Cryptanalysis of An Authentication and Key Distribution Protocol. IEEE Commun. Lett., Vol. 3 (1999) 7–8
4. Krawczyk, H., Bellare, M., Canetti, R.: HMAC: Keyed-Hashing for Message Authentication. IETF RFC 2104 (1997)

A Study of Security Association Management Oriented to IP Security

Wonjoo Park, Jaehoon Nah, and Sungwon Sohn

Information Security Technology Division,
Electronics and Telecommunications Research Institute
Daejeon, Korea, 305-350
{wjpark,jhnah,swsohn}@etri.re.kr

Abstract. The security architecture of the Internet Protocol known as IP Security (IPsec) is the most advanced effort in the standardization of Internet security. IPsec provides secure communication between two IP nodes, called IPsec peers. These types of communications are essentially sets of Security Associations (SAs) and define which protocol should be applied to packets. To implement IPsec cooperation with key management protocol– either manual or automated-, it is required prior to the security association management tool and database(SADB). This paper proposes the design and implementation of SA management method oriented to IP security.

1 Introduction

Apart from increased connectivity and a broad range of new services, the Internet has also given technically advanced intruders the opportunity to carry out a variety of attacks, thereby threatening the integrity of its infrastructure and violating the privacy of its users. A widely adopted solution consists of physically separating networks from the rest of Internet as firewalls. The current cryptographic security offers a viable alternative to packet by preserving a strongly connected global network.

Initial cryptographic security focused on application-level protocols and software, such as Secure Sockets Layer(SSL), which is used mainly for securing Web traffic; Secure Shell (SSH), which is used for securing TELNET sessions and FTP; and Pretty Good Privacy (PGP), which is used for securing e-mail. These forms of security can be limiting, as the application itself needs to support these. Another place to support internet security is at the network layer, as the applications are secured, even if they are not themselves aware of the security mechanisms. IPsec is based on this model. IPsec uses the approach that no matter what applications is used, all packet level information must pass the network layer. By securing the network layer, all application packets can benefit from the security offered by that layer. IPsec has strength that it allows organizations to implement strong security without the need to change any of their applications. As with IP, IPsec is completely transparent from the end-user's perspective.[1]

IPsec uses two protocols to provide traffic security--Authentication Header(AH) and Encapsulation Security Payload(ESP)[2][3]. These protocols are applied to traffic to provide a desired security services. These types of communications are essentially

I. Chong (Ed.): ICOIN 2002, LNCS 2344, pp. 381–388, 2002.
© Springer-Verlag Berlin Heidelberg 2002

sets of security associations (SAs) and define which protocols should be applied to IP packet, as well as the keying between the two endpoints. The concept of a 'Security Association' is fundamental to IPsec.

The management of SAs is required prior to the provision of security services between communicating entities, either manual or automated. In this paper, we describe the design and implementation of manual and automated solution for management of SAs.

2   Background and Related Works

2.1   IP Security, IPsec

IPsec provides security services at the IP layer by enabling a system to select required security protocols, determine the algorithms to user for the services, and put in place the requested services. IPsec has two main functions: (1) Authentication only, provided through the Authentication Header(AH) protocol, (2) Authentication and confidentiality(encryption), provided through the Encapsulation Security Payload (ESP) protocol. The AH offers connectionless integrity, data origin authentication, and optional anti-replay services. And The ESP protocol provides confidentiality (encryption), and limited traffic flow confidentiality. It also may provide connectionless integrity, data origin authentication, and an anti-replay service. [1][2][3]

IP AH and IP ESP may be applied alone or in combination with each other. Each protocol can support two modes: transport mode and tunnel mode. In transport mode, the security mechanisms of the protocol are applied only to the upper layer data and the information pertaining to IP layer operation as contained in the IP header is left unprotected. In tunnel mode, both the upper layer protocol data of the IP header of the IP packet are protected or tunneled through encapsulation.[1]

2.2   Security Associations, SAs

IPsec Security Associations define how two or more IPsec peers will use security services in the context of a particular security protocol, AH, ESP, or both, to communicate securely on behalf of a particular flow. An SA is unidirectional in that connection. In the case when both AH and ESP services are required, two or more different SAs should be created to afford protection to the traffic stream. When the traffic requires multiple SAs for two hosts to communicate securely, the collection of SAs is called *SA bundles*. [1]

The SA of a node are stored in the Security Association Database(SADB) and each SA is uniquely identified by three parameters, a Security Parameter Index(SPI) , an IP destination address, and a protocol(AH or ESP) identifier. SAs specify whether IPsec is used in transport or tunnel mode. A transport mode SA is a security association between two host. The security provided by IPsec is end-to-end. A tunnel mode SA is used when the IPsec peer is not the final destination of the IP traffic. This mode is typically used when either end of an IPsec security association is a security gateway or when both ends are security gateways. Thus, tunnel mode is used between two gateway or between a host and a gateway.

SAs can be created manually or automated by using Internet Key Exchange(IKE)[4]. In manual SA management, security association parameters should be predefined by the authorized manager. These parameters should include IP destination address, SPI, the security protocol, and keying material. The authorized SA manager can delete a specified SA and clear SADB, as well.

3 Design and Implementation

3.1 C-ISCAP Architecture

We propose the C-ISCAP(Controlled Internet Secure Connectivity Assurance Platform), which is an internet information security system based on IPsec. It offers a IPsec solution for end-to-end security in the distributed network. It includes Universal IPsec engine(UGINE), Automatic Internet Key Exchange (AUTOKEM), Universal Internet Key Management System (UKEM), Security Policy System (SEPS)[5], Internet Security Management System,(SEMS)[6], and the Internet Security Evaluation System(ISE).

The UGINE supports the functions of IPsec Engine, interacting with the AUTOKEM and the SEPS, and the AUTOKEM is in the charge of key exchange and SA negotiation required for C-ISCAP security information services. SEPS is a distributed database of security policy information and provides the mechanisms for processing security policy information of secure hosts or network domain. SEMS offers control functions to security manager and ISE estimates the system safety and finds the threat factors before it occurs. Figure 1 shows the C-ISCAP architecture and their interfaces.

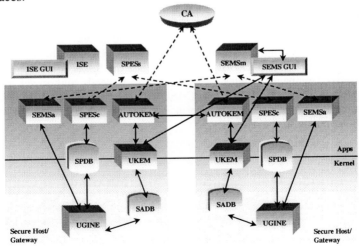

Fig. 1. C-ISCAP architecture

Secure host and gateway communicate with other host or gateway by applying security policy and security association securely. When an application sends a packet through IPsec engine, engine refers the SPD with selector-source address, destination address, source port and destination port. In case of applying the specified security protocol(AH or ESP), SEPS triggers key exchange system to establish key material and SA information between two IPsec peers(host or gateway). After the establishment of key material and SA information by AUTOKEM negotiation, it passes them key management system (UKEM). UKEM receives key material and SA information and put the SA information included secret key in SADB at the kernel.

When negotiation of security policy and security association is done, user packet is ready to apply IPsec engine processing routine. An outbound packet just refers and applies the SA, which is searched SA identifier-IP destination address, Security protocol, SPI. Inbound packets are performed in reversed processing of the outbound. C-ISCAP has Security Policy System(SEPS) to decide hash or encryption algorithm, IPsec action(bypass, apply, or drop). Also, SPDB has IPsec mode and IPsec SA bundle information. Internet Security Management system(SEMS) offers control mechanism to security manager and has GUI(Graphic User Interface). The GUI produces configuration and management other system of C-ISCAP. Internet Security Evaluation system(ISE) estimates the system safety and finds the threat factors in system.

3.2 Security Association Database, SADB

Security services are afforded to an SA by the use of AH, or ESP, but not both. In C-ISCAP implementation, there is a Security Association Database(SADB), in which each entry defines the parameters associated with one SA. Each SA has an entry in the SADB.

The two most important tasks of SA management are creation and deletion. SAs in C-ISCAP can be managed either manual or automated methods. The SA management has an interface to communicate with SADB in the kernel.

Manual managing is mandatory to support and it was used during the initial development and testing of IPsec engine. The two sides should agree on the parameters of the SA by other offline means. Once the SA is agreed, it never expires until authorized manager deletes it. In this case, it needs authorized manager and interface and it is just SEMS-GUI in C-ISCAP. The GUI for manual SA add-on is shown in Figure 2. There are more interfaces to delete the specified SA and to clear SADB of a peer.

Fig. 2. Manual 'new SA ADD' interface

Also Manual keying of UKEM is a useful for debugging the base IPsec engine when the key establishment protocols are failing. However, the use of manage keying is unsure of stableness.

In an environment where C-ISCAP is deployed, the SAs are established through the AUTOKEM. It is invoked by the SEPS that the connection should be secure and there is not the SA. AUTOKEM negotiates the SA with the destination and more peers, depending on the policy, and establishes the SAs. Once the SAs are established, AUTOKEM should pass the SA information and key material to UKEM. While IKE is application -level protocol, IPsec engine is located in kernel. PF_KEY protocol is used to pass key material and other configuration information from user space to kernel space, and to report asynchronous events such as invoking for expiration of lifetime from kernel space to a user-level keying daemon. The PF_KEY is a new socket protocol family used by trusted privileged key management applications to communicate with a kernel's key management internals. It is derived in part from of BSD routing socket, PF_ROUTE. In this system. PF_KEY is used for communicating AUTOKEM or GUI applications with SADB in kernel. Figure 3 shows the PF_KEY protocol to communicate with UKEM internals.[7]

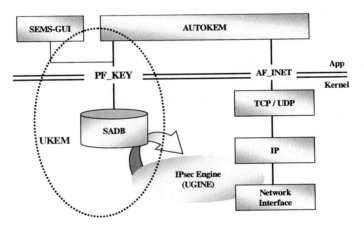

Fig. 3. PF_KEY protocol to communicate with UKEM internals

The SA is deleted for lifetime expiration, compromised key, requests to delete by end host, and so on. The SAs can be deleted manually. When existing SA is deleted manually, manager has to know the specified SA identifier<destination address, SPI, security protocol>. Once lifetime of the SA is expired, the SA should be deleted and negotiated a new SA. When soft lifetime is expired, kernel warns to create new SA. And hard lifetime is expired, kernels deletes existing SA and invoke to negotiate new SA.[1]

It is necessary to have SADB structure in kernel for supporting the interoperability and functionality. So we propose SADB-it contains parameters that associated with each SA- in kernel. The SADB structure is shown in Figure 4. It has SA identifier, AH/ESP algorithm-AH:MD5, SHA1,ESP: DES,3DES,etc which is required the policy-, anti-replay window to reject old or duplicated packet that could be used as an attack, soft lifetime, soft and hard lifetime, SA source address, SA destination address, authentication key, encryption key, initial vector value, and so on.

0	7	15	23	31	39	47	55	63
destination_address				SPI				
Security protocol				sadb sequence number				
pid of process				transformation type				
transform flags				name(readble)			algorithm errors	
auth alg	enc alg	replaywin	state	replaywin_last_seq				
replaywin_bitmap								
replaywin_maxdiff				sadb_flags				
lifetime_allocation_current				lifetime_allocation_soft				
lifetime_allocation_hard				tunnel/transport mode				
lifetime_bytes_current								
lifetime_bytes_soft								
lifetime_bytes_hard								
lifetime_addtime_current								
lifetime_addtime_soft								
lifetime_addtime_hard								
lifetime_usetime_current								
lifetime_usetime_soft								
lifetime_usetime_hard								
sadb_source_address				sadb_destination_address				
src_addr_size		dst_addr_size		sadb_key_bits_a		sadb_auth_bits		
sadb_key_bits_e		sadb_iv_bits		iv_size		padding		
sadb_key_a_size				sadb_key_e_size				
sadb_key_authentication								
sadb_key_encryption								
sadb_initial vector								

Fig. 4. SADB structure

Once authentication or encryption key material is established, UKME can create a key applying security protocol algorithms. When new SA is added linked list, destination address, SPI, protocol values are hashed and its value is identifier to get or put a SA in SADB structure. Therefore, when it knows all identifier, it can add or delete SA. IPsec protocols can be used either in tunnel or transport mode. So mode filed is set to tunnel mode, transport mode, or wild card. When this field is set to wild card, it implies that SA can be used either for tunnel or transport mode.

Also, we can display a SA or all SAs of DB by the SEMS GUI. SA information in kernel is registered in proc file system to show to applications.

When the PF_KEY is initiated, pfkey_sadb, pfkey_registerd_socket, and pfkey _supported_socket proc file are registered. Pfkey_sadb is used for showing the SADB information, and pfkey _registerd_socket and pfkey_supported_socket is used for invoking listening socket of application by the kernel. A SA that is gotten from a peer is shown in Figure 5.

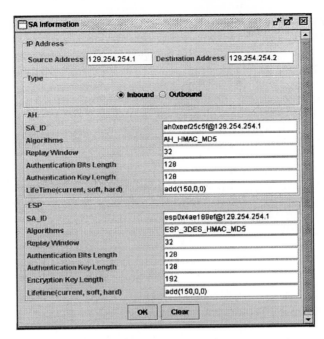

Fig. 5. SA information

UKEM in C-ISCAP supports also preshared-key method and public key method to authorize among peers. UKEM cooperates with CA(Certificate Authority) which offers the X.509 certificate standard.[8] If the key management want to authenticate others without CA, key's numbers grow exponentially as it adds peers. So, the CA offers a good solution to such an environment. UKEM has more functions to store and manage the preshared-key , public and private key. The keys from CA is stored in file encrypted.

4   Interoperation with Other SA Management

When security is addressed as a global network problem, a major issue is the management of security services, because of the complexity of interaction between various security mechanisms implemented in the protocols and the need for automatic configuration of the mechanisms. UKEM and AUTOKEM - is based on RFC 2407- in C-ISCAP offers a suitable solution for the management of security associations and the exchange of shared keys with IPsec protocols. To ensure security services, different IPsec engine from the C-ISCAP should have security policies and security associations. If a packet is received without SPI – namely, there is no SA to apply – in its header, it may be dropped, and internet security services are not supported. They may support security services, whether the other has Key Exchange(e.g.IKE)and when they are interoperable each other, or our or its manual interface can set SA and key information to be same with our SADB.

5   Remarks and Future Works

The need for security is much stronger for network control and management functions that are responsible for maintaining the connectivity over the global network. In this paper, we have described and implemented the C-ISCAP and key management system. The implementations are for the Linux kernel. C-ISCAP is a solution to produce security services at the IP layer by enabling a system to select required security protocols. Especially, we have designed and implemented Internet Key Management system and Security Association Database structure that is part of C-ISCAP.

Also, multicast has the potential to be very useful, but it suffers form guaranteeing security. No support for multicasts is currently provided, and the IPsec supports only unicast IP datagram. The future works focused on extension IPsec to secure multicasting services. Also it is necessary to study group key and SA managements for multicasting[9][10].

References

1. S. Kent, R. Atkinson, "Security Architecture for the Internet Protocol", RFC2401, Nov. 1998
2. S. Kent, R. Atkinson, "IP Authentication Header", RFC 2402, Nov. 1998
3. S. Kent, R. Atkinson, "IP Encapsulating Payload", RFC 2406, Nov. 1998
4. D. Harkins, D. Carrel, " The Internet Exchange(IKE)", RFC 2409, Nov.1998
5. M. Blaze, M. Richardson, "IPSP requirements", Internet draft, Jul. 2001
6. T. Jenkins, J.Shriver, "IPsec Monitoring MIB", Internet draft, Feb. 2001
7. D. McDonald, C. Metz, PF_KEY Key Management API, Version 2", RFC 2367, Jul. 1998
8. C. Adams, S. Farrell, "Internet X.509 Public Key infrastructure Certificate Management Protocols", RFC 2510, Mar. 1999
9. http://www.ietf.org/html.charters/msec-charter.html
10. http://www.ipmulticast.com/community/smug/

Design of Network Security Control System for Cooperative Intrusion Detection

Byoung-Koo Kim[1], Jong-Su Jang[1], and Tai M. Chung[2]

[1] Network Security Department, Electronics and Telecommunications Research Institute,
161 Gajeong-Dong, Yuseong-Gu, Daejeon, 305-350, KOREA
{kbg63228, jsjang}@etri.re.kr
[2] School of Electrical and Computer Engineering, Sungkyunkwan University,
300 Chunchun-Dong, Changan-Gu, Suwon, 440-746, KOREA
tmchung@ece.skku.ac.kr

Abstract. As intrusions and other attacks become more widespread and more sophisticated, it becomes more difficult to detect them at a single intrusion detection system(IDS). Therefore, IDSs have become focused on various intrusions (and/or attacks) in large scale network environments. But, it is not easy to detect various intrusions, since the design of early IDSs are based on analyzing the audit trails supported just a single host. Here we have made effort to design and implement IDS which can detect more complex attacks as well as support security management through cooperating each others. In this paper, we present the architecture of our system that detects various intrusions in large scale network environments as well as supports flexibility, portability, and extensibility for policy based security management.

1 Introduction

The fast extension of inexpensive computer networks has increased the problem of unauthorized access and tampering with data. As a response to increased threats, many IDSs have been developed. These IDSs have two major approaches; misuse intrusion detection and anomaly intrusion detection[4]. The first is based on the detection of intrusions that follow well-defined patterns of attack exploiting known system's and application's software vulnerabilities. The second is based on the detection of anomalous behavior or the abnormal use of the computer resource. These approaches all depend on what is being audited. Most of previous auditing mechanisms catch and analyze only the audit source generated by a single host. However, as intrusions and other attacks become more widespread and more sophisticated, it becomes more difficult to detect them at single IDS. With the widespread use of the Internet, IDSs have become focused on various intrusions in large scale network environments. But, it is not easy to detect various intrusions, since the design of early IDSs are based on analyzing the audit trails generated by host operating systems, and monitoring just a single host. Therefore, we have made effort to design and implement IDS which can

I. Chong (Ed.): ICOIN 2002, LNCS 2344, pp. 389–398, 2002.
© Springer-Verlag Berlin Heidelberg 2002

detect more complex attacks as well as support integrated security management through cooperating each others. However, managing IDSs scattered over widespread network is not an easy task.

As an example of effort to integrate various kind of IDSs, a working group was created under the auspices of the defense advanced research projects agency(DARPA) to develop a common intrusion detection framework(CIDF)[8]. They discussed the efforts of the CIDF working group to design a framework on which intrusion detection and response(ID&R) systems can cooperate with one another. As a result, they can achieve some extended goal which have not been achieved by a single IDS. Most of all, the most important contribution of CIDF is they could introduce reusability of components in intrusion detection with defining interfaces by which the different kinds of IDSs can communicate. Additionally, there are NIDES, EMERALD, AAFID, GrIDS, etc. as studies about cooperative security monitoring among intrusion detection modules[3],[5],[6],[7]. But, they have problems such as bottleneck, overhead, and so forth in collecting and analyzing data in a specific component. Consequently, while basic intrusion detection is performed independently, more effective detection strategies must be investigated.

In this paper, we present the architecture of our system designed to detect intrusions in large scale network environments as well as to support flexibility, portability, and extensibility for policy based security management.

2   Design Principle

Our IDS, named 'Network Security Control System' (shortly, NSCS) is designed to detect intrusions in widespread global network environments. In early studies, IDSs have focused on attacks to the several hosts or restricted area of network, e. g. a sub-network of LAN environment. As network grows more and more and composes a part of the Internet, various kind of attacks we haven't even named yet are emerging. We focused on the intrusions which involve several origins(or just looks like, but actually one origin) in attacking one target. We coined the term "multi-attack" to describe these attacks[1]. We classified multi-attack into three types as following;

1. Multi denial of service attack: The attacks are observed when multiple-attacker IP address are working denial of service together toward a common target. Multi denial of service attack includes multi sync flooding, multi smurf, multi new teardrop, and so forth.
2. Multi IP spoofing toward TCP session: The attacks are observed when multiple-attacker IP address are working denial of service and IP spoofing toward a common session. Multi IP spoofing toward TCP session includes IP spoofing, TCP slicing, and so forth.
3. Attacker hiding using redirect attack: The attacks are observed when attacker IP address is working behind the identity of indirect target toward a intended target. Attacker hiding using redirect attack includes redirect finger attack, redirect attack using anonymous ftp, and so forth.

Besides, we focused on the anomalous behavior or abnormal use of network re-source on each protected area. Namely, we employed anomaly detection approach for detecting anomalous behavior of each protected area. Sometimes these anomaly de-tection methods tend to be computationally expensive because of several maintained metrics that are updated after every system activity. However, we have limitation that is difficult to define every attack types in large scale network environments. There-fore, we tried to take benefit of the best practices of both misuse and the lowest anom-aly detection approaches.

We designed NSCS to detect anomalous behavior of each protected area and spe-cial intrusions such as multi-attacks described above. Also, we tried to support secu-rity management based on detection information in large scale network environments.

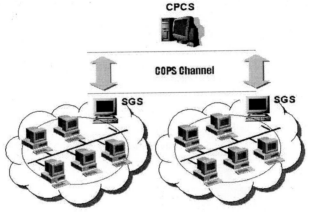

Fig. 1. The architecture of NSCS

3   The Architecture of NSCS

In this section we introduce the architecture of the Network Security Control Sys-tem(NSCS). The architecture consists of two main components; Security Gateway System(SGS) for low-level intrusion detection and response, Cyber Patrol Control System(CPCS) for policy based security management, and high-level intrusion detec-tion and response. First, SGS analyzes network packet data collected on a single SGS for simple intrusions. Second, CPCS performs integrated analysis on suspicious data collected from SGSs, and manages each SGSs. In other words, CPCS analyzes suspi-cious data which is can't clearly determined by SGS. Communication channel be-tween CPCS and SGSs use COPS protocol in order to guarantee availability to support policy based security management in the future. As shown in the figure 1, this is over-all architecture of our system; For effective intrusion detection and response in large scale network environments, the results of intrusion detection operation of SGSs are passed to their parent CPCS, and integrated analysis in CPCS on collected information can detects intrusions difficult to determine in a single system.

3.1 Component of the Architecture

CPCS.

CPCS analyzes suspicious data which is can't clearly determined whether it is intrusion or not by SGSs, and performs integrated analysis on it. CPCS receives and processes requests from security manager for configuring detection policies, and managing each SGSs. CPCS collects additional event log data from SGSs to produce statistical information to security manager and maintains these data periodically. Specially, CPCS also maintains independently TCP session information with UDP/ICMP packet information and formatted event log profile using maintained event log data. This information are combined to intrusion information from SGSs, and analyzed for detecting intrusions difficult to be determined by SGSs. Consequently, CPCS performs detecting operations independently using suspicious data such as event log data and intrusion information data from SGSs.

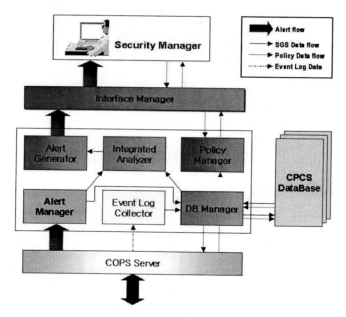

Fig. 2. Detailed CPCS architecture

As shown in the figure 2, we can divide CPCS into several sub-modules; alert generator(AG), integrated analyzer(IA), policy manager(PM), database manager(DBM), event log collector(ELC), alert manager(AM), interface manager(IM), COPS server and so forth. The summary of the internal modules is following;

- **Security Manager**: Security manager is responsible for whole system management. Most of all, security manager receives intrusion information from IM, and indicates proper response.
- **Interface Manager (IM)**: IM processes all incoming request messages and error messages from security manager. Also, intrusion information data from AG sends to security manager for security alarm.
- **Alert Generator (AG)**: AG generates intrusion alert message about analysis results from IA.
- **Integrated Analyzer (IA)**: IA is the major module of CPCS to detect multi attacks and anomalous behavior. IA performs detecting operations and sends the results of intrusion detection operation to AG.
- **Policy Manager (PM)**: PM handles pattern policy about SGSs. In other words, PM receives whole system policy from security manager, and applies to lower SGSs. Therefore, PM is the major module of CPCS to support policy based security management.
- **Database Manager (DBM)**: DBM handles not only database such as UI table, agent table, policy table, system log table and mapping table (needed system management), but also database such as event log profile table, alert log profile table, TCP session table and UDP/ICMP packet information table (needed IA's operations).
- **Event Log Collector (ELC)**: For integrated analysis, ELC collects event log data periodically from SGSs by timer. Then, ELC maintains additional information such as TCP session information and UDP/ICMP packet information, event log profile, and so forth.
- **Alert Manager (AM)**: Intrusion information data from SGSs sends to IA for multi attack detection, and maintains alert log profile.
- **COPS Server**: COPS Server is responsible for communication with NSCS.

With collected data from SGSs, CPCS make detection results from SGSs more trustworthy as well as can detect multi-attacks; for instance, redirect attack, ip spoofing, distributed scanning attack, and so forth. Also, CPCS can detect anomalous behavior or abnormal use of network resources on each protected area. Precisely, the major functionality of CPCS is to detect these types of attack. In the prospect of management, the functionality of CPCS can increase flexibility, portability, and scalability as well as supports effective maintenance. Finally, we designed NSCS can be managed through Web environments in favor of human manager. The main functionality of Web-interface is to display intrusion information to security managers.

SGS.
IDS is classified into host-based IDS and network-based IDS. Audit sources discriminate the type of IDSs based on the input information they analyze. Host- based IDS analyzes host audit source, and detects intrusion on a single host. With the widespread use of the Internet, IDSs have become focused on network attacks. Therefore, we employed network-based IDS, named SGS.

SGS is a substructure of CPCS aimed at real-time network-based intrusion detection based on misuse detection approach. SGS analyzes data packets as they travel across the network for signs of external or internal attack. Then, SGS that find signs of attack immediately send warnings to CPCS. Precisely, the major functionality of SGS is to detect network attacks and communicate with parent CPCS for applying policy information. Additionally, SGS sends to CPCS suspicious data, that is, event log data about which SGS can't determine whether it is intrusion or not as well as the information of detected intrusion.

As shown in the figure 3, SGS consists of four parts; COPS Client for communication channel with CPCS, CP-Agent for database management and intrusion response, network-analyzer for intrusion detection operation, network-sensor for network packet capturing. Again, we can divide network-sensor into several sub-modules; packet collector(PC), packet decoder(PD), packet reducer(PR), event log generator(ELG). Also, we can divide network-analyzer into several sub-modules; event log manager(ELM), database manager(DBM), event log analyzer(ELA), alert generator(AG). The summary of the internal modules is following;

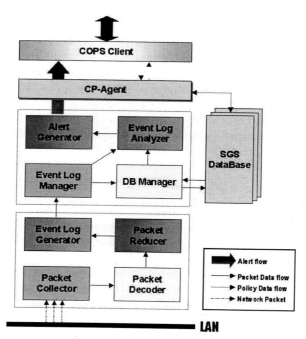

Fig. 3. Detailed SGS architecture

- **Alert Generator (AG):** AG generates intrusion alert message about analysis results from ELA.
- **Event Log Analyzer (ELA):** ELA is the major module of SGS to detect network attacks. ELA performs detecting operations and sends the results of intrusion detection operation to AG. ELA analyzes event log data from

ELM, and finds signs of attack immediately that matches an attack profile.

- **Event Log Manager (ELM)**: ELM receives event log data from ELG. Then, ELM sends event log data to ELA for intrusion detection operations. Also, ELA sends event log data to DBM for maintaining event log data.
- **Database Manager (DBM)**: DBM handles not only database such as system configuration table, system log table and mapping table (needed system management), but also database such as attack profile table, alert log table, event log table (needed ELM's operations and upper CPCS's operations).
- **Event Log Generator (ELG)**: ELG receives reduced packet data from PR, and converts reduced packet data to event log data format. Then, ELG sends generated event log data to ELM of network-analyzer as audit source for intrusion detection operation.
- **Packet Reducer (PR)**: PR receives decoded packet data from PD, and reduces decoded packet data as remove unnecessary data. Then, PR sends reduced packet data to ELG for data format conversion.
- **Packet Decoder (PD)**: PD receives network packet data from PC, and decodes received network packet data. PD consists of two parts; protocol decoder for each protocol and service decoder for each application. Then, PD sends decoded packet data to PR for data reducing.
- **Packet Collector (PC)**: PC simultaneously monitors numerous packet over network. PC uses network filter (supported by libpcap library) to reduce the processing load to the minimum. PC sends raw network packet data to PD for data decoding.

3.2   Integrated Analysis Mechanisms

In our design of NSCS, the major functionality is to detect multi attacks and anomalous behavior of each protected area. Therefore, role of CPCS is very important. CPCS uses maintaining event log information and alert log information for detecting intrusions difficult to be determined by SGSs. IA, the major sub-module of CPCS, has integrated analysis mechanism to detect these intrusions, and consists of two parts; rule based multi-attack detector(RB-MAD) for misuse detection, periodic anomaly detector(PAD) for anomaly behavior of each protected area. Consequently, IA combines anomaly and misuse intrusion detection mechanism that tries to take benefit of the best practices of both misuse and anomaly detection approaches. The description of operation to detect multi attacks and anomaly behavior by using sub-modules of IA is as follows:

As shown in the figure 4, IA uses event log data and alert log data for integrated analysis. ELC of CPCS collects event log data that generated from SGSs periodically, and these information are converted to event log profile and additional information needed for integrated analysis. AM of CPCS also receives alert messages from SGSs, and converts these messages to alert profile and alert log data. Event log profile and alert log profile are needed for detecting anomalous behavior, and additional information from event log data and alert log data are needed for multi-attack detection.

First, RB-MAD is based on misuse detection approach for detecting multi-attack. Therefore, RB-MAD seeks to discover multi-attack by precisely defining them ahead of time and watching for their occurrence. The underlying idea behind the misuse detection approach is to combine the present alert message received from SGSs with additional information generated from event log data and alert log data, and match multi-attack pattern rule. In other words, RB-MAD analyzes the alert message from AM, and if the message is involved in multi attack pattern, RB-MAD receives additional information needed for multi attack pattern analysis from DBM and analyzes suspicious data combined with intrusion detection information and event log information. If the result of analysis proves multi attack, RB-MAD delivers multi attack information to AG. The last step of the processing is that AG sends multi attack alarm message to CPCS. Besides, RB-MAD sends simple attack alarm message from SGSs to AG.

Second, PAD is based on statistical anomaly detection approach for detecting anomalous behavior or abnormal use of network resource. Therefore, PAD operates on the assumption that misuse or intrusive behavior deviates from normal network use. With a statistical approach to anomaly detection, PAD is constantly measuring the deviance of the present event log profile or alert log profile from the original. In other words, if the present profile generated by timer deviates from the original profile, PAD delivers anomalous information to AG. The last step of the processing, in the same way as RB-MAD, is that AG sends anomalous behavior alarm message to CPCS. Finally, the present profile information is added to the original profile information, and the original profiles are updated to new event log profile and alert log profile.

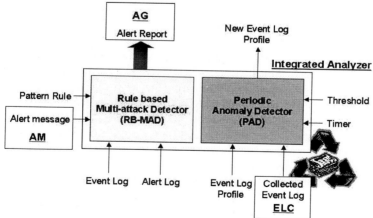

Fig. 4. Integrated Analysis Mechanisms

As description in above, integrated analysis mechanisms are willing to detect multi-attacks and anomalous behavior in large network environments, but it is necessary to manage effectively much maintaining information such as event log profile, alert log profile, TCP session information and so forth.

4 Discussion and Future Work

The main advantages of the NSCS are making detected intrusion information more trustworthy as well as detecting various intrusions by processing integrated analysis on intrusion information which is collected from sub systems. Also, in the prospect of management, the functionality of NSCS can increase flexibility, portability, and scalability as well as supports effective maintenance. We have demonstrated the feasibility of the architecture by the implementation of working prototype. We analyzed the functions of various intrusion detection systems in our testbed network and have developed the prototype of our IDS. For resolving the problem derived from the verification of implemented system prototype, we will go and consider on system performance, availability, fault tolerance test with prototype. And now, we are defining SNMP MIB and XML DTD in order to integrate various security systems to NSCS. Finally, we will implement our designed system and give more effort to demonstrate effectiveness of our system. Also, we will guarantee secure the transmission of messages among communication of external entities.

5 Conclusion

In this paper, we propose the architecture for Network Security Control System called NSCS, which detects various intrusions in large scale network environments as well as supports flexibility, portability, and extensibility for policy based security management by using COPS protocol. Also, we designed and prototyping the architecture of Cyber Patrol Control System called CPCS that allows data to be collected from multiple sub systems, thus allowing us to combine the best characteristics of misuse and anomaly intrusion detection approaches for detecting various attacks and intrusions in global network environments. Furthermore, the modular characteristics of the architecture allow it to be easily extended, configured and modified, ether by adding new components, or by replacing components when they need to be updated.

References

1. Byoung Koo Kim, D.S. Kim, and Tai M. Chung, "A Design of Integrated Intrusion Detection Systems in a Large Scale Network Environment", *APNOMS 2000*, pp. 187–197, Nara, Japan, Oct., 2000.
2. D. E. Denning, "An Intrusion-Detection Model", In *Proceedings of the IEEE Symposium on Security and Privacy*, pp. 118–131, 1986.

3. D. Anderson, T. Frivold and A. Valdes, "Next-gene ration intrusion detection expert system (NIDES)", Technical Report SRI-CLS-95-07, May, 1995.
4. S. Kumar, "Classification and Detection of Com puter Intrusions", Phd, Purdue University, 1995.
5. S. Staniford-Chen, S. Cheung, R. Crawford, M. Dilger, J. Frank, J. Hoagland, K. Levitt, C. Wee, R. Yip and D. Zerkle, "GrIDS-A Graph based intrusion detection system for large networks", In *Proceedings of the 19th National Information Systems Security Conference*, 1996.
6. P. A. Porras and P. G. Neumann, "EMERALD: Event monitoring enabling responses to anomalous live disturbances", In *National Information Systems Security Conference*, pp. 353–365, Baltimore, MD, Oct., 1997.
7. J. S. Balasubramaniyan, J. O. Garcia Fernandez, D. Isacoff, E. Spafford and D. Zamboni, "An Architecture for Intrusion Detection using Autonomous Agents", COAST Technical Report 98/05, Jun., 1998.
8. C. Kahn, P. A. Porras, S. Staniford-Chen and B. Tung, "A Common Intrusion Detection Framework-data formats", Internet draft-ietf-cidf-data-formats-00.txt, Mar., 1998.
9. H. Debar, M. Dacier and A. Wespi, "Research Report Towards a Taxonomy of Intrusion Detection Systems", Technical Report RZ 3030, IBM Research Division, Zurich Research Laboratory, Jun., 1998.

Policy-Based Intrusion Detection and Automated Response Mechanism

Jinoh Kim, Kiyoung Kim, and Jongsoo Jang

ETRI (Electronics & Telecommunications Research Institute)
Tel: +82-42-860-5350 Fax: +82-42-860-5611
zyno21@etri.re.kr

Abstract. Automated response to intrusions has become a major issue in defending critical systems. Because the adversary can take actions at computer speeds, systems need the capability to react without human intervention. Policy-based network simplifies the many tasks associated with coordinating the resources and capabilities of the network with the business-level goals of the network administrator. This paper provides policy-based security management architecture enabling network-wide intrusion detection and automated response. And this paper provides required functionality to realize the automated response mechanism. This paper also presents security policies to facilitate security management functions in policy-based networks.

1 Introduction

Today's information systems in government and commercial sectors are distributed and highly interconnected via local area and wide area computer networks. While indispensable, these networks provide potential avenues of attack by hackers, international competitors, and other adversaries. To monitor and protect against such threats, organizations increasingly deploy intrusion detection systems and network boundary control devices (i.e., firewalls, filtering routers, and guards). When suspicious activities are detected, intrusion detection systems alert human administrators or automated processes that undertake corrective action [1].

Automating intrusion analysis and response is critical for two reasons. First, human analysis of a single intrusion can take hours, days, or longer. During this period, the costs and damage caused by an intrusion, e.g., "down time" or corruption of data, can rapidly mount. Second, the analysis is complex; it requires expert network administrators, who are chronically in short supply. If such experts are not immediately available when an intrusion occurs, determining how to respond may be delayed, allowing damage to increase further [2].

To provide consistently high service quality and to ensure mission success, an automated system is required to counter these threats since humans are not fast enough to react to high speed or broad scale attacks effectively. With appropriate automation, it may be possible to respond more quickly and at times when expert administrators are not available [3].

I. Chong (Ed.): ICOIN 2002, LNCS 2344, pp. 399–408, 2002.
© Springer-Verlag Berlin Heidelberg 2002

Several vendors have developed products that support intrusion response. However these products all use proprietary protocols and are limited by an architecture that requires all response decisions to be made at a central controller. Devices such as firewalls are simply response mechanisms and not full participants in the response decision making process.

While useful, these automated mechanisms have important limitations, especially when applied to large internetworked environments or the information infrastructure. That is intrusion detection systems detect local intrusion symptoms and can only react locally. Because an attacker may cross many network boundaries, a response local to the target can't identify or mitigate the true source of the attack, which may be several networks removed [1].

Policy-based network begins where traditional systems end. It originates from a different set of assumptions, practical needs, and philosophies. It assumes a functioning network, and proactively attempts to tune that network towards achieving pre-defined organizational objectives and goals. Expressing human goals as a set of rules that can be comprehended by network nodes is no easy task – in fact, that is one area where consensus among experts is hard to come by. However, it is commonly agreed upon that such sets of rules are known as "policy".

PBN simplifies the many tasks associated with coordinating the resources and capabilities of the network with the business-level goals of the network administrator. Desired network behavior can be expressed by the user in high-level, human-friendly terms, i.e. "goals", the PBN system automatically manages the translation of these terms into commands and configuration parameters that can be understood by the network devices that provide the behavior, i.e. "procedures" [4].

This paper provides a discussion of how security management functions can be provided in policy-based networks. That is this paper adopts security management functions in the policy-based networks to enable better security control and response. Section 2 describes the overview of policy-based network. Section 3 provides the policy-based intrusion detection and automated response mechanism. Section 4 presents security policies to facilitate security management in the policy-based networks. Section 5 provides a summary.

2 Policy-Based Networks

This section describes policy-based network (PBN) concept and architecture.

2.1 PBN Concept

Distributed systems may contain a large number of objects and potentially cross organizational boundaries. New components and services are added or removed from the system dynamically, thus changing the requirements of the management system over a potentially long lifetime. There has been considerable interest recently in policy-based management for distributed systems

A Policy is information which can be used to modify the behavior of a system. Separating policies from the managers which interpret them permits the modification of the policies to change the behavior and strategy of the management system without recoding the managers. The management system can then adapt to changing requirements by disabling policies or replacing old policies with new ones without shutting down the system [5].

Policy can be defined from two perspectives. The one is a definite goal, course or method of action to guide and determine present and future decisions. The other is a set of rules to administer, manage, and control access to network resources [6] [7].

PBN signals a shift in the way networks are controlled and managed. Traditional network management originated from an engineering need to detect and debug network problems. It therefore provides network administrators with many knobs to "tweak", and many failure detection conditions to set (using mechanisms such as SNMP and CLI) [4].

PBN enables the coordination of network information, and dynamically maps it to configuration information, including queuing mechanisms, packet treatment methods, link capacity based on service class, etc. Configuration information of this type will vary more often, depending on the specific services the network supports. Different devices may also require different configurations to support the same service. PBN can match this resource information with user and application needs that may change frequently – daily, or even hourly.

2.2 PBN Components

PBN employs policy servers as the PDP (Policy Decision Point) to control the network devices that enforce the policy. And the network devices that receive and enforce the decisions from the PDP are refered to as PEPs (Policy Enforcement Points). PBN also offers a policy repository for storing policy information accessed by the PDPs in the system. The COPS (Common Open Policy Service) policy protocol is used to communicate policy information between PDPs and PEPs, while the LDAP (Lightweight Directory Access Protocol) accesses the policy repository. Figure 1 shows PBN components and their relationships.

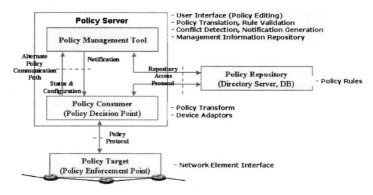

Fig. 1. PBN components and relationship

The PDP is the PBN component that directly controls the network devices or policy enforcement points. Functionally, the PDP takes policy information that has been entered into the PBN management system and processes it, together with network resource information, to make policy decisions, and send directives to the network devices. The policy data used by the PDP can either be obtained in real-time as it gets entered at the management console, or from the policy repository on an as-needed basis.

The network devices that receive and enforce the decisions from the PDP are referred to as PEPs. In both outsourcing and provisioning policy management models, PEPs receive policy decisions and enforce them at the packet level as data passes through the devices.

The PBN architecture assumes that multiple policy systems may need to interoperate within a single domain, and share the same policy information. A central repository can be used to store, distribute, and coordinate policy information among such systems. Directories and the Lightweight Directory Access Protocol (LDAP) are the industry choice for interoperable standard policy storage [8]. A multi-organizational effort involving the DMTF (Distributed Management Task Force), IETF, and some of the major vendors, has been defining both information model and LDAP schemas to represent standard policy information. The policies in the directory are typically at a business-centric level of abstraction. In addition, this information is typically static, relative to the policy rules that need to be installed, and changed, in the PEPs, to implement the business policies.

The IETF Resource Allocation Protocol (RAP) WG has developed the Common Open Policy Service (COPS) as a policy protocol for use in PBN management systems. COPS represents a revolutionary approach to the proactive management of network devices. It was developed in contrast to traditional network management protocols such as SNMP, which was found to be incapable of efficiently supporting PBN, The COPS protocol stack can be conceptually divided into three distinct layers: the base protocol, client-type usage directives, and policy data representation. Together, these three layers, along with the other advantages COPS offers over traditional network management protocols, makes COPS especially well-suited for the PBN environment [9].

3   Policy-Based Security Management

This section describes policy-based security management architecture and required functionality enabling automated responses in the architecture.

3.1   Security Management Architecture

A distributed security system would comprise some security agents and a communication system. Security agents would be IDSs, firewalls, and other management systems, which work on information collection, sharing, and correlation.

Figure 2 shows the concept of our policy-based security management architecture. Security agents are deployed on the links of an inter-network, at an appropriate point for security and traffic monitoring. A security zone is an administrative domain controlled by a security manager with same policy strategy. Each security zone consists of a couple of security agents and local networks that can be one or more autonomous systems and small-scale access networks. Like a security manager cares a security zone, a security agent cares a local network. To do this, each agent maintains its care-of-network IP address spaces of corresponding local network.

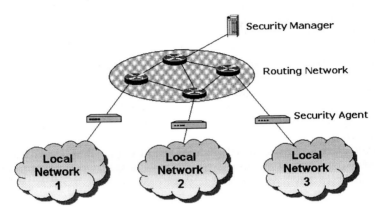

Fig. 2. Policy-based security management architecture

From the policy-based network model, a security manager plays a role as a PDP and security agents are PEPs that enforce security policies. A security manager monitors security information such as alert and traffic data from security agents, and controls the zone security appropriately based on the gathered information and the results of the analysis. A security agent performs security actions such as intrusion detections and packet filtering, and reports alerts to the security manager whenever events occurred. By collecting security information at the security manager, it has a more complete picture of the overall security state.

3.2 Functionality for Automated Responses

The main purpose of gathering information from security agents is to respond appropriately with network global view. And the responses can be formed as several ways. This paper divides response levels by activeness and intelligence as below.

Notification – Just notifies the event thus there's no need intelligence.
Information – Notifies with related information of the corresponding event.
Recommendation – Recommends responses available against the event
Automated Response – Performs automated responses against the event

Although the automated response mechanism has a lot of benefits, it should be to take a response only if the response causes the system less harm than the attack. This means the security manager should know current security condition correctly and should be able to make response decisions against the attacks according to the current security condition. Because mission objectives and administrator preferences affect the determination of the desired response to a given situation, the security manager has a number of controls available to administrators to ensure that only the desired responses are taken. This allows the administrator to trade the response benefits in terms of blocking the attack against the cost to the organization in terms of lost services caused by the response. To facilitate the automated response mechanism, some functionality is required for the security manager. This paper classifies the functionality as follows: (1) high-level detection function, (2) security observation function, (3) response decision function, and (4) administrator assistant function.

The high-level detection function monitors alert data and traffic flows, detects network-wide intrusion, and detects mid-term and long-term intrusion trial using cumulated alert data. And it also checks certainty and severity of the alert received. Certainty represents the likelihood that the reported event is an attack and severity represents the potential damage that may be caused by the attack.

The security observation function grasps the security situation of the network. To represent current security condition, it uses security level. The security level reflects the security condition of the zone and each local network. If security level changes, response policies may be changed. This paper provides 5 levels of security conditions as below.

Security Level 4 – Green state; The security condition is clear.
Security Level 3 – Yellow state; This state requires more strict watch.
Security Level 2 – Orange state; The security condition is serious because some services are being attacked.
Security Level 1 – Red state; The security condition is panic because most services could not be going on.
Security Level 0 – Dark state; The current security condition could not be decided because the security manager fails or it could not keep communicating with the security agents.

The response decision function provides the response strategy based on the current security condition. It makes possible responses against the triggered events, and selects the best one among them on the security level. If there are needed to add new policy rules, it generates new one, and adopts instantly. Also it changes the rules that are performed at the security agents if needed.

The security manager also provides the administrator assistant function. The function provides (1) preferences configuration such as threshold, (2) security information notification such as security level change and threshold exceed, and (3) statistics and log information. Especially, if the security manager is running the automated response mode, it logs response actions. Thus the administrators know how to be responded and what rules have been newly adopted.

Figure 3 illustrates how the traceback and the ingress filtering[10] are accomplished in the automated response mode. If a security agent detects an attack, the agent sends an alert message to the security manager as well as performs the local actions against the attack. If the security manager decides to do ingress filtering as a response, it sends the corresponding policy to the security agent that cares the source domain. The security manager refers security level, attack severity, care-of-network table and more security information to decide responses.

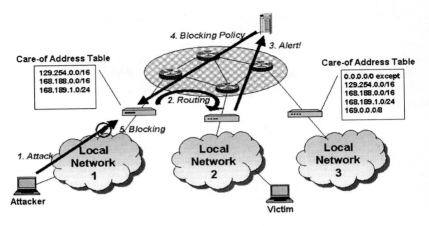

Fig. 3. Traceback and ingress filtering procedure

4 Security Policies

This section describes security policies that enable intrusion detection and automated responses. As mentioned above, policy rules are typically expressed as condition/action pairs (or if/then clauses). Security policy rules are also expressed as same manner. This can be generated from the administrators or the security manager and enforced at the security agents.

This paper classifies the security policy into two categories as "static" and "instant" from its behavior. Static policy rules stored in repository servers and loaded to each security agent. It is built up by administrators to reflect their volition. On the other hand, instant policy rules are generated instantly by resulting of automated analysis. It is adopted temporally and recovered when the situation is over. That is instant policy rules are not kept in repository servers and used temporally if needed.

The security policy comprises multiple sections. Different policy sections cover different concerns, such as intrusion detection, alert control, or filtering control. The following sections describe security policy functionality and mechanisms.

4.1 Intrusion Detection Policy

This policy provides to detect intrusions. This is the most general and basic policy for security management and can be defined within security signatures that are provided by the security agent.

The intrusion detection policy rules define the attack types. That is the rules understood as detection requirements by security agents. For the same attack type, administrators direct different actions as basis of the importance of the asset. This means this policy defines intrusion detection condition and value of network resources by describing different actions.

Figure 4 shows the intrusion detection policy rule examples. The example 1) defines DOS Teardrop attack and the example 2) and 3) define FTP password retrieval attempt attack. From the example 2) and 3), the actions can be described differently even though the attack is same as the FTP password retrieval attempt. Against the attack, alert message sending is just directed in general. For the particular host, however, the corresponding actions are defined as blocking the packets and sending alert messages.

---

1) If *Protocol is UDP and Id is 242 and FragBit sets 'M'',*
   Then *Alert and Block;*

2) If *Protocol is TCP and TCPFlags set 'ACK' and*
   Content includes "RETR" and "passwd",
   Then *Alert;*

3) If *Protocol is TCP and TCPFlags set 'ACK' and*
   Content includes "RETR" and "passwd" and
   Target is 129.254.1.1/32,
   Then *Alert and Block;*

Fig. 4. Example intrusion detection policy

4.2 Alert Control Policy

Alert control policy provides alert traffic reduction mechanisms to protect resources such as network bandwidth, processing power and so on. This policy consists of two parts of alert suppression and aggregation.

The alert suppression mechanism reduces repetitive alerts that can cause administrator overload. In some instances, security agents may generate a large number of alerts that are all related to the same event. For example, a full port scan of a subnet can cause detectors to emit thousands of essentially identical port scan reports. These alerts can be used to flood the alert messaging and thereby inhibit automated response. These alerts also flood the administrator's console with unnecessary information. The alert control policy provides alert suppression and this can be accomplished by the administrators and the result of the automated responses. This mechanism can also be used for friendly vulnerability scanning. Sometimes organizations often perform friendly scanning to help identify and close security vulnerabilities. For this friendly vulnerability scanning, this mechanism can summarize the port scanning alerts from the friend devices.

The alert aggregation mechanism enables an administrator to aggregate a set of alerts into one alert message. To protect valuable resources, the administrators command to perform alert aggregation when alerts are generated exceeding threshold within a period. Then the security agent received the policy rule performs aggregation whenever it faces that situation.

Figure 5 shows alert suppression and alert aggregation examples. For the first example, the security agent sends the first port scan alert, and then sends an suppressed alert after 30 times or 30 seconds, whichever comes first. This would reduce the number of alerts in a full port scan by a factor of 30. The second policy rule commands to perform alert aggregation by every minute against the port probing of the particular host.

---

1) **If** *attack is PortProbing and Target is 129.254.1.1/32,*
   Then *AlertSuppress(times=30, time=30);*

2) **If** *attack is PortProbing and Target is 129.254.1.1/32,*
   Then *AlertAggregate(time=60);*

Fig. 5. Example alert control policy

4.3   Filtering Control Policy

This policy deals with packet handling rules such as permitting and blocking packets. This filtering control policy has two aspects. The one is the network-level packet control and the other is service-level control. The network-level control provides to permit and block packets to/from the particular hosts, subnets, and domains. The service-level control provides to permit and block for the particular services such as FTP, HTTP, and so on. The applications can be configuration control, ingress filtering, service control, and others.

The packet filtering policy can be used for configuring the care-of network of security agents, access control list, and blocking packets from the blacklist members. When automated responses are triggered and the responses include packet filtering, packet filtering rules are generated instantly. From the traceback mechanism, ingress filtering can be accomplished by the filtering control policy.

Using this policy, administrators also can set some constraints to the network. For instance, it can be used to disallow IPX service in the network or IP packet fragment. This means it is possible to provide service control as well as blocking some packets from adversaries using the filtering control policy.

5   Summary

It has been increased the needs of automated responses against network-based attacks. To provide consistently high service quality and to ensure mission success, an automated system is required to counter these threats since humans are not fast enough to react to high speed or broad scale attacks effectively. Automation of responses is critical to keeping pace with the speed of the intrusion trials and attacks.

In this paper, we discussed policy-based security management architecture to provide the intrusion detection and automated responses. We described security levels as the security condition measurement and required functionality to perform the automated responses appropriately. And we also presented the security policies concerned intrusion detection, alert control, and filtering control in order to manage and control the network security in efficient.

References

[1]   D.Schnackenberg, K. Djahandari, and D. Sterne, "Infrastructure for Intrusion Detection and Response", Proceedings of the DARPA Information Survivability Conference and Exposition, Hilton Head, SC, Jan. 2000.

[2]   D.Schnackenberg, H. Holliday, R. Smith, K. Djahandari, and D. Sterne, "Cooperative Intrusion Traceback and Response Architecture (CITRA)", DISCEX'01, Anaheim, California, June. 2001.

[3]   S. M. Lewandowski, D. J. Van Hook, G. C. O'Leary, J. W. Haines, and L. M. Rossey, "SARA: Survivable Autonomic Response Architecture", DISCEX'01, Anaheim, California, June. 2001.

[4]   IPHIGHWAY, Inc., "Introduction to Policy-based networking and Quality of Service", http://www.iphighway.com.

[5]   E. Lupu and M. Sloman, "Conflicts in Policy-based Distributed Systems Management", IEEE Transactions on Software Engineering, Vol. 25, No. 6, Nov. 1999.

[6]   A. Westerinen, J. Schnizlein, J. Strassner, M. Scherling, B. Quinn, S. Herzog, A. Huynh, M. Carlson, J. Perry, and S. Waldbusser, "Terminology for Policy-Based Management", IETF <draft-ietf-policy-terminology-04.txt>, July 2001.

[7]   B. Moore, E. Ellesson, J. Strassner, and A. Westerinen, "Policy Core Information Model – Version 1 Specification", IETF RFC3060, Feb. 2001.

[8]   M.Wahl, T.Howes, S.Kille, "Lightweight Directory Access Protocol (v3)", IETF RFC 2251, Proposed Standard, Dec. 1997.

[9]   J. Boyle, R. Cohen, D. Durham, S. Herzog, R. Rajan, A. Sastry, "The COPS (Common Open Policy Service Protocol", IETF <draft-ietf-rap-cops-07.txt>, Aug. 1999.

[10] P. Ferguson, D. Senie, "Network Ingress Filtering: Defeating Denial of Service Attacks which employ IP Source Address Spoofing", IETF RFC2827, May. 2000.

A Power-Sum Systolic Architecture in GF(2^m)*

Nam-Yeun Kim[1], Hyun-Sung Kim[2], Won-Ho Lee[1], and Kee-Young Yoo[1]

[1]Department of Computer Engineering, Kyungpook National University,
Daegu, Korea 702-701
knyeun@hanmail.net purmi@purple.knu.ac.kr yook@knu.ac.kr
[2] Kyungil University, Kyungsansi, Kyungsangpookdo, Korea
hskim@purple.knu.ac.kr

Abstract. Finite field arithmetic operations have been widely used in the areas of data communication and network security applications, and high-speed and low-complexity design for finite field arithmetic is very necessary for these applications. This paper presents a new algorithm and architecture for the power-sum operation (AB^2+C) over GF(2^m) using the standard basis among these finite field operations. The proposed algorithm is based on the MSB-first fashion, plus the architecture has a low hardware complexity and small latency compared to conventional approaches. In particular, the hardware complexity and latency of the proposed array are about 19.8% and 25% lower than Wei's over GF(2^m), respectively. In addition, since the proposed architecture incorporates simplicity, regularity, modularity, and pipelinability, it is well suited to VLSI implementation and can be easily applied to inversion architecture.

1 Introduction

Finite field arithmetic operations have been widely used in the areas of data communication and network security applications such as error-correcting codes [1], switching theory and cryptosystems [2,3,4]. A high-speed and low-complexity design for finite field arithmetic is very necessary for meeting the demands of wider bandwidths, better security, and higher portability for personal communication [5]. Addition, multiplication, exponentiation, and inversion/division are the most important finite-field arithmetic operations. Besides these operations, the power-sum (AB^2+C) operation has been used as an efficient basic operation in decoding error-correcting codes. For example, the computation of $(S_1)^6 + S_3(S_1)^3 + S_5(S_1)^1 + (S_3)^2$ is required in a triple-error-correcting binary BCH step-by-step decoder, where S_1, S_3, and S_5 are the syndrome values calculated from the received words. The computation may be performed using only three power-sum operations and one multiplication; that is, $[(S_1)^3 + S_3]^2 + S_1[S_3(S_1)^2 + S_5][6]$. In addition, the power-sum(AB^2+C) can be used as an efficient method for public-key cryptosystems including RSA, ElGamal cryptosystem, and Elliptic Curve Cryptosystem(ECC). For example, when designing an ECC, the sum of two points on the elliptic curve requires a number of divisions,

* This work was supported by grant No. 2000-2-51200-001-2 from the Korea Science & Engineering Foundation

I. Chong (Ed.): ICOIN 2002, LNCS 2344, pp. 409–417, 2002.
© Springer-Verlag Berlin Heidelberg 2002

which can be computed using multiplication and a multiplicative inverse ($A/B = AB^{-1}$). An inverse can also be regarded as a special case of exponentiation because $B^{-1} = B^{2^m-2} = (B(B(B\cdots B(B(B)^2)^2\cdots)^2)^2)^2$. The following equation can be computed as [6]:

```
Step1 :    R = B;
Step2 :    For i = m - 2 downto 1
Step3 :       R = B·R²;
Step4 :    R = R²;
```

In this case, the result $R = B^{-1}$ and power-sum operations can be used to compute the operations in steps 3 and 4. However, since these operations are rather time-consuming, they require the design of high-speed and low-complexity circuits.

Systolic architecture is commonly used when designing high speed and low-complexity circuits for implementing arithmetic operations over GF(2^m). For a high-speed multiplier over GF(2^m), Jain [7] proposed the LSB and MSB-first algorithm for multiplication, squaring, and exponentiation using semi-systolic arrays. However, the broadcasting problem becomes more difficult when the bit length m becomes larger. To reduce the cell complexity, Lee [8] developed the AB^2+C bit-parallel systolic architecture in GF(2^m) based on irreducible all one polynomials(AOP) of degree m, plus many other bit-parallel low-complexity multipliers have been proposed for cryptographic applications, for example, those in [9 and 10]. Recently, Wei [10] and Wang et al.[11] both proposed systolic multipliers for AB^2+C computing over GF(2^m), however, none are very suitable for cryptographic applications, due to the complexity and latency involved.

In addition, many architectures over GF(2^m) have been developed using different bases; normal, dual, and standard. Although normal and dual architectures have their own distinct features, they still require a basis conversion, whereas standard basis architectures do not.

Accordingly, the current paper presents a new algorithm for a power-sum operation over GF(2^m) using the standard basis. The proposed algorithm is then used as the foundation to introduce a parallel-in parallel-out systolic architecture and supports the MSB-first scheme with pipelinability. Moreover, the proposed architecture has a hardware complexity of $m^2(3\text{AND}+3\text{XOR})-m(2\text{AND}+2\text{XOR})+(9m^2-5m-3)$Latches and latency of $3m-1$, both of which are lower than conventional architectures, plus it is well suited to VLSI implementation and can be easily applied to inversion architecture.

2   Algorithm

A finite field GF(2^m) has 2^m elements and, in this paper, all the (2^m-1) non-zero elements of GF(2^m) are represented using the standard basis. Consider three elements A, B, and C of GF(2^m), represented as polynomials of x, that is $A(x) = a_{m-1}x^{m-1} + a_{m-2}x^{m-2} \cdots + a_1x + a_0$, $B(x) = b_{m-1}x^{m-1} + b_{m-2}x^{m-2} \cdots + b_1x + b_0$, and $C(x) = c_{m-1}x^{m-1} + c_{m-2}x^{m-2} \cdots + c_1x + c_0$, where a_i, b_i, and $c_i \in$ GF(2) ($0 \leq i \leq m-1$). A finite field of GF(2^m) elements is generated by a primitive polynomial of degree m over GF(2). Let $F(x)$ be an irreducible polynomial that generates the field and is expressed as $F(x) = x^m + f_{m-1}x^{m-1} + \cdots + f_1x + f_0$. If α is the root of $F(x)$, then $F(\alpha) = 0$, and $F(\alpha) \equiv \alpha^m = f_{m-1}\alpha^{m-1} + \cdots + f_1\alpha + f_0$, $F'(\alpha) \equiv \alpha^{m+1} = f'_{m-1}\alpha^{m-1} + \cdots + f'_1\alpha + f'_0$ where $f_i, f'_i \in$ GF(2) ($0 \leq i \leq m-1$). To compute the $AB^2 + C$ operation, the proposed algorithm commences with the following equation:

$$P = AB^2 \bmod F(x) + C \tag{1}$$

$$= A(b_{m-1}\alpha^{2m-2} + b_{m-2}\alpha^{2m-4} + \cdots + b_1\alpha^2 + b_0) \bmod F(x) + C$$

$$= (Ab_{m-1}\alpha^{2m-2} + Ab_{m-2}\alpha^{2m-4} + \cdots + Ab_1\alpha^2 + Ab_0) \bmod F(x) + C$$

$$= (\cdots(\cdots((Ab_{m-1})\alpha^2 \bmod F(x) + Ab_{m-2})\alpha^2 \bmod F(x) + \cdots + Ab_{m-i})\alpha^2$$

$$\bmod F(x) + \cdots + Ab_1)\alpha^2 \bmod F(x) + Ab_0 + C$$

A new recursive relation is derived that is suitable for efficient power-sum systolic-array implementation. Here, $c_i = 0$ ($0 \le i \le m-1$), because the current interest is computing an inverse. Beginning with the first term of eq.(1), $Ab_{m-1}\alpha^2$, the subsequent terms in the above equation are accumulated until reaching the end. The procedure of the new algorithm is as follows.
First,

$$P_1 = Ab_{m-1}\alpha^2 \bmod F(x) \tag{2}$$

$$= [\sum_{k=0}^{m-1} a_k b_{m-1}\alpha^k]\alpha^2 \bmod F(x)$$

$$= [\sum_{k=0}^{m-1} a_k b_{m-1}\alpha^k \alpha^2] \bmod F(x)$$

$$= \sum_{k=0}^{m-1} d_k^1 \alpha^{k+2} \bmod F(x)$$

$$= d_{m-1}^1 \alpha^{m+1} + d_{m-2}^1 \alpha^m + d_{m-3}^1 \alpha^{m-1} + \cdots + d_1^1 \alpha^3 + d_0^1 \alpha^2 \bmod F(x)$$

$$= d_{m-1}^1(f_{m-1}'\alpha^{m-1} + \cdots f_1'\alpha + f_0') + d_{m-2}^1(f_{m-1}'\alpha^{m-1} + \cdots + f_1'\alpha + f_0') + \cdots$$

$$+ d_1^1 \alpha^3 + d_0^1 \alpha^2$$

$$= \sum_{k=0}^{m-1} p_k^1 \alpha^k$$

where,

$$\begin{cases} d_k^1 = a_k b_{m-1}; \\ \\ p_k^1 = d_{m-1}^1 f_k^{'} + d_{m-2}^1 f_k + d_{k-2}^1 \ (k = 2, \cdots, m\text{-}1); \\ \\ p_k^1 = d_{m-1}^1 f_k^{'} + d_{m-2}^1 f_k \ (k = 0, 1); \end{cases}$$

In general,

$$P_i = (P_{i-1} + Ab_{m-i})\alpha^2 \text{ mod } F(x) \tag{3}$$

$$= [\sum_{k=0}^{m-1} (p_k^{i-1} + a_k b_{m-i})\alpha^k \alpha^2] \text{ mod } F(x)$$

$$= [\sum_{k=0}^{m-1} d_k^i \alpha^{k+2}] \text{ mod } F(x)$$

$$= \sum_{k=0}^{m-1} p_k^i \alpha^k$$

where,

$$\begin{cases} d_k^i = p_k^{i-1} + a_k b_{m-i}; \\ \\ p_k^i = d_{m-1}^i f_k^{'} + d_{m-2}^i f_k + d_{k-2}^i \ (k = 2, \cdots, m\text{-}1); \\ \\ p_k^i = d_{m-1}^i f_k^{'} + d_{m-2}^i f_k \ (k = 0, 1); \end{cases}$$

Finally,

$$P_m = P_{m-1} + Ab_0 \tag{4}$$

$$= \sum_{k=0}^{m-1} p_k^{m-1} \alpha^k + \sum_{k=0}^{m-1} a_k b_0 \alpha^k$$

$$= \sum_{k=0}^{m-1} (p_k^{m-1} + a_k b_0)\alpha^k \quad = \sum_{k=0}^{m-1} d_k^m \alpha^k$$

where,

$$\begin{cases} d_k^m = p_k^{m-1} + a_k b_0; \\ \\ p_k^m = d_k^m; \end{cases}$$

Thus the product P for $AB^2 + C$ in GF(2^m) can be computed efficiently using the above new recursive algorithm.

3   Power-Sum Systolic Architecture in GF(2^m)

From the new power-sum algorithm, a corresponding parallel-in parallel-out systolic array architecture can be obtained by following the process in [12 and 13]. Fig.1 shows the proposed systolic power-sum circuit over GF(2^4), where D^n denotes an n-unit delay element.

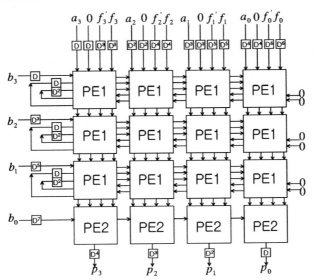

Fig. 1. Systolic architecture for power-sum operation in GF(2^4)

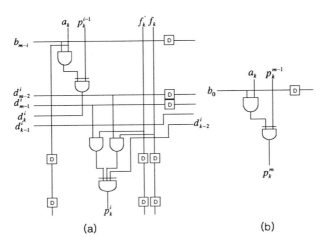

Fig. 2. Basic architecture for power-sum circuit (a) PE1 (Processing Element1) (b) PE2

The inputs A, F, and F' enter the array in parallel from the top row, while B enters from the leftmost column. The output P is transmitted from the bottom row of the array in parallel. Inputs A and B are both read into the system in the same order with the most significant bit first (MSB), plus the output is also produced in the same order as the inputs. The MSB-first scheme supports pipelinability in the architecture more easily compared to the LSB-first scheme. In Fig.1, there is a traverse line in the (i, k) cell. This is required to pass the signal d^i_{k-1} from the $(i, k\text{-}1)$ cell to the $(i, k\text{+}1)$ cell. When the cell is located in the first row; $k=m\text{-}1$, d^i_k must be connected to the d^i_{m-1} signal line, while d^i_{k-1} must be connected to the d^i_{m-2} signal line. In Fig. 2 (a), PE1 denotes the logic circuits of the basic cells, and in Fig. 2 (b), PE2 denotes the logic circuits of the bottom cells of the array. Fig 2(a) shows the basic cell for the general case where the circuit's function is primarily governed by the following two equations:

$$d^i_k = p^{i-1}_k + a_k b_{m-i};\tag{5}$$

$$p^i_k = d^i_{m-1} f_k + d^i_{m-2} f_k + d^i_{k-2};\tag{6}$$

where the k-th bit (p^i_k) of P_i is the partial product. If the cell is located in the lowest two columns ($k=0$, $k=1$), d^i_{k-2} is zero.

Note that the cells in the bottom row only have to calculate d^m_k, as shown in Fig. 2(b). As such, the bottom cell circuit is very simple and reduces the total cell complexity compared to previous architectures. Since the vertical path of each cell only requires two delay elements, except for the cells in the bottom row, the latency is $3m\text{-}1$.

4   Simulation and Performance Analysis

The new array was simulated and verified using an ALTERA MAX+PLUSII simulator.

4.1 Simulation

Fig.3 shows the simulation results for AB^2+C in GF(2^4). The inputs of the proposed power-sum systolic multiplier were $A(x)$, $B(x)$, $C(x)$, $F'(x)$, and $F(x)$, while the output was $P(x)$. The input data $A(x) = x^3 + x^2 + x + 1 = (\,1\ 1\ 1\ 1\,)$, $B(x) = x^3 + x^2 + x + 1 = (\,1\ 1\ 1\ 1\,)$ and $C(x) = (\,0\ 0\ 0\ 0\,)$ were the three elements in GF(2^4), while $F(x) = x^4 + x + 1$ and $F'(x) = x^5 + x^2 + x$ were the two modulo polynomials in GF(2^4). If α is the root of $F(x)$, then $F(\alpha) = 0$, $F(\alpha) \equiv \alpha^4 = \alpha + 1 = (\,0\ 0\ 1\ 1\,)$, and $F'(\alpha) \equiv \alpha^5 = \alpha^2 + \alpha = (\,0\ 1\ 1\ 0\,)$ were entered as the input data. The output data $P(x)$ was $AB^2+C \bmod F(x)$. Fig.3 shows the result $P(x) = x^3 + x^2 = (\,1\ 1\ 0\ 0\,)$ obtained after $3m\text{-}1 = 11$ clock cycles.

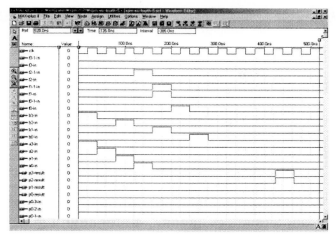

Fig. 3. Simulation results for AB^2+C in GF(2^4)

4.2   Analysis

Tables 1 and 2 present some comparisions between the proposed architecture and the related circuit described in [10]. Wei [10] proposed systolic arrays for performing power-sum operations. As mentioned in Wang [11], the systolic power-sum circuit proposed in [10] is not accurate: to ensure its proper operation, three 1-bit latches need to be added to each cell of the circuit. Accordingly, the assumption in [11] was used to make an accurate comparison. In the comparisons, the following assumptions were made: 1) T_{AND2} and T_{XORi} denote the propagation delay through a 2-input AND gate and i-input XOR gate, respectively; 2) The 3-input gate was constructed using two 2-input XOR gates. [14]; 3) The 2-input AND gate, 2-input XOR gate, and 1-bit latch consisted of six, fourteen, and eight transistors, respectively [15]; 4) The 2-input AND gate and 2-input XOR gate had 2.4ns and 4.2ns gate delays, respectively[15].

The cell complexity of the Wei [10] architecture was m^2 (3AND + 3XOR+13Latches), whereas that of the proposed architecture was m^2(3AND+3XOR)-m(2AND+2XOR) +($9m^2$-5m-3)Latches. Therefore, the proposed architecture reduced the cell complexity by m(2AND+2XOR)+($4m^2$+5m+3)Latches. The latency of the architecture of Wei [10] was $4m$, whereas that of the proposed architecture was $3m$-1. As such, the hardware complexity and latency were approximately 19.8% and 25% lower than that for [10] in GF(2^m), respectively.

Table 2 compares the proposed architecture with Wei's [10] based on the area-time product. The results showed that the proposed array was more efficient in terms of area and time complexity than [10].

5   Conclusion

This paper presented a new algorithm and parallel-in parallel-out systolic architecture for performing the power-sum operation (AB^2+C) over GF(2^m) using the standard

Table 1. Comparison of two systolic architectures for Computing AB^2+C in GF(2^m)

Circuit Item	Wei[10]		Proposed	
Number of cells	m^2		m^2	
Function	AB^2+C		AB^2+C	
Throughput	1		1	
Latency	$4m$		$3m-1$	
Computation time per basic cell	$T_{AND}+T_{XOR3}$		$T_{AND}+T_{XOR3}$	
Cell complexity	3	2-input AND	PE1	3 2-input AND
	1	2-input XOR		1 2-input XOR
	1	3-input XOR		1 3-input XOR
	13	1-bit latches		9 1-bit latches
			PE2	1 2-input AND
				1 2-input XOR
				1 1-bit latch
Algorithm	LSB		MSB	

Table 2. Comparison of Area-Time Product (ϕ : Number of transistors Δ: Gate Delay(ns) by Daniel D. Gajski [15])Circuit

Item	Wei[10]	Proposed
Area Complexity	$164m^2\phi$	$(132m^2-80m-24)\phi$
Time Complexity	$10.8\ m^2\Delta$	$10.8\ m^2\Delta$
AT Product	$1771.2m^4\phi\Delta$	$(1425.6m^4-864m^3-259.2\ m^2)\phi\Delta$

basis. The algorithm is based on the MSB-first scheme. Furthermore, since the proposed architecture has a low hardware complexity and small latency, it is efficient for computing inversion/division in GF(2^m) and well suited to VLSI implementation due to its simplicity, regularity, modularity, and pipelinability.

References

1. W.W. Peterson and E.J. Weldon, *Error-correcting codes*, MIT Press, MA, 1972.
2. D.E.R. Denning, *Cryptography and data security*, Addison-Wesley, MA, 1983.
3. IEEE P1363/D9(Draft Version 9), *Standard Specifications for Public Key Cryptography*, IEEE standards Draft, USA, 1999.
4. T. ElGamal, "A public key cryptosystem and a signature scheme based on discrete logarithms," *IEEE Trans. on Info. Theory*, vol. 31(4): 469–472, 1985.
5. W.C.Tsai and S.J.Wang, "Two systolic architectures for multiplication in GF(2m)", IEE Proc. –Comput. Digit. Tech, Vol. 147:375–382, 2000.
6. S.W. Wei, "VLSI architectures for computing exponentiations, multiplicative inverses, and divisions in GF(2m)," *Proc. IEEE Trans. Circuits and Systems,* 44: 847-855, 1997.
7. S.K. Jain, L. Song, and K.K. Parhi,, "Efficient semisystolic architectures for finite field arithmetic," *IEEE Trans. VLSI Syst.*: 101–113, 1995.

8. C.Y. Lee, E.H. Lu, and L.F. Sun , "Low-complexity Bit-parallel Systolic Architecture for Computing AB^2+C in a Class of Finite Field GF(2^m)," *IEEE Trans. On Circuits and Systems*, Vol. 48: 519–523, 2001.

9. C.H. Liu, N.F. Huang, and C.Y. Lee , "Computation of AB^2 Multiplier in GF(2^m) Using an Efficient Low-Complexity Cellular Architecture," *IEICE Trans. Fundamentals*, Vol. E83-A: 2657–2663, 2000.

10. S.W. Wei, "A Systolic Power-Sum Circuit for GF(2^m)," *IEEE Trans. Computers*, 43: 226–229, 1994.

11. C.L. Wang and J.H. Guo , "New systolic arrays for $C+AB^2$, inversion, and division in GF(2^m)," *IEEE Trans. Computers*: 49, 1120-1125, 2000.

12. S. Y. Kung, *VLSI Array Processors*, Prentice-Hall, 1987.

13. K. Y. Yoo, *A Systolic Array Design Methodology for Sequential Loop Algorithms*, Ph.D. thesis, Rensselaer Polytechnic Institute, New York, 1992.

14. N. Weste and K. Eshraghian, *Principles of CMOS VLSI Design*, A System Perspective. Reaking, Mass. Addison-Wesley, 1985.

15. Daniel D. Gajski, *Principles of Digital Design*, Prentice-hall international, INC, 1997.

Design and Analysis of Hierarchical VPN Based on SCG

Chan-Woo Park[3], Jaehoon Nah[1], JongHyun Jang[2], and Chimoon Han[3]

[1]Internet Security Research, ETRI
161 Gajeong, Yuseong, Daejeon, Korea 305-503
jhnah@etri.re.kr
[2] Open System Architecture Platform, ETRI
161 Gajeong, Yuseong, Daejeon, Korea 305-503
jangjh@etri.re.kr
[3]Communication Networks Lab, Hankuk University of F.S.
89 WangSan(San), MoHyun, YoungIn, KyungKi, Korea 449-850
{cwpark,cmhan}@hufs.ac.kr

Abstract. Most of VPN (Virtual Private Network) protect the information only through the public network. There is need to develop the various forms of VPN. Therefore, various types of VPN are going to be studied in the view of SCG (Secure Communication Group). In this paper, the problems of path-definition method and the area-definition method to construct SCG are analyzed, then possible models among VPNs based on SCG technology are reviewed. This paper proposed the hierarchical VPN based on SCG number that apply the area-definition method to the internet, and analyzed the characteristics of the proposed VPN model on the point of the authentication frequency and the number of management keys.

1 Introduction

Most of company and public offices leased the line for the communication between the head and the branch offices, and were going to extend the private network by the line. This type of communication has to cost the expensive network equipment, software investment, and also the fare is high. In order to solve these difficulties, the VPN that uses the public internet infrastructure, has been studied.

VPN affords secure channel and access control for the inside or outside users of the company. VPN is logically same with the CUG (Closed User Group). These days, the commercial VPNs are categorized into three kinds: the intranet VPN, Remote Access VPN, extranet VPN. These types of VPNs cover only the public network security. There can be several types of data in the company. According to the co-report of FBI and CSI (Computer Security Institute), the occurrences of hacking have happened at the inner part of company. So the construction of VPN is needed to secure the data according to the class. And the concept able to build a small VPN of each department in a company is requested [1], [2], [3].

The research on the VPNs based on the SCG (Secure Communication Group) has been ongoing [4]. The ways of the SCG construction are categorized into two groups: path-definition method, and area-definition method. Path-definition of IPsec (IP Security Protocol) & IKE (Internet Key Exchange) has the strong security

I. Chong (Ed.): ICOIN 2002, LNCS 2344, pp. 418–429, 2002.
© Springer-Verlag Berlin Heidelberg 2002

and wide coverage, but the heavy management traffic. The other hand, the area-definition of the GSP (Group Search Protocol) is a proper way to be applied to the intranet and not a good one to be adjusted to the internet [5]. In this paper, the concept of SCG on the intranet environment is introduced. To build the SCG, the characteristics of the path-definition and the area-definition methods are analyzed, the hierarchical VPN that can be applied by the area-definition of SCG number is proposed.

The introduction is followed by chapter 2; the VPN construction methods using the cryptography. In chapter III, the construction method of the VPN by SCG number is proposed. In chapter IV, the construction way for the hierarchical intranet VPN based on SCG number be proposed. And the conclusions are at the end.

2 VPN Construction by Cryptography

The very important requirement in case of VPN is security. The security can be done through the encryption of sending data. If the encrypted data are sent, someone catches the data but cannot decrypt the data without the key. If a group with the same key can be seen as a VPN, this makes various VPNs available.

2.1 SCG Definition [4]

The SCG means that the each group node has peer keys fully meshed to every nodes or symmetric key shared by all nodes.

Figure 1 shows the SCG built by the path-definition method. The CE (Communication Entity) can be one either a terminal equipment or a sub-network with encrypting function. MGE (Management Entity) authenticates each CE and manages the keys. MGE manages the list of all CE and distributes the keys to the CEs on the whole paths. A CE communicates securely by one key with the peer.

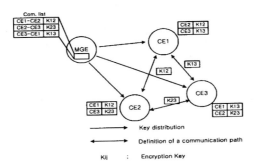

Fig. 1. SCG construction based on the Path-Definition method

There are two key distribution means from MGE to CE in SCG. One is periodic key distribution, and the other on-demand key distribution. Figure 2 shows the periodic key distribution method. This has a disadvantage of traffic increase because the number of keys increases linearly in the case of a huge system.

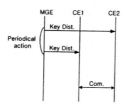

Fig. 2. Periodical key distribution method on the Path-Definition SCG

Figure 3 shows the on-demand key distribution method. There is an advantage in the view of the key management traffic but the initiation processing of key distribution has the delay and decreases the efficiency of other systems. Whenever there is a demand between MGE and CE, and between CEs, the authentication has to be processed.

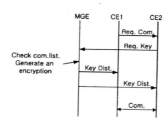

Fig. 3. On-demand key distribution method on the Path-Definition SCG

Figure 4 shows SCG built by the area-definition method. This is a way to distribute the same key after defining the area in contrast to the path-definition. The set of CEs with the same key forms a SCG. That is, CE1, CE2, CE3 have the same key K to communicate securely. There are two key distributions: periodic and on-demand.

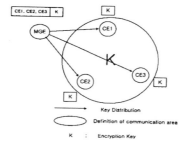

Fig. 4. SCG construction based on the Area-Definition

Figure 5 shows the periodic key distribution. In the area-definition method, although the number of CE increases, the number of keys does not increase linearly. Because of that, the periodic key distribution method is mainly chosen. Before the key distribution, the authentication is needed between MGE and CE. KEYDIST includes the authentication header and encrypted key. In the case of the periodic key distribution, there is a problem that the new key and the old one can exist simultaneously. This problem can be solved by the stop/restart command.

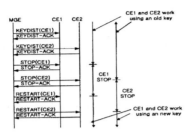

Fig. 5. Periodical key distribution method on the Area-Definition SCG

2.2 Methods Constructing SCG

Figure 6 shows the way that constructs a SCG by IPsec and IKE [6]. In the initial processing, MS (Management Server) distributes the public key to a CE, then the key is used to authenticate the peer CE on the communication path. And SPD (Security Policy Database) sends the key to the every EE (Encryption Entity). SPD has the parameter values for EE's operation, e.g. CE1 should communicate with CE2, CE3. After the communication begins, each peer CE establishes a SA (Security Association). During the SA processing, the authentication between CEs is done by the public key. And the created encryption key will be shared. After creating SA, the packet is being transfered in the form of IPsec. In order to operate smoothly, the size of SPD needs to be very big. When the initiation of a SA be established, the delay and the size variation of the packet, cause to degrade the efficiency.

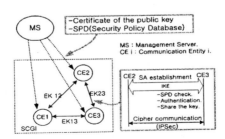

Fig. 6. SCG construction by IPSec & IKE

Figure 7 shows the way that constructs the SCG by GSP (Group Search Protocol) [5]. After MS makes authentication with each CE in the Group, the key will be distributed. For a CE would like to communicate with a peer CE, the CE can transfer the packet that be encrypted by the key from MS.

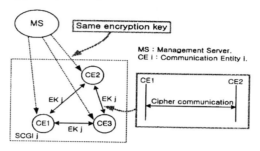

Fig. 7. SCG construction by GSP

2.3   Comparison and Analysis of Two Methods

In the path-definition way, the configuration information is calculated from SCG's definition and data related to IP address at MS. The set of these calculated values and SCG configuration information can be downloaded initially from MS. Whenever the physical variation of the configuration happens, IP address and data related to IP address should be registered, and the new parameters should be downloaded. So the traffic is going up and there are heavy delays to be authenticated by every CE.

The other hand, in the area-definition way, the SCG can be built in the intranet by means of GSP. This method has advantage against the one that is built by IPsec and IKE in the view of key's number and authentication's delay. However this is operating independently with the IP. It is impossible to apply this method to the public internet.

Table 1. Comparison of IPSec & IKE with GSP

	IPSec & IKE	GSP
Key allocation	Path-based	Area-based
Key sharing	On-demand	Periodic
Management	Complex -Complex registration parameter -IP dependent -Location dependent	Simple -Simple Parameters -IP independent -Location free
Performance	Overload factor -IKE negotiation -Encapsulation -Encryption	Overload factor -Encryption
Security	Strict	Loose
Application	Open network (Internet)	Close Network (Intranet)

Table 1 shows the comparison of the SCGs of the IPsec & IKE and GSP. The IPsec & IKE has the feature of complex registration procedure and degradation of the performance, and the other hand, the GSP has less factors of decreasing performance. So in the environment of the physical intranet, it is good to build the SCG by GSP. The GSP operates independent on the IP and is impossible to be applied at the internet. In order to reduce the number of keys and the authentication delay, the solution should be found to apply the area-definition method to the internet. There would be a solution that SCG be constructed by GSP at the intranet and the SCG numbers are shared by MS in each individual intranet.

3 VPN Constructed by SCG Number

The SCG number is binding the one intranet SCG and the other intranet SCG and, then makes a new VPN. In the inner part of intranet, the SCG is built by the area-definition of the GSP and the SCG number can be used for constructing a SCG with the CE of another intranet. In this case, the path-definition method would be used for sharing the SCG topology information and transferring the new SCG configuration's parameters.
The below explains the procedures of binding the SCGs by the SCG number.
1. Construct the SCG by GSP at the intranet
2. CE sends its own SCG number to MS (intranet is secure or sends the encrypted number encoded by the intranet key)
3. MGE catches the intranet SCG topology
4. Each MS exchanges the SCG topology information (when there is a change at the topology, the information should be exchanged: synchronization) ➔ Initialization process: constructing SCG at an intranet
5. On building the new SCG with an intranet's SCG, the init CE sends the request message of building SCG to the init MS
6. The init MS asks the target MS
7. The target MS transfers the request to the representative CE of the target SCG
8. The CE returns either acceptance or rejection message to the target MS
9. The target MS sends the parameters for creating the SCG
10. The init MS and the target MS distribute the SCG parameters to the CEs, the SCG topology information is going to be changed
11. Then the new Secure Communication Group be constructed

After Init CE initialize, the CE executes the GSP protocol to build the SCG. If there is response from the intranet, the SCG be built. If there is no response from the intranet, the CE asks the MS to build the SCG and then waits. If the CE receives the parameters from the MS, the new SCG will be built with other intranet.

During the initialization of the Init MS, the parameters of SCG configuration will be distributed to the CEs. Each CE builds the SCG by GSP protocol and sends the SCG number to the MS. Without creating and calculating the SCG number for building the SCG, MS receives the SCG number and has the SCG configurational information of the intranet.

MS synchronizes and gets the whole SCG configurational information. Because the configurational information only be shared, the overall system's performance is not affected severely.

After receiving the SCG configurational information from the other CE, the CE searches the SCG topology information and sends that to the MS related to the intranet. On receiving the key and algorithm, the MS sends response message to the target MS and distributes the key and algorithm to the CEs. After confirming the construction of the SCG, the MS changes the configurational information.

The MS has the SCG configurational information at the initialization of the target MS. If the MS receives the SCG request from the init MS, it searches the configurational information and sends the message to the representative of the CEs. If the target MS receives the message from the target CE, it creates the parameters for constructing the SCG and sends the message to the init MS. When the target MS receives the response from the init MS, it sends the parameters of the SCG configuration to the each CE of the target SCG. If the new SCG configuration is changed, the SCG configurational information should be updated.

When the target CE receives the SCG construction request from the MS, it sends the response to the target MS. If the parameters for constructing SCG be received, the new SCG will be constructed.

4   Constructing the Hierarchical VPN

In this chapter, the intranet VPN is focused. Figure 8 shows the concept of the hierarchical VPN by SCG. The hierarchical VPN is composed of the 3 units: Secure Family Group (SFG), Secure Group (SG), Communication Entity (or sub-network). That is, the VPN is divided into smaller VPNs applied by SCG recursively.

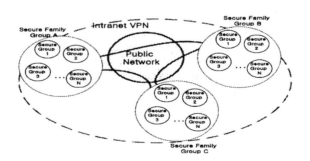

Fig. 8. Hierarchical VPN design by SCG

4.1 Hierarchical VPN by the Traditional Method

In the table 2, there are possible VPN methods in the view of the basic constructing ways of figure 8. From the table 2, we can see that only 4 methods can be applied. Although in the inner part of SFG, the path-definition way and/or area-definition way can be applied for constructing SCG, there can't be applied by the area-definition way between SFGs. At this time, the path-definition way can build a SCG by IPsec & IKE, and the area-definition way can build a SCG by GSP.

Table 2. Possible methods based on the hierarchical VPN by SCG

	Inside SG	Bet. SGs	Bet. SFGs	Possibility of application
Method 1	Path	Path	Path	O
Method 2	Path	Path	Area	X
Method 3	Path	Area	Path	O
Method 4	Path	Path	Area	X
Method 5	Area	Path	Path	O
Method 6	Area	Area	Area	X
Method 7	Area	Area	Path	O
Method 8	Area	Path	Area	X

4.2 Hierarchical VPN by SCG Number

The hierarchical VPN based on SCG number is proposed. The VPN builds the SCG by GSP at the intranet, and the SCG by IPsec & IKE in the Internet. By using these, a new SCG is made. This method uses the path-definition between SFG and MS. Because in case of CE-based encryption key, only one key per SCG should be maintained, it is the same as the area-definition is applied. Because in the public internet, each MS shares the SCG number, on receiving the request of building the SCG, the CE references the SCG number and requests the other MS to build the SCG.

Figure 9 shows the hierarchical VPN constructed by the SCG number proposed in this paper. If the initialization assumed to be already done, it shows that the new SG is going to be built between the SGs at each different side. The SG1 of SFG A and the SG2 of the SFG L create the new SCG. The CE1, the representative CE of the SG1, sends the SG request message to MS A (①), MS A searches the SG configurational information and sends the request message to MS L(②). Then MS L sends this message to CE4, the representative of SG2 (③), then CE4 sends the accept message to MS L (④), MS L creates the key and encryption algorithm and sends the MS A (⑤). If MS A receives the SCG configurational information, it sends the response message to MS L (⑥), each MS sends the configurational information to every CE (⑦) and updates the SCG topology.

4.3 Comparison of the Methods

In order that the individual methods at the section 1,2 of chapter IV would be analyzed, some conditions are assumed as follows:

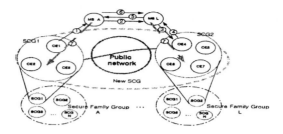

Fig. 9. Hierarchical VPN by SCG number

- Each CE has to join a party with VPN
- A new SCG can be created only between SGs
- Every SG has connections with two SGs
- A new SCG is composed of all SGs
- The number of SFG (L), the number of SG (M), the number of CE
  per SG (N), the number of the total CE (LMN)

Table 3 shows the maximum number of management keys at CE. Among those, method 3 and 5 based on the area-definition and the path-definition increase the number of redundant keys. The proposed method doesn't make any affection to the number of CE. It means that there are a few keys to be handled at CE.

If under the condition of 5 SFGs, and 10 SGs, the number of CE is going to be changed, the maximum number of management keys at CE be shown in figure 10. Comparing the methods 3 and 7 with the methods 1 and 5, the management keys are a few.

Table 3. Maximum number of management keys at CE

	Max number of management key at CE
Method 1	$(N-1)+N(M-1)+(L-1)MN+1$
Method 3	$(N-1)+(M-1)+(L-1)MN+1$
Method 5	$1+N(M-1)+(L-1)MN+1$
Method 7	$1+(M-1)+(L-1)MN+1$
Proposed	$1+(M-1)+(L-1)M+1$

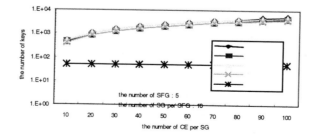

Fig. 10. Maximum number of management keys at CE

Table 4 shows the maximum number of authentication at CE. The result is depicted in figure 11. Among those, method 3 and 5 based on the area-definition and the path-definition increase the number of authentication. The method 7 that area-definition is applied at intranet and the path-definition is applied at internet, has the property of a few authentication. In the proposed method, CE has to authenticate only one with MS. There is no need to authenticate for constructing a new SCG because CE has authenticated with MS and every MS has authenticated with each other. That is, if CE has got only to authenticate with its own intranet's MS, it's enough and nothing's needed.

Table 4. Maximum number of authentication at CE

	Max number of authentication at CE
Method 1	$(N-1)+N(M-1)+(L-1)MN+1$
Method 3	$(N-1)+(L-1)MN+1$
Method 5	$N(M-1)+(L-1)MN+1$
Method 7	$(L-1)MN+1$
Proposed	1

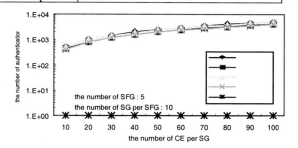

Fig. 11. Maximum number of authentication at CE

Table 5 shows the maximum number of management keys at MS. Among those, method 3 and 5 based on the area-definition and the path-definition increase the number of redundant keys. The method 7 that the area-definition is applied at intranet and the path-definition is applied at internet, has the fewest number of management keys according to the growth of CE. In the proposed method, the number of management keys is steady because the number doesn't have any affection of the number of the CE. So the proposed method is efficient in terms of performance.

Table 6 shows the maximum number of authentication at MS. Among those, method 3 and 5 based on the area-definition and the path-definition increase the number of authentication. In the proposed method, MS has to authenticate every CE of its own intranet's and every MS of other intranet's. There is no need to authenticate for constructing a new SCG because CE has authenticated with MS and every MS has authenticated with each other. That is, there is no affection by the number of CEs and the proposed method has a fewer frequency to authenticate than the others.

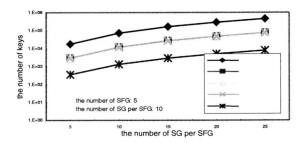

Fig. 12. Maximum Number of management keys at MS

Table 5. Maximum number of management key at MS

	$(\because A = 1 + \sum_{i=2}^{N-1} K)$
Method 1	$MA + \binom{M}{2}A + \left(\binom{LM}{2} - \binom{M}{2}\right)A + MN$
Method 3	$MA + \binom{M}{2} + \left(\binom{LM}{2} - \binom{M}{2}\right)A + MN$
Method 5	$M + \binom{M}{2}A + \left(\binom{LM}{2} - \binom{M}{2}\right)A + MN$
Method 7	$M + \binom{M}{2} + \left(\binom{LM}{2} - \binom{M}{2}\right)A + MN$
Proposed	$M + \binom{M}{2} + \left(\binom{LM}{2} - \binom{M}{2}\right) + L + MN$

Table 6. Maximum number of authentication at MS

	Max number of authentication at MS
Method 1	$MN + \binom{M}{2}N + \left(\binom{LM}{2} - \binom{M}{2}\right)N + MN$
Method 3	$MN + \left(\binom{LM}{2} - \binom{M}{2}\right)N + MN$
Method 5	$\binom{M}{2}N + \left(\binom{LM}{2} - \binom{M}{2}\right)N + MN$
Method 7	$\left(\binom{LM}{2} - \binom{M}{2}\right)N + MN$
Proposed	$L + MN$

5   Conclusions

In order to construct the varieties of VPNs, the hierarchical VPN was designed on the concept of SCG and the possibility of the every model was reviewed. On building the

VPNs at the intranet, the area-definition method was more efficient than the path-definition one. Because it was not proper to apply the intranet VPN to the internet, the path-definition method should be used. It was proposed that the design way of the hierarchical VPN was applied by the concept of SCG number, and the characteristics of the VPN were analyzed.

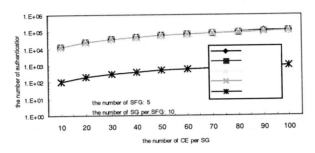

Fig. 13. Maximum number of authentication at MS

It was shown that the area-definition method in the intranet was superior. In the case of the mixed usage of the path-definition and the area-definition, it could be found that the maximum numbers of authentication at CE and MS were less when the area-definition method was applied between the SGs. Because the area-definition could not be applied in the Secure Family Group, the path-definition had to be applied. Without relating to the SG constructed in advance, the new SG was constructed by the path-definition. When the area-definition method was applied in the intranet, the CE should handle all of the two methods. Also there was no big difference between the area-definition and mixed area-definition methods among SGs. If the hierarchical VPN was constructed by SCG number, the number of keys and the number of authentication were less than ones in case of the traditional methods.

By the proposed method, it was shown that the authentication delay and traffic efficiency of the hierarchical VPN was more improved than ones of the traditional VPNs.

References

1. KISA, "Secure VPN Standard between IP and Application (Plan)", March 1999.
2. B. Gleeson, A. Lin, J. Heinanen, G. Armitage, A. Malis, "A Framework for IP Based Virtual Private Networks", RFC 2764, February 2000.
3. P. Rapalus, "Issues and Trends: 1998 CSI/FBI Computer Crime and Security Survey", CSI Press Release, March 1998.
4. A. Watanabe, S. Seno, Y. Koui, T. Ideguchi, M. Yabe, "Realization Method of Secure Communication Groups using Encryption", Transactions of Information Processing Society of Japan V.38 N.4, April 1997.
5. A. Watanabe, T. Inada, T. Ideguchi, I. Sasase, "Proposal of Group Search Protocol Making Secure Communication Groups for Intranet", ICC 2000, June 2000.
6. S. Kent, R. Atkinson, "Security Architecture for the Internet Protocol", RFC 2401, November 1998.

Design and Analysis of Role-Based Security Model in SNMPv3 for Policy-Based Security Management

HyungHyo Lee[1] and BongNam Noh[2]

[1] Div. of Information and EC , Wonkwang University, Iksan, Korea 570-749
hlee@wonkwang.ac.kr
[2] Dept. of Computer Science, Chonnam National University,
Gwangju, Korea 500-757
bongnam@chonnam.ac.kr

Abstract. Policy-Based Network Management(PBNM) architecture is to meet various needs of network users, and to provide effective management facilities in distributed and large-scaled networks to network administrators. In PBNM, network administrators perform network management operations by stipulating a set of rules rather than control each network component. On the other hand, by providing security services such as authentication, privacy of messages as well as a new flexible and extensible administration framework, SNMPv3 enables network administrators to monitor and control the operation of network components more secure than its predecessors. Despite of its enhanced security services, SNMPv3 has deficiencies in managing distributed, large-scaled network because it does not provide centralized and policy-based security management facilities. In this paper, we propose a new security model, named Role-based Security Model(RSM) with security management policy, to support scalable and centralized security management for SNMP-based networks. Also, the analysis of the SNMPv3 security system extended by RSM is also described.

1 Introduction

As the number of network elements connected to computer network increases, it is getting difficult for network administrators to configure and control each component in an effective way. Policy-Based Network Management (PBNM) is to solve those network management problems by intelligently configuring and executing the software on them using some well-defined policies or rules [15,16]. In other words, PBNM provides a way of performing network management functions via a set of high-level rules or business policies that should be abided by similar devices rather than managing device by device. There are a number of research efforts on the design and implementation of PBNM framework and Internet Drafts from IETF working group [3,4,5,6].

On the other hand, SNMPv3(Simple Network Management Protocol), a general framework for all three versions of SNMP and future development in SNMP management with minimum impact on existing operations, provides security facilities in managing network management information more secure than its previous versions. View-based access control to MIB(Management Information Base) data, authentication, encryption and integrity check of protocol data are some of security services newly provided by SNMPv3 [10,11,17,18].

I. Chong (Ed.): ICOIN 2002, LNCS 2344, pp. 430–441, 2002.
© Springer-Verlag Berlin Heidelberg 2002

But, some security features of SNMPv3 such as user-oriented authentication/encryption pass-phrases management, replication of security related MIB like users' password among managers and agents make it difficult to administrate networks security management in a centralized manner. The lack of central and scalable security management of SNMPv3 may cause a critical security hole in management of security MIB in a large-scaled network. Therefore, in order to provide a centralized and scalable management view to network administrator at a single point, policy-based security management in network management arena is one of the challenging research approaches.

In this paper, we reviewed the components and characteristics of SNMPv3 security services and the shortcomings from the security management perspective. And, we describe a need and an importance of centralized security management. Finally, a new security management model, *Role-based Security Model*(RSM), for specifying and enforcing security management policy is presented and analyzed in terms of security management efficiency. Also, the structure and operation of SNMPv3 security system extended by RSM is described.

In section 2, a brief review of policy based system and PBNM are described. And, the characteristics and shortcomings of SNMPv3 security services in terms of security management are presented in section 3. In section 4, the components and operation of RSM to supplement the security management facilities of SNMPv3 are described. Also, the analysis of the SNMPv3 security system extended by RSM in terms of security management is also presented. Section 5 concludes the features of the proposed security management model and further research areas.

2   Policy-Based Network Management

2.1   Objectives

As high-availability requirements, new applications, and multimedia increase pressure on networks, IT industry realizes that adding bandwidth is not enough to solve traffic-related problems. Network administrators must also find ways to manage networks resources effectively and efficiently. Fig. 1 illustrates the simple premise of policy-based network management applied to quality of service, or security, or even provisioning and configuration(Fig. 1. a) [1,15]. A network manager creates policies to define how resources or services in the network can or cannot be used. A policy is a representation of a business objective to be implemented in the management domain and is applied using a set of rules, which define how those business functions will be met (Fig. 1. b). The PBNM system transforms these polices into configuration changes and applies those changes to the network.

2.2   Policy-Based Management Architecture

Fig. 2 depicts a general architecture for a policy-based system [2,3,4,5,6]. This architecture includes the following components:

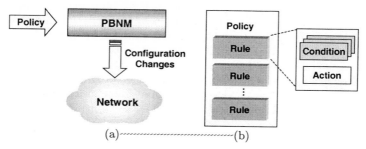

Fig. 1. PBNM Architecture and Policy Components

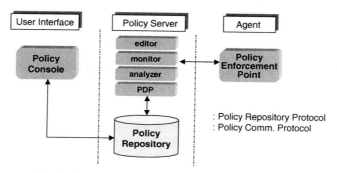

Fig. 2. General Architecture for Policy-Based System

Policy Decision Point(PDP) is a process that makes decisions based on policy rules and the state of the services those policies manage. The PDP is responsible for policy rule interpretation and initiating deployment. Its responsibility may include trigger detection and handling, rule location and applicability, network and resource-specific rule validation and device adaptation functions. *Policy Enforcement Point* (PEP) is an agent running on or within a resource (e.g. device) that enforces a policy decision and/or makes a configuration change. Network devices can be categorized into groups: policy-aware and policy-unaware. For policy-aware devices, the PEP is a component residing in those devices. But policy-unaware devices are those that are unable to interpret any portion of the policy information. However, they can still participate in a policy-based network if an application (PEP on an agent) can translate the policies into a form that the policy-unaware device can implement. *Policy Repository* is a directory and/or other storage service where policy and related information is stored. *Policy communication protocols* is a protocol to read/write data from the repository, and a protocol to communicate between PDP and PEP (e.g. COPS).

In addition to PDP, there are modules for editing, monitoring and analyzing policy in a policy server. *Policy editor* is the user interface to construct policies, deploy policies, and monitor status of the policy-managed environment. The policy editor provides the policy editing, policy presentation, rule translation, and the rule validation. In this component, many rules are translated from abstract or human readable forms to the syntax of the policy information model of the repository.

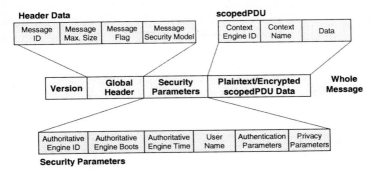

Fig. 3. The SNMPv3 Message Format

Policy monitor receives dynamic information from the network to validate or adjust policy to ensure users receive required level of service. *Policy analyzer* performs both pre-deployment and post-deployment analyses. The former determines which policies need to be deployed, the latter determines their effectiveness.

3 SNMPv3 Security System

3.1 Characteristics

One of the main objectives in developing SNMPv3 was the addition of security features for SNMP-based network management. Authentication and privacy of information, as well as authorization and access controls, have been addressed in SNMPv3 specifications [11,12,13].

The SNMPv3 architecture permits flexibility to use any protocol for the authentication and privacy of information. However, the SNMPv3 working group has specified for its security subsystem a *User-based Security Model* (USM). It is called user-based because it follows the traditional concept of a user, identified by a user name with which security information is associated. In addition, SNMPv3 provides more secure and flexible access control model, called *View-based Access Control Model* (VACM), than community-based access policy. As shown in Fig. 3, the SNMPv3 message format, drastically revised to accommodate security services, consists of four groups of data as shown in Fig. 3.

Configurable Security Services through Multi-level Security Functions. Depending on the sensitivity of management data, network administrator can configure the strength of security by leveraging the security level ('noAuthNoPriv', 'authNoPriv', 'authPriv') and view type('READ', 'WRITE', and 'NOTIFY') of the data.

User-based Security Model (USM) [12]. In order to ensure the data integrity and data origin authentication, the authentication module in USM provides HMAC-MD5-96 or HMAC-SHA-96 authentication protocols [8,12]. The privacy module of USM provides data confidentiality service using CBC-DES encryption algorithm. And the timeliness module provides a message timeliness check preventing message delay or relay based on the concept of authoritative engine. Three fields in SNMPv3 message, 'snmpEngineID', 'snmpEngineBoots', and 'snmpEngineTime', are used to check the timeliness of the message (Fig. 3).

View-based Access Control Model (VACM) [13]. VACM addresses access control, which deals with who can access network management information and what they can access. It defines a set of services that an application in an agent can use to validate command requests and notification generators, and consists of 5 elements; groups, security level, contexts, MIB views and families, and access policy. Once the SNMPv3 message is arrived, VACM decides whether the access request is granted or not by examining the security parameters such as security model and security level information in header data, user name, authentication and privacy values in security parameters, and context name, access mode (read/write/notify), and the object instance to be accessed in scopedPDU data. With VACM, network administrator can specify very fine-grained access control rules.

3.2 Shortcomings of SNMPv3 Security System

Despite its provision of enhanced security services, however, SNMPv3 has some disadvantages from the security management point of view [9].

Problems in Sharing Authentication and Encryption Pass-phrases. In USM, the authentication and the privacy key of each network administrator are generated from authentication and privacy pass-phrases by hash function, respectively. These pass-phrases should be maintained in secure manner and shared by a set of SNMP managers and agent to be used both in authentication and encryption/decryption processes.

In order to share the pass-phrases for authentication and encryption between the SNMP managers and agents, some proprietary configuration steps are needed in the initial phase of the SNMP network management. Since the initial setup procedure is not standardized in SNMPv3, there can be many different implementations of initial setup and management procedures for sharing the authentication and encryption pass-phrases.

This pass-phrase sharing problem stems from the USM's adoption of the one-way hash function to derive a key from the pass-phrase. Though the problem can be relieved by the public key cryptography system, we do not deal with it in this paper.

Deficiency in Centralized Security Management. Despite flexible and configurable security services, there exists a security management shortcomings in SNMPv3. Assume that the SNMPv3 network management environment of a corporate which has two different management domains (Fig. 4). Each management domain is composed of a set of SNMP managers, agents and network administrators, and it is managed under the security management policy specified by its network administrators. The SNMP agents maintain the security information on network administrators who have an access rights to their local MIBs. Some SNMP agents, 3 and 4, are common members of both management domains.

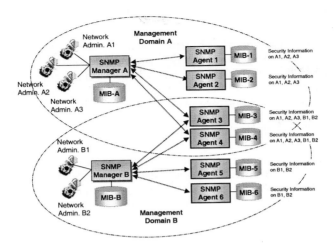

Fig. 4. Network Management Environment Example

In Fig. 4, the security information such as user name, authentication and encryption keys of network administrator A1 is stored and managed by the SNMP agents 1 through 4. If the security information on network manager A1 is modified or deleted, the local MIBs of the SNMP agents 1, 2, 3, 4 should be updated in atomic way. This might be a great overhead in managing large-scaled networks in which a large numbers of network managers are involved. This problem is caused by the USM's user-centric information management property.

In addition, there is no global view of the security management for entire network. Each network administrator has a local security management scope for some parts of an entire network. This can cause a security breach if conflicting security management policies are applied to SNMP agents without any coordination. In the case of the SNMP agents 3 and 4, uncoordinated security management policy update by network managers from different domains may leave a network in an inconsistent or an insecure state.

4   Role-Based Security Model

Role-Based Access Control (RBAC) model is getting widely used as an access control model in commercial environments due to its effective permission management using the inheritance relationship between roles [7,14]. RBAC supports several well-known security principles and policies including the least pricilege principle and the specification and enforcement of conflict-of-interest tasks, etc.

In this section, we briefly review the security properties of RBAC that can be useful for supplementing the drawbacks of SNMPv3 security management, and then propose a *Role-based Security Model* (RSM).

4.1 Characteristics of RBAC Model

The most distinguishing property of RBAC model is the entity entitled to have privileges is a role. Therefore, the access permissions to resources are given to roles, not to users [7,14]. Under RBAC, if a user wants to perform a privilege, he/she at first should be a member of a role that has the privilege. This greatly simplifies the permission management in security systems that consist of a large number of users and privileges. Adding or removing a user is done by granting or revoking a role membership to and from the user, without having to deal with the complex privilege assignment to the user.

Inheritance relationship between roles also relieves a security administrator from the security management burden. Due to permission inheritance property, the privileges assigned to junior role can be performed by a member of senior roles. So, he/she does not need to assign common privileges to all senior roles.

In addition, RBAC enables a security administrator to configure and manage the security policy in a centralized way. Because privilege assignment can be done only by a security administrator and the security policy can be changed by modifying the components of RBAC model. Therefore, RBAC provides useful characteristics for configuring a security policy and enforcing that policy in a controllable manner.

4.2 Policy-Based Security Management

The major functions of security management include maintaining of security information, controlling access request to sensitive data and managing encryption process. But the user-centric and distributed management of security information of SNMPv3 makes it difficult for network administrators to perform scalable network security management.

But in RSM, security management policy is described and managed at a single and centralized system, and security information is managed on a network management role basis. By describing a policy, network administrators can easily stipulate the security management rules to be applied to every network component, instead of specifying and managing network components individually. Also, role-based security information management simplifies the permission maintenance process in network management.

Components of Security Management Policy. A network security management policy in RSM is an abstract and human readable rules that specify the elements of security management as following:
• *Management Domain*: identifiers for network components in managed network(e.g. IP address, DNS name)

rsmUserToRoleTable		
rsmUserName	SnmpAdminString	read-create
rsmRoleName	SnmpAdminString	read-create

rsmRoleHierarchyTable		
rsmJuniorRoleName	SnmpAdminString	read-create
rsmSeniorRoleName	SnmpAdminString	read-create

rsmEngineTypeTable		
rsmEngineID	SnmpEngineID	read-create
rsmEngineType	INTEGER	read-create

rsmRoleAccessTable		
rsmRoleName	SnmpAdminString	not-accessible
rsmAccessContextPrefix	SnmpAdminString	not-accessible
rsmAccessSecurityModel	SnmpSecurityModel	not-accessible
rsmAccessContextMatch	INTEGER	read-create
rsmAccessReadViewName	SnmpAdminString	read-create
rrsmAccessWriteViewName	SnmpAdminString	read-create
rsmAccessNotifyViewName	SnmpAdminString	read-create
rsmAccessStorageType	StorageType	read-create
rsmAccessStatus	RowStatus	read-create

Fig. 5. Structure of Security Management MIB Tables

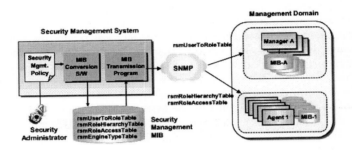

Fig. 6. Operation of Security Management

- Administrator-to-Role Assignment
- Management Privileges-to-Role Assignment
- *Role Hierarchy* among network management roles
- Functions of each network component(manager, agent or both)

Although we defined the syntax for the security management policy, we do not describe it due to the space limitation.

MIB Structure of Security Management. The security management policy should be transformed into concrete format and stored in MIB tables to be enforced by SNMP systems. The managers and agents in a management domain enforce the security management policy by accessing information from MIB tables and performing their own functions according to that information. Some important MIB tables are shown in Fig. 5.

rsmUserToRoleTable' is a MIB table to store network administrator-to-role assignment information, and role hierarchy relations among roles are stored in 'rsmRoleHierarchyTable'. And 'rsmRoleAccessTable', like 'vacmAccessTable' in VACM, provides 'READ', 'WRITE' and 'NOTIFY' MIB views associated with each role. Finally, the table 'rsmEngineTypeTable' determines which network management function that each SNMP system performs, such as manager or agent. Also, the entries of 'rsmEngineTypeTable' are the members of management domain.

Since, the tables accessed by network managers and agents are different, all the MIB tables are not transferred to every SNMP system. Depending on the value of 'rsmEngineType' field in 'rsmEngineTypeTable', some tables need not be transferred to the SNMP systems. For example, while the table 'rsmUserToRoleTable' is copied to the system if the type of the SNMP system is an SNMP manager, 'rsmRoleHierarchyTable' and 'rsmRoleAccessTable' tables are transferred to the system in case of the type of the SNMP system is an SNMP agent, because they are needed in the access control procedure. Since the information on the network administrators and their role membership is stored in the SNMP manager's MIBs, it is easy to add or remove the network administrator by manipulating the MIB table 'rsmUserToRoleTable' without modifying the MIB's of the SNMP agents.

Operation. Before the SNMP network management systems get in action, security administrator analyzes all the components pertinent to the security management such as network management privileges, network administrators, network management roles and their inheritance relationship according to the regulation of the corporate. And then, he/she stipulates the security management policy by assigning network management privileges to network management role(s) as well as network administrators to network management role(s).

After finishing the description of the security management policy, security administrator converts it to MIB tables(Fig. 5) using policy-to-MIB conversion software as shown in Fig. 6. Finally, the MIB tables are sent to the SNMP managers and agents by the security MIB transmission program. MIB transmission program refers 'rsmEngineTypeTable' to determine which tables should be transferred to each SNMP system.

Modified Access Control Procedure. In order to support the modified access control process of the RSM, the security framework of SNMPv3 needs slight modification(Fig. 7). In RSM, a network administrator is identified by the role of which he/she is member, instead of his/her user name. A network administrator is authenticated in the SNMP manager and then mapped to a network management role according to the 'rsmUserToRoleTable' table. Also, the access control process in the RSM of the SNMP agents refers the table 'rsmRoleAccessTable' to determine whether the MIB object specified within a PDU can be accessed by the given network management role. The interesting behavior of the modified access control procedure is that even if an access is denied, it tries to check to see if any of its junior roles obtained from 'usmRoleHierarchyTable' has an access privilege. If exists, the access request is finally granted.

4.3 Analysis

The purpose of the proposed RSM is to supplement the shortcomings of SNMPv3 security system in the management of security information, not to replace USM. In fact, RSM makes use of USM's message authentication and encryption services. With the help of characteristics of RBAC model, however, RSM has advantages in

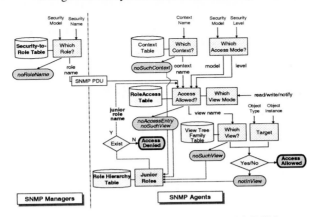

Fig. 7. Modified VACM Procedure with RSM

managing security management information including pass-phrases for network administrators, compared with USM.

To compare RSM with USM in terms of security management efficiency, we choose two parameters: *the number of security data stored and managed in SNMP systems* and *the degree of security data distribution*. Since, if a large number of security data are stored among SNMP systems and managed in a distributed manner, it is getting difficult to configure and administrate security data in a consistent way.

Table 1 shows the comparison of the number of security data in SNMP systems USM and RSM. Assume that the number of network administrators, network administrators stored in managers and agents as N_{admin}, $N_{admin'}$, $N_{admin''}$, respectively. Similarly, the number of roles, roles stored in manager and agents as N_{role}, $N_{role'}$, $N_{role''}$. And $N_{group''}$ and $N_{perm''}$ represent the number of groups and network management privileges stored in agents. As to groups and roles, there is no general rule for how groups are mapped into roles. So, in this paper, we assume that the number of $N_{group''}$ and $N_{role''}$ are at the same magnitude.

Table 1. Comparison of the number of Security Management Data between USM and RSM

	Security Information	USM	RSM
Managers	Pass-phrases for administrators	$2 * N_{admin'}$	$2 * N_{admin'}$
	Pass-phrases for roles	~	$2 * N_{role'}$
	Administrator-role assignment	~	$N_{role'} + N_{admin'}$
Agents	Pass-phrases for administrators	$2 * N_{admin''}$	-
	Pass-phrases for roles	~	$2 * N_{role''}$
	Group-privileges assignment	$N_{group'} + N_{priv'}$	-
	Role-privileges assignment	-	$N_{role''} + N_{priv''}$
	Role hierarchy	-	$N_{role''}$

In general, since one network administrator performs management operations on several agents and one or more administrators are assigned to the same role, (eq. 1), (eq. 2) and (eq. 3) hold.

$$N_{admin} \leq N_{admin'} < N_{admin''} \tag{1}$$

$$N_{role} \leq N_{role'} < N_{role''} \tag{2}$$

$$N_{role} < N_{admin}, N_{role'} < N_{admin'}, N_{role''} < N_{admin''} \tag{3}$$

From Table 1, we can analyze that the overhead for managing security data entailed by adopting RSM is (eq. 4), while eliminating the overhead for storing security data for network administrators in agents, $2 * N_{admin''}$.

$$N_{role'} + N_{admin'} + 2 * N_{role'} + 2 * N_{role''}$$
$$< 2 * N_{admin'} + 2 * N_{role'} + 2 * N_{role''} \tag{4}$$

$$< 4 * N_{admin'} + 2 * N_{role''} \tag{5}$$

Considering the real network management environments in which one administrator performs management operations on many agents ($N_{admin'} \ll N_{admin''}$) and several administrators are assigned to a singe role ($N_{role''} \ll N_{admin''}$) , we can conclude that (eq. 5) is smaller than $2 * N_{admin''}$, the number of security data managed by agents in USM.

5   Conclusions and Future Work

The SNMPv3 network management standard provides a reliable and evolutionary network management environment by defining an enhanced security services as well as a flexible and extensible network management framework. The security subsystem provides services such as authentication and privacy of messages and potentially contains multiple security model including USM or other user-defined ones. With the USM and VACM provided by the SNMPv3 security subsystem, the individual network manager can setup and enforce the security management policy. But, the user-centric feature of the USM and no provision of the global security management view can make the SNMPv3 security subsystem unsuitable for the management of large-scaled networks.

To supplement those drawbacks of SNMv3 security management, we propose a Role-based Security Model(RSM), which has the following features; the effective management of network managers using RBAC model, and the centralized policy-based security management The role-based management of network administrators greatly reduces the complexity of permission management in large network management systems. Also, the security management policy specified and managed by a designated security administrator provides a unified security management view at a single point as well as consistent security enforcement function.

In addition, the implementation of the RSM module in the SNMPv3 security subsystem is another future work.

Acknowledgements. This work is supported in part by the Ministry of Information & Communication of Korea, under the "Support Project of University Information Technology Research Center(ITRC)" supervised by KIPA.

References

1. Y. S. Shin, Policy-based Security Management Network Architecture, NETSEC-KR2001, April, 2001.
2. Wang Changkun, "Policy-based Network Management," Communication Technology Proceedings, 2000.
3. Policy Framework Core Information Model, draft-ietf-policy-core-info-schema-02.txt, Internet Draft, February 1999.
4. The COPS(Common Open Policy Service) Protocol, draft-ietf-rap-cops-06.txt, Internet Draft, February 1999.
5. Policy Framework, draft-ietf-policy-framework00.txt, Internet Draft, September 1999.
6. Policy Framework Core Information Model, draft-ietf-policy-core-info-model-02.txt, Internet Draft, October 1999.
7. David F. Ferraiolo, Janet A. Cugini, D. Richard Kuhn, "Role-Based Access Control(RBAC): Features and Motivations," Proceedings of the 11th Annual Computer Security Applications Conferences, December 1995, pp. 241–248.
8. Warwick Ford, *Computer Communications Security: Principles, Standard Protocols and Techniques*, Prentice-Hall, 1994.
9. HyungHyo Lee, DongIk Lee, BongNam Noh, "Policy-based Security Management in SNMPv3: Role-based Approach," Workshop on Information Security Applications, November 2000.
10. RFC 2571, An Architecture for Describing SNMP Management Frameworks, May, 1999.
11. RFC 2572, Message Processing and Dispatching for the Simple Network Management Protocol(SNMP), May 1999.
12. RFC 2574, User-based Security Model (USM) for version 3 of the Simple Network Management Protocol (SNMPv3), April 1999.
13. RFC 2575, View-based Security Model (VACM) for the Simple Network Management Protocol(SNMP), April 1999.
14. Ravi S. Sanhdu, Pierangela Samarati, "Access Control: Principle and Practice," IEEE Computer, September 1994, pp.40–48.
15. Susan J, Shepard, "Policy-Based Networks: Hype and Hope," IT Pro,_ January-February 2000.
16. Morris Sloman, Network and Distributed Systems Management, Addison-Wesley, 1994.
17. Stallings, W. SNMP, SNMPv2, SNMPv3 and RMON1 and RMON2, Third Edition, Addison-Wesley, 1998.
18. Mani Subramanian, Network Management: Principles and Practice, Addison-Wesley, 2000.

A Taxonomy of Spam and a Protection Method for Enterprise Networks

Tohru Asami, Takahiro Kikuchi, Kenji Rikitake, Hiroshi Nagata,
Tatsuaki Hamai, and Yoshinori Hatori

KDDI R&D Laboratories, Inc.
2-1-15 Ohara, Kamifukuoka-shi, Saitama 356-8502, Japan
{asami, kick, kenji, hi-nagata, hamai, hatori}@kddilabs.jp
http://www.kddilabs.jp/index.html

Abstract. This paper presents a basic design and principles to protect enterprise networks from spam mail after considering a taxonomy
of spam mail based on their delivery schemes as well as their message
envelope formats. The analysis of MTA mail logs of a mail gateway at a
sample enterprise network provides a correlation between senders' mail
addresses and their IP addresses. Thus, a mail filter for gateways of these
networks is proposed to prevent them from receiving spam mail from the
Internet. Finally, the effectiveness of this filter is demonstrated based on
the results of mail log analysis.

1   Introduction

The Internet mail system consists of individual mailboxes, MUA (Mail User
Agent) where users operate and refer mailboxes and MTA (Mail Transfer Agent)
where mail is distributed, as shown in Figure 1. One of the MTA's basic functions is to receive mail sent from local users or other MTAs and send the received
mail to local users' mailboxes or other MTAs as typified by Sendmail[1] by E.
Allman. In this paper, a network consisting of an MTA and MUAs accessing
the MTA's mailboxes is called the MTA's management network. Internet service providers' networks and corporate LANs generally correspond to MTAs'
management networks.

Communication among MTAs has been peer-to-peer and symmetrical since
the initial Internet design concept. Considering a personal computer's capability,
the protocols of POP3[2] or IMAP4[3] were developed for the present mail system
to download the contents from the mailboxes on a POP server or an IMAP server,
respectively, when local users refer to contents of these mailboxes.

The e-mail system, the most basic communication tool in existence since the
initial stage of the Internet, is still one of the most important communication
tools because of its push-type information distribution (SMTP among MTAs),
which is not with the Web. Especially, the latest trend is that a growing number
of companies and services are running mailing lists for many clients or members. Of course, the more users they have, the more damage they suffer from

I. Chong (Ed.): ICOIN 2002, LNCS 2344, pp. 442–452, 2002.
© Springer-Verlag Berlin Heidelberg 2002

email with computer viruses or spam mail, which could create a headache for corporate management as well as an ISP's help desk services. Spam mail, which were initially known to trouble NetNews, generally indicated the sending of a considerable amount of unsolicited mail such as advertisements, and they inflict even more damage on today's email systems[4]. Moreover, since spamming is found even in BBSs, or bulletin board systems scattered on the Internet, countermeasures for this have been discussed for these systems as well.

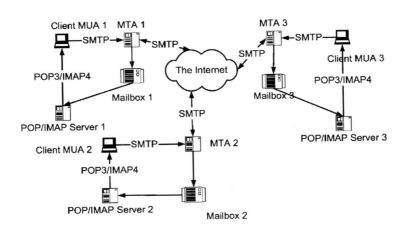

Fig. 1. Structure of mail delivery in the Internet

In this paper, among a variety of e-mail misuses, countermeasures focusing on spam mail that are in the RFC2505[5] are described. Hereafter, a classification of methods for sending spam mail and general countermeasures against them are described in Section 2, the e-mail log analysis at a mail gateway against the Internet in a sampled company is described in Section 3 and a proposal of possible protection at the MTA, where spam mail is received, and its effectiveness are described in Section 4.

2   Classification of Sending Spam Mail and Common Anti-spam Countermeasures

In this paper, the term "host name" or "domain name" is used, which should be interpreted as a Fully Qualified Domain Name, FQDN, and returned from the DNS in response to a PTR query (.IN-ADDR.ARPA). Also, this refers to a name with a DNS A or MX record associated to an IP address when it is translated to a name. Reference is also made to the local part of a mail address by "username," and the domain part of a mail address by "mail domain name."

When a mail transfer protocol SMTP[6], as shown in Figure 1, is functionally equivalent either in MUA or in MTA as far as the sending function is concerned.

As a result, users have to set up their e-mail address in MUA, which is a fundamental cause of spam mail. This annoys Internet beginners very much when they start using e-mail, and also annoys Internet Service Providers having to deal with the beginners' problems. As shown in Figure 2, spam mail has two distribution patterns. Class A sends mail via MTA with illegal third party SMTP relaying. Class B sends mail directly to MTA (MTA 2). With Class A, mail is sent via an MTA 1 server with third party SMTP relaying to block their recipients to identify the sender. Therefore, Class A is the standard method for sending spam mail. The spammer, the sender of the spam mail, is sometimes a provider dedicated to the spam mail but mostly a user with dial-up or an equivalent environment.

With Class B, many of the spam messages are directly sent from senders in a dial-up environment, but some are from a provider with anonymously paid fees such as prepaid cards or from a charge-free provider with fake registration of personal information. In these cases, concealing the senders' identity is based on the fact that senders' IP addresses are dynamically allocated by DHCP when connecting to a provider.

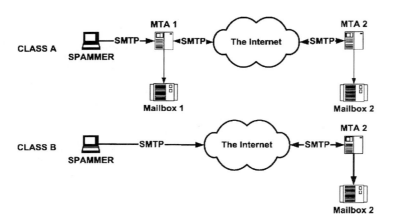

Fig. 2. Taxonomy of sending spam mail

Next, spam mail is categorized into the following two types based on the form of the message envelope: Message Type 1 where the actual mailbox corresponds to the "MAIL From:" (envelope originator) address and Message Type 2 where it does not. If the distribution system is in Class A, Message Type 1 is categorized again into the following two types: Message Type 1.1 whose sender address or sender's mailbox is on a steppingstone MTA, and Message Type 1.2 where the mailbox was imitated as another mailbox on another MTA on the Internet. If the distribution system is in Class B, Message Type 1 is categorized again into the following two types: Message Type 1.1 where the sender's mailbox was under management of the MTA to which the spammer belongs, and Message Type 1.2 where the mailbox was imitated as another mailbox on another MTA on the Internet.

Furthermore, Message Type 2 is categorized into the following two types: Message Type 2.1 where the mail domain name of the sender mail address can be identified by DNS but the username or local-part of the mail address does not exist, and Message Type 2.2 where it cannot be identified by DNS. If the distribution system is in Class A, Message Type 2.1 is categorized again into the following two types: Message Type 2.1.1 where the domain name of the mailbox was the same as that of a steppingstone MTA, and Message Type 2.1.2 where the mailbox was imitated as another mailbox on another MTA on the Internet. If the distribution system is in Class B, Message Type 2.1 is categorized again into the following two types: Message Type 2.1.1 where the domain name of the mailbox was the same as that of the spammer's MTA and Message Type 2.1.2 where it was imitated as another MTA on the Internet.

Both Message Type 1 and 2 can be sent from ordinary Internet users. However, if those messages are spam, actual Internet users of good will, who are imitated by spammers, suffer terrible social damage in the case of Message Type 1. While for Message Type 2, it is difficult for receivers to cope with the spam mail because of unidentifiable senders.

In the meantime, the way to prevent spam mail at MTA is detailed in RFC 2505[5]. At the same time, it's effectiveness is limited because there is no sure way to authenticate the sender with SMTP that is the standard Internet mail sending protocol. Therefore, the SMTP authentication in RFC 2554 [7] is stipulated as the standard track, which is the way to authenticate the sender with SMTP. This expansion is supported by mail server systems such as qmail and sendmail 8.10/8.11[8] and has become supported by client software programs such as Outlook Express 4.0 and 5.0[9]. The SMTP authentication, however, is not used in all MTAs or MUAs because of the operational concern over the lower compatibility of the Internet mail system. This is the biggest obstacle to eradicating spam mail.

Accordingly, the general countermeasure is not to receive messages from a spammer, who is forced to be identified by the sender's mail address or envelope originator address received at MTA, comparing it with those in the address table. In this case, MTA has such an address table with the admitted senders' or clients' mail addresses (in fact, the domain names), and their IP addresses (network addresses), registered in advance. This filter is already installed into many MTAs such as sendmail[10],[11]. Table 1 shows the effectiveness of the above measure with message types and distribution systems as the parameters. In this table, 'O' indicates effective, 'X' indicates ineffective and '$\star$' indicates effective on condition.

In Table 1, the problem of $\star_1$ can be solved in principle when the management network of MTA 1 is operated precisely because the spammer is with a provider or in a corporate network whose MTA 1 is used as a steppingstone.

Table 1. Effectiveness of the limitation of third party SMTP relaying for spam

	Class A	Class B
Message Type 1.1	$\star_1$	X
Message Type 1.2	O	X
Message Type 2.1.1	$\star_1$	X
Message Type 2.1.2	O	X
Message Type 2.2	O	O

1. Ineffective if the spammer belongs to the same manage-
    ment network whose MTA 1 is used as a steppingstone
    by him.

One of the typical measures for MTA 2 receiving the mail messages is MAPS
Relay Spam Stopper (RSS)[12]. It excludes the mail messages received from the
MTA if they are found to belong to the MTA database, which allows third party
SMTP relaying. In reality, however, even spam mail such as Class A that can be
dealt with comparatively easily cannot be expected to be eradicated in the near
future, because 6% of the MTAs on the Internet were still allowing third party
SMTP relaying as of January 2001 [13].

The third party SMTP relaying restriction is not effective enough to cope
with spam mail corresponding to Message Type 1 and Message Type 2.1 of Class
B sent by a mock user or host from dial-up terminals. Consequently, measures
such as MAPS DUL (DialUp List) for MTA 2 that prevents the receiving of mail
masages from pre-registered dial-up addresses have been proposed [14]. The basic
idea, though, is similar to the restriction with the third party SMTP relay.

Another proposed measure is MAPS RBL (Realtime Blackhole List) that
prevents the receiving of mail messages by MTA 2 from pre-registered spammers or spam services such as Web servers [15]. Operations at MTA 2 are quite
bothersome, but MAPS is effective for encouraging MTAs and services, advantageous to spammers, to cope with spam mail by releasing its database to the
public.

Class B is also difficult to deal with in terms of the convenience of Internet services. For instance, if taking a roaming service into account, only the
restrictions on senders' domain names or IP addresses will make it difficult to
send mail from the original domain name when connecting MUA to another ISP.
Therefore, some ISPs and companies adopt POP before SMTP[16] to authenticate the sender of SMTP. This mechanism is such that the sender has been
authenticated by POP before sending a mail with SMTP from the sender's host.
The installations of this system, however, have not been as widespread as SMTP
AUTH[17].

As for SMTP transmission within earned hours, ten minutes, for example,
after POP authentication, POP before SMTP has the following security holes
because of being checked only by IP addresses:

1. If the user connects to an ISP in the dial-up environment, the next user to the same dial-up port can use the line within the former user's earned hours even after the former one has finished. (Not all ISPs announce the information of their earned hours.)
2. If the user connects from their office via NAPT, his or her colleague can also use the line.

3 Analysis of Mail Log at an Mail Gateway on Senders' Addresses

Table 2 shows a mail log recorded at a mail gateway (sendmail-8.11.6.) of a Japanese company. The period of this log was from 2001/07/15 03:41:16 to 2001/09/10/11:56:40. The operation principle of this server (gateway) was "All local Mail allowed and Mail from the Internet was filtered with MAPS Spam Filter." The analysis of this log suggested that the MAPS Spam Filter rejected 2.8% of the 72,648 mail messages (70,653+1,995) that were sent from the Internet outside the firewall.

Table 2. Results of the mail log analysis

	Total	Itemized Details	
SMTP connections	177,818	accept	175,823
		reject	1,995
Received Mail (accept)	175,823	Mail Sent from Local Hosts$_1$	105,170
		Mail Sent from the Internet outside the Firewall$_2$	70,653
Mail Sent from Local Hosts	105,170	null$_3$	2,228
		normal$_4$	102,942
Mail Sent from the Internet outside the Firewall	70,653	null	3,508
		normal	67,145
Rejected Mail by MAPS Filters (reject)	1,995	RSS$_5$	1,932
		DUL$_6$	54
		RBL$_7$	9

1. Mail sent from inside the firewall
2. Mail sent from the Internet outside the firewall of this MTA
3. Mail with null envelope originator address ("MAIL From: <>")
4. Mail with non-null envelope originator address
5. Mail rejected by MAPS RSS[12]
6. Mail rejected by MAPS DUL[14]
7. Mail rejected by MAPS RBL[15]

Among the 67,145 mail messages that had usual or non-null envelope originator ("MAIL From:") addresses sent from the external host (Class A MTA 1 or

Class B MUA in Figure 2), the number of unique pairs of the external host IP address and its corresponding mail domain name "<External Host IP Address, Mail Domain Name>" was 7,454, where the external host IP is defined as the IP address of a host outside the firewall, and also the corresponding mail domain as the domain name part of the envelope originator address excluding empty addresses, i.e. "MAIL From: <>", thrown to this server.

A three-step operation was devised on these host IP addresses.

1 Send DNS PTR queries for these host IP addresses to obtain host names or FQDNs for these addresses.
2 Send A queries to these FQDNs to obtain the corresponding IP addresses.
3 Compare the produced IP addresses with the original host IP addresses.

The results of this are shown in Table 3.

Table 3. Results of address matching for external host addresses

	Total	Itemized Details	
External Host IP Address	4,684	Reliable Hosts	3,980
		Unreliable Hosts	704
Unreliable Hosts	704	DNS PTR Query Failures	559
		DNS A Query Failures	80
		No Match between the original IP and that provided by DNS A Query	65

A host whose original IP address is equal to the produced address is defined as a reliable host. There are 6,523 <External Host IP Address, Mail Domain Name> unique pairs for these 3,980 reliable hosts. The other 931 mail messages (=7,454-6,523) are indicated as the "irregular" pairs from the unreliable hosts. This result suggested that about 15% of the 4,684 external hosts accepted by the MAPS filters were not reliable because those host IP addresses did not correspond to proper FQDNs. Thus, the MTAs can be run to reject mail from these unreliable hosts.

Applying the following procedure to 6,523 < External Host IP Address, Mail Domain Name> pairs corresponding to 3,980 "reliable" external hosts gives Table 4.

(1) Obtain an FQDN (FQDN1) for each external host IP address.
(2) Match FQDN1 with the Mail Domain Name (FQDN2) using the following domain matching rules.
(3) If it fails, match FQDN1 with the MX record (FQDN21) for FQDN2 using the domain matching rules.

The domain matching rules are shown in regular expressions as follows:

Table 4. Organization domain matching between external host IP address and corresponding envelope originator address ("MAIL From:")

	Total
Unique Pairs	6,523
Organization Domain Matches	4,684
Organization Domain Matches after Consideration of MX	704
Not Matches	1,135

i *.??? : Match at the second level domain (SLD) for a 3-letter TLD (ex. kddi.com, ieee.org)

ii *.??.??: Match at the third level domain for a 2-letter TLD and SLD (ex. kddlabs.co.jp, dion.ne.jp)

iii ???*.??: Match at the second level domain (SLD) for a 2-letter TLD with an SLD with more than or equal to 3 letters (ex. kddilabs.jp, inria.fr,···)

The domain name derived by these rules is referred to as the "organization domain." Organization matching can be done for a fairly wide range of domain names with just these 3 rules (except net.th, etc). Organization domain matches were obtained for 69% of the 6,523 < External Host IP Address, Mail Domain Name> unique pairs

An examination of the remaining 31% of the pairs reveals that there were quite a few mailing list mail messages resent from the mailing list server, retaining the original message senders' addresses, instead of the list servers' addresses, as envelope originator addresses. These mailing list mail messages are operated assuming the existence of a third party SMTP relaying MTAs. This shows: (1) there are still a number of MTAs which allow third party SMTP relaying even though they are reliable and not rejected by MAPS filters; (2) the management of current Internet mailing lists often assumes third party SMTP relaying at MTAs; and (3) there is still the possibility of spam mail being sent directly from MUAs connected to ISPs.

4 Possible Protection Methods at MTAs Receiving Spam Mail and Their Effectiveness

In the previous section, it was shown that it is not enough to apply MAPS filters as a spam mail protection method, as well as to reject the receiving from unreliable MTAs. In this section, a more strict spam protection method is proposed using the following filtering rules on receiving mail.

(1) Deny the third party SMTP relaying
(2) Deny receiving mail messages with MAPS filters (2.8%)
(3) Deny receiving mail messages from unreliable external hosts (5.3%)
(4) Deny receiving mail messages which fail in the domain matching (23.0%)

Table 5. Details of effectively received mail messages

	Total	Itemized Details	
Effectively Received Mail	69,140 **(100%)**	Rejected Mail by MAPS Filters	1,995 **(2.8%)**
		Normal Mail Sent from the Internet outside the Firewall	67,145
Normal Mail Sent from the Internet outside the Firewall	67,145	Mail from Reliable Hosts	63,452
		Mail from Unreliable Hosts	3,693
Mail from Reliable Hosts	63,452	**Mail that Organization Domain Matches Succeeded in**	47,535**(68.8%)**
		Mail that Organization Domain Matches Failed	15,917**(23.0%)**
Mail that Organization Domain Matches Succeeded in	47,535	Organization Domain Match	44,411
		Organization Domain Match after Consideration of MX	3,124
Mail from Unreliable Hosts	3,693**(5.3%)**	No Reverse look-up	2,796
		No Forward look-up	384
		No Match between the original IP and that provided by DNS re-consultation	513

Here, the new rule (4) has been added and the ratio of the corresponding mail messages to the total number of mail messages is attached at the end of each rule. Table 5 is a rewrite of the results in the previous section from the perspective of message numbers, where mail messages other than error mail messages, which correspond to "MAIL From: <>", are referred to as "effectively received mail messages," the number of which is the sum of the number of normal mail messages sent from outside the firewall and that of mail messages rejected by MAPS filters.

This table shows that MAPS filters can manage 2.8% out of 31.2% of the possible spam mail in total. Adding organization domain matching to the mail filters can provide more reliable anti-spam environments with the current mail server management. It will be safer to use this new rule for B to B communications, which usually require more trustworthy management

The rejected mail messages may contain those from people in good faith even though they may contain spam mail messages or third party SMTP relayed mail messages. It has been noted that a growing number of Internet users under "always-on" Internet connectivity with static routable IP addresses will use ISP name servers for their DNS PTR queries even though they have their own name servers. ADSL services have made it possible even for individuals as well as corporate organizations to enjoy "always-on" Internet connectivity with static routable IP addresses. They can obtain their own DNS domain names and run their own mail servers or name servers on their computers. As for DNS PTR queries, they have to use the FQDNs, which are pre-assigned by ISPs, or often pay extra fees for allowing ISPs to modify their DNS systems. Services without DNS PTR look-up are becoming increasingly popular even for dedicated line services with static IP address for corporate users. In each of the above two cases, DNS PTR query provides us with FQDNs, which ISPs have pre-assigned. All the other Internet environments are the same as regular users'. It's sufficient to delegate this service with an NS record to the static IP address users. However, many ISPs with this delegation right are unwilling to do so because of operational costs. One of the other possibilities is for users to login to the ISPs' DNS systems to set up this kind of delegation through the Internet after some customer authentication. This kind of service has not entered into wide use.

5   Conclusions

A new principle for protecting enterprise networks from spam mail has been presented after considering a taxonomy of spam mail based on their delivery schemes as well as their message envelope formats. The new filtering rule called organization domain matching is based on the match between the MTA sending the mail and its envelope originator mail address with the organization domain names. It has been shown that 68.8% of the observed mail messages fall into this category. Assuming the management of ISPs and enterprise networks is conducted as shown by the procedure from (1) to (3) in section 3, it has been shown that very powerful filtering can be achieved, based on the analysis of MTA mail logs of a mail gateway at a sampled enterprise. It was also shown that the current MAPS filters are only effective for 9% (2.8% out of 31.2%) of the possible spam mail in total.

References

1. Costales, B., Allman, E.: sendmail. Second Edition. O'Reilly and Associates, Cambridge Köln Paris Sebastopol Tokyo (1997)
2. J. Myers, J., Rose, M.: Post Office Protocol – Version 3. RFC1939 (1996)
3. Crispin, M.: Internet Message Access Protocol – Version 4rev1. RFC2060 (1996)
4. Schwartz, A., Garfinkel, S.: Stopping Spam. O'Reilly and Associates, Cambridge Köln Paris Sebastopol Tokyo (1998)
5. Lindberg, G.: Anti-Spam Recommendations for SMTP MTAs. RFC 2505 (1999)

6. Postel, J.: Simple Mail Transfer Protocol. RFC0821 (1982)
7. Myers, J.: SMTP Service Extension for Authentication. RFC2554 (1999)
8. http://members.elysium.pl/brush/smtp-auth/server.html
9. http://members.elysium.pl/brush/smtp-auth/client.html
10. MAPS: Basic Mailing List Management Guidelines For Preventing Abuse. http://www.mail-abuse.org/manage.html (2001)
11. MAPS Transport Security Initiative Team: MAPS Transport Security Initiative. http://www.mail-abuse.org/tsi/ (2001)
12. http://work-rss.mail-abuse.org/rss/index.html
13. Hoffman, P.: Allowing Relaying in SMTP: A Series of Surveys. Internet Mail Consortium Report IMCR-015, http://www.imc.org/ube-relay.html (2001)
14. http://www.mail-abuse.org/dul/
15. http://www.mail-abuse.org/rbl/
16. Gellens, R., Klensin, J.: Message Submission. RFC 2476 (1998)
17. http://www.emaillab.org/win-mailer/table-otherspec.html

Security Policy Deployment in IPsec

Geon Woo Kim, So Hee Park, Jae Hoon Nah, and Sung Won Sohn,

Network Security Department
Electronics and Telecommunications Research Institute
161 Kajong-Dong , Yusung-Gu, Taejon, KOREA, 305-350
{kimgw, parksh, jhnah, swsohn}@etri.re.kr

Abstract. Security policy system defines the basic rules and strategies for information security systems, e.g., IPsec, IKE, and Firewalls. But due to the heterogeneous network security systems and the corresponding applications of security policy system, it is difficult to build up a unified policy database and deploy it efficiently. In this paper, we propose a unified security policy database schema to be applied to IPsec system and a synchronous methodology of security policy deployment with controlling network schedule. By these, we can guarantee more efficient security services in the policy-based distributed network.

1 Introduction

Network security technologies are quickly being deployed over the Internet to support electronic commerce and virtual private networking services. In particular, security gateways, commonly known as firewalls, are installed along the perimeters of enterprise internets to enforce data origin authentication and access control, and security protocols, e.g., IP security (IPsec) protocols and Transport Level security (TLS/SSL) protocols, are used to provide end-to-end and hop-to-hop confidentiality, integrity and authentication protection to the Internet traffic[1].

Though Security policy is often regarded as a solution for efficient network security management, the definition is various [2]. A policy is a set of rules that controls the behavior of a network under different conditions. Rather than offering an uniform or best-effort service to all users, a policy-enabled network can take into account priorities or other user-level characteristics, and dynamically determine the treatment to give the each packet [3].

The Security Policy System (SPS) is a distributed database of security policy information, It provides the mechanisms needed for discovering, accessing and processing security policy information of hosts, subnets or networks of a security domain[4].

The Security Policy Protocol (SPP) defines how the policy information is exchanged, processed, and protected by clients and servers. The protocol also defines what policy information is exchanged and the format used to encode the information[5]. While there are a lot of security policy systems attached to various security applications to control them, each has a heterogeneous database schema and deployment mechanism different from others.

I. Chong (Ed.): ICOIN 2002, LNCS 2344, pp. 453–464, 2002.
© Springer-Verlag Berlin Heidelberg 2002

Deployment mechanism describes that how we should access security policy information and solve synchronization problem between policy systems and applications

IPsec system resides in IP layer in kernel and could be regarded as a policy client. As a client should wait server's response in the client-server system, an ipsec engine in the kernel should wait. But it is subject to cause serious problems

Therefore, in this paper, we suggest transparent and effective mechanism that can be used for applying security policy, moreover, solve the synchronization problem between security policy system and IPsec engine

2 SPS in C-ISCAP

Controlled Internet Secure Connectivity Assurance Platform (C-ISCAP) offers a universal IPsec solution for end-to-end security in the distributed network.

C-ISCAP is composed of IPsec Engine (i-Gine), IKE Server (AutoKem), Security Policy System(SPS), Security Management System (SMS), and Security Evaluation System (SES).

The i-Gine applies cipher and/or authentication algorithms to each outbound/inbound packet, and references Security Association (SA) negotiated by the AutoKem. The i-Gine locates in the IP Layer of system kernel and there are three ways for installation: Integration, Bump in the Wire, Bump in the Stack.

The AutoKem handles SA generation, management, and certificate issue by communicating with the CA server. When receiving SA request message from the i-Gine, the AutoKem searches SA Database (SADB), and return the corresponding SAs based on the selector values. In case of absence, the AutoKem references security policy through the SPS, and then negotiates the new SAs with the peer IKE server.

Negotiation process needs two steps: phase1 and phase2. During the SA negotiation, security policy information is used for both the secure communication channels and SAs being negotiated.

2.1 i-Gine (IPsec Engine)

The i-Gine supports the functions of IPsec Engine, interacting with the AutoKem and the SPS. In the terminal systems, security host functions are installed. In the systems located along the perimeters of security domains, having multiple network interfaces, commonly known as routers, firewalls, and VPN servers, security gateway functions are installed.

Regardless of installed systems, the i-Gine establishes the secure communication between source host and destination host, between source host and gateway, between gateway and destination host, and between gateway and gateway in compliance with security policy information [6][7][8][9][10].

After the new security policy is founded, SAs between two systems are set up using autonomous key negotiation. These new SAs are used for exchanging secure messages among systems(hosts or gateways).

Figure 2 shows the correlation between the i-Gine units.

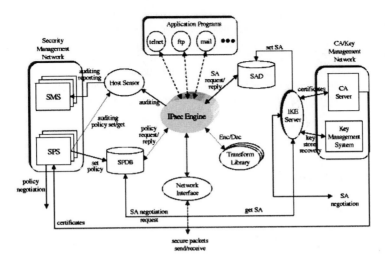

Fig. 1. C-ISCAP Architecture

2.2 AutoKem (IKE System)

The AutoKem is in charge of key exchange and SA negotiation required for C-ISCAP security information services. Exchanged key is used as cryptographic library key, assuring confidentiality, authentication, and integrity for the packets. Also, as is applied as a basis for SA. We can support the negotiation process both manually and autonomously.

The AutoKem negotiation process is composed of two steps for secure key exchange: phase1 and phase2.

In phase 2, IPsec SAs and key materials being used for the i-Gine, are negotiated based on the new key and SA which are negotiated between peer IKE servers for secure communication in Phase 1. That is, phase 2 negotiation channel is under the protection of IKE SAs and key negotiated in phase 1. The use of Perfect Forward Secrecy (PFS) in phase 2 is optional.

The Autokem is divided into 4 units: Main Mode Processing Unit (MMPU), Aggressive Mode Processing Unit (AMPU) for the phase 1, Quick Mode Processing Unit(QMPU) for the phase 2, New Group Mode Processing Unit (NMPU) not used for key negotiation but for negotiating group for the future use. By the SPS, the Autokem negotiates SAs, generates security key, and stores those in SADB. In need of new group, NMPU is closely connected with either MMPU or AMPU [11].

ISAKMP SA and IPsec SA generated during the phase 1/phase 2 are stored in the ISAKMP SADB and the IPsec SADB respectively.

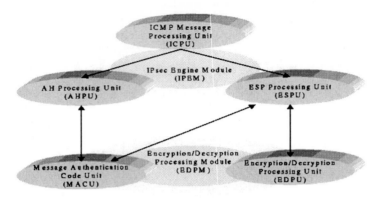

Fig. 2. Correlation between i-Gine units

2.3 SPS(Security Policy System)

The Security Policy System (SPS) is a distributed database of security policy information. It provides the mechanisms needed for discovering, accessing and processing security policy information of hosts, subnets or networks of a security domain.

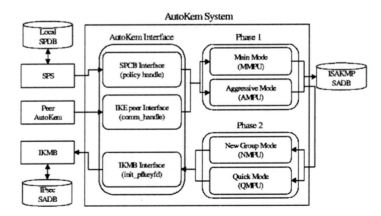

Fig. 3. AutoKem Structure

In SPS, each security domain has a master file that uniquely defines a security domain by its network resources (hosts, subnets, networks) and policies to access them. These policies reside in a database local to the security domain, The Policy Server (PS) provides access to these policies to client applications requesting policy information for a particular host, subnet or network.

Policy Client (PC) providing interface between application/IPsec system and policy server generates messages corresponding to the security policy.

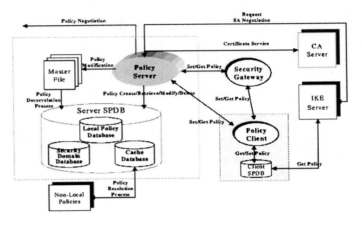

Fig. 4. SPS Components and Interactions

2.3.1 SPS Message Flow

♦ Message *Q1*
The policy client A of security domain A transmit query message to the policy client B of security domain B. The selector values, such as source ip address, destination ip address, source port number, destination port number and protocol, are contained in this message *Q1*. The SPP defines the format.
The policy Server A, receiving message *Q1*, searches a SPDB and returns corresponding policy information through message *R2*.

♦ Message *Q1*
The policy client A of security domain A transmit query message to the policy client B of security domain B. The selector values, such as source ip address, destination ip address, source port number, destination port number and protocol, are contained in this message *Q1*. The SPP defines the format.
The policy Server A, receiving message *Q1*, searches a SPDB and returns corresponding policy information through message *R2*.

♦ Message *Q2*
If the policy information doesn't exist, the policy server A transmit message *Q2* with signature for negotiating new policy information.

♦ Message *R2/R3*
The policy server B, receiving message *Q2*, either searches the new policy information and return it to policy server A through message *R2*, or return policy client B's policy information through message *R3*.

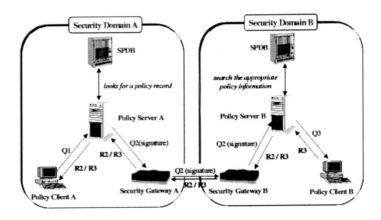

Fig. 5. SPS Flow

2.3.2 IPsec-SPS Interactions

The i-Gine references policy information for every inbound/outbound packets. The difference between inbound packet handling and outbound packet handling is following. In case of outbound packets, the i-Gine references the policy information before applying SA, but in case of inbound packets, references the policy information for deciding the correctness after applying SA.

The SPS client installed in each network node resides through the application area and the kernel area. The reason why an extra SPDB exists in the kernel area is to promote the performance by minimizing the interactions between kernel and application. All the policies of the application SPDB are not loaded in the kernel SPDB. There are two critical reasons. First, As some of SPDB are needed to I-Gine and others are only for IKE materials, we can save the kernel memory. Second, by reducing searching routine, we can improve the network performance. In C-ISCAP, policies and encryption/decryption are applied to each inbound /outbound packet, it is very important to optimize this processing. This kernel security policy database is a selDB.

During the application programs, such as ftp and telnet, transmit packets to other network nodes, ip socket interface captures the message. Using source ip address, destination ip address, source port number, destination port number and protocol from message as a default selector value, it inquires of application SPDB about the policy information. If there is no corresponding policy information, the source host negotiates a new security policy with the peer security policy server. Some of the negotiated policies are stored in the selDB, and returns a confirm message to I-Gine. In this ways, this system can maintain the consistency between the application SPDB and the selDB.

We use two character-devices for communication between kernel and application. One is for one-way download from application, the other for communicating. As kernel has a higher priority than application, it is important to support robust synchronization.

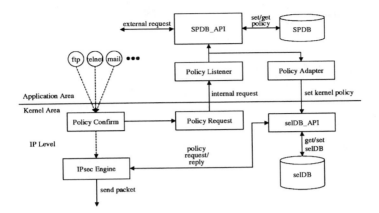

Fig. 6. IPsec-SPS Interactions

To send a request message to sps in application, kernel wakes up a character-device module by signal and sleeps on the wait_queue. But in some cases, it needs to sleep on wait_stack. In the next section, I'll explain.

The i-Gine receiving confirm message continues normal flow and establishes secure communication channel with it.

3 Security Policy Deployment

3.1 Adaptive Secure Sessions for Policy Negotiation

When a host is to communicate with other host/network and there are no policies found, then it needs to generate a new security policy. The initiator policy server is passive and the responder policy server is responsible for choosing the policy information to be used among them.

The PS_A receiving query message from PC_A transmits the request message(message 1) to the PS_B through the SG_A. The message 1 is the security policy negotiation request message for corresponding session.

The intermediate routers receiving message 2 inspect their own SPDB. If a policy exists, then the intermediate routers return the found policy to PS_A using message 2-1. Else, they forward message 3. The SG_B inspects the policy information and replies to PS_A with the negotiated security policy message(message 5).

A security policy system provides the security systems, such as IPsec/IKE, with the security strategies, and must not have their own security holes

From a viewpoint of SPP, messages for a policy negotiation are transmitted as a plain text, though some documents only define that these messages should be protected by other security mechanisms.

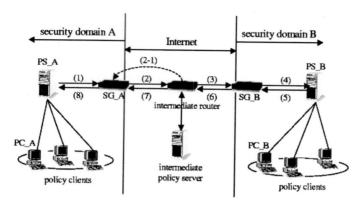

Fig. 7. Security Policy Negotiation

But, it is a dilemma that SPS for providing IPsec and SSL with the basic rules should be protected by IPsec or SSL for security policy negotiation. And it is impossible to establish SAs among policy servers all over the world. Therefore, it comes to a conclusion that we must establish secure negotiation channel by security mechanisms other than IPsec or SSL.

During the process of new security policy negotiation between end-points, the policy information of intermediate routers is applied, and generated policy is applied to them. Also, we can use multiple tunnelings through SA bundle. By only one policy negotiation flow, we can distribute keys protecting all sessions, and remove own security holes.

For example, a PS_A want to negotiate a new policy with a PS_D and there are intermediate routers, a PS_B and a PS_C between the two.

The following tables show the contents of each policy server's SPDB.

Table 1. Contents of PS_B's SPDB

source address	destination address	source end point	destination end point	mode
A	D	A	B	Tunnel
A	D	B	C	Tunnel
A	D	B	D	Tunnel

Table 2. Contents of PS_C's SPDB

source address	destination address	source end point	destination end point	mode
A	D	B	C	Tunnel

Table 3. Contents of PS_D's SPDB

source address	destination address	source end point	destination end point	mode
A	D	A	D	Transport
A	D	B	D	Tunnel

The following figure shows the Diffie-Hellman messages and control messages for secure sessions. When the initiator policy server PS_A sends control message with own signature to the responder policy server PS_D, the message is modified in each hop. Each intermediate router attaches its own signature on the message, and forwards it to the destination. This enables the chain of trust. The PS_D receiving message from the PS_A, inspects it's own SPDB for corresponding policy, and extracts the needed information for secure sessions from it. In each secure session, The PS_D establishes secure channel with Diffie-Hellman key exchange protocol. This is performed recursively in all nodes from the PS_D to the PS_A. After this, we can guarantee that all channels for policy negotiation from the PS_A to the PS_D are secure.Consider n is the number of nodes, and T the number of secure sessions needed. Then the needed number of rounds, R is $2n + 3T -2$. So, we can say that the number of rounds for security policy negotiation between end-points is $2n + 3T -2$.

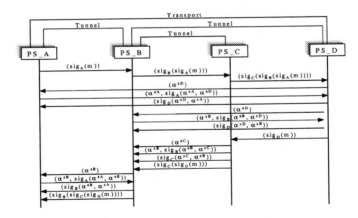

Fig. 8. Adaptive Secure Sessions for Negotiation

The format of message m is following.

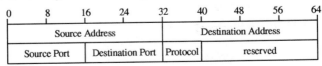

Fig. 9. The format of message m

3.2 Enhanced ASPNS

In the previous section, we introduce the mechanism for establishment of secure negotiation sessions. However, we can reduce the round number by transmitting signature message with random number used as Diffie-Hellman key exchange material. While previous mechanism operates sequentially, this improved mechanism operates in parallel. Each node uses Diffie-Hellman key exchange protocol for secure sessions after forwarding control message m, if any.

For this, it is needed that random number be attached to the control message m.

Every intermediate nodes divide message into forwarding resources and Diffie-Hellman resources. Consider n is the number of nodes, T the number of secure sessions needed and d the number of dummy SPD sessions. Then the needed number of rounds, R is $n + 2T + d - 1$.

The term 'dummy' means that a source host in each secure session doesn't have any policy information for the session. The source host can't send its random number for the session, so 3 rounds are needed for Diffie-Hellman key exchange. Otherwise only 2 rounds are needed.

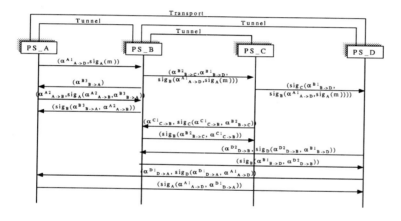

Fig. 10. Enhanced Adaptive Secure Sessions for Negotiation

In the previous example, the dummy SPD session is a session between the PS_A and the PS_B.

3.3 Kernel Scheduling for Security Policy

In the previous section, we introduce a mechanism for applying security policy information to i-Gine in the kernel. This character-device adapts client-server model and kernel should wait until the corresponding application replies. In the kernel, common method for process scheduling is FIFO using wait-queue. whereas, in the IPsec system, we must use wait_stack. For example, when a application program want to send TCP packets without SA, then ip process should wait until the sps replies with the corresponding SA.

For this SA, IKE should negotiate the new SA with peer IKE server using UDP with port number 500. But if this IKE packet has no SA, ip process also should wait sps's reply. In this flow, wait_stack is needed.

However, i-Gine wakes up the sleeping character-device read module and begins to sleep on the write_wait_stack waiting for character-device write module to wake up.

In the IPsec system, we use the LILO for kernel scheduling.

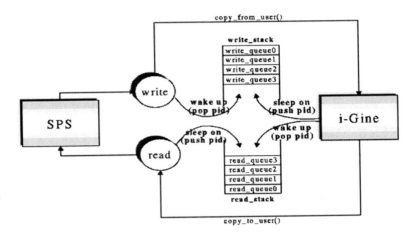

Fig. 11. Kernel Module Scheduling

4   Conclusion

It is not reasonable to establish a security policy negotiation channel using other security mechanism such as IPsec, moreover can be a security hole.

We introduce that it is more desirable to dynamically classify a type of session and set up secure communication channel through Diffie-Hellman key exchange than to know all the needed secure sessions before security policy negotiation. Besides, with Diffie-Hellman random number attached to the signature message, each intermediate node can establish multiple secure sessions in parallel. Consequently, not only the needed number of rounds is definitely reduced, but also we can remove system security holes and provide with a high-performance security policy system.

When the kernel is a client, it is difficult to setup a robust synchronization between kernel and application. To solve this problem, we propose a kernel scheduling mechanism using two wait_stacks. By this, it is possible to control the IPsec system.

The security policy system in the distributed environment can cause many overheads to systems and networks, so the researches on safety and efficiency are needed more.

References

1. Zao J, Sanchez L, Condell M, Lynn C, Fredette M, Helinek P, Krishnan P, Jackson A, Mankins D,   Shepard M, Kent S, „Domain Basd Internet Security Policy Management", Proceedings DARPA Information Survivability Conference and Exposition, DISCEX'00, IEEE Comput, Soc, Part vol.1, 1999, pp.41–53 vol.1, Las Alamitos, CA, USA.
2. Young-Ju Lee, Nam-Kyung Um, Ji-In Lee, Sang-Ho Lee, Geon-Woo Kim, „Design of A Working Mechanism for Hierachical Security Policy", Proceedings of The 12'th KISS Fall Conference, vol.1, 2000, pp.36–40, Taejon, KOREA.
3. Baltatu M, Lioy A, Mazzicchi D, „Security policy system: status and perspective" Proceedings IEEE International Conference on Networks 2000 (ICON 2000), Networking Trends and Challenges in the New Millennium. IEEE Comput, Soc, 2000, pp.278–284, Los Alamitos, CA, USA.
4. L.A. Sanchez, M.N. Condell, „Security Policy System", Internet Draft, Network Working Group, IETF, November 18, 1998.
5. L.A. Sanchez, M.N. Condell, „Security Policy Protocol", Internet Draft, Network Working Group, IETF, July 17, 2000.
6. S. Kent, R. Atkinson, „Security Architecture for the Internet Protocol", RFC2401, Network Working Group, IETF, November, 1998.
7. S. Kent, R. Atkinson, „IP Authentication Header", RFC2402, Network Working Group, IETF, November, 1998.
8. C. Madson, R. Glenn, „The Uses of HMAC-SHA-1 within ESP and AH", RFC2404, Network Working Group, IETF, November, 1998.
9. C. Madson, N. Dorawamy, „The ESP DES-CBC Cipher Algorithm with Explicit IV", RFC2405, Network Working Group, IETF, November, 1998.
10. S. Kent, R. Atkinson, „IP Encapsulating Payload", RFC2406, Network Working Group, IETF, November, 1998.
11. D. Harkins, D. Carrel, „Internet Key Exchange", RFC2409, Network Working Group, IETF, November, 1998.

V. Multimedia Application

Mobile Agent-Based Adaptive Multimedia Communication

Koji Hashimoto[1], Yoshitaka Shibata[1], and Norio Shiratori[2]

[1] Faculty of Software and Information Science, Iwate Prefectural University,
152-52 Sugo, Takizawa, Iwate, Japan 020-0193
{hashi, shibata}@iwate-pu.ac.jp
[2] Research Institute of Electrical Communication, Tohoku University,
2-1-1 Katahira Aoba Sendai, Japan 980-8577
norio@shiratori.riec.tohoku.ac.jp

Abstract. On interconnected computer networks, we can communicate with each other using realtime media such as audio and video by distributed multimedia system that can integrate various realtime and non-realtime media data. The multimedia system is required to have end-to-end quality of service (QoS) guarantee functions. When users communicate with each other on interconnected different bandwidth networks, if the system can use translator or mixer functions that are defined by RTP dynamically, it will be able to guarantee more flexible QoS considering to wide band through narrow band networks. We have proposed Flexible Multimedia System (FMS) that is able to guarantee end-to-end QoS according to priority of parameters and consensus policy. In addition, the system is able to organize various multimedia services dynamically. The FMS is a mobile agent based system, therefore the system will be able to organize translator or mixer dynamically. In this paper, we survey the translator and mixer in RTP, discuss about QoS guarantee functions using mobile agent and re-design our prototyped system to use mobile transcoding functions.

1   Introduction

The advent of high performance computers that can process and integrate audio, video, graphics and text on high speed networks make it possible for us to communicate with each other by realtime and stored media, even if using reasonable price personal computers.

At ordinary times, we actually can use multimedia applications such as IP telephone, IP radio, Video-on-Demand system, multimedia teleconference system and so on. We can communicate with remote places easier than before by using these multimedia applications.

These multimedia applications have various convenient functions for multimedia communication, but many applications don't have QoS(Quality of Service) functions that consist of QoS mapping, QoS adaptation and resource management. The QoS functions are required to guarantee of quality of service requested

I. Chong (Ed.): ICOIN 2002, LNCS 2344, pp. 467–478, 2002.
© Springer-Verlag Berlin Heidelberg 2002

from users. On the other hand, there are a number of works concerned with QoS guarantee for media data flow on mainly ATM networks[1]. For example, there are control methods of end-to-end delay and jitter, bandwidth allocation schemes and so on. In addition, there are some applications that implement RSVP[2] or RTP[3] protocol to use realtime media data.

However it is difficult for the system to realize end-to-end QoS guarantee or adaptation functions on interconnected network environment that is constructed by different bandwidth networks. For example, suppose that remote teleconference is being held on 100Mbps networks, and bidirectional DV(Digital Video)[4] streams are used in the teleconference. Now, some participants on a business trip have mobile computer, and the computer is connected to narrow bandwidth network about 1.5Mbps. The participants want to join the teleconference, but the mobile computer and narrow bandwidth network can't process DV streams in realtime, because about 35Mbps network bandwidth is required to process a DV stream. If translator and mixer functions defined by RTP are available on the environment, it is able to translate DV stream into another stream such as MPEG or M-JPEG. By using translator function, the participants can join to the teleconference and communicate with each other even if they are on the business trip. Well, if the communication environment is fixed, it is able to prepare the translator node in advance. However, the preparation is very difficult, because the environment is not always fixed. Therefore we consider that if required transcoding functions such as translator and mixer are able to move into suitable intermediate node, the communication system will have more flexible QoS guarantee functions.

We have proposed Flexible Multimedia System (FMS)[5] that is able to organize various multimedia services dynamically. The system has end-to-end QoS functions[6] based on priority of QoS parameters and consensus policy. Various functions in the system are based on mobile agent. Therefore if the above mentioned transcoding functions are implemented as a mobile agent in this system, the agent will be able to move itself into suitable intermediate node.

In this paper, we introduce transcoding functions such as translator and mixer into our FMS, discuss how to use transcoding functions according with requirements from users. Moreover we describe a implementation of prototype system.

2   Transcoding Functions

RTP[3] defines translator and mixer as media transcoding functions in intermediate node. The translator translates format of media stream according to available bandwidth. It is able to make realize video communication within available network bandwidth that the translator node translates each video stream into another format, regulates actual frame rate and color depth.

On the other hand, mixer integrates some input media streams into one output media stream. For example, if voice data streams of some senders are simple PCM data streams, mixer adds each byte data of all streams to integrate some inputs into one output. As a result, required bandwidth decreases without degradation of voice quality.

By using these transcoding functions, the communication system offers high quality media stream to users if the users can use wide bandwidth networks, and even if the users can only use narrow bandwidth networks, the users can communicate with each other.

2.1   Transcoding Node

When available bandwidth is $N[bps]$ and another network bandwidth is $M[bps](N > M)$, to communicate using multimedia, transcoding functions should be in an intermediate router or on the $N[bps]$ network. Figure 1 show the three type of intermediate nodes that can perform transcoding functions.

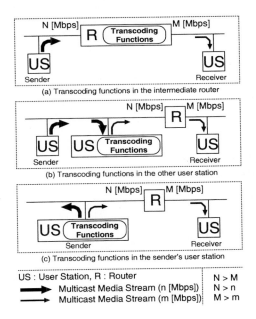

Fig. 1. Transcoding Node

In the figure 1(a), when the transcoding functions are in the router, stream data will be transcoded effectively without some extra traffic, but the router may get heavy load. On the other hand, the transcoding functions can also perform in a node that doesn't associate with media data stream. In the figure 1(b), the left side network will get extra traffic. When there are enough computer power and available bandwidth in the left side network, this transcoding is effective. Moreover transcoding functions perform in the sender node too. The figure 1(c) shows this situation. In this case, sender node has one input source and two outputs media stream for each receiver. When there are enough computer power in the sender node and available bandwidth in the left side network, this transcoding is effective.

3   To Introduce into FMS

In order to use transcoding functions dynamically, suitable node must be selected according to target node's computing power and available network bandwidth. Our proposing agent oriented FMS is based on concepts of flexible[7]. By applying agent technology[8,9], the system defines a framework to realize multimedia communication in various user environments.

The FMS consists of User Stations and Repositories. It is able to organize[10] some required mobile agent based functions into User Station from Repositories according to service and QoS requirements of users.

3.1   Agent Configuration

After agent organization phase, organized agents start on each media processing in user station. Figure 2 shows one of agent configurations in user station to communicate using live audio and video data.

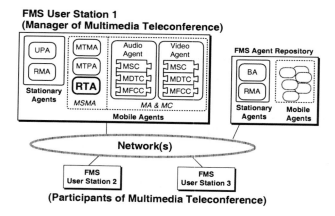

Fig. 2. Agent Configuration for Multimedia Teleconference Service

User Partner Agent (UPA) receives service or QoS requirements from user, and Resource Management Agent (RMA) performs reservation, allocation and release required resources. These functions depend on the fixed environment of each user station. Therefore, these agents are realized by stationary agent.

On the other hand, Multimedia Service Management Agent (MSMA) realizes management functions that depend on requested multimedia service. In case of offering multimedia teleconference service to users, Multimedia Teleconferencing Management Agent (MTMA) is organized in the user station of conference manager, and Multimedia Teleconferencing Participant Agent (MTPA) is organized in each user station of participant. In addition, required Media Agents (MAs) and Media Components (MCs) are organized in all user stations. MAs perform monitoring and control of media processing, and MCs process required media data. Media Synchronization Component (MSC) performs inter-media and intra-media synchronization, Media Data Transform Component (MDTC) performs codec (M-JPEG, MPEG1/2/4, H.261, etc.), format conversion, mixing of audio data and so on. Media Flow Control Component (MFCC)[5] regulates of packet interval, performs variable bit rate transmission to control of packet loss rate. In figure 2, each user station organizes MAs and MCs to process realtime audio and video. these agents are mobile agent that is transferred from the repository to user station by Broker Agent (BA).

In addition, Round Trip Agent (RTA) goes around each user station, collects information about QoS requirements and media processing, and informs each user station of adapted QoS[6].

3.2 Transcoding Agent

Figure 3 shows an example of transcoding agent configuration when the type of transcoding node is figure 1(b).

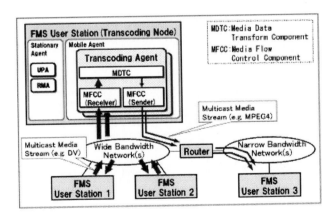

Fig. 3. Transcoding Agent

Transcoding node is realized by organized transcoding agent into user station in FMS. Transcoding agent is a type of MA. The agent has two MFCCs and one MDTC. Therefore the agent can transcode input media stream into output media stream. Of course, there are both UPA and RMA in the transcoding node, and the each agent performs communication among user stations and manages resource for transcoding process. If hardware codec is available, organized transcoding agent may use the functions.

Moreover in order to control media transmission rate and packet loss rate, transcoding agent performs variable bit rate transmission and packet interval regulation in MFCC[5]. Transcoding agent is receiver and sender for a media stream. Two MFCCs in a transcoding agent perform media flow control independently. Therefore end-to-end QoS is guaranteed even if there are something extra loads.

4   Multicast Session

When some users want to communicate with each other using live media such as audio and video, each sender and each receiver may have each QoS. Therefore the communication system must consider both sender and receiver side QoS requirements. Especially, if multicast communication is used, some methods to adapt QoS requirements from different receivers are required.

Here, we define multicast session as a media transmission using multicast that consists of a sender and some receivers. In this section, transcoding functions are applied to multicast session.

Figure 4 shows about a set of Multicast Session QoS(MSQ). One multicast session has one MSQ to reflect QoS requirements of users.

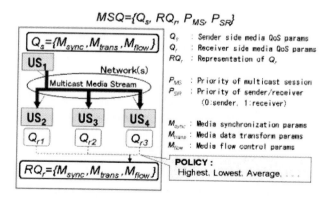

Fig. 4. A Set of Multicast Session QoS Parameters

A MSQ includes both sender side QoS(Q_s) and representation receiver side QoS(RQ_r). RQ_r is calculated by some receiver side QoSs(Q_r) and consensus policy($POLICY$) decided in advance.

Q_s, RQ_r and Q_r include M_{sync}, M_{trans}, M_{flow}. These are each set of QoS parameters of media synchronization, media data transform and media flow control.

$$M_{sync} = \{S_F, R_F, T_{START}\}$$
$$M_{trans} = \{CODEC, FORMAT\}$$
$$M_{flow} = \{S_{MDU}, S_{PEAK}, R_{MDU}, R_{LOSS}\}$$

In M_{sync}, S_F[byte] shows frame size, R_F[1/sec] shows frame rate and T_{START} is start time of media processing. M_{trans} has parameters associated with codec and format type. In M_{flow}, S_{MDU}[byte] is average size of media data unit (MDU) to send and receive media data in FMS. S_{PEAK}[byte] is the peak size. R_{MDU}[1/sec] shows MDU rate, R_{LOSS}[%] shows MDU loss rate. Quality of media data decided by these media QoS parameters.

FMS can adapt different receiver side QoS requirements by using consensus policy. However if the system adapts wide difference QoS requirements from receivers, this method isn't suitable. For an example, there are two type of users. One user can use 100Mbps network and requires sending DV stream, other users can use same network and requires receiving DV stream, the other users can only use 1.5Mbps network and require receiving MPEG4 stream. It isn't realistic that the system adapts these two type of receiving QoS requirements to representation QoS. Here, we introduce transoding node into intermediate node in multicast session.

Figure 5 shows an example of using transcoding function in intermediate node of wide bandwidth network and narrow bandwidth network. The system can offer DV multicast stream to users on the wide bandwidth network, at the same time, the system can offer MPEG4 multicast stream to users on the narrow bandwidth netwok.

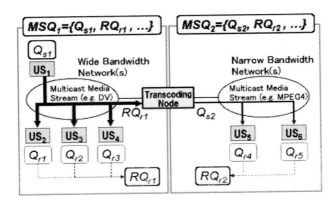

Fig. 5. Transcoding Node and MSQs

5   Dynamic Positioning

FMS is one of mobile agent system considering multimedia communications. Therefore implemented transcoding agents can position into suitable intermediate node dynamically. In order to position transcoding agent to the intermediate node (figure 1), the system must find out FMS User Stations in multicast session paths.

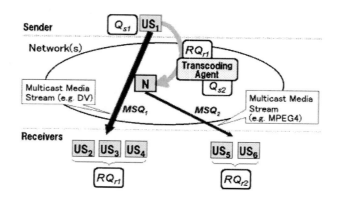

Fig. 6. Dynamic Positioning of Transcoding Agent

Following is a flow of positioning of transcoding agent.

1. Some representation receiver side QoS parameter sets are decided by requested QoS from user, available network bandwidth, media data format and so on. (RQ_{r1} and RQ_{r2} in figure 6)
2. Gets paths to receivers without media transcoding. (from US_1 to US_2, US_3, US_4 in figure 6)
3. Gets paths to the other receivers that require media transcoding. (from US_1 to US_5, US_6 in figure 6)
4. Finds out branch node (N).
5. If node (N) is FMS User Station and the results of admission test are all right to transcode media data, the transcoding agent moves to the node (N).
6. If node (N) isn't FMS User Station or can't transcode required media stream, the system finds out FMS User Station in the path of US_1 and node (N).

By using this procedure, the system finds out suitable transcoding node. FMS realizes more flexible QoS guarantee schemes by dynamic positioning of transcoding agent.

6   System Design

When the system uses transcoding functions in intermediate node on multimedia communication paths, it is considered that the node doesn't need to play-
out stream data such as audio and video. In addition, the system should be
lightweight in ordinary situation, because transcoding functions are only used
when the functions are required. So far, our prototyped FMS considers that each
User Station is used by only end users. Therefore the basic modules of FMS has
many GUI parts, so the prototyped system is heavy system in ordinary situation.

Here, in order to construct lightweight system and to introduce transcoding
functions, FMS is re-designed. Figure 7 shows the Java class structure of re-
designed system using UML.

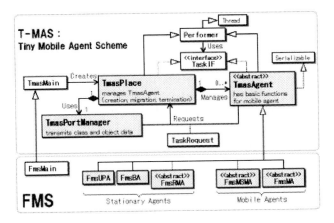

Fig. 7. Class Structure of Flexible Multimedia System

First of all, we picked up required minimum functions that are agent creation,
migration and termination from prototyped system, next asynchronous message
delivery scheme was re-designed. This basic scheme is named T-MAS (Tiny Mobile Agent Scheme). Core classes of T-MAS are TmasPlace, TmasPortManager
and TmasAgent that are shown in figure 7.

TmasPlace is working environment of mobile agent and performs agent creation, migration and termination. TmasPortManager realizes transmission of
required class and object data demand on agent migration. TmasAgent is a super class of mobile agent, has basic functions to realize migration. In FMS, all
agents extend TmasAgent. TmasAgent can make byte data from object data of
itself, because TmasAgent implements Serializable class.

Both TmasPlace and TmasAgent extend Performer class. Performer class
extends Thread class in Java (Ver.1.3), has a queue to receive asynchronous
requests using TaskRequest class objects. The request messages in the queue are
processed by one of classes that implement TaskIF class.

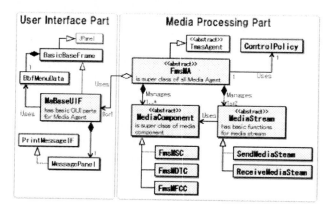

Fig. 8. Class Structure of Media Agent

Media Agent manages media processing components in FMS. In figure 8, FmsMA class is super class of all Media Agents. There are two type of classes associated with FmsMA. The one is associated with media processing and another is associated with user interface. In order to realize media processing, FmsMA has some MediaComponent classes and MediaStream classes, and manages these class objects.

On the other hand MediaStream class is extended to realize media transmission. Each SendMediaStream and ReceiveMediaStream includes some Media Components that are FmsMSC, FmsMDTC and FmsMFCC.

When Media Agent is created or migrated, the system gives media control policy to the Media Agent. ControlPolicy class has the control policy that is a set of Media QoS Parameter such as MSQ and the priority value of the parameters.

In addition FmsMA can use MaBaseUIF class to realize graphical user interface. The MaBaseUIF class has GUI parts such as base frame, menu bar, message output panel. FmsMa and MaBaseUIF is separated completely. Therefore when the agent performs transcoding functions without GUI parts, processing loads associated with GUI do not arise.

Figure 9(a) shows that ReceiveMediaStream is constructed from three MediaComponents (FmsMSC, FmsMDTC, FmsMFCC) . Each MediaComponent has InputMediaStream and OutputMediaStream, and Media Agent connects MediaStreams of MediaComponents to realize media processing.

Figure 9(b) shows the configuration of ReceiveMediaStream and SendMediaStream in transcoding agent. Media Agent connects OutputMediaStream of FmsMDTC to InputMediaStream of FmsMFCC to send transcoded media data to destination.

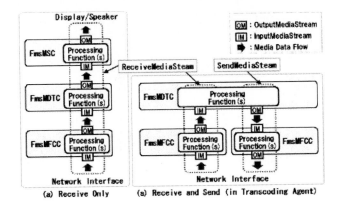

Fig. 9. MediaComponent and MediaStream

7 Conclusion

By realizing transcoding functions by mobile agent, distributed multimedia communication system is able to use the functions such as translator and mixer dynamically. The system will be able to guarantee more flexible QoS when users communicate with each other on interconnected different bandwidth networks. In this paper, we introduced transcoding functions into our proposed FMS, discussed how to use transcoding functions according with requirements from users. Moreover we described re-design of prototype system.

In order to perform transcoding functions in suitable intermediate node, the system must be able to manage cpu power of transcoding node, available devices for multimedia streams, network configurations and traffic load and so on. As future work, we will design protocols to find out suitable intermediate node, transcoding agent, lightweight migration protocols in detail. Then, we will construct prototyped system to hold multimedia teleconference on FMS, and will evaluate the mobile transcoding functions on both high speed network such as JGN (Japan Gigabit Network) and relatively narrow bandwidth network such as wireless LAN environment.

References

1. S.Okubo, S.Dunstan, G.Morrison, M.Nilsson, H.Radha, Dale L. Skran and Gary Thom.: *ITU-T Standardization of Audiovisual Communication System in ATM and LAN Environments*, IEEE Journal on Selected Areas in Communications, Vol. 15, No. 6, pp. 965–982, Aug. 1997.
2. Tsipora P. Barzilai, Dilip D. Kandlur, Ashish Mehra and Debanjan Saha.: *Design and Implementation of an RSVP-Based Quality of Service Architecture for an Integrated Services Internet*, IEEE Journal on Selected Areas in Communications, Vol. 16, No. 3, pp. 397–413, Apr. 1998.

3. H. Schulzrinne, S. Casner, R. Frederick and V. Jacobson.: *RTP: A Transport Protocol for Real-Time Applications*, RFC 1889, January 1996.
4. K.Sugiura, T.Sakurada, A.Ogawa, *Transferring High Quality Media Using Internet*, Information Processing Society of Japan (IPSJ) Magazine, Vol. 41, No. 12, pp. 1321–1326, Dec. 2000.
5. K.Hashimoto, Y.Shibata and N.Shiratori, *Flexible Multimedia System Architecture with Adaptive QoS Guarantee Functions*, Proc. of the 7th International Conference on Parallel and Distributed Systems (ICPADS'00), pp. 119–126, Jul., 2000.
6. K.Hashimoto, Y.Shibata and N.Shiratori, *MOBILE AGENT BASED MULTIMEDIA COMMUNICATION SYSTEM*, Proc. of the 2001 IEEE International Conference on Multimedia and Expo (ICME2001) CD-ROM, pp. 697–700, Aug., 2001.
7. Shiratori N., Sugawara K., Kinoshita T. and Chakraborty G.: *Flexible Network:Basic Concepts and Architecture*, IEICE Trans. Communication, Vol. E77-B, No.11, pp. 1287–1294, 1994.
8. Pauli Misikangas and Kimmo Raatikainen.: *Agent Migration between Incompatible Agent Platforms*, Proceedings of the 20th IEEE International Conference on Distributed Computing Systems (ICDCS2000), pp. 4–10, Apr. 2000.
9. Shinji Tanaka, Hirofumi Yamaki, and Toru Ishida *Mobile-Agents for Distributed Market Computing*, Proceedings of the 1999 International Conference on Parallel Processing, pp. 472–479, Sep. 1999.
10. Koji Hashimoto, Yoshitaka Shibata and Norio Shiratori.: *The System Organization and QoS Functions for Flexible Multimedia System*, Proc. of the 6th International Conference on Distributed Multimedia Systems (DMS'99), pp. 209–216, Jul. 1999.

An Alternative Multicast Transmission Scheme for VCR Operations in a Large VOD System⋆

Seung-Won Lee[1], Ki-Wan Kim[2], Jung-Min Choi[1], and Ki-Dong Chung[1]

[1] Dept. of Computer Science,
Busan National University,
Kumjeong-Ku, Pusan 609-735 Korea
{bluecity,choijm,kdchung}@melon.cs.pusan.ac.kr
http://apple.cs.pusan.ac.kr
[2] Dept. of Computer Science,
KOREA THIRD MILITARY ACADEMY,
Changha-Dong, Yongchon 770-849 Korea
wan1434@kornet.net

Abstract. Multicast delivery is one of solutions to reduce the cost in large VOD systems. However, it is difficult to implement the interactive operations for an individual user in multicast transmission system. In this paper, we propose the Alternative Multicast Transmission Scheme (AMTS) to serve a large number of users and reduce the initial delay while supporting VCR operations. For hot video users, this scheme uses the Frame-Based Multicast Transmission (FBMT) to support both normal playback and VCR playback via the identical channels. The proposed FBMT is to deliver frames via an appropriate channel based on frame type (I, P, B) of MPEG video. The AMTS has better performance than other multicast systems in terms of both initial blocking probabilities and VCR blocking probability.

1 Introduction

The past few years have shown the dramatic growth of multimedia applications such as videoconference and video-on-demand system, which involve video streaming over the Internet. Server and network resources, in particular server I/O bandwidth and network bandwidth, have proved to be major limiting factors in the widespread usage of video streaming over the Internet. More recently, researches have concentrated on the multicast transmission techniques that efficiently utilize both server and network resources, to support a large number of clients in VOD system.

Multicast transmission scheme is different from one-to-one connection scheme, which each customer is served by exclusive unicast channel. In multicast delivery system, one stream is used to serve a number of concurrent customers. So multicast transmission can greatly reduce server I/O and network bandwidth.

⋆ University Research Program supported by Ministry of Information & Communication in South Korea

I. Chong (Ed.): ICOIN 2002, LNCS 2344, pp. 479–490, 2002.
© Springer-Verlag Berlin Heidelberg 2002

But it is expensive for multicast scheme to support VCR functions such as pause, fast-forward, since an exclusive channel is not allocated to each user.

In previous studies, various transmission schemes providing VCR functionality in a multicast VOD system were proposed. Almeroth [2] assumes that the set-top box has some video frame buffering capability and explores the use of multicast communication to address the issue of providing scalable and interactive VOD service. Other schemes [3],[4],[5] have been proposed to deal with the problem of providing interactive functions in multicasting of follows near VOD systems. However, such schemes have addressed interactive functionality in the environment of an on-demand batching video delivery system and depended on both client buffering and "interactive" channels to provide VCR functions.

Liao and Li [3] tried that customers can join back to multicast streams after VCR operations, Ho Kyun [5] proposed a scheme that can support an interactivity for all users requesting the same video stream with batching.

This paper suggests a multicast transmission scheme named Alternative Multicast Transmission Scheme (AMTS) that can support an interactive VCR services in a Huge VOD system. We used two transmission methods based on the popularity of video. Our transmission scheme transmits hot videos by Frame based Multicast Transmission (FBMT)[10]. Cold videos exploit the general catching method and use unicast channel for VCR services. The proposed multicast transmission methods use basically a catching scheme to reduce initial latency and provide nearly instantaneous service for a large number of clients.

The paper is organized as follows. Section 2 will describe the Alternative Multicast Transmission Scheme. In section 3, we explain the Frame Based Multicast Transmission (FBMT). And, section 4 describes scheme for providing VCR functions with the proposed FBMT based on play mode. In section 5, a formulation process is described to decide what transmission method is used. A simulation result of the proposed system is given in section 6 and some conclusions are drawn in section 7.

2   Alternative Multicast Transmission Scheme (AMTS)

Several techniques to support VCR services in multicast transmission had been proposed. But, most researches had not reduced the cost for supporting VCR services. Also, No consideration about access pattern of customers had leaded to the inefficiency of network utilization. Due to these points, it is difficult to support VCR services in multicast VOD.

On the whole, the access pattern of customers in VOD forms the Zipf's distribution. Namely, the access pattern of customers tends to concentrate in popular videos. On the other hand, there is no way to measure accurately the occurrence frequency because VCR service requests occur randomly. However, if VCR requests occur regularly, hot videos will have more VCR requests than cold videos. Accordingly, in this assumption VCR support has need of alternative method exploiting the access frequency of customer.

By and large, the support of VCR service in multicast leads to much waste of network resource. In a multicast VOD, the separate user requests of VCR service require more network channels. For this reason, it is expensive to support VCR services in multicast VOD system. Therefore, minimum network cost is a big issue for supporting VCR service.

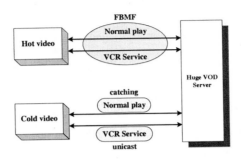

Fig. 1. The overview of AMTS

In this paper, we propose Alternative Multicast Transmission Scheme (AMTS) which incorporates both the user access pattern and the efficient utilization of network resource. AMTS applies the access pattern of customers to transmission policy; it uses two methods alternatively. For hot videos, we suggest Frame-Based Multicast Transmission (FBMT). And, for cold videos, we use a catching method for normal playback and additional unicast channel for VCR service. Fig. 1 shows the overview of AMTS. Because the method for cold videos is a general method for supporting VCR service, we will not explain in detail.

In this paper, what we emphasize is how FBMT reduces the waste of network resource in VCR service. Also, we will show that AMTS can reduce unnecessary channels when exploiting FBMT in cold videos.

3   Frame-Based Multicast Transmission (FBMT)

The goal of Frame-Based Multicast Transmission (FBMT) is to serve a large number of users and to reduce the initial delay and the network cost while supporting VCR operations. In this paper, we assume that the video is stored and transmitted in MPEG format. FBMT consists of two-level transmission layer. The first level layer uses a general delivery technique that transmits a whole video with catching. And the second level layer uses a transmission technique of a segment and policy to deliver frames via an appropriate channel based on frame type. Fig. 2 shows two-level transmission layer. As shown in figure, a video object is partitioned into three segments, A, B and C of equal length d, each segment is separately delivered via different channels (I, P, B channel) based on frame type (I, P, B frame).

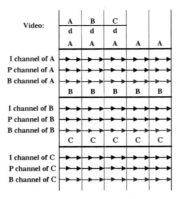

Fig. 2. An overview of two-level transmission scheme

3.1   First Level Transmission Layer

In this level layer, we utilize the catching delivery technique that a video object is served with multicast transmission[1]. Compared with other video transmission schemes, batching and patching, this catching must have superior performance over those in both channel usage efficiency and service latency.

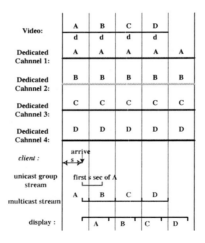

Fig. 3. First level transmission layer scheme

Fig. 3 illustrates that a video object is partitioned into four segments, A, B, C and D of equal length d, each segment is broadcasted periodically. A client who arrives s second after the beginning of the current broadcast cycle of segment 'A' can immediately watch a video instead of waiting for next broadcast cycle.

The number of exclusive channels should affect network resource usage efficiency as well as buffer. Therefore, in order to achieve superior performance, we must derive the optimal allocation of exclusive channels. The optimal number is as follows (1).

$$Length of video i : L_i$$

$$Request rate of video i : \lambda_i \qquad (1)$$

$$Optimal number of channel = O(\log(\lambda_i L_i))$$

Since the arrival rate means the popularity of video, the expected number of channels become more as the request rate gets larger.

3.2 Second Level Transmission Layer

The second level layer describes an inner transmission scheme of a first level layer segment. A segment is delivered by frames via independent channels on a basis of frame type. I frames use two channels that transmit frames at the normal playback rate. P frames and B frames are delivered at double playback speed via a channel. We allocate odd I frames to first I channel, even I frames to second I channel. But P frame and B frame is sequentially delivered at double playback rate via one channel.

This technique can avoid the waste of channel, since normal users and interactive users do not use each allocated channels but share the same channel.

Fig. 4 shows the expanded transmission scheme that delivers a segment (consists of k GOP). Several I frames that are included in a segment are delivered via first I channel (treat odd frame) and second I channel (treat even frame).

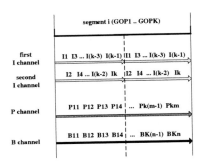

Fig. 4. Second level transmission scheme

4 VCR by Frame-Based Multicast Transmission

When we use the proposed FBMT, the playback mode can be classified into four modes: normal, fast-forward, pause and switching mode. For a segment of length d and size S, we will describe the transmission technique and the buffering schemes of four modes.

4.1   Normal Mode

As frames are received at double playback rate, the receiver in normal mode plays back the half of received data while temporarily storing the remaining data.

Hence, the buffer size of the client receiver gets increased at the beginning of the current cycle and points out to the half of total segment size $\frac{S}{2}$ at the end of cycle $\frac{d}{2}$. When the current segment is completely delivered, the client does not join to the on-going broadcast cycle of the next segment but waits for the next broadcast cycle.

As the buffered data begins to play back at time $\frac{d}{2}$, the total size is decreased. So the buffer is empty at time d that is played back overall data of current segment. The trace of buffer size is repeated as the segment proceeds to next cycle. Fig. 5 shows the buffering size based on the time. The maximum buffer size is the half of the segment size S.

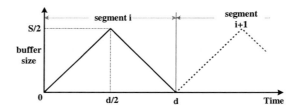

Fig. 5. Trace of buffer size on the normal mode

4.2   Fast-Forward Mode

In the case of fast-forward mode, users join to the first I stream and the second I stream, except for P stream and B stream. And they orders and plays only I frame. Since the transmission rate is the same as the fast-forward rate, data is seldom stored in buffer. Since it does not need additional channels as well as buffering, the FBMT scheme has superior performance over existing techniques in fast-forward mode.

4.3   Pause Mode

If user issues a pause operation at time t_0, it stops playing and stores the received data in buffer. This buffering is to resume the current playbacks immediately without delay.

For pause mode, a receiver needs smaller buffer than the segment size S. The worst case is that a client's pause requests arrive immediately after the beginning of the cycle of segment.

The necessary buffer size is S, since each segment is broadcasted periodically. If the data to be played back remains in buffer, the customer does not join to the current cycle of segment $i+1$ at time $\frac{d}{2}$ that the current segment i is completely received.

Since the customer receives the data by joining to the next cycle, they only consume the buffer from time $\frac{d}{2}$ to time d. Fig. 6 shows that the size of buffer

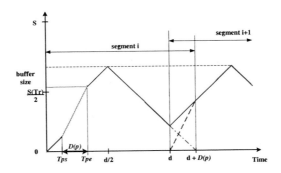

Fig. 6. Trace of buffer size on the pause mode

is changed as the time proceeds. The customer who is playing back segment i request a pause operation at the pause starting time T_{ps} and stops pausing at the pause ending time T_{pe}.

5   Formulation of Channel's Number

Based on the popularity of video. AMTS transmits hot videos by Frame based Multicast Transmission (FBMT). Cold videos use the general catching method and use unicast channel for VCR services - we call this scheme as adaptive catching. In other words, FBMT is much more efficient for hot video and adaptive catching for cold video. For more optimal number of channel, we formularize the result of the analysis.

We assume that there are n videos and m users in the server. The probability of requesting any video is assumed to follows Zipf's distribution, therefore probability which user will select video i is defined as:

$$P_i = \frac{F_i}{\sum_{j=1}^{n} F_j} \qquad where \qquad F_i = \frac{1}{i^{1-\theta}}$$

And a function is similar to real access pattern, when θ is equal of 0.271. Due to this probability, we can calculate the number of channels for a video of length L_i and request rate of λ_i.

In case of hot video, we transmit video through FBMT scheme. Because overall video is transmitted based on catching, optimal number of segment is equal to $O(\log(\lambda_i L_i))$[1]. A segment is delivered via different channel (two I channel and one P, B channel), and so the total number of channel for hot video is as $4\,O(\log(\lambda_i L_i))$. But, FBMT makes many unnecessary channels in case of cold video, so we transmit cold video using adaptive catching method. Given the VCR request rate of video i as δ_i, we formulate the number of channel for transmitting cold video.

Overall video is separated to several segment, one segment is transmitted through one channel. When a VCR request arrives, we create a new unicast channel for VCR service.

On transmitting one video, server needs the number of segments $O(\log(\lambda_i L_i))$ added to the number of unicast channels for VCR service. The number of unicast channel for VCR request is formulized with considering Zipf distribution and VCR request rate. By multiplying the total video number m and probability which select Video i as P_i, we have the number of selected video i and then multiple this result by VCR request rate δ_i.

$$O(\log(\lambda_i L_i)) + (\delta_i P_i m) \tag{2}$$

On transmitting all video, a little more channel must be required for enhancing performance. Hot video and cold video are transmitted by each different transmission scheme. So we decide efficiently point γ to classify hot and cold video.

On n videos, we consider videos satisfying following state as hot videos. And then, the maximum satisfied value is γ.

$$\sum_{i=1}^{\gamma}\{4\,O(\log(\lambda_i L_i))\} \le \sum_{i=1}^{\gamma}\{O(\log(\lambda_i L_i)) + (\delta_i P_i m)\} \tag{3}$$

The total number of channels that are required to transmit all video is sum of channels of hot video and channels of cold video. A value of γ calculated in (3). In result we calculate number of channel of γ hot videos and $(n - \gamma)$ cold videos.

$$\left[\sum_{i=1}^{\gamma}\{4\,O(\log(\lambda_i L_i))\}\right] + \left[\sum_{i=n-\gamma}^{n}\{O(\log(\lambda_i L_i)) + (P_i \delta_i m\}\right] \tag{4}$$

6   Simulation Model

This section presents the simulation results of the proposed Alternative Multicast Transmission Scheme (AMTS) for a multicast VOD system.

To develop a simulation model, we assume that there are 200 movies in the server. The characteristics of user requests are modeled as a Poisson arrival process. The probability of requesting any video is assumed to be q_i which follows

Zipf's distribution, i.e., if $q_1 \leq ... \leq q_M$, where M is the number of movies, $q_i = \dfrac{c}{i}$, where c is a normalizing constant such that $\displaystyle\sum_{i=1}^{M} \dfrac{c}{i}$.

We further assume a customer is in one of two states, the normal and the interaction states. He/she starts in the normal state that the video is being played at normal speed. He stays in this state for a period of time that follows an exponential distribution with parameter α. Then, he/she issues an interactive operation. p_1 and p_2 are the pause and fast forward probabilities such that $\displaystyle\sum_{i=1}^{2} p_i = 1$. He/she stays in the interactive state for another period of time that follows an exponential distribution with parameter β_{VCR}. He/she shuffles between two states several times until the end of the movies. In the simulation model, we assume that the number of hot videos is 100. Table 1 and 2 show the parameters of the system and the customer's behavior.

Table 1. Parameter of the simulation model

Parameter	Range of Values	Nominal Value
Simulation Time	6 hours	-
No of Movies	200	-
Bitrate of Movies	1.5Mbps	-
Movie's Length	120 minutes	-
Arrival Rate	0.01 - 0.3 arrivals/s	0.15 arrivals/s
Bandwidth	40 - 100 Mbps	70 Mbps

Table 2. Parameter of the customer behavior

Parameter	Range of Values	Nominal Value
Pause Probability	1/3	-
FF Probability	1/3	-
$1/\alpha$	10 - 50 minutes	30 minutes
$1/\beta_{Pause}$	5 minutes	-
$1/\beta_{FF}$	0.5 minutes	-

In order to compare the performance of different multicast system, six different schemes were simulated using the same set of parameters.

Fig. 7 and 8 show the initial blocking probability and the VCR blocking probability of different systems when the arrival rate is increased. We find that the initial blocking probability is increased as more customers arrive within six hours. When the arrival rate is low, AMTS shows higher initial blocking

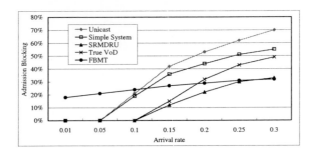

Fig. 7. Initial blocking probability vs. number of requests

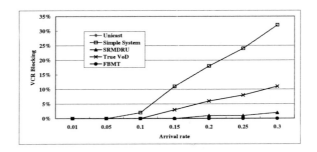

Fig. 8. VCR blocking probability vs. number of requests

probability than only SRMDRU. The AMTS shows better performance as the arrival rate is increased. And, for arrival rate is more than 0.15 the AMTS shows equal performance with SRMDRU. But, the VCR blocking probability is about 0% in AMTS when the arrival rate is high.

In addition, we see that the Initial blocking probability of the AMTS is decreased as the arrival rate is increased further. Because normal mode player and interactive mode player share multicast channel in hot videos, VCR blocking probability is about 0% to admitted users.

Fig. 9 and 10 show the blocking probability as a function of the bandwidth. As we predicted, both the initial blocking probability and the VCR blocking probability are decreased when the total bandwidth is increased. It is because more bandwidth is available to serve the new customers and interactive operations. The initial blocking probability of the AMTS is less than 7% when bandwidth is 90 Mbps in the system. However, the increase of the total bandwidth has less influence on AMTS in VCR service. That is because most users concentrate on hot videos. Therefore, many users are served by FBMT. In FBMT, normal and VCR users share channels. As a result, the increase of the total bandwidth has no influence on FBMT.

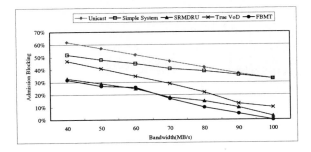

Fig. 9. Initial blocking probability vs. bandwidth

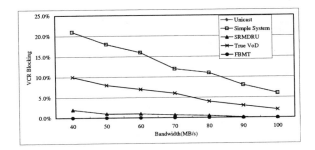

Fig. 10. VCR blocking probability vs. bandwidth

7   Conclusions and Future Works

In this paper, we have presented an Alternative Multicast Transmission Scheme (AMTS) to inexpensively provide VCR functionality. The goal of this scheme is to serve a large number of users and reduce the initial delay and channel cost while supporting VCR operations.

This scheme has two methods based on the access pattern of users: Frame-Based Multicast Transmission (FBMT) and normal catching. The FBMT is divided into two levels transmission layer. The first level layer uses a general delivery technique that a whole video is transmitted with catching. And the second level layer describes the inner transmission technique of a segment and policy to deliver frames via an appropriate channel based on frame type that organizes the segment.

Because of AMTS, we can reduce the number of channels and support both normal playback and VCR playback via one channel.

The performance of our scheme is superior over existing techniques in terms of both network resource, VCR blocking probability and initial blocking probability. When the total bandwidth is increased, both the initial blocking probability and the VCR blocking probability are decreased. And the initial blocking probability of AMTS is decreased as the arrival rate is increased further, VCR blocking probability is about 0% to admitted users because normal players and interactive players share the multicast channels.

FBMT is effective for "hot" (i.e., frequently accessed) video objects, not for "cold" video objects. So we use normal catching and unicast for cold videos. But, because this method is inefficient, we are to develop a new technique to efficiently delivery "cold" video objects in the future.

References

1. L. Gao and Z.-L. Zhang and D. Towsley, "Catching and Selective Catching : Efficient Latency Reduction Techniques for Delivering Continuous Multimedia Streams". ACM Multimedia' 99 Preceedings.
2. K. C. Almeroth and M. H. Ammar, "The Use of Multicast Delivery to Provide a Scalable and Interactive Video-on-Demand Service". IEEE Journal on Selected Areas in Communications, Vol. 14, August 1996.
3. W. Liao and V.O.Li, "The split and merge protocol for interactive video-on-demand," IEEE Multimedia, vol. 4, October-December 1997. Also in Proceedings of Infocom'97
4. E. L. Abram-Profeta and K. G. Shin, "Providing unrestricted VCR functions in multicast video-on-demand servers". in Proceedings of IEEE international Conference on Multimedia Computing and Systems(ICMCS'98), 1998.
5. H. K. Park and H. B. Ryou, "Multicast Delivery for Interactive Video-On-Demand Service". Information Networking, 1998. (ICOIN-12) Proceedings, Twelfth International Conference on, 1998, Page(s): 46–50
6. Wing-Fai Poon and Kwok-Tung Lo and Jian Feng, "Multicast video-on-demand system with VCR functionality". Communication Technology Proceedings, 1998. ICCT '98. 1998 International Conference on, 1998, Page(s): 569–573 vol.1
7. Wing-Fai Poon and Kwok-Tung Lo and Jian Feng, "Design and analysis of multicast delivery to provide VCR functionality in video-on-demand systems". ICATM' 99. 1999 2nd International Conference on, 1999, Page(s): 132–139.
8. Poon, W.-F. and Lo, K.-T. and Feng, J, "Design of multicast delivery for providing VCR functionality in interactive video-on-demand systems". Broadcasting, IEEE Transactions on Volume: 451, March 1999, Page(s): 141–148.
9. Z. Fei and I. Kamel and S. Mukherjee and M. H. Ammar, "Providing Interactive Functions for Staggered Multicast Near Video-On-Demand Systems," IEEE Multimedia, 1999
10. Jungmin Choi, Seungwon Lee, Kidong Chung, "A Frame-Based Multicast Transmission (FBMT) Scheme for Efficient Service in a VOD System", ICME 2001 Page 156–159.

Enhancing Information Quality for Web Pages

Wallapak Tavanapong, Wenlu Li, and Ki-Hwan Kim

The MiΔea Laboratory, Department of Computer Science
Iowa State University, Ames, IA 50010-1040, U. S. A.
{tavanapo,wli,khkim}@cs.iastate.edu

Abstract. Recent years have witnessed an explosive growth of the
World-Wide Web with an increasing number of images corporated in
a Web page. Images was mentioned to constitue upto 70% of the Internet traffic and are currently used on the Web for various purposes such
as text-related content, decoration, navigation, and advertisement. Unfortunately, not all images contain valuable information for most users.
For instance, advertisements, banners, or decorative images are typically
not the reason for most users to visit a particular Web site. While the
information value of these images appear small, they consume network
bandwidth and increase service delays. This paper discusses an application service that enhances the quality of information on a Web page.
The service allows a user to specify desirable categories of images to
retain in a page. The service automatically detects image purposes associated with the requested Web page and renders the page with only the
desirable images.

1   Introduction

Recent years have witnessed the explosive growth of the World-Wide Web. It
has evolved from a simple information-sharing mechanism offering only static
text to a rich assortment of images and dynamic and interactive services such as
video/audio conferencing, electronic commerce, and distance learning. The success of the Web has imposed a heavy demand on networking resources and Web
servers. Users often experience long and unpredictable delays when retrieving
Web pages.

A Web page typically does not contain only desirable information for users.
It can be seen as a combination of two types of components: content components and non-content components. Content components convey information to
the users and are often found in the form of text or images. Examples of content components are HTML files, thumbnail news stories on news Web sites,
thumbnail book covers on online bookstores, etc. Non-content components do
not convey information relevant to the objectives of the sites and are found
mostly in the form of images such as advertisements, bullets, and background
images. While images in Web pages form up to 70% [1] of Internet traffic, many
of the images can be considered non-content components. While they do not
provide useful information to Web users, these components consume networking
resources and contribute to long service delays.

I. Chong (Ed.): ICOIN 2002, LNCS 2344, pp. 491–500, 2002.
© Springer-Verlag Berlin Heidelberg 2002

To measure the quality of information that a Web page carries, we define *information quality* of a Web page as the percentage of bytes of content components over the total number of bytes needed to render the page[1]. For instance, if a Web page contains five inlined images and only two of them are content images, the information quality of this page is only 50% assuming that the sizes of the HTML file and the images are the same, and the HTML file is always considered as a content component.

Information quality of a page varies depending on the purpose of a Web site and a user. For instance, an image of a book cover posted on a news Web site can be considered an advertisement for most users whereas the same image posted on an online bookstore such as www.amazon.com is considered content. For most users, advertisement images are considered non-content whereas for advertising agencies, advertisements may be considered as content when the agency performs a survey on how its advertisements are used. Unlike the available programs that block popup advertisements or the feature of Web browsers that allow users to blindly disable all the images, our approach offers users a finer granularity of control on the information that they wish to see on a Web page. In the proposed approach, images are automatically divided into five categories: (i) **content images**, such as thumbnail images of headline news for news Web sites, or book covers for online bookstores; (ii) **advertisement images**; (iii) **decorative images**, such as background, balls, masthead, or bullet images; (iv) **informational images**, such as text images (e.g., 'Under Construction', 'What's New', show logos from news sites, or company logos); and (v) **navigational images** that help users navigate through the site quickly such as arrows, 'back to home', links to video and audio files, or image maps.

A user can receive the service by visiting our Web site and specifying the desired image categories. After the user inputs the URL of their desired Web page, he/she is presented with the requested page and only the images in the chosen categories. The image classification is done on the fly based on a set of rules generated from prior semi-automatic analyses of Web sites. In general, users will receive better quality pages. For users with a limited network bandwidth and battery power, the service will also reduce the amount of transferred data, which could result in less battery power consumption and a quicker response time.

The remainder of this paper is organized as follows. Section 2 discusses related work. Our approach is presented in detail in Section 3. Finally, concluding remarks are given in Section 4.

2   Related Work

Recently, several work (e.g., [2,3,4,5]) have been proposed to modify Web pages to make it more suitable for a limited network bandwidth or for a limited display capability of personal digital assistants, handheld computers, smart phones, etc.

---

[1] Our definition is different from the definition of information quality based on whether the Web content is credible.

At a Web server or a proxy server, an image is typically either transcoded into a lower quality version or entirely removed to satisfy the limited bandwidth or the capability of the display device. No automatic detection of image purposes are done or used in these techniques.

AdEater [6] learns rules that determine whether an image is an advertisement image or not using several features extracted from the image's URL as well as phrases within anchored text enclosed in $< A >$ tag and alt. Paek et al. proposed a technique to detect image purposes for cataloging and indexing images for better retrieval [1]. Their technique is a decision-tree learning algorithm. Twenty-one attributes are extracted from both HTML files and visual properties of the inlined images such as the number of colors in the images are used to classify the images into two categories: content and non-content images. Content images are then cataloged and indexed.

3 Enhancing Information Quality

We present the design and implementation of our service in this section.

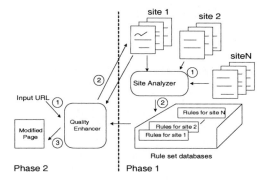

Fig. 1. Software Architecture

Fig. 1 depicts the two main software modules: **Site Analyzer** and **Quality Enhancer**. Site Analyzer downloads Web pages from popular sites (see Label 1 in Fig. 1 Phase 1) and analyzes these pages to obtain rules suitable for classifying images into the aforementioned categories (see Label 2 in Fig. 1 Phase 1). The rules are associated with the hostname of the site, and stored in a **rule database** implemented as a relational database. We query the rules from the database instead of hard coding them in Quality Enhancer so that existing rules can be modified since characteristics of images change over time and new rules can be easily added without any modification to the program.

While Site Analyzer works as a background process and does not interact with users, Quality Enhancer takes an input URL from the user and the selected image categories to retain. The program issues a request to the Web site according to the URL and retrieves the rule set associated with the Web site from the database. The rules are then used to determine whether the images referenced in the requested page are desirable for the user. If the images are not in the desirable categories, the HTML page is modified to discard the reference to these images. Only the modified HTML file and the images in the user-selected categories are passed along to the user's browser. If no rule sets are found in the database for the requested site, Quality Enhancer uses a default rule set and triggers Site Analyzer to download the requested site for further analysis. We note that caching of pages requested previously for users who subsequently ask for the same page can also be done to further reduce the service time.

In order for our service to provide good information quality for Web pages, the rules must be sufficiently accurate. Although the technique proposed by Paek et al. [1] can be modified to produce the rules, it requires twenty-one attributes to be extracted for each image, making it less attractive to provide fast service. Furthermore, it seems possible to improve the classification accuracy of the rules by selecting a smaller and better set of attributes. This motivates us to manually analyze Web sites to gain some insights on characteristics of images and their purposes in order to select good attributes. We present some statistics of two Web sites in this paper. First, *wget* program was executed to retrieve the HTML files and most of the embedded objects through markups such as <href> and <img src> from www.abcnews.com and www.espn.com. These two sites are referred to hereafter as ABCNEWS and ESPN, respectively. They were chosen because of their popularity and the fact that Web pages on these sites are quite complex and consist of a wide variety of images. If good rule sets can be obtained from these sites, it is likely that we can obtain good sets for simpler sites. Since some images and videos in a popup window cannot be downloaded using wget, to ensure the accuracy of our analysis, the HTML files were parsed to detect the missing images which were then retrieved manually.

Table 1 demonstrates the percentage of files of each category based on the total number of files. For these sites, the majority of the images are content images, followed by informational images, advertisements, and navigational images, with the number of decorative images being the smallest. The statistics also showed that there were only slight changes in the percentages of images in different categories between two consecutive days. Thus, the rule sets need not be updated often. Table 2 depicts the frequency of reference of each category in an HTML page. If the same image was referenced several times on the same page, the image was counted as being used only one time since it was transferred only once to the user. On average, there are ten images per page and nine were non-content images. For ABCNEWS, a content image was found in nearly all of the HTML pages, and there were about two advertisement images per HTML page. This statistical information suggests that on average, the information quality of a Web page on these sites can be improved significantly if only the content image is desired.

To estimate the average improvement in information quality, Table 3 summarizes the descriptive statistics for each category. On average, for ABCNEWS,

Table 1. Counts of various categories

Counts	ABCNEWS		ESPN	
	Day 1	Day 2	Day 1	Day 2
Total	3388	3199	3843	3321
HTML files	41.15%	43.54%	61.15%	59.92%
All Images	58.85%	56.46%	38.85%	40.08%
Content images	38.75%	37.45%	23.42%	21.17%
Advertisement	3.36%	3.53%	8.38%	11.08%
Decoration	1.18%	1.25%	1.43%	1.63%
Information	12.28%	10.97%	4.66%	4.55%
Navigation	3.28%	3.25%	0.96%	1.65%

Table 2. Frequency of reference of each category per HTML page

Counts	ABCNEWS		ESPN	
	Day 1	Day 2	Day 1	Day 2
Content images	0.98	0.91	0.60	0.58
Advertisement	2.24	2.13	1.54	1.73
Decoration	2.70	2.74	4.05	3.46
Information	4.79	4.83	1.91	1.82
Navigation	2.98	3.13	0.26	0.17

the information quality of a page if a user is only interested in the HTML file and the content image is only 50% or $\frac{34906.04}{68882.39} \cdot 100$ calculated according to the definition using the sum of all the products of the median of image size and the frequency of reference for each category. Similarly, the average quality of a page for ESPN is 62.6% or $\frac{27976.4}{44675.53} \cdot 100$. Thus, on average, the information quality can be increased by 48-50% for these two sites. Note that the increase in information quality will be different for different Web sites.

3.1   Web Site Analyzer

Since available traces (e.g., [7,8]) do not contain enough information to identify the five image purposes, Site Analyzer invokes *wget* program [9] to download the entire sites under consideration. Repeated Incremental Pruning to Produce Error Reduction (RIPPER) program [10] is then used to classify images site by site. We chose RIPPER since it produces rule sets that are relatively easy to understand. Rule learning systems also perform better than decision-tree learning systems on many problems [10]. After the rule sets for the Web site have been obtained, they can be used to automatically classify images for the same site.

Since RIPPER generates the rule sets based on training examples, the examples must first be constructed from the downloaded pages. A training example consists of a set of attributes describing an image to be classified and a category of the image determined manually. Since the selection of appropriate attributes is very important to obtain accurate rules, we selected the set of attributes based

Table 3. Statistics for various categories

Category	Statistics	ABCNEWS		ESPN	
		Day 1	Day 2	Day 1	Day 2
All file	Mean	15559	15722	17193	18621
	Median	13803	13833	12032	20864
	Stdev.	11648	11598	15403	14329
Image	Mean	8375	8451	4567	5262
	Median	7401	7507	2332	3464
	Stdev.	7274	7266	4405	4723
HTML	Mean	25845	25153	25215	27556
	Median	25475	25157	26684	28364
	Stdev.	8630	9110	14492	11335
Content images	Mean	10967	10994	4336	5297
	Median	9948	9964	2154	2480
	Stdev.	7356	7323	4460	5123
Advertisement	Mean	8232	8105	7095	7009
	Median	9256	8752	7122	6001
	Stdev.	3799	3688	3894	3955
Decoration	Mean	212	208	241	866
	Median	44	44	192	663
	Stdev.	416	417	187	1044
Information	Mean	2782	2797	3268	4058
	Median	2257	2228	2509	3151
	Stdev.	2518	2641	3531	3586
Navigation	Mean	1797	1835	923	866
	Median	761	746	627	663
	Stdev.	2862	2948	1260	1044

on the observations gained during the manual analysis of the two news sites. We observed that i) more than 80% of the JPEG (with extension jpg) files are content images; ii) content images are usually larger than others both in file size and in the width and height; iii) most of the advertisement images are animated gif files; and iv) most of the content images tend to have a similar width and height, while some advertisement and decorative images have a larger width than height. These observations lead to the use of the following six attributes being extracted from the images. The extraction of these attributes can be done very easily by parsing the results of a program called *identify* on the image. The program comes with Linux distribution.

– Format - Image format: gif(0), jpg(1)
– Filesize - Size of a file in bytes
– Width - Image width in pixels
– Height - Image height in pixels
– WHratio - A ratio of the image width to the height
– Animation - This attribute describes whether an image is likely an animated-gif image. Since an animated-gif file typically contains several images, the

number of images contained in the file is used as the value for this attribute. If this number is greater than one, the image is an animated-gif image.

Given the training data, RIPPER forms a list of rules to describe different classes of images: advertisement (adv), content images (cont), decorative images (dec), informational images (info), and navigational images (nav). The resulting rules are in the form:

classA: attrib1=x, attrib2=y.
classB: attrib3=z.
default classC.

That is, an image belongs to classA if attributes 1 and 2 of the image are x and y, respectively. The image belongs to classB if it does not belong to classA and attribute 3 is z. Otherwise, the image belongs to the default class, classC. Multiple conditions are taken to mean that all conditions must hold.

To check whether the chosen attributes are indeed good attributes, the rule sets were used to classify images from a different day and the accuracy of the classification was measured. The classification accuracy of a category is defined as the percentage of images that are automatically classified correctly over the total images classified manually to be in the same category. The accuracy of the automatic classification is illustrated in Table 4. For advertisement, content, and decoration categories, the accuracies of the automatic classification are greater than 93% whereas the classification accuracies for informational and navigational images vary according to sites. The rule sets for ABCNEWS and ESPN are given in Tables 5. We observe that the rules were simple and the number of rules is small. This reduces the time to determine the category of each image. Other sites are analyzed similarly. We also found that using only 10% of the images on these sites as training samples still yields classification accuracies of about 82.76% and 87.3% for five image categories for ABCNEWS and ESPN, respectively. We note that our technique outperforms the decision-tree learning algorithm by Paek and Smith [1], which has more complicated rules and has an overall classification accuracy of 84% for classification of two categories.

Table 4. Classification accuracy for five categories

Category	ABCNEWS	ESPN
Content	93.60	99.67
Advertisement	94.74	97.83
Decoration	97.50	94.55
Information	91.59	78.77
Navigation	59.46	89.19
Overall	91.42	96.32

Table 5. Rule sets

ABCNEWS	ESPN
ID conditions	ID conditions
1 dec : filesize $\leq$ 44	1 nav : height $\leq$ 22, filesize $\geq$ 249, $\;\;\;\;$ 4 $\leq$ whratio $\leq$ 7
2 dec : filesize $\leq$ 411, width $\geq$ 259	2 nav : 35 $\leq$ height $\leq$ 37
3 dec : filesize $\leq$ 310, height $\geq$ 31	3 nav : height $\leq$ 17, width $\leq$ 47, $\;\;\;\;$ filesize $\geq$ 190
4 dec : filesize $\leq$ 310, width $\geq$ 14, $\;\;\;\;$ height $\leq$ 13	4 nav : 75 $\leq$ filesize $\leq$ 115
5 dec : width $\geq$ 440, filesize $\leq$ 1410	5 dec : filesize $\leq$ 366
6 nav : filesize $\leq$ 339, height $\leq$ 14	6 info : 39 $\leq$ height $\leq$ 49, width $\geq$ 115
7 nav : filesize $\leq$ 413, height $\leq$ 20	7 info : whratio $\geq$ 1, animation $\leq$ 1, $\;\;\;\;$ width $\geq$ 115, format = 0, height $\geq$ 62
8 nav : width = 110, format = 0	8 info : height $\leq$ 25
9 nav : 88 $\leq$ width $\leq$ 96, filesize $\leq$ 919, $\;\;\;\;$ height $\geq$ 13	9 info : whratio $\geq$ 1, height $\geq$ 90, $\;\;\;\;$ width $\leq$ 200
10 nav : height = 98, filesize$\geq$ 9577	10 info : 76 $\leq$ height $\leq$ 78, animation $\leq$ 1
11 adv : height = 60	11 info : height $\leq$ 80, animation $\leq$ 1, $\;\;\;\;$ format = 0, 108 $\leq$ width $\leq$ 188
12 adv : 30 $\leq$ height $\leq$ 35, $\;\;\;\;$ 115 $\leq$ width $\leq$ 125	12 info : width $\leq$ 64, 1368 $\leq$ filesize $\leq$ 2482
13 info : filesize $\leq$ 3813, format = 0	13 info : filesize $\leq$ 1187, height $\leq$ 34
14 info : format= 0, width $\geq$ 104, $\;\;\;\;$ 56 $\leq$ height $\leq$ 98	14 info : 72 $\leq$ width $\leq$ 76
15 info : width $\leq$ 96, filesize$\leq$ 2532, $\;\;\;\;$ height $\geq$ 85	15 info : height $\geq$ 130, width $\leq$ 120, $\;\;\;\;$ 3033 $\leq$ filesize $\leq$ 3809
16 info : 84 $\leq$ width $\leq$ 123, format= 0, $\;\;\;\;$ height $\leq$ 162	16 adv : whratio $\geq$ 2
17 info : width $\leq$ 67	17 adv : animation $\geq$ 2
18 default cont.	18 default cont.

3.2   Quality Enhancer

The rule database consists of two relations depicted in Figs. 2 and   3. The primary key of the *rule* relation consists of two attributes h, a foreign key to hostmap.hostid and *id* denoting the rule number. Fig. 3 demonstrates an example of how rules 1 to 4 from Table 5 are actually stored in the rule database.

hostid	hostname
1	www.abcnews.com
2	www.espn.com

Fig. 2. Rule Database - (hostmap relation)

h	id	type	format	lb_size	ub_size	lb_w	ub_w	lb_h	ub_h	lb_a	ub_a	ub_wh	lb_wh
1	1	dec	NULL	NULL	<= 44	NULL	NULL	NULL	NULL	NULL	NULL	NULL	NULL
1	2	dec	NULL	NULL	<= 411	>= 259	NULL	NULL	NULL	NULL	NULL	NULL	NULL
1	3	dec	NULL	NULL	<= 310	NULL	NULL	>= 31	NULL	NULL	NULL	NULL	NULL
1	4	dec	NULL	NULL	<= 310	>= 14	NULL	NULL	<= 13	NULL	NULL	NULL	NULL

Fig. 3. Rule Database - (rules relation)

Quality Enhancer is a Common Gateway Interface (CGI) program written in Perl language using the CGI, HTTP, LWP, HTML, URL, and DBI modules. The program extracts the hostname from the user's URL and issues a SQL query to retrieve all rules relevant to the hostname from the rule database. Concurrently, Quality Enhancer retrieves the page as well as images associated with the page from the original site. Once the page and the image files are received, Quality Enhancer categorizes the images into the five categories according to the rules. For images in the categories not selected by the user, Quality Enhancer eliminates <img src> tags associated with these images. The modified HTML page is sent to the user along with the selected images. To allow the user to click on the hyperlinks and enhance the quality of the pages referred to by the hyperlinks, these links are modified to call Quality Enhancer with the URL of the original links as the parameter.

4   Concluding Remarks

We have presented the design and implementation of a novel multimedia application service. The goal of the service is to enhance quality of information on a Web page. The service categorizes images based on the rule set associated with the requested Web site and filters out images in the undesirable categories specified by the users. The rule sets are automatically obtained and are quite accurate in detecting image categories. The implementation of the service is also flexible allowing easy addition or deletion of rules without changing the filtering code. Future investigations will focus on exploiting the rules to provide a better Quality of services for Web servers serving wireless personal digital assistance.

Acknowledgment. This material is based on work supported by the National Science Foundation under Grant CCR-0092914. Any opinions, findings and conclusions or recommendations expressed in this material are those of the author(s) and do not necessarily reflect those of the National Science Foundation.

References

1. Paek, S., Smith, J. R.: "Detecting image purpose in world wide web documents," in *IST/SPIE Symposium on Electronic Imaging: Science and Technology Document Recognition*, Janunary 1998.

2. Smith, J. R., Mohan, R., Li, C-S.: "Scalable multimedia delivery for pervasive computing," in *Proc. of ACM Multimedia'99*, Orlando, FL, October 1999, pp. 131–140.
3. Han, R., Bhagwat, P., LaMaire, L., Mummert, T., Perret, V., Rubas, J.: "Dynamic adaptation in an image transcoding proxy for mobile web browsing," in *IEEE Personal Communications*, December 1998, vol. 5, pp. 8–17.
4. Liu, H., Fang, H.: "Real-time image on qos web," in *Proc. of Performance Analysis of Systems and Software*, 2000, pp. 70–75.
5. Chandra, S., Ellis, C-S., Vahdat, A.: "Differentiated multimedia web services using quality aware transcoding," in *Proc. of IEEE Infocom 2000*, 2000, pp. 961–969.
6. Kushmerick, N.: "Learning to remove internet advertisements," in *Proc. of Int'l Conf. on Autonomous Agents*, Seattle, WA, 1999.
7. Digital Equipment Cooperation, "Digital's web proxy traces," in *ftp://ftp.digital.com/pub/DEC/traces/proxy/webtraces.html*.
8. National Lab for Applied Network Research, "NLANR proxy traces," in *ftp://ircache.nlanr.net/Traces/*.
9. "wget": in *http://sunsite.auc.dk/wget/index.html*.
10. Cohen, W., "Fast effective rule induction," in *Proc. of the Twelfth Int'l Conf. on Machine Learning*, Lake Tahoe, California, 1995.

A Replacement Policy with Growth-Curve on the Continuous Media Proxy Cache (CMPC)

Seongho Park[1], Yongwoon Park[2], Hyeonok Hong[2], and Kidong Chung[1]

[1] Department of Computer Science, Pusan National University, Rep. of Korea
{shpark, kdchung}@melon.cs.pusan.ac.kr
[2] Department of Computer Science, Dongeui Institute of Technology, Rep. of Korea
{ywpark, hohong}@dit.ac.kr

Abstract. With rapid proliferation of the internet, lots of content providers make their market debuts on the internet for internet broadcasting systems, VoD service and advertizement using moving pictures; this eventually leads to heavy traffic on the internet. Proxy caching is one of the widely known techniques to solve this problem but so far it has been told not to show satisfiable performance for continuous media.

So, in this paper, we propose a replacement policy called LNGV for continuous media files on the proxy cache. For this, we frist analyze the log file of iMBC to see what characteristics the continuous media objects have. In LNGV, the growth-curve function is established to check how each media object's access frequency varies after $n(n \geq 1)$th period and then is used to choose the victim objects for replacement so as to reserve cache space for newly requested object.

Simulation results show that our proposed algorithm shows the average 11.5% performance improvement compare with LFU and the performance gap between LNGV and LFU can be reduced as the cache capacity increases.

1 Introduction

With the rapid proliferation of the Internet, we have seen that many sites servicing continuous media objects including VoD and AoD dominate the Internet and so far they have made efforts to provide more qualified service to the users. Although, both computing and networking technologies have been developed significantly, their development speed can not catch up with the speed of the volume increase of contents and users on the Internet [1]. With Gigabit Ethernet and Fiber Channel technology, the cost of installing and running a local area gigabit network becomes increasingly cheaper. Therefore, reducing the total bandwidth requirement of the backbone network should be an important objective in the design of a real time continuous media data delivery system [2].

Based on this, researchers have made their efforts to find solutions of reducing the network traffic overhead caused to send objects from central servers to end users [3,4]. Among these, proxy caching policy, which is intermediary storage archives existing between the servers and end users, is one of the most attractive policies in terms of reducing network overhead [5]. Recently, several proxy

I. Chong (Ed.): ICOIN 2002, LNCS 2344, pp. 501–511, 2002.
© Springer-Verlag Berlin Heidelberg 2002

servers have been designed to service Web requests consisting of textual and image objects in general [7]. However, due to the characteristics of continuous media objects, which require much more storage space and network bandwidth than conventional data, they will degrade the performance of the proxy server if they are not properly controlled. As shown in Fig. 1, the proxy caches for the continuous media objects are located at network connection points among Access Network, Distribution Network and Core Network so that they could service users' requests. However, research on continuous media objects is still in its infancy that it is desirable intense researches should be done to provide more qualified service.

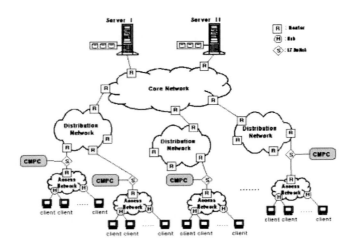

Fig. 1. Proxy cache configuration on the internet

To address problems in caching continuous media objects, the access frequency, recency and size of the requested objects should be taken into consideration but so far research about those characteristics has not made much progress. So, in this paper, we analyze log files of continuous media file server of iMBC corporation in Korea and then analyze each object's access pattern. Using the access patterns of each object, we create a function called growth-curve function per each object and then propose a replacement algorithm called LNGV (Least Next Growth Value) using the output of the function. Moreover, we do trace-driven experiment to check the efficiency of our proposed algorithm.

This paper is composed as follows. In Section 2, we will refer to the related researches and in section 3, we will show how we analyze log files of continuous media file server and create the growth-curve function. Consequently, in section 4, we propose a new caching policy for continuous media files called LNGV algorithm afterward. In section5, we do simulations and then do analyze whether or not our proposed algorithm show better performance than other caching algorithms. Finally, we draw conclusions about our research and refer to the further research.

2   Related Researches

2.1   Web Caching

Generally, web caching has focused on the discrete media such as images and text files and is different from the conventional file caching in that the size of objects not always the same as is the case of file caching. For the efficient cache space management, researches have made their effort on the replacement algorithms so as to evict the least likely accessed documents or objects. Generally, three major factor are taken into consideration for the replacement policies: frequency, recency and size of objects and the most widely known algorithms are told to be LRU, LFU, LTU-SIZE, LRU-MIN, LRFU and GD-SIZE [6]. However, these policies are designed for the conventional objects with small size that they are not appropriate for continuous media objects whose size is large and requires time constraint delivery.

2.2   Proxy Caching for Continuous Media Objects

Continuous media objects are larger in size and require higher transfer bandwidth than conventional media objects that it is very likely that both server and network are overloaded leading to packet loss and jitter. It is only recently that researchers have been working to improve the performance of proxy server in dealing with continuous media objects[7] because it has been known that continuous media files do not go well with the existing caching algorithms. [7] applied the concept of interval caching, which is one of the widely known continuous media file caching policy.

In this policy, how much the system resources such as cache bandwidth and space required is taken into consideration for cache management. [4] proposed the prefix proxy caching where initial portions of archived objects are cached so that their initial latency otherwise occur to access the objects from the central server could be hided and the burstiness of VBR streams could be smoothed. [10] proposed a caching policy based on the layered encoding technology so that caching could be done on the layer basis and thus transfer the stream adaptively depending on the available network bandwidth. [6] proposed a perfetching policy in that the initial portion of the objects are cached and thus the network bandwidth to access the remaining portions of streams could be smoothed on the heterogeneous network environment. [8,9] proposed a partial caching policy to cache either the initial portion or some part of the objects step by step so as to reduce the initial latency and manage cache space efficiently. In this policy, the size of the cached portion of an object is determined on its popularity.

Each research shows it effectiveness in both backbone network bandwidth performance and the central server but no research refers to the subject on how to guarantee the QoS when caching is applied.

3 Trace Analysis

In this paper, we analyze 21 days of log files of a continuous media server in iMBC (http://www. imbc.com) to find out the characteristics of continuous media files. Based on the results of our analysis, we show how to anticipate the future access patterns of the objects and propose a new proxy space replacement policy.

Table 1 shows the analysis environment. In iMBC server, about 20–30 objects are created everyday. The log data of Window Media server includes both start and random access requests. The start request is a request to playback the requested object from the beginning while the random access one is to choose the playback point arbitrarily at user's own will in the current stream. The percentage of the created objects out of all objects after we started the analysis reaches 11.3% and the percentage of the start and random access requests over all objects are 46.8% and 48.5 respectively. In this paper, we analyze and test objects created after we start our analysis

Table 1. Parameter of the simulation model

Classification	Characteristics
Server's O/S	Windows NT
Server's streaming S/W	Windows Media Server
Total number of objects	2,527
Total number of objects created for period of observation	325
Average transmission rate	300k bps
Play time of objects	20–70 min
Total number of requests	2,661,243
Total number of start requests	1,233,789
Total number of requests for objects created during the obse	1,246,293
Total number of start requests for objects created during th	598,217

3.1 The Life Cycle of Continuous Media Objects

The life cycle of continuous media objects indicates how users' access pattern varies as time passes after creation. Fig. 2 shows the life cycle of the continuous media objects created 5 times per week (the total playback time of each objects is 20 min and we analyzed the access pattern every 2 hours). As shown in Fig. 2, the access patterns of the objects show the following characteristics.

◇ The number of start requests varies with contents of the requested objects.

◇ The number of accesses for the object shows its peak point during the first 1 or 2 days after its creation.

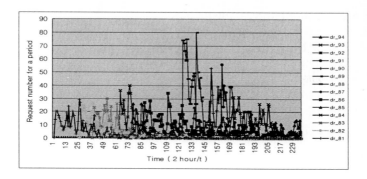

Fig. 2. The life cycle of the objects created 5 times per week

3.2   Growth-Curve of Continuous Media Objects

Fig. 3 shows the growth-curve of the continuous media objects for the objects created 5 times per week. The growth-curve in this paper means the accumulated numbers of start requests for all objects after their creation. As shown in Fig. 3, the growth-curves of the objects show the following characteristics.

◇ The number of access frequency for the objects increases abruptly during first couple of days after their creation but remains dormant after those days.

◇ Although the access frequencies of the objects in the same SG (Series Group) show some differences, in general, they are similar in their fashion.

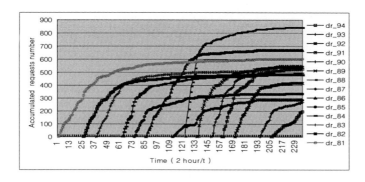

Fig. 3. The Growth-curve of the objects created 5 times per week

So, in this paper, we have derived a Eq. (1) from the growth-curve distribution. we call Eq. (1) growth- curve function. Using this function, we can anticipate the appropriate cache allocation and replacement time after $n(n \geq 1)$ rounds.

$$GF_i(t) = MC_i(1 - e^{-\frac{t}{s_i}}) \tag{1}$$

i : Identification number of media objects
t : Time period passed after object i is created
$GF_i(t)$: Growth-curve function of object i
MC_i : The maximum number of accumulated users of object i
S_i : Skew value of Life Cycle of object i
* period : An interval of the observation

Fig. 4 shows the accumulated value of growth-curve distribution and growth-curve function's distribution for the continuous media objects created five times per week. We use 's appropriateness experimentation to check whether or not the accumulated growth-curve distribution follows the estimated growth-curve function. The average tested statistics is, which is smaller than, that we can not reject the hypothesis. So, we can tell that the accumulated growth-curve distribution based on the real objects follows growth-curve function's distribution within the significant level of 0.005.

$$T = \frac{(O_i - E_i)^2}{E_i} \tag{2}$$

The slope of growth-curve function's distribution gets steeper as the value of gets bigger while gets flatter as the value of gets smaller. However, when the accumulated access frequency of object stops growing, the optimal value of and can be determined.

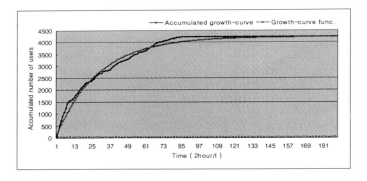

Fig. 4. Growth-curves of observed and approximated values

4   Cache Replacement Policy

In the replacement policy of continuous media objects, other than those of conventional objects, the available disk and network bandwidth must be taken into consideration to service user's requests issued at random point in time keeping their real time constraints in terms of data delivery. Therefore, we can determine the time of replacing an object by anticipating the growth value of each object after the nth period. However, we could know the optimal value of MC_i and S_i only after an access frequency of object i does not increase any more. Therefore, instead of using MC_i and S_i, we use the adjusted values of $\overline{MC_i}$ and $\overline{S_i}$ to provide the adjusted growth-curve function $\overline{GF_i}(t)$.

$$\overline{GF_i} = \overline{MC_i}(t-1)(1 - e^{-\frac{t}{S_i(t-1)}}) \tag{3}$$

$\overline{GF_i}(t)$: Adjusted growth-curve function of $i$object
$\overline{MC_i}$: Adjusted MC_i to compute $\overline{GF_i}(t+n)(n \geq 1)$ at t time period
$\overline{S_i}$: Adjusted S_i to compute $\overline{GF_i}(t+n)(n \geq 1)$ at t time period

4.1 Adjusted Algorithm for MC_i and S_i

MC_i and S_i parameters for the growth-curve function of continuous media object i created in real time could be adjusted based on $\overline{GF_i}(t)$ at $t-1$ time period and $O_i(t)$ for t time periods. Eq. 4 is an equation to compute the value of tested statistics to get the error using $\overline{GF_i}(t)$ and $O_i(t)$.

$$T_i(t) = \frac{O_i(t) - \overline{GF_i}(t)^2}{\overline{GF_i}(t)} \tag{4}$$

$O_i(t)$: Observed access frequency of object i for t time periods.
$T_i(t)$: The tested statistics of χ^2 distribution of $O_i(t)$ and $\overline{GF_i}(t)$ when access frequency distribution is $\chi^2(0.05, 1)$.

Fig. 5 and Fig. 6 represent the adjusted algorithms of $\overline{MC_i}$ and $\overline{S_i}$ respectively. $\overline{MC_i}$ and $\overline{S_i}$ could be adjusted at t time period if the value of tested statistics, $T_i(t)$ is greater than $\chi^2(0.05, 1)$. $\overline{MC_i}$ and $\overline{S_i}$ of the just created object i are set to the average values of the objects in the same series.

---

if$(t = 0)$ $MC_i(t) = Avg(MC_k)$, (when $i \in SG_p, \forall k \in SG_p$)
  if$(T_i(t) \leq \chi^2(0.05, 1), \chi^2(0.05, 1) = 3.841)$
    $MC_i(t) = (MC_i(t-1))$
  else$\{$
    if$(LC_i(t) < O_i(t))$
      $MC_i(t) = (MC_i(t-1) + (O_i(t) - LC_i(t))/log_{3.841} * chi_i^2(t)$
    else
      $MC_i(t) = (MC_i(t-1) - (O_i(t) - LC_i(t))/log_{3.841} * chi_i^2(t)$
  $\}$
  if$(T_i(t) < O_i(t))$ $MC_i(t) = O_i(t) + 1$

Fig. 5. Algorithm to adjust $\overline{MC_i}(t)$

SG : Objects Series Group.
p : Identification number of media objects series.

$$
\begin{aligned}
&\text{if}(t=0)\ S_i(t) = Avg(S_k),\ (\text{when } i \in SG_p, \forall_k \in SG_p) \\
&\text{if}(T_i(t) \leq \chi^2(0.05,1), \chi^2(0.05,1) = 3.841) \\
&\quad S_i(t) = (S_i(t-1)) \\
&\text{else}\{ \\
&\qquad K_i(t) = K_i(t-1) + \frac{-t}{log(1 - \frac{O_i(t)}{MC_i(t)})} \\
&\qquad S_i(t) = \frac{K_i(t)}{Revision_i} \\
&\qquad Revision_i + + \\
&\}
\end{aligned}
$$

Fig. 6. Algorithm to adjust $\overline{S_i}(t)$

4.2 LNGV (Least Next Growth Value) Algorithm

LNGV algorithm computes the value of $R_i(t)$, which is used to determine whether or not to replace object i, using the growth-curve function $\overline{GF_i}(t+n)$ at $t+n(n \geq 1)$ and $O_i(t)$ at t (See Eq. 5). Particularly, as shown in Fig. 10, the load factor of proxy cache fluctuates in a day, so replacements could be done at proxy's idle time.

In LNGV algorithm, n will be chosen so that the cache shows as light traffic as possible at $t+n(n \geq 1)$ time period. We could explain the LNGV algorithm more detail in Fig. 7.

$$R_i(t) = \overline{GF_i}(t+n) - O_i(t),\ (n \geq 1) \tag{5}$$

5   Performance Evaluation

In this section, we test our proposed algorithm based on the simulation using log files of iMBC. The log files used in this simulation includes information about 1,246,293 times of users' requests for both start and random access requests. We designed the proxy cache to cache the streaming data progressively at the object level.

To analyze the impact of the cache size on the cache's performance, we changed the cache size from 4 GB to 19 GB by 1 GB unit using LNGV, LFU and LRU.

Figure 8 shows the cache hit ratio changes by the algorithms. In general, LNGV shows better performance than LFU by 3% to 15% (the average is 11.5%). Particularly, smaller cache size result in bigger performance gap between two algorithms.

Figure 9 and 10 presents the number of replacement and point at time. When the cache size is smaller than 15 GB, LNGV shows more replacement counts than LFU. However, When the cache size is larger than 15 GB, LNGV shows the same replacement counts as LFU. Particularly, as shown in Fig. 10, the LNGV algorithm replaces objects considering the overall traffic from/to proxy. Therefore the overall the number of replacement itself does not matter even if its number of replacement is higher than that of LFU.

```
do_replace(measured replace value(R_i(t))){
        Cache_original-remaind = Cache_remaind
    while(1){
      if(Size_i < Cache_remaind)
        Eject Victim Objects from Cache,
        Cache i.
        return;
        Obj_k = min(Obj_j), Obj_j ∈ Cache, j = 1, m
        if(R_i(t) ≥ R_k(t)){
          if( Obj_k is not finished caching) continue
          Set Obj_k as victim object.
          Cache_remaind+ = Size_Obj_k
        }else{
          Release the expected victim Objects
          Cache_remaind = Cache_original-remaind
          return
        }
    }
}
```

Fig. 7. Replace algorithm of LNGV

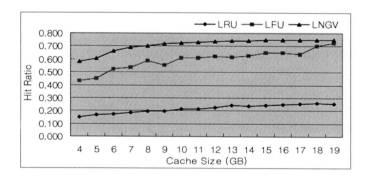

Fig. 8. Cache hit ratios

6   Conclusion and Next Research

In this paper, we have proposed a replacement policy called LNGV for continuous media files on the proxy cache. For this, we frist analyzed the log file of iMBC to see what characteristics the continuous media objects have. In LNGV, the growth-curve function is exploited to analyze how each media object's access frequency changes. Using the growth-curve function, the expected growth-curve value at $(t+n)(n \geq 1, t : currentperiod)$ time period for each object is computed,

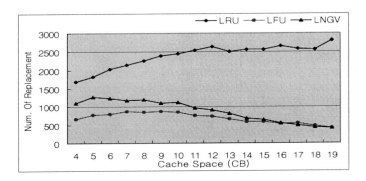

Fig. 9. Cache space vs. Replacement counts

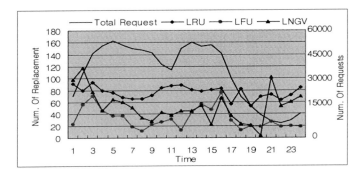

Fig. 10. User's requests and replacement point when the size of proxy cache is 6 G Bytes

so the object with the highest expected growth-curve value will be chosen as the caching object after time periods.

By the simulations of log files of iMBC that include information about 1,246,239 times of users' requests, we know that our proposed algorithm shows the average 11.5% performance improvement compare with LFU, and the performance gap between LNGV and LFU will becomes bigger when the cache capacity is small. The replacement counts of our proposed algorithm are larger than a case of LFU but LNGV algorithm replaces considering the overall traffic from/to proxy. Therefore the overall number of replacement itself does not matter even if its number of replacement are larger than that of LFU. Our proposed algorithm shows higher hit ratios than that of LFU but higher hit ratio itself does not guarantee QoS for the requested objects. However we are beyond the scope of this research and remain our future work.

References

1. Garth A. Gibson, Jeffrey S. Vitter, John Wilkes, Strategic Directions in Storage I/O Issues in Large-scale Computing, ACM Computing Surveys pp. 779-793 vol.28, No. 4, 1996.
2. Zhi-Li Zhang, Du, D.H.C., Dongli S, Yuewei Wang, A Network conscious Approach to End-to-End Video Delivery over Wide Area Networks Using Proxy Servers, in Proc. of INFOCOM'98 pp. 660-667 vol.2, 1998.
3. D. Eager, M. Ferris, and M. Vernon, Optimized Regional Caching for On Demand Data Delivery, in Proc. Multimedia Computing and Networking, January 1999.
4. S. Sen, J. Rexford, and D. Towsley, Proxy Prefix Caching for Multimedia Streams, in Proc. IEEE INFOCOM'99 pp.1310-1319 vol.3, April 1999.
5. C. Maltzahn, K. Richardson and D. Grunwald, Performance Issues of Enterprise Level Web Proxies, in Proc. of the SIGMETRICS Conference on Measurement and Modeling of Computer System, June 1997.
6. Y. Wang, Z.-L. Zhang, D. Du, and D. Su, A Network-Conscious Approach to End-to-End Video Delivery over Wide Area Networks Using Proxy Servers, in Proc. IEEE Infocom'98, April 1998.
7. R. Tewari, H. M. Vin, A. Dan, D. Sitaram, Resource-based Caching for Web Servers, in Proc. SPIC/ACM Conference on Multimedia Computing and Networking, January 1998.
8. E. J. Lim, S.H. Pakr. H.O. Hong. K.D. Chung, A Proxy Caching Scheme for Continuous Media Streams on the Internet, in Proc. ICOIN'15 , 2001.
9. S.H. Park, E.J. Lim K.D. Chung, Popularity-based Partial Caching for VoD Systems using a Proxy Server, in Proc. 15th International Parallel & Distributed Processing Symposium, 2001.
10. Reza Rejaie, Haobo Yu, Mark Handely, Deborah Estrin, Multimedia Proxy Caching Mechanism for Quality Adaptive Streaming Applications in the Internet, In Proc. of IEEE Infocom'2000 , Tel-Aviv, Israel, March 2000.

Hierarchy and Region Based Source-Channel Combined Coder for Image Transmission over Mobile Links

ByungGil Lee[1] and KilHoum Park[2]

[1] Electronics and Telecommunications Research Institute
161 Gajeong-dong, Yuseong-gu, Daejeon, KOREA,
BGLee@etri.re.kr, http://www.etri.re.kr
[2] Kyungpook National University, 1370 Sankyuk-dong,
Buk-gu, DaeGu, KOREA

Abstract. This paper presents a new combined source channel coding approach to an unequal error protection method based on hierarchical and regional sensitivity to error propagation utilizing wavelet transform for mobile channels is presented. The proposed scheme performed in a very robust manner, when tested in a mobile multimedia system.

1   Introduction

The wireless multimedia industry is experiencing a growing demand for efficient image/video transmission, particulary in applications such as real-time mobile video phones, mobile video conferencing and wireless image multicasting. The transmission of high bandwidth real-time image sequences over a mobile communication system, such as IMT-2000, presents several challenging problems that remain to be resolved. A major problem relates to the source bandwidth in a costly wireless channel. This can be reduced by source coding it with a number of algorithms such as discrete cosine transform(DCT)-based coding and wavelet-based coding. However, these algorithms are degraded by burst erroneous mobile channels. The mobile channel impedes the correct transmission and reception of highly compressed video sequences by introducing burst errors, which are caused by the multi-path fading of the transmitted signal. In worst cases, Rayleigh fading occurs whereby no direct line of sight exists between the transmitter and the receiver in a moving object. To reduce the number of errors introduced by the wireless channel, channel-coding techniques combined with source coding are frequently employed [1,2,3,4,5]. Generally, a channel code with different rates is applied non-uniformly by the error sensitivity-weighting factor of source coded data prior to transmission [6]. However, this is relatively inefficient because different source coded blocks require different processing for error propagation protection. To overcome this problem, we have modified the source coder so as to provide information relative to the significance of the source bits, which can then be utilized to adjust the level of error protection applied by the channel coder. In wavelet source image coding, every coding scheme has an

I. Chong (Ed.): ICOIN 2002, LNCS 2344, pp. 512–520, 2002.
© Springer-Verlag Berlin Heidelberg 2002

error propagation problem. Error propagation may cause more severe damage in compression structure. For protecting against error propagation, we propose the use of non-uniform size sub-blocks by error sensitivity. This paper describes a source-channel combined system structure and simulation results in a W-CDMA environment. We describe source coding and a concept of UEP source coding in Chapter 2 and the structure of a combined coder for UEP in Chapter 3. Chapter 4 describes the simulation model and the results of the proposed scheme. Finally, in Chapter 5 we offer some concluding remarks.

2   Source Coding

2.1   Wavelet Transformation of Source Image

Wavelet transformation indicates functions which occur as a result of dilation and translation from an arbitrary underlying function $\Psi(x)$ and represents a linear combination of wavelet underlying functions having a simultaneous locality in a time-frequency zone. It can be numerically expressed as follows:

$$H_{ab}(t) = |a|^{-1/2} h\left(\frac{t-b}{a}\right) \quad a, b \in \mathrm{R}. \tag{1}$$

With regard to a, the large value, the underlying function, a form expanded from the mother wavelet, serves as a low frequency underlying function, and with regard to a, the small value, the underlying function contracted from the mother wavelet, serves as a high frequency underlying function. Wavelet transformation is defined as follows:

$$X_w(a, b) = \frac{1}{\sqrt{a}} \int_{-\infty}^{+\infty} h\left(\frac{t-b}{a}\right) x(t) dt. \tag{2}$$

The Wavelet transformed sub-band tree structure is shown in Fig. 1. Four types of sub-bands are gained, namely sub-bands HH, HL, LH and low frequency sub-band LL.

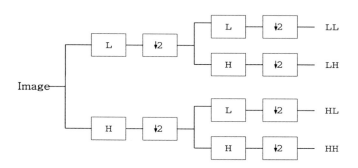

Fig. 1. Wavelet Decomposed Tree

The highest layer coefficient of the sub-bands has many more statistical properties of the original image than other layers.

2.2 Error Sensitivity Analysis

For image source coding, Fig. 2 shows a typical decoded error sensitivity at each bit position in the SPIHT (Set Partitioning Hierarchical Trees)[7] algorithm which, when divided by the wavelet coefficient using the zerotree, is partitioned according to the wavelet coefficient's importance. In the case of a compressed image, an analysis of the PSNR value of the reconstructed image revealed a bit error. In terms of sensitivity, Fig. 2 illustrates the difference in error sensitivity with the position of a bit error, which can be up to 20 dB in terms of PSNR performance.

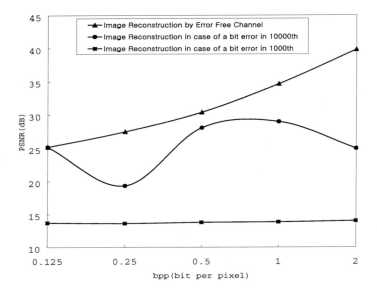

Fig. 2. Error Sensitivity

The reconstructed image quality for each layer is dependent on the lowest frequency band (LL) of the highest layer 0, as shown in Table 1.

2.3 Wireless Robust Image Coding

Highly compressed wavelet-transformed data of images are extremely sensitive to burst errors, which are commonly encountered in wireless fading channels. To resolve this problem, different sized sub-blocks are divided into sub-banded

Table 1. PSNR of the Reconstructed image using the remaining coefficients, except for a specific band

Layer	0	1			2			3		
Omitted band	LL3	HL3	LH3	HH3	HL2	LH2	HH2	LH1	LH1	HH1
PSNR(dB)	6.1	26.8	30.9	31.1	28.9	32.2	32.4	31.0	33.4	33.8

images. The blocks are then channel coded independently to give the individual block property. In order to minimize image distortion, an efficient bit allocation process and optimization of a quantization parameter is used. Using this independent sub-block processing, the influence of errors can be kept as short as the block, because the decomposed sub-blocks are encoded independently for error sensitivity. The next erroneous block can be decoded perfectly, thus minimizing error spreading.

We adapt a 7/9 -taped antonini filter to achieve wavelet transform, and we generate 10 types of sub-band images (LL3, LH3, HH3, LH2, HH2, HL1, LH1, HH1). The images of the LL3, HL3, LH3 and HH3 bands into 8x8 block sizes LL3, HL3, LH3 and HH3 bands are decomposed into 16x16 block sizes, and the HL1, LH1 and HH1 bands into 32x32 block sizes. On the other hand, the decomposed image is retained as the same number of sub-blocks and an EOB (End of Block) is inserted at the end of each block.

To optimize block bit allocation, the Largrangian method is used for the Rate-Distortion Curve. The whole target bit rate is also defined independently, and is then re-scaled by adding the target bit rate of the block. In this process, the optimum quantization parameter of each block is quantized uniformly and the adaptive arithmetic coding method is used at the end of the source coding process, as shown in Fig. 3.

3 Channel Coding

3.1 A Channel Coding for Unequal Error Protection

In a typical W-CDMA fading channel, bit errors occur in bursts. The bit error rate in an erroneous data frame is much higher than the time-averaged channel BER. We analyzed the source coder to provide information concerning the significance of the source bits, which can then be utilized to adjust the level of error protection applied by the channel coder. It is assumed that the total source coded data rate is limited to 57.6Kbps of asynchronous data transformation. The highest layer of each decomposed frequency band has the capability of a 3 byte error correction using the additional RS coder rate. In addition, the other layer has the capability of 2 byte, 2byte, 1byte, symbol error corrections as shown in Fig. 4.

In Fig. 4, the RS coder [8] uses an 8 bit-symbol so each layer can use (16, 10), (32, 28), (48, 44), (64, 62) code by the signification of the layer. The wavelet

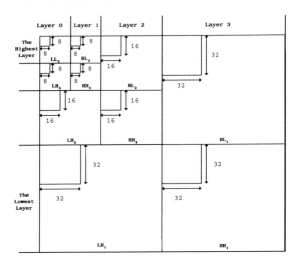

Fig. 3. The wavelet layer decomposition scheme divided into sub-blocks using the band property

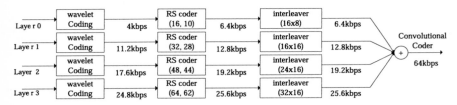

Fig. 4. Channel coding scheme using the UEP for each layer

decomposed band images are partitioned into different sub-blocks, and each sub-block in each layer does not affect the others, thus permitting more uncorrupted information to reach the highly compressed wavelet source which has intra and inter-band images. In the wavelet pyramid decomposition stage, the source image efficiently concentrates most of the subjectively important information in the lowest frequency band. As a result, the wavelet and RS codes, in combination with convolutional codes and proper interleaving techniques, can be more efficiently performed. This arises because the decoding of these combined codes is more efficient due to the simultaneous correction of error propagation, location and value.

3.2   Retransmission Scheme

Initially, an ARQ [9] buffer in a transmitter contains the data frames that have been transmitted but not yet acknowledged as well as data-frames that have not yet been transmitted. The sub-block of the error frame in the receiver buffer contains the non-error data sub-block in a frame which has been transmitted as an error frame. The process of selecting the error-free parts from error packet data is accomplished using the combined coder. The performance of this scheme was compared with that of an existing conventional hybrid Type I and was found to provide improvement in throughput of the error frame. The advantage of the proposed ARQ is that the error frame is not discarded and is used for improving the image quality of each sub-block.

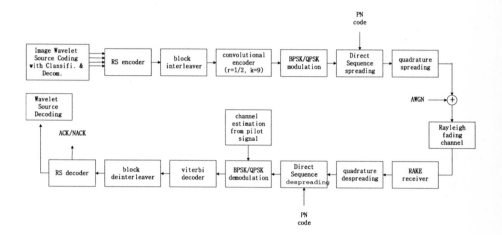

Fig. 5. Simplified diagram of the mobile communication system model for image transmission

4   Discussion of the Simulation Results

4.1   Requirements Related to Basic Wireless Internet Roaming Service

A traffic channel of a power-controlled mobile wideband CDMA simulation environment was used as a model to examine the performance of the source-channel combined coding method. The control channel is also affected by the mobile multi-path environment. Simulations were performed for the source combined coder as shown in Fig. 5, and the following processing procedures and parameters were used. As indicated above, the image is wavelet-transformed and the band images are decomposed. To improve the transmission efficiency, an RS coder is used as a UEP channel coder. After RS coding, the total transmission rate is 64Kbps and interleaved. The convolution coding rate is set at r=1/2,k=9 and is inner interleaved as well. In order to spread the CDMA a direct sequence spread spectrum was performed. In the case of the PN code, spreading codes of 4.096 Mcps were used. The communication channel of the Rayleigh fading channel is assumed to use a Doppler frequency of 5 Hz. In order to accommodate a data transmission link, the transmission and receiving links are independent and the receiver structure assumes that 2 antennas, with the same path, are used for diversity.

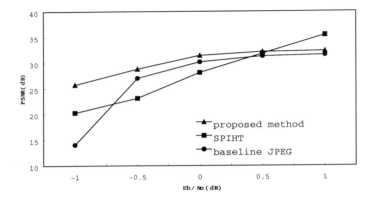

Fig. 6. Reconstructed LENA image of a near average PSNR at the same Eb/No

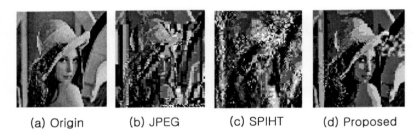

(a) Origin   (b) JPEG   (c) SPIHT   (d) Proposed

Fig. 7. Average PSNR Comparison of the reconstructed LENA images

Each simulation run was performed using channel estimation and power control and the parameter assumes optional conditions. In order to simulate an asynchronous CDMA system, the received signals are despreaded and QPSK demodulated with channel estimating from a pilot signal. The demodulated signal is deinterleaved and viterbi decoded using a soft viterbi decoding scheme. The decoded signal is wavelet-decomposed after RS decoding and the received signal is then reconstructed as an image. The image used for the test was a 256 gray level image and a 256x256 LENA image. The PSNR (Peak Signal to Noise Ratio), an objective method for evaluating the quality of an image, was used to examine the performance of the source coder. Between the original and restored image of x(m,n), y(m,n), we define the MSE (Mean Square Error) using equation (3); PSNR is represented by equation (4).

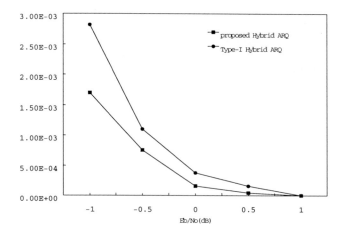

Fig. 8. BER comparision of the Type-I Hybrid ARQ and proposed Type-I Hybrid ARQ

$$MSE = 1/M\Sigma\Sigma[x(m,n) - y(m,n)] \qquad (3)$$

$$PSNR = 10\log(255/MSE) \qquad (4)$$

We assume a difference in the basic image reconstruction performance in a noise-free channel. The PSNR and compression rates of the LENA image, as reconstructed by the noise-free channel, are shown in Table 2 [10].

Table 2. PSNR comparison of reconstructed LENA image in noiseless channel

Measure \ Method	JPEG(baseline)	SPIHT	Proposed
PSNR(dB)	31.75	35.4	32.83
Comp. rate(bit/pixel)	0.84	0.84	0.84

For image coding in a wireless environment, Fig. 6 shows the SPIHT algorithm, which has worse quality than the other method when the channel error is due to walking mobility. In addition, the JPEG algorithm has decoding problems in every server erroneous channel(-1 Eb/No value below).

Fig. 7 shows reconstructed LENA images in the near average PSNR when transmitted from the original image at the same Eb/No(-1dB)

Therefore, the proposed source-channel combination encoding method more effective for image transmission in mobile communication systems. In addition, the proposed method using the subband algorithm is a more acceptable method for human visual systems [11] and the blocking effect of the existing algorithm based on the block is not seen.

5 Conclusion

In this paper, we describe an efficient image transmission system using a designed source-channel combined coder in a W-CDMA mobile communication system. In the proposed schemes, the wavelet transformed hierarchical band-images are decomposed into different size blocks that have different properties in terms of error sensitivity. An RS (Reed-Solomon) coder with a different coding rate is used for each decomposed source block which has a different importance. In addition, the retransmitted error frames in the Truncated Hybrid Type I ARQ are combined. The proposed algorithm shows efficient image transmission methods in an erroneous wireless channel because it is not significantly degraded in the PSNR compared with the existing source-channel coder, which is not combined.

References

1. Charles D. Creusere.: A New Method of Robust Image Compression Based on the Embedded Zerotree Wavelet Algorithm. IEEE Trans. on Image Processing vol. **6**, no. **10**, Oct. (1997) 1436–1442
2. A. J. Viterbi.: Convolutional Codes and Their Performance in Communication systems. IEEE Trans. on Comm. vol. COM-**19**, (1981) 751–772
3. Man H., Kossentini F., and Smith MJT.: A family of efficient and channel error resilient wavelet/ subband image coders. IEEE Trans. on Circuits and Systems for Video Technology, vol. **9**, no. **1**, Feb. (1999) 95–108.
4. Liyanage C. De Silva and Hari K. Garg.: Performance Comparison of Source-Channel Codecs for Image Transmission in Noisy Mobile Channels using CDMA, IEEE APCC/ICCS. Nov. (1998) 79–82
5. Po-Rong.: Spread Spectrum CDMA Systems for Subband Image Transmission. IEEE Transactions on Vehicular Technology, vol. **46**, no. **1**, Feb. (1997) 80–95
6. J. E. Kleider, G. P. Abousleman.: Adaptive-rate image compression for wireless digital data transmission systems. IEEE International Conference on Acoustics, Speech, and Signal Processing, vol. **5**, Number **4**, May (1998) 2629–2632
7. A. Said and W. A. Pearlman.: A new, fast, and efficient image codec based on set partitioning in hierarchical trees. IEEE Trans. Circuits Syst. Video Technol, vol. **6**, Number **4**, June (1996) 243–250
8. S. B. Wicker, V. K. Bhargava.: Reed-Solomon Codes and Their Applications. IEEE Press, (1994)
9. S. Lin, D. J. Costello, Jr.: Automatic Repeat Request Error Control Schemes. IEEE Comm. Magazine. vol. **22**, no. **12**, Dec. (1984) 7–16
10. G. K. Wallac.: The JPEG Still Picture Compression Standard. Comm. ACM. vol. **34**, Number **4**, Apr. (1991) 30–44
11. Oh-Jin Kwon, Seok-rim Choi.: Motion Compensated Subband Video Coding with Arbitrarily Shaped Region Adaptivity. ETRI journal, vol. **23**, Number **4**, Dec. (2001) 190–198

Development of "High Presence" Video Communication System – Trial Experiment of the Next Generation Real-Time Remote Lecture –

Takahiro Komine[1], Akihiko Machizawa[1], Shin-ichi Nakagawa[1], Fumito Kubota[1], and Yasuo Tan[2]

[1] Communications Research Laboratory, Information and Network Systems Division, 4-2-1, Nukui-Kitamachi, Koganei-shi, Tokyo 184-8795, Japan
{komine, machi, snakagaw, Kubota}@crl.go.jp
[2] Japan Advanced Institute of Science and Technology, School of Information Science, 1-1, Asahidai, Tatsunokuchi, Ishikawa 923-1292, Japan
ytan@jaist.ac.jp

Abstract. We have been developing the Next Generation remote lecture system that real-time transmits "high presence" audio/visual images (that are, as real as it feels no distance or screen is ever existing), without having any disturbance in a long distance transmission. We figured out what functions are required to perform the real-time remote lectures with multi-interaction between an instructor and students. We also reconfirmed the technical difficulties by executing the remote debate trial experiments and the distributed cardiac surgery workshop trial experiments and evaluating them. This paper presents introduces the Next Generation remote lecture system, which is developed based on the results and analysis of our past experiments, and reports the results of our first trial for the real-time remote lecture. It also introduces our ongoing plan to further develop this system.

1 Introduction

The history of remote conference including the remote lecture began with tele-conferences, which were later developed into TV conferences through ISDN. The transmission video quality of those systems was never sufficient in many ways, and the realization of "high presence" communications was once considered almost impossible.

With the spread of the Internet, the concept of remote conference over the Internet came into public mind. In fact, this turned out as the ongoing development of Internet videoconference systems, such as VIC and VAT, which gave birth to several commercial TV conference systems, such as Polycom ViewStation. On the other hand, a system that provides each student with different contents has been proposed in order to take care of the ability gaps among individuals. Commercial servers for remote communications have already become ready and available.

The problem is, although conventional systems are all applicable for an individual use, that they cannot manage group discussions. A remote lecture with a large group

I. Chong (Ed.): ICOIN 2002, LNCS 2344, pp. 521–528, 2002.
© Springer-Verlag Berlin Heidelberg 2002

requires higher quality of video images and less audio delay than what existing systems can offer.

There are already several remote lecture examples in Japan such as the distance-learning project called School of Internet (SOI) [1] presented by Widely Integrated Distributed Environments (WIDE). With its on-demand style lectures, the system provides the whole public with equal opportunity to learn. However, as a one-way communication system, it does not let instructors and students interact with each other; hence, it is impossible for instructors to change the contents of his lecture program according students' action/reaction during a lecture.

In order to achieve real-time video communications with multi-interaction, we have developed several communication systems and evaluated them in our trial experiments. The actual functions required in remote lectures with multi-interaction are introduced in the chapter 2. Chapter 3 describes our actual experiments implemented so far, while the chapter 4 mentions the Next Generation remote lecture system, which is based on the results and analysis of our past experiments, and reports the results on our first trial experiment. Finally, the chapter 5 concludes this paper by informing our future improvement plans and evaluation methods.

2   Actual Functions Required in Remote Lectures

The realization of "high presence" audio/visual images is very important for successful real-time video communication with multi-interaction between an instructors and students. In the case of real-time remote lectures, it is important to enable an instructor and students to interact with have each other as well as to change the lecture contents according students' action/reaction during the lecture.

The following three actual functions are required in the real-time remote lectures with multi-interaction. One is to transmit high quality audio and video with shortens time-delay, while another is to carefully arrange the composition of video screens and cameras for people to make mutual eye contacts. The other function is for an instructor to be able to listen various low sounds in the lecture room to follow students' little actions/reactions through these sounds.

3   Past Trial Experiments

3.1   Remote Debate Trial Experiments

We have challenged the Remote Debate trial experiments connecting far apart three sites, including two debaters' sites and a judge's site [2][3]. The debate requires a "high presence" communication system that enables the debaters to appeal their presentations to the judge and to change their tactics in response to subtle expression changes of the opponent.

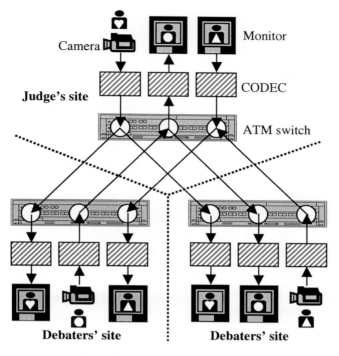

Fig. 1. System configuration on remote debate

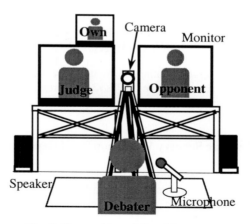

Fig. 2. Composition of first debate stage

Figure 1 shows the system configuration on the first trial experiment. We connected each sites through ATM networks and use the NTSC CODEC as the real-time communication system, which use the MPEG1 audio format and the MPEG2 video format and spent about 0.3 seconds for the encoding and decoding.

In this experiment, experienced debaters had a debate trial with the composition of debate stage shown in Figure 2. The video camera was placed closer to the center of three TV monitors displaying a judge, opponent, and own picture. However it was rather difficult for the debaters to look at both the camera and all the monitors at the same time, and this caused them frustration.

Fig. 3. Composition of second debate stage

In the second trial experiment, we adopted the multiplexed video monitor that displays the images at both the opponent's site and a judge's site while checking the remaining time (shown in Figure 3). We verified that this method functioned effectively and that it was important to arrange the video screens and cameras into the composition where the participants can make mutual eye contacts for realizing "high presence" communication.

We also found out important technical difficulties caused by the lack of the video quality and transmission time-delay in MPEG2 compression as well as audio echo back, by filling out a questionnaire about these experiments and evaluating it.

3.2   Distributed Cardiac Surgery Workshop Trial Experiments

Communications Research Laboratory (CRL) has produced the Distributed Cardiac Surgery Workshop trial experiments connecting CRL in Japan and Communications Research Centre (CRC) in Canada, where the cardiac surgeons discussed on recent technical issues for cardiac operations [4]. They are in need of "high presence" video images in which they can detect even the fine thread for the operation.

Figure 4 shows the system configuration. The experimental network was set up between CRL and CRC via an ATM commercial network in Japan, an INTELSAT circuit of 45 Mbps, and the CA*netII in Canada and spent about 0.3 seconds for transmission delay. We used the HDTV CODEC as the real-time communication system, which use the MPEG1 audio format and the MPEG1 video format and spent about 0.7 seconds for the encoding and decoding.

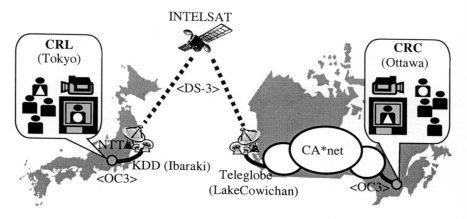

Fig. 4. System configuration on distributed cardiac surgery workshop

The focus was on the real-time discussion using "high presence" video image transmission. We asked the surgeons to compare the video quality by HDTV CODEC with that by NTSC CODEC. Then we all discussed the availability of this workshop using HDTV CODEC as well as recent technical issues for cardiac operations. As a result, we verified that the real-time communication system using this HDTV CODEC had enough performance to display the "high presence" video images.

On the other hand, we found out important technical difficulties that audio echo back and long transmission time-delay (about 1 seconds) caused frustrating conversation, by filling out a questionnaire about these experiments and evaluating it. Because the video quality was mach better than what we had expected, we could not help being disappointed with the audio deficiency relatively.

4   Next Generation Remote Lecture Trial

4.1   Summary

Japan Advanced Institute of Science and Technology (JAIST) and CRL have been conducting joint research on remote lecture using a "high presence" real-time video communication system called "the Next Generation Remote Lecture system".

A trail experiment using the Next Generation remote lecture system was performed as a JAIST intensive lecture program during two weeks on September 2000 and some students took this lecture for two credits. In these experiments, some instructors are located at CRL and gave a remote lecture to JAIST graduate students in a JAIST lecture room. Several methods were adopted to improve technical difficulties, such as effective screen arrangements on dual screens and reducing audio echo back. We examined the efficiency of this method by collecting comments and impressions especially on the aspects of "high presence" from all instructors and several graduate students. And furthermore, the regular remote lecture program between JAIST and CRL began in October 2001.

4.2   First Trial Experiment

Figure 5 shows the system configuration of the first trial experiment. CRL and JAIST are connected via Japan Gigabit Network (JGN), which is a high-speed nationwide network testbed for R&D [5][6]. The nucleus of this system is ATM-IEEE1394 Link Unit (Link Unit) [7], which carries out real-time mutual signal conversion between ATM cells and DV packet signals on IEEE1394.

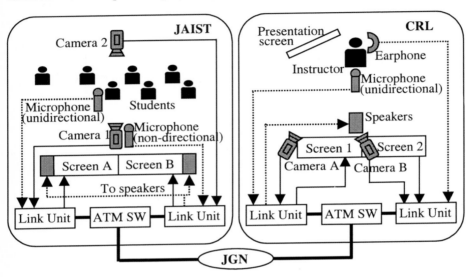

Fig. 5. System configuration on first remote lecture

Table 1 shows the frame of Link Unit. Although no compression format equipped, this system excels in maintaining the quality of video images and shortening time-delay without any compression process, such as MPEG2.

We placed one unidirectional microphone in front of an instructor and one speaker just behind the microphone so that the instructors could interact with JAIST graduate students despite being at CRL. While watching the high quality video images, he could also hear various low sounds in the JAIST lecture room with his earphone to follow the students' little reactions.

4.3   Results

We verified that the method, which uses the Link Unit and displays dual video images of other site, have a practical use for bilateral "high presence" video communication in a remote lecture. One incomplete thing was the use of the earphone: it created huge echo back that made the instructor uncomfortable, while it gave certain reality in the mock-up situation.

Table 1. Frame of Link Unit

IEEE1394 (DV) part	

Chroma format
Vertical size
Frame rate
Image compression ratio
Stream rate About
	4:1:1
	525 lines
	30 frames/s interlace
	1/5
	28 Mbps
	   (Image 25.9 Mbps)
	   (Voice 1.7 Mbps)

ATM part	
ATM interface Transport rate	AAL5 About 35 Mbps
Exchanging part	
ATM buffer size DV buffer size	512 AAL-PDU 1024 packets

5   Conclusion

We figured out the actual functions required in the real-time remote lectures with multi-interaction and reconfirmed the technical difficulties through the remote debate trial experiments and the distributed cardiac surgery workshop trial experiments and evaluating them.

This paper presented the Next Generation remote lecture system, which was developed based on the results and analysis of our past experiments. Also it reported the results of our first trial experiment for the real-time remote lecture.

We could not solve the technical difficulty with echo back through all the experiments; although adopting a commercial echo-canceling device at each site was expected to reduce echo back, we could not observe any improvement in our past experiments.

For the future, we intend to analyze the reason why the echo-canceling devices keep failing to reduce echo back so that we may find the practical solution. When successful with this, we will perform another regular remote lecture program between JAIST and CRL and evaluate its availability.

Furthermore, we intend to develop the Next Generation remote lecture system to make it available on the Next Generation Internet and ATM networks, by using the Digital Video Transport System (DVTS) [8], which carries out real-time mutual signal conversion between IP packets and DV packet signals on IEEE1394.

Acknowledgement. The authors would like to thank their staffs at JAIST and CRL for their helpful comments and support. The authors also appreciate the tremendous effort of all participants in these experiments.

This research is the joint research with TAO using JGN, JGN-G11027.

References

1. http://www.soi.wide.ad.jp/aboutsoi/aboutsoi_e.html
2. M. Itou, H. Higaki, T. Komine, K. Nagaoka, "Comfortable Environment for Remote-Debaters – Display Effect on Remote-debaters –", JSiSE, Transaction of Japanese Society for Information and Systems in Education Vol. 14 No.3, August 1997
3. G. Hamada, K. Ebina, S. Matsumoto, T. Komine, R. Suzuki, F. Kubota, "Multimedia Distance Education Experiments in Japan – Tree site Debate Trial by MPEG2 over ATM system –", 20th Annual Pacific Telecommunications Conference (PTC '98), 1998
4. T. Komine, G. Hamada, R. Suzuki, E. Tsang, F. Kubota, "GIBN Multimedia Network Experiments – ATM Satellite Communication Experiments between Japan and Canada –", 20th Annual Pacific Telecommunications Conference (PTC '98), 1998
5. http://www.jgn.tao.go.jp/english/index_E.html
6. T. Komine, F. Kubota, S. Nakagawa, A. Amemiya, "Development of Japan Gigabit Network", Joint IEEE ATM Workshop 2000 and 3rd International Conference on ATM (ICATM '2000)
7. Y. Tan, T. Nomura, H. Tamori and K. Koshiba, "Plug and Play Campus Digital Video Network with IEEE1394 and ATM", International Conference on Computer Communication 99, 1999
8. http://www.sfc.wide.ad.jp/DVTS/

SRTP: TCP-Friendly Congestion Control for Multimedia Streaming

Byunghun Song[1], Kwangsue Chung[1], and Yongtae Shin[2]

[1] School of Electronics Engineering, Kwangwoon University, Korea
byungh@adams.gwu.ac.kr, kchung@daisy.gwu.ac.kr
[2] School of Computing College of Information Science, Soongsil University, Korea
shin@computing.soongsil.ac.kr

Abstract. Multimedia streaming applications are becoming increasingly popular on the Internet. Most of these applications do not share the available bandwidth fairly with applications built on TCP, such as web browsers, FTP- or email-clients. The Internet community strongly fears UDP-based multimedia applications, because the UDP traffic could lead to a congestion collapse and starvation of TCP traffic. For this reason, TCP-friendly protocols are being developed that behave fairly with respect to co-existent TCP flows. In this paper, we propose a new TCP-friendly protocol, called *Smart RTP(SRTP)*, for multimedia streaming. Our proposed protocol has two salient features. One is TCP-friendliness and the other is rate smoothness. As a result, the end-to-end quality of service (QoS) is improved. The evaluation results demonstrate the effectiveness of our proposed schemes.

1 Introduction

There has been increasing interest in multimedia streaming over the Internet recently. However, efficient delivery of streaming media over the Internet presents many challenges. Especially, the current Internet only provides best-effort service and it does not provide the Quality of Service (QoS) guarantee or provision for multimedia services. While data applications such as Web and FTP are based on TCP, multimedia applications will be based on UDP due to its characteristics of real-time. However, UDP does not support congestion control. For this reason, wide usage of multimedia application in the Internet might lead to overload situation. Furthermore, the UDP causes the starvation of congestion controlled TCP traffic which reduces its bandwidth share during overload situation. To avoid such a situation, UDP-based application must be enhanced with congestion control algorithms. This algorithms collect, exchange and process information about the network congestion status and adjust the behavior of the end systems based on this information [1].

In this paper, we consider the aspects of friendliness of UDP traffic towards competing TCP connections. Therefore, we propose novel TCP-friendly congestion control schemes for UDP-based multimedia communications. The efficiencies

I. Chong (Ed.): ICOIN 2002, LNCS 2344, pp. 529–538, 2002.
© Springer-Verlag Berlin Heidelberg 2002

of the proposed schemes are proven through implementations as well as through simulations comparisons to similar approaches.

For the collection and exchange of information about the network load, losses and delays, the proposed schemes use the Real-time Transport Protocol (RTP). Recently, the RTP widely used for multimedia communication in the Internet, because it offers the necessary mechanisms for collecting and exchanging information about losses and delay [2]. But, it only supports the exchange of information in intervals of a few seconds. Therefore, our proposed schemes are designed to achieve an optimal streaming transmission rate adaptation behavior.

The rest of this paper is organized as follows: We first present the background and related work in Section 2. In Section 3, we are described the design options for rate adaptation schemes. In Section 4, a TCP-friendly adaptation scheme for multimedia streaming is proposed. Section 5 gives implementation and evaluations. Our conclusions and future work are presented in Section 6.

2   Background and Related Work

Recently, there has been several proposals for TCP-friendly adaptation schemes [3,4,5]. In this paper, TCP-friendliness is the terminology used for non-TCP flow. Non-TCP flows are defined as TCP-friendly when "their long-term throughput does not exceed the throughput of a conformant TCP connection under the same conditions".

We use the scheme shown in Fig. 1 to classify the different approaches. This classification distinguishes between single-rate and multi-rate congestion control at the top-level and rate-based versus window-based congestion control at the second level [6].

2.1   Rate-Based Approaches

Many rate-based congestion control protocols mimic TCP's AIMD behavior to achieve TCP-fairness. This approach is used to adjust the rate according to a analytical model of TCP traffic. Early work in this area was presented in Jacobs's research [7].

2.2   Window-Based Approaches

The domain of window-based congestion control is well covered by TCP. Golestani and Sabnani propose to use a window-based approach where each receiver keeps a separate congestion window adjusted similar to the congestion window of TCP [8]. From the size of the window and the number of outstanding packets, each receiver calculates the highest sequence number.

2.3   Modeling Bandwidth Share of TCP

Padhye et al. present an analytical model for the available bandwidth share (T) of TCP connection with s as the segment size [9], p as the packet loss rate, t_{RTT} as

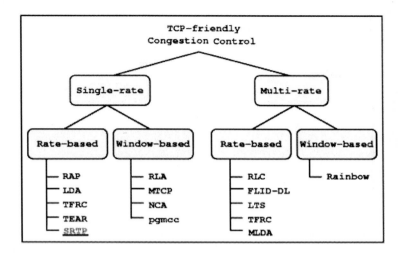

Fig. 1. Classification Scheme for TCP-friendly Protocols.

the round trip time,t_{RTO} as the retransmission timeout. The average bandwidth share of TCP depends mainly on $t_{RTT}$and p.

$$T = \frac{s}{t_{RTT}\sqrt{\frac{2bp}{3}} + t_{RTO}\ \min\left(3\ \sqrt{\frac{3bp}{8}}\right)\ p(1 + 32p^2)} \tag{1}$$

3   Design Options for Rate Adaptation Schemes

The TCP-friendly congestion control (the increase and decrease action) indicates that competing TCP and UDP flows with similar loss and delay values. Therefore, the increase and decrease action of multimedia streaming should be designed in a manner as to follow for smooth and stable reactions to the network congestion state. Additionally, as congestion control schemes needs to operate over wide variety of networks with different loss and delay. Therefore, the increase and decrease actions should be dynamically adapted to the environment they are used in.

To achieve a stable behavior the resource share some flow is utilizing needs to converge to a steady state. However, with no information about the explicit share to use, adaptation schemes do not converge to a single steady state. Instead, the system reaches an equilibrium in which it oscillates around the optimal value. The time taken to reach this equilibrium (responsiveness) and the size of the oscillations (smoothness) jointly determine the convergence [10]. Ideally, the response time as well as the oscillation should be small as shown in Fig. 2.

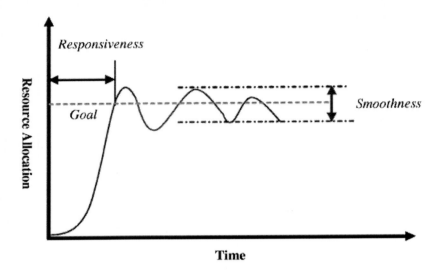

Fig. 2. Responsiveness and Smoothness

4   TCP-Friendly Adaptation Schemes for Multimedia Streaming

Most of the TCP-friendly adaptation schemes present that the sender adapts its transmission behavior based on frequent feedback messages, such as TCP. This is particularly important for the case of reliable transport where the sender needs to retransmit lost packets.

On the contrary, our proposed scheme, Smart RTP (SRTP), was designed to use the RTP for exchanging feedback information about the round trip time and the losses at the receiver. RTP is currently being proposed as an application-level protocol for multimedia services over the Internet. RTP defines a data and a control part. For the data part, RTP specifies an additional header to identify the sender and type of data. With the control part called Real-time Transport Control Protocol (RTCP), each member of a communication session periodically sends control reports to all other members. This report contains information about sent and received data state. However, the report interval between sending two RTCP messages is usually around five seconds. The infrequency of the RTCP feedback messages dictates that an RTP sender can not benefit fast enough from rapid changes in the network conditions. Thus, the goal of RTCP-based adaptation is to adjust the sender's transmission rate to the average available bandwidth. This might be actually more appropriate for multimedia streaming than other schemes where the transmission rate is rapidly changed on the basis of very frequent feedback messages.

The process of optimal control of appropriate transmission rate using SRTP consists of three stages: (1) estimating packet loss ratio and round trip time, (2) estimating available bandwidth share, (3) adjusting transmission rate.

4.1 Packet Loss Ratio and RTT Estimation

RTCP messages include information about the losses and delays noticed in the network. Losses (p) are estimated at the receiver by counting the gaps in the sequence numbers included in the RTP header of the data packets. N_{real} is number of actually received packet, N_{max} is maximum number of received packet, N_{first} is number of first received packet [11].

$$p = \frac{N_{real}}{(N_{\max} - N_{first})} \tag{2}$$

The round trip delay (t_{RTT}) is estimated by including a timestamp in the sender reports indicating the time the report was generated. In its reports, the receiver includes the timestamp of the last received sender report (T_j) and the time elapsed in between receiving that report and sending the receiver report (T_i). Knowing the arrival time (t) of the RTCP receiver report the sender calculates the round trip time (t_{RTT}) as follows:

$$t_{RTT} = (\alpha \cdot \overline{t_{RTT}}) + (1 - \alpha)(t - T_i - T_j) \tag{3}$$

$\overline{t_{RTT}}$ is the current round trip time, α is a weighting parameter that is set to 0.80 in our work. This weighting parameter is set to 0.875 in the traditional TCP congestion control algorithm. Considering the real-time requirement, we choose a smaller weighting value so that the recent RTT value has a higher impact on the RTT estimation.

4.2 Available Bandwidth Share Estimation

After the procedure mentioned, the sender estimates available bandwidth share. Therefore, the sender limits its transmission rate to the available bandwidth share calculated using Eqn. 1.

4.3 Transmission Rate Adjustment

Having got the available bandwidth share, sender can adjust its transmission rate based on the estimated value. The SRTP is a kind of AIMD algorithm with the addition and reduction values determined dynamically on the basis of the current network state [10].

We define two thresholds to determine the network state as *congested, loaded* or *uncongested* according to the classification in Fig. 3. There are a strong resemblance between this scheme and the AIMD approach. With losses above a certain threshold called the upper loss threshold (λ_c) the sender can decrease its transmission rate multiplicatively. On the other hand, with losses below a second threshold called the lower loss threshold (λ_u) the sender can increase its transmission rate additively.

For the case of *congested state* the transmission rate (T_{SRTP}) is determined as Fig. 4, where $\overline{T_{SRTP}}$ is current transmission rate, β is the weighting parameter for

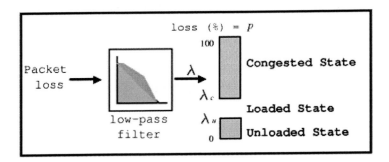

Fig. 3. Network State Classification

rate smoothing. During congestion situations, SRTP estimates the transmission rate to be minimally the available bandwidth share determined with Eqn. 1. Form Eqn. 1, it is evident that the overall TCP bandwidth share (or throughput) is inversely proportional to the square root of the loss values. Hence, we propose to reduce the rate of the SRTP flow in a similar manner to TCP approach. Thus, after receiving a loss notification from the receiver, the sender can reduce its transmission rate follows:

$$T_{SRTP} = \overline{T_{SRTP}} \cdot (1 - \sqrt{p}) \qquad (4)$$

For the case of *unloaded state* the transmission rate (T_{SRTP}) can be increased by additive increase value. This value does not exceed the increase of the available bandwidth share of a TCP connection with the same RTT and packet size. Thus, the SRTP sender should make the transmission rate to be smaller than that of competing TCP connection, in order to ensure the bandwidth fairness. I_{TCP} is additive increase value of TCP determined with Eqn. 1. I_{SRTP} is additive increase value of SRTP.

5   Implementation and Evaluation

We implement our proposed schemes. The purpose of this Section is to demonstrate that: (1) bandwidth share of SRTP; (2) comparison of SRTP with a rate-based TCP-friendly scheme. A prototype MPEG-4 streaming system, which we have developed in Linux environments, is utilized in our experiments.

5.1   Description on Experimental System

Fig. 5 shows the architecture of the system that we have developed for MPEG-4 delivery that serves as a testbed for our experiments. It consists of a video server and a video client.

The server listens for requests on an Real Time Streaming Protocol (RTSP) port and streams requested data to the video client via an implementation of

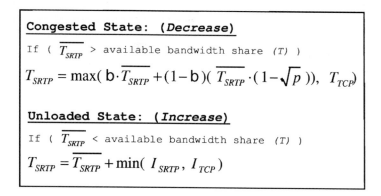

Fig. 4. Transmission Rate Adjustment Schemes

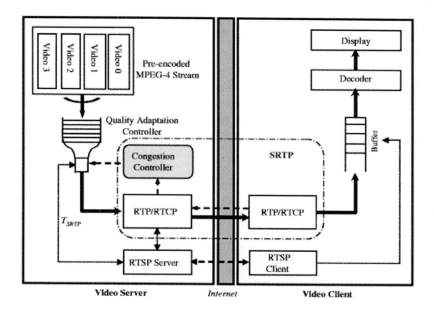

Fig. 5. A System Architecture

RTP (over UDP) [12]. Feedback is provided to the video server via RTCP receiver reports and is used to adjust the increase/decrease schemes. The decoder drains data from buffers and feeds the display.

We conducted experiments using the testbed shown in Fig. 6. The video server (running on a Pentium Linux 2.4 machine) streamed data to the receiver (also a Pentium Linux 2.2 machine) across a 1.5 Mbps link with 100 ms latency, configured using Dummynet [13]. Background Web cross-traffic was introduced using the SURGE toolkit [14] to emulate the varying network conditions. An actual streaming server might see in the face of competing Web traffic on the Internet. For testing the performance of SRTP we used a simple topology, consisting of bottlenecked link connecting l SRTP-based streaming connections, m TCP-based FTP connections and n TCP-based WWW connections.

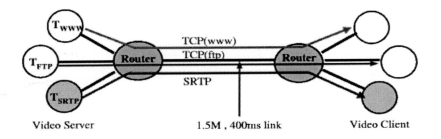

Fig. 6. Experimental Testbed

5.2   Bandwidth Sharing

Fig. 7 shows the bandwidth distribution for the case of Web traffic competing with SRTP flows ($l = 20$) on the one hand and with long lived FTP connections ($m = 20$) on the other. The SRTP flows receive a much higher bandwidth share than the Web traffic ($n = 20$).

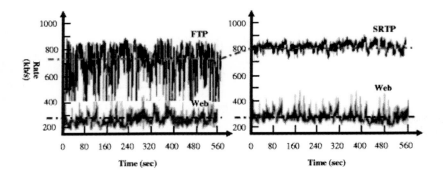

Fig. 7. Bandwidth Sharing ($l = m = n = 20$)

Also the share of the SRTP flows is similar to that of the TCP connections under the same conditions. The SRTP flows does not affect the performance of the Web traffics.

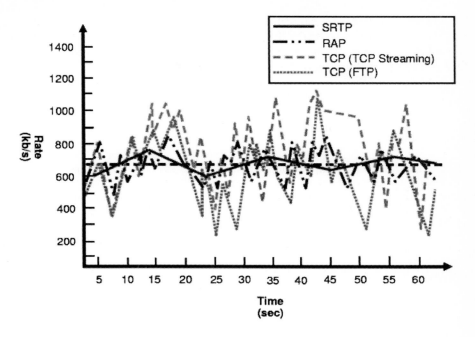

Fig. 8. Comparison of SRTP and RAP and TCP

5.3   Comparison of SRTP with a Rate-Based TCP-Friendly Scheme

We used the network simulator (NS) version 2 to compare the SRTP with a rate-based TCP-friendly adaptation scheme [15]. The network topology in the simulation consists of a single shared bottleneck link (dumbbell model), as shown in Fig. 5. All links are sufficiently provisioned to ensure that any drops and delays occurred are only caused by congestion at the bottleneck link. In the simulations, the background traffic consists of infinite-duration TCP-based connections (FTP), and infinite-duration real-time adaptive protocol (RAP) connections [5]. The RAP is an end-to-end rate-based congestion control mechanism for multimedia streaming. It deploys an AIMD algorithm for coarse grain rate adjustment to emulate window adjustment in TCP.

Transmission rates for different connections are illustrated in Fig. 8. We can see that our proposed SRTP provides smoother transmission rate than TCP and RAP.

6   Conclusion and Further Work

It is very important for multimedia streaming to be "TCP-friendly", because a dominant portion of today's Internet traffic is TCP-based. The multimedia streaming systems are expected to react to congestion by adapting their transmission rates and maintain the inter-protocol's fairness, in order to be efficiently transported over the Internet.

The advantage of our scheme is very smoothly rate adaptation according to the available bandwidth share. In other words, the SRTP has less variation in the transmission rate. In summary, our proposed SRTP has two salient features. One is TCP-friendliness and the other is rate smoothness.

In the future, we will further enhance the estimation function of available network bandwidth based on the media characteristics.

References

1. S. Floyd and F. Kevin, "Router mechanisms to support end-to-end congestion control," Technical Report, LBL-Berkeley, Feb. 1997.
2. I. Busse, B. Deffner, and H. Schulzrinne, "Dynamic QoS control of multimedia applications based on RTP," *IEEE Computer Communications*, Jan. 1996.
3. S. Floyd, M. Handley, J. Padhye, and J. Widmer, "Equation-based congestion control for unicast appli- cations," *In Proceedings of SIGCOMM 2000*, 2000.
4. D. Sisalem and H. Schulzrinne, "The Loss-Delay Based Adjustment Algorithm: A TCP-Friendly Adaptation Scheme," *In Proceedings of NOSSDAV 1998*, 1998.
5. R. Rejaie, M. Handley, and D. Estrin, "An end-to-end rate-based congestion control mechanism for realtime streams in the Internet," *In Proceedings of NFOCOMM 1999*, 1999.
6. J. Widmer, R. Denda, and Martin Mauve, "A subvey on TCP-Friendly congestion control," *IEEE Network*, May/June 2001
7. S. Jacobs and A. Eleftheriadis, "Providing video services over networks without quality of service quarantees," W3C Workshop, Oct. 1996.
8. S. Golestani and K. Sabnani, "Fundamental observations on multicast congestion control in the Internet," *In Proceedings of INFOCOMM 1999*, 1999.
9. J. Padhye, V. Firoiu, D. Towsley and J. Kurpose, "Modeling TCP throughput: A simple model and its empirical validation," *In Proceedings of SIGCOMM 1998*, 1998.
10. D. Chiu and R. Jain, "Analysis of the increase/decrea se algorithms for congestion avoidance in computer networks," *IEEE Computer Networks*,June 1989.
11. H. Schulzrinne, S. Casner, R. Frederick, and V. Jacobson, "RTP: A Transport Protocol for Real-Time Applications,". *IETF*, RFC 1889, Jan. 1996.
12. B. Song, K. Chung, H. Lee, and S. Oh, "A study on the interworking platform supporting adaptive stream QoS," *In Proceedings of PV 2001*, 2001.
13. Dummynet, http://www.iet.unipi.it/~luigi/.
14. Scalable URL Reference Generator(SURGE), http://www.cs.bu.edu/home.html.
15. The Network Simulator (NS), http://wwwmash.cs.Berkeley.edu/ns/ns.html.

A Method for Functional Testing of Media Synchronization Protocols

Makoto Yamada, Takanori Mori, Atsushi Fukada, Akio Nakata, and Teruo Higashino

Department of Informatics and Mathematical Science,
Graduate School of Engineering Science, Osaka University,
Toyonaka Osaka, 560-8531, Japan

Abstract. In this paper, we propose a functional testing method of media synchronization protocols, which control the synchronization between audio and movie, described in concurrent synchronous timed I/O automata. In order to trace all test sequences (I/O event sequences) with synchronization on the model, we need to execute each I/O event at an adequate timing which satisfies the whole timing constraint for all the given test sequences. However, the outputs are given from the IUT and uncontrollable. Also each output/synchronization timing may affect executable timing for its succeeding I/O events in the test sequences. In this paper, we propose a technique to derive a set of time intervals which make all the given test sequences executable, and propose a method for functional testing using the technique.

1 Introduction

Recently many multimedia systems, which make use of various information media such as audio, video, images and text, have been developed and utilized. Multimedia systems usually can be modeled as real-time systems and timing constraints on I/O events are imposed upon such systems in order to guarantee their QoS. Especially, multimedia synchronization protocols can be modeled as a concurrent model in which multiple real-time modules work cooperatively [4,5]. Timed automata or timed Petri nets have been taken into account for specifying such real-time systems[1] and used for specifying multimedia systems[3,6]. Functional testing[7], which tests whether a given IUT (Implementation Under Test) possesses the functions designated in its specification, is an effective way for improving the reliability of such multimedia systems.

In this paper, we propose a functional testing method for media synchronization protocols, which control the synchronization among audio and video media. Here, for simplicity of discussion, we assume that each media synchronization protocol consists of three modules, (1) a module receiving and playing out audio data, (2) a module receiving and playing out video data and (3) a module controlling the synchronization between the above two modules. The modules (1) and (2) receive audio and video data through the network. The control module (3) synchronizes audio and video every suitable intervals. As the testing items, we consider the following items, (a) whether the synchronization among audio and video media works properly and (b) whether the IUT can execute an alternative operation when the IUT cannot receive data because of the delay in the network.

I. Chong (Ed.): ICOIN 2002, LNCS 2344, pp. 539–550, 2002.
© Springer-Verlag Berlin Heidelberg 2002

We assume that each module of the media synchronization protocol is described as a timed I/O automaton[2]. In the timed I/O automaton model, a clock and registers are available. And each timing constraint on state transitions consists of a logical product of linear inequalities composed of the variables that keep the values of the clock and registers. We generate test sequences that consist of a series of I/O events for each timed I/O automaton. However, the generated sequences may not be always executable because executable timing constraints may be imposed on state transitions. Therefore we need to check the executability of the test sequences, and if they are executable, we also need to find suitable values of the variables (such as I/O event timing) that enable the sequence execute. Generally each input to the IUT can be given at the specified timing. On the other hand, the timing of each output from the IUT is uncontrollable. Accordingly, when we check the traceability of a test sequence, it is desirable not only finding an input timing set that makes the whole test sequence executable but also finding feasible input/output interval set.

In [2], we proposed a method for checking the traceability of a state transition sequence on a single timed I/O automaton automatically. If the sequence is traceable, we can derive values of such parameters as input timing automatically. In this paper, combining this method and an idea of [4] for scheduling event timing on concurrent EFSMs, we propose a method for (1) checking the traceability of concurrent sequences of state transitions on concurrent timed I/O automata, in which multiple timed I/O automata work cooperatively, and for (2) deriving a series of intervals of I/O event timing that enables the concurrent sequences.

The rest of this paper is organized as follows. In Section 2, we show our concurrent timed I/O automata model. In Section 3, the traceability of concurrent state transition sequences is defined. In Section 4, we propose a method for generating test sequences and deriving a set of executable intervals if the sequences are executable. And in Section 5, application results of our method are shown. Section 6 concludes this paper.

2   Concurrent Timed I/O Automata

2.1   Timed I/O Automata

Definition 1. Timed I/O automaton is defined as a 10-tuple $M = <S, A, I/Otype, t, V, Pred, Def, \delta, s_{init}, \{x_{1init}, x_{2init}, \ldots, x_{kinit}\}>$, *where*

- $S = \{s_0, s_1, \ldots, s_n\}$ is a finite set of states.
- A is a finite set of I/O events.
- $I/Otype = \{!, ?\} \cup \{?_\mathbf{v} | \mathbf{v} \in V\}$ is a finite set of I/O types, where the symbols ? and ! represent input and output, respectively. The symbol $?_\mathbf{v}$ represents that the input value is assigned to the variable $\mathbf{v}$, whose value may be used in the transition conditions of its succeeding events.
- t is a global clock variable which holds the current time.
- $V = \{x_1, x_2, \ldots, x_k\}$ is a set of variables.
- $Pred$ is a set of transition conditions, each of which is a logical conjunction $P[t, x_1, x_2, \ldots, x_k]$ of linear inequalities consisting of variables.

- Def is a set of assignments. An assignment is a function which maps variables $x_i \in V$ to a linear expression $f(t, \mathbf{v}, x_1, x_2, \ldots, x_k)$, denoted by $x_i \leftarrow f(t, \mathbf{v}, x_1, x_2, \ldots, x_k)$.
- δ is a transition function. $S \times A \times I/Otype \times V \rightarrow S \times V$.
- $s_{init} \in S$ is the initial state of M.
- $\{x_{1init}, x_{2init}, \ldots, x_{kinit}\}$ is a set of the initial values of variables $x_1, x_2, \ldots, x_k \in V$.

A transition on M is denoted by $s \xrightarrow{a\$[P]D} s'$, where $s, s' \in S$, $a \in A$, $\$ \in I/Otype$, $P \in Pred$ and $D \subseteq Def$. $\qquad\qquad\square$

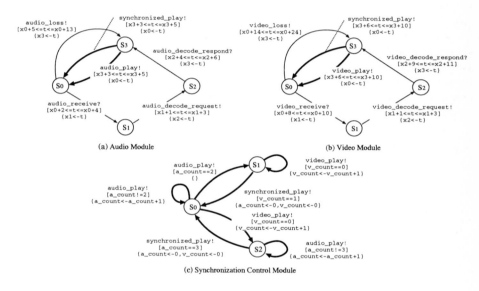

Fig. 1. Media Synchronization Protocol

For example, Fig.1 is a system which receives audio and video data from senders, synchronizes and plays out these media. The system is described as the concurrent timed I/O automata model whose formal semantics is described in Section 2.2. Modules (a) and (b) receive audio and video data, ask external decoders to decode them and play out them, respectively. They repeat these behavior. In this system, the fourth frame of audio and the second frame of video must be synchronized with module (c).

Intuitively, the semantics of our timed I/O automaton is as follows. At state s_0 in Fig.1(a), if the initial value of x_0 is zero, input event audio_receive? is executable between time 2 and 4 since the transition condition $x_0 + 2 \le t \le x_0 + 4$ becomes true. By executing this transition, module (a) moves to state s_1. In addition, the current value of clock t is assigned to variable x_1 according to the assignment $x_1 \leftarrow t$. If the execution time of the input event is 3.5, then $x_1 = 3.5$ and output event audio_decode_request!

becomes executable between time 4.5 and 6.5. If the output event is executed, module (a) moves to state s_2.

Module (a) receives an audio frame, asks a decoder to decode it and plays out it simultaneously with a video frame or independently. If module (a) cannot receive a frame (input event audio_receive? cannot be executed) at state s_0, module (a) assumes that the frame is lost, and it executes output event audio_loss! and moves to s_3.

In general, the behaviour of a timed I/O automaton is formally defined as a transition relation between *concrete states*. A concrete state is a pair of a control state and an value-assignment for all variable including the clock variable t. The transition relation between concrete states are defined as follows.

Definition 2. A *value assignment* σ is a mapping from the set of state variables $V \cup \{t\}$ into the set of real-numbers R. For any value assignment σ and any non-negative real-value d, let $\sigma + d$ denote the same value assignment as σ except that the value of the clock variable t is increased by d (i.e. $(\sigma + d)(t) = \sigma(t) + d$). For any value assigment σ and any set of assignment statements $D \subseteq Def$, let σD denote the same value assignment as σ except that the assigned value of variables x_i appeared in the left hand of each assignment statements is the corresponding right hand expression, i.e. if $(x_i \leftarrow e_i) \in D$, then $\sigma D(x_i) = e_i$. We write $\sigma \models P$ if the predicate P holds when the value-assignment σ is applied. □

Definition 3. The transition relation between concrete states of a timed I/O automaton M is defined as the minimum relation derived by the following rules:

- For each transition $s \xrightarrow{a\$[P]D} s'$,
  - for any value assignment σ and non-negative real-number d such that $(\sigma + d) \models P$, $(s, \sigma) \xrightarrow{d} (s, \sigma + d)$ holds (transition by time passage).
  - if $\$ \neq ?_\mathbf{v}$, i.e. the action $a\$$ is not an input action with a variable, for any σ such that $\sigma \models P$, $(s, \sigma) \xrightarrow{a\$0} (s, \sigma D)$ holds[1] (transition by I/O actions without input values)
  - if $\$ = ?_\mathbf{v}$, for any σ such that $\sigma \models P$, $(s, \sigma) \xrightarrow{a\$v} (s, \sigma(D \cup \{\mathbf{v} \leftarrow v\}))$ (transition by an input action with an input value). □

Note that our timed I/O automaton can simulate the classical Alur's timed automaton [1]. The details are shown in [2].

2.2   Concurrent Timed I/O Automata

In this paper, each media synchronization protocol is modeled as concurrent timed I/O automata which work cooperatively. Here, we assume that all the automata refer to the same clock variable t and that all the I/O events with the same name must be executed at the same time synchronously.

---

[1] For technical convenience, we attach a dummy value 0 for both input actions without input values and output actions.

For example, on the system shown in Fig.1, when synchronized_play! is executed on module (a), synchronized_play! is also executed on both modules (b) and (c). The transition condition for the synchronous events are the logical conjunction of the transition conditions corresponding to synchronized_play! on modules (a), (b) and (c). So, the synchronous event is executable if and only if the three transition conditions hold. Formally, it is defined as follows.

Definition 4. A *concurrent timed I/O automaton* $\mathcal{M}$ is a finite vector of timed I/O automata $(M_1, \ldots, M_k)$, whose behavior is defined by the minimum transition relation between concrete states derived by the following rules:

- for any $i \in \{1, \ldots k\}$, if $(s_i, \sigma_i) \xrightarrow{d} (s'_i, \sigma'_i)$, then $((s_1, \ldots, s_k), (\sigma_1, \ldots, \sigma_k)) \xrightarrow{d}$ $((s'_1, \ldots, s'_k), (\sigma'_1, \ldots, \sigma'_k))$ holds (synchronized time passages).
- for any I/O action $a\$$, let I be the set of indices of timed I/O automata such that the transition $(s_i, \sigma_i) \xrightarrow{a\$v} (s'_i, \sigma'_i)$ is executable. Then $((s_1, \ldots, s_k), (\sigma_1, \ldots, \sigma_k)) \xrightarrow{a\$v}$ $((s'_1, \ldots, s'_k), (\sigma'_1, \ldots, \sigma'_k))$, where $s'_j = s_j$ and $\sigma'_j = \sigma_j$ for $j \notin I$ (synchronized execution of the same I/O actions). $\qquad\square$

3   Executability of Sequences

In testing of real-time protocols, tester can usually control input timing to IUT. On the other hand, tester cannot control output timing from IUT. Each output timing may affect executable timing for its succeeding I/O events in sequences. Hence, it is desirable that we can find I/O event sequenses which can be executed whenever the IUT produces outputs. In this section, we define executability of sequences on our concurrent I/O timed automata.

3.1   Symbolic Trace

Since values of variables may be updated by executing transitions in a timed I/O automaton, we have to consider how to change the values of those variables for deciding executability of sequences. We translate values of variables and transition constraints in given sequences into expressions(symbolic traces) consisting of initial values of variables, execution time of transition or input values[2].

Formally symbolic traces are defined as follows.

Definition 5. For a path $\alpha = s_1 a_1 \$_1 s_2 \ldots s_k a_k \$_k s_{k+1}$ on a timed I/O automaton M (each $s_i, a_i, \$_i (i \in \{1, \ldots, k+1\})$ can be repeated), sequence $w = (a_1 \$_1, t_1, P_1) \cdots$ $(a_k \$_k, t_k, P_k)$ is said to be a symbolic trace of α, if the following conditions hold.

- $t_1, \ldots, t_k$ are n different variables.
- Assume that Iin denotes the set of subscripts $\{i \mid \$_i =?_{v_i}\}$. $v_i (i \in Iin)$ are n different variables.
- Each predicate P_i contains only the initial values of state variables V on M (x_{1init}, $\ldots, x_{kinit}$), input variables before i-th event ($\{v_j \mid j \in Iin \cap \{1, \ldots, i\}\}$) and variables ($t_1, \ldots, t_i$).

- If M can execute a sequence $(s_1, \sigma_1) \xrightarrow{d_1} (s_1, \sigma_1') \xrightarrow{a_1 \$_1 v_1} (s_2, \sigma_2) \xrightarrow{d_2} \cdots \xrightarrow{a_i \$_i v_i}$ (s_{i+1}, σ_{i+1}), then $P_1 \wedge \cdots \wedge P_k$ is satisfied, where $\{v_i \mid \$_i =?_{v_i}\}$ are input values v_i and t_j $(j \in \{1, \ldots, i\})$ are execution time of event a_j $(t_j = \Sigma_{m=1}^{j} d_m)$. In this case, we say that the execution sequence above *satisfies* the symbolic trace w. □

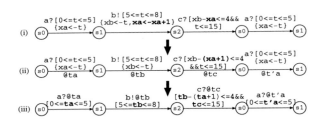

Fig. 2. Symbolic Trace

Let us construct a symbolic trace for a sequence shown in Fig.2 (i). First, we express each variable by expressions containing some of the preceding execution time, initial values of variables and input variables. Next, for the obtained sequence(ii), we prepare variables (t_a, t_b, t_c) for execution time of events and express constraints of transitions by using these variables. As a result we obtain a symbolic trace for the sequence $a?@t_a[0 \leq t_a \leq 5], b!@t_b[5 \leq t_b \leq 8], c?@t_c[t_b - (t_a + 1) \leq 4 \leq t_c \leq 15], a?@t_a'[0 \leq t_a' \leq 5]$.

We say that a symbolic trace w is traceable, if for some output timing there exists some input timing such that the succeeding sequence can be executed. We formalize the traceability as follows.

Definition 6. For symbolic trace $w = (a_1 \$_1, t_1, P_1) \cdots (a_k \$_k, t_k, P_k)$, $T(w)$ denotes a tuple of conditions $(P_1', \ldots, P_k')$ for the initial values of variables $(x_{1init}, \ldots, x_{kinit})$, execution times t_i of events $a_i \$_i$ $(1 \leq i \leq k)$ and input values v_i (if they exist) $(1 \leq i \leq k)$, where P_i' is equivalent to P_i or stricter condition than P_i $(P_i' \Rightarrow P_i)$. We call $T(w)$ as an I/O timing interval. □

Intuitively, we say that a symbolic trace w is traceable by an I/O timing interval $T(w)$ if the corresponding action sequence of w is executable at any I/O timing which satisfies $T(w)$.

Definition 7. A symbolic trace $w = (a_1 \$_1, t_1, P_1) \ldots (a_k \$_k, t_k, P_k)$ is *traceable* with respect to I/O timing interval $T(w) = (P_1', \ldots, P_k')$, if for any solution $(t_1, \ldots, t_k, \mathbf{v_1}, \ldots, \mathbf{v_k})$ of the condition $P_1' \wedge \cdots \wedge P_k'$, there exists some concrete trace $(s_1, \sigma_1) \xrightarrow{d_1} (s_1, \sigma_1') \xrightarrow{a_1 \$_1 v_1} (s_2, \sigma_2) \xrightarrow{d_2} \cdots \xrightarrow{a_i \$_i v_i} (s_{i+1}, \sigma_{i+1})$ such that $d_1 = t_1$ and $d_i = t_i - t_{i-1}$ $(i \in \{2, \ldots, k\})$. □

3.2 Executability of Concurrent Sequences

The concrete trace of concurrent timed I/O automata is generally defined as a concrete trace of its composed automaton, as defined in Definition 4. Intuitively, we say that such a concrete trace can *trace* the given set of symbolic traces if the *projection* of the trace to each consisting timed I/O automaton satisfies the corresponding symbolic trace. Moreover, if there exists such a concrete trace for a given set of symbolic traces, we say the set of symbolic traces is *traceable*. Formally, these notions are defined as follows.

Definition 8. For sequences $((s_1, \ldots, s_k), (\sigma_1, \ldots, \sigma_k)) \xrightarrow{d_1} \xrightarrow{a_1\$_1} \ldots \xrightarrow{d_n} \xrightarrow{a_n\$_n}$ $((s'_1, \ldots, s'_k), (\sigma'_1, \ldots, \sigma'_k))$ on concurrent timed I/O automata $\mathcal{M} = (M_1, \ldots, M_k)$, $Proj_i(d_1 a_1 \$_1 \ldots d_n a_n \$_n)$ denotes a projection for timed I/O automaton M_i.

$$Proj_i(d_1 a_1 \$_1 \alpha) = \begin{cases} d_1 a_1 Proj_i(\alpha) & if\, a_1 \in A_i \\ Proj'_i(\alpha, d_1) & otherwise. \end{cases}$$
$$Proj_i(\epsilon) = \epsilon$$
$$Proj'_i(d_1 a_1 \$_1 \alpha, d) = Proj_i((d + d_1) a_1 \$_1 \alpha))$$
$$Proj'_i(\epsilon, d) = d$$

Then, for a given set of symbolic traces $w_1, \ldots, w_k$ of a concurrent timed I/O automaton $\mathcal{M} = (M_1, \ldots, M_k)$ and a concrete trace α of $\mathcal{M}$, if $Proj_i(\alpha)$ satisfies w_i for each $i \in \{1, \ldots, k\}$, we say that α *satisfies* the set of symbolic traces $w_1, \ldots, w_k$. If such a concrete trace exists, we also say that $w_1, \ldots, w_k$ is *traceable*. □

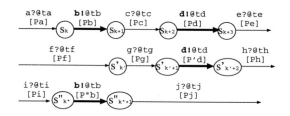

Fig. 3. Symbolic Traces for Concurrent Timed I/O Automata

Sequences on three concurrent timed I/O automata are shown in Fig.3. By definition, actions whose names are the same (output events $b!$ and $d!$ in the figure) are executed synchronously (simultaneously). For deciding the executability of concurrent sequences, we also construct I/O timing intervals from the tail action to the head one of the sequence by similar method described in Section 3.1.

Since tail events of each sequence ($e?$, $h?$ and $j?$) are input events and they are executed independently, a condition which denotes that they are executable independently is $\exists t_e[P_e] \wedge \exists t_h[P_h] \wedge \exists t_j[P_j]$. Assuming that each timing variable of some action (such as t_e) never appears in the transition conditions of other actions (such as P_h or P_j), the expression is equivalent to $\varphi_1 = \exists t_e \exists t_h \exists t_j[[P_e] \wedge [P_h] \wedge [P_j]]$. Next, for synchronous output event $d!$ that is executed most recently before these three input actions, we consider a condition which expresses that "there exists some timing t_d such that these output actions can be executed simultaneously and all the succeeding actions can also be executed." The condition can be described as $\varphi_2 = \exists t_d[[P_d] \wedge [P'_d] \wedge \varphi_1]$. $e?$, $h?$ and

j? are executable if and only if the condition holds. For input events c? and g? which are executed independently, we obtain a condition $\varphi_3 = \exists t_c \exists t_g [[P_c] \wedge [P_g] \wedge \varphi_2]$. We construct conditions for the rest of sequences. For a synchronous output event b!, we obtain a condition $\varphi_4 = \exists t_b [[P_b] \wedge [P_b''] \wedge \varphi_3]$. Finally, by considering input events a?, f? and i?, we obtain a condition $TrCond = \exists t_a \exists t_f \exists t_i [[P_a] \wedge [P_f] \wedge [P_i] \wedge \varphi_4]$. Three sequences shown in Fig.3 are executable simultaneously if and only if $TrCond$ is true.

4   Functional Testing

In this paper, since we consider systems which communicate through networks, each input timing to IUT may have jitter (receive timing of frames and acknowledgment etc.) It is required that there exists some margin for input timing. Furthermore, the test sequence might not be executable for some output timings, since output timings are uncontrollable. Therefore, we apply a static scheduling proposed in [4] in order to obtain suitable I/O timing intervals. By using the scheduling method, for given concurrent sequences we decide whether there exists some execution timing(intervals) for I/O events such that all the sequences are executable. If there exists such timing intervals for sequences, we obtain them and determine the sequences are executable. Although the sequences may not be executed if the IUT does not produce outputs in the expected intervals, we minimize the influence of output timings on the executability of test sequences by making the output timing intervals as wide as possible. Since we can get information about the output timing corresponding to the executability of sequences by applying the scheduling method, we can quickly detect that we fail to execute the given sequences on testing.

4.1   Proposed Testing Method

In the proposed method, we first generate test sequences for checking the correctness of the functions which we intend to test. Then, we derive time intervals that enable the test sequences execute for the I/O events in the generated test sequences.

The problem of deriving time intervals for the I/O events is formally defined as follows.

Definition 9. Given concurrent timed I/O automata $\mathcal{M} = (M_1, \ldots, M_k)$ and a set of symbolic traces $w_1, \ldots w_k$, the problem of deriving time intervals for I/O events is deriving a set of time intervals $T(w_1), \ldots, T(w_k)$ for the I/O events, which makes all of $w_1, \ldots, w_k$ traceable. □

For the timing of input event, we can choose several timing covering the derived interval for the input event, when we carry out functional testing actually. In [7], a useful method for choosing adequate input timing for testing multimedia systems has been proposed. Ref. [7] recommends that we should test more at timing close to the boundary in the input timing interval than at timing close to the middle of the interval as shown in Fig.4.

On the other hand, it is impossible to control the timings of output events. So, in order to make the output timing earlier or later intentionally, we adopt the following

Fig. 4. Input Timing

way. In [8], we have proposed and implemented a method to execute existing program codes in multi-threaded environments where a given program can be converted into a thread and it can be executed with other threads cooperatively. Here, we assume that each IUT is given as a program (object coeds). By using the method in [8], we can convert a given IUT into a thread as shown in Fig.5. In Fig.5, we add the load-control thread (load controller) as the cooperative thread. If we want to make the output timing later, the tester asks the load-controller to use CPU time more. As the result, the IUT cannot use enough CPU time and the output timing from the IUT becomes late. By adjusting CPU time used by the load-controller, we can make the output timing of the IUT earlier or later intentionally.

Fig. 5. Testing Environment

Here, we will apply the following functional testing method to the IUT. First, we generate a set of test sequences for functional testing. Then, we derive time intervals which make the given test sequences executable by using the method described in the next sub-section. We give the inputs to the IUT at some timing in the derived time intervals and observe the outputs from the IUT. If the outputs are observed at an earlier time than the expected time intervals, then we make the load-conroller consume much CPU time so that we can observe the outputs at some timing in the derived time intervals. We repeat this functional testing process and give inputs at different timing as shown in Fig.4. If we can observe the corresponding outputs at adequate timing for several times by adjusting the CPU load for the load-controller, we conclude that we succeed the functional testing. If we cannot observe the I/O events at adequate timing even if we repeat the above process, then we conclude that there may have some errors for the implementation of the function to be tested.

4.2   Method for Deriving I/O Timing Intervals

In the proposed method, first, we generate test sequences corresponding to the functions that we want to test and transform them into symbolic traces, which reflect the changes of the values of variables. Based on the scheduling method in [4], we derive executable time intervals for the I/O events in the given traces if they exist.

When we test a synchronized output between audio and video for the media synchronization protocol in Fig.1, we generate test sequences which make the synchronized event "synchronized_play!" executable for three modules (a), (b) and (c), and transform them into symbolic traces. Since module (c) in Fig.1 controls only the synchronization by counting the number of repetition and no time constraint is imposed on the events in it, we omit module (c) from the discussion below. Here, we consider two symbolic traces in Fig.6. These are traces that synchronized_play! is executed synchronously in modules (a) and (b) in Fig.1 when they repeat the cyclic transitions $s_0 \rightarrow s_1 \rightarrow s_2 \rightarrow s_3 \rightarrow s_0$ for 4 times and 2 times, respectively.

Hereafter, we propose a method for deriving I/O timing intervals of given test sequences.

sequence in module (a)	sequence in module (b)
audio_receive?@t_{a1}	video_receive?@t_{v1}
audio_decode_request!@t_{a2}	video_decode_request!@t_{v2}
audio_decode_respond?@t_{a3}	video_decode_respond?@t_{v3}
audio_play!@t_{a0}	video_play!@t_{v0}
audio_receive?@t'_{a1}	video_receive?@t'_{v1}
audio_decode_request!@t'_{a2}	video_decode_request!@t'_{v2}
$\cdots$	video_decode_respond?@t'_{v3}
audio_decode_respond?@t'''_{a3}	**synchronized_play!**@t'_{v0}
synchronized_play!@t'''_{a0}	video_receive?@t''_{v1}
audio_receive?@t''''_{a1}	

Fig. 6. Test Sequences for Synchronized Events (Symbolic Traces)

The condition for each transition $a@t_a$ is supposed to be represented as a logical product of linear inequalities consisting of variable t_a representing execution time of a and variables used in the preceding events. Therefore, the conditions are represented as logical products of expressions whose forms are $\alpha \leq t_a$ or $t_a \leq \beta$. In order to derive the I/O event intervals that make the two traces executable, we give the following constraints.

(1) Each event is never executed ahead of the preceding events. So, the expressions below should be satisfied. We add them as the constraint.
$$t_{a1} \leq t_{a2} \leq t_{a3} \leq t_{a0} \leq t'_{a1} \leq \cdots \leq t'''_{a0} \leq t''''_{a1}$$
$$t_{v1} \leq t_{v2} \leq t_{v3} \leq t_{v0} \leq t'_{v1} \leq \cdots \leq t'_{v0} \leq t''_{v1}$$

(2) We introduce two new variables t_{xmin} and t_{xmax} that represent the earliest and latest executable time of t_x, respectively, and replace the original constraint $\alpha \leq t_x \leq \beta$ with $\alpha \leq t_{xmin} \leq t_{xmax} \leq \beta$.

As for the timing of the synchronized event (synchronized_play! in the sequences in Fig.6), we replace the execution time t'''_{a0} and t'_{v0} of two synchronized events with the same variable pair (t_{0min}, t_{0max}). Then the constraint on synchronized_play! in the audio module is replaced $[t'''_{a3} + 3 \leq t_{0min} \leq t_{0max} \leq t'''_{a3} + 5]$, and that in the video module is replaced $[t'_{v3} + 6 \leq t_{0min} \leq t_{0max} \leq t'_{v3} + 10]$.

(3) As the objective function for solving the above linear programming problem, we give the weighted sum of time intervals $\Sigma w_i(t_{imax} - t_{imin})$, where adequate weights (w_i) are given. If we want to derive wider time intervals for output events, we can give larger weights w_i to the output time intervals.

(4) We solve the linear programming problem.

Thus, the solution tells executable time intervals for each event. For each output event, we judge that the rest of the test sequence is executable if the output is observed in the derived time interval. On the other hand, we decide that given test sequences are not executable if we cannot obtain a solution.

5   Example

We apply our method to two test sequences for checking the correctness of the functions of synchronized events and timeout handling on the media synchronization protocol shown in Fig.1.

5.1   Functional Test Sequences for Synchronized Events

We apply our method to two symbolic traces and obtain the following solutions[2].

$$2.4 \leq t_{a1} \leq 2.8 \quad 3.8 \leq t_{a2} \leq 5.4 \quad 9.4 \leq t_{a3} \leq 9.8 \quad 12.8 \leq t_{a0} \leq 14.4$$
$$16.4 \leq t'_{a1} \leq 16.8 \quad 17.8 \leq t'_{a2} \leq 19.4 \quad 23.4 \leq t'_{a3} \leq 23.8 \quad 26.8 \leq t'_{a0} \leq 28.4$$
$$30.4 \leq t''_{a1} \leq 30.8 \quad 31.8 \leq t''_{a2} \leq 33.4 \quad 37.4 \leq t''_{a3} \leq 37.8 \quad 40.8 \leq t''_{a0} \leq 42.4$$
$$44.4 \leq t'''_{a1} \leq 44.8 \quad 45.8 \leq t'''_{a2} \leq 47.4 \quad 51.4 \leq t'''_{a3} \leq 51.8 \quad 55 \leq t_0 \leq 56.4$$
$$58.4 \leq t''''_{a1} \leq 59$$
$$8 \leq t_{v1} \leq 9 \quad 10 \leq t_{v2} \leq 11 \quad 20 \leq t_{v3} \leq 21 \quad 27 \leq t_{v0} \leq 28.4$$
$$36.4 \leq t'_{v1} \leq 37 \quad 38 \leq t'_{v2} \leq 39.4 \quad 48.4 \leq t'_{v3} \leq 49 \quad 55 \leq t_0 \leq 56.4$$
$$64.4 \leq t''_{v1} \leq 65$$

If we give inputs to IUT at adequate timing in obtained intervals and observe outputs in obtained intervals, the sequence is executable. So we can check the behavior of IUT and test whether IUT synchronizes correctly.

5.2   Functional Test Sequences for Timeout Handling

We consider that if a synchronized event is executed correctly after executing a timeout event, the timeout event (timeout handling) is also executed correctly. We apply our method to the sequences which are obtained by replacing a part of the sequence on module (a) shown in Fig.6 (third appearance of audio_receive?@t''_{a1}, audio_decode_request!@t''_{a2}, audio_decode_respond?@t''_{a3}) with timeout handling (audio_loss!@t''_{a3}). The result is as follows.

$$2 \leq t_{a1} \leq 2.4 \quad 3.4 \leq t_{a2} \leq 5 \quad 9 \leq t_{a3} \leq 9.4 \quad 12.4 \leq t_{a0} \leq 14$$
$$16 \leq t'_{a1} \leq 16.4 \quad 17.4 \leq t'_{a2} \leq 19 \quad 23 \leq t'_{a3} \leq 23.3 \quad 26.3 \leq t'_{a0} \leq 28$$
$$37 \leq t''_{a3} \leq 38 \quad 41 \leq t''_{a0} \leq 42$$
$$44 \leq t'''_{a1} \leq 45 \quad 46 \leq t'''_{a2} \leq 47 \quad 51 \leq t'''_{a3} \leq 52 \quad 55 \leq t_0 \leq 56$$
$$58 \leq t''''_{a1} \leq 59$$
$$8 \leq t_{v1} \leq 9 \quad 10 \leq t_{v2} \leq 11 \quad 20 \leq t_{v3} \leq 21 \quad 27 \leq t_{v0} \leq 28.4$$
$$36.4 \leq t'_{v1} \leq 37 \quad 38 \leq t'_{v2} \leq 39.4 \quad 48.4 \leq t'_{v3} \leq 49 \quad 55 \leq t_0 \leq 56$$
$$64 \leq t''_{v1} \leq 65$$

---

[2] If we want to derive wider intervals for output events, the interval widths for some input events may become zero. However, it is practically difficult to execute input events at exact timing. To cope with the problem, we specify the minimum interval width for each input event. In this example, for input events audio_receive?, audio_decode_respond?, video_receive? and video_decode_respond?, we specify the minimum widths 0.4, 0.3, 0.6 and 0.5, respectively.

6 Conclusion

In this paper, we specify a media synchronization protocol as a model in which multiple timed I/O automata work in parallel and cooperatively. Then, we propose a method for functional testing on the model. We also propose a technique for deriving wide executable time intervals of I/O events in a given set of test sequences by using linear programming techniques. As the future work, we are planning to develop an actual environment for functional testing with the multi-threaded programming method in [8].

References

1. R. Alur and D. L. Dill: "A theory of timed automata", Theoretical Computer Science, Vol. 126, pp. 183–235 (1994).
2. T. Higashino, A. Nakata, K. Taniguchi and A. R. Cavalli: "Generating test cases for a timed I/O automaton model", Proc. of 12th IFIP Workshop on Testing of Communicating Systems (IWTCS'99), pp. 197–214 (Sept. 1999).
3. C. M. Huang and C. Wang: "Synchronization for Interactive Multimedia Presentations", IEEE MULTIMEDIA, Vol. 5, No. 4, pp. 44–62 (Oct.-Nov. 1998).
4. H. Katagiri, M. Kirimura, K. Yasumoto and T. Higashino and K. Taniguchi: "Hardware Implementation of Concurrent Periodic EFSMs", Proc. of Joint International Conference on 13th Formal Description Techniques and 20th Protocol Specification, Testing, and Verification (FORTE/PSTV2000), pp. 285–300 (Oct. 2000).
5. D. Lee and M. Yannakakis: "Principles and Methods of Testing Finite State Machines - A Survey", Proc. of the IEEE, Vol. 84, No. 8 (1996).
6. T. D. C. Little and A. Ghafoor: "Synchronization and storage models for multimedia objects", IEEE Journal of Selected Areas in Communications, Vol. 8, No. 3, pp. 413–427 (Apr. 1990).
7. V. Misic, S. T. Chanson and S. C. Cheung: "Towards a Framework for Testing Distributed Multimedia Software Systems", Proc. International Symposium on Software Engineering for Parallel and Distributed Systems (PDSE98) (Apr. 1998).
8. K. Abe, T. Matsuura, K. Yasumoto and T. Higashino: "A Method to Execute Existing Program Codes in Multi-theread Environments and Its Implementation" Journal of Information Processing Society of Japan, Vol. 41, No. 9, pp. 2603–2613 (Sep. 2000) (In Japanese).

Measurements to Validate Optimised Bandwidth Usage by the Distributed Network Entities Architecture for Multimedia Conferencing

R. Sureswaran and O. Abouabadalla

Network Research Group
School of Computer Sciences
Univ. Sains Malaysia
11800 Penang, Malaysia
sures@cs.usm.my

Abstract. Real-time Multimedia conferencing is becoming an important part of today's communications architecture. Such applications are gaining popularity for use within both the Internet and Corporate Intranets. Some of the reasons include the following: 1. Multimedia applications can be easily seen, heard and understood. 2. Multimedia communications are set to become the future generation of communications. 3. Multimedia conferencing will allow meetings to be held in a virtual manner. This means office and critical staff can have meetings with others without even leaving their office. 4. Video streaming will allow instantaneous access to video files scattered around the globe. In order to allow the current and future system and network structures to support such high bandwidth and resource hungry applications, a form of distributed processing is necessary. This focus of this paper is to discuss the design of such a distributed system and to validate the bandwidth claims by using certain mechanisms to obtain measurement results. This paper will take us initially through the theory, design stages and implementation of this distributed architecture. It will then discuss methods of testing and validating the testing measurement results obtained. *The above research is being funded in parts by IRPA (Ministry of Science, Malaysia), Multimedia Research Labs and Network Research Group (NRG) University of Science Malaysia.*

1 Introduction

Practically all multimedia conferencing systems try to provide real-time connections as well as on-line receive and transmit capabilities. This means all or most participants are receiving and transmitting at the same time [6,7,8,9,10]. This would be considered an ideal situation if not for these two factors:
– The confusion generated when everyone tries to speak at the same time.
– The tremendous amount of network traffic generated by all these participating sites.
The RSW Control Criteria [1,2,18] resolves the bandwidth requirements of multimedia conferences as well as to create a system of order for these conferences. The RSW Control Criteria is discussed in greater detail under section 4.0
   Using this control criteria, we were able to implement a multimedia conferencing system using a highly developed form of distributed architecture. This distributed

I. Chong (Ed.): ICOIN 2002, LNCS 2344, pp. 551–562, 2002.
© Springer-Verlag Berlin Heidelberg 2002

architecture was called "distributed network entities" [16,17,19] and based on the following design principles:

-    Each network entity is an independent of each other. Its only dependency is upon the server, which will provide it directions and control information.
-    Each network entity will communicate with each other using the universally supported standard communication protocol, the Internet protocol (IP).
-    Each network entity should be implemented in a black box style format, where the inputs, outputs and requirements of the black box are clearly defined.
-    Each network entity can be plugged into a network and unplugged without crashing the system.

The implementation principles include the following:
-    Open Systems Architecture (Requirement of non-proprietary hardware).
-    The ability to implement within any OS that can support the specifications.
-    The ability to reside anywhere on the network, including sharing the same host as other network entities.

Using these fundamental design principles and implementation principles as a guide, we have defined and implemented four such distributed network entities to support this multimedia conferencing application. They are:

◆   The server entity
◆   The client entity
◆   The multilan IP converter (MLIC) entity
◆   The data compression (DC) entity

This paper will discuss the actual implementation of the above distributed network entities to support the Multimedia Conferencing System (MCS) Ver 4.0 as well as the measurements used to validate the claims of advantages offered by its architecture.

The flow of this paper covers research into the multimedia conferencing system and how the distributed network entities environment ties into it. It starts with the site status and control criteria (2.0), the distributed environment (3.0), the measurements (4.0) and ending with the conclusion (5.0).

2   Site Status

In order to resolve the first issue, one has to simply observe how an actual conference is conducted around a meeting table. Following the systems set within these meetings as a guideline, the control criteria was designed.

A multimedia conferencing system using the RSW Control Criteria is a server based multipoint system and not a peer to peer system [14, 15]]. The system encompasses one server per LAN segment (separated by IP routers) [11, 12] and as many sites as the LAN can support.

Any workstation (client) involved in a multimedia conference is called a client site. It is assumed that all sites have full network connectivity. Independent of its site status, a workstation at a site may be fully equipped to transmit and receive multimedia communications (inclusive of video and audio).

The RSW Control Criteria defines six different types of client sites status and one server site involved in a multimedia conference. *Figure 1* shows the relationship between the different sites.

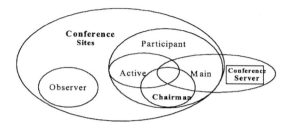

Fig. 1. Site Status Relations

The Conference sites are all the sites involved in the conference, that is the main site, active site, participant site, chairman site and observer site.

The Observer site is a site that is allowed to only view the conference and is not allowed to take active part in the conference.

The Participant site is like an upgraded observer who can participate and contribute towards the conference. In other words, a participant can request and become an active site.

The Chairman site is the site that initiates the conference. The chairman site will have main site status if the RSW Control Criteria used is the option in which the chairman gets to decide who the next active site will be (*Chairman Main Site Option*). The chairman site, however, is the only site that will always have the following features:

➢ The ability to cut short a lengthy active site, that is, to "kill" a currently active site.
➢ To become the default site if there is no other active site requests.
➢ To have the ability to change the status of a participant to observer and vice-versa.
➢ The ability to end or terminate the conference.

There can only be one chairman site per conference. The chairman site is considered a static site and cannot be reassigned to another participant site during that particular conference.

The main site is the main controlling site for the conference. The main site is automatically assigned to the conference chairman site during the start of most conferences. The main site is the only site that can decide who the next active site will be.

An active site is the only site in the conference that is allowed to transmit its multimedia stream onto the network. The active site is dynamic and when a site is designated to become the active site, it is the only site that is allowed to make any changes to the conference's audio, video and document.

➢ A participant becomes an active site by making an active site request, then awaiting his or her turn to be given the active site status.

The Conference server is not really a site, but a server that will manage the conference. A client will have to first connect (log into) a conference server before it can begin a conference. The server will establish a virtual point to point TCP link to all clients to handle control communications and another link to handle document communications.

The RSW Control Criteria is used to address the following two issues with multimedia conferencing:
- The confusion generated when everyone tries to speak at the same time.
- The tremendous amount of network traffic generated by all these participating sites.
In order to resolve the first issue, one has to simply observe how an actual meeting is conducted. Following the parameters set within these meetings as a guideline, the control criteria was designed.

The second issue could be resolved based on the method of control used to resolve the first issue. The issue of bandwidth cannot be removed completely, but an optimal amount can be found. Once again, observing the regular conference table, we notice that only one person needs to talk at any one time to hold an effective conference, while the chairman is usually coordinating the meeting. Thus, translating this into a virtual multimedia conferencing system, we can cut down the network traffic to only one station to transmit at any one time. Figure 2 show the GUI dialog box that is used by the control criteria to control the conference.

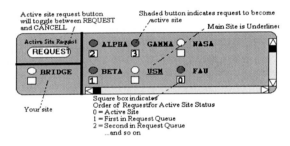

Fig. 2. The Dialog Box

Some of the other benefits from using the RSW Control Criteria include:
➢ The ability to hold real-time multimedia conferences like Desktop Video Conferencing over long delay networks like satellite links.
➢ The ability to coordinate changes that are being done to a shared document because these changes must be done sequentially in order to avoid conflicts or errors (data integrity problem).
➢ The ability to support a distributed processing environment for multimedia conferencing.

The following subsections discuss the numerous options proposed within our control criteria.

Option 1: Equal Privileges

This option is the most open option, where all conference sites have an equal opportunity of becoming active sites, that is, all conference sites have participant status. The user that gets active site status is also given main site status and the privilege of choosing the next active site by simply clicking the mouse on one of the marked sites (in Fig. 2, the sites Alpha, Beta, Gamma and FAU are marked, i.e. requesting main site privilege).

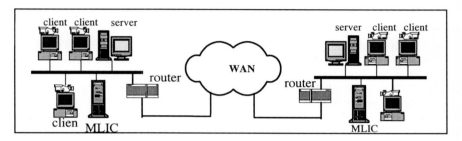

Fig. 3. General MCS network architecture.

Option 2: First come first serve
This time the decision for choosing the active sites resides within the server. The application will assign active site status to the sites in the order the request comes in.

Option 3: First come first serve, with time-out
This option is similar to option 2, except that there is a time-out feature incorporated within the application itself. Each site is only allowed a certain maximum time limit, reaching which, the site is warned that they will automatically be cut off within the next **x** seconds.

Option 4: Organizer Main site
From our feedback and beta site testing, we have found this option to be the most popular. This option gives the privileges of choosing the active site to the site that organizes the conference, which will then hold the status of main site. Thus now, the conference organizer will decide which site is to be the next active site and for how long.

Option 5: Restricted Active sites
In this option, the organizing site will restrict the sites allowed to participate in the conference. The participating sites may be that of a mutual prearrangement made earlier by the organizers of the conference. Other sites may be allowed to participate only as observers and never given active site privileges. This option is not used alone, but is used to enhance one of the existing options.

Option 6: Restricted active sites, upgradeable observer sites
This option is very similar to that of option 5, except that we will try to resolve the main limitations of option 5, which is the disability to upgrade observer sites to active sites in real-time. If the main site control, which is the only site that can permit the

upgrade, resides within the application, then the application will have to decide if the observer site requesting active site should be upgraded. If the application was to decide, then a decision taking algorithm will have to be used to help process the decision

3 The Distributed Environment

When the RSW Control Criteria is used to Control Multimedia Conferences, it has to be applied in a distributed processing client/server manner.

The basic modules are the client workstations and the conference server. The client workstations are completely independent of the server. They may use any operating system that has good graphic support. Similarly, the server may run any multitasking Operating System.

This form of distribution can be further extended by adding the multilan IP address conversion unit and the WAN data compression unit. Again, each of these units is a separate independent unit. They function like black boxes. Fig. 3 shows the general network architecture of the MCS entities and Fig. 7 shows an example output of the MCS system.

The input parameter of each of these black boxes are defined, its job function is defined and its output parameters are also clearly defined. The information style, language and operating systems are transparent to the user and even the entire system. This section will describe the following four units as distributed network entities. They are

➢ The server entity
➢ The client entity
➢ The multilan IP converter (MLIC) entity
➢ The data compression (DC) entity

3.1 The Server Entity

The main purpose of the server entity is to maintain and control the conference according to the RSW control criteria used. The functions of the server are as follows.

➢ Control the Conferencing using the RSW control criteria chosen.
➢ To reflect the document conferencing TCP/IP packets to all participants.
➢ Establishes inter-server links (only during multi server conferences).

3.2 The Client Entity

Whenever a workstation uses the RSW control conferencing to hold a multimedia conference, it cannot run the conference independently. It must use a server to establish and start the conference.

The advantages of using the distributed network architecture for the clients include:

➢ It is OS independent, thus different operating systems (Mac, Win 95, Win 3.1, X Win) can all participate in the same conference.
➢ Message passing is done using standard forms, i.e. TCP/IP and UDP.

➤ The workstation CPU only has to concentrate on Multimedia (Audio, Video and Document) capture, packaging, transmission and vice-versa.

➤ Implementation on different platforms (operating system and hardware) becomes easy as the clients main functions only involve multimedia capture, transmission and playback.

Figure 4 shows the client objects connectivity.

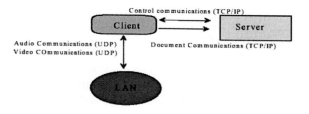

Fig. 4. The Client Object (Active Site Status)

The functions of the client are as follows:

➤ Multimedia capture (capture of audio, video and document information).

➤ Multimedia playback (display of audio, video and document information).

➤ Data packetization (for transmission) and data reconstruction (for receiving/playback).

➤ To communicate with the server, especially to maintain the RSW Control Criteria.

3.3   The Multilan IP Converter (MLIC) Object

The MLIC is needed when more than one IP LAN is involved in the multimedia conference. This is because the UDP packets transmitted by the client object is an IP multicast packet. Most routers will drop multicast packets, thus the Multicast audio and video UDP packets will never get to cross over a router. The job of an MLIC is to function as a bi-directional tunneling device that will encapsulate the multicast packets so as to be able to transport them across routers, WANs and the Internet.

Figure 5 shows the configuration of 2 IP LANs using the MLICs. An MLIC has two interfaces, the LAN interface and the router interface.

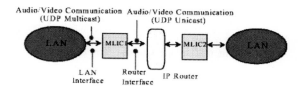

Fig. 5. Two IP LANs using the MLICs

The UDP multicast from the active site client in LAN1 will be picked up by MLIC1. MLIC1 will then encapsulate the packet into a unicast packet, and then pass this packet to the router.

The router receives the UDP unicast packet and will transmit it to LAN2. At LAN2, the UDP unicast packet is picked up by MLIC2 and will de-capsulate it back into a UDP multicast packet for LAN2.

All MLICs are bi-directional and can provide the functionality of receiving and transmitting at the same time. MLICs can also handle more than one conference at a time.

The functions of the MLIC can be defined as follows:
➢ Detect Audio/Video UDP multicast at the LAN interface.
➢ Change to Audio/Video UDP unicast and transmit via router interface.
➢ Receive Audio/Video UDP unicast from router interface.
➢ Change to Audio/Video UDP multicast and transmit via LAN interface.
➢ Unicast transmission must be sent to all MLICs that are defined for that conference.
➢ All other packets arriving at both interfaces are transparently forwarded to the opposite interface.

The MLIC has been implemented in the following platforms:
-   DOS (using 80x86 Assembly code)
-   Win95 (using Visual C++)
-   Linux (Using C++)

3.4   The Data Compression (DC) Object

The DC unit has two interfaces. One interface is connected to the MLIC router interface (MLIC interface) while the other interface is connected directly to the router (router interface). Figure 6 shows the DC unit in the distributed multimedia conferencing environment.

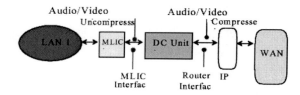

Fig. 6. The Data Compression Unit

The DC's first function is to first detect and intercept the multimedia conferencing UDP unicast packets that are being sent out by the MLIC to the router. The data section of these packets are then compressed using the compression algorithm implemented. Both, runtime or fixed frame compression may be used. The resulting

UDP unicast packet is now transmitted to the router with a compressed data section, which should be significantly smaller than its original size. This will save the much treasured WAN bandwidth. The DC's second function is to receive the compressed UDP unicast from another LAN's MLIC and decompress it.

It is possible to incorporate the DC unit into the MLIC. This may even be advantageous as the both the units' first layers perform the same function. The functions of the DC unit are as follows:

➢ Detect Audio/Video UDP unicast at the MLIC interface.
➢ Compress Audio/Video data and transmit via router interface.
➢ Receive compressed Audio/Video UDP unicast from router interface.
➢ Uncompress Audio/Video data and transmit via LAN interface.
➢ All other packets arriving at both interfaces are transparently forwarded to the opposite interface.

The DC Entity has been implemented for the following compression standards:
- *YUV411*
- *Indio*
- *MJPEG*
- *MPEG*
- *H.261*
- *VDO*

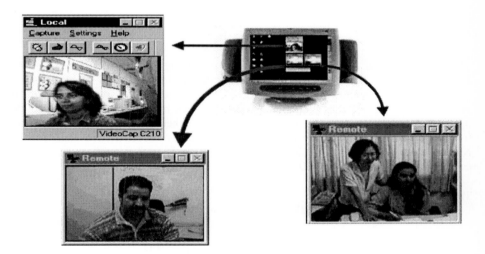

Fig. 7. MCS Ouput

4 Measurements

The following experiments were carried out for the control criteria and the distributed network entities architecture to evaluate the capabilities and to match the practical outcomes with the theoretically predicted values.

4.1 The Control Criteria

The tests carried out are to evaluate the bandwidth usage with and without the control criteria. With the control criteria, transmission is from one site only, while without the control criteria, transmission is generated from n sites ($n=2$ to 4). There is no theoretical upper limit to n, but 4 was chosen as a practical value and due to hardware availability for testing.

Theoretical Values:
Given transmission from one site is V Bytes per second, then the following is true:

$$BW_1 = V \text{ Bytes per second .}$$

Given a contention-less and collision free network, with unlimited bandwidth, then the following holds true:

$$BW_n = n \ x \ V \text{ Bytes per second .}$$

Given the 10Mbit condition placed on Ethernet, we can then narrow down this formula a little more:

$$BW_n = n \ x \ V \text{ Bytes per second .}$$
[for values of n where $BW_n =< 10$Mbits]

Experimental Values

These experimental results were obtained from running experiments on an isolated Ethernet network. This reduced background traffic to almost 0%. The compression used is VDO Wavelet. This codec tends to transmit different amounts of data if movement takes place thus the camera was positioned to point to a constant motion image during the tests to enable more accurate data gathering.

All the readings in the table below were taken in a controlled environment using Netmon Ver 3.0, a real-time Ethernet bandwidth analysis tool, with an average of transmission over a period of 10 seconds.

Transmission from 1 site
$$BW_1 = 245 \text{ Kbits per second}$$

Table 1. Transmission from n sites (Kbits)

No of Sites (*n*)	With Control Criteria		Without Control Criteria	
	Theoretical Value	Actual Reading	Theoretical Value	Actual Reading
1	245	245	245	245
2	245	250	490	475
3	245	254	735	716
4	245	258	980	940

It is interesting to note that a linearly increasing value is not maintained. This can be explained as theoretical assumptions are based on a perfect Ethernet network condition where there is no contention or collisions. However, in reality, there will be contention and collision as more than one site compete for transmission. This is why the experimental results do not follow an exact linear increase (See Table above).

This experiment proves that the control criteria optimises the bandwidth to a constant, while without the control criteria, bandwidth increases almost linearly.

The main advantage of using the distributed multimedia conferencing environment are bandwidth optimisation and better processor utilisation.

5 Conclusion

As multimedia conferencing is become more popular, proper methods of control of such conferences are needed. One method to successfully provide this form of control is the RSW control criteria.

By using the RSW Control Criteria, long haul and high delay network links like satellite links can now get involved in real-time multimedia conferences. Combined with its the powerful distributed network entity architecture, this system can create a global mesh of servers with world wide links, allowing unlimited numbers to join into any multimedia conference at the same time, in real-time.

The distributed network entity architecture allows power hungry, high bandwidth multimedia applications like interactive multimedia conferencing to be run even over shared Ethernet networks on non-dedicated PCs. This paper proves this that the combination of implementing the RSW Control Criteria by using the Distributed Networks Entities Architecture optimises the Bandwidth requirements for multimedia conferencing.

MCS Ver.4.0, which fully utilises this technology, can transmit video up to 40 fps, CD quality audio and yet manage to keep bandwidth low and provide smooth, full motion video.

Future work includes the study of migrating H.323 architecture to an architecture based on the system presented within this paper.

References

[1]   R. Sureswaran and R.K. Subramaniam.: "A control Criteria to optimize Collaborative Document and Multimedia Conferencing Bandwidth Requirements." IEEE SICON, Singapore (1995).

[2]   R. Sureswaran. "A Control Criteria for Document and Multimedia Conferencing". Proceedings of IASTED conference on Multimedia and Distributed Systems, Honalulu (1994).

[3]   NCR Corporation. "NCR Telemedia: Setup and User Guide". NCR Corporation, Dayton, Ohio (1993).

[4]   V. Anupam and C.L.Bajaj. "Shastra: Multimedia Collaborative Design Environment". IEEE Multimedia, Summer (1994) 39–49.

[5]   V. Anupam and C.L.Bajaj. "Collaborative Multimedia Scientific Design in SHASTRA". ACM Multimedia Conference (1993) 447–456.

[6]   AT&T Technical Education Center. "Multimedia Communications Technology, Systems, and Application DC2001". ver 2.1 (1994)

[7]   J.F.K. Buford. 2Multimedia Systems". ACM Press. Addison-Wesley (1994)

[8]   E. Craighill et al. "CECED: A System For Informal Multimedia Collaboration". ACM Multimedia Conference (1993) 437–445

[9]   H. Ishii and N. Miyake. "Towards An Open Shared Workspace: Computer and Video Fusion Approach of TeamWorkStation". Communications of the ACM, Vol.34, No.12 (1991) 37–50.

[10] M. Altenhofen et al. "The BERKOM Multimedia Collaboration Service". ACM Multimedia Conference (1993) 447–456.

[11] A.S. Tanenbaum. "Computer Networks". 2nd Edition (1988)

[12] W. Stallings. "Data And Computer Communications". 4th Edition (1994).

[13] S. Casner. "Are You on the Mbone". IEEE Multimedia, Summer (1994) 76–79.

[14] Intel. "Intel Proshare Personal Conferencing, Product Information". Intel Corporation.

[15] Creative Labs. "SharedVision User Guide." (1995)

[16] R.Sureswaran. "Using the RSW Control Criteria to Create a Distributed Environment for Multimedia Conferencing" REDECs '97, Malaysia (1997)

[17] R.Sureswaran "Using the RSW Control Criteria to Support Desktop Video Conferencing Using Satellite Links", Nasonet, Malaysia (1997)

[18] R.Sureswaran "Using the RSW Control Criteria to Support Ethernet Based Multimedia Conferencing" SEACOMM'96, Malaysia (1996)

[19] R.Sureswaran & S. Madria "A method to Support Multimedia Conferencing Using satellite Links" IWTS98, Shah Alam, Malaysia (1998)

VI. Distributed Systems

Consistent and Efficient Recovery for Causal Message Logging

JinHo Ahn, Sung-Gi Min, and ChongSun Hwang

Dept. of CS & Eng., Korea University
5-1 Anam-dong, Sungbuk-gu, Seoul 136-701, Republic of Korea
{jhahn,hwang}@disys.korea.ac.kr, sgmin@korea.ac.kr

Abstract. To reduce the number of stable storage accesses and impose no restriction on the execution of live processes during recovery, Elnozahy proposed a recovery algorithm based on causal message logging. However, the algorithm with independent checkpointing may force the system to be in an inconsistent state when processes fail concurrently. In this paper, we identify these inconsistent cases and then present a recovery algorithm to perform consistent recovery by allowing the recovery leader to collect recovery information from the other recovering processes as well as all live ones. Our recovery algorithm requires no additional message compared with Elnozahy's algorithm.

1   Introduction

Log-based rollback recovery is a well-known technique to provide fault-tolerance for distributed systems. It requires that each process periodically saves its local state with or without synchronizing with other processes, and logs each received message [2]. Message logging protocols are classified into three categories: pessimistic, optimistic and causal [2]. Among the three approaches, causal message logging approach has the following advantages. Firstly, each process piggybacks its log of determinants [2] in the volatile storage on every sending message in order to prevent *orphan* processes. The *orphan* processes are live processes whose states are inconsistent with the recovered states of failed processes. Thus, the causal message logging approach restricts rollback of each failed process to the most recent checkpoint on the stable storage without rolling back any live process. It can also handle concurrent and multiple failures by utilizing volatile storages of the dependent processes. Therefore, the approach has the advantage of pessimistic message logging approach. Secondly, each process saves the determinant of each received message and the determinants piggybacked on the message in the stable storage asynchronously. Therefore, the causal message logging approach has the advantage of optimistic message logging one. Finally, as the technologies of processors and networks have more rapidly been developed than those of stable storage, the causal message logging approach is very attractive for distributed systems as a low-cost fault-tolerance technique due to the desirable properties [1].

I. Chong (Ed.): ICOIN 2002, LNCS 2344, pp. 565–574, 2002.
© Springer-Verlag Berlin Heidelberg 2002

To recover the system to be a consistent state even in case of concurrent failures, the recovery algorithm of Family Based Message Logging protocols ($FBML$) prevents live processes from continuously executing their computations during recovery [2]. This undesirable property may increase the relative cost of a process failure and significantly degrade system performance [1]. $Manetho$'s recovery algorithm [3] requires each live process to write a list of determinants for recovery into the stable storage after receiving a recovery message from each recovering process. Moreover, the algorithm delays the delivery of every received message, which may potentially create any inconsistency with the state of the recovering process, until the recovering process completes its recovery. Thus, the algorithm may require unnecessary delays in delivering application messages that would not create any inconsistency and frequent stable storage access. Elnozahy proposed an efficient recovery algorithm (ERA) based on causal message logging tolerating concurrent failures [1]. ERA reduces the number of stable storage accesses and allows live processes to progress their computations even in concurrent failures. But, it requires more additional messages than the traditional recovery algorithms such as $FBML$'s and $Manetho$'s algorithms [1].

However, if the causal message logging approach were integrated with independent checkpointing [2], ERA might force the system to be in an inconsistent state in case of concurrent failures. Thus, in this paper, we identify these inconsistent cases and then present an efficient recovery algorithm ($CERA$) ensuring the system consistency even in case of integrating the causal message logging approach with independent checkpointing. $CERA$ requires the small number of stable storage accesses and enables each non-failed process to perform its execution continuously in concurrent failures. Compared with ERA, $CERA$ can satisfy these goals without any additional message.

The rest of the paper is organized as follows. In section 2 and 3, we introduce the system model and identify ERA's inconsistent cases. In section 4, we present our recovery algorithm and prove its correctness. Section 5 compares our work with related works with respect to various aspects. Finally, section 6 concludes this paper.

2   System Model

The distributed computation on the system, N, consists of $n(n > 0)$ processes concurrently executed on hosts. Processes have no global memory and global clock. The system is asynchronous: each process is executed at its own speed and communicates with each other only through messages at finite but arbitrary transmission delays. We assume that the communication network is immune to partitioning, there is a stable storage that every process can always access and hosts fail according to the fail stop model [2]. Events of processes occurring in a failure-free execution are ordered using Lamport's *happened before* relation [2]. The execution of each process is *piecewise deterministic* [2]: at any point during the execution, a state interval of the process is determined by a *non-deterministic* event, which is delivering a received message to the appropriate application. The

k-th state interval of process p, denoted by $si_p^k (k > 0)$, is started by the delivery event of the k-th message m of p, denoted by $dev_p^k(m)$. Therefore, given p's initial state, si_p^0, and the non-deterministic events, $[dev_p^1, dev_p^2, \cdots, dev_p^i]$, its corresponding state s_p^i is uniquely determined. Let p's state, $s_p^i = \sum_{k=0}^i si_p^k$, represent the sequence of all state intervals up to si_p^i. s_p^i and $s_q^j (p \neq q)$ are *mutually consistent* if all messages from q that p has delivered to the application in s_p^i were sent to p by q in s_q^j, and vice versa. A set of states, which consists of only one from each process in the system, is *a globally consistent state* if any pair of the states is mutually consistent [2].

The information needed for replaying the delivery event of a message during recovery is called *determinant* of the event. The determinant consists of the identifiers of its sending process(sid) and receiving process(rid), send sequence number(ssn) and receive sequence number(rsn). In this paper, the determinant of message m and the set of determinants in process p's volatile storage are denoted by $det(m)$ and $d\text{-}set_p$.

3   Inconsistent Cases of *ERA*

ERA allows live processes to progress their computations during recovery and reduces the number of stable storage accesses compared with the traditional recovery algorithms. To satisfy the desirable properties, each recovering process performs the following four phases in *ERA*.

(**Phase 1.**) Every recovering process restores its latest checkpoint and increments its incarnation number, which is an integer that is incremented by one whenever each process recovers from a failure. Then, it obtains from the other processes the information about whether the processes are live or recovering respectively.

(**Phase 2.**) Recovering processes elect a recovery leader among them.

(**Phase 3.**) The leader collects the incarnation number from every recovering process and saves the number into a vector, called *incs*.

(**Phase 4.**) The leader sends every live process a recovery message for requiring the determinants needed for all the recovering processes, which includes the vector. When each live process receives the request message from the leader, it updates its *incs* using the *incs* included in the request message. Then, the live process sends the recovery leader the determinants. Afterwards, even if each live process receives an stale application message sent from any recovering process before the failure of the recovering process, the live process can immediately know that the application message creates inconsistency by using its *incs*. This ensures that after sending the determinants to the recovery leader, *ERA* will not acquire any application message that may force the system to be in an inconsistent state. If a live process fails before replying to the leader, the leader restarts the phase 4 after updating its *incs*. This step ensures that the leader obtains a consistent set of the determinants despite of concurrent failures during recovery. After obtaining the determinants from every live process, the leader distributes the determinants to the other recovering processes and then every recovering process recovers to be a consistent state. If the leader fails during recovery, a new leader is elected among the recovering processes and re-executes the phase 2 through 4.

However, if the causal message logging approach were integrated with independent checkpointing, ERA might force the system to be in an inconsistent state in case of concurrent failures. Figure 1 illustrates the inconsistency problem of ERA. In this figure, process q takes its local checkpoint, $Chk_q^{~j}$, receives message m_1 and then m_2 from process p and r, and then sends message m_3 to r. Process r receives the message m_3 from q and then takes its local checkpoint, $Chk_r^{~k+1}$. Then, suppose that process q and r fail concurrently like in figure 1. In ERA, q and r perform phase 1 and then elect q as a recovery leader in phase 2. In phase 3, q obtains r's incarnation number, inc_r. In phase 4, it requires the determinants needed for the recoveries of q and r from a live process p. However, q cannot obtain any determinant from p because p has no determinant for the recoveries of q and r. In this case, q may replay message m_2 and then m_1 in phase 4 because it has no information about receipt sequence numbers of m_1 and m_2, and message transmission delay is arbitrary in distributed systems. Therefore, q may generate another message m_x, not m_3, in phase 4. However, process r restarts from its latest checkpoint, $Chk_r^{~k+1}$, saved after it has received message m_3 from q. Thus, the recovered state of q is inconsistent with that of r after q and r have completed ERA respectively. In conclusion, ERA has the inconsistency problem because the recovery leader collects the determinants for all recovering processes from only live processes in phase 3.

4   The Proposed Recovery Algorithm

4.1   Basic Idea

In this section, we propose a *C*onsistent and *E*fficient *R*ecovery *A*lgorithm ($CERA$) to solve the inconsistency problem of ERA. $CERA$ enables the recovery leader to collect determinants, needed for the recoveries of all the recovering processes, from not only live processes but also the other recovering processes without any additional message compared with ERA. In other words, $CERA$ is informally described as follows: $CERA$ consists of four phases like ERA. The first and the second phases of $CERA$ are similar to those of ERA. However, in the third phase, the recovery leader collects the incarnation number of every recovering process and the determinants in *d-set* of the recovering process, and then saves the number into a vector, called *incs* and includes the determinants in *d-set* of the leader. The fourth phase of $CERA$ is similar to that of ERA.

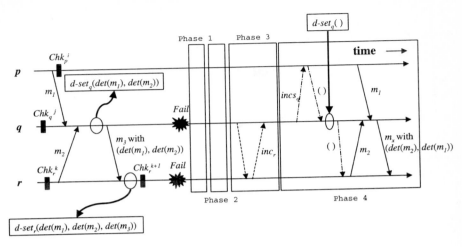

Fig. 1. An inconsistency case of ERA

To illustrate how to perform consistent recovery in $CERA$, consider the example shown in figure 2. After q and r concurrently fail like in this figure, they execute phase 1 respectively and then in phase 2, elect a recovery leader. In this case, q is elected as the leader. Thus, in phase 3, q collects r's incarnation number, inc_r, and the determinants in d-set_r, $(det(m_1), det(m_2))$, from r. Then, q includes $(det(m_1), det(m_2))$ in d-set_q like in this figure. In phase 4, q requires the determinants for q and r from live process p. In this case, p cannot provide q with any determinant because p has no determinant in d-set_p. However, q can replay message m_1 and then m_2 in phase 4 because it has the determinants of m_1 and m_2 in d-set_q. Therefore, q regenerates the message m_3 in phase 4. Therefore, after q and r has completed $CERA$ respectively, the recovered state of q is consistent with that of r. Additionally, in phase 3, every recovering process except the leader, for example r, includes its incarnation number and determinants in a reply message and then sends the message to the leader. Therefore, we can see that $CERA$ requires no additional message compared with ERA.

The data structures and procedures for process p in $CERA$ are formally given in figure 3, 4, 5 and 6. Procedure RECOVERY() in figure 4 is the entire procedure executed when each process fails and recovers. In this procedure, ELECT_LEADER() allows all recovering processes to elect a recovery leader among them and know whether every process in the system is live or recovering. In $CERA$, the recovery leader elected among recovering processes executes procedure LEADER() in figure 5 and the other recovering processes perform RECOVERING() in figure 6. In LEADER(), the leader p obtains the incarnation number and determinants of every recovering process q by sending message $Req\_Inc\_Dets()$ to and receiving message $Rep\_Inc\_Dets(inc_q, dets_q)$ from q. Then, p collects determinants from every live process r by sending message $Req\_Dets(incs_p)$ to and receiving message $Rep\_Dets(status_r, inc_r, dets_r)$ from r. When r receives from p a recovery message, $Req\_Dets(incs_p)$, for re-

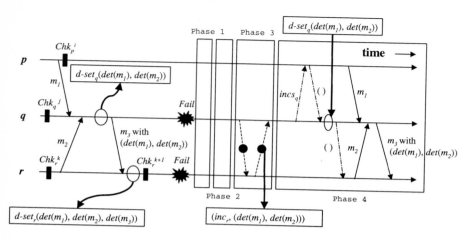

Fig. 2. An example of consistent recovery of $CERA$

- $d\text{-}set_p$: It is a set that records determinants of messages delivered by p and determinants piggybacked on the messages. It is initialized to $\emptyset$.
- $incs_p$: It is a vector which records the incarnation number of every process in the system. Each element of the vector is initialized to 0. p piggybacks its incarnation number, $incs_p[p]$, on every sending message. Thus, when p receives an application message from process q, it discards the message if q's incarnation number piggybacked on the message is less than $incs_p[q]$. Otherwise, it accepts the message.
- $procstatus_p$: It is a vector which records the status of every process in the system. The value of the status of each process is 'L' if the process is live or 'R' if the process is recovering.

Fig. 3. Data Structures for process p in $CERA$

quiring the determinants needed for recoveries of all recovering processes, it performs procedure PROVIDE_DETS() in figure 6. In this procedure, r provides its $d\text{-}set$ for p. If r fails before replying to p, p restarts from label T after updating $dets_p$ and $procstatus_p[r]$. After obtaining the determinants from every live process, it distributes the determinants to q by sending message $Deliver\_Incs\_Dets(incs_p, temp\_dets)$ to q. Then, p and q replay the messages received beyond the latest checkpoint during failure-free operation by using their determinants in RECOVERY(). If p fails during recovery, q re-executes ELECT_LEADER() and the next steps.

4.2   Correctness

In this section, we prove the correctness of $CERA$.

Lemma 1.   When the causal message logging approach is integrated with independent checkpointing, the system can perform globally consistent recovery

procedure RECOVERY()

    get its latest checkpoint, $d\text{-}set_p$ and inc_p **from** the stable storage ;

    restore its state using the latest checkpoint ; $inc_p \leftarrow inc_p + 1$;

    save inc_p **into** the stable storage ; $incs_p[p] \leftarrow inc_p$;

    send $Recovery\_Msg()$ **to** the other processes ; $leader\_fail \leftarrow false$;

    do {

        $(i, procstatus_p) \leftarrow$ ELECT_LEADER$(p,' R')$;

        if $(p = i)$ **then**   LEADER() ;

        else $leader\_fail \leftarrow$ RECOVERING(i) ;

    } **while**$(leader\_fail)$;

    replay the messages received beyond the latest checkpoint by using $d\text{-}set_p$;

Fig. 4. Recovery Procedure for process p in $CERA$

in case of $f(1 \leq f \leq n)$ concurrent failures only if the recovery leader among recovering processes collect determinants from not only all live processes but also the other recovering processes.

Proof : We prove this lemma by contradiction. Assume that consistent recovery is impossible even if the recovery leader among recovering processes collect determinants from all live processes and the other recovering processes. Let C be the set of all the failed processes in N. Suppose that message m_1 depends on message m, $q(q \in C)$ is the receiver of m, and $r(r \in N)$ is the receiver of m_1. There are two cases to consider as follows:

Case 1: The determinant of m is piggybacked on m_1.

In this case, there are two sub-cases to consider:

Case 1.1: $r \in (N - C)$.

In this case, there are two sub-cases to consider:

Case 1.1.1: r receives from the recovery leader a recovery message for requiring the determinants after r has received m_1.

In this case, r can provide the determinant of m for the recovery leader because m's determinant is included in $d\text{-}set_r$. Thus, r never becomes an orphan process.

Case 1.1.2: r received from the recovery leader a recovery message for requiring the determinants before r receives m_1.

When r receives the recovery message, it updates its incarnation vector, $incs_r$, to the latest using the incarnation vector included in the message. However, in this case, r cannot provide the recovery leader with m's determinant. Afterwards, if the incarnation number of m_1's sender in r's incarnation vector is more than the incarnation number piggybacked on m_1 when r receives m_1, r discards m_1 because q may not obtain the determinant of m and replay m. Thus, the current state of r is consistent with the recovered state of q. Therefore, r never becomes an orphan process.

Case 1.2: $r \in C$

In this case, there are two sub-cases to consider:

Case 1.2.1: r received m_1 and then took its latest checkpoint before r fails.

procedure LEADER()
    $dets_p \leftarrow \emptyset$;
    for all $q \in N$ **st** $((p \neq q) \wedge (procstatus_p[q] =' R'))$ **do** {
        send $Req\_Inc\_Dets()$ **to** q ;
        receive $Rep\_Inc\_Dets(inc_q, dets_q)$ **from** q ;
        $dets_p \leftarrow dets_p \bigcup dets_q$; $incs_p[q] \leftarrow inc_q$;
    }
    T: $d\text{-}set_p \leftarrow \emptyset$;
    for all $q \in N$ **st** $(procstatus_p[q] =' L')$ **do** {
        send $Req\_Dets(incs_p)$ **to** q ;
        receive $Rep\_Dets(status_q, inc_q, dets_q)$ **from** q ;
        $incs_p[q] \leftarrow inc_q$;
        if$(status_q =' L')$ **then** $d\text{-}set_p \leftarrow d\text{-}set_p \bigcup dets_q$;
        else $dets_p \leftarrow dets_p \bigcup dets_q$;
        $procstatus_p[q] =' R'$;   **goto T** ;
    }
    $d\text{-}set_p \leftarrow d\text{-}set_p \bigcup dets_p$;
    for all $q \in N$ **st** $(procstatus_p[q] =' R')$ **do** {
        $temp\_dets \leftarrow \emptyset$;
        for all $e \in d\text{-}set_p$ **st** $(e.rid = q)$ **do**
            $temp\_dets \leftarrow temp\_dets \bigcup \{e\}$;
        send $Deliver\_Incs\_Dets(incs_p, temp\_dets)$ **to** q ;
    }

Fig. 5. Procedure for recovery leader p in $CERA$

If r receives from the recovery leader a recovery message for requiring the determinants, r can send the determinant of m to the leader because the latest checkpoint includes the determinant of m. Thus, r never becomes an orphan process.

Case 1.2.2: r took its latest checkpoint and then received m_1 before r fails.
If r receives from the recovery leader a recovery message for requiring the determinants, r can send no determinant of m to the leader. However, r never becomes an orphan process because the latest checkpoint includes no delivery event of m.

Case 2: The determinant of m is not piggybacked on m_1.
In this case, r can provide no determinant of m for the recovery leader. However, as at least one process among $f + 1$ processes storing the determinant of m never fails even in f concurrent failures, it can always provide the determinant of m for the recovery leader. Thus, r never becomes an orphan process.

Thus, consistent recovery is possible in all the cases. This contradicts the hypothesis.   □

Theorem 1. $CERA$ enables the system to recover to be a globally consistent state in case of f concurrent failures.

procedure RECOVERING(*leader*)
    while(true) {
        receive m **from** q ;
        if((m is of the form $Recovery\_Msg()$) $\wedge$ $q = leader$) **then**
          return *true* ;
        else if(m is of the form $Req\_Inc\_Dets()$) **then**
          send $Rep\_Inc\_Dets(inc_p, d\text{-}set_p)$ **to** q ;
        else if(m is of the form $Req\_Dets(incs_q)$) **then**
          send $Rep\_Dets('R', inc_p, d\text{-}set_p)$ **to** q ;
        else if(m is of the form $Deliver\_Incs\_Dets(incs_q, dets_q)$) **then** {
          $d\text{-}set_p \leftarrow d\text{-}set_p \cup dets_q$; $incs_p \leftarrow incs_q$;   **return** *false* ;
        }
    }

procedure PROVIDE_DETS($incs_q$)
    for all $k \in N$ **st** $(p \neq k)$ **do**
        $incs_p[k] \leftarrow \max(incs_p[k], incs_q[k])$;
    send $Rep\_Dets('L', inc_p, d\text{-}set_p)$ **to** q ;

Fig. 6. Procedures for recovering or live process p in $CERA$

Proof: From lemma 1, in order to perform globally consistent recovery in case of f concurrent failures, the recovery leader among recovering processes should collect determinants from not only all live processes but also the other recovering processes. $CERA$ follows the same rule. Thus, $CERA$ enables the system to recover to be a globally consistent state in case of f concurrent failures. $\square$

Theorem 2. $CERA$ terminates within a finite time.
Proof: Whenever a live process fails before the leader collects the requested determinants from the live process, the leader discards the determinants received so far, and then recollects the determinants from the other processes again. As no more than f failures occur, it re-executes the recovery algorithm at most up to f times. Additionally, $CERA$ does not block the execution of any live process and recovering processes except the leader are blocked only until the leader finishes gathering the determinants needed for the recoveries of all the recovering processes. Thus, $CERA$ terminates within a finite time. $\square$

5   Related Works

$FBML$'s recovery algorithm blocks the execution of live processes during recovery [1]. This undesirable property may increase the relative cost of a process failure and highly degrade system performance.

In *Manetho* [3], if each live process receives a recovery message from a recovering process, the live process writes all the determinants related to the recovering process into the stable storage and then sends the determinants to the recovering process. If the live process receives application messages which may potentially create any inconsistency, it should delay the delivery of the messages to the corresponding application until the recovering process completes recovery. Therefore, this protocol may prevent live processes from executing their computations after they have received any recovery message from recovering processes, and require extra accesses on stable storage.

Elnozahy's recovery algorithm [1] reduces the number of stable storage accesses and allows live processes to progress their computations even in concurrent failures whereas requiring more additional messages. The algorithm performs consistent recovery in case of integrating the causal message logging approach with coordinated checkpointing [2]. However, if the causal logging approach is combined with independent checkpointing, the algorithm may force the system to recover to be an inconsistent state.

Mitchell and Garg [5] presented a non-blocking recovery algorithm where every process creates the $Recovery$ $Reply$ $State(RRS)$, a global state for which every local state has the same incarnation vector. In this algorithm, each recovering process autonomously executes its recovery procedure without any election among recovering processes. However, the recovery may not be completed [4].

6   Conclusion

In this paper, we identified that ERA has the inconsistency problem in case of being integrated with independent checkpointing because the recovery leader collects the determinants for all recovering processes from only live processes. Thus, we presented $CERA$ to perform consistent recovery by allowing the recovery leader to obtain the determinants from the other recovering processes as well as the live ones. $CERA$ can satisfy these goals without any additional message compared with ERA.

References

1. E. N. Elnozahy. On the Relevance of Communication Costs of Rollback Recovery Protocols. *In Proc. the 15th ACM Symposium on Principles of Distributed Computing*, 1995, pp. 74–79.
2. E. N. Elnozahy, L. Alvisi, Y. M. Wang and D. B. Johnson. A Survey of Rollback-Recovery Protocols in Message-Passing Systems. Technical Report CMU-CS-99-148, Carnegie-Mellon University, 1999.
3. E. N. Elnozahy and W. Zwaenepoel. Manetho: Transparent rollback-recovery with low overhead, limited rollback and fast output commit. *IEEE Transactions on Computers*, 41 (1992) 526–531.
4. B. Lee, T. Park, H. Y. Yeom and Y. Cho. On the impossibility of non-blocking consistent causal recovery. *IEICE Transcations on Information Systems*, E83-D(2) (2000) 291–294.
5. J. R. Mitchell and V. Garg. A non-blocking recovery algorithm for causal message logging. *In Proc. of the 17th Symposium on Reliable Ditributed Systems*, 1998, pp. 3–9.

Server-Based Dynamic Server Selection Algorithms

Yingfei Dong[1], Zhi-Li Zhang[1], and Y. Thomas Hou[2]

[1] Dept. of Computer Science and Engineering, University of Minnesota, Minneapolis,
MN 55455
{dong,zhang}@cs.umn.edu,
Tel:612-626-7526, FAX: 612-625-0572.
[2] Fujitsu Labs of America, 595 Lawrence Expressway, Sunnyvale, CA 94085
thou@fla.fujitsu.com

Abstract. Server selection is an important problem in replicated server systems distributed over the Internet. In this paper, we study two server selection algorithms under a server-based framework we have developed. These algorithms utilize server load and network performance information collected through a shared passive measurement mechanism to determine the appropriate server for processing a client request. The performance of these algorithms is studied using simulations. Comparison with two naive server selection algorithms is also made. The initial simulation results show that our dynamic server selection algorithms have superior performance over the two naive ones, and as a result, demonstrate the importance of dynamic server selection mechanisms in a replicated server system.

Keywords: Server replication, server-based dynamic server selection, QoS, passive measurement

1 Problem Formulation and Related Work

Server replication (or mirroring) is a common technique that has been used to provide scalable distributed service over the Internet. If done appropriately, server replication can avoid server overload and significantly reduce client access latency. In addition to the issues such as where to place replicated servers in the global Internet and how clients can locate a server with the desired service, an important and challenging problem in server replication is how to select a server to process a client request so as to provide the "best service" for the client. This problem, referred to as the *server selection* problem, is the focus of our paper.

In addressing the server selection problem, two procedures are typically involved. First, statistics about server/network performance need to be collected. Based on the collected statistics, a server selection algorithm then selects the "best" server among a pool of eligible servers for processing a client request. Depending on where the statistics are collected and the server selection decision is made, existing server selection approaches [5,7,8,11,14,16,19,23] can be classified as either *client-based* or *server-based*.

I. Chong (Ed.): ICOIN 2002, LNCS 2344, pp. 575–584, 2002.
© Springer-Verlag Berlin Heidelberg 2002

Under the client-based approach, statistics about the network and server performance is typically collected using the *active probing* method [7,8,11,16,19]: a client [8] or its "proxy" (e.g., the service resolver in [11]) sends probe packets to measure the network performance.[1] As an exception, Seshan *et al.*[23] employs a novel *passive measurement* method where clients share their network performance discovery through performance monitoring at the client side. The major drawback of the client based server selection approach is that it is not *transparent* to clients: either clients or their proxies (e.g., service resolvers) need to know the name/location of all the servers providing a given service. Furthermore, the client-based approach requires modification of client browser software and installation of network measurement tools at every client side, and in some cases, it may rely on the deployment of an Internet-wide measurement infrastructure (e.g., modification of DNS to incorporate service resolvers). In the case where active probing measurement techniques are used, the extra traffic introduced by probe packets can lead to network bandwidth wastage or even congestion. In the long term the client-based approach may have its appeal, especially when an Internet-wide measurement infrastructure [15,21,22] is in place. However, in the immediate future, server selection mechanisms using the client-based approach are unlikely to be widely available to many clients to take advantage of replicated server systems.

In contrast, the server-based approach relies on the cooperation of servers providing a given service to determine the "best" server for a client and inform the client the location of the server via, say, HTTP redirect mechanism. An example of the server-based approach is HARVEST system [5] which uses a simple network metric, *hop count*, as the criterion to select the "best" server for clients.

In [14], we propose and develop a server-based measurement and server selection framework which employs passive network measurement techniques and performance information sharing among servers to dynamically select the "optimal" server for processing client requests. The proposed framework contains three major components: (1) a server-based passive network measurement mechanism based on tcpdump [18] and a prototype of which called *Woodpecker* is described in [10], (2) a metrics exchanging protocol for sharing server load and network performance information among the servers, which will be described in a future paper, and (3) a dynamic server selection mechanism for selecting an "optimal" server to process a given client request, which is the focus of this paper.

In this paper we describe two server-based *dynamic* server selection algorithms. These two algorithms utilize both the server load and network performance information collected by each server and shared among them. Based on these performance statistics, a server decides whether itself or another server would be the appropriate server to process a client request it receives. In the latter case, the client request is redirected to the selected server (see Fig. 1).

---

[1] In the case of [11] servers also collaborate with service resolvers by "pushing" their performance statistics to service resolvers periodically.

The performance of these algorithms is studied using simulations. Comparison with two naive server selection algorithms is also made. The initial simulation results show that our dynamic server selection algorithms have superior performance over the two naive ones, and as a result, demonstrate the importance of dynamic server selection mechanisms in a replicated server system.

Our server-based approach to the problem of server selection has the following salient features. First, it is client-transparent. Second, it does not require any modification to any client software or relies on the availability of an Internet-wide measurement infrastructure. Therefore it is readily deployable. Third, by using shared passive network performance measurement, we avoid the wastage of network resources of active probing measurement techniques. The network bandwidth overhead incurred by metrics exchanges among the servers is relatively low, as the number of servers in a replicated system is generally small, in particular, when compared to the number of clients (or the sub-networks). Fourth, by taking both the server load and network congestion status of paths between servers and clients, we can employ relatively sophisticated dynamic server selection algorithms to choose the "optimal" server to service client requests without incurring too much latency and overhead. Our approach is particularly amenable to an enterprise-based replicated server system, where a number of servers connected through an enterprise network provide certain services over the Internet to "regular" clients who need access to the services frequently.

The remainder of this paper is organized as follows. In Section 2, the two dynamic server selection algorithms are presented. The performance results are shown in Section 3. In Section 4, we conclude the paper and discuss the future work.

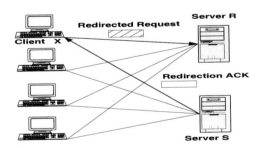

Fig. 1. An example system

2   Server-Based Dynamic Server Selection Algorithms

In this section, we present two dynamic server selection algorithms. Before describing our algorithms, we define two major performance metrics used in our server selection algorithms: *server loads* and *fast paths*, which indicate the status of servers and different network paths from servers to clients.

```
1. if (a client X is a regular client)
2.  if (server load is light) and (the path from S to X is not slow)
3.    accept the request
4.  else
5.    if (all other servers are heavily loaded)
6.      accept the request
7.    else
8.      randomly redirect the request to R other than S
9.  else
10. if (server load is light) or (all other servers are heavily loaded)
11.   accept the request and generate metrics about the client
12. else
13.   redirect to a light-loaded server
```

Fig. 2. Random Redirection Server Selection Algorithm

2.1   Performance Metrics

Server Load: For given server S, its load, denoted by L_S, is defined as the ratio of T_{total} to R_{max}. Here, T_{total} is the time period required to serve all requests in the waiting queue on S, and is related to the capability of S. It is computed based on the current and historical information of S including the processing delay on S and the transferring delay from S to a client. R_{max} is the maximum response delay that a typical client can accept for a request of average size, such as 10 KBytes [6]. L_S is called *light* if $L_S \leq 80\%$; otherwise, it is called *heavily loaded.*

Fast Path: As shown in Figure 1, if the transferring delay of a reply from a server S to a client X is k times longer than the transferring delay from another server R to X plus the cost of redirecting the request from S to R, the path from S to X is called *slow*, and the path from R to X is called *fast*. Here, k is an experimental parameter, such as 2 or 4. The transferring delay depends on the size of the reply and the bottleneck bandwidth of the path. For simplicity, the bottleneck bandwidth is approximated through the TCP-friendly formula [17,24]. The redirection cost is the delay of an acknowledgment from S to X plus the delay of the redirected request from X to R. Both transferring delays are estimated through the RTTs of the paths which are obtained through the passive measurement on servers.

2.2   Server Selection Algorithms

Our two simple server selection algorithms are given here. When a client X sends a request to server S, there are two situations in which this request may be redirected: (1) server S is overloaded, or (2) the path from S to X is *slow*. We give two algorithms to choose another server R which may provide better service under both situations. The first algorithm is called *Random Redirection (RR)* which randomly redirects a request to another server R other than S. The

```
1. if (a client X is a regular client)
2.  if (server load is light) and (the path from S to X is not slow)
3.    accept the request
4.  else
5.    computing equivalent classes
6.    if (no better server)
7.      accept the request
8.    else
9.      redirect the request to the ''best'' server R
10. else
11.  if (server load is light) or (all other servers are heavily loaded)
12.    accept the request and generate metrics about the client
13.  else
14.    redirect to a light-loaded server
```

Fig. 3. Best-guess Redirection Server Selection Algorithm

second algorithm examines all servers and selects the "best" one as R based on the current metrics on S. Because the current metrics on server S may not be accurate due to the delay of metrics exchange among servers, the second method is called *Best-guess Redirection (BR)*. RR is the simplest method which doesn't required too much calculation. BR, however, is one of the most complicated methods, in which server must know the status of all servers and related network paths to clients and compare all the possible choices.

Figure 2 and Figure 3 gives the basic frame of algorithms RR and BR. When a server S receives a request from a client X, S checks whether accepting this request into its waiting queue or redirecting this request to another server R. As shown from line 1 to 3, if X is a regular client, and S's load is light and the path from S to X is not slow, S accepts this request. Otherwise, in RR method in Figure 2 from line 5 to 8, S randomly redirect the request to another server R other than S; or, in BR method in Figure 3 from line 5 to 9, S computes equivalent classes to find whether a better server exists. *The equivalent class* of server S respective to a specific request is the set of servers whose current response time for this specific request is within a threshold compared with that of the server S. If all the servers are in the same equivalent class (i.e., there is no better server), S has to accept the request. Otherwise, S redirects the request to R which could provide "best" service.

Lines 10 to 14 shows the case that there is no information available about this client. If S's load is light, it accepts this request. Otherwise, it redirects this request to a light loaded server. In order to avoid the request oscillating among the servers, S has to accept this request if all other servers are heavily-loaded.

3   Simulation

3.1   System Models

In our simulation, a system consists of clients, servers and network paths. Figure 1 shows an example system with 2 servers, 4 clients and 8 network paths.

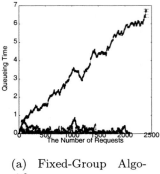

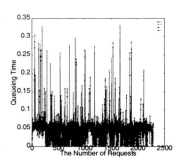

(a) Fixed-Group Algorithm.

(b) Random-Redirection Algorithm.

Fig. 4. Queueing delays of all 4 servers under FG and RR algorithms

Client Model: A client generates requests and receives replies from servers. For generating requests, the size of a request is Pareto-distributed with an average request size of 10 KBytes [6] and a shape parameter 1.66 [9]; the inter-arrival of requests is exponential distributed. In addition, we assume a new request can be generated even though previous requests have not been served. For receiving a reply, if the reply is a redirection acknowledgment, the client resends the request to the redirected server; if the reply is data, the client records the response time of the request.

Server Model: When a server receives a request, it first checks whether to accept this request or to redirect it to another server using the server selection algorithms presented above. If a redirecting decision is made, the server sends a redirecting acknowledgment back to the client; otherwise, the server accepts this request into a FIFO queue which holds all waiting requests when the server is serving a previous request.

A server always processes the request at the head of the waiting queue if the queue is not empty. The cost of processing a request on a server consists of its queuing cost, its operating system cost for starting and ending a service, and its storage accessing cost. Queuing cost is the time period between the request's arriving and leaving the queue. Operating system cost is counted using an average value (e.g., 0.1ms). Storage accessing cost are computed through a storage model [13] with the parameters, such as an average seek time (e.g., 10ms), an average rotation delay (e.g., 1ms), a controller overhead (e.g., 1ms), and the transferring cost from the storage to the memory with a 20 Mbytes/second I/O bus. The effect of disk cache and memory cache will be considered in our future study.

Network Model: Each client has a network path to each server, while the bottleneck bandwidth of the path can be approximated using the formula in [17,24] through the Maximum Transfer Unit (MTU), the Round-Trip Time (RTT) and the loss rate of the path.

$$Bandwidth = \frac{1.3 \cdot MTU}{RTT \cdot \sqrt{LossRate}}$$

From the results in [4,6,9,17,24], the typical values of loss rates, MTUs, and RTTs of network paths are: a loss rate is between 0.01% and 5%, a MTU is between 576 bytes to 1500 bytes, and a RTT is between 20 ms and 500 ms. In our simulation, the loss rate, MTU and RTT of a path are generated within the above ranges to simulate the dynamic status of the network paths.

Table 1. Response times among all clients

Method	AVG	MAX	MIN	STD	# Requests Served
FS	4.048	14.16	0.007	3.953	6955
FG	1.189	6.932	0.009	1.851	7949
RR	0.142	3.489	0.003	0.196	8341
BR	0.140	2.647	0.003	0.210	8343

3.2   Initial Simulation Results

In this section, we present our initial simulation results. Besides the proposed two server selection algorithms, RR and BR, two other static server assignment approaches are also tested in our simulation to investigate the effect of the server selection algorithms. The first one is called *Fixed-Server (FS)* assignment method: a client keeps requesting services from a fixed server which is initially randomly chosen by the client, and a server has a fixed number of clients during a simulation. This is the most naive situation. The second method is called *Fixed-Group (FG)* assignment approach: all clients are divided into equal size groups, and each group of clients only contacts with a single default server. In this method, each server has the same number of clients during a simulation. FS and FG are similar to the methods used in existing systems.

Our initial simulation results are given as follows. In this group of simulations, 16 clients require services from 4 servers during a period of 1 minute. In Table 1, the average response time of all clients under four approaches are given, as are the maximum, minimum and standard deviation of the response time. The total number of requests served are listed in the last column of Table 1. It is easy to see that RR and BR served more requests than FS and FG.

Table 2 shows the queuing times on the simulated servers. Similarly, the average value, the maximum, the minimum and the standard deviation are given.

Choosing the average results of BR as the base 1.00, Table 3 clearly shows that RR and BR are significant better than FS and FG both in the average response time among clients and the average queuing time among servers. An interesting finding is that RR is only slightly worse than BR. BR requires much more computing cost than RR, but RR did more redirections than BR. In our simulation, 51664 times of redirections took place in the RR approach, while

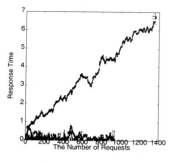

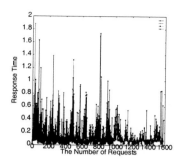

(a) Fixed-Group Algorithm.

(b) Random-Redirection Algorithm.

Fig. 5. Response times of the first 4 clients under FG and RR algorithms

Table 2. Queuing delays among all servers

Method	AVG	MAX	MIN	STD
FS	4.028	14.17	0.0	3.956
FG	1.166	6.749	0.0	1.850
RR	0.049	0.342	0.0	0.028
BR	0.044	0.327	0.0	0.028

47686 times of redirections happened in the BR approach. A large scale simulation will show even greater differences between them.

Table 3. Summary of simulation results

	FS	FG	RR	BR
Avg Response Time over all clients	28.91	8.49	1.01	1.00
Avg Queuing Time over all servers	85.70	24.80	1.04	1.00

The dynamic methods, such as RR and BR, distributed the requests more evenly among servers than the static methods such as FS and FG. The FG method is better than the FS method in static methods while the RR method is slightly worse than the BR method in dynamic methods. Therefore, we choose the FG and RR methods as representatives to compare static methods with dynamic methods. Figure 4 shows the queuing times of requests on four server in the FG and RR methods. The x axis is the number of requests which have been served on servers, and the y axis is the queuing time. Figure 4(a) shows that one waiting queue keeps building up in the FG method; however, Figure 4(b) shows that all the queuing times of requests are distributed in a small range in the RR method. Figure 5 shows response times of 4 clients under the FG and RR methods. The x axis is the number of requests which clients have received their replies, and the y axis is the response time. In Figure 5(a), the response time of one client keeps increasing in the FG method; on the other hand, Figure 5(b) clearly shows that, the response time of all client stay within a small range in the RR method.

4   Concluding Remarks

In this paper, we study two server selection algorithms under a server-based framework that we have developed in [14]. Both algorithms employ server load and network performance metrics collected through a shared passive measurement mechanism to determine the appropriate server for a client request. Simulations show that both dynamic algorithms are significantly better than static methods in terms of response times and queuing delays. Work is under way to implement this mechanism and investigate its performance over the Internet.

References

1. H. Balakrishnan, V. N. Padmanabhan, S. Seshan, M. Stemm, and R. H. Katz, "Analyzing stability in wide-area network performance," *in Proc. ACM SIGMET-RICS'97,* June 1997.
2. H. Balakrishnan, V. N. Padmanabhan, S. Seshan, M. Stemm, and R. H. Katz, "TCP behavior of a busy Internet server: analysis and improvements," *in Proc. IEEE INFOCOM'98,* San Francisco, CA, March 1998.
3. P. Barford and M. E. Crovella, "Generating representative web workloads for network and server performance evaluation," *Technical Report BU-CS-97-006,* Dec. 1997.
4. J. Bolot, "Characterizing end-to-end packet delay and loss in the I Internet," *Journal of High Speed Networks,* vol.2, no.3, pp. 305–323, Dec. 1993.
5. C. M. Bowman, *et. al.,* "Harvest: a scalable, customizable, discovery and access system," *Technical Report CU-CS-732-94,* Dept. of Computer Science, University of Colorado at Boulder, 1994.
6. N. Cardwell, S. Savage, and T. Anderson "Modeling the performance of short TCP connections," Oct. 1998.
7. R. L. Carter and M. E. Crovella, "Measuring bottleneck link speed in packet-switched networks," *Performance Evaluation,* vol. 27 & 28, 1996.
8. R. L. Carter and M. E. Crovella, "Dynamic server selection using bandwidth probing in wide-area networks," *in Proc. IEEE INFOCOM'97,* Kobe, Japan.
9. C. R. Cunha, A. Bestavros, and M. E. Crovella, "Characteristics of WWW client-based traces," *in Proc. ACM SIGMETRICS'96.*
10. Y. Dong, Y. T. Hou, and Z.-L. Zhang, "A novel server-based measurement infrastructure for Enterprise Networks ", *Technique Report TR 98-031,* Dept. of Computer Science, University of Minnesota, 1998.
11. Z. Fei, S. Bhattacharjee, E. W. Zegura, and M. Ammar, "A novel server selection technique for improving the response time of a replicated service," *in Proc. IEEE INFOCOM'98,* San Francisco, CA, 1998.

12. J. Guyton and M. Schwartz, "Locating nearby copies of replicated Internet services," *in Proc. ACM SIGCOMM'95,* Cambridge, MA, 1995.

13. J. Hennessy and D. Patterson, *Computer Architecture: A Quantitative Approach,* Second Edition, Morgan Kaufmann Publishers, Inc., 1996.

14. Y. T. Hou, Y. Dong, and Z.-L. Zhang, "Network performance measurement and analysis - part 1: a server-based measurement infrastructure", *Technical Memorandum FLA-NCRTM98-01,* Fujitsu Laboratories of America, CA, July, 1998.

15. Internet Distance Maps Project, `http://idmaps.eecs.umich.edu/`.

16. V. Jacobson, `pathchar` - A Tool to Infer Characteristics of Internet Paths, `http://ee.lbl.gov/nrg-talks.html`, April 1997.

17. J. Mahdavi and Sally Floyd, "TCP-Friendly Unicast Rate-Based Flow Control", The TCP-Friendly Website.

18. S. McCanne, C. Leres, and V. Jacobson, `tcpdump` , Lawrence Berkeley National Laboratory.

19. K. Moore, J. Cox, and S. Green, "SONAR - A Network Proximity Service," *Internet-Draft,* `http://www.netlib.org/utk/projects/sonar/`, Aug. 1996.

20. A. Myers, P. Dinda, and H. Zhang, "Performance characteristics of mirror server on the Internet," *In Proc. IEEE INFOCOM'99,* March 1999.

21. V. Paxson, "Measurements and analysis of end-to-end Internet dynamics," *Ph.D. Dissertation,* U.C. Berkeley, 1997.

22. V. Paxson, J. Mahdavi, A. Adams, and M. Mathis, "An architecture for large-scale Internet measurement," *IEEE Commun. Magazine,* pp. 48–54, Aug. 1998.

23. S. Seshan, M. Stemm, and R. Katz, "SPAND: shared passive network performance discovery," *in Proc. USENIX'97,* 1997.

24. M. Yajnik, J. Kurose, and D. Towsley "Packet loss correlation in the MBone multicast network," *Technical Report UMASS CMPSCI 96-32,* Dept. of Computer Science, Univ. of Mass. at Amherst, 1996.

Designing a Virtual Environment for Large Audiences

Jori Liesenborgs, Peter Quax, Wim Lamotte, and Frank Van Reeth

Expertisecentrum Digitale Media
Limburgs Universitair Centrum
Wetenschapspark 2, B-3590 Diepenbeek, Belgium
{jori.liesenborgs, peter.quax, wim.lamotte, frank.vanreeth}@luc.ac.be

Abstract. In this paper we describe our approach in creating a truly scalable distributed virtual environment. The resulting system should not only support large amounts of users, but it should also be able to work with a general purpose network like the Internet and it is intended to work with typical personal computers or game consoles. To realize this, we propose a distributed client-server approach in which most of the responsibilities are concentrated at the client side. The servers will only be used for control purposes – e.g. access control. Direct communication between clients helps preventing servers from becoming bottlenecks and the network load is kept low by distributing only the actions of a client instead of continuously distributing position updates. Finally, we also address some security related issues which arise from this client-oriented approach.

1   Introduction

Distributed virtual environments have already been the topic of several research projects. The resulting systems were usually meant to run on powerful computing platforms, often with a dedicated network where much bandwidth was available. These systems usually allowed only a limited number of users and were often only useful for specific application domains (mostly military).

Currently, we are exploring the feasibility of a truly large-scale distributed virtual environment, to which users can connect over a general-purpose network like the Internet. In this situation, there are several differences with previously realized projects.

First of all, one of the key interests of our research is to make it possible for the system to support huge amounts of users, all occupying the same virtual world. Previously designed systems often had difficulties with large amounts of users (over one thousand). With our approach, we aim to provide access to the virtual world for a much larger number of users (over ten thousand).

Furthermore, since we are not working with a dedicated network, there are no guarantees about network delay or jitter. The accessibility of the virtual environment through the Internet also raises security-related issues: measures must be taken to avoid malicious users from interfering with the virtual environment.

I. Chong (Ed.): ICOIN 2002, LNCS 2344, pp. 585–595, 2002.
© Springer-Verlag Berlin Heidelberg 2002

Finally, the system is not intended to be useful only with powerful computing platforms. Instead, typical personal computers and possibly game consoles are the main target platforms.

This paper describes our ongoing work concerning distributed virtual environments for large audiences. While some of the proposed techniques have not yet been implemented, most of our early testing suggests that the principles are sound and applicable. In the next section, we will discuss some related work. Afterwards, our general architecture will be explained and some security related issues are discussed. The paper concludes with the current status of our work.

2   Related Work

Support for multiple users in networked virtual environments emerged primarily from real-time military simulation systems. One of the first real multi-user virtual environments was SIMNET, later followed by its successor DIS (Distributed Interactive Simulation Protocol) [1][16][17].

In these frameworks, packets containing position and velocity information were continuously distributed among the participants to create the virtual world experience. To reduce the amount of bandwidth required for this information, a technique called dead-reckoning was introduced. With this technique, all participants derive the current position of an object from position and velocity information which was previously received from that object. When the difference between actual and predicted position becomes too large, the object distributes new information.

Early versions of these frameworks used broadcast communications to distribute data between clients; later versions included multicast protocols for use on LAN and WAN's. Another military framework that was implemented for real-time simulations was NPSNET [7][5], which divides the world in hexagonal cells, whereby every cell is associated with its own multicast group. When the users move through the world, they dynamically leave and join other multicast groups. Most of these military systems were dependent on a more or less static world.

Later efforts to implement large networked virtual environments included Spline [14], which introduced the idea of locales [2]. These locales represent specific parts of the world and each locale has a specific set of multicast groups associated with it. The general idea behind this was that the user was probably not interested in something that happened at the other end of the (large) world. Therefore, in Spline the user did not receive data from locales that were not currently nearby. Spline also featured local world models, whereby each of the participants kept its own database of objects in the world (or locale). The Diamond Park [14] virtual world was based on Spline.

Experimental collaborative virtual environments also included MASSIVE and MASSIVE 2 [1] [18] [19], which respectively used unicast and multicast distribution models. They were however still relatively limited in the number of users that could participate in one session.

Finally, the DIVE [10] platform included a number of the techniques described above, but it also included peer-to-peer communications, whereas many of the other examples (heavily) rely on a client-server model.

Our architecture combines ideas from the above frameworks, and extends them to enable the creation of very large scale virtual environments.

3   Architecture

In this section we discuss the general approach that we are following in our design of a large scale distributed virtual environment. First, some design considerations are explained. Afterwards, the client side responsibilities are presented and the section is concluded with an overview of the tasks of a server in our system.

3.1   Design Considerations

Before explaining the architecture we designed, we will highlight some observations which led to this architecture. First of all, it is important to recognize the necessity of some kind of client-server mechanism. The server side is needed to provide access control to the virtual environment, to perform security related tasks, etc.

Furthermore, to be able to scale to the large amount of users that we envision, the architecture should be distributed: the total load of the system should be distributed over the available servers to prevent a server from becoming a bottleneck.

Even with this distributed approach it is still possible for the servers to become bottlenecks when too much network traffic has to go through them. For this reason, in our approach we concentrate most of the responsibilities at the client side and keep the interaction with the servers to an absolute minimum.

3.2   Client Side

To reduce the load on the servers in our architecture, clients will communicate with each other in a direct way. To make efficient use of the available network resources, communication will be done using multicasting. Even though our goal is to create a distributed virtual world that works on a *general purpose* network, we believe the use of multicasting is justified since multicast supporting network devices are becoming more widely employed every day.

Clients use different multicast groups in a hierarchical way. Those which are nearby in the virtual world will receive the necessary messages for a detailed view of each other. Clients will also periodically send an update of their information to a more general multicast group, which will be monitored by clients far away. This way, a client can still get a general idea of what is happening at a long distance. Some distinction will have to be made based upon the properties of an object. For example, large objects should be visible from a greater distance than smaller ones.

Previously designed distributed virtual worlds usually depended on continuous updates of position and velocity. Dead-reckoning was employed to reduce the required bandwidth, but it still required much of the network's capacity. In our architecture, we use a more high-level approach: instead of sending positional updates, action descriptions are sent. For example, an action could be *start walking forward* or *stop walking*. This obviously reduces the required network capacity. The efficiency of this approach depends on the actual messages sent and on the way these messages are encoded, but it clearly requires less bandwidth.

To make sure that each client knows exactly when an action occurred, the action is accompanied by a timestamp. The internal timestamp clocks of all clients are synchronized by the servers, as will be explained in the next section. Using this timestamp-action based scheme, all clients will have the same idea of what goes on in the virtual world.

A client has full responsibility for what it sees of the world. As will be explained in the next section, the virtual world will be divided into smaller pieces, managed by different servers. To get information about different parts of the world, a client has to connect to servers which manage those parts. This way, for example, the clients can determine themselves 'how far' they can see.

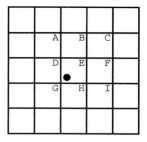

Fig. 1. Example world grid

Let us illustrate this concept. Consider the situation depicted in figure 1. Each square represents a part of the world that is managed by a different server. The client – represented by the small black dot – is in the region managed by server E. Because the client is quite close to the borders of this world, it will probably make a connection to servers D, G and H. If the client should be able to 'see' even further, it could connect to other servers as well to obtain their client information. To make this technique possible, a server must know all of its neighbors and a client must be able to access this information.

It is important to stress the fact that the only purpose of the servers is to provide information about the clients in their respective regions. All information regarding changes in the world – e.g. an avatar which starts to move – is distributed peer-to-peer. This way, the client does not have to do anything special when it moves into a region managed by another server.

Each client is also responsible for collision detection of other clients with its own avatar. By only detecting collisions with itself and spreading the collision response in the form of an action, the collision detection scheme becomes distributed: each client only has to check against the other clients. This has an algorithmic complexity of $O(n)$, where n is the number of clients nearby. This obviously scales very well with increasing n.

There are some subtle difficulties however. Network related issues like delay or packet loss combined with the action based scheme make it possible that, for example, collisions are detected which actually did not occur. This could be possible when a *stop walking* message was not received in time.

Currently, we are still working out how to resolve such problems. Depending on the situation, it is possible that resolving the inconsistency – i.e. backtracking and performing the correct action – is more disturbing to the user than simply ignoring it. Further tests need to be done to decide what is best in a specific situation.

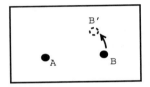

Fig. 2. Discontinuous motion

Network delay is also responsible for other phenomena: discontinuous motion and late collision responses. Consider the situation in figure 2 where client A observes client B. At time t, client B distributes a *start walking* message. However, because of the network delay δ, this message will only reach client A at time $t + \delta$. When client A has to display the actual position of B, a certain distance will already be traveled and B will seem to jump to a different position (B' in the figure).

Two techniques can help to resolve this problem. First of all, by adjusting the timestamp of an action, we can make the action take place in the future. If the difference we introduce in the timestamp is larger than the network delay δ, it is clear that there will not be any discontinuous motion. However, to keep the experience realistic, we cannot introduce too much delay this way: as described in [11], only interaction delays below 150 ms are not noticeable.

The following example will clarify this. Suppose that at a given time a user instructs its client to move forward. The client will then build a packet containing the *start walking* message. However, instead of setting the timestamp to the current time, the timestamp is set to a time in the future, e.g. 150 ms later. Locally, this will cause the action to take place after a small but hardly noticeable delay. When the action is distributed, this increased timestamp value makes sure that network delays below 150 ms do not affect the synchronization between clients.

The second technique copes with the delay in a different way: instead of displaying the actual position jump, a smooth transition from one state to the other will be calculated and displayed. At a certain time, the displayed position will converge with the actual position. As an example, on the reception of a *start walking* message the corresponding avatar would start walking with a slightly higher velocity until the actual and displayed positions are the same.

Late collision responses can originate from the fact that each client distributes its own collision response. Again, because of the network delay, this collision response will reach other clients only after a specific time, resulting in a late collision response.

Currently, we are still exploring possible solutions to this problem. One approach would be the following: when a client detects a collision with its own avatar, it not only distributes its own collision response, but it also predicts and displays a collision response for the other client involved. When the actual collision response of the other party is received, the client will make the predicted path converge with the real one.

3.3   Server Side

In our proposed architecture, the servers act as managers for parts of the large virtual world. The general idea is to severely limit the amount of data that has to be processed by the servers, thereby leaving the responsibility of data distribution to the clients. To achieve this goal we limited the server's tasks, as will be shown in detail below.

A first task of a server is distribution of participant information. When a new user joins an ongoing session, the server distributes information about this new participant to the interested clients. Information that is sent includes the type of avatar that the new user has chosen and communication information, like the the user's network address (for peer-to-peer communications between clients).

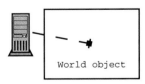

Fig. 3. World object managed by a single server

The server also provides synchronization information to its clients. As was explained before, the clients in our system use a system based on timestamps and action descriptions. Because of this, there is need for some sort of synchronization between participants in a session. In our model, a client application has an internal timestamp clock which is synchronized by a server. To stay synchronized, clients regularly contact the server to check what the difference is between their own clocks and the server's clock, taking into account the network delay.

In the previous section, we explained how multicasting techniques are employed to make data distribution as efficient as possible. Ideally, each participant in the virtual world should have its own multicast group, but this may not be realistic. The most restraining factor will probably be the network resources: for each multicast group that is joined, memory is needed in network devices (network cards, routers, etc).

This brings us to another task of the servers: client grouping. Instead of allocating a multicast group for each client, it is more efficient to group several clients which are nearby and allocate a single multicast group for all of them.

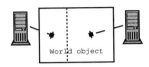

Fig. 4. World object managed by two servers

An important fact in our architecture is that there is no static allocation of servers to different parts of the virtual world. Instead, there is only a pool of servers which dynamically manage different parts of the world in a hierarchical way. Let us clarify this concept:

First, it is essential to note that all objects in our world model are represented as clients. Even objects that act as scenery or static objects will themselves connect to servers. Considering this fact, when a session is started we have to introduce an initial object that represents the world. Initially, this world object is managed by a single server, as shown in figure 3. When more and more clients become part of the world, the management of the world will be divided among several servers, as depicted in figure 4.

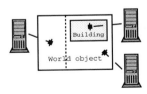

Fig. 5. Building object managed by a server

When an object with an interior is introduced, a building for example, this interior can in turn be managed by a server. This is illustrated in figure 5. When this interior gets crowded, management can be divided over several servers, just like with the world object. A large object inside the building – e.g. a room – can again be managed by a different server etc. Furthermore, the actual area managed by each server can be adapted according to the distribution of clients to balance the load on the servers.

It should be clear from the description of the server's tasks that they will be able to handle a large number of clients in a virtual world. There is no responsibility like maintaining a continually changing 'world database' that is accessed by all of the clients, thereby generating a lot of network traffic and load on the server itself. There is also no need for the servers to act as distributors of data between the clients, as they only use multicast and/or peer-to-peer communications. Load on the server will only temporarily be high when for example, management of areas has to be rearranged or when a large number of users joins the world at the same time.

4   Security Issues

In the previous sections we explained the basic ideas behind our client-oriented architecture. Putting most of the responsibilities at the client side reduces the possibility of servers becoming bottlenecks for the system. However, allowing a client to do so much on its own also has an impact on the security of the system.

For example, because a client is responsible for detecting collisions with its own avatar, a malicious user could make modifications to the client in such a way that no collisions are detected at all. This would enable this user to do things that should not be possible, like walking through walls. Similarly, a user could be able to teleport himself to an arbitrary location in the virtual environment. The definition of illegal or unwanted behavior is of course application-dependent.

A possible solution to such problems is to introduce some 'trusted objects' into the virtual world. These objects could then monitor other clients and when inappropriate behavior is observed, the server could be instructed to disconnect the involved client. Because of our client-oriented approach, this does not only require the control connections with the servers to be broken, but information must also be spread to make sure that other clients know they should ignore the malicious client.

This monitoring functionality could be incorporated in objects which do not have to do much on their own, like buildings or trees for example. Alternatively, some invisible clients could be connected to the world to perform this task. One can even imagine some sort of virtual police force to monitor for 'illegal' behavior.

Another security related issue is privacy. For example, a client could join the multicast channel of another client's speech data and hear everything that is said. Since only the appropriate multicast group needs to be joined to do this, this is nearly impossible to detect. The only way this kind of abuse can be avoided is by making a client encrypt its own speech data. The key to decode this data could then either be distributed to the right set of clients by a server, or supplied on demand by the client itself. The downside of encryption however, is that it slows things down and consumes more bandwidth.

5   Project Status & Future Work

Currently we are focusing on the client side issues of our architecture. For this purpose, a lightweight server has been implemented that allows clients to connect to a prototype world. At the moment, the communication channels are unicast instead of multicast and clients communicate with each other through the server, which distributes messages if necessary.

The prototype of the client application implements many of the techniques that were described in the client-side section such as action distribution, synchronization, elementary collision detection and action processing. Extensions to this prototype will support multicasting, detection of invalid movements and extended collision detection. Voice communication in the virtual world is made possible using JVOIPLIB[15]. Below, a screenshot of the client application is shown.

When the client side issues are resolved, this one-server implementation will be tested to see how many clients it can support. From this testing process we intend to obtain the necessary information concerning bandwidth use, thereby enabling us to make more accurate predictions about the supported numbers of simultaneous users. It is important to get an idea of the scalability of our approach before starting work on the multiple servers approach.

The 'pool of servers', as described above, has to be managed somehow. In the initial stages of the research, they will be managed by a centralized server application to be able to concentrate on the core issues. However, it is clear that some distributed protocol should be used instead or else the central server itself will become a very vulnerable point. This protocol will be defined and implemented in the final stages of the project.

Fig. 6. Screenshot of a client's view

6   Conclusion

In this paper we described our approach in designing a scalable virtual environment for a large number of users. To achieve this scalability, we use several techniques which are summarized below.

First of all, the world is divided into smaller pieces in a hierarchical way. This way, the total load can be divided among the available servers. Furthermore, a client can decide on its own which parts of the world it receives information from, effectively only monitoring the regions it is interested in.

To reduce the likelihood of a server becoming a bottleneck, interaction with the server is reduced to an absolute minimum. Efficient usage of the available bandwidth is achieved by using multicasting combined with a high level action-based mechanism.

Each client only interacts with a relatively small number of clients: those which are close enough to be of interest. In general this will only be a small fraction of the total amount of clients in the entire virtual environment. It is this localized interaction that will make very large virtual environments possible. As stated before, we hope to support this claim by further development of our prototype framework.

Although the project is still under development, we believe that with the approach discussed in this paper, a very scalable virtual environment can be constructed.

Acknowledgments. This research is funded by EFRO (European Fund for Regional Development) and the Flemish government.

References

1. Matijasevic, M.: A Review of Networked Multi-User Virtual Environments. World Wide Web, http://citeseer.nj.nec.com/matijasevic97review.html (1997)
2. Barrus, J., Waters, R., Anderson, D.: Locales and Beacons: Precise and Efficient Support for Large Multi-User Virtual Environments. Proceedings of VRAIS'96, Santa Clara CA (1996).
3. Waters, R.C.: Time Synchronization in Spline. World Wide Web, http://www.merl.com/reports/TR96-09/, 1996.
4. Lea, R., Raverdy, P.-G., Honda, Y., Matsuda, K.: Scaling a shared virtual environment. World Wide Web, http://www.csl.sony.co.jp/person/rodger/ICDCS/icdcs2.html, 1996.
5. Macedonia, M.R., Brutzman, D.P., Zyda, M., Pratt, D.R., Barham, P.T. Falby, J., Locke, J.: NPSNET: A Multi-Player 3D Virtual Environment over the Internet. In: Symposium on Interactive 3D Graphics (1995) 93–94, 210
6. Roehl, R.: Distributed Virtual Reality – An Overview. World Wide Web, http://ece.uwaterloo.ca/~broehl/distrib.html, 1995.
7. Macedonia, M. R., Zyda, M. J, Pratt, D.R., Barham, P.T., Zeswitz, S.: NPSNET: A Network Software Architecture for Large-Scale Virtual Environment. Presence, **3** (1994) 265–287
8. Department Of Computer Science and Engineering, The Chinese University of Hong Kong. Issues in a Very Large Scale Distributed Virtual Environment. World Wide Web, http://citeseer.nj.nec.com/202052.html, 1997.
9. Pryce, N.: Group Management and Quality of Service Adaptation in Distributed Virtual Environments. World Wide Web, http://www.brunel.ac.uk/faculty/tech/systems/groups/vvr/vrsig97/proceed/007/group-qos.html, 1997.

10. Frécon, E., Stenius, M.: DIVE: A Scaleable network architecture for distributed virtual environments. Distributed Systems Engineering Journal **5** (1998) 91–100
11. Vaghi, I., Greenhalgh, C., Benford, S. Coping with Inconsistency due to Network Delays in Collaborative Virtual Environments. In: Proceedings of the ACM symposium on Virtual reality software and technology. (1999) 42–49
12. Macedonia, M., Zyda, M., Pratt, D., Brutzman, D., Barham, P.: Exploiting Reality with Multicast Groups: A Network Architecture for Large-scale Virtual Environments. Proceedings of VRAIS'95 (1995)
13. Roehl, B.: Channeling the Data Flood. IEEE Spectrum **34** (1997) 32–38
14. Waters, R., Anderson, D., Barrus, J.: Diamond Park and Spline: A Social Virtual Reality System with 3D Animation. World Wide Web, http://www.merl.com/reports/TR96-02a/ (1996)
15. Liesenborgs, J., Lamotte, W., Van Reeth, F.: Voice over IP with JVOIPLIB and JRTPLIB. The 26th Annual IEEE Conference on Local Computer Networks (LCN) (2001)
16. Calvin, J., Dicken, A., Gaines, B., Metzger, P., Miller, D., Owen, D.: The SIMNET virtual world architecture. Proceedings of VRAIS'93, (1993) 450–455,
17. DIS Steering Committee: The DIS Vision: A Map to the Future of Distributed Simulation, (1993)
18. Greenhalgh C., Benford, S.: MASSIVE: A distributed virtual reality system incorporating spatial trading. In: International Conference on Distributed Computing Systems. (1995) 27–34
19. Greenhalgh, C., Benford, S.: A multicast network architecture for large scale collaborative virtual environments. In European Conference on Multimedia Applications, Services and Techniques. (1997) 113–128

Optimal k-Coteries That Maximize Availability in General Networks

Eun Hye Choi[1], Tatsuhiro Tsuchiya[2], and Tohru Kikuno[2]

[1] Corporate Research & Development Center, TOSHIBA Corporation
1 Komukai Toshiba-Cho, Saiwai-Ku, Kawasaki 212-8582, Japan
[2] Graduate School of Information Science and Technology, Osaka University
1-3 Machikaneyama, Toyonaka-City, Osaka 560-8531, Japan
{choi, t-tutiya, kikuno}@ics.es.osaka-u.ac.jp

Abstract. We consider a distributed system where there are k indistinguishable shared resources that require exclusive access by different computing nodes. The k-mutual exclusion problem is the problem of guaranteeing the resource access request in such a way that no more than k computing nodes enter a critical section simultaneously. The use of a k-coterie, which is a special set of node groups, is known as a robust approach to this problem. In this paper, we address the problem of constructing optimal k-coteries in terms of availability in general topology networks with unreliable nodes and links. We formulate the problem of constructing an optimal k-coterie as a combinatorial optimization problem with a linear objective function. To solve the formulated problem, we propose a branch-and-bound method based on a linear programming relaxation. The feasibility of the proposed method is shown through an experiment conducted for several networks.

Keywords. k-mutual exclusion, k-coterie, availability, combinatorial optimization problem, branch-and-bound

1 Introduction

In a distributed environment, a number of different resources are shared by computing nodes. The resource could be, for example, a database or other data structure that requires exclusive access in order to avoid interference among the operations of different computing nodes. The distributed mutual exclusion problem is the problem of guaranteeing that only one computing node will access a shared resource at a given time, and is recognized to be one of the most fundamental distributed problems.

The distributed *k-mutual exclusion* is a generalization of the classical mutual exclusion and is specifically the problem of guaranteeing that no more than k computing nodes enter the critical section (CS), in other words, access to shared resources simultaneously. The solution to the k-mutual exclusion problem is useful for various applications in a distributed environment. For example, the solution can be used to restrict the number of broadcasting nodes for congestion control. It can also be useful in replicated databases that allow bounded ignorance[6]. In such databases, simultaneous updates are allowed in order to increase concurrency.

The use of a *k-coterie* is known as a promising approach for solving the k-mutual exclusion problem[1,4,5,7]. A k-coterie is a special set of subsets of nodes. Each element

I. Chong (Ed.): ICOIN 2002, LNCS 2344, pp. 596–608, 2002.
© Springer-Verlag Berlin Heidelberg 2002

in a k-coterie is called a *quorum*. For any $k+1$ quorums in a k-coterie, there is always a node that is shared by at least two of the $k+1$ quorums. Each node has to gain permissions from all nodes of a quorum before it is allowed to enter the CS, and thus it is guaranteed that no more than k nodes enter the CS simultaneously.

In the presence of node and link failures, k-coterie-based mutual exclusion mechanisms provide desirable fault-tolerance capability. For example, as long as all nodes of a quorum are connected via operational paths when failures occur, there still exists a node that can enter the CS by obtaining permission from the quorum. The availability of a k-coterie-based mechanism clearly depends on the k-coterie adopted. In order to achieve high reliable k-mutual exclusion, therefore, construction of k-coteries that provide high availability is indispensable.

Several constructions for k-coteries have been proposed[1,4,5]. Among them, a k-*majority coterie*[5,9] was shown to be optimal in the sense of maximizing availability. However, this holds only when all nodes are fully connected by completely reliable links[5]. No construction of optimal k-coteries has been suggested for general topology networks so far.

In this paper, we address the problem of constructing an optimal k-coterie for a given arbitrary topology network with unreliable nodes and links. We first propose a new notion called a k-*acceptance set* as an alternative representation of a k-coterie and show that the problem of constructing an optimal k-coterie is equivalent to that of finding an optimal k-acceptance set. We next formulate the problem of finding an optimal k-acceptance set as a combinatorial optimization problem with a linear objective function. In order to solve the formulated problem, we also propose a method using the branch-and-bound technique. The search space for the branch-and-bound is reduced by estimating an upper bound of the objective function. For the upper bound estimation, we derive a linear programming problem as a relaxation of the formulated problem. Through an experiment conducted for several networks, we show the feasibility of the proposed method.

2 Preliminaries

2.1 System Model

We consider a distributed system modeled by an undirected graph $G = (V, E)$, where $V = \{v_1, v_2, \cdots, v_n\}$ is a set of vertices and E is a set of edges incident on vertices in V. Each vertex $v_i \in V$ represents a computing node (*node*, for short) and each edge $e_{i,j} \in E$ represents a communication link (*link*, for short) between node v_i and node v_j. We use the term *component* to indicate a node or a link.

We assume that failures of components are mutually independent and each component has two states: operational and failed. Let $p_i(0 < p_i \leq 1)$ and $p_{i,j}(0 < p_{i,j} \leq 1)$ denote the probability that v_i is operational and the probability that $e_{i,j}$ is operational, respectively. We refer to these probabilities as *reliabilities* of components.

Based on the above assumptions, we introduce some definitions. We call a tuple (Vc, Ec) with $Vc \subseteq V$ and $Ec \subseteq E$ a *configuration* of G. The *current configuration* is defined as a configuration (Vc, Ec) such that Vc and Ec are the set of all currently operational nodes and links, respectively. The current configuration represents the current status of the network. The universe U of all possible configurations over G is defined as follows: $U = \{(Vc, Ec)|Vc \in V \text{ and } Ec \in E\}$.

Nodes in a configuration are divided into some isolated groups which we call *partition groups*. Formally, given a configuration $(Vc, Ec) \in U$, a subset N of Vc is called a partition group iff all nodes in N are connected by links in Ec and N is maximal with respect to this property. We use the notion $\mathcal{P}(Vc, Ec)$ to represent a set of all partition groups in configuration (Vc, Ec).

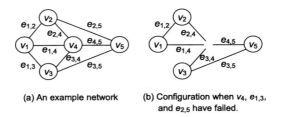

(a) An example network

(b) Configuration when v_4, $e_{1,3}$, and $e_{2,5}$ have failed.

Fig. 1. An example of a network.

Example 1. Consider the network $G = (V, E)$ shown in Figure 1(a). Figure 1(b) sche-matically illustrates the current configuration when v_4, $e_{1,3}$, and $e_{2,5}$ have failed. The current configuration (Vc, Ec) is then $(\{v_1, v_2, v_3, v_5\}, \{e_{1,2}, e_{1,4}, e_{2,4}, e_{3,4}, e_{3,5}, e_{4,5}\})$. All partition groups in (Vc, Ec) are $\{v_1, v_2\}$ and $\{v_3, v_5\}$; that is, $\mathcal{P}(Vc, Ec) = \{\{v_1, v_2\}, \{v_3, v_5\}\}$.

2.2 k-Coteries

This section introduces the formal definition of a k-coterie and explains how to achieve k-mutual exclusion using a k-coterie. In the following, we refer to a nonempty subset of V as a *node group* to avoid confusion.

Definition 1   (k-coterie and quorum[3]). A nonempty set $\mathcal{C}$ of node groups is called a k-coterie iff the following two conditions hold:

- Intersection property: For any $k + 1$ elements $Q_1, \cdots, Q_{k+1} \in \mathcal{C}$, there exists a pair Q_i and Q_j such that $Q_i \cap Q_j \neq \emptyset (1 \leq i, j \leq k + 1)$.
- Minimality property: For any $Q \in \mathcal{C}$, there is no $Q' \in \mathcal{C}$ such that $Q' \subset Q$.

Each element of a k-coterie $\mathcal{C}$ is called a *quorum*.

Example 2. Suppose $V = \{v_1, v_2, v_3, v_4, v_5, v_6\}$ and $k = 3$. Now, consider the following sets of node groups $\mathcal{C}_1, \mathcal{C}_2$, and $\mathcal{C}_3$.

$$\mathcal{C}_1 = \{\{v_1, v_2\}, \{v_1, v_3\}, \{v_4\}, \{v_5\}\}.$$
$$\mathcal{C}_2 = \{\{v_1, v_2\}, \{v_3\}, \{v_4\}, \{v_5, v_6\}\}.$$
$$\mathcal{C}_3 = \{\{v_1, v_2\}, \{v_1, v_3\}, \{v_4\}, \{v_5\}, \{v_4, v_5\}\}.$$

Among the above sets, C_1 is a 3-coterie. C_2 and C_3 are not 3-coteries either because they do not satisfy the intersection property and the minimality property, respectively.

k-Mutual exclusion is achieved using a k-coterie as follows: Before entering the CS, a node has to gain permission from every node in at least one quorum. Each node is allowed to give permission to at most one node. If a node has attained permission, it enters the CS and holds the permission until it leaves the CS. By the intersection property, it is always guaranteed that no more than k nodes enter the CS simultaneously.

2.3   Availability of k-Coterie

Availability Measure. As an availability measure for a k-coterie, we introduce a *computation availability*[2]. Computation availability has been used for evaluating gracefully degradable systems in terms of both availability and performance. Here we define the computation availability of a k-coterie as follows. (In the rest of the paper, we refer to computation availability as availability for short.)

$$\text{Availability} = \sum_{r=1}^{k} \alpha(r) \times r\text{-availability}$$

where $\alpha(r)$ denotes the computation capability when $r(1 \leq r \leq k)$ nodes can enter the CS simultaneously, and an r-availability is the probability that the maximum number of nodes that can enter the CS simultaneously is exactly r[5].

Assuming that α_r is proportional to r, we set α_r to r/k. The implication of this assumption is that the computation capability increases proportionally in the amount of available resources, and it has often been used in the literature. We then obtain the following equation.

$$\text{Availability} = \frac{1}{k} \sum_{r=1}^{k} r \times r\text{-availability}$$

Availability in Network. In the presence of node and link failures, the CS can be entered iff there is a partition group that contains a quorum in the current configuration. We say that a quorum is *available* in a configuration if there exists a partition group that contains the quorum in the configuration. Thus r nodes can enter the CS simultaneously iff r disjoint quorums are available in the current configuration. The r-availability of a k-coterie is then equal to the probability that the maximum number of disjoint quorums in the k-coterie that are available in the current configuration is equal to r.

In the following, we present the formal definition of an availability of a k-coterie C in a network $G = (V, E)$. Given k-coterie C and integer $r(1 \leq r \leq k)$, let $U_{C,r}(\subseteq U)$ be the set of all configurations in which at least r disjoint quorums in C are available. For example, consider the network shown in Figure 1(a) and the 3-coterie C_1 discussed in Example 2. In the configuration (Vc, Ec) illustrated in Figure 1(b), two disjoint quorums $\{v_1, v_2\}$ and $\{v_5\}$ are available because the quorums are contained in partition groups $\{v_1, v_2\}$ and $\{v_3, v_5\}$, respectively. Thus $(Vc, Ec) \in U_{C,1}$, $(Vc, Ec) \in U_{C,2}$, and $(Vc, Ec) \notin U_{C,3}$. Based on configurations, the availability of a k-coterie is then defined as follows.

Definition 2 (availability). Let $\mathcal{C}$ be a k-coterie. The availability of $\mathcal{C}$, $A(\mathcal{C})$, is defined as the following formula:

$$A(\mathcal{C}) = \tfrac{1}{k} \sum_{r=1}^{k} \sum_{(Vc,Ec) \in U_{\mathcal{C},r}} prob(Vc, Ec)$$

where $prob(Vc, Ec)$ is the probability that the current configuration is (Vc, Ec).

3 k-Acceptance Set

3.1   Definition

In this section, we introduce a new notion called a k-acceptance set and show that an optimal k-coterie is obtained directly from an optimal k-acceptance set.

Definition 3 (k-acceptance set). A nonempty set of node groups is called a k-acceptance set iff it satisfies the intersection property of the k-coterie in Definition 1.

A k-acceptance set differs from a k-coterie only in that the minimality property is not required for a k-acceptance set. For example, $\mathcal{C}_3$ shown in Example 2 is a 3-acceptance set.

We say that a k-acceptance set S is *maximal* iff for any element N in S, a superset of N is also an element of S. Consider a k-coterie $\mathcal{C}$, and a set S of node groups that consists of every quorum in $\mathcal{C}$ and every node group containing a quorum in $\mathcal{C}$; that is, $S = \{N \subseteq V | N \supseteq N'(\in \mathcal{C})\}$. Obviously, S is a maximal k-acceptance set.

Let $h(N)$ be the probability that a node group N forms a partition group in the current configuration, i.e.,

$$h(N) = \sum_{\{(Vc,Ec) \in U | N \in \mathcal{P}(Vc,Ec)\}} prob(Vc, Ec).$$

Additionally, let $F_{S,r} : 2^V \rightarrow \{0, 1\}$ be a function that is determined from a k-acceptance set S and an integer $r (1 \leq r \leq k)$ as follows: For every $N \subseteq V$, (1) $F_{S,1}(N) = 1$ iff $N \in S$, and (2) $F_{S,r}(N) = 1$ with $2 \leq r \leq k$ iff there exist r disjoint elements $Q_1, Q_2, \cdots, Q_r$ in S such that $N = Q_1 \cup Q_2 \cup \cdots \cup Q_r$.

Definition 4 (availability of k-acceptance set). Let S be a k-acceptance set. The availability of S, $\hat{A}(S)$, is defined as follows:

$$\hat{A}(S) = \tfrac{1}{k} \sum_{r=1}^{k} \sum_{N \subseteq V} h(N) F_{S,r}(N).$$

We say that a k-acceptance set S is optimal iff the availability of S, $\hat{A}(S)$, is the highest among all possible k-acceptance sets.

Consider a maximal k-acceptance set S and a set $\mathcal{C}$ that contains only minimal elements in S. Obviously, $\mathcal{C}$ is a k-coterie. As will be shown in Lemma 1, the availability of $\mathcal{C}$ is equal to the availability of S.

Lemma 1. Let S be a maximal k-acceptance set and a k-coterie $C = \{N \in S | \forall N' \in S, N' \not\subset N\}$. The availability of C is then equal to the availability of S.

Proof. For a set $\mathcal{D}$ of node groups, let $d(\mathcal{D}, N)$ denote the maximum number of disjoint elements in $\mathcal{D}$ that are contained in a node group N. Then $d(C, N) = d(S, N)$, for any node group N. Since $F_{S,r}(N) = 1$ only if $1 \leq r \leq d(S, N)$, $d(C, N) = d(S, N) = \sum_{r=1}^{k} F_{S,r}(N)$. Thus,

$$A(C) = \tfrac{1}{k} \sum_{(Vc, Ec) \in U} \sum_{N \in \mathcal{P}(Vc, Ec)} d(C, N) \times prob(Vc, Ec)$$

$$= \tfrac{1}{k} \sum_{(Vc, Ec) \in U} \sum_{N \in \mathcal{P}(Vc, Ec)} \sum_{r=1}^{k} F_{S,r}(N) \times prob(Vc, Ec)$$

$$= \tfrac{1}{k} \sum_{r=1}^{k} \sum_{N \subseteq V} h(N) F_{S,r}(N) = \hat{A}(S).$$

$\square$

Lemma 2. If S is an optimal k-acceptance set but is not maximal, there exists a maximal k-acceptance set S' such that $\hat{A}(S') = \hat{A}(S)$ and $\{N \in S' | \forall N' \in S', N' \not\subset N\} = \{N \in S | \forall N' \in S, N' \not\subset N\}$.

Proof. Let $Q = \{N \subseteq V | N \notin S, \exists N' \in S(N \supset N')\}$ and $S' = S \cup Q$. Then S' is a maximal k-acceptance set and $\{N \in S' | \forall N' \in S', N' \not\subset N\} = \{N \in S | \forall N' \in S, N' \not\subset N\}$. For any node group N and $r (1 \leq r \leq k)$, $F_{S',r}(N) \geq F_{S,r}(N)$, and thus $\hat{A}(S') = \hat{A}(S)$.

$\square$

Theorem 1. If a k-acceptance set S is optimal, a k-coterie $C = \{N \in S | \forall N' \in S, N' \not\subset N\}$ is optimal.

Proof. Assume that there is a k-coterie C_{opt} that has a higher availability than C. Let a k-acceptance set $S_{opt} = \{N \subseteq V | N \supseteq N' (\in C_{opt})\}$. S_{opt} is then maximal. By Lemma 1, $A(C_{opt}) = \hat{A}(S_{opt})$. If S is maximal, $A(C) = \hat{A}(S)$. $\hat{A}(S_{opt}) > \hat{A}(S)$ since $A(C_{opt}) > A(C)$. This is a contradiction of the optimality of S. If S is not maximal, by Lemma 2, there is a maximal k-acceptance set S' such that $\hat{A}(S') = \hat{A}(S)$ and $C = \{N \in S' | \forall N' \in S', N' \not\subset N\}$. By Lemma 1, $A(C) = \hat{A}(S')$. Hence $\hat{A}(S_{opt}) > \hat{A}(S)$ since $\hat{A}(S) = \hat{A}(S') = A(C)$. This is a contradiction of the optimality of S.

$\square$

By theorem 1, an optimal k-coterie C that maximizes $A(C)$ can be obtained directly from an optimal k-acceptance set S that maximizes $\hat{A}(S)$ by removing all nonminimal elements in S.

3.2 Finding Optimal k-Acceptance Set

In this section, we show that the problem of finding an optimal k-acceptance set for a given network is formulated into an optimization problem with a linear objective function.

In the formulation, we first number all possible node groups of V. Next, we encode each node group N as an integer i in such a way that the binary representation of i is equal to $b_n b_{n-1} \cdots b_1$ where b_j is 1 iff N contains v_j. Each node group is then encoded to each of integers $\{1, 2, \cdots, m(= 2^n - 1)\}$. For example, when $V = \{v_1, v_2, v_3\}$, a node group $\{v_1, v_2\}$ is encoded as $3(= 011_2)$.

We call the node group encoded to integer i simply the ith node group and denote it by N_i. A k-acceptance set S can be represented by a 0-1 vector of length m, where the ith element of the vector is equal to 1 iff N_i is an element of S. Thus it could be natural to formulate the problem with m 0-1 variables, but we consider an $m \times k$ 0-1 variables. By doing so, we can treat the problem as having a linear objective function.

Let A be an $m \times k$ matrix of 0-1 variables, and let $A_r (1 \leq r \leq k)$ denote the rth column vector of A. Let $a_{i,r} (1 \leq i \leq m, 1 \leq r \leq k)$ denote the ith element of A_r, and $a_{i,r} = 1$ iff N_i is an union of r disjoint elements in S. A_1 is then exactly the column vector that represents S, and A_r with $2 \leq r \leq k$ is determined from A_1. Let $\mathcal{M}(i, r)$ denote a set of all sets of r integers $x_1, x_2, \cdots, x_r$ such that the ith node group is equal to the union of the the x_yth node groups for $1 \leq y \leq r$ and those node groups are disjoint. Each $a_{i,r}$ is represented by the following sum of Boolean variables.

$$a_{i,r} = \bigvee\nolimits_{\{x_1, x_2, \cdots, x_r\} \in \mathcal{M}(i,r)} a_{x_1,1} \wedge a_{x_2,1} \wedge \cdots \wedge a_{x_r,1}. \tag{1}$$

Example 3. Suppose $V = \{v_1, v_2, v_3\}$ and a 2-acceptance set $S = \{\{v_1, v_2\}, \{v_3\}, \{v_1, v_3\}, \{v_2, v_3\}, \{v_1, v_2, v_3\}\}$. Then, A corresponding to S is a 0-1 vector with two columns and seven$(= 2^3 - 1)$ rows as follows.

$$A^T = \begin{bmatrix} 0011111 \\ 0000001 \end{bmatrix}.$$

For example, $a_{3,1} = 1$ and $a_{4,1} = 1$ because the third node group $\{v_1, v_2\}$ and the forth node group $\{v_3\}$ are elements of S. (Note that $\{v_1, v_2\}$ and $\{v_3\}$ are encoded as $3(=011_2)$ and $4(=100_2)$, respectively.) Then, $a_{7,2} = 1$ because $a_{3,1} = 1$, $a_{4,1} = 1$, and $\{3, 4\} \in \mathcal{M}(7, 2)$. This means that the 7th node group $\{v_1, v_2, v_3\}$ is the union of two disjoint node groups $\{v_1, v_2\}$ and $\{v_3\}$ in S.

Let H be a row vector of length m, where the ith element is a real number equal to the probability that N_i becomes a partition group, i.e., $h(N_i)$. ($h(N_i)$ is calculated by the algorithm that we proposed in [8]. We omit the detail of the algorithm in this paper.) If A_1 represents a k-acceptance set S and $A_r (1 \leq r \leq k)$ meets the above condition, then

$$\hat{A}(S) = \tfrac{1}{k} \sum\nolimits_{r=1}^{k} \sum\nolimits_{i=1}^{m} h(N_i) \times a_{r,i} = \tfrac{1}{k} \sum\nolimits_{r=1}^{k} H A_r.$$

Now we define a *partition* of V as a set of disjoint node groups such that any node in V appears in exactly one of the node groups. Each partition of V can be represented by a 0-1 row vector of length m, where the ith element is a 1 iff N_i exists in the partition. Let $T(\mathcal{R})$ denote this row vector corresponding to a partition $\mathcal{R}$. Thus the following theorem holds.

Theorem 2. A set S represented by a column vector V of length m is a k-acceptance set iff $T(\mathcal{R})V \leq k$ for every partition $\mathcal{R}$.

Proof. ($\leftarrow$) Assume that there exist $k + 1$ disjoint elements $Q_1, \cdots, Q_{k+1}$ in S. Since $Q_1, \cdots, Q_{k+1}$ exist in some partition $\mathcal{R}$, then $T(\mathcal{R})V \geq k + 1$. ($\rightarrow$) Assume that $T(\mathcal{R})V > k$. Then S contains more than k disjoint elements, and thus S is not a k-acceptance set. $\qquad \square$

By regarding A as a matrix of 0-1 variables to be computed, the problem of constructing an optimal k-acceptance set S is formulated into the following problem P of finding an optimal solution of A_1 and an optimal value z_P of the objective function:

$$\text{Maximize } \sum_{r=1}^{k} H A_r \text{ subject to}$$

(1) holds for every $a_{i,r}(1 \leq i \leq m, 1 \leq r \leq k)$ of A, (C1)

$$T(\mathcal{R})A_1 \leq k \text{ for every partition } \mathcal{R}. \qquad\qquad (C2)$$

3.3 Illustrative Example

In this section, we illustrate the formulation using an example network. Consider a network $G = (V, E)$, where $V = \{v_1, v_2, v_3\}$ and $E = \{e_{1,2}, e_{1,3}, e_{2,3}\}$. Concerning reliabilities of components, assume $p_1 = 0.7, p_2 = 0.8, p_3 = 0.9$ and $p_{1,2} = p_{1,3} = p_{2,3} = 0.95$. In this case, there are seven node groups. By the encoding scheme, they are ordered as follows: $\{v_1\}, \{v_2\}, \{v_1, v_2\}, \{v_3\}, \{v_1, v_3\}, \{v_2, v_3\}, \{v_1, v_2, v_3\}$. All partition of V are as follows: $\{\{v_1, v_2, v_3\}\}, \{\{v_1\}, \{v_2, v_3\}\}, \{\{v_2\}, \{v_1, v_3\}\}, \{\{v_3\}, \{v_1, v_2\}\},$ $\{\{v_1\}, \{v_2\}, \{v_3\}\}$.

Any partition $\mathcal{R}$ can be represented by a 0-1 vector $F(\mathcal{R})$ in length seven. For example, a vector corresponding to a partition $\mathcal{R} = \{\{v_1\}, \{v_2\}, \{v_3\}\}$ is $F(\mathcal{R}) = [1101000]$.

Vector H has seven elements, each of which is equal to $h(N)$ for one node group N, that is, the probability that N appears as a partition group. In this case,

$$H = [h(\{v_1\})h(\{v_2\})\ h(\{v_1, v_2\})\ h(\{v_3\})h(\{v_1, v_3\})\ h(\{v_2, v_3\})\ h(\{v_1, v_2, v_3\})]$$
$$= [0.024360\ 0.038860\ 0.054397\ 0.072360\ 0.120897\ 0.206397\ 0.500346].$$

For example, $\{v_1\}$ becomes a partition group when both of the following two conditions hold, provided that v_1 is operational. The first condition is that v_2 or $e_{1,2}$ has failed, while the second one is that v_3 or $e_{1,3}$ has failed. Thus $h(\{v_1\})$ is $0.7 \times (1 - 0.8 \times 0.95) \times (1 - 0.9 \times 0.95) = 0.7 \times 0.24 \times 0.145 = 0.02436$.

Let $k = 2$ and let A be a 7×2 matrix of 0-1 variables. The problem of finding an optimal k-acceptance set is formulated into the following problem P:

$$\text{Maximize } \sum_{r=1}^{k} H A_r \text{ subject to}$$
$$a_{3,2} = a_{1,1} \wedge a_{2,1}, \quad a_{5,2} = a_{1,1} \wedge a_{4,1}, \quad a_{6,2} = a_{2,1} \wedge a_{4,1},$$
$$a_{7,2} = (a_{1,1} \wedge a_{6,1}) \vee (a_{2,1} \wedge a_{5,1}) \vee (a_{3,1} \wedge a_{4,1}),$$
$$[1000010]A_1 \leq 2, \quad [0100100]A_1 \leq 2, \quad [0011000]A_1 \leq 2,$$
$$[1101000]A_1 \leq 2, \quad [0000001]A_1 \leq 2.$$

Using the proposed branch-and-bound method, which will shown in the next section, we can solve this problem P and obtain the following optimal solution of A_1 and z_P:

$$A_1^T = [0111111] \text{ and } z_P = 1.7.$$

This result means that the optimal 2-acceptance set S is $\{\{v_2\}, \{v_1, v_2\}, \{v_3\}, \{v_1, v_3\}, \{v_2, v_3\}, \{v_1, v_2, v_3\}\}$, and its availability is equal to $0.85(=1.7/k)$. Consequently, we obtain the following optimal 2-coterie C by removing all nonminimal elements in S.

$$C = \{\{v_2\}, \{v_3\}\}.$$

4   Branch-and-Bound Method

In order to solve the formulated optimization problem P, we propose a branch-and-bound method. The branch-and-bound algorithm proceeds by traversing a search tree in which each node represents a subproblem of the initial problem P in order to find a feasible leaf node with an optimal value, z_P, of the objective function of P. The search space for the branch-and-bound tree is reduced by estimating an upper bound of the objective function. To estimate the upper bound, we derive a linear programming problem LP from P by relaxing the constraints of P.

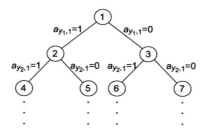

Fig. 2. An example search tree.

The problem P is to find an optimal solution A corresponding an optimal k-acceptance set. As mentioned previously, $A_r (2 \leq r \leq k)$ is uniquely determined from A_1. Thus we search A_1 that represents an optimal k-acceptance set. The branch-and-bound search space then forms a binary tree of depth $m(= 2^n - 1)$, because A_1 contains m elements each of which is 0 or 1.

Figure 2 illustrates a simple example of a search tree. The root node of the tree, node 1, represents the original problem P. We adopt depth-first search since its storage requirement is only linear in the depth of the search tree unlike best-first search. Then the search algorithm proceeds as follows. Assume that $a_{y_1,1}$ is first chosen as a branching variable, P is then split into two subproblems associated with the son nodes 2 and 3 by fixing $a_{y_1,1} = 1$ and $a_{y_1,1} = 0$, respectively. This procedure will be iterated on these two nodes. After solving the relaxed problem of subproblem 2, $a_{y_2,1}$ is next chosen as a branching variable. Fixing its value to 0 or 1 produces subproblems 4 and 5, respectively. For example, subproblem 4 has both $a_{y_1,1}$ and $a_{y_2,1}$ fixed to 1. In this way, branching operation is iterated on non-leaf nodes of the search tree.

Let t be a node of the search tree, and let P_t and LP_t denote the subproblem associated with t and its relaxed problem, respectively. LP_t must have the property that $z_{LP_t} \geq z_{P_t}$. Additionally, let z be the maximal value of the objective function obtained so far. Branching on node t is not needed, that is, P_t needs not be further subdivided into smaller subproblems in the following two cases: (1) $z_{LP_t} \leq z$, and (2) LP_t has no feasible solution. In the former case, there is no hope to find a better value than z by branching on t because z_{LP_t} is the upper bound on z_{P_t}. In the latter case, P has no solution either that satisfies the constraints. The branch-and-bound algorithm terminates when all the generated nodes either produced their son nodes or not needed to branch.

Note that we must be able to solve LP_t much faster than P_t to achieve speed up by space reduction. In the rest of this section, we describe how we determine LP_t. Let B be an $m \times k$ matrix whose element x is a real number with $0 \leq x \leq 1$. Let $B_r(1 \leq r \leq k)$ denote the rth column vector of B. We define the relaxed problem LP of P as follows:

$$\text{Maximize } \sum_{r=1}^{k} H B_r \text{ subject to}$$
$$T(\mathcal{R}) B_r \leq \lfloor k/r \rfloor \text{ for every partition } \mathcal{R} \text{ and } r(1 \leq r \leq k).$$

For the relaxed problem LP_t of subproblem P_t, the constraints resulting from fixing branching variables are added to the above problem. Since all variables of LP_t are real numbers and the constraints are linear, LP_t is a linear programming problem. For the problems $\text{P}(=\text{P}_1)$ and $\text{LP}(=\text{LP}_1)$ and their subproblems, the following theorem holds.

Theorem 3. $z_{\text{LP}_t} \geq z_{\text{P}_t}$.

Proof. Let $\text{P}(r)$ with $1 \leq r \leq k$ be the problem defined as follows:

$$\text{Maximize } H A_r \text{ subject to } T(\mathcal{R}) A_1 \leq k \text{ for every partition } \mathcal{R}.$$

The subproblem $\text{P}(r)_t$ of $\text{P}(r)$ is also determined similarly. Since constraint C1 of P_t is not required for each $\text{P}(r)$,

$$z_{\text{P}_t} \leq \sum_{1 \leq r \leq k} z_{\text{P}(r)_t}. \tag{2}$$

Additionally, let $\text{LP}(r)$ with $1 \leq r \leq k$ be the problem defined as follows:

$$\text{Maximize } H B_r \text{ subject to } T(\mathcal{R}) B_r \leq \lfloor k/r \rfloor \text{ for every partition } \mathcal{R}.$$

We denote by $\text{LP}(r)_t$ the subproblem of $\text{LP}(r)$ that is associated with t. We then obtain the following equation:

$$z_{\text{LP}_t} = \sum_{1 \leq r \leq k} z_{\text{LP}(r)_t}. \tag{3}$$

Assume that $T(\mathcal{R}) A_1 \leq k$ and $T(\mathcal{R}) A_r = i > \lfloor k/r \rfloor$ for some partition $\mathcal{R}$ and integer $r(1 \leq r \leq k)$. There then exist i elements $a_{x_1,r}, a_{x_2,r}, \cdots, a_{x_i,r}$ of A_r such that $a_{x_j,r} = 1$ and $N_{x_j} \cap N_{x_k} = \emptyset$ for any $1 \leq j, k \leq i$. By definition, for each $a_{x_j,r}(1 \leq j \leq i)$, there are r elements of A_1 each of which corresponds to one of r disjoint node groups that are all subsets of N_{x_j}. Thus $T(\mathcal{R}) A_1 \geq r \times i > k$. This is a contradiction. Hence, for every $\mathcal{R}$ and $r(1 \leq r \leq k)$, $T(\mathcal{R}) A_r \leq \lfloor k/r \rfloor$ if $T(\mathcal{R}) A_1 \leq k$.

Since the constraint of $\text{LP}(r)$ is that $T(\mathcal{R}) B_r \leq \lfloor k/r \rfloor$ for every $\mathcal{R}$, $\text{LP}(r)$ is exactly the problem that relaxes the constraints of $\text{P}(r)$, and thus,

$$z_{\text{LP}(r)_t} \geq z_{\text{P}(r)_t} \text{ with } 1 \leq r \leq k. \tag{4}$$

From formulae (2), (3), and (4), $z_{\text{LP}_t} \geq z_{\text{P}_t}$. $\square$

LP_t is a linear programming problem. Thus it can be solved in polynomial time. For the initialization, we solve $\text{LP}(1)(= \text{P}(1))$ and set the initial values of Z and z to solution A of $\text{LP}(1)$ and $\sum_{r=1}^{k} H A_r$, respectively. Branching variables are chosen according to the number of nodes in its corresponding node group. The search space is then explored by iterating branching and bounding using LP in the way described before.

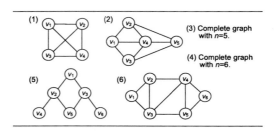

Fig. 3. Sample networks.

5 Experimental Results

We implemented the proposed algorithm in C++ language, and conducted an experiment for the networks shown in Figure 3. In the experiment, we used the linear programming package lp_solve3.0 for solving the linear programming problem LP. The experiment is conducted on a PC/AT compatible workstation with PentiumIII 700MHz running LINUX.

Table 1 shows the optimal k-coteries obtained and their availabilities when $k = 2$ and 3. In the experiment, we assumed $p_i = 0.90$ for every node v_i and $p_{i,j} = 0.95$ for every link $e_{i,j}$ for simplicity only. For the comparison, we also show the availabilities of k-majority coteries[1], which are sets of all node groups that contain exactly $\lceil (n+1)/(k+1) \rceil$ nodes[5]. In the case of networks 2 and 3, the optimal 2-coterie is identical with the 2-majority coterie. In other cases, the availability of the optimal k-coterie is 12-26% better than that of the k-majority coterie.

Table 2 shows the running times and the numbers of explored nodes in the branch-and-bound search tree. As expected, the running time increases exponentially with the value of n and linearly with the value of k. We also show the number of all nodes in the search tree for comparison. From the result, one can see that the search space is efficiently reduced by the proposed relaxation LP.

6 Conclusion

In this paper, we proposed the method for obtaining an optimal k-coterie in a general topology network with unreliable nodes and links. We first introduced the new notion of a k-acceptance set, and showed that an optimal k-coterie is obtained directly from an optimal k-acceptance set. We then formulated the problem of finding the k-acceptance set into the optimization problem P with the linear objective function. In order to solve the problem P, we next proposed the branch-and-bound algorithm based on the linear programming relaxation LP.

Using the proposed method, we obtained optimal k-coteries for several networks in the experiment. To the best of our knowledge, this is the first time that optimal k-coteries

---

[1] A k-majority coterie exists only when $\lceil (n+1)/(k+1) \rceil \leq n/k$[9].

Table 1. The obtained optimal k-coteries.

Network $(n, \|E\|)$	k	Optimal k-coterie	Availability	Availability of a k-majority coterie
1(4,6)	2	$\{\{v_1, v_2\}, \{v_1, v_3\}, \{v_2, v_3\}, \{v_4\}\}$	0.934855	0.824803
2(5,8)	2	$\{\{v_1, v_2\}, \{v_1, v_3\}, \{v_1, v_4\}, \{v_1, v_5\}, \{v_2, v_3\},$ $\{v_2, v_4\}, \{v_2, v_5\}, \{v_3, v_4\}, \{v_3, v_5\}, \{v_4, v_5\}\}$	0.957015	0.957015
	3	$\{\{v_1\}, \{v_2\}, \{v_3, v_4\}, \{v_3, v_5\}, \{v_4, v_5\}\}$	0.923173	-
3(5,10)	2	$\{\{v_1, v_2\}, \{v_1, v_3\}, \{v_1, v_4\}, \{v_1, v_5\}, \{v_2, v_3\},$ $\{v_2, v_4\}, \{v_2, v_5\}, \{v_3, v_4\}, \{v_3, v_5\}, \{v_4, v_5\}\}$	0.958751	0.958751
	3	$\{\{v_1, v_2\}, \{v_1, v_3\}, \{v_2, v_3\}, \{v_4\}, \{v_5\}\}$	0.923870	-
4(6,15)	2	$\{\{v_2, v_3\}, \{v_2, v_4\}, \{v_2, v_5\}, \{v_2, v_6\}, \{v_3, v_4\},$ $\{v_3, v_5\}, \{v_3, v_6\}, \{v_4, v_5\}, \{v_4, v_6\}, \{v_5, v_6\}\}$	0.958990	0.765031
	3	$\{\{v_1, v_4\}, \{v_1, v_3\}, \{v_1, v_4\}, \{v_1, v_5\}, \{v_2, v_3\},$ $\{v_2, v_4\}, \{v_2, v_5\}, \{v_3, v_4\}, \{v_3, v_5\}, \{v_4, v_5\}, \{v_6\}\}$	0.939327	0.838474
5(6,6)	2	$\{\{v_1\}, \{v_2, v_3\}, \{v_2, v_5\}, \{v_3, v_5\}\}$	0.924175	0.764317
	3	$\{\{v_2\}, \{v_6\}, \{v_1, v_3\}, \{v_1, v_5\}, \{v_3, v_5\}\}$	0.916117	0.793207
6(6,9)	2	$\{\{v_2, v_3\}, \{v_2, v_4\}, \{v_2, v_5\}, \{v_2, v_6\}, \{v_3, v_4\},$ $\{v_3, v_5\}, \{v_3, v_6\}, \{v_4, v_5\}, \{v_4, v_6\}, \{v_5, v_6\}\}$	0.951775	0.756160
	3	$\{\{v_1\}, \{v_2, v_3\}, \{v_2, v_4\}, \{v_2, v_5\}, \{v_2, v_6\},$ $\{v_3, v_4\}, \{v_3, v_5\}, \{v_3, v_6\}, \{v_4, v_5\}, \{v_4, v_6\}, \{v_5, v_6\}\}$	0.934513	0.830988

(- means that the k-majority coterie does not exist.)

Table 2. The running times and the numbers of explored nodes.

Network	k	Running time (in seconds)	# of explored nodes	# of all nodes in a search tree
1	2	4.64	108	2^{16}-1
2	2	26.64	583	2^{32}-1
	3	39.29	807	//
3	2	27.11	581	//
	3	39.69	807	//
4	2	656.47	12624	2^{64}-1
	3	1458.06	29921	//
5	2	600.90	12424	//
	3	1435.56	29921	//
6	2	621.42	12627	//
	3	1452.07	29921	//

have been obtained for general topology networks. The experimental results show that the proposed method is feasible up to $n = 6$. Development of an branch-and-bound method that can handle larger networks should be included in future research. For instance, a more efficient relaxation will reduce the search space further, and parallelization will also speed up the search.

References

1. D. Agrawal, Ö. Eğecioğlu and A. E. Abbadi, "Analysis of quorum-based protocols for distributed $(k+1)$-exclusion," *IEEE Trans. Parallel and Distributed Systems*, Vol. 8, No. 5, pp. 533–537, 1997.
2. M. D. Beaudry, "Performance-related reliability measures for computing systems," *IEEE Trans. Computers*, Vol. 27, No. 6, pp. 540-547, 1978.

3. S. T. Huang, J. R. Jiang, and Y. C. Kuo, "k-Coteries for fault-tolerant k entries to a critical section," *Proc. 13th IEEE International Conference on Distributed Computing Systems*, pp. 74–81, 1993.
4. J. R. Jiang, S. T. Huang and Y. C. Kuo, "Cohorts structures for fault-tolerant k entries to a critical section," *IEEE Trans. Computers*, Vol. 46, No. 2, pp. 222–228, 1997.
5. H. Kakugawa, S. Fujita, M. Yamashita, and T. Ae, "Availability of k-coterie," *IEEE Trans. Computers*, Vol. 42, No. 5, pp. 553–558, 1993.
6. N. Krishnakumar and A. J. Bernstein, "Bounded ignorance: a technique for increasing concurrency in a replicated system," *ACM Trans. Database Systems*, Vol. 19, No. 4, pp. 586–625, 1994.
7. M. L. Neilsen and M. Mizuno, "Nondominated k-coteries for multiple mutual exclusion," *Inform. Process. Lett.*, Vol. 50, pp. 247–252, 1994.
8. T. Tsuchiya, "Design and Evaluation of Coteries for Distributed Mutual Exclusion," *Ph.D. thesis*, Dept. Informatics and Mathematical Science, Osaka University, 1998.
9. S. M. Yuan and H. K. Chang, "Comments on availability of k-coterie," *IEEE Trans. Computers*, Vol. 43, No. 12, pp. 1457, 1994.

The Fast Bully Algorithm: For Electing a Coordinator Process in Distributed Systems

Seok-Hyoung Lee[1] and Hoon Choi[2]

[1] Information System Research Department, Korea Institute of Science and Technology Information, DaeJeon, Korea ,305-806
skyi@kisti.re.kr

[2] Department of Computer Engineering, Chungnam National University, DaeJeon, Korea , 305-764
hchoi@ce.cnu.ac.kr

Abstract. Many of software in distributed systems require a special process called the coordinator or the leader. The coordinator is elected among the processes participating a particular distributed service when the service begins operation or the existing coordinator fails. The procedure of selecting the coordinator is called the election. One of the most popular election algorithms is the bully algorithm. However, this algorithm is costly because it generates $O(n^2)$ messages for n processes. Also the time delay till the completion of the election is considerably long. This paper proposes a new election algorithm called the fast bully algorithm. The algorithm is described and performance of the algorithm has been analyzed. The fast bully algorithm shows a clear performance improvement. It works with less messages, $O(n)$ messages, and with shorter time delay than the bully algorithm. Also the problem of having two coordinators at the same time that appears by the bully algorithm can be avoided by the FBA.

1 Introduction

Many of software in distributed systems require a special coordinator process among the distributed process. For instance, a manager process controls allocation and deallocation of the shared resources to guarantee mutual exclusion when multiple distributed processes access shared resources [1]. The election algorithm is performed to elect the coordinator process when a distributed system begins operation or the existing coordinator fails. The key concept of the election algorithm is to find the highest priority process among the active processes in process group and to notify other processes of it.

Several election algorithms have been studied [2,3,4,5,6,7,8,9]. In the bully algorithm (BA) suggested by Garcia-Molina [2], the process that firstly detects the coordi-

---

* Hoon Choi was supported by the Korea Science and Engineering Foundation under the Grant 981-0925-134-2.

I. Chong (Ed.): ICOIN 2002, LNCS 2344, pp. 609–622, 2002.
© Springer-Verlag Berlin Heidelberg 2002

nator's failure makes both itself and all other processes begin the election procedure. In the ring algorithm suggested by Chang and Robert [3] and improved by Tanenbaum [4], the election messages circulate in the logical ring of the processes group. The local leader election algorithm [5] elects the coordinator whenever the logical partitions are created at a cost of heavy message traffic because the coordinator must send election messages periodically to other processes.

Among these algorithms, the BA is commonly used because it is simple and fault-tolerant. However, this algorithm is costly. $O(n^2)$ messages are generated for n processes [1]. In result, not only is the delay occurred on communication channels due to traffic overflowing, but also does the queuing delay at the system buffer become long. Also, an inconsistency may occur by the BA. Suppose that the coordinator process fails and a process with higher priority number than the coordinator joins the process group. This new process will become the new coordinator by the algorithm. When the previous coordinator process recovers from failure, it starts the election procedure and notifies other processes that itself is the coordinator without knowing of newly joined, higher priority process in the group. So there exist two coordinators in the group. This situation is not resolved until another election procedure is started.

To solve these problems, we suggest a new algorithm, named the fast bully algorithm (FBA). We implemented and evaluated the performance of the algorithm using the simulation tool, COVERS [10]. The FBA in this paper is the improved version of the one suggested in [11]. The FBA shows a superior performance compared with the BA and ring algorithms. The number of generated message is decreased from $O(n^2)$ to $O(n)$ and response time is also decreased by about half. Also the proposed algorithm has a good tolerance against various faults including the case that a new process is added in the group during other process's failure.

We describe the BA introduced in [4] in Chapter 2 and the proposed algorithm in Chapter 3. After explaining the implementation of the algorithm in Chapter 4, we analyze the performance of the FBA by comparing with the BA in Chapter 5.

2   Bully Election Algorithm

In the BA, each process is assumed to have its own priority number. A process that firstly detects the coordinator's failure makes itself and other processes begin the election procedure. All active processes perform the algorithm independently. If a process notices that it does not have the highest priority, it waits until the new coordinator is elected and is notified of [2]. Although a process fails during the election procedure, the BA can cope with this exceptional situation. If a recovered process has a higher priority than the existing coordinator, it regains the coordinator position. The basic assumptions for running the BA are as follows.

- Each process in the group knows the priority number, address of other processes and can send messages one another.
- The delivery of messages on communication channel is guaranteed but a process may fail.
- A process does not know in advance whether other processes failed or not.

If a process *Pi* sends a message to the coordinator but the coordinator does not respond within the time T1, *Pi* regards the coordinator to have failed and it starts the BA shown in Fig. 1. All the processes have the same value set for timers T1, T2 and T3 of the algorithm.

The Bully Algorithm

BEGIN
1. A process *Pi* does not receive a response within T1 from the coordinator.
   1.1   *Pi* sends an *election* message to every process with higher priority number.
   1.2   *Pi* waits for *answer* messages for the interval T2.
   1.3   If no answer within T2, /**Pi* is the new coordinator*/
      1.3.1   *Pi* sends a *coordinator* message to other processes with lower priority number.
      1.3.2   *Pi* stops its election procedure.
   1.4   If the *answer* messages are received,
      1.4.1   *Pi* begins the interval T3 to wait for a *coordinator* message.
   1.5   If the *coordinator* message is received,
      1.5.1   Admit the sender as the new coordinator.
      1.5.2   *Pi* stops its *election* procedure.
   1.6   If no *coordinator* message within T2, *Pi* restarts the election procedure.
2. A process *Pj(i<j)* may receive an *election* message from *Pi*.
   2.1   *Pj* sends an *answer* message to *Pi*.
   2.2   *Pj* begins its *election* procedure (step 1).
3. A process *Pj(i>j)* may receive a *coordinator* message from *Pi*.
   3.1   Admit *Pi* as the new coordinator.
   3.2   *Stops* the election procedure.
END

Fig. 1. The Bully Algorithm

3   Fast Bully Election Algorithm

In the FBA, the process that firstly detects the coordinator's failure controls the election procedure. Not only does it make other processes perform the election algorithm, but does it find out a process that has the highest priority number based on other processes' response and then it informs a process to be a candidate for the coordinator. And the candidate process announces itself to other processes as the coordinator and the election algorithm ends.

The FBA is also fault-tolerant. It can cope with various situations where processes fail during the election procedure. The recovered process can regain the coordinator's position immediately if its priority number is higher than the current coordinator's. Besides, by receiving the view message, it can figure out whether any process has been added in the group and which is the new coordinator. The basic assumptions for running the FBA are the same as the BA. Six types of messages are used in the FBA.

(1) election: the request message to start election procedure
(2) answer: the response message to the election message
(3) nomination: the message sent to the highest numbered process to notify that it is a candidate for the coordinator
(4) coordinator: the message that claims that the sender is the coordinator

(5) IamUp: the message that is sent by the recovered process

(6) view : the response message to the IamUp message containing a list of the processes in the group

The algorithm is described in Fig. 2.

The Fast Bully Algorithm

BEGIN
0   A process Pi recovers from failure :
  0.1 Pi sends an *IamUp* message to every process.
  0.2 Pi waits for *view* messages for the interval T2.
  0.3   If no *view* messages within T2, Pi stops the procedure. // Pi is the coordinator
  0.4   If the *view* messages are received within T2 :
    0.4.1 Pi compares its view with the received views.
    0.4.2   If the received view is different from the Pi's view, Pi updates its view.
    0.4.3   If Pi is the highest priority numbered process,
      0.4.3.1 Pi sends a *coordinator* message to other processes with lower priority number.
      0.4.3.2 Pi stops the procedure.
    0.4.4   Otherwise,
      0.4.4.1 Admit the highest priority numbered process as the coordinator.
      0.4.4.2 Pi stops the election procedure.
1   A process Pi may receive an *IamUp* message from Pj,
  1.1 Pi sends a *view* message to Pj.
2   A process Pi does not receive a response within T1 from the coordinator:
  2.1 Pi sends an *election* message to every process with higher priority number.
  2.2 Pi waits for *answer* messages for the interval T2.
  2.3   If no answer within T2,
    2.3.1 Pi sends a *coordinator* message to other processes with lower priority number.
    2.3.2 Pi stops its election procedure.
  2.4   If the *answer* messages are received within T2,
    2.4.1 Pi determines the highest priority number of the answering processes.
    2.4.2 Pi sends a *nomination* message to this process.
    2.4.3 Pi waits for a *coordinator* message for the interval T3.
  2.5   If the *coordinator* message is received,
    2.5.1   Admit the sender as the new coordinator.
    2.5.2 Pi stops its *election* procedure.
  2.6   If no *coordinator* message within T3,
    2.6.1   Repeat step 2.4.2 for the next highest priority numbered process.
    2.6.2   If no process left to choose, Pi restarts the election procedure.
3   If a process $Pj(i<j)$ receives an *election* message from Pi,
  3.1 Pj sends an *answer* message to Pi.
  3.2 Pj waits for either a *nomination* or a *coordinator* message for the interval T4.
  3.3   If no *coordinator* message or *nomination* message within T4.
    3.3.1 Pj restarts the procedure.
  3.4   If a process $Pj(i<j)$ receives the *nomination* message from Pi,
    3.4.1 Pj sends a *coordinator* message to all the processes with lower priority numbers.
    3.4.2 Pj stops its election procedure.
4   If a process $Pj(i>j)$ receives a *coordinator* message from Pi,
  4.1   Admit Pi as the new coordinator.
  4.2   Stops its election procedure.
END

Fig. 2. The Fast Bully Algorithm

When a process Pi in the group sends a service request to the coordinator but the coordinator does not respond within the time T1, Pi initiates the FBA (Fig. 2). Values of T1,T2 and T3 in the algorithm are the same for all processes, but values of T4 are different each other depending on the process priority number. The value of T4 is in inverse proportion to the processes' priority number as the equation (1). Therefore, a candidate for the coordinator (the process with the highest priority number) expires T4 earlier than other processes. The values of a and b in equation (1) are determined so that the value of T4 of each process is sufficiently different from each other. For example, if the average message transmission and propagation delay is 10 msec, the value of T4 of the process 4 may be 100 msec, that of the process 3 is 200 msec and that of the process 2 is 300 msec etc. Although transmission and propagation delays of messages may differ from message types and processes, this difference is negligible and the operation of the algorithm is not affected. Furthermore, which process's T4 expires first has nothing to do with electing a correct coordinator. The reason we set the smallest value to the T4 of the highest priority numbered process is only to improve the performance of the election algorithm in an abnormal situation as shown in the Fig. 6. Even though T4 of the lower priority numbered process expires first, the highest priority numbered process will be elected as the coordinator.

$$T4 = a/\{ b*(process's\ priority\ number)\} \quad where\ a, b : \text{some constants} \tag{1}$$

The operating procedure of the FBA is depicted in Fig. 3 assuming that there are five processes in the group and process 1 detects the coordinator's failure.

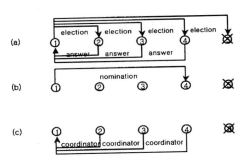

Fig. 3. The operating procedure of the FBA in case of no process failure during the election

Process may fail even during the election procedure. The FBA can cope with all the possible exception cases that are listed as follows.

(1) A candidate of the coordinator fails before declaring itself.

(2) A process that initiates the election procedure fails before completing the algorithm.

(3) Processes in the group other than either the candidate or the initiating process fails during the election.

(4) Another election is initiated by another process during the election.

Fig. 4 shows a case where the candidate for the coordinator fails before notifying other processes that it is the coordinator. Fig. 4(a) is the same situation as Fig. 3(a). Process 1 nominates the process 4 for a candidate of the coordinator and sends the nomination message to the process 4. If the process 4 also fails and does not respond within the time T3, then the process 1 nominates the process 3 for the new candidate. The process 3 notifies the process 1 and 2 that it is the coordinator.

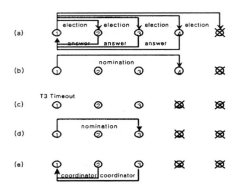

Fig. 4. A scenario of the exception case (1)

Fig. 5 shows a case where the process initiating the election fails before completion. Assume that the process 1 fails after Fig. 5(a). Processes 2,3,4 start the timer T4 after sending the answer message to the process 1. Because the value of T4 of the process 4 is smallest as explained in equation (1), the process 4 expires T4 first and restarts the FBA by the step 3.3.1 of the algorithm. The process 4 sees itself being the highest priority numbered process among active processes, then it sends a coordinator message to the processes 2,3 and completes the algorithm. The process 2 and 3 complete the algorithm by the step 2.5.

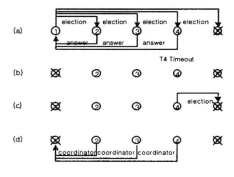

Fig. 5. A scenario of the exception case (2)

The third case is that the processes other than a candidate or the process initiating the election fail. Although these processes fail, it has no influence on the election procedure. Since the process initiating the election waits for the answer messages only during T2, it just ignores any process that fails and does not respond (step 2.4).

Fig. 6 shows the last case where multiple election procedures are on going. This can occur when the timer T1 of the lower priority numbered process expires during the election. Suppose the process 2 starts the election after detecting the process 5's failure. Also suppose T1 of the process 1 expires and the process 1 initiates another election. Once the process 2 receives an election message from the process 1, it ceases the controlling role because it replies an answer message and waits for a new coordinator (step 3).

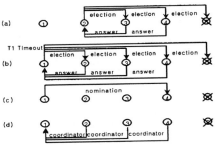

Fig. 6. A scenario of the exception case (4)

Whenever a process recovers from failure, it runs the FBA. The recovered process sends an IamUp message to every other process in the group. It receives view messages as a response. The view message shows a list of processes in the group, and the recovered process checks whether any process has been added in the group during its failure. As a result, it can figure out which process has the highest number, i.e. the coordinator in the group without election procedure. Note that whether this highest priority process is currently alive or not does not matter. The process that receives an IamUp message performs normal operations after it responses with a view message. If the priority number of the recovered process is the highest, the recovered process regains the coordinator by sending out a coordinator message(step 0.4.3.1, step 4), otherwise the coordinator is not changed.

In the BA, when a new process is added and elected as the coordinator during the failure of the former coordinator, the former coordinator does not know about it after it recovers. Thus, two coordinator processes may exist in the group. This problem can be solved only when another election is performed by a process except the former coordinator. This problem does not occur by the FBA.

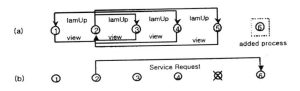

Fig. 7. Operation of the algorithm in case that a new process is added in the group during the failure of other process

4   Implementation

We may implement the algorithm on a real network of computers. But, the real environment is not appropriate to raise various exception conditions and to verify correctness of the algorithm. Therefore, we implemented the algorithms on a simulated network environment using the tool, COVERS[10].

We grouped the states of a process into Failure state and Active state. The Active state consists of five substates (Fig. 8). A description of each state is as follows.

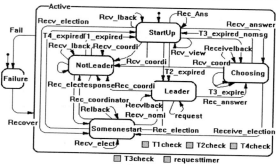

Fig. 8. State transition diagram of the FBA

(1) Failure : A process enters this state when it fails. When the model begins execution, all the processes starts from this state. The time taken for a process to enter the Active state is exponentially distributed with the mean of 2,000 seconds.

(2) Active : In this state, the process is operational. A process in the Active state must be in one of the StartUp, Choosing, SomeoneStart, Leader and NotLeader states. The time for a process in any Active state to enter the Failure state is exponentially distributed with the mean of 100,000 seconds.

a) StartUp : When the process is recovered from failure, or it detects the coordinator's failure and starts the election procedure, it enters this state.

b) Choosing : The process enters this state after receiving an answer message from other processes during T2 (step 2.4 of Fig. 2).

c) SomeoneStart : If the process receives an election message from other process, it enters the SomeoneStart state (step 3).

d) NotLeader : The normal operating state of a process that is not the coordinator.

e) Leader : The normal operating state of a process that is the coordinator.

5   Performance Evaluation

5.1   Case That the Election Is Due to the Failure of the Coordinator

We get the average response time and the number of messages of the two algorithms as in Table 1. Case (a) represents that the process 1 first detects the coordi

> - Number of processes in the group: 5
> - Request rate of service to the coordinator per one process: 0.002 per second
> - T1: 15 second, T2: 20 second, T3: 35 second, T4: *30 second + ((number of processes in the group -1)-(priority number of the process+1)) x 6*
> - Transmission and propagation delay of a message: 1 second (constant)
> - Channel's transmission capability: multicast
> - Simulation time: 30,000,000 second (about 1 year),
> - Number of simulation runs : 10
> - Assumptions: When a process thinks that it is highest numbered, it does not send out the election message and the larger priority number has higher priority.

Fig. 9. Simulation parameters. We simulated various failure/recovery situations and compared the performance of the FBA and the BA using the same parameters.

nator's failure and starts the election procedure, cases (b) (c) (d) represent the process 2, process 3 and process 4 initiates the election, respectively. The response time in Table 1 considers only the normal situation that no other process fails during the election procedure. Failures during the election will be discussed in section 5.5.

Table 1. Performance of the case that the election is due to the failure of the coordinator

	Response Time (sec)		Number of Messages	
	BA	FBA	BA	FBA
(a)	23.0	4.0	19	11
(b)	23.0	4.0	12	9
(c)	23.0	4.0	7	7
(d)	1.0	1.0	3	3

In the case (a) of the BA, when the process 1 starts the election, it sends an election message to processes 2,3,4 and 5 and then starts T2 timer. It takes one second for the process 1 to transmit an election message. The processes 2,3 and 4 take one second to reply with answer messages, and then take another one second to send election messages to the higher numbered processes and start the T2 timer individually. The process 4 will not receive an answer message from the process 5 in T2 time. After T2 expires, the process 4 sends a coordinator message taking one second. Thus the response time of the BA in case (a) is 23 seconds in total.

In the case (a) of the FBA, the process 1 takes one second to send an election message to the higher numbered processes and starts T2 timer. All the processes except the process 5 will reply in one second. At this time, the process 1 does not need to wait until T2 expires for the message from the process 5 because it knows that the process 5 has failed. Process 1 takes one second to send a nomination message and the new coordinator takes one second to multicast a coordinator message. Therefore, in the FBA, the response time of the case (a) is the sum of the transmission time of the election, answer, nomination and coordinator messages in sequence, 4 seconds in total. In the case (b) and (c), the response times of the BA is 23 seconds as the case (a), and those of the FBA is 4 seconds as the case (a).

In the case (d), because the process 4 knows that the process 5, the coordinator, failed, it just notifies other active processes in the group that it is the new coordinator. So, the response time is one second, only the transmission time of a coordinator message. The number of messages is totally 3; three coordinator messages that the process 4 sends to the processes 1,2 and 3.

5.2   Case That the Election Is Due to a Process's Recovery

We get the response time and the number of messages in case (a) of Table 2 that process 1 recovers and then starts the election. Likewise (b),(c),(d),(e) is for the process 2, process 3, process 4 and process 5, respectively. As in section 5.1, no process failure is assumed during the election. But we assume that a new process is added in the group during the failure of one of the existing processes.

In the case (a), (b), (c) and (d) of the BA, the recovered process does not know of the added process. When it starts the election, process 6, the newly added process, receives election messages from processes in the group except the recovered one. Because the process 6 is the coordinator, it will send a coordinator message to all the lower priority numbered processes including the recovered one. The response time becomes 3 seconds.

But, in the case (e), the recovered process, process 5, does not know of the added process and it still thinks that itself is the highest ranked. Thus it immediately sends out a coordinator message. Other lower numbered processes think the process 6 has failed, they regard the process 5 as the new coordinator, resulting two coordinators in the group. This problem can be fixed only when another election is started by a process other than the former coordinator. So the delay time is unpredictable in the case (e). In the FBA, however, the recovered process knows whether the new process is added and which process is the coordinator by the view messages. The delay time of the FBA in all cases is 2 seconds, and the number of messages is 8, four IamUp messages and four view messages.

Table 2. Performance of the case that the election is due to a process recovery

	Response Time (sec)		Number of Messages	
	BA	FBA	BA	FBA
(a)	3.0	2.0	33	8
(b)	3.0	2.0	23	8
(c)	3.0	2.0	15	8
(d)	3.0	2.0	9	8
(e)	N/A	2.0	33 or more	8

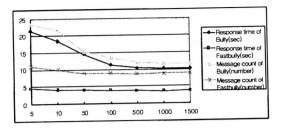

Fig. 10. Performance with respect to the service request interval

5.3 Performance with Respect to Mean Service Request Interval

We analyze the response time and the message count of the algorithms with respect to time values of Fig. 9. When the mean service request interval is short, each process communicates frequently with the coordinator. If the coordinator fails, the chance is high that multiple processes detect the failure and initiate the elections independently. As we have shown in Table 1, the time to complete the election procedure is longer in the BA than the FBA. Thus the probability that other elections are overlapped when one election is in progress becomes larger in the BA, resulting longer response time and more messages.

5.4 Performance with Respect to the Number of Processes

Because messages are transmitted in multicast mode, the average response time of the election is comparatively uniform regardless of the number of processes. The number of messages increases quadratic as the number of processes increases in the BA. On the other hand, the number of messages increases linearly in the FBA. This is because only one process sends an election message and other processes reply an answer message and wait for the coordinator message.

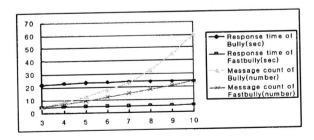

Fig. 11. Performance with respect to the number of processes

5.5 Performance in Exception Situation

As explained in section 3.2, exceptional situations can occur during the election procedure. In Table 3, (a) is the case that the process 1 and process 3 detect the failure of
the coordinator and start the election individually, (b) is the case that after the process
3 starts the election, T1 of the process 1 expires after some time, for instance in 2
seconds, and the process 1 initiates another election and (c) shows the overall performance results considering all the normal, exceptional situations during the simulation.

In the case (a), although the two processes start the election in parallel, process 3
goes into the state of waiting for a coordinator message because it receives an election
message from the process 1. As a result, the response time is the same as the case
where the process 1 alone starts the election. The response time of the FBA is shorter
than that of the BA. Now we check the message count. In both algorithms, process 3
sends two election messages to processes 4,5 and receives an answer message from
the process 4. Hence three more messages are needed compared with the case where
election is begun by one process. Again, the total message count of the FBA is smaller
than that of the BA.

In the case (b), the response time became 2 seconds longer in both algorithms. This
is because the process 3 becomes silent after the process 1 initiates its election, and
two seconds spent by the process 3 is added to the response time of the election by the
process 1. The number of messages increases by 3 in the BA, two election messages to
the processes 4, 5 and one answer message from the processes 4. The response time of
the FBA also gets two seconds longer. Process 3 has spent two seconds when the process 1 starts the election. Regardless of which state the process 3 is in at that time, it
will answer to the election message from the process 1, and the process 1 will take
control of the election. If we assume two seconds gap, the process 3 may have sent the
nomination message to the process 4 by that time. The FBA requires four more messages in this case compared with the single process case. Even if we consider the exceptional situations, the FBA is superior to the BA.

Table 3. Performance of exception situations

	Response Time (sec)		Number of Messages	
	BA	FBA	BA	FBA
(a)	23.0	4.0	22	14
(b)	25.0	6.0	22	15
(c)	22.6992	3.92457	11.9206	8.60701

5.6 Number of Messages

Let n be the number of processes in the group. The best case of the BA with respect to
the number of messages to complete the algorithm is that the process having the next
highest priority number starts the election algorithm. This process sends an election
message to the dead coordinator and sends a coordinator message to the other (n-2)

processes. Therefore, *n-1* messages are required. The worst case is that the process having the lowest priority number starts the BA. In this case, $O(n)$ processes perform the algorithm and each process causes $O(n)$ message. Consequently, $O(n^2)$ messages are required[1].

The number of messages in the FBA is only *1/n* of that by the BA. The best case is the same as the BA. The worst case is that the process having the lowest priority number initiates the election algorithm. In this case, *(n-2)* election messages, *(n-2)* answer messages, one nomination message and *(n-2)* coordinator messages, totally $O(n)$ messages are required. Therefore, the FBA is better than the BA.

When an exception case occurs during the election, the procedure may be restarted and the cost may be doubled. This applies both to the BA and FBA.

Table 4. Number of messages by each algorithm (*n*: the number of processes)

	BA	FBA
Number of messages	$O(n^2)$	$O(n)$
The best case	*n-1*	*n-1*
The worst case	$O(n^2)$	$O(n)$

6 Conclusion

The bully algorithm (BA) is a commonly used election algorithm, but it is not efficient with respect to the number of messages and the response time. In this paper, we proposed the fast bully algorithm (FBA) that is based on the BA but is superior to the BA. The number of messages is reduced from $O(n^2)$ to $O(n)$ for *n* processes and the response time is reduced by 35%~80%. Also the problem of having two coordinators at the same time that appears by the BA can be avoided by the FBA. The time values used in the simulation should be adjusted in order to fit delay characteristics of a real network.

The FBA needs three more types of messages and implementation cost may be a little higher than the BA. But this overhead is negligible considering the performance improvement. We expect this FBA to be used in wide range of distributed applications such as the server process of network printer, which is a manager process of shared resources therefore requires the mutual exclusion and concurrency control.

References

1. G. Coulouris, J. Dollimore and T. Kindberg, *Distributed Systems, Concepts and Design (3rd Ed.)*, Addison-Wesley, 2001.
2. H. Garcia-Molina, "Elections in a distributed computing system," IEEE Trans. on Computers, vol. 31, no. 1, pp. 48–59, Jan. 1982.

3. E. G. Chang and R. Robert, "An improved algorithm for decentralized extreme-finding in circular configurations of processors," Communications of ACM, vol. 22, no. 9, pp. 281–283, 1979.

4. A. S. Tanenbaum, *Modern Operating Systems*, Prentice Hall, 1992.

5. Christof Fetzer and Flaviu Cristian, "A Highly Available Local Leader Election Service," Proc. of Dependable Computing and Fault Tolerant Systems, vol. 11, pp. 83–102, 1997.

6. Y, Huang and P.K. McKinley, "Group Leader Election under Link-State Routing," Proc. of International Conference on Network Protocols, pp. 95–104, 1997.

7. Riccardo Gusella and Zatti Stefano, "An Election Algorithm for a Distributed Clock Synchronization Program," Proc. of the 6th International Conference on Distributed Computing System, pp. 364–371, 1986.

8. T.W. Kim, E.H. Kim and T.Y. Kim, "A Leader Election Algorithm in a Distributed Computing System," Proc. of the IEEE Computer Society Workshop in Future Trends of Distributed Computing System, pp. 481–487, 1995.

9. S. Singh and J. Kurose, "Electing 'Good' leaders," Journal of Parallel and Distributed Computing, vol. 21, pp. 184–201, May 1994.

10. Andrei V. Borshchev, Yuri G. Karpov and Victor V. Rouadkov, "System Modeling, Simulation and Analysis Using COVERS Active Object, "http://www.xjtek.com/reference /articles/monterey/index.html, 1997.

11. Hoon Choi, "An efficient algorithm for electing a coordinator process in distributed systems," Journal of Korea Information Science Society, vol.25, no 9, pp. 926–936, 1998.

An Enhancement of SIENA Event Routing Algorithms*

Ki-Yeol Ryu and Jung-Tae Lee

Division of Information and Computer Engineering
Ajou University
Suwon 442-790, Republic of Korea
{kryu,jungtae}@madang.ajou.ac.kr

Abstract. A new class of applications based on event interactions is emerging for the wide-area network such as Internet, which is characterized as loose coupling, heterogeneity, and asynchrony. Content-based event routing has been studied to implement an event notification service for wide-area networks. In this paper, we analyze some anomalies of the event routing algorithm in SIENA, a recently developed as a representative event notification service architecture, and develop enhanced routing algorithms. The proposed algorithms take the advantage of SIENA while resolving the anomalies.

1 Introduction

Event service has been widely studied and used as a concept of messaging for event-driven systems. Recently, the event notification service (also referred to as event publish/subscribe service) is considered a mechanism to support interaction among distributed components or applications on a wide area network such as Internet [1,2,3,4,5]. An event notification service can be implemented in various ways in terms of event brokering which is responsible for the delivery of event subscriptions and event notifications between event subscribers and event publishers. A way of traditional event brokering is group-based systems, usually adopted in the centralized, hierarchical or client-server networks in various forms such as channel, subject, or topic [6,7,8] etc.

As an alternative to group-based, a new class of the event notification service called content-based, is recently emerging for distributed applications on a wide area network. In the content-based event notification, a client subscribes events to a network of event servers (or brokers) and the client also publishes events to the network. The main difference in the content-based event notification service is that event servers deliver events to clients in a way of content-based addressing without a pre-determined address like an IP address. The content-based event notification service has been implemented and experimented in various experimental systems such as Elvin [9], Keryx [10], Gryphon [1,2], and SIENA [3,4,5]

---

* This work was supported by Korea Research Foundations (Grant KRF-2000-EA0079)

I. Chong (Ed.): ICOIN 2002, LNCS 2344, pp. 623–633, 2002.
© Springer-Verlag Berlin Heidelberg 2002

systems. Among them, Gryphon and SIENA are based on general peer-to-peer network architectures on which we focus in this paper.

The kernel of the content-based event notification service on a peer-to-peer architecture is the distributed event routing algorithm that routes an event notification from an event publisher to the event subscribers that subscribed to the event. The event routing algorithm in SIENA uses an elegant strategy for optimizing performance called *the most general subscription (or filter) principle* in which if an event broker delivers a more general subscription to its neighbors, it does not have to deliver less general subscriptions to its neighbors without loss of events. This principle reduces the traffic of both event subscriptions and event notifications considerably. However, in the SIENA routing algorithm, we have found two anomalies called *the multi-path anomaly* and *the incomplete subscription anomaly*, which will be described in detail in Section 2.

In this paper, we propose a distributed event routing algorithm preserving the major traffic advantage in SIENA as well as resolving the two anomalies, while breaking the most general subscription principle in part. Our approach increases the event subscription traffic, in general small amount compared to that of event notification traffic. The event notification traffic, however, are basically same in both approaches.

Section 2 briefly describes the concept of the event and the event routing algorithm in SIENA and the two anomalies. An enhanced routing algorithm is proposed in Section 3 and it is analyzed and compared with the related work in Section 4. We conclude the paper in Section 5.

2   SIENA

In this section, we summarize the important concepts in SIENA [3,4,5] such as the event, event filter, poset structure, and event routing algorithms, on which our approach is based, and explain the two anomalies of SIENA.

2.1   Events and Event Filters

An event in SIENA is represented as a set of typed attributes. An attribute has a type, a name, and a value. And an event subscription is represented as an event filter, simply a filter, which specifies a set of attribute constraints. An attribute constraints is a tuple of a type, a name, an operator, and a value.

An attribute $\alpha = (type_\alpha, name_\alpha, value_\alpha)$ matches an attribute constraints $\varphi = (type_\varphi, name_\varphi, operator_\varphi, value_\varphi)$ if and only if $type_\alpha = type_\varphi \wedge name_\alpha = name_\varphi \wedge operator_\varphi(value_\alpha, value_\varphi)$, and it is denoted as $\alpha \prec \varphi$.

An event n matches an event filter f or equivalently f covers n ($n \prec f$ for short):

$$n \prec f \Leftrightarrow \forall \varphi \in f : \exists \alpha \in n : \alpha \prec \varphi$$

The coverage relation between two filters can be derived. A filter f covers[1] another filter f' :

$$f' \prec f \Leftrightarrow \forall n : n \prec f' \Rightarrow n \prec f$$

To implement the concept of the most general filter principle, SIENA introduces a poset structure at each event server that stores subscription filters and their relations. Fig. 1 shows the poset structure between two filters f and f' in which f covers f'. The subscriber set attached by a filter denotes the set of neighbors that subscribe the filter immediately and the forward set denotes the set of neighbors to which the server delivers f.

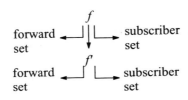

Fig. 1. Poset structure($f \prec f'$)

2.2   Event Routing Algorithms

SIENA provides three algorithms: event subscription, event unsubscription, and event notification (event delivery). An event subscription is basically delivered to all nodes (potential event publishers) in the event server network through a spanning tree. For routing the event subscription, each event server (more strictly, an access point from clients) has its own spanning tree through whose paths the subscription is delivered from the subscriber to the potential event publishers.

An original subscriber (client) creates an event subscription and transfers it to its neighbor in its spanning tree. The receiving server records the subscription information in its poset and then delivers it to its neighbors following the paths in the spanning tree of the original subscriber. The original subscriber of a filter is distinguished from other servers that broker the filter directly to their neighbors. We call them the immediate subscribers or just subscribers.

SIENA introduced an optimization technique to reduce the traffic of subscription delivery, called the most general filter principle. When a server receives a filter f, if the same or more general filter f' ($f \prec f'$) from its neighbor has been already subscribed, then f does not need to be subscribed and thus it stops the delivery of f to its neighbors.

---

[1] The covering relation symbol $\prec$ is shared for the three relations without any confusion.

Event notifications are delivered through the reverse paths by using the subscriber information maintained on the poset in the event servers. Because only the most general filter is stored for a group of related filters in some covering relations, the amount of event notification delivery can be decreased dramatically. (Refer to [3,4,5] for the detailed algorithms).

2.3   Two Anomalies

SIENA event routing algorithms explained above have two anomalies, *the multipath anomaly* and *the incomplete subscription anomaly*. The multi-path anomaly means that one event notification may be delivered to a subscriber through more than one route.

Consider an exemplary network in Fig. 2-a. Fig. 2-b and Fig. 2-c show the resulting posets of all servers after b and a subscribe filter f on the network, respectively. Fig. 2-b and Fig. 2-c also show the spanning trees of b and a, respectively. Suppose that an event e ($e \prec f$) is published at node 4. The event can be routed to the subscriber a through more than one path as follows: 4-1-2-a and 4-1-3-2-a. Moreover, the path may contain a cycle e.g. 4-1-2-3-2-3...

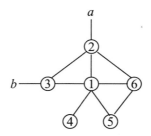

(a) An exemplary event server network

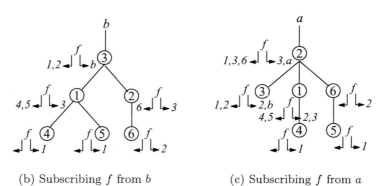

(b) Subscribing f from b (c) Subscribing f from a

Fig. 2. An exemplary server network and event subscriptions

This anomaly is raised because the subscriptions of a same filter from various subscribers come cross at some server directing to different publishers and thus the server can have more than one subscriber for the filter. However, the server cannot differentiate which subscriber intends to subscribe the filter to which publisher only by seeing its poset. When the server receives an incoming event for the filter, it sends the event to all subscribers for the filter that cover the event.

For example, in Fig. 2-c the filter f in node 2 has the subscriber set $\{3, a\}$ in which node 3 subscribed it for node 6 and node a subscribed it for node 1 and node 3. But node 2 receiving an event e that matches f sends it to both 3 and a no matter where it came from.

The incomplete subscription anomaly, originated from the most general filter principle, is that some subscription information can be lost at some nodes and thus event notifications for the subscriptions cannot be delivered to some original subscribers.

For example, at node 2 the subscription f from a in Fig. 2-c is not propagated to node 6, therefore not to node 5, because the same filter from b has already been delivered to node 6 through node 2. Instead, node 5 has received the filter via node 1 when b subscribed it. But an unsubscription of f from b raises a critical problem that removes the filter f at node 5, although it is still necessary for the subscription from a. The main reason is that the subscription from a was not delivered to node 5 through its own spanning tree, but b's spanning tree.

3   Enhanced Routing Algorithms

We basically adopt the concept of events and event filters and the data structures such as poset in SIENA. In our approach, a subscription from a client is broadcast to all the nodes in the network through its own spanning tree to resolve the incomplete subscription anomaly. Subscriptions from the clients that use a same spanning tree still follow the most general filter principle without incurring the anomaly. And an event notification from a publisher is delivered to the subscriber via the exact reverse path of the subscription to resolve the multi-path anomaly.

While breaking the principle in SIENA and thus resulting in increasing the subscription traffic, our approach preserves the advantage of reducing event notification traffic in SIANA by introducing the merge and split operations on filters, which will be explained in next two sections. Fortunately, the traffic of event notification is in general considerably heavier than that of event subscription. In this section we describe the modified event routing algorithms for the event subscription, event unsubscription, and event notification.

3.1   Event Subscriptions

When a server X receives a subscription message subscribe(U, f, a) from the subscriber U (an immediate subscriber or original subscriber), the subscription is processed in X as follows:

algorithm subscribe(U: subscriber, f: filter, a: f's original subscriber)
begin
(1) **if** $\exists f' \in P_s : f \prec f'$ and $U \in subscribers(f')$
(2)   **if** $U_a \in tagged\_subscribers(f')$
(3)   **return**
(4)   **else**
(5)   Add U_a into $tagged\_subscribers(f')$ — (merge)
(6)   **endif**
(7) **elseif** $\exists f' \in P_s : f' = f$ and $U \notin subscribers(f')$
(8)   Add U_a into $tagged\_subscribers(f')$
(9) **else**
(10)   Insert a node f in P_s in the correct position according to
        the covering relation and add U_a into $tagged\_subscribers(f)$
(11) **endif**
(12) **if** $\exists f' P_s : f' \prec f$ and $\exists t : U_t \in subscribers(f')$
(13)   Remove U_t from $tagged\_subscriber(f')$ and add it into
        $tagged\_subscribers(f)$ — (merge)
(14)   **if** $subscribers(f') = \phi$
(15)   Remove f' from P_s
(16)   **endif**
(17) **endif**
(18) Send subscribe(X, f, a) to each node in $children(a, X)$
end

A *tagged subscriber* denotes a subscriber subscripted with the original subscriber (e.g. U_a). In our approach, the subscriber set in the poset structure is the set of tagged subscribers (denoted as $tagged\_subscribers(f)$). The $subscribers(f)$ denotes the subscriber set that all tags are removed from $tagged\_subscribers(f)$. And $children(a, X)$ denotes the set of the child nodes of X node in the spanning tree of a. The children set corresponds to the forward set in SIENA.

In the above algorithm, step 1-6 deals with the situation that X has received a more general filter f' from U. The situation can have two cases: one is that f' was originally sent from client a, and the other is not the case. In the former case, like SIENA, f is not necessary to be delivered without causing the incomplete subscription anomaly. In the latter case, however, after we merge the two filters into the more general filter f', unlike SIENA we record the tagged subscriber (U_a) of f to the subscriber set of f' and then deliver f to the neighbor servers of X (step 18). Step 7-10 handles the case that X has not received any that f' covers f. After inserting f into the poset, if there exists f' which is covered by f and whose subscriber is U, then it should be merged to the more general filter f. This is shown in step 12-17.

The merge operation merges more than two filters into the most general filter according to the covering relation as in the step 5 and 13 in the above algorithm (see Fig. 3). After merging of such filters, the poset structure in the node where the merging is performed is basically same as that of SIENA. But the difference is that even the hidden filter due to the merging can be propagated to child nodes, while, in SIENA, such filter is not propagated to the neighbors due to the most general filter principle, resulting in the incomplete subscription anomaly.

But this merged filter must be split into the original poset structure when one of the merged filters needs to be unsubscribed (see Fig. 3). The unsubscription can be performed after splitting. This operation will be described in detail in next section.

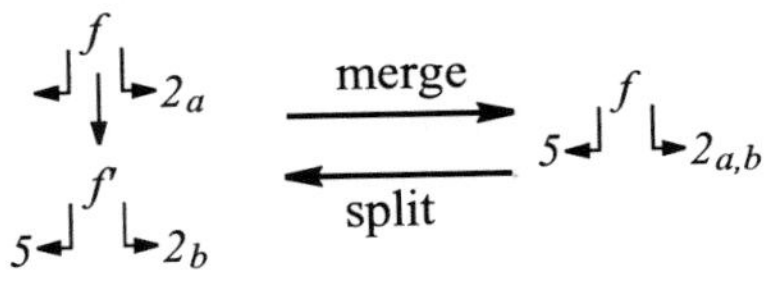

Fig. 3. The merge and split operations

Fig. 4 shows posets after a filter f from b and then f' ($f' \prec f$) from a are subscribed on the network in Fig. 2. In node 6, f' delivered from 2, originated from a, is merged into the more general one f.

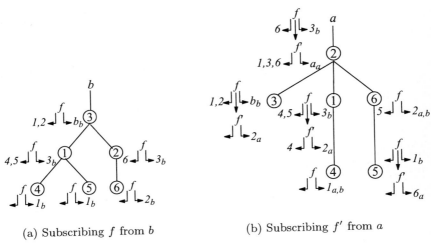

(a) Subscribing f from b (b) Subscribing f' from a

Fig. 4. Event subscriptions ($f \prec f'$)

3.2   Event Unsubscriptions

An event unsubscription is the reverse action to an event subscription. Thus the unsubscription from a client is also broadcast to all the nodes through its own spanning tree. One thing to note is that unsubscription at a merged filter node can be performed after the merged filter is splitted into the original poset structure. When a server X receives a message unsubscribe$(U,\ g,\ a)$, for each filter f in its poset that matches g and contains U_a in its subscriber set, one of the following three cases is executed.

1. For the most general filter $f' \in P_s$ satisfying $f' \prec f$, if there exists $b \in$ $original\_subscribers(f')$ such that $children(a, X) \cap children(b, X) \neq \phi$, then send the split$(f,\ f',\ X,\ b)$ message to each element in the intersection. And then remove U_a from $tagged\_subscribers(f)$.
2. For the most general filter $f' \in P_s$ satisfying $f \prec f'$, if there exists $b \in$ $original\_subscribers(f')$ such that $children(a, X) \cap children(b, X) \neq \phi$, then send the split$(f',\ f,\ X,\ b)$ message to each element in the intersection. And then remove U_a from $tagged\_subscribers(f)$.
3. Otherwise, remove U_a from $tagged\_subscribers(f)$.

If the subscriber set of f remains an empty set after removing U_a, the filter itself is removed from the poset. Finally, the unsubscribe message unsubscribe$(U,\ g,\ a)$ is propagated to each node in $children(a, X)$.

The unsubscription algorithm is very simple compared to that of SIENA because a unsubscription from an original subscriber is not affected by subscription from other subscriber except with the case of splitting filters. First two cases in the above algorithm deal with the case of splitting filters.

One thing to note here is that the server in which a merging was occurred does not know whether it happened or not. Fortunately, the subscriber of the merged filter can find out the fact that its neighbor merged some filters, one of which was delivered from the subscriber. For example, in Fig. 4-b, server 2 knows that server 6 has a filter that merges f and f' . Therefore when node 2 receives an unsubcription message g that matches f from node 3, it must send node 6 a split message that allows to split f and f' into the original poset structure before merging. Similarly, when node 2 receives an unsubscription message g that matches f' from client a, it must send a similar split message to node 6.

The server that receives split$(f,\ f',\ X,\ b)$ message creates a new node f' with the subscriber of X_b in its poset and removes X_b from the subscriber set of f node. Fig. 5 shows the poset structure after client b unsubscribes f in Fig. 4-b. A split operation is delivered from node 2 to node 6 before the unsubscription of f is sent to node 6. Similarly, node 1 sends a split message to node 4.

3.3   Event Notifications

To route an event notification from a publisher to a subscriber through only one path, unlike SIENA we use two kinds of data structure, the poset and an inverted tree (a spanning tree for a publisher). The inverted tree is made of all the paths from each client (potential subscriber) to the publisher in the subscribing spanning tree of the client. We can make such a set of the spanning trees of clients that the set of all subscription paths from the clients to a publisher is an

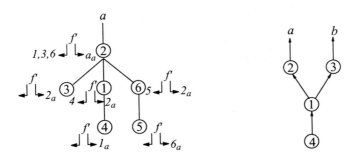

Fig. 5. Unsubscription of f from b **Fig. 6.** Inverted spanning tree for node 4

inverted spanning tree for the publisher. Fig. 6 shows the spanning tree for node 4 (and thus all the clients that has node 4 as the access point) in Fig. 2-a.

The spanning trees for all publishers can be kept in a form of local routing information (child nodes) in each node. For example, the local routing information for node 2 in Fig. 2 is as follows:

$$children(1, 2) = \{a\}, \; children(2, 2) = \{a, 3\},$$
$$children(3, 2) = \{a\}, \; children(4, 2) = \{a\},$$
$$children(5, 2) = \{a\}, \; children(6, 2) = \{a, 3\}$$

In the equations, $children(P, X)$ denotes the set of child nodes of X in the spanning tree of publisher P.

When node X receives an event notification e from a publisher P from its neighbor, the next nodes, $next(P, e, X)$, to which X should deliver the event, can be calculated as follows:

$$next(P, e, X) = children(P, X) \cap subscribers(X, e) \tag{1}$$

In the formula, the term $subscribers(X, e)$ denotes the union of the subscriber sets of all the filters which cover e in the poset of X. For example, an event e ($e \prec f$) delivered to node 2 from publisher 4, can be delivered only to node a, not to node 3.

4   Analysis

If the set of the subscription paths from potential subscribers to a publisher makes an inverted tree that has the publisher node as the root, that is, the spanning tree for the publisher, it is easy to prove that the equation (1) guarantees that an event from the publisher is always delivered to all the subscribers only through one path. Event notification using only subscription information in posets, denoted as $subscribers(X, e)$ in equation (1) can deliver the event to the all subscribers at least once and the set of delivery paths essentially includes the reverse path of the subscription. And the publisher's spanning tree, represented by $children(P, X)$ in equation (1), consists of only one path, the reverse path of subscription, to each potential subscriber from the publisher. Thus, equation (1) calculates one and only one path from the publisher to a subscriber, the exact reverse path of the subscription. Therefore, the multiple-path anomaly can be avoided.

One problem is whether we can make the spanning trees for subscribers so that there exists the spanning tree for each publisher. Fortunately, there exists an easy way to make such spanning trees for clients that there always exists a spanning tree for each publisher. If each client chooses a tree with minimum depth as its spanning tree, the spanning tree for every publisher can be made of subscription paths of the spanning trees of clients.

In our approach, an event subscription is propagated to neighbor nodes in the network through the original subscriber's spanning tree even though a more general subscription has already been sent to them. Though some filters can be merged in the poset, they are recovered when unsubscribing one of the filters. Thus the incomplete subscription anomaly cannot occur since the problem occurs when some subscription information is not delivered to some nodes.

Finally we compare the routing traffic of our approach with SIENA with respect to both subscriptions and notifications. First, the subscription traffic would be much more than that of SIENA because, in our approach, event subscriptions are broadcast to all nodes while SIENA follows the most-general subscription principle. Second, the notification traffic is basically same in both systems (even assuming that SIENA does not raise the two anomalies) because our merge/split operations make the same poset structure with what SIENA does. Fortunately, the frequency of subscriptions would be in general much less than that of event notifications and the size of a subscription quite smaller than that of a notification. The extensive comparisons of SIENA with other related work can be found in [3], most of which can be adopted as the comparison of our approach with them because the basic concept and structure of our approach and SIENA are virtually same.

5   Conclusion

SIENA is a highly expressive and scalable event notification service for wide area networks. It provides a strategy called the most general filter principle to optimize performance. But its content-based event routing algorithms have two non-trivial anomalies, the multi-path anomaly and the incomplete subscription anomaly. In this paper, we have revealed the two anomalies and proposed a solution for the problems. The proposed algorithm adopts the strategy of the most general filter principle in the event notification part, dominant in traffic, while breaking it in the subscription part.

References

1. M. K. Aguilera, R.E. Strom, D.C. Struman, M. Astley, T.D. Chandra, "Matching Events in a Content-Based Subscription System," Proceedings of 18th ACM Symposium on Principles of Distributed Computing(PODC'99), Atlanta, GA, May 1999, pp. 53–61.
2. G. Banavar, T. Chandra, B. Mukherjee, J. Nagarajaro, "An Efficient Multicast Protocol for Content Based Publish-Subscribe Systems," Proceedings of 19th IEEE ICDCS'99, Austin, TX, May 1999, pp. 262–272.
3. A. Carzaniga, D.S. Rosenblum, and A.L. Wolf, "Design and Evaluation of a Wide-Area Event Notification Service," ACM Transactions on Computer Systems, 19(3), Aug. 2001, pp. 332–383.
4. A. Carzaniga, D.S. Rosenblum, and A.L. Wolf, "Achieving Expressiveness and Scalability in an Internet-Scale Event Notification Service," Proceedings of 19th ACM Symposium on Principles of Distributed Computing (PODC2000), Portland OR July, 2000.
5. A. Carzaniga, D. S. Rosenblum, and A. L. Wolf, "Interfaces and Algorithms for a Wide-Area Event Notification Service," Technical Report CU-CS-888-99, Department of Computer Science, University of Colorado, October, 1999 (revised May 2000).
6. G. Cugola, E. Di Nitto, and A. Fuggetta. "Exploiting an Event-based Infrastructure to Develop Complex Distributed Systems." Proceedings of the 20th ICSE98, Kyoto, Japan, Apr. 1998.
7. M. Hapner, et.al. Java Message Service Version 1.0.2, Sun Microsystems, Nov. 9. 1999.
8. Object Management Group, "Notification Service," Technical Report, Nov. 1998.
9. B. Segall and D. Arnold. Elvin has left the building: A publish/subscribe notification service with quenching. Proceedings of AUUG97, Brisbane, Queensland, Australia, Sept. 1997.
10. M. Wray, R. Hawkes, Distributed Virtual Environments and VRML: an Event-based Architecture," Computer Networks and ISDN Systems 1-7, 30, 1998. pp. 43–51.

Dynamic Behavior of Multiagents with Subjective Cooperative Relations

Takashi Katoh[1], Tetsuo Kinoshita[2], and Norio Shiratori[1]

[1] Research Institute of Electrical Communication, Tohoku University, Sendai, Japan
{p-katoh,norio}@shiratori.riec.tohoku.ac.jp
[2] Information Synergy Center, Tohoku University, Sendai, Japan
kino@riec.tohoku.ac.jp

Abstract. The aim of this paper is to analyze scaling behavior of the system based on subjective cooperative relations. In our previous works, we have analyzed the efficiency of the system based on subjective cooperative relations and behavior of the system under several settings of agents as basic characteristics of the system. According to the results of further experiments, we show several interesting characteristics of the system based on subjective cooperative relations.

1 Introduction

To organize an effective multiagent system existing large amount of agents, it is needed to compose organizations consisting of several agents and to cooperate with each other. Therefore agents are required to know other agents' trait or current status via inter-agent communication and assign tasks based on the information.

There are several works regarding cooperation among agents, e.g. contract net protocol [1] and multistage negotiation protocol [2]. They try to decide a role of an agent to do some fixed tasks (goals). These methods can be regarded as a way to compose organizations consisting of several agents. However, the amount of communication required for composing organizations or cooperated problem solving will increase if the number of agents on the system increases. As a result, the efficiency of the system may be degraded. On this account, the contract net protocol may cause much traffic because the agent issues its task announcements by broadcasting to all agents and has no way to select the most suitable agent based on its own sense of values.

There are several works on formation of coalitions [3,4]. These works try to improve the efficiency by grouping all agents on the system into several groups and allocating each task to each group. But this method divides the whole system into the number of tasks and processes each task in each group when several tasks exist. It is efficient when it is enough just to process only those tasks. However, it may not work well with a real multiagent system if tasks occur irregularly. Also the overheads of this method may become too large because this method is needed to form coalitions whenever it is given new tasks to the system. Therefore, this is not always a suitable method in a real multiagent system.

I. Chong (Ed.): ICOIN 2002, LNCS 2344, pp. 634–643, 2002.
© Springer-Verlag Berlin Heidelberg 2002

The organizational self-design [5] is aimed at improving the performance of a system by changing organization of agent using 'composition' and 'decomposition'. The organization of this method, however, does not reflect the other agents' status.

On the other hand, methods of the agents' dynamic organization formation based on the concept social awareness [6,7] or other task allocation methods [8,9] are being suggested in recent years. In these methods, the benefit is introduced explicitly and agents decide their behavior based on it. In case of agents which process actual tasks, however, it is difficult to apply this idea to the system because the time needed to perform the task, in general, is not known a priori.

In our previous papers, we proposed *subjective information* and a concept of *subjective cooperative relations*[10,11]. Each agent determines its behavior based on its subjective information, which are internal information of agents to be used as criteria for decision making. And the subjective cooperative relation, which represents relations between an agent and the neighboring agents, is defined on the basis of subjective information. This relationship may change from time to time and an agent can know a situation of the peer's surroundings by exchanging information about subjective cooperative relations. Therefore, an agent can grasp information from a wider range by communicating only with a specific peer agent.

In this paper, we carried out a set of experiments to study the behavior of multiagent system based on subjective cooperative relations, and show that our proposed method is useful especially when the system is almost balanced in terms of an amount of incoming tasks and ability of the system. We also present and discuss the characteristics in other cases, i.e., overloaded and sufficient situations of multiagent systems.

The rest of the paper is organized as follows. We give an outline of the subjective information of agents and subjective cooperative relations based on subjective information in Sect. 2. We show results obtained from our experiments in Sect. 3, and conclude our work in Sect. 4.

2   Subjective Information and Subjective Cooperative Relations

In this section, we give a brief definitions of (1) subjective information which is the internal information of agents and reflects the agents' cooperative behavior of the past, and (2) subjective cooperative relations which represent relations of an agent with the neighboring ones, and is defined on the basis of subjective information [11].

2.1   Subjective Information

In multiagent systems, agents are required to behave flexibly not only to realize a desirable behavior of a multiagent system as a whole but also to work effectively.

Thus each agent should have criteria of the decision making and determine its behavior based on the criteria.

Subjective information is used as criteria of decision making. Each agent possesses, for example, the opponents' reliabilities based on the results of the previous negotiations.

Definition 1. Subjective information *is information maintained by an agent i internally and is used when an agent deals with subject S, which is a matter that an agent deals with (e.g. problems, concerning agents). Subjective information is denoted as follows, i.e.,*

$$m_i(S) \ ,$$

where m is an identifier to specify a subjective information.

Each subjective information is assigned its value val *and referred as:*

$$m_i(S) = \text{val} \ .$$

□

For example, a subjective information $Reliability_A(\text{Agent } B) = 10$ means "agent A has a subjective information regarding $Reliability$ of agent B and its current value is 10."

Each agent updates its subjective information based on the received messages. Assume that agent j asks for processing a task to agent i. The behavior is defined as:

$$Reliability_j(i) \leftarrow \begin{cases} Reliability_j(i) + \alpha & \text{if ACCEPTed} \\ Reliability_j(i) - \beta & \text{if REFUSEd} \end{cases} , \qquad (1)$$

where α and β are terms defined with respect to the agents. An agent j increases its $Reliability_j(i)$ to agent i when the result is an ACCEPT and decreases its $Reliability_j(i)$ when the result is a REFUSE. This implies that the subjective information of agents reflect the agent's behavioral history.

The $Reliability_j(i)$ can be utilized as a criterion to select the most relevant opponent to send a request because a rate of returning an ACCEPT decreases when an agent become busy and the $Reliability_j(i)$ will also be gradually decreased. By changing agents' own subjective information depending on the results of repetitive communication, the agents can know the opponents' current status or situation of a group of agents.

2.2   Subjective Cooperative Relations for Multiagent Systems

Next, we describe a concept of a group of agents, called *subjective cooperative relations,* to define a method for obtaining information of a multiagent system and selecting suitable agents to cooperate with.

Definition 2. Subjective cooperative relation *of an agent i is a set of agents C_i that satisfy the following condition:*

$$C_i = \{j \mid F_i(\{m_i(j)\}) > \theta\} \ ,$$

where $\{m_i(j)\}$ is a set of agent i's subjective information to agent j, F_i is a function of subjective information and θ is a threshold of $F_i(\{m_i(j)\})$. A function F_i and threshold θ is defined with respect to the agents. □

F_i is a function to evaluate an agent whether it should cooperate or not. We denote the subjective cooperative relation as 'SCR' here after.

The SCR represents relations between an agent and the neighboring agents. This relationship may change from time to time and by exchanging information about a SCR, an agent can know a situation of the peer's surroundings. Therefore, an agent can grasp information from a wider range by communicating only with a specific peer agent and SCRs help each agent to determine the appropriate partners to cooperate with.

3 Experiments on Dynamics of Multiagent System Based on Subjective Cooperative Relations

In this section, we demonstrate the results obtained from our experiments and discuss the dynamics of the multiagent system based on SCRs.

In our previous works [10,11], we showed the efficiency of system based on SCRs and the effects of the settings of agents' behavior and so on, to forming of SCRs. Hence, the aim of experiments in this paper is to analyze scaling behavior of the system based on SCRs.

The agents of multiagent system in our experiments have been implemented using Perl, and executed on AT compatible computers of FreeBSD 2.2.8R / 4.3R and Sun Ultra SPARCStations of Solaris 2.5.1.

3.1 Properties and Functions of Agents

In our experiments, a multiagent system consists of many task agents. A user provides a task T to this system via a task agent. A task agent divides a task received from a user into subtasks and processes subtasks if the agent can perform the subtasks by itself, otherwise tries to assign them to other agents.

Task agents behave as follows: (T1) When a task agent receives a request to perform a subtask from another task agent (requester agent), (T1-1) the agent returns an ACCEPT if it is currently free, or returns a REFUSE if it is not free (dealing with another subtask). (T1-2) Along with the reply message (ACCEPT or REFUSE), the agent discloses information about its SCR and its members' abilities to the requester agent.

(T2) When a task agent receives a request of task T from a user, (T2-1) the task agent divides the task into subtasks and distributes the subtasks to other

Table 1. Abilities of agents ('t_1-20' denotes an agent which can deal with subtask t_1 and takes 20 clocks)

Task \ Process time	20	40	60	80	100
t_1	t_1-20	t_1-40	t_1-60	t_1-80	t_1-100
t_2	t_2-20	t_2-40	t_2-60	t_2-80	t_2-100
t_3	t_3-20	t_3-40	t_3-60	t_3-80	t_3-100
t_4	t_4-20	t_4-40	t_4-60	t_4-80	t_4-100
t_5	t_5-20	t_5-40	t_5-60	t_5-80	t_5-100
t_6	t_6-20	t_6-40	t_6-60	t_6-80	t_6-100

task agents (candidate agents) considering their abilities. An assignment of subtasks ends up in a failure if there exists no candidate agents with required ability of performing the given subtask. (T2-2) The task agent updates its subjective information depending on the reply message as described in Section 2. Here we define α and β appeared in (1) as follows:

$$\alpha = \beta = 1 \ ,$$

i.e. when the result is an ACCEPT, the agent increases the *Reliability* of the candidate agent by +1, and when the result is a REFUSE, it decreases the *Reliability* of the candidate agent by −1. (T2-3) The task agent updates a list of known agents according to the information given by candidate agents. The default value of *Reliability* to a new agent is 0.

3.2   System Configuration

We have carried out several experiments with the multiagent systems consisting of 30, 36, 42, 48, 54, 60, 90 task agents respectively. A user gives a task T to a randomly selected task agent in every 5 clocks. Each task T is a combination of randomly selected (one or more) subtasks $(t_1, \ldots, t_6)$ and each subtask cannot be subdivided anymore.

The abilities of agents are defined as follows: For 30 agents system, there exist five agents which can perform each of the subtasks $(t_1, \ldots, t_6)$, and the agents take 10, 20, 30, 40 and 50 clocks to deal with the subtasks respectively (Table 1).

For 36 agents system, we add six agents of 20 clocks (t_1-20, ..., t_6-20) to 30 agents system. For 42 agents system, we add further six agents of 40 clocks (t_1-40, ..., t_6-40) to 36 agents system, and so on.

We have set up the following initial conditions: (1) A task agent knows only two other randomly selected agents and their abilities, (2) There exists no SCRs of agents in the initial state.

In order to compare efficiency of our method with an ideal system, we observed the behavior of the ideal multiagent system. Here, the ideal multiagent system is that the subtasks are assigned by an omniscient task assigner, in other

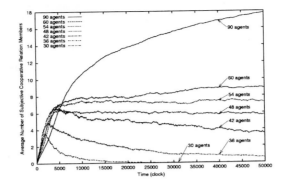

Fig. 1. Transitions of average sizes of subjective cooperative relations

words, the subtasks are assigned to the agents which have the shortest process time to deal with the subtask among the free agents, which don't have any assigned tasks.

3.3   Results of Experiments

We have carried out several sets of experiments with different task sequences and the following results show the average values.

The transitions of average sizes of SCRs, i.e. the average numbers of SCR members, are depicted in Fig. 1.

The left-hand side, middle and right-hand side of Fig. 2 show the transitions of the average sizes of the SCRs, the system performance values of both the SCR-based system and the ideal system. These figures show the typical behaviors of a overloaded system ($n = 30$), a balanced system in terms of an amount of incoming tasks fit to system's ability ($n = 42$) and a system in which incoming tasks are fewer than system's ability ($n = 90$). Here, n denotes the number of agents on the system.

Table 2 and Table 3 show the performance values of both the SCR-based system and the ideal system respectively.

3.4   Analysis and Evaluation

From these results, we can derive the following conclusions:

1. The transitions of the average sizes of SCRs (Fig. 1) reflect the situations of the whole system, i.e., (1) when a system becomes the overloaded situation, the average sizes of SCRs are going to decrease after the respective peak values (i.e. almost all agents have assigned subtasks), (2) when the system becomes the balanced situation in terms of an amount of incoming tasks fit to system's ability, the average sizes of SCRs is almost constant, (3) when the system can deal with the incoming tasks sufficiently, the average sizes of SCRs increase monotonously.

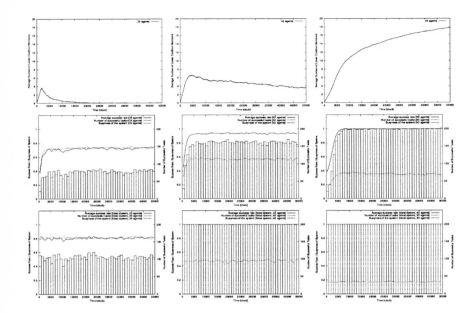

Fig. 2. Transitions of average sizes of SCR (top), system performance indicators of SCR-based system (middle) and system performance indicators of ideal system (bottom). LHS is results of $n = 30$ (an overloaded state), middle is results of $n = 42$ (balanced system in terms of an amount of incoming tasks and ability of system), and RHS is result of $n = 90$ (incoming tasks are fewer than ability of system)

The situations explained above can be used as an indicator to grasp the system's performance as a whole.

2. The proposed method is useful for making good use of a multiagent system especially when the system is almost balanced in terms of an amount of incoming tasks and system's ability.

The elapsed time per subtasks for $n = 42$ (i.e. a system which is almost balanced), is shorter than those of other settings as shown in Table 2. We therefore infer that each agent can behave adaptively by making its decision using subjective information, which is updated according to the results of its requests to reflect the system's situation.

In the above setting ($n = 42$), the performance values of the system are a little worse than those of the ideal systems (Table 2 and 3). It may come from the simplicity of the behavior functions of each agent. As shown in our previous papers [11], this performance degradation can be overcome by improving the behavior functions of the agent, i.e. modifying a method to convey informa-

Table 2. Performance indicators of SCR-based systems for each 10000 clocks

Period (clock)		$n = 30$	$n = 36$	$n = 42$	$n = 48$	$n = 54$	$n = 60$	$n = 90$
~10000	average of success rates	0.6590	0.7846	0.8553	0.8204	0.8388	0.8284	0.8257
	number of successful tasks	658.3	996.0	1350.0	1183.7	1319.3	1281.7	1406.3
	elapsed time per subtasks	44.03	38.59	37.56	41.16	40.95	46.91	48.14
~20000	average of success rates	0.7131	0.8377	0.9393	0.9186	0.9457	0.9439	0.9995
	number of successful tasks	750.0	1141.7	1626.7	1481.0	1664.7	1638.7	1996.0
	elapsed time per subtasks	43.87	38.18	36.55	40.01	40.12	45.82	46.50
~30000	average of success rates	0.7278	0.8430	0.9378	0.9146	0.9464	0.9461	0.9996
	number of successful tasks	798.0	1145.7	1618.3	1472.7	1666.3	1650.7	1995.3
	elapsed time per subtasks	43.61	38.04	36.28	39.73	39.98	45.29	45.65
~40000	average of success rates	0.7265	0.8410	0.9358	0.9134	0.9520	0.9457	0.9997
	number of successful tasks	765.7	1137.3	1608.0	1480.7	1695.7	1644.3	1998.0
	elapsed time per subtasks	43.43	37.78	36.26	39.61	39.83	45.43	45.11
~50000	average of success rates	0.7321	0.8462	0.9355	0.9159	0.9443	0.9465	0.9999
	number of successful tasks	812.0	1172.3	1611.7	1491.3	1655.7	1651.3	2000.0
	elapsed time per subtasks	43.34	37.85	36.23	39.69	39.78	45.21	45.01

Table 3. Performance indicators of ideal systems

Period (clock)		$n = 30$	$n = 36$	$n = 42$	$n = 48$	$n = 54$	$n = 60$	$n = 90$
1~50000	average of success rates	0.8064	0.9850	0.9995	1.000	1.000	1.000	1.000
	number of successful tasks/5	1051.3	1897.9	1997.1	2000.0	2000.0	2000.0	2000.0
	elapsed time per subtasks	34.61	31.05	28.30	27.93	27.92	27.86	21.88

tion of SCR ((T1-2) in Section 3.1) and / or a method of updating subjective information ((T2-2) in Section 3.1).

3. **A member of a multiagent system should change its behavioral functions according to situation to maintain the whole system's performance.**

(1) It is desirable to exchange the explicit information of agent's ability when a system is not too busy as a whole (i.e. an amount of incoming tasks are fewer than system's ability). In case of $n = 90$ (a system which is not too busy), the values of both the average of success rates and the number of successful tasks are by no means inferior to an ideal system (Table 2 and Fig. 2). The elapsed time per subtasks for $n = 90$, however, is worse than the ideal system. The reason is that the system is not busy as a whole and almost all requested tasks are accepted without retry operations. Therefore, the agents' subjective information don't reflect the ability (the required time to deal with a subtask) of the other agents. This phenomenon implies that it is required to exchange the explicit information of the agent's ability when many agents which do not have assigned subtasks exist in the system.

However, in other cases (overloaded and balanced situations), exchanging the explicit information may cause the degradation of agent's ability to adapt to the situation because the other agents' abilities have also been changed simultaneously. Still, our method is useful even if the agents don't exchange the explicit information when the whole system is not too busy. Because a SCR-based system needs a small amount of communication (as shown in [12]) than a broadcast-based system such as a system using contract net protocol.

(2) Under the overloaded situation, it should avoid rapid decrement of SCR members. Since the formation of SCRs become slow (Fig. 1, LHS of Fig. 2) in an overloaded situation ($n = 30$), exchanging information of SCRs should be hindered. As a result, an agent cannot get information of unknown (i.e. new) agents by using information of SCRs sent by the other agents, and also cannot issue a request for a subtask.

Summarizing the above discussions, in order to improve the performance of multiagent systems based on SCRs, it is important to (1) design a suitable way to manipulate the subjective information, and (2) select and replace the agents' behavioral strategies with respect to the changes of runtime situation of the system.

4 Conclusion

To organize an effective agent-based system where a large number of agents exist, it is required to compose organizations consisting of agents which cooperate with each other. To do so, the agents have to know surrounding situations such as other agents' trait and current status via inter-agent communication. We have proposed the concept of subjective information of agents and subjective cooperative relations (SCRs) based on subjective information.

In this paper, we carried out a set of experiments to analyze the scaling behavior of multiagents based on SCRs in order to design effective multiagent systems with SCRs. And then, we showed the behavioral characteristics of multiagents based on SCRs. As a result, we showed that the proposed method is useful especially when the system is almost balanced as a whole in terms of incoming tasks and ability of the system, and also discussed characteristics in other cases, i.e. overloaded and sufficient situations of multiagent systems.

The future works include experimenting with other different settings, e.g. considering the coexistence of agents of various settings on the same system.

References

1. Reid G. Smith. The contract net protocol: High-level communication and control in a distributed problem solver. *IEEE Transactions on Computers*, C-29(12):1104–1113, December 1980.
2. Susan E. Conry, Kazuhiro Kuwabara, Victor R. Lesser, and Robert A. Meyer. Multistage negotiation for distributed constraint satisfaction. *IEEE Transactions on Systems, Man, and Cybernetics*, 21(6):1462–1477, November/December 1991.
3. Onn Shehory and Sarit Kraus. Task allocation via coalition formation among autonomous agents. In Chris S. Mellish, editor, *Proceedings of the International Joint Conference on Artificial Intelligence, 1995 (IJCAI-95)*, volume I, pages 655–661, Denver, Colorado, August 1995. International Joint Conferences on Artificial Intelligence, Inc., Professional Book Center.
4. Onn Shehory and Sarit Kraus. Methods for task allocation via agent coalition formation. *Artificial Intelligence*, 101:165–200, 1998.

5. Toru Ishida, Les Gasser, and Makoto Yokoo. Organization self-design of distributed production systems. *IEEE Transactions on Knowledge and Data Engineering*, 4(2):123–134, April 1992.
6. L. M. Hogg and N. R. Jennings. Socially rational agents. In *Proc. AAAI Fall symposium on Socially Intelligent Agents*, pages 61–63. AAAI, November 1997.
7. S. Kalenka and N. R. Jennings. Socially responsible decision making by autonomous agents. In *Proceedings of Fifth International Colloquium on Cognitive Science (ICCS-97)*, May 1997. San Sebastian, Spain.
8. Martin R. Andersson and Tuomas W. Sandholm. Contract types for satisficing task allocation: II Experimental results. In *Proc. AAAI 1998 Spring Symposium on Satisficing Models*. AAAI, March 1998.
9. Tuomas W. Sandholm. Contract types for satisficing task allocation: I Theoretical results. In *Proc. AAAI 1998 Spring Symposium on Satisficing Models*. AAAI, March 1998.
10. Takashi Katoh, Tetsuo Kinoshita, and Norio Shiratori. A model of coalition formation based on agents' mental states. In *The Second Pacific Rim International Workshop on Multi-Agents (PRIMA99)*, pages 149–163, January 1999.
11. Takashi Katoh, Tetsuo Kinoshita, and Norio Shiratori. Dynamic properties of multiagents based on a mechanism of loose coalition. In Chengqi Zhang and Von-Wun Soo, editors, *Design and Applications of Intelligent Agents*, volume 1881 of *Lecture Notes in Artificial Intelligence (LNAI) 1881*, pages 16–30. Springer, August 2000. Third Pacific Rim International Workshop on Multi-Agents, (PRIMA 2000).
12. Takashi KATOH, Tetsuo KINOSHITA, and Norio SHIRATORI. A method of group formation of multi-agents. *The Transactions of the Institute of Electronics, Information and Communication Engineers D-I*, J84-D-I(2):173–182, February 2001. (in Japanese).

A File Migration Scheme for Reducing File Access Latency in Mobile Computing Environments

Moon Seog Han[1], Sang Yun Park[2], and Young Ik Eom[2]

[1] Dept. of Administration and Computer Science,
Wonju National College, San 2-1, Heungup-Ri,
Heungup-Myun, Wonju-Si, Kangwon-Do, Korea
mshan@sky.wonju.ac.kr
[2] School of Electrical and Computer Engineering,
Sungkyunkwan Univ., Chunchun-Dong 300,
Jangan-Gu, Suwon-Si, Kyounggi-Do, Korea
{bronson, yieom}@ece.skku.ac.kr

Abstract. In mobile computing environment, mobile support station provides a wireless communication link to mobile clients and help them to access remote file servers that reside on a fixed network. Mobile hosts send requests to the mobile support station to access files on remote file servers. No longer this environment requires a user to stay at a fixed position in the network and enables almost unrestricted user mobility. The user mobility affects the file access latency. Mobile hosts have severe resource constraints in terms of limited size of non-volatile storage, and the burden of computation and communication load raises file access latency. In this paper, we propose a scheme for reducing the file access latency by using file migration scheme. We develope an on-demand file migration scheme which determines when the file server should migrate files to another server, and when it should transfer files to mobile hosts. Using simulation, we examine the effects of the parameters such as file access frequency, file size, and mobility rate on the file system performance. We present simulation results, providing useful insight into file system latency.

1 Introduction

The recent mobile computing environment is growing rapidly due to the development of wireless mobile communication technologies and the generalization of portable terminals. Mobile computing environment consists of fixed hosts that reside in wired networks and mobile hosts communicating through wireless networks. Of the fixed hosts, some servers provide various information services and the mobile support stations not only provide a wireless interface to the mobile hosts, but also control a cell, which is a geographical area in the mobile computing environment:[1].

I. Chong (Ed.): ICOIN 2002, LNCS 2344, pp. 644–655, 2002.
© Springer-Verlag Berlin Heidelberg 2002

Mobile computing environment must help mobile clients to access remote information at any time and at any place. Mobile clients send requests to mobile support stations to access files which reside on remote servers. Mobile support stations receive the files from the remote servers and deliver them to the mobile clients. The high mobility of mobile hosts and the narrow bandwidth in the wireless network may cause some delay in accessing files. Also, mobile clients may request file services too often and give rise to bandwidth contention among themselves, which can cause latencies on the whole:[2].

In this paper, to solve the file access latency problem, we propose a file migration scheme. The scheme entails moving the files that the mobile host with limited resources asks for from the home server to the remote server where the mobile host resides. Using the proposed migration scheme, we can reduce the file access latency time by distributing the load to the fixed hosts who provide the stable services. This is to say that we propose a dynamic file migration scheme that decides whether to move a file or not when an access to the file is requested. For our file migration scheme, the file migration criteria were set by taking into consideration the file size, the mobility rate of the mobile host, and the file access frequency. We analyzed the performance of our scheme according to the file migration criteria by comparing the cost of transferring the file that the mobile host has requested to the mobile support station and the mobile host, with the cost of migrating the file to the server in the area the mobile host currently resides and providing service.

In Section 2, we investigate the related works. We introduce the system model in Section 3, and in Section 4, we propose a file migration scheme based on the migration criteria. In Section 5, we describe the evaluation results acquired from the simulation experiments. Finally, in Section 6 we evaluate the proposed scheme and present the future research directions.

2   Related Works

Usually, Mobile hosts are under massively changing conditions. In order to operate efficiently in such conditions, they must adapt their behaviors to the dynamically changing conditions. Resource management under the mobile computing environments is done in similar way to that of the distributed systems in a fixed network, except that the wireless network environments have some restrictions. For traditional distributed system environments, there have been many research attempts to improve the efficiency of the resource usage by distributing the load to all available servers through file assignment/allocation, file migration, and file replication, but there has not been much research on the topic for mobile computing environments:[5],[6],[7].

Also, researches on load sharing and balancing in distributed systems were conducted for moving processes from heavily loaded servers to lightly loaded servers:[3]. Pope proposed a method that could reduce the load of the mobile host and prevent service latency by migrating the computation-bound application programs of a mobile host so that it could be run on a fixed server:[4]. Such researches on migration of application programs have not completely solved the problems of system performance, and still are not easy to implement due to the heterogeneity of the hosts and the mobility of the hosts.

Unlike the situation in existing distributed environments, the characteristics such as wireless communication, mobility, and scarce resources should be considered for constructing mobile computing environments. Up to now, most researches for increasing the performance of mobile computing systems have been focused on caching strategies. Caching mechanism for mobile computing environments should be very different from that of distributed systems because mobile hosts in mobile computing environments have poor resources and low bandwidth communication links that may be disconnected frequently:[8],[2]. Among the various caching methods, prefetching and hoarding are typical methods that cache the files in advance with high possibility of usage:[1],[9],[10].

In mobile computing environments, mobile hosts change their location frequently while connected to the network:[11]. Each mobile host will try to connect to the nearest server from his location to acquire the information that he wants:[1]. In this paper, we propose a file migration scheme that can reduce file access latency and also reduce the overhead that arises due to the consistency or recency of the files during caching.

3   System Model

The proposed system model as shown in Fig. 1 consists of a fixed network and several mobile wireless networks. In the fixed network, there are many file servers and mobile support stations. Each mobile support station supports his cell. There are several cells in the wireless network, and mobile hosts are in one of the cells. It is assumed that there are large number of files in the remote file servers that reside in the fixed network. File migration occurs only among the servers and not between the server and the mobile support station or the mobile host. Now, we describe some assumptions on our system model.

- There is no file sharing among the servers.
- Network partition does not occur in the fixed network.
- There is only one replica of a file in the servers.
- Mobile host does not have enough resources and is a kind of personal handheld computers like diskless PDAs.

We use the following notations to describe the file migration scheme proposed in this paper.

- S : a set of file servers
- S_n : the n-th file server
- SM : a set of mobile hosts
- mh_i : the i-th mobile host
- C : a set of wireless cells
- C_k : the k-th wireless cell

- MSS : a set of mobile support stations
- MSS_k : the k-th mobile support station
- SF : a set of all files stored in the file servers
- f_m : the m-th file in the SF
- $SM\,(C_k, t)$: a set of mobile hosts staying in the cell C_k at time t
- $SF\,(S_n, t)$: a set of files maintained in the server S_n at time t

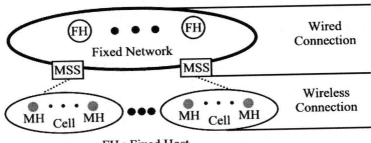

FH : Fixed Host
MSS : Mobile Support Station
MH : Mobile Host

Fig. 1. System model

So, S, SM, C, MSS, and SF can be defined as follows.

- $S = \{S_1, S_2, \ldots, S_n\}$
- $SM = \{mh_1, mh_2, \ldots, mh_l\}$
- $C = \{C_1, C_2, \ldots, C_k\}$
- $MSS = \{MSS_1, MSS_2, \ldots, MSS_k\}$
- $SF = \{f_1, f_2, \ldots, f_m\}$

A set of mobile hosts in a specific cell at time t is can be expressed as follows.

- $SM(C_1, t) = \{mh_{11}, mh_{12}, \ldots, mh_{1l_1}\}$
- $SM(C_k, t) = \{mh_{k1}, mh_{k2}, \ldots, mh_{kl_k}\}$
- $l = \sum_{i=1}^{k} l_i$: total number of mobile hosts

A set of files in servers at time t can be expressed as follows.

- $SF(S_1, t) = \{f_{11}, f_{12}, \ldots, f_{1m_1}\}$
- $SF(S_n, t) = \{f_{n1}, f_{n2}, \ldots, f_{nm_n}\}$
- $m = \sum_{i=1}^{n} m_i$: total number of files

By taking into consideration the wired network which has high bandwidth, and the wireless network which has low bandwidth, data transmission units are different for each network. In the fixed network, the servers exchange data by transferring them in the whole file. This is because the fixed network is connected through a high-speed wired network and it is assumed that network partition does not occur in the fixed network. We assume that the data transmission through the wireless network is performed in blocks.

4   File Migration Scheme

4.1   Criteria for File Migration

File size and file access frequency are important factors that affects the file migration scheme in distributed systems. Our scheme considers the mobility of the mobile hosts as well as the file access frequency and file size in determining file migration. Especially, considering the response time, our scheme can dynamically determine whether to migrate a file or not using some threshold values of the factors we used, which resulted in reduction of the latency time in the overall system and improved the system performance. Next, we discuss the criteria for file migration that are considered in our scheme.

File access frequency.

- $\Delta = t - t_i$: the length of the time interval $[t_i,\ t]$
- $a_{ij}(t,\ \Delta)$: the access frequency (at time t) that mobile host mh_i requests file f_j during Δ
- $A_{ij}(t_0,\ t)$: the total access frequency by mobile host mh_i to file f_j during $[t_0,\ t]$
- $A_{total}^i(t,\ \Delta) = \sum_{j=1}^{m} a_{ij}(t,\ \Delta)$: the total access frequency on the files in file servers that mobile host mh_i requests

Mobility rate.

- $N_{a,b}^i(t,\ \Delta)$: the number of handoffs of the mobile host mh_i from cell a to cell b during the time interval Δ (measured at time t)
- $N_b^i(t,\ \Delta) = \sum_{a=1}^{k} N_{a,b}^i(t,\ \Delta)$: the total number of handoffs that mh_i moves to the specific cell b
- $N^i(t,\ \Delta) = \sum_{a=1}^{k} \left(\sum_{b=1}^{k} N_{a,b}^i(t,\ \Delta) \right)$: the total number of handoffs of the mh_i during Δ $\left(assume\ N_{a,a}^i(t,\ \Delta) = 0\ for\ all\ C_a \right)$
- $N_{avg}^i(t,\ \Delta) = N^i(t,\ \Delta)/\Delta$: the mean mobility rate of mobile host mh_i during Δ

When a mobile host requests a specific file, the server applies the migration criteria to decide whether to migrate the file or not. The migration time it takes to migrate from the source server to the destination server and the response time the mobile host gets are the main concerns of our scheme. The factors for evaluating the cost of file migration scheme are as follows.

- M : the message size for requesting a file
- L_j : the size of the file f_j
- α : the block size
- $k = \frac{L_j}{\alpha}$: the total number of blocks of the file f_j
- B_{fixed} : the bandwidth of the fixed network
- $B_{wireless}$: the bandwidth of the wireless network

- RT^i_{kj} : the response time a mobile host mh_i gets when he requests a file f_{kj} in the server S_k
- MT^l_{ij} : the migration time for moving the file f_{ij} in the server S_i to the server S_l
- $C = [c_{ij}]$, $1 \le i \le k$, $1 \le j \le n$: the communication cost matrix between the mobile support station MSS_i and the server S_j (if $i = j$, $c_{ij} = 0$)
- $D = [d_{ij}]$, $1 \le i \le n$, $1 \le j \le n$: the communication cost between the server S_i and the server S_j (if $i = j$, $d_{ij} = 0$)

Response time includes the time it takes to send the request message to the server when the mobile host asks for a file and the time it takes for the server to transfer the file to the mobile host. Migration time is the time it takes to send the file between file servers. The factors to be considered for evaluating the response time and migration time are the bandwidth of each network, the message size, and the file size.

The following shows the formula for evaluating the response time and the migration time based on the factors mentioned above.

$$Message\ Transmission\ Time: \ MTT = \frac{M}{B_{fixed}} + \frac{M}{B_{wireless}}. \qquad (1)$$

$$File\ Transmission\ Time: \ FTT = \frac{L_j}{B_{fixed}} + \frac{k \times \alpha}{B_{wireless}}. \qquad (2)$$

$$Response\ Time: \ RT^i_{kj} = \frac{L_j + M}{B_{fixed}} + \frac{k \times \alpha + M}{B_{wireless}}. \qquad (3)$$

$$Migration\ Time: \ MT^l_{ij} = \frac{L_j}{B_{fixed}}. \qquad (4)$$

4.2   Migration Scheme

The scheme proposed in this paper considers two different cases in determining whether to migrate a file. The cost model considers the relative cost of the migration and non-migration schemes according to the file access frequency and mobility. Fig. 2 shows our model for migration scheme.

When a mobile host mh_i, staying in the cell C_a, requests the mobile support station MSS_a for a file f_j, the server S_k, which has received the request, decides whether to migrate the file or not based on the migration decision criteria. Now, the total cost for accessing a file by the mobile host when the file migration scheme is used is as follows.

$$COST_{FM} = \frac{L_j + M}{B_{fixed}} \times d_{kl} \times N^i\,(t,\ \Delta)$$

$$+ \left(\frac{L_j + M}{B_{fixed}} \times c_{al} + \frac{k \times \alpha + M}{B_{wireless}} \right) \times a_{ij}\,(t,\ \Delta). \qquad (5)$$

On the other hand, the total cost for accessing a file by the mobile host when the file migration scheme is not used is as follows.

$$COST_{NM} = \left(\frac{L_j + M}{B_{fixed}} \times c_{ak} + \frac{k \times \alpha + M}{B_{wireless}} \right) \times a_{ij}\,(t,\ \Delta)\,. \qquad (6)$$

The first scheme decides to migrate the file or not after comparing the response times based on the file access frequency. The server that has received the request from the mobile host at time t calculates the file access frequency during the previous time interval Δ. If the response time is longer than the migration time and the file access frequency is greater than or equal to the threshold value, the file should be migrated. The threshold value is the file access frequency at the point when the response time of the non-migration scheme becomes longer than that of the migration scheme. The equation below is the determinant equation of the migration scheme that uses the file access frequency.

$$a_{ij}\,(t,\ \Delta) \geq \left(MT^l_{kj} + a_{ij}\,(t,\ \Delta) \times RT^i_{lj} \right) \,/\, RT^i_{kj} \text{ and}$$
$$a_{ij}\,(t,\ \Delta) > \Phi, \text{where } \Phi \text{ is a threshold value.} \qquad (7)$$

The second scheme determines to migrate the file or not according to the file access frequency and the relative ratio of the mobility. Although the mobile host may have requested the file, if the mobility of the mobile host is relatively greater than the file access frequency, the cost of frequent file migrations between servers may be much greater than the cost of simple file service without migration. Therefore, when a mobile host with high mobility asks for a file, it may be more efficient for the server to simply service the file request instead of migrating the file. This scheme migrates the file to the destination server when the relative ratio of mobility and the file access frequency are larger than the predefined threshold value.

$$\Lambda = \frac{a_{ij}\,(t,\ \Delta)}{N^i_{avg}\,(t,\ \Delta)} \;>\; \xi, \text{ where } \xi \text{ is the threshold value.} \qquad (8)$$

5   Evaluation and Analysis

5.1   Simulation Model

We used SIMLIB for simulation of the file migration schemes:[12]. We assumed five cells for our simulation environments and evaluated how much the file migration scheme can reduce the file access latency in the mobile computing environments. The mobile host can move randomly to the neighboring cell, and can ask for file services on a random basis during the time it stays in the cell. It is assumed that each fixed host or file server is always connected to the fixed network and is able to communicate each other.

The purpose of our experiments is to compare the costs of the migration scheme and non-migration scheme. Additionally, we can find out the necessary

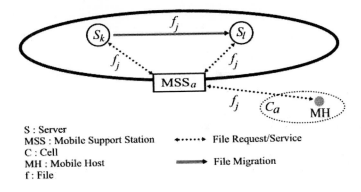

S : Server
MSS : Mobile Support Station   •◄·····► File Request/Service
C : Cell
MH : Mobile Host   ══════► File Migration
f : File

Fig. 2. System model for file migration

threshold values with our experiments. We simulated the system with the system parameters such as the arrival rate and the departure rate of the mobile host in each cell, file service time, the mobility rate of the mobile host, the file access frequency, and the file size. Also, we used exponential and Poisson distribution for random variables in the simulation environments.

The simulation was done for 1440 minutes. The simulation was performed 5 times and produced its final results by averaging all the simulation results. Table 1 shows the simulation factors that we have considered in our simulation experiments.

Table 1. Simulation parameters and its values

Parameter	Value
Wired bandwidth	10 Mbps
Wireless bandwidth	2 Mbps
Simulation duration	24 hours
Simulation count	5 times
Inter-arrival time of the mobile host	Exponential Distribution with mean of 10 minutes
Message size	512 Kbyte
File size	Exponential Distribution
Mean time staying at a cell	Exponential Distribution
File access rate	Poisson Distribution

5.2   Simulation Results

We analyzed the performance of the migration and the non-migration scheme according to mobility, file size, and file access frequency. The results of the experiments mainly shows the effects of mobility on the file access latency.

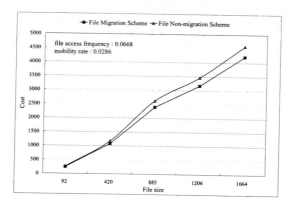

Fig. 3. Comparison of $COST_{FM}$ and $COST_{NM}$ as file size increases

Fig. 3 illustrates the changes in costs for the two different schemes according to the file size. As the size of the file requested by a mobile host increases, so do the file migration and the non-migration costs in a similar ratio. But, the cost for the migration scheme is always smaller than that of the non-migration scheme and the gap between the two costs increases, too. That is because, when the size of the requested file increases, both the cost of transferring the whole file in the wired network and the cost of transferring the blocks of the file several times in the wireless network increase every time the file service is provided. In other words, the migration scheme seems to be able to reduce the file access latency through migration of the file in the fixed network and ultimately improves the system performance.

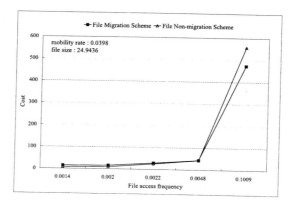

Fig. 4. Comparison of $COST_{FM}$ and $COST_{NM}$ as file access frequency increases

The file access frequency means the average number of accesses for a file per unit period of time(Δ hours) by a mobile hosts. The result in Fig. 4 shows that, as the file access frequency by mobile hosts increases, the cost of the non-migration scheme is slightly smaller than that of the migration scheme in some interval. However, the cost of the two schemes increase rapidly as the file access frequency passes the threshold value($\Phi = 0.0048$). Φ can be set as the threshold value because the costs of the migration scheme and the non-migration scheme cross at this point. As the file access frequency increases, the cost of both the wired and wireless communication increases. But, if we migrate the file to the destination server near the cell the mobile host resides when the file access frequency is greater than Φ, we can reduce the file access latency and improve the service transparency for the mobile hosts.

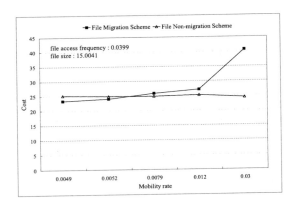

Fig. 5. Comparison of $COST_{FM}$ and $COST_{NM}$ as mobility rate increases

The mobility rate implies the rate at which the mobile host moves to other cells during the unit period of time(Δ), that is, the average rate of handoffs. As the mobility rate increases, the migration scheme migrates the files more frequently to the server near the area the mobile host currently resides, which causes increase in the communication traffic and cost. On the other hand, the non-migration scheme may not be affected by the mobility since the mobile host always receives the file service directly from the server without file migration. Fig. 5 indicates the cost changes according to mobility. The lower the mobility rate, the smaller the cost of the migration scheme. When the mobility rate becomes high, the non-migration scheme is more advantageous.

Another factor to be considered is the correlation between mobility and file access frequency. As we have mentioned above, for the reduction of the file service latency, it is desirable to migrate the file to the server near the cell the mobile host currently resides when the mobility is low and not to migrate the file when the mobility rate is high. Fig. 6 shows this situation. As we have defined earlier, the cost would be low if we migrate the file when $\Lambda > \xi$. In the result of this experiment, the cost of the non-migration scheme gradually increases as the ratio of the file access frequency to the mobility rate became 1.61 and rapidly increases as it passes 1.90. Therefore, the value of the relative ratio $\xi = 1.61$ can be the threshold value determining the file migration.

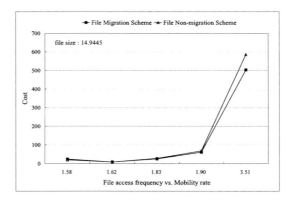

Fig. 6. The cost by the ratio of file access frequency to mobility rate

Mobility does not have a great effect on the non-migration scheme, but, in the migration scheme, since the cost rapidly increases as the mobility increases, the non-migration scheme is quite attractive when the mobility rate is high. As the relative ratio of the mobility rate to the file access frequency increases, the cost of the non-migration scheme gradually increases and rapidly increases from the point of 1.90.

6   Conclusions

In this paper, we proposed a file migration scheme that can reduce the file access latency and the system load by distributing the load of the mobile host to the fixed hosts capable of providing stable file services. For our schemes, we set the file size, the mobility rate of the mobile host, and the file access frequency as the system parameters to be considered in deciding file migration. According to these criteria, we presented the file migration scheme that could determine whether to migrate the file or not at the time the request is made.

We analyzed the performance according to the migration criterion by comparing the cost of servicing a file requested by a mobile host without migration, with the cost of servicing the file through migrating the file to the server near the area of the cell at which the mobile host currently resides. As the results indicate, unlike the existing distributed system environments, the file access latency due to mobility and file access frequency can occur in mobile computing environments. In this paper, we focused on the migration scheme that can increase the access transparency by reducing the file access latency for the mobile hosts.

Future research will be necessary for finding the correlation among the file replication scheme, caching scheme, and the file migration scheme for reducing the access latency and the load of mobile hosts in mobile computing environments.

References

1. Forman, G. H., Zahorjan, J.: The Challenges of Mobile Computing. IEEE Computer, Vol. 27 , No. 4 (1994) 38–47
2. Bright, L., Raschid, L.: Efficient Remote Data Access in a Mobile Computing Environment. International Workshop on Parallel Processing, Toronto (2000) 57–64
3. Griffioen, J., Anderson, T. A., Breitbart, Y.: A Dynamic Migration Algorithm for a Distributed Memory-Based File Management System. Proc. of the IEEE 7th International Workshop on Research Issues in Data Engineering, Birmingham (1997) 151–160
4. Pope, S.: Application Migration for Mobile Computers. Proc. 3rd International Workshop on Services in Distributed and Networked Environments, Macau (1996) 20–26
5. Hac, A: A Distributed Algorithm for Performance Improvement Through File Replication, File Migration, and Process Migration. IEEE Trans. Software Engineering, Vol. 15, No. 11 (1989) 1459–1470
6. Hurley, R. T.: File Migration and File Replication : A Symbiotic Relationship. IEEE Trans. on Parallel and Distributed Systems, Vol. 7, No. 6 (1996) 578–587
7. Hurley, R. T., Black, J. P., Wong, J. W.: Limited Effects of Finite Storage on a Beneficial File Migration Policy. Proc. 19th IEEE Conf. Local Computer Networks, Minneapolis (1994) 432–439
8. Khurana, S., Kahol, A., Gupta, S. K. S., Srimani, P. K.: An Efficient Cache Maintenance Scheme for Mobile Environment. Proc. of the 20th International Conference on Distributed Computing Systems, Taipei (2000) 530–537
9. Froese, K. W.: File Cache Management for Mobile Computing. MS Thesis, Dept. of Computer Science, Univ. of Saskatchewan, Canada (1996)
10. Kroeger, T. M., Long, D. D. E.: Predicting File System Actions from Prior Events. USENIX Conference Proceedings, California (1996) 319–328
11. Rajagopalan, S., Badrinath, B. R.: An Adaptive Location Management Strategy for Mobile IP. Proc. of the 1st Annual International Conference on Mobile Computing and Networking, California (1995) 170–180
12. Law, A. M., Kelton, W. D.(2-ed): Simulation Modeling and Analysis. 2nd edn. McGraw-Hill (1991)

Fresh Information Retrieval Using Cooperative Meta Search Engines

Nobuyoshi Sato, Minoru Uehara, Yoshifumi Sakai, and Hideki Mori

Department of Information and Computer Sciences, Toyo University
2100 Kujirai, Kawagoe City, Saitama 350-8585 Japan
jju@ds.cs.toyo.ac.jp, {uehara,sakai,mori}@cs.toyo.ac.jp

Abstract. In web page retrievals, search engines are usually used. However, conventional search engines have a problem in that their update intervals are very long because they are based on centralized architecture, which gathers documents using robots. So we proposed the Cooperative Search Engine (CSE) in order to reduce the update interval. CSE is a distributed search engine, which integrates small local search engines into a large global search engine by using local meta search engines. A local meta search engine hides a local search engine in each web site. Although CSE can reduce the update interval, the retrieval performance is not enough. So, we proposed several speed up techniques. In this paper, we describe the structure and behavior of CSE and its efficiency.

1 Introduction

It has been a long time since the Internet became popular. Today, we often use the Internet as a large-scale database. In web page retrievals, search engines are indispensable. Many search engines have indexes (or inverted files) of terms for the purpose of fast retrieval. Naturally, each index must be updated periodically. We call that period the update interval. An index is updated as follows; web pages are gathered, index information is extracted and stored into the index file. In a conventional search engine based on centralized architecture, a robot traverses hyperlinks to gather web pages. Generally, a robot wastes a very long time gathering them. For an example, Waseda University's Senrigan requires almost a month to gather 500,000 documents. Even if distributed robots are used, it is hard to make the update interval very short. Therefore, we developed Cooperative Search Engine (CSE) [1] [2], which is based on distributed architecture.

By the way, recently, data mining and knowledge management have become popular in business. For such applications, fresh information retrieval is very important. Therefore, a search engine for fresh information retrieval is needed in the intranet. Many commercial search engine vendors sell their search engines for the purpose of intranet information retrieval. Furthermore, several small-scale search engines such as Namazu [4] and SGSE [5] are developed. These small-scale search engines are distributed based on GPL and other free licenses. However, because these search engines are based on centralized architecture, it is

I. Chong (Ed.): ICOIN 2002, LNCS 2344, pp. 656–670, 2002.
© Springer-Verlag Berlin Heidelberg 2002

difficult to reduce the update interval because there is a lack of a fast documents gathering method. The number of web servers in intranets is far fewer than that of the Internet's, but the update interval of intranets should be much shorter than the Internet's. Therefore, the update interval is relatively problem in the intranet. CSE is suited for intranet information retrieval in point of the update interval.

As mentioned above, CSE is a distributed search engine that is developed in order to reduce the update interval. A local search engine, which is installed in each web server, searches only site local documents. These local search engines are integrated by local meta search engines. They communicate with each other and behave as a global search engine. In CSE, an index can be updated locally. CSE can greatly reduce the update interval. However, the retrieval performance is not good enough because of communication delays. So, it is important for CSE to improve the retrieval performance. In this paper, we propose several techniques to improve the retrieval performance.

The organization of this paper is as follows; in Sect. 2, we describe related works in distributed information retrieval. In Sect. 3, we describe the outline and behavior of CSE. In Sect. 4, we describe the evaluation of CSE. Finally, we summarize the feature work and conclusions.

2   Related Works

First, there are the following kinds of approaches to reduce the update interval; (1) using distributed robots, (2) gathering a lot of documents at once, (3) increasing parallelism, and (4) improving gathering methods.

As distributed robots, there are Gatherer of Harvest [6], WebAnts [7], and PRS [8]. Harvest requires gathering targets to be set manually. WebAnts sets gathering targets automatically, while PRS makes gathering more efficient by using a concept of distance.

PRS is a distributed robot that aims to gather all documents in Japan in 24 hours. PRS Master (PRSM) allocates gathering targets automatically with consideration to distances in the network between the robots and gathering targets. PRS gathers documents on 103 servers using 7 robots, and 5 to 20 times performed. This result seems to be the best for situations that HTTP is used as the method to access documents. However, the hidden meaning of this result is that accessing to documents by HTTP has limitations.

The behaviors of the robots are restricted by the regulation "A Standard for Robot Exclusion" [9]. This regulation recommends the interval time to gather documents from the same site. In the case of a robot keeping this regulation strictly, only 288 pages can be gathered from one site in a day, and this is not suited for practical use. Because of this reason, many robots ignore this regulation. There is also the approach of increasing parallelism rather than distributed robots. According to a paper by Noto and Takeno [10], 10 Mbps of bandwidth is not filled up even if 8000 parallel processing is used. Increasing parallelism seems to be effective in the Internet. However, it is difficult to apply increasing parallelism to intranets.

Infoseek [11] and FreshEye [12] reduce their update intervals by improving gathering method. Infoseek uses a list of all files or modified files on each web server. However, the administrator's cooperation is needed to use this method. Infoseek also proposed a distributed information retrieval method using a meta index, as the same as CSE. However, the update interval is not very short.

Although FreshEye is based on centralized architecture, it realizes the update interval shorter than 10 minutes by limiting some domains in which documents are gathered, or by gathering in the order of modification frequencies of documents. Therefore, FreshEye cannot gather all documents in all domains within a specified period.

In principal, it is extremely difficult to gather all documents on the Internet using these approaches. And it is also difficult to realize the short update intervals required in intranets.

Meta search engines such as MetaCrawler [13] and SavvySearch [14], have a problem in ranking when different types of search engines are used. Especially, in the case of the $tf \cdot idf$ scoring method, the value idf cannot be calculated locally. NECI's Inquirus [15] downloads hit documents in the meta search, and calculates their score again using these downloaded documents. However, this method spends time calculating scores. In CSE, the LSE searches documents based on tf, and the LMSE calculates scores again using the idf.

Now, there are the following approaches to distributed information retrieval.

Whois++ [16] collects Forward Knowledge (FK) as centroid. Each Whois++ server transfers the query based on centroid, and requests retrieval on information which is not known by the server. This is called query routing.

Harvest [6] consists of two main components; Gatherer and Broker. Gatherer gathers documents, and transfers their summaries to Broker as Summary Object Interchange Format (SOIF). SOIF is the format for common use of attributes and summaries of documents, and it includes information such as the author's name, title of document, keywords etc. In Harvest, the full-text must be included in SOIF to do a full-text retrieval, so Gatherer transfers entire documents and their attribute information to Broker. Broker makes an index, accepts retrieval request, and retrieves documents in cooperation with other Brokers. Broker makes indexes and retrieves using search engines such as Glimpse and Nebula. Glimpse has a compact index. Nebula searches rapidly. In Harvest, Gatherer can access document files directly through the file system. However, Harvest needs to transfer entire documents because Gatherer does not make an index. Therefore, Harvest is more difficult to reduce the update interval than CSE.

Ingrid [17] is an information infrastructure that aims to realize searching and browsing at the topic level. Ingrid makes links between gathered resources, and constructs original topology. The Forward Information (FI) server manages this topology, and the Ingrid navigator retrieves the documents by communicating with other FI servers, and searching the route. Although Ingrid is flexible, Ingrid may have large delays at the retrieval phase because Ingrid navigator searches the routes sequentially. CSE has minimal delay at the retrieval because LS searches the routes only once.

3   Cooperative Search Engine

We developed Cooperative Search Engine in order to reduce the update interval. In this section, we describe the structure and behavior of CSE.

3.1   CSE Principles

In CSE, an index is made locally in each web site to reduce the update interval. Then, the information about "who (site) knows what (keyword)" is sent to only one manager. This information is called Forward Knowledge (FK). FK is meta knowledge which means who knows what, and it's the same information as Ingrid's FI. At the retrieval phase, the manager answers who "knows" the keyword. As a result, a communication delay occurs at the retrieval phase in CSE. However, overall, CSE can reduce the update interval greatly. Therefore, it is important for CSE to keep the delay at the retrieval phase small.

CSE consists of the following components shown as Fig.1.

Location Server (LS): LS is the server which manages FK. Only one LS exists in the whole of CSE.

Cache Server (CS): CS is the server which caches FK and retrieval results. LS is thought to be a top level CS, but LS does not cache retrieval results. CS realizes fast continuous retrieval by caching the results. Here, continuous retrieval is defined as the method to show the retrieval results divided into 10 or 20 items in a page. CS calls LMSE in parallel.

Local Meta Search Engine (LMSE): LMSE is a local meta search engine which hides the difference between the LSEs. LMSE accepts a user's request, and transfers it to CS. LMSE manages their local LSE, and LMSE retrieves locally using LSE.

Local Search Engine (LSE): LSE gathers local documents, makes index, and searches them locally. CSE uses Namazu and SGSE as LSE.

Both Namazu and SGSE are full-text Japanese search engines for the small-scale environments. Namazu, especially, is widely used in the open source community. CSE can use them without any changes.

Figure 2 shows the outline of caches in Cache Server. In Fig. 2, Q_{sorted} is the sorted cache; Q_1 and Q_2 are unsorted caches corresponding to $LMSE_1$ and $LMSE_2$. Retrieval results received from $LMSE_n$ are stored in Q_n temporarily. The contents of Q_n are sorted by $LMSE_n$. The contents of every Q_ns are merge-sorted by CS and stored into Q_{sorted}. The rest that couldn't be merge-sorted stays in Q_ns.

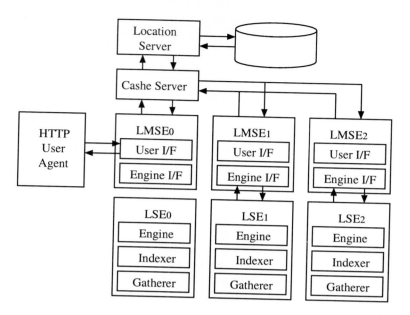

Fig. 1. The overview of CSE

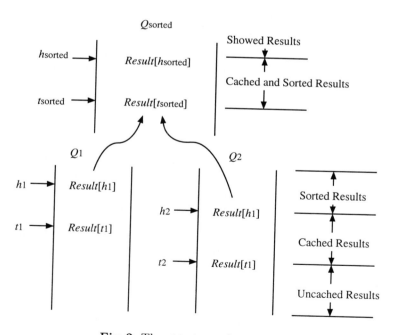

Fig. 2. The structure of cache in CS

3.2   Behavior of CSE at Updating

The definitions of index update and its order are as follows:

1. **Gathering Documents** is the step to gather documents from the web server.
2. **Making Index** is the step to make the index.
3. **Transferring Index** is the step to transfer indexes made in Step 2 to search engine. Generally, centralized search engines do not need to transfer because they make indexes locally.

Gathering documents and making index can be processed in parallel. Generally, gathering documents requires the most time, followed by making index, and transferring index.

The methods to gather documents and the terms are as follows:

Direct Access: The method to gather documents through the file system such as NFS and local mass storage devices.

Archived Access: The method to gather documents using a probe CGI that archives document files. Archived file is transferred.

HTTP Access: The method to transfer documents in each file through HTTP. This method is generally used.

In these three methods, direct access is the fastest method, the next is archived access, and the slowest is HTTP access. This reflects the distance to gathering target, and the communication delay.

The CSE's algorithm at updating is as follows:

1. Gatherer of LSE gathers documents in target web sites. At this time, gatherer uses the fastest method as possible. That is, direct access is used if available, if not, archived access is used if available. HTTP access is used if these two methods are not available.
2. Indexer of LSE makes indexes.
   a) Engine I/F of $LMSE_i$ extracts N_i, K_i, $k \in K_i$, $n_{k,i}$. Here, N_i is the number of all documents in $LMSE_i$, K_i is the set of keywords in $LMSE_i$, k is the keyword, $n_{k,i}$ is the number of documents which includes k in $LMSE_i$. Engine I/F sends this data to LS with Engine I/F's URL.
   b) LS records $n_{k,i}$ and N_i from $LMSE_i$.

CSE supports all three methods to gather documents as described above. Therefore, CSE needs to install the probe CGI program into each web server to use archived access. The probe CGI program archives SERVERROOT and each user's documents on the web server, and transfers them.

3.3   Behavior of CSE at Retrieval

Here, we describe the behavior of CSE at the retrieval phase. CSE sacrifices the retrieval phase efficiency to accelerate the updating phase efficiency. In CSE, a lot of communications happen at the retrieval phase. So, saving communications is very important. CSE reduces communication delays by using the following four techniques.

The first technique is a query based site selection. CSE allows Boolean retrieval based on Boolean formulas. The operators "and", "or", and "not" are available in CSE's query. Here, the "not" operation is not the negation; is a binary operator used in a a field of information retrieval, which means the difference of two operands. It is also known as $A - B$. Let S_A, S_B be a set of sites selected for query A, B:

- $S_A \cap S_B$ is a set of sites selected for query A and B
- $S_A \cup S_B$ is a set of sites selected for query A or B
- S_A is a set of sites selected for query A not B

According to this, the number of messages could be reduced.

The second technique is look-ahead caching. Most search engines display the retrieval results as divided into several items in a page, typically 10 or 20 items. We call this way continuous retrieval. Through experience, we know that many users are satisfied with the first 1 or 2 pages of the results of continuous retrieval. CSE uses look-ahead caching to minimize the response time at continuous retrieval. In look-ahead caching, the next page is also pre-fetched into CS at every continuous retrieval.

The third technique is a score based site selection. In this technique, each site gives the maximum score that will be retrieved at the next continuous retrieval. As a result, unnecessary queries are not sent to sites having lower scores. At most, only N_p sites need to be searched to display the result page that includes N_p items. However, in CSE, scores of each document are not known until once searched in LMSE. In this way, at the first page retrieval, CS must send retrieval requests to all sites which may have documents matching the query. However, at the second and later page retrieval, by using the maximum scores received from the LMSEs at the previous retrieval, CS can know which site will supply the results needed to display the next page. Accordingly, the target sites are limited to N_p sites.

The last technique is a global shared cache. In CSE, LMSE receives a user's query and asks to a close CS. Therefore, there are many caches related to the same query in each CS, and the same query is repeatedly sent. So, we decided to share the caches by requesting retrievals to the CS which is given the query and starts the retrieval in the first place.

However, the load balancing becomes a problem if the cache is simply shared. If the Internet scale retrievals rush into a CS, the CS will become similar to when CS is under DoS attack. Therefore, to cope with both cache sharing and load balancing, a copying cache is needed. CS has a retrieval result cache, and LS has a site selection cache. The set of LMSEs and CS that sends the query first is recorded in the site selection cache.

CS mainly carries out the retrieval process in CSE. CS has the following three kinds of queues: Sorted Queue (SQ), Sorting Queue (Q), and Q_is for every $LMSE_i$s. Each queue stores the tuples (n, u, w) ordered according to score w.

When $n = 0$, the tuple (n, u, w) is called a data, and u is the URL of a document. When $n \neq 0$, the tuple (n, u, w) is called a continuation, where n denotes the number of documents read from the LMSE whose URL is u. The value of w is the score of the document indicated by n and u. In addition, a data (n, u, w) with $w = 0$ represents EOF. The queues SQ and Q are persistent.

The algorithm for the retrieval process is as follows.

1. The user sends a search query E with the rank N_s of the first site displayed, and the number N_p of the sites displayed in each page, to $LMSE_0$.
2. $LMSE_0$ sends $Search(E, N_s, N_p)$ to close CS, CS_0.
3. If E is a new query (that is, CS_0 has no results for the query in cache), the followings are carried out:
   a) CS_0 requests to LS the list of continuations for the query.
   b) LS checks the site selection cache. If E is in the site selection cache, LS decides CS' from the cache, and returns it. Otherwise, LS binds CS_0 and E, and finds the set of $LMSE_i$s using Query based Site Selection, and calculates an expected score w_i for each i, and makes the list L of continuations $(0, LMSE_i, w_i)$s.
   c) LS returns N for the number of all the documents, and $idf(k)$s for all the words k in E, to CS_0.
   d) If CS_0 receives CS', CS_0 sends E to CS'. Q in CS' is transferred to CS_0. (goto 6). Otherwise, goto 3e.
   e) CS_0 adds to Q the continuations by score order.
4. CS generates the two threads, $Display(N_s, N_p, SQ)$ and $Lookahead(N_s, N_p, 2, SQ, Q)$, and carries out the following procedures.
5. $Display(h, l, SQ)$: wait until $|SQ| \geq h + l - 1$, and returns the sequence of l tuples of SQ starting from the h-th one.
6. $Lookahead(h, l, p, SQ, Q)$: does the following to each tuple in Q from the first one to the $(h + (p+1)l - 1 - |SQ|)$-th one while $|SQ| < h + (p+1)l - 1$ holds:
   a) if Q is empty, then quit,
   b) if the tuple is a data, then delete it from Q and add it to SQ,
   c) if the tuple is an unmarked continuation, then
      i. mark it (i.e. make n_i negative), make Q_i if not existent, insert $(n_i, LMSE_i, w_i)$ into Q_i, generate $LMSE_i(Q_i)$ thread for each unmarked continuation up to the $(h + (p+1)l - 1 - |SQ|)$-th one, and
      ii. wait for the same continuations as marked ones, and continue the process.
   d) if the tuple is a marked continuation $(-n_i, LMSE_i, w_i)$, then
      i. if Q_i is empty, then wait,
      ii. if the head of Q_i has score $w = 0$, then delete Q_i,
      iii. if the head of Q_i is a continuation, insert it into Q without deleting from Q_i,
      iv. if the head of Q_i is a data, then wait until $|Q_i| \geq 2$ holds, delete it from Q_i and insert it into Q. Furthermore, delete Q_i, if the resulting head is EOF, and insert into Q the tuple $(n_i, LMSE_i, w_k)$ for the score w_k of the resulting head of Q_i, otherwise, and

     v. delete marked continuations from Q, and wait until the terminations of all the $LMSE_i$ threads, and transfer Q_i to Q.

7. In $LMSE_i(Q_i)$:

    a) if the tail of Q_i is EOF, halt immediately,

    b) if the tail $(n_j, LMSE_i, w_j)$ of Q_i satisfies $n_j - 1 \leq h + (p+1)l - 1$, halt immediately,

    c) delete the tail of Q_i, and

    d) send $Search(E, idf(k)|k \in E, n_j, h + (p+1)l - 1 - n_j + 1)$ to $LMSE_i$.

    e) In $LMSE_i$, $LMSE_i$ asks LSE_i to search for each word included in E, and makes a list D_k of documents including the word k listed according to *tf* scoring. Furthermore, it processes Boolean operations based on E, and calculates the score $w(d, E)$ for document d. Score $w(d, E)$ is defined recursively as follows.

      i. $w(d, k) = tf(d, k) \cdot idf(k)$,

      ii. $w(d, A \text{ and } B) = \min(w(d, A), w(d, B))$,

      iii. $w(d, A \text{ or } B) = \max(w(d, A), w(d, B))$,

      iv. $w(d, A \text{ not } B) = w(d, A)$,

    where k is a single word, and A and B are any Boolean formulas. Finally, it makes as the search result a list D_E of documents listed according to $w(d, E)$.

    f) The $LMSE_i$ thread stores the result into Q_i. Since $LMSE_i$ packs the result in packets, the number of the returned documents may exceed the requirement. Moreover, if possible, some extra continuations also packed in the same packets.

8. $LMSE_0$ transforms the result into a HTML form and returns it.

The threads of *Display*, *Lookahead* and $LMSE_i$ are executed in parallel.

4   Evaluations of CSE

4.1   Evaluation of Update Phase

There are the following kinds of updating:

Simple Update: newly created or modified files are added to indexes.

Full Update: Indexes are completely remade. Information of removed files is removed from indexes.

A full update is equivalent to a simple update after removing all index files. Thus, a full update requires more time than a simple update. We consider simple update is important for CSE because the simple update is run as a daily routine while full update is run as a maintenance.

    In CSE, LSE performs almost all parts of the index update, and spends a lot of time. So we have developed an LSE which performs document gathering and index making in parallel. At the beginning, as a preliminary experiment, we experimented on index update at the Media Computing Center of Faculty of Engineering, which is the largest site at Toyo University (See Table 1). In

Table 1. The preliminary evaluation on test machines

	Gathering [min:sec]	Indexing [min:sec]	Transfer index [min:sec]	Total [min:sec]
Namazu full update	25:50	20:32	00:00	46:22
Namazu simple update	19:51	01:27	00:00	21:18
CSE simple update	00:09	01:27	00:18	01:54
Parallel CSE simple update	00:02	00:37	00:11	00:50

this experiment, two PCs (Celeron 333 MHz, 128 MB of memory and Celeron 400 MHz × 2, 128 MB of memory) are used.

As the main experiment of the Media Computing Center, we experimented on index updating (See Table 2). Used 17 workstations at the Media Computing Center are Sun SPARCstation 20, SuperSPARC II 75 MHz × 2, 224 MB of memory. We compared a sequential update and a 17 parallel update in CSE's simple updating. In each case of the main experiment, more times was spent than in the preliminary experiment due to differences in CPU performance. And, a high enough effect on the CPU number was not archived, this is because other making was creating a bottleneck. However, as a result of main experiment, CSE can update all documents in the Media Computing Center in 15 to 30 minutes even if the fluctuation of loads is considered.

Table 2. The evaluations on real machines

The number of processes	Execution time [min:sec]
1	34:39
17	11:15

Next, we performed an experiment of archived access method targeting ToyoNet, a web server of other schools of Toyo University. Disk capacity adjustment and installing program by the user is restricted because the operation of ToyoNet is trusted to a commercial Internet service provider. So we could not use the direct access method not to mention installing a LMSE. Therefore, we had to make and manage a index on other computers. We decided to use archived access at ToyoNet because CGI programs written in Perl could be used. Table 3 shows documents gathering times for HTTP and archived access from ToyoNet. Here, we used wget for HTTP access. All the files of several users are archived in a file and transferred. When archived access is used, we confirmed the documents gathering time is kept less than 1/3 compared with HTTP access. Table 4 shows updating times of ToyoNet in case of archived access is used and archived in 50 users. The condition of the experiment is the same as the preliminary experiments.

Table 3. Gathering time by both HTTP and archived access

Access method	The number of users	Gathering time [h:min:sec]
HTTP access		1:35:32
Archived access	1	0:52:17
	10	0:28:01
	20	0:26:57
	50	0:24:10
	100	0:24:19

Table 4. Update time of ToyoNet

	Gathering [min:sec]	Indexing [min:sec]	Transfer index [min:sec]	Total [min:sec]
Namazu full update	1:35:32	1:16:25	0:00:00	2:51:57
Namazu simple update	1:35:32	1:01:03	0:00:00	1:36:35
CSE simple update	0:24:10	0:01:03	0:01:07	0:26:20
Parallel CSE simple update	0:06:51	0:00:23	0:00:34	0:07:48

4.2   Evaluation of Retrieval Phase

In compensation for the reduction of updating times, CSE has the communication delay at the retrieval phase. Table 5 shows the response time of local searches. SGSE is slower than Namazu because SGSE is written in Perl. CSE is slower than SGSE due because CSE needs communication.

Table 5. The performance comparison in local keyword search

	Namazu [sec]	SGSE [sec]	CSE [sec]
Execution time	0.046	0.448	1.267

CSE retrieves documents in parallel in order to reduce this overhead. Table 6 shows response times for a 24 sites retrieval where Namazu is used sequentially and CSE is used in parallel. The slow response of Namazu is because Namazu does not have a parallel retrieval feature. There is a difference between "and operation" and "or operation". This is why the communication time is different by their number of retrieved items.

We discussed about narrowing the search target sites at the paper[3], and forecast that the number of retrieval target sites will be reduced to 40 %. In our experiment, we targeted 27 sites in Toyo University, the number of elements of set of LMSEs S_k was 1.6, here, k is the all of keyword registered in LS. And was

Table 6. Namazu (seq.) vs CSE (par.) in Boolean search

	Namazu (sequential) [sec]	CSE (parallel) [sec]
AND operation	5.6	1.5
OR operation	5.6	5.0

proved the number of sites is reduced to 6 %. This is because a few number of sites such as Media Computer Center have almost all documents. Narrowing the search target sites is more effective in case of distribution of the documents is uneven.

Next, we describe the evaluation of continuous retrieval. We analyzed a web proxy server's logs at our university. As the result, the ratio of continuous retrieval reaches 70 % of all search requests. So, we developed CS in order to make continuous retrievals more efficient. Table 7 shows the performance of CS. Here, a medium frequency keywords are used in this experiment because medium frequency keywords are relatively important. When CS is not used, all of the results are transferred at every retrieval. In the case of CS being used, only the results needed to show a page are transferred. As this result, response times are reduced to about 20–45 % at the first page retrieval, and to 10–20 % at the second and later pages. These second and later page retrieval times are especially faster than SGSE's. As mentioned above, the retrieval time of CSE is not inferior to retrieval times of centralized search engines, so CSE is equal to the task of practical use.

Table 7. The performance of look ahead cache

Hit items	Without CS [sec]	With CS 1st page [sec]	2nd page [sec]
903	2.97	0.80	0.31
544	1.75	0.66	0.32

In order to increase efficiency at continuous retrieval of the second and later pages, we introduced a score based site selection. Here, the second and later page retrievals are run by CS in the background, while users are seeing retrieval results. Consequently it seems that the effects of score based site selection do not generally appear. However, score based site selection reduces the time of continuous retrieval spent by CS, and other communication. Where the number of LMSE is increased, a score based site selection is necessary. Figure 3 shows this result. Here, two PCs are used. Retrieval is performed at a maximum of 10 parallel for per a PC. In the second page retrieval where a score based site selection is used, the retrieval time in case of 20 LMSEs are retrieved is the same as in case of 10 LMSEs are retrieved. Even if the number of LMSEs are increased, we believe that the response time will not increase on a large scale.

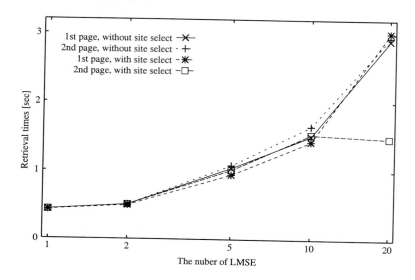

Fig. 3. The evaluation of score based site selection

Next, We evaluated global shared cache. When the cache is shared, the response time is reduced because only the contents of cache are transferred instead of sending queries to many LMSEs by CS, and waiting for their responses.

Table 8 shows the evaluation of the global shared cache. Here, as a comparison target, we also evaluated the response time of CS_0 which receives the query from a LMSE when the retrieval result is cached. Each of 5 LMSEs were installed into two PCs and each CS was installed into two PCs. The query form $LMSE_0$ is first 10 items. Furthermore, the contents of the prepared indexes in the LMSEs are exactly the same; it is the worst case for the behavior of CS.

Table 8. The comparison of the response times

	Cache is not shared		Cache is shared
	Hit in cache [sec]	No hit in cache [sec]	[sec]
Response time	0.45	4.67	0.66

Where the case of cache was shared, transferring the cache and showing the results was finished in 1 second, and is much faster than when the cache is not shared. However, as a matter of course, it spends longer than where the result is cached in CS_0. Also, if the result is cached in CS_0, the response time will be almost the same whether the cache is shared or not.

5   Conclusions

In this paper, we describe the structure, behavior and efficiency of CSE. CSE realizes the short update interval time by integrating distributed local search engines using meta search engines. CSE can update the indexes of all documents in the Media Computing Center of Toyo University. This means that web pages made in student laboratory classes can be successfully retrieved in the same class time. We believe that this has a positive effect on education. Furthermore, we think CSE can update all documents in 1 minute by using PC servers for the exclusive use of CSE. Additionally, we proved that by using a read forward cache, score based site selection and a global shared cache, CSE's long retrieval phase response time problem can be solved.

As future works, we investigate that CSE is not useful for only intranets but also for the Internet. And, we will do a long term running test of CSE for practical use.

Acknowledgment. This research is performed cooperatively as part of a Distributed Computing for Wide Area Network project at Toyo University. The INOUE ENRYO Memorial Foundation provided support for this project. Finally, we would like to thank the office of the Foundation at Toyo University.

References

1. Nobuyoshi Sato, Minoru Uehara, Yoshifumi Sakai, Hideki Mori: Fresh Information Retrieval in Cooperative Search Engine. Proc. of 2nd Software Engineering Artificial Intelligence, Networking & Parallel / Distributed Computing 2001 (SNPD'01). pp.104–111 (2001)
2. Nobuyoshi Sato, Minoru Uehara, Yoshifumi Sakai, Hideki Mori: On Updating in Very Short Time by Distributed Search Engines. Proc. of The 2002 Symposium on Applications and the Internet (SAINT2002). pp.176–183 (2002)
3. Minoru Uehara, Nobuyoshi Sato, Takashi Yamamoto, Yoshihiro Nishida, Hideki Mori: Minimizing Query Targets in CSE. Proc. of 7th Workshop on Multimedia Communication and Distributed Processing Systems (DPSWS'99). pp.85–90 (1999) (in Japanese)
4. Namazu Project: Namazu: a Full-Text Search Engine. http://www.namazu.org/
5. Sony Corp.: SonyDrive Search Engine. http://www.sony.co.jp/sd/Search/
6. C. Mic Bowman, Peter B. Danzig, Darren R. Hardy, Udi Manber, Michael F. Schwartz: The Harvest Information Discovery and Access System. Proc. of 2nd International WWW Conference. http://www.ncsa.uiuc.edu/SDG/ IT94/Proceedings/Searching/schwartz.harvest /schwartz.harvest.html (1994)
7. Lycos Inc.: WebAnts. http://polarbear.eng.lycos.com/

8. Hayato Yamana, et al.: Experiments of Collecting in WWW Information using Distributed WWW Robots. Proc. of 21st Annual International ACM SIGIR Conference on Research and Development in Information Retrieval (SIGIR'98). pp.379–380 (1998)
9. M. Koster: A Standard for Robot Exclusion. http://info.webcrawler.com/mak/projects/robots/norobots.html (1994)
10. Tokiharu Noto, Hiroshi Takeno: Design and Implementation of a Scalable WWW Information Collection Robot. Proc. of 8th Workshop on Multimedia Communication and Distributed Processing Systems (DPSWS 2000). pp.7–12 (2000) (in Japanese)
11. Steve Kirsch: Infoseek's approach to distributed search. Report of the Distributed Indexing / Searching Workshop (DISW'96). http://www.w3.org/Search/9605-Indexing-Workshop/Papers/Kirsch@Infoseek.html (1996)
12. FreshEye Corp.: FreshEye. http://www.fresheye.com/
13. Info Space Inc.: MetaCrawler. http://www.meracrawer.com/
14. CNET Networks, Inc.: SavvySearch. http://www.savvysearch.com/
15. S. Lawrence, C. L. Gills: The NECI Metasearch Engine. http://www.neci.nec.com /~lawrence/inquirus.html
16. C. Weider, J. Fullton, S. Spero: Architecture of the Whois++ Index Service. RFC1913 (1996)
17. Nippon Telegraph and Telephone Corp.: Ingrid. http://www.ingrid.org/

AMC: An Adaptive Mobile Computing Model for Dynamic Resource Management in Distributed Computing Environments

Dong Won Jeong and Doo-Kwon Baik

Software System Laboratory, Korea University
1, 5-ka, Anam-dong, Sungbuk-ku Seoul 136-701, Korea
{withimp, baik}software.korea.ac.kr

Abstract. The structure of systems is needed to be rearranged and reorganized for extension or load-balancing. In this situation, one or more new systems are added into the existing system, and then a part of the roles taken charge by the previous systems are distributed to new sub-systems added later. However the resource management like this has some problems. The worst disadvantage is that the services provided by some systems or the whole system should be suspended for a few days to be reconstructed. Thus, users can't use the services and it also causes a lot of problems. In this paper, we propose a Mobile Computing Model that is able to add a new sub-system and balance the load of the system automatically and adaptively in variable environments that need dynamic resource management.

1 Introduction

We know well what is a server. In this paper, we define the server system as a group of systems providing specific service to users. Services including mail service, FTP service, web service, map service and other application services are sometimes established in one system, or distributed and located in several systems. In any case, we should distribute them to a few or many systems and provide the same services.

In any special case, we create a server system that divides a large problem into several small sub-problems and distributes each sub-problem to the corresponding server systems. The server distributing sub-problems to the sub-server systems collects the results of the sub-problems and generates the final result from them. Sometimes sub-server systems collaborate or communicate to cooperate among them. All of the cases we described are defined as the distributed computing. Each case will be described with more details later.

Now, Let's suppose the following situation.

"We hope to extend our server system for high-performance or load-balancing."

In general, most of systems' extension has been achieved through the following process.

I. Chong (Ed.): ICOIN 2002, LNCS 2344, pp. 671–678, 2002.
© Springer-Verlag Berlin Heidelberg 2002

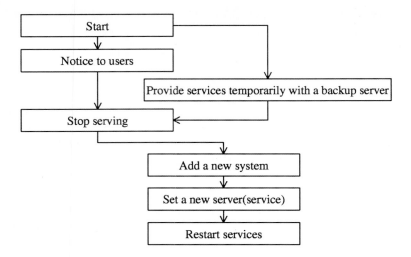

Fig. 1. Generic extension process: As you know through this figure, we must stop services to extend a server. It is very important and critical problems in any cases. More than all problems, suspension of services cause very critical problem.

Figure 1 above is visualizing two types, but both of them have some problems. The following is the summaries of the problems:

1. Users can't use the services for a few hours or days. So it weakens the reliability of the services.
2. They should pay high cost to manage a backup server system.
3. They will waste a few systems' resources according to the situation. For example, some server systems can be useless owing to the reduction of users and it demands unnecessary costs from the management.
4. It is impossible to optimize and re-construct a system in the dynamic and real-time environments.

In this paper, we propose an AMC (Adaptive Mobile Computing) model to solve the problems already described. This model provides a framework for optimization and reconstruction of the resources of a server system. It also improves the performance of the system.

This paper is organized as the following. Chapter 2 describes some types of distributed computing, and the similarities and the differences between them. In chapter 3, we describe what is the mobile computing, and then we inspect how to help dynamic reconstruction and performance improvement of a server system. Also we describe the Adaptive Mobile Computing (AMC) model that we propose. In chapter 4, we describe three types of AMC model and evolution processes of the AMC model based systems, and finally chapter 5 includes future direction of our research.

2 Mobile Agent Computing and Distributed Computing Types

In this chapter, we overview the mobile computing paradigm simply. Also, we define and describe some types of the distributed computing.

2.1 Types of Distributed Computing

There are some types of distributed computing and we examined and summarized about the types of distributed computing. The classified types are shown in Table 1. It's very practical and actual. It is assorted using some factors like the number of server systems, service type, division and collaboration.

Table 1. Classification of Distributed Computing.

Type	The num. of server	Service type	Division	Collaboration
Type 1	1	Not equal	Not created	Not needed
Type 2	n	Not equal	Not created	Not needed
Type 3	n	Equal	Not created	Not needed
Type 4	n	Just the same Problem	Created	Not needed
Type 5	n	Just the same Problem	Created	Needed

The number of servers is the number of sub-server systems distributed. The service type indicates whether sub-server systems process the same job or not. The division column's "Not created" means one job or problem is not divided into several sub-jobs or sub-problems. On the other hand, "Created" means that a problem is divided into several sub-problems and the top server system gathers all the results later, and then get the final result from them. Collaboration column's field will be filled with "Needed", if sub-servers cooperate and communicate to get a solution of each given sub-problem.

The following is the summary of each type:
1. Type 1: This is not a true distributed computing but just a centralized client-server system. It is used to construct a small-scale and simple system usually.
2. Type 2: Type 2 has many different services and each service is provided through different server systems. That is, Services are distributed to several servers and provided to users.
3. Type 3: Type 3 is similar to type 2, but, in this type, many sub-server systems provide a same service independently. The number of servers is n and there is no communication among sub-server systems.
4. Type 4: In type 4, A large and complex problem is divided into several independent sub-problems. Sub-problems are distributed into proper sub-server systems to process. The communication or collaboration between sub-server systems is not needed, because each sub-problem can be solved independently. The root server collects the results of them and merges together to make the final result. However this type's not practical in the real world.
5. Type 5: Type 5 is similar to Type 4. However the collaboration or communication among sub-server systems is essential, because each sub-problem cannot be solved independently.

The distributed types described in previous can be classified by the important characteristics(factors) and it is visualized in Figure 2.

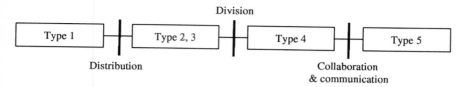

Fig. 2. Factors and Classification: The boundary between Type 1 and others is distribution. In other words, Type 1 indicates centralized client/server systems. Division is related with divide and conquer algorithm. Therefore, in case of Type2 and 3, services in each sub-server system are operated and ran independently, but Type 4 is needed to collection and mergence step indispensably. Collaboration and Communication is more complex mechanism for solving a problem. In this case, each service or process of sub-server systems must share data or cooperate each other to solve their sub-problem.

The goal of this paper is to provide a system model for automatic reconstruction and management. We define the ability of automatic reconstruction as the adaptability of the system. We use the mobile agent paradigm to achieve the goal.

In the next section, we will consider the definition and the advantages of the mobile agent paradigm.

2.2 Mobile Agent Paradigm

The mobile agent paradigm has evolved from the traditional paradigm called the client/server paradigm. A more generic and similar concept is a mobile object. It encapsulates data along with the set of operations, and can be transported from a remote node to another [1,2,3,4].

In tradition, services in distributed systems have been constructed using client/server paradigm. In this paradigm, a client send a request to a server in server systems, and then the corresponding server processes the request and send results that the client hope to receive. Therefore, all requests is concentrated on a server system.

The mobile agent paradigm has many advantages for solving problems of the traditional paradigm, client/server paradigm. [8] identified the advantages of the mobile agent paradigm.

Figure 3 shows explicitly how the mobile agent paradigm differs from the client-server paradigm[5].

The property of mobile agent paradigm like this is the key of our research and a model we propose. Also it makes a server system to be reconstructed and extended automatically and dynamically.

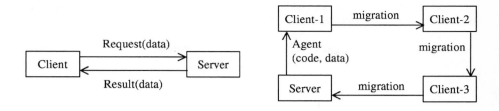

Fig. 3. Traditional C/S Paradigm and Mobile Agent Paradigm

3 Adaptive Mobile Computing Model

3.1 AMC Model and Evolution Theory

The AMC(Adaptive Mobile Computing) model we propose is very similar to Darwin's evolution theory. Charles Darwin is considered as the father of evolution as we well recognize and he proposed a viable mechanism for evolution, namely natural selection[5,6].

We have thought that the same theory should be applied to the next generation systems, because they must adapt to new, variable and various circumstances, and change their environments autonomously and then provide services consistently.

We propose the AMC(AMC: Adaptive Mobile Computing) model applying Darwin's theory. The AMC model supports server systems' dynamic and autonomous extension depending on external environments. It can optimize resources of a system.

The AMC model supports all types of server systems based on the traditional paradigm and the distributed mobile computing paradigm we described in the previous section. Someone can use the model to construct and manage selectively as the type they want.

In addition, The AMC model supports way to use resources of the client system not the server system.

3.2 Architecture of AMC Model

The AMC model has three layers-AMC layer, Server layer and Client layer. First, AMC layer is a main server system and it controls the whole system and their services. It plays the most important role and its roles are the following:

1. It monitors the internal and external factors and determines whether to evolve or not.
2. It reconstructs a system when it should be evolved. At that time, it generates agents and then they are moved into sub-systems. Some agents may be server programs and others may be just simple programs to process a given sub-job or sub-problem.
3. It provides a collaboration pool (communication pool) that enables sub-systems to communicate.

The root server system in the AMC layer distributes one or more tasks(services, processes) to several other systems when it couldn't process all requests. They, which are called as sub-server system, take the second layer, namely server layer. Sometimes agents in the server layer can be servers or simple programs solving sub-problems given by AMC server. The final layer is the client layer. This layer is formed when requests are too many for resources of server side to provide responses. In this situation, programs and data required to solve the requests are moved into a client system temporarily, and then all requests are processed using resources in the client system.

Figure 4 shows three layers of the AMC model.

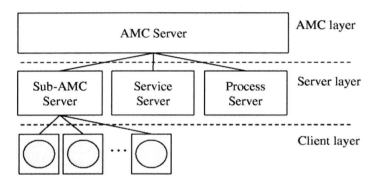

Fig. 4. Three layers of AMC model

The AMC model has the following characteristics:

1. Continuance: Services is being provided to users continuously.
2. Intelligence: The Server system in the AMC layer monitors and identifies various and variable environments, and distributes tasks to other systems intelligently.
3. Mobility: Operations and data is migrated from the root server system to sub-server systems. They are installed in a sub-server system and provide a given service or solve any problem. This property is derived from the Mobile Agent paradigm[1,2].
4. Dynamic Extensibility: Providing services continuously, we can extend or reorganize a system.
5. Optimization: Client systems can be used as sub-server systems. It is derived from the Mobile Agent paradigm [7,8].

3.3 Evolution of AMC Model-Based System

The basic concept of AMC model is based on Evolution theory created by Darwin. Like the procedure of evolution, AMC model-based systems evolve gradually according to internal and external factors, namely circumstances.

In general, an AMC model-based system evolves and reconstructs through five steps. Figure 5 shows all of the evolution steps and the system architecture of each step.

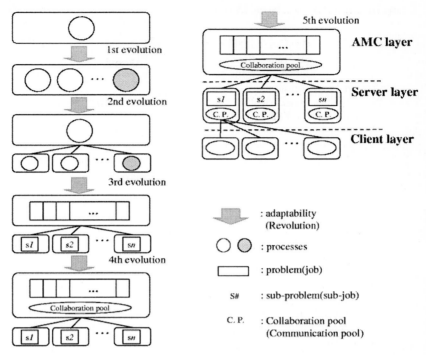

Fig. 5. General evolution steps of the AMC model based systems.

The AMC model-based system has six system architectures as shown in Figure 8. All architectures are mapped to one of all types in Table 1 except the sixth architecture. Table 2 shows the relationship between types in the Table 1 and architectures in Figure 4.

Table 2. Relationship between AMC model based system architecture and distributed computing type

Architect.	Type	Remarks
1st	Type 1	Provides just one service. Not practical.
2nd	Type 1	Provides various services-ftp, web, map services and etc.
3rd	Type 2, 3	Has many server systems and always need not communicate between AMC layer and Server layer.
4th	Type 4	Divides a large problem into several independent sub-problems and distributed them into sub-system for processing. Not practical.
5th	Type 5	Be similar to 4th but Collaboration is needed.
6th	None	Use resources of a client system and can reduce loads of server side. Security is indispensable.

4 Conclusions and Future Works

The current server resource management process has many problems, because it is extended or reorganized manually not automatically and dynamically. So we proposed the AMC model to solve the problems and to evolve a system according to internal or external environments.

The AMC model is based on the types of the current distributed computing, the Mobile Agent paradigm and Darwin's theory.

An AMC model-based system evolves according to circumstances. It generates agents including operation and data. They are moved to the corresponding system and then provide specific services. AMC model based systems have a lot of advantages-autonomy, intelligence, mobility, extensibility, adaptability, continuance, optimization and so on.

In this paper, we didn't consider various external and internal factors. However we will deal with them in our next work and will study a well-defined evolution algorithm with factors.

References

1.   Robert Gray, David Kotz, George Cybenko and Daniela Rus, "Mobile agents: Motivations and state-of-the-art systems", Dartmouth colledge, TR2000-365, 19, April 2000
2.   N. C. Lobley, "Intelligent mobile systems", BT Technology Journal, Vol. 13, No. 2, April 1995
3.   H. S. Nwana, L. Lee and N. R. Jennings, "Co-ordination in software agent system", BT Technology Journal, Vol 14, October 1996
4.   L. C. Lee et al, "The stability, scalability and performance of multi-agent systems", BT Technology Journal, Vol. 16, No. 3, July 1998
5.   http://www.aboutdarwin.com/
6.   Charles Darwin, "The Origin of Species", Greg Suriano (Editor), New York: Grammercy, 1998
7.   Mitsubishi Electric ITA, "Mobile Agent Computing", 19, January 1998.
8.   Danny B. Lange and Mitsuru Oshima, "Seven Good Reasons for Mobile Agents, Vol. 42, No. 3, Communications of the ACM, March 1999
9.   M. Wooldridge, N. R. Jennings, "Intelligent Agents: Theory and Practice", The Knowledge Engineering Review, Vol 10, No. 2, January 1995

A Hybrid Fault-Tolerant Scheme Based on Checkpointing in MASs

YunHee Kang[1], HoSang Ham[2], and ChongSun Hwang[3]

[1] Department of Computer and Communication Engineering, Cheonan University,
115, Anseo-dong,
Cheonan 330-180, Choongnam, Republic of Korea
yhkang@infocom.cheonan.ac.kr
[2] Electronics and Telecommunications Research Institute
161, Gajeong-dong,
Yusong-Gu, Daejon, 305-350, Republic of Korea
hsham@etri.re.kr
[3] Department of Computer Science and Engineering, Korea University
5-1, Anam-dong, Seongbuk-ku,
Seoul, 136-701, Republic of Korea
hwang@disys.korea.ac.kr

Abstract. The paper introduces a hybrid fault-tolerant scheme of MAS. The scheme is composed of facilitator based recovery scheme and coordination based recovery scheme. To discuss the fault-tolerance of working agents in MAS, we propose a novel approach to fault-tolerance of agent's execution in MAS based on multiple facilitators architecture. In coordination based recovery scheme, the facilitators maintain a task plan in a stable storage to preserve consistency by using checkpoints. In addition, facilitator based recovery scheme is appropriate for supporting autonomous as well as fault-tolerant characteristics of MAS.

1 Introduction

MAS(Multi-Agent System) focuses on the collaborative resolution of global problems by a set of distributed entities. Agents attempt to satisfy their own local goals as well as the collaborative global goals [1][2][3]. In MAS, agents become suddenly unavailable cause of typical node or communication link failures. In [4], we consider the fault-tolerance of MAS model to tolerate the failure of working agents based on task delegation and observation which can be independent of an arbitrary number of node failures.

The agents continuously sense the external environment based on their internal state and take actions that affect the environment. Agents lack a clear notation of a transaction that can be used to log changes to its internal state onto a permanent store. The agents are embedded in a continuously changing environment and often interact with other agents and/or humans to carry out a variety of tasks. As a result, even if a failed agent was capable of being recovered in finite time, the recovered agent would still be out from the external environment. These differences require us to modify existing technique and adapt it for

I. Chong (Ed.): ICOIN 2002, LNCS 2344, pp. 679–690, 2002.
© Springer-Verlag Berlin Heidelberg 2002

agent recovery [5]. Since the worker agents in MASs have independence strategies to maintain its states, we consider the suitable protocols designed for these environments save the checkpoints in multiple facilitators [4]. The autonomous characteristics of worker agents in MASs can make traditional checkpointing and message logging unsuitable [6].

This paper proposes a new hybrid approach composed of a facilitator based recovery scheme and a coordination based recovery scheme to fault-tolerance of worker agent's execution in MAS. If any agent is to be failure in the long time, then the facilitator takes over the failure by using a facilitator based recovery scheme. In the coordination based recovery scheme, we describe the *checkpointing and rollback recovery* mechanism depend on control messages piggybacked.

The advantage of the scheme is more reliable due to collaborative failure handling mechanism that the result of task is maintained by the facilitator only and the group of agents only hold the overall status of the task by messages piggybacked. Each agent takes a local checkpoint when one receives the request of checkpointing for synchronization with other agents in the group of worker agents. We apply the checkpoints of a facilitator to maintains the checkpoints on stable its storage by using coordination with worker agents.

The rest of paper is organized as follows: Section 2 describes related works. Section 3 describes an extended MAS model including system model. Section 4 describes a hybrid fault-tolerant scheme and observations. We conclude in section 5 with some future directions.

2   Related Works

The MAS concept is largely based on the idea that complex activities are the outcome of interactions between relatively independent software entities called agents [7]. MAS may be defined as a set of agents that interact with each other and with the environment to solve a particular problem in a coordinated manner [2].

MAS dynamically changes concern, location and its heterogeneity of available resources and content as well as different user communities and societies of agents each of which is pursuing its goals that may conflict with the goals of others. The essential capabilities of any kind of middle-agents are to facilitate the interoperability of appropriate services, agents and systems, building trust, confidence and security in a flexible manner [8].

There are some MASs that have middle agent such as facilitators in Genesereth's federated systems [1], SRI's Open Agent Architecture [9], and matchmakers and brokers in RETSINA [8][10]. As a MAS with middle agents becomes more complex and as the number of the agents in the system increases, the number of middle agents in the system will also increases due to reasons such as domain specialization and efficiency that in a large multi-agent system with multiple middle agents, the middle agents could potentially serve as backups for each other.

An agent requesting the information to the facilitator, a kind of middle agent, receives information about the agents having the capability corresponding to the task. Facilitators are used to accept and respond routing request from the agent, to locate services and to register capability [1][9][10].

In distributed AI, it assumes that agents execute under the fault-free condition. In MAS, the multiple brokers are proposed to be used as middle agent for fault-tolerant mechanism [11]. There are few works to discuss fault handling in MAS in the viewpoint of distributed system [11][12]. It needs an additional recovery mechanism from worker agent failures [4].

Typically, checkpointing is one of the techniques for providing fault tolerance in distributed systems. Saving a copy of the system state during system execution is usually referred to as checkpointing. Checkpointing is an important operation for many functions in distributed systems, such as rollback recovery, debugging, and detection of stable properties [13][14]. After a failure is occurred, the application rolls back to the last state that was saved and starts its re-execution.

Communication-induced checkpointing [6] is a kind of coordinated checkpoint, the coordination for checkpointting is done in an agent group by piggybacking control information on application messages.

3   The Extended MAS Model

MASs provide clean separation between individual agents and their interactions in overall systems. The separation is natural for the design of multi-agent architecture [1][2]. Coordinating, the behavior of intelligent middle agents, is a key issue in MAS [15]. Computation and coordination are separate and orthogonal dimensions of all useful computing. The environment changes and multi-agent must evolve to adapt to the changes.

3.1   Assumptions

Let us assume the following general model of a MAS: the system consists of a set of *agents* connected through communication *links* that are used by the agents to exchange *messages*, which are typed. Agents will be separate processes which may be running in the same address space or on separate nodes. MAS consists of a group of agents that can take specific roles within an organizational structure. A facilitator, the supervisor of agents, takes a checkpoint when receiving the messages correspond to worker agents having delegated tasks by saving its state on stable storage. We propose the MAS based on multiple facilitators interact with agents and other facilitators.

In the proposed MAS, the facilitators maintain current state of a task plan in a stable storage by using checkpoints taken when either completing a task or receiving a response of a subtask within the task plan [4]. The facilitator plays a role of an observer that monitors the agents as a role of *name* and *resource* server as kind of a directory server.

The facilitator defines a task and its constraint. In order to execute the task you have to set actions, fulfill constraint and return the result. Actions are done by the agents at each host. The facilitator is responsible that the defined criteria are met and finally the facilitator takes the responsibility for the whole task and only it decides when the whole task is finished and the results are returned [9].

3.2   Failure Semantics

In MAS, agents may be up and down at any time. Agents need to orient their behaviors to deal with a dynamic environment. A basic robustness behavior is to detect and recover from facilitator failure [4][11][12]. Either the current request action is aborted, or the requester waits and tries to contact the facilitator again [11]. This behavior depends on the assumption that facilitator services are themselves robust and that if a facilitator goes down, it will be back as soon as the underlying hardware and software can recover.

The failure in MAS may lead to a partial or complete loss of the agent. For instance, a failure of the device on which the agent is currently executing causes all information about the agent node held in stable storage to be lost. A simple checkpointing based execution of an agent, it ensures that the agent is not lost, is prone to blocking. Whereas adding a simple timeout based failure detection mechanism solves the blocking problem, it also leads to the violation of the exactly-once execution property, which is fundamental to many applications [16].

In facilitator based recovery scheme, we consider a simple system of two entities which are a failure detector fd and an agent w which connected through a communication link. w may fail by crashing, and the link between fd and w may delay or drop messages. fd monitors w and determines either "I suspect that w has crashed" or "I trust that w is up".

The agent w sends a heartbeat message to the failure detector fd; when fd receives a heartbeat message, it trusts w and starts a timer with a fixed timeout value TO. In the following we describe the assumptions of MAS operations for fault-tolerant computing:

- The agents communicate with facilitators, and the facilitators send the messages and route them to the appropriate places.
- The facilitators provide location service and resource management service to a group of agents.
- The MAS exploits that the facilitator has responsibility for a manager by taking checkpoints.

In this Fig. 1, f_k and f_l represent facilitators. p, q and r represent agents. p is registered into f_k received information of its capability and preference; f_k takes care of p. Then f_k multicasts information message of the registered agents to a group of facilitators. And f_k has responsibility of task execution of p which has been failure by using failure detector. In this paper, failure detector is out of scope.

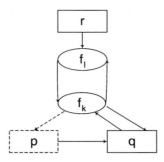

Fig. 1. Worker agent failure

As shown in Fig. 2, if f_k goes down in MAS, the agents which are registered into the f_k are adopt into "godfather". The godfather means one of other facilitators which play a role of the alternative observer keeping track the orphan agents. f_k will be back as soon as the underlying hardware and software can recover. These problems can be reduced by using multiple, distributed facilitators, but at the cost of increased system complexity. The proposed MAS follows the assumption of *consistency preservation* for faulty facilitator [11]. In this paper, we consider agent failure with such as failure the loss of an agent results in a temporary loss of some functionality of the overall system.

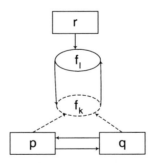

Fig. 2. Facilitator failure

3.3 System Model

We consider MAS computation consisting of n agents which are interacted by message passing via facilitators. Each pair of agents is connected by a two-way reliable channel whose transmission delay is unpredictable but finite. Agents are registered into facilitators consisting of m facilitators, $F=\{f_1, f_2, \ldots, f_m\}$, which coordinate a set of agents.

Agents are autonomous in the sense that they do not share memory, do not share a common clock value, and do not have access to private information of other agents such as clock drift, clock granularity, clock precision, and speed. We assume, finally, agents follow a fail-stop behavior. Agents can be inactive in a finite bound time due to a failure.

Communication from the components of MAS is said to satisfy First-In-First-Out(FIFO) delivery if for all messages m and m' of agent or facilitator i and j.

$$send_i(m) \rightarrow send_i(m') \Rightarrow deliver_j(m) \rightarrow deliver_j(m')$$

The coordination of F and configuration of A in MAS at any time t is :

$$\text{Coordintion(F,t)} \Leftrightarrow \text{Configuration(A,t)}$$

We define **Coordintion(F, t)** as the same problem that "any facilitator f_i in F multicasts m to F" which can be represented by :

$$(multicast\ F\ m) \equiv \exists\ [\text{t, t} + \triangle],\ e_{f_i,h} \rightarrow e_{f_i,k}$$

- $k = succ.h$
- $e_{f_i,h} = send_G(m)$
- $e_{f_i,k} = receive_G(m')$
- $\exists\ f_k \notin G : \neg\ (\ e_{f_k,k} = receive_{f_i}(m))$

where m represents a message for $PROF$ which is information of agent status , m' represents a message for ACK. G represents a group of facilitators which are available currently. In order to distinguish control messages, we use e to denote events that satisfy synchronous ordering.

We define **Configuration(A, t)** as the same problem that "for all p in AG, the facilitator f_i in F checkpoints the state of p" which can be represented by :

$$(checkpoint\ W\ m) \equiv \exists\ [\text{t, t} + \triangle],\ e_{f_i,h} \rightarrow e_{f_i,k}$$

- $k = succ.h$
- $e_{f_i,h} = send_{AG}(m)$
- $e_{f_i,k} = receive_{AG}(m')$
- $\exists\ p \notin AG : \neg\ (\ e_{p,k} = receive_p(m))$

where AG represents the group of agents, m represents a message for *Request-ForCheckpoint*, m' represent a message for ACK.

4   The Proposed Hybrid Fault-Tolerant Scheme

In this section, we classify the recovery scheme into *facilitator based recovery scheme* and *coordination based recovery scheme*.

4.1   Facilitator Based Recovery Scheme

Fault-tolerant distributed systems are designed to provide reliable and continuous service despite the failures of some components. Fault tolerance is crucial from real-world applications of MAS. The facilitator based recovery scheme allows the facilitator to remove the failure agent when it detects the timeout delay beyond a given deadline caused by node crash.

Facilitator based recovery algorithm. We consider high availability system to provide adaptivity by using **FD**. To take over single point of failure of MAS, we design the ring structure linked logically for MAS to detect a facilitator has been failure and to eliminate the failure facilitator from the ring element.

The failure detector **FD**, the major subsystem of facilitator, is assume to build a sequence of view $v_0(ag)$, $v_1(ag)$, ... $v_i(ag)$ ... corresponding to the succession composite of the group ag as proceed by **FD** [17]. The sequence of views satisfies the following conditions:

- if an agent $p \in ag$ has crashed, then **FD** will eventually detect it and define a new view from which p is excluded (if p recovers after the crash, it comes back with a different agent information, and is thus considered as an agent);
- if p joins the group ag, **FD** will eventually detect it, and define a new view including p;
- if a view $v_{i(ag)}$ is defined and $p \in v_{i(ag)}$, or $\exists\, k > 0$, such that $p \notin v_{i+k(ag)}$;
- $p, q \in ag$, if p and q receive views $v_{i(ag)}$ and $v_{j(ag)}$ ($i \neq j$), then p and q both receive $v_{i(ag)}$ and $v_{j(ag)}$ in the same order.

Theorem 1. $fd_{vc} < q_{vc}$ *iff q is incarnated agent and $fd_{vc} = q_{vc}$ iff q is right agent, where vc is view number*

When any worker agent register, unregister and delete, we consider multi-cast m_{view} having current configuration if message m_{view} has been received by any facilitator(faulty or not), then message m_{view} is received by all facilitators that reach a decision. In the facilitator recovery algorithm of Fig. 3, a facilitator detects failure agents and multicasts gathered information to a group of facilitators.

---

for every **FD** of f_k
      for all $q \in$ all_agents(FD_{f_k}) do
      $q =$ all_agents(FD_{f_k})
      repeat periodically
         1. issue a heart beat message $\mathcal{HBM}$ to q in f_k.
         2. send $\mathcal{HBM}$ to q.
         3. receive alive messages m_{ack} from q.
         4. determine the faulty agents and update suspicion lists.
         5. multicast relevance workers related with faulty agents to F.
         6. update view count.
      until all_worker(FD_{f_k}) got(m_{ack})

Fig. 3. Algorithm of the facilitator based recovery scheme by using the failure detector in a facilitator

4.2   Coordination Based Recovery Scheme

The coordination based recovery scheme enables each agent to participate in the execution of task to recover a failure agent by forcing the agent to piggyback only an additional n-size vector, where n is the number of agents.

Facilitator initiated checkpointing. We apply the checkpointing mechanism based on communication-induced checkpointing to coordination based recovery scheme as follows: each agent in AG is forced to take checkpoints based on some application messages such as *complete_subtask*, *complete_overall_task*, *suspend_task* and *resume_task*. Information such as task execution vector is piggybacked on each agent so that the receiver can examine the information prior to processing the message.

Let us consider that each agent p has a variable sn_p which represents the sequence number of its last checkpoint. Each time a checkpoint $C_{p,sn}$ is taken by an agent p to start an explicit coordination with a facilitator f_k.

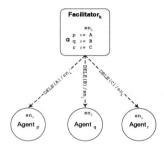

Fig. 4. Coordination between a facilitator and the agents in checkpointing: Initial task delegation

The request agent p transmits the request of service G to the facilitator f_k. As shown in Fig. 4, f_k delegates G to the agents p, q and r after decomposing into the subtasks of A, B, and C. At the same time, the facilitator f_k piggybacks its sn of task execution in p, q and r. p, q and r save their sn numbers in their stable storage and respond to the sn numbers. After that, the facilitator f_k takes checkpoints to store the status of G in its stable storage. p, q, r related to G have the same sn value sn_i. Because the current state of the G is maintained by the facilitator f_k, the current local state maintains the same sn number and is consistent by checkpointing algorithm [4].

The values of $sn_{p,i+2}$, $sn_{q,i+2}$, $sn_{r,i+2}$, $sn_{f_k,i+2}$ are all the same. The state of sn_{i+2}, which are related to G, is maintained by f_k. The agent r transmits *RESP(C)* which is the execution result of the task C delegated to it and the new sn value sn_{i+3} to the facilitator f_k after piggybacking it.

Rollback algorithm. The facilitator f_k informs the agents p, q, and r to change the sn number to sn_{i+3}. As is shown in Fig. 5, in the execution state where the sn value is sn_{i+2}, the agent q is blocked by a failure because of system c rash. Following the fail-stop model, after some time passes, the agent q starts recovery using sn it has saved. The agent q realizes that its sn number sn_{i+2} and the sn number of the facilitator f_k sn_{i+3} are not the same. In order to maintain the consistency of the system, the agent q receives the task plan which is checkpointed from the facilitator f_k, replays it, reflects the execution result of the task B' within f_k (checkpoint sn_{i+3}), changes its sn number to sn_{i+3}, and saves it.

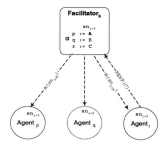

Fig. 5. Failure at worker agent: dotted circle represents failure agent

In the algorithm of Fig. 6, a recovering agent q rolls back to its checkpoint based on sn on its stable storage of f_k. In step 1) of recovery algorithm, q restores a state of an agent and knows that $TEG_{sn}^{f_k}$ of G is completed. In step 2), 3), the agent p sequentially receives t holding a partial result of a subtask from a provider agent based on sn. And it compares its sn and sn of a facilitator f_k. It then installs its checkpointed state from f_k and resumes execution until its sn is the same with the sn of a facilitator f_k. In the final step of the recovery algorithm, the sn of q is set to sn_k of a facilitator f_k. For recovery, we assume that if an agent q fails, it restores its last checkpoint sn_{i+2} as shown in Fig. 5.

Global Consistency. We assume that the system can ensure the situation that there is no missing or orphan message in the concerned ith checkpointing interval C_{sn_i}, then the set of all the ith checkpoints taken by its consistent. Our observation means that the sending as well as receiving of any message have to be recorded in the same checkpoint of the sender and the receiver accordingly. In any time, The worker agents are to be failure state, then those take over a failure state based on piggybacked information autonomously.

In the viewpoints of communication, it improves the number of messages to preserve the consistency and enhances the autonomous property of agents. Each

Data Structure at agents such as p, q and r.

      sn : sequence number

      t: task assigned by worker agent

Whenever failure agent p restart with a local checkpoint number sn

Procedure $Do\_Recovery$(message m) {

    1. restore sn from stable storage of agent

    2. request sn to a facilitator f_k

    3. receive t with m_{sn} from f_k

    4. **if** $(sn < m_{sn})$

        replay t corresponding m_{sn}

    5. update sn /* $sn \leftarrow m_{sn}$ */

}

Fig. 6. Algorithm of rollback-recovery based on sn

agent takes a local checkpoint when one receives the request of checkpointing for synchronization with other agents in a group.

Whenever failures are occurred, the facilitator do not consider the worker's status of task. In recovery time, the autonomous worker takes over the failures based on *active restart* by interaction with facilitator, then it takes rollback recovery to preserve global consistency in MAS.

Theorem 2. *The group of agents, AG, is consistent iff : each agent in AG has the same number of sn on stable storage of the agent.*

Proof : *The sequence of events related to execution of agents can be consistent with the causal order. Causal order ensures the guarantee of causally ordered message delivery to the group of agents. Let m_{Chk_p} represent the piggypack message with a request for checkpoint related with an agent p and m_{ACK_p} represents the reply message after checkpointing by p. If two message m for m_{Chk_p} and m' for m_{ACK_p} are received and sent by p, then m must be delivered everywhere before m'. m' causally depends on m(m happens before m').*

   $- k = h + 1$,

   $- \forall p : e_{f,h} = send_{f_k}(m_{Chk_p})$ and $e_{f_k,k} = receive_{f_k}(m_{ACK_p})$,

   $- \exists e_{r,z} : (e_{f_k,h} \rightarrow e_{r,z}) \wedge (e_{r,z} \rightarrow e_{f_k,k}) \Rightarrow |AG| = |m_{ACK_p}|$

Informally, the facilitator f_k maintains optimal agents in AG after looking up its $PROF_k$. The facilitator preserves the consistency of the agents in AG which are the components of $TEG_{sn}^{f_k}$. If an agent is a member of AG, its sn is preserved by piggybacked information via facilitator. The agent increases the sn by receiving the synchronization message m_{sync} from the facilitator.

The additional information required for such checkpoint synchronization is appended at the end of the piggybacked synchronization message. The information is utilized by the agents that then synchronize their checkpoints with each other. The paper describes that any global checkpoint taken in AG is consistent and hence the system has to be rollbacked to the last saved state in case of a failure.

> **Theorem 3.** *The changed task plan is preserved the* consistent *state in the group of agents AG.*
>
> **Proof** : *Any long failure is derived to change the task plan in order to isolate the faulty agents. If a facilitator f_k finds a faulty agent p, then p is excluded from AG, and reconstructs a task plan $TEG_{sn}^{f_k}$. On the other hand, if the faulty agents are in AG_i, the faulty agents are recovered from the last checkpoint based on local sn and interact with f_k. In both cases, we can observe that the faulty agent can be either replaced by other worker agent or restart from the last checkpoint. Then system is globally consistent with its sound state.*

5   Conclusion and Future Works

In this paper, we have introduced a scheme for fault-tolerance of agent execution in MAS. The multiple facilitator covers the avoidance of blocking in a MAS and improves availability. The fault-tolerant scheme improves the blocking avoidance of MAS based on alternative in worker pool and recovery scheme of agents based on checkpointing. We apply the sequence number of a checkpoint to preserve consistence of states for tasks execution between agents and a facilitator.

In section 3 and 4, we showed an appropriate MAS model and applied multiple facilitator to provide a fault-tolerant scheme and checkpointing mechanism in a MAS. We presented a recovery mechanism to preserve consistence based on coordination with a facilitator. We observed some cases of facilitators or agents failure and showed that the proposed MAS is well-formed and recoverable. The facilitator checkpoints consist of a set of the independent AGs. Both facilitator and coordination based recovery scheme rely on a facilitator. In addition, the facilitator based recovery scheme is appropriate for supporting autonomous as well as fault-tolerant characteristics of MAS.

In this paper, we do not cover performance evaluation of the algorithm in failure-free and recovery. In order to assess feasibility and applicability, we have developed a prototypical instance of our MAS model for agent services. In the future, we apply fault-tolerant scheme to mobile environment such as e-commerce transaction requiring "exactly-once" property and non-blocking execution of mobile agents.

References

1. M. Genesereth: An agent based framework for interoperability, In: J. M. Bradshaw (eds.): Software Agents, AAAI Press, (1997) 317–345
2. M.R. Genesereth and S.P. Katchpel: Software agents, Communications of the ACM. **37** No. 7, (Jul. 1994) 1029–1036

3. Sycara, K., Decker, K., Pannu, A., Williamson, M. and Zeng, D.: Distributed Intelligent Agents. IEEE Expert. (Dec. 1996)
4. YunHee Kang, HoSang Ham and ChongSun Hwang.: A Fault-Tolerant Scheme of Multi-agent System for Worker Agents, In: J.Liu et al.(Eds.): AMT 2001, LNCS Vol.2252, Springer-Verlag (2001) 171–181
5. Arvind K. Bansal, Kotagiri Ramohanarao, and Anand Rao: Distributed Storage of Replicated Beliefs to Facilitate Recovery of Distributed Intelligent Agents, In: Munindar P., Singh Anand Rao and Michael J. Wooldrideg (Eds.): Intelligent Agents IV, ATAL'97,LNCS Vol.1356, Springer-Verlag (1997)
6. R.Baldoni, J.H.Helary, A. Mostefaoui, and M.Raynal: A Communication-Induced Checkppointing Protocal that Ensures Rollback-Dependency Trackability. Proc. IEEE Int'l Symp.Fault Tolerant Computing. (1997) 68–77
7. Charles J. Petrie: Agent-Based Engineering, the Web, and Intelligence. *IEEE Expert,* **11** No.6 (Dec.1996) 24–29
8. Decker, K., Sycara, K. and Williamson, M.: Matchmaking and Brokering. Proceedings of the Second International Conference on Multi-Agent Systems (ICMAS-96), (Dec. 1996)
9. D. Martin, A. Cheyer, and D. Moran: Building Distributed Software Systems with the Open Agent Architecture. Applied Artificial Intelligence. **37** No.12, (1999) 92–128
10. Hao-Chi Wong and Katia Sycara : A Taxonomy of Middle-agents for the Internet, In: *Proceedings of the Fourth International Conference on Multi-Agent Systems (ICMAS'2000),* (2000)
11. S. Kumar, P. R. Cohen, and H. J. Levesque : The Adaptive Agent Architecture: Achieving Fault-Tolerance Using Persistent Broker Teams, Proceedings of the International Conference on Multi-Agent Systems (ICMAS2000), Boston, MA USA, (2000)
12. Johansen, D., Marzullo, K., Schneider, F.B., Jacobsen, K., Zagorodnov, D.: NAP: practical fault-tolerance for itinerant computations. Distributed Computing Systems. (1999) 180–189
13. O. Babaoglu and K. Mazullo: Consistent Global States of Distributed Systems: Fundamental Concepts and Mechanisms. University of Bologna, Italy, (1993)
14. K.M. Chandy and L. Lamport: Distributed Snapshots: Determining Global State of Distributed Systems. ACM Transaction on Computer Systems. **3**, No.1 (1985) 63–75
15. H.V.D. Parunak. : Characterizing multi-agent negotiation. In International Workshop on Multi-Agent Systems(IWMAS-98), (1998)
16. Tushar Deepak Chandra and Sam Toueg: Unreliable Failure Detectors for Reliable Distributed Systems. Journal of the ACM, **43** No.2 (Mar. 1996) 225–267
17. E. Lotem, I Keidar and D. Dolev: Dynamic voting for consistent primary components. in Proc. of the 16th Annual ACM Symposium on Principles of Distributed Computing. Santa Barbara, CA, August (1997) 63–71

A Fast Commit Protocol for Distributed Main Memory Database Systems

Inseon Lee and Heon Y. Yeom

Seoul National University, Seoul,151-742 , Korea,
{inseon,yeom}@arirang.snu.ac.kr

Abstract. Although the distributed database systems has been studied for a long time, there has been only few commercial systems available. The main reason for this is that the distributed commit processing costs too much which results in little or no performance gain compared with single node database system. In this paper, we note the difference in the update and logging policy between disk based database and main memory database in the distributed environment, and presents a fast distributed commit protocol for the main memory database. In the proposed protocol, instead of sending and receiving two sets of messages one after the other as in two phase commit, only one set of messages are sent after the coordinator completes committing a distributed transaction. The main idea of this fast commit processing is to send all the redo-logs to the coordinator so that the coordinator alone can make the decision to commit or abort when the time comes. As a result, the frequency of the communication and the disk access related to the commit processing can be significantly reduced. Our simulation study shows that the proposed commit protocol achieves the high performance as we expect.

Keywords – Distributed database system, Transaction Processing, Main memory database, Commit Processing.

1 Introduction

The main memory database with its very high speed, has been gaining popularity as the price of the main memory continues to drop. Likewise, computer clusters and networked workstations begin playing an important role as the performance of off-the-shelf personal computers reaches a new height. The memory resident database or main memory database (MMDB) has a potential for significant performance improvement over conventional database where the data reside in the disk[4, 1]. Since there is virtually no I/O overhead for reading and writing the database, the MMDB shows faster response time and improved throughput[6, 7]. With the falling price of RAM and the rising demand of high performance system, MMDB has become a real alternative for database systems. However, for the database residing in the main memory, the occurrence of failures can be

I. Chong (Ed.): ICOIN 2002, LNCS 2344, pp. 691–702, 2002.
© Springer-Verlag Berlin Heidelberg 2002

more fatal, and hence, logging and checkpointing become the most important part of the MMDB design. Moreover, the only disk access in MMDB is due to logging and checkpointing activity and the efficient implementation of logging and checkpointing is essential to avoid the performance bottleneck of the system. As the idea of main memory database has been materialized, the need to utilize the main memory database in distributed environment is arising. Although the distributed database systems has been studied for a long time, there has been only few commercial systems available. The main reason for this is that the distributed commit processing costs too much which results in little or no performance gain compared with single node database system. This paper identifies the problems of the commit protocol for the distributed database systems, and presents a fast distributed commit protocol for the main memory database. In the proposed protocol, instead of sending and receiving two sets of messages one after the other as in two phase commit, only one set of messages are sent after the coordinator committed a distributed transaction. The main idea of this fast commit processing is to send all the redo-logs to the coordinator so that the coordinator alone can make the decision to commit or abort when the time comes. As a result, the frequency of the communication and the disk access related to the commit processing can be significantly reduced. To compare the performance of the proposed protocol with the existing two-phased distributed commit protocol, we have conducted extensive simulation. The simulation results match our expectation that the proposed protocols perform well in heavy traffic situation where the MMDB system would be most useful.

The rest of this paper is organized as follows: We first review some previous works including distributed commit protocols in section 2. We also identified the problems with current protocols. The distributed MMDB system model is given in section 3. Our proposed fast commit protocol is presented and the recovery process for the commit protocol is outlined in Section 4. The performance of the proposed protocol is studied in depth using simulation in Section 5 and we conclude the paper in Section 6.

2   Related Works

2.1   Distributed Database

The work on distributed database systems can be divided into several classes. First, there is shared disk environment where each site has some portion of the database and a shared disk has the whole database. The work on ARIES-SD[9] proposed that each site have separate logs and pages are shipped between them as needed. Another work[11] suggested a recovery scheme with NOFORCE strategy, entry logging and on-request invalidation scheme for coherence control with a direct exchange of modification between nodes. Nonvolatile log was assumed. Second, there is client-server environment where the clients as well as the server participate for the checkpointing and recovery process. The client-server version of EXODUS[2] also describes recovery scheme for distributed database system. Both of these are data sharing architecture while our scheme is based on shared nothing environment. The amount of main memory needed for the processing should be a lot less in shared nothing environment compared with the data sharing environment.

2.2 Distributed Commit Protocols

There has been a lot of research to mitigate the performance degradation from the two-phase commit generally used for the distributed database systems[3, 5]. Some of the research have also been applied to the main memory database systems[10]. These can be classified into three groups. The first is the Presumed Abort/Presumed Commit which try to reduce the number of messages passed during the commit process as well as the amount of log. The second is Precommit/Optimistic commit which reduces the lock contention by releasing the locks earlier. Lastly, there is group commit which minimizes the number of disk writes to enhance the overall system throughput.

The Presumed abort[8] is to assume that the transactions abort if they are not explicitly committed to reduce the messages such as acknowledgement message from the cohorts to the coordinator and the disk write for the abort log record while the Presumed Commit[8] assume the transacions commit if they are not explicitly aborted. Since transactions usually commit under the normal circumstances, it has the advantage if we omit the messages related to the commit processing as well as the log write. However, it has the disadvantage that all the logs should be sent to the coordinator before two phase commit process begins. Optimistic commit[5] concentrates on reducing the lock waiting time by lending the locks the committing transactions hold. However, in the main memory database systems where all the database resides in the main memory, lock waiting and processing time are relatively small compared to the commit time and optimistic commit has no or little effect.

When the precommit is applied to the distributed main memory database system, it is possible for the global transactions to commit in the inconsistent order. To prevent such an inconsistency, it is proposed to propagate the transaction dependency information[10]. It is also noted that precommit can also result in the cascading abort if there is a site failure in the middle of commit process. It can be catastrophic to the system performance.

Even though the group commit can enhance the system performance by reducing the number of disk writes, it also increases the lock holding time of finished transactions.
Hence, group commit is usually applied in conjunction with the precommit[10].

3 System Model

The distributed main memory system we consider in this paper is composed of N sites connected through a communication network, and each site of the system contains the database residing in the main memory and the system components; the *transaction manager*, the *memory manager*, the *scheduler* and the *recovery manager*. The transaction manager handles the execution of the transactions including their initiation and termination. A transaction accessing only the database of its initiating site is called a *local transaction*, and a *global transaction* consists of a set of subtransactions accessing the database at remote sites. The transaction manager of the initiating site is called the *coordinator* of a global transaction and the ones managing the subtransactions are called the *cohorts*. The coordinator and the cohorts communicate by message passing and the messages are managed by the log message manager while the log manager is responsible for the disk logging of the necessary information. The checkpoint manager periodically backs up the database into the two backup disks.

To preserve the serializable precedence order between the conflicting transactions, we assume that the scheduler follows the strict two-phase locking protocol for the concurrency control.

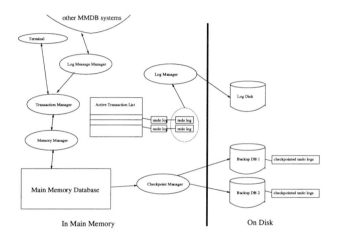

Fig. 1. The architecture of the distributed main memory database system

For the main memory database update, the in-place update policy is assumed; when a transaction updates a data item, it directly writes on the database. Due to this update policy, an 'undo'log has to be appended to the log before the write operation and the 'redo'log has to be appended to the log after the write operation[6]. These logs stay in the volatile memory until the transaction is ready to commit at which time only the 'redo'logs are flushed to the disk. Upon successful completion of the disk write, the transaction manager can safely commit the transaction. Note here that the 'undo'log is not needed after the transaction commits.

We assume the fuzzy checkpointing in this paper. Under fuzzy checkpointing, the checkpointed database states may not be consistent, and hence the undo logs of the transactions active during the checkpointing also have to be saved. Also, two backup databases are needed to alternatively checkpoint ('ping-pong'update[12]) to preserve the write-ahead-logging rule. In case of a system failure, the recovery manager first reloads the checkpointed database and makes a consistent database state by undoing the transaction operations in the saved undo log. Then, the recovery manager performs the redoing of the transaction operations saved in the stable system log, in order to reflect the effects of the transactions which have been committed after the checkpointing.

In the next section, we present our fast commit protocol, which requires only one set of messages from a coordinator and only one disk write per transaction if we do not consider the checkpointing activity.

4   A Fast Distributed Commit Protocol

We first look at the difference between the distributed database system based on the disk and the distributed main memory database system whose data resides on the memory. Finally, the fast distributed commit protocol based on the main memory database is given.

4.1 Comparison between Disk Resident DBMS and Main Memory DBMS in Distributed Environment

In a distributed database system, databases are stored on the disks at each site and a global transaction is split into subtractions and submitted to each site. After each subtransaction is finished, the two phase commit protocol is initiated. Disk write as well as two passes of message exchange are needed before the transaction is committed. Two phase commit is necessary to preserve the ACID property of the transactions in distributed environment. Another characteristic of distributed database system is the use of deferred update. When using deferred update, actual database update on the disk is performed during the first phase of two phase commit. Each subtransction, upon receipt of 'PREPARE'message from the coordinator, updates the database on the disk after writing the undo/redo log on the log disk. It then sends the 'READY'message to the coordinator and concludes the first phase. When all the subtransactions finish the first phase, the coordinator can safely decides to commit the global transaction since each subtransaction can be re-done in case of failure. The coordinator writes 'COMMIT'record on its log disk and sends 'COMMIT'message to all the subtransactions. When the subtransactions receive the 'COMMIT'message, they can write 'COMMIT'on the log disk, release the locks, and conclude the two phase commit. For a global transaction with n subtransactions, the number of messages needed for the two phase commit is 3n and the number of disk writes is 2n+1.

There are certain differences when it comes to the main memory database system. Since the database resides on the main memory, in-place update is usually used instead of deferred update. Hence, only the redo log needs to be written to the disk. This redo log writing is the only disk write in the main memory database system while the disk resident database incurs several more disk writes including data read/write and undo log write. We note that we do not have to use the two phase commit when the only disk write is for the redo logs.

4.2   Protocol Description

In the proposed protocol, the commit process is completed in one phase without incurring costly disk writes on each site. Instead of logging the redo log after receiving 'PREPARE'message from the coordinator, each subtransaction sends the redo log to the coordinator when the subtransaction finishes processing. The coordinator, after receiving the redo logs from all the subtransactions, decides to commit by writing the logs all at once and sends 'COMMITDONE'message to all the cohorts. What follows is the formal protocol description.

[Cohort's action when a subtransaction is finished]
1. If a subtransaction successfully completes processing, its cohort sends its 'redo'log as well as 'WORKDONE'message to its coordinator.
2. Otherwise, the subtransaction is involved in a deadlock due to global/local inconsistent lock request, it then releases all the locks, sends 'ABORT'message to the coordinator immediately and aborts the transaction.

[When the coordinator receives 'WORKDONE'messages from all its subtransactions]
1. The coordinator writes the 'redo'logs from the subtransactions and 'COMMIT'record on its log disk.
2. The coordinator sends 'COMMITDONE'messages to all the subtransactions after finishing disk write.
3. Each cohort releases all the locks and ends the subtransaction.

[When the coordinator receives any 'ABORT'message]
1. The coordinator sends 'ABORT'message to all the other subtransactions which sent their 'redo'log already and abort the transaction itself.
2. Each cohort, upon receiving 'ABORT'message, performs 'undo' using the 'undo' log, releases all the locks, and aborts the subtransaction.

[When the coordinator fails during commit processing]
1. When recovering, if the 'COMMIT'log is the last log, it re-sends 'COMMITDONE'message to all the cohorts although it could be duplicate. If there is no 'COMMIT'log for some transaction, it must have failed before writing the logs. In this case, it just sends 'RECOVERING'messages to all the sites.
2. If the cohort does not receive 'COMMITDONE'message and it receives 'RECOVERING'message, the cohort aborts the subtransaction which is managed by the recovering coordinator.

[When a cohort fails during commit processing]
1. When the cohort recovers, it broadcasts the 'RECOVERING'message to all

the other sites since it does not have any idea regarding the transactions committed after its latest checkpoint.

2. Each site, upon receiving the 'RECOVERING'message, sends all the 'redo'logs of the committed sub transactions that had been executed in the crashed site after the checkpoint.

Theorem 1 : The fast commit ensures a safe commit.

Proof : It is trivial from the protocol description. Since the commit process is performed only at the coordinator site as soon as all the redo logs are received from the cohorts and the commit process at the coordinator site is the same as a single DBMS's, durability as well as atomicity is guaranteed. The consistency is also preserved, since the aborted transactions can be 'undone'using the undo log of each site. □

The proposed fast commit protocol has the following characteristics.

1. When a global transaction is composed of n subtransactions, the number of messages needed is only n which is 'COMMITDONE'message from the coordinator to each n cohorts because the 'PREPARE'and 'READY'messages are not necessary where $3n$ messages are needed in the two-phase commit protocol. Also, the subtransaction blocking time is dramatically reduced.

2. For n subtransactions, our protocol incurs only 1 disk write compared with $2n + 1$ disk writes needed for the 2PC which results in tremendous performance improvement.

3. In case of a failure, the recovering site has to gather 'redo'logs from other sites since it only has the 'redo'logs of the transactions which it managed.

4. To reduce the amount of log on the disk , each site needs to broadcast the checkpointing information so that unnecessary logs can be truncated.

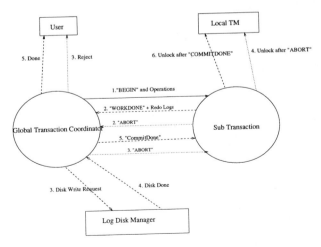

Fig. 2. Inner working of the fast Commit Protocol

4.3   Checkpointing

We assume fuzzy checkpointing with two backup databases. We need two disk backup databases since the checkpointed image of the main memory database is not consistent until the checkpointing is finished. If there is only one backup database and there is a failure during the checkpointing process, it is impossible to restore the consistent database state because the fuzzy checkpointing is neither transaction consistent nor action consistent. Also, the active transaction list(ATL) and the undo logs of the transactions in ATL are written to the disk while checkpointing.

After the checkpointing, some of the redo logs become useless. However, with fast commit, the redo logs are stored on the disk of the coordinator of the transaction. Each site needs to inform the other sites of their checkpointing progress so that other sites can perform garbage collection.

4.4   Failure-Recovery

For a system site, S_i, to recover from a failure, the recovery manager of the site, say RM_i, first restores its database state from the most recently checkpointed backup database of the two. The RM_i should have the active transaction list, undo logs of the active transactions, and the redo logs after the checkpoint. Since we assume the fuzzy checkpointing, the restored database state may not be consistent and the recovery manager RM_i has to perform the undo operations using the undo log saved with the checkpoint. Then, the redo operations for the transactions which have been committed after the checkpointing are performed to make the database state of S_i up-to-date. Some redo logs of the committed transactions can be found in the stable log of S_i, and the others can be found in the stable log of other system sites. Hence, RM_i first broadcasts the *recovery_message* to the recovery managers of the other sites in the system. Upon the receipt of the *recovery_message*, each of the recovery managers sends the redo logs for S_i back to RM_i. RM_i then arranges the redo log entries collected from other recovery managers. For the efficient arrangement, a sequential log number assigned to each redo log entry generated by a system site can be used. With the arranged redo log entries, RM_i performs the redo operations and the site S_i can resume the normal transaction service when the redo procedure is completed.

5   Performance Study

5.1   Simulation Model

A distributed main memory database system consisting of 8 sites was simulated. At each site, a global transaction is initiated with a rate 300 transactions per second on the average with exponential distribution and each global transaction consists of 3 subtransactions. The sites executing the subtransactions are selected randomly and each subtransaction accesses 6 pages and the proportion of the write operation is specified as 0.6. For each transaction-related message, 5 milliseconds of communication delay is assumed including sending and receiving event processing. Since we assume the main memory resident database, the only reason for disk access is the stable logging and we assume that it takes 5 milliseconds for the logging of one data page. The disk delay consists of rotational delay and the data transfer time. Since we assume that the log disk is dedicated only for log storage, there is no seek time involved with log writes when there is no failure.

Here, we have some additional assumptions for the simulation model.

- A closed queueing model is used.
- Global transactions are generated according to the MPL at the terminals.
- The only stable storage is the hard disk. There is no non-volatoile memory available.
- Strict twho phase locking is used.
- The database is composed of pages which is the locking granularity.
- There is no transaction abort due to user request or system failure. The only transaction abort is to resolve deadlock.
- When there is a lock reqest leading to the deadlock, it is identified using the distributed deadlock detection algorithm and the transaction issued the offending lock request is terminated.
- A transaction begins with 'BEGIN' operation, performs 'READ' and 'WRITE' operations and ends with 'END' operation.
- The grnularity of each operation a word, the log is composed of pages, and a transaction only read/write a page only once.
- If a transaction reads and then writes a page, we assume there is only one 'WRITE' operation and one 'WRITE' operation is further divided into 'locate the word' + 'read' + 'write'
- A transaction is managed by the transaction manager of the site it was generated and aborted transaction is re-submitted until it is finished.
- We assume fuzzy checkpointing. However, we did not consider checkpointing time since we are mainly interested in the commit processing.
- Also, we do not consider system failure and database reload.

For the simulation, we have used CSIM[13]and the performance metric is the average response time of the transactions. We have run 35,000 transactions to the completion and measure the values from the 30,000 transactions excluding the first 5,000.

5.2   Simulation Results

We have compared the performance of 2 phase commit(2PC) protocol and the proposed fast commit(FC) protocol. The performance metric for the comparison are the average response time, throughput, log disk utilization and CPU utilization.

Figure 3 shows the response time of transactions under 2PC and FC with varying MPL, respectively. We can see from the figure that the proposed FC protocol reduces the response time dramatically compared with 2PC.

If we look at the composition of response time, we can see that the long commit delay has a huge effect on the operation time since the commit delay directly affects the lock holding time. Even though all the operations are performed in the memory, transactions running 2PC protocol show long operation time compared with the FC running transactions as shown in figure 4 and figure 5.

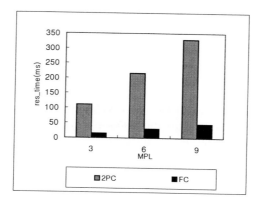

Fig. 3. The Effect of MPL on Response Time

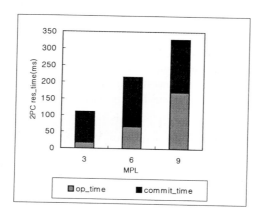

Fig. 4. Response Time composition of transactions with 2PC

Overall, the proposed fast commit protocol shows better performance compared to the two-phase commit protocol, mainly due to the reduced frequency of the commit-related communication and disk accesses.

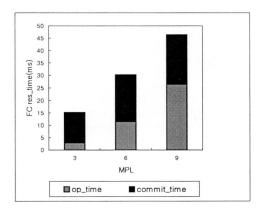

Fig. 5. Response Time composition of transactions with FC

6   Conclusions

The rapidly growing main memory size and the price have made the main memory database a reality. Also, the high speed network enables the main memory databases to be distributed over the network when the main memory of one system is not enough to store the entire database. However, when the system is extended to the distributed environment, we need a consistent distributed protocol. One of the widely used such protocol, two-phase commit protocol has a very high communication overhead in terms of messages exchanged as well as the blocking time. This delay has a more profound effect on the main memory database where all the operations are performed on the main memory with very fast response time.

To solve this problem, we note that only the redo log is to be written to the disk in the main memory database system and propose a one pass distributed commit protocol which we call fast commit protocol. In the fast commit, each cohort sends its redo logs when it reports completion of the subtransaction and wait for the coordinator to decide the fate of the transaction. Other than sending the redo log, the cohort does not do anything until it receives the decision of the coordinator. The coordinator can unilaterally commit the transaction after it receives all the redo logs from its cohorts and the logs are safely written on the log disk. Afterwards, the coordinator informs the decision to the cohorts so that the cohorts can release the locks the subtransactions hold and commit.

Our proposed protocol guarantees the atomicity and durability of the global transactions while reducing the number of messages exchanged into 1/3, and needing only one disk write operation. Hence it shows superior performance compared with two phase commit. The only drawback of the fast protocol is the recovery process is a little more complicated. A site needs to broadcast its recovering state and fetch necessary redo logs from other sites for the transactions not originated from that site. However, we believe this is a small cost compared with the huge performance gain as shown using the simulation.

References

1. David J. DeWitt, Randy H. Katz, Frank Olken, Leonard D. Shapiro, Michael R. Stonebraker, and David Wood. Implementation techniques for main memory database systems. *SIGMOD Record (ACM Special Interest Group on Management of Data)*, 14(2):1–8, 1984.
2. M.J. Franklin, M.J. Zwilling, C.K. Tan, M.J. Carey, and D.J. DeWitt. Crash recovery in client-serverEXODUS. In *Proceedings of ACM-SIGMOD International Conference on Management of Data*, pages 165–174, June 1992.
3. A. Citron G. Samaras, K. Britton and C. Mohan. Two-phase commit optimizatins in a commercial distributed environment. *Journal of Parallel and Distributed Databases*, 3(4), 1995.
4. H. Garcia-Molina and K. Salem. Main memory database systems: An overview. *Trans. on Knowledge and Data Engineering*, 4(6):509–516, December 1992.
5. Ramesh Gupta, Jayant Haritsa, and Krithi Ramamritham. Revisiting commit processing in distributed database systems. In *Proceedings of the 1997 ACM SIGMOD*, pages 486–497, 1997.
6. H. V. Jagadish, Avi Silberschatz, and S. Sudarshan. Recovering from main-memory lapses. In *International Conference On Very Large Data Bases*, pages 391–404, San Francisco, Ca., USA, August 1993. Morgan Kaufmann Publishers, Inc.
7. Tobin J. Lehman and Michael J. Carey. A recovery algorithm for A high-performance memory-resident database system. In Umeshwar Dayal and Irving L. Traiger, editors, *Proceedings of the Association for Computing Machinery Special Interest Group on Management of Data 1987 Annual Conference*, pages 104–117, San Francisco, California, 27–29 May 1987.
8. C. Mohan, B. Lindsay, and R. Obermarck. Transaction management in the r* distributed database management system. *ACM Trans. on Database Systems*, 11(4), 1986.
9. C. Mohan and I. Narang. Recovery and coherency-control protocols for fast inter-system page transfer and fine-granularity locking in a shared disks transaction environment. In *Proceedings of 17th International Conference on Very Large Databases*, pages 193–207, September 1991.
10. Taesoon Park and Heon Y. Yeom. A consistent group commit protocol for distributed database systems. In *Parallel and Distributed Computing Systems*, August 1999.
11. E. Rahm. Recovery concepts for data sharing systems. In *Proceedings of the 21st International Conference on Fault- Tolerant Computing (FTCS-21)*, pages 109–123, June 1991.
12. Kenneth Salem and Hector Garcia-Molina. System M: A transaction procrressing testbed for memory resident data. *tkde*, 2(1):161–172, March 1990.
13. H. Schwetman. Csim user's guide for use with csim revision16. In *MCC*, 1992.

Mobile Agent Model for Transaction Processing in Distributed Database Systems

Takao Komiya, Hiroyuki Ohshida, and Makoto Takizawa

Dept. of Computers and Systems Engineering
Tokyo Denki University
{komi, ohsida, taki}@takilab.k.dendai.ac.jp

Abstract. In database applications, application programs are performed on clients and issue requests to object servers. Then, the object servers send responses to the clients. On the other hand, programs named agents move to object servers where the agents manipulate objects in a mobile agent approach. If agents complete manipulating objects in the servers, the agents move to other servers. If an agent conflicts with other agents on an object server, the agents negotiate with each other to resolve the confliction under various constraints among servers like atomicity and consistency.

1 Introduction

In client-server database applications, application programs are performed on clients, which issue SQL [2] requests to object servers. The object servers send responses to the clients on completion of the requests. Requests and responses are exchanged among clients and servers in networks. The more number of requests are issued to object servers by applications and the more number of responses are sent back to the applications, the more communication overheads are increased. In the three-tier client-server architecture [4], applications move to application servers from clients in order to decrease the communication overheads between clients and servers.

In database applications, transactions are required to manipulate objects in object servers so as to satisfy ACID (atomicity, consistency, isolation, and durability) properties. [4]. For example, objects in multiple object servers are required to be atomically manipulated and transactions are serializable. In the traditional systems, objects are locked to realize the serializability [4] of transactions. In the locking protocol, multiple accesses to an object are coordinated based on a principle that only one transaction is a winner which can hold the object and the others are losers. There is another way like timestamp ordering [4]. Here, transactions are totally ordered in their timestamps. Transactions manipulate objects according to the timestamp order, i.e. the elder, the earlier. The locking protocol implies deadlock but no deadlock occurs in the timestamp ordering protocol.

In another computation paradigm, programs named mobile *agents* [1] move around data servers. First, an agent lands at a server and then is performed to

I. Chong (Ed.): ICOIN 2002, LNCS 2344, pp. 703–713, 2002.
© Springer-Verlag Berlin Heidelberg 2002

manipulate data objects in the server. If the agent finishes manipulating the data objects in the server, the agent moves to another server which has data to be manipulated. Here, agents manipulate objects in object servers without exchanging messages in a network. Compared with traditional process-based applications like client-server applications, mobile agents have following characteristics;

1. Agents are autonomously initiated and performed.
2. Agents negotiate with other agents.
3. Agents are moving around computers.

In this paper, we discuss how to manipulate multiple object servers by using agents. Agents move around object servers without exchanging messages in the network. On the other hand, application programs and object servers are exchanging messages in the network. In addition, an agent negotiates with other agents if the agents manipulate objects in a conflicting manner. Through the negotiation, each agent autonomously makes a decision on whether the agent continues to hold the objects or gives up to hold the objects.

In section 2, we present object servers. In section 3, we present an agent model for processing transactions. In section 4, we discuss how agents negotiate with other agents. In section 5, we discuss consensus conditions on which agents make an agreement in negotiation.

2   Object Servers

A system is composed of object servers $D_1, \ldots, D_m$ ($m \geq 1$), which are interconnected with reliable, high-speed communication networks. Each object server supports a collection of objects and methods for manipulating the objects. Objects are encapsulations of data and methods. Objects are manipulated only through methods supported by the objects.

Applications in clients initiate transactions in application servers. A transaction manipulates objects in one or more than one object server. A *transaction T* is an atomic sequence of methods for manipulating objects in object servers. A subsequence T_i of methods in T to manipulate objects in one object server D_i is referred to as *subtransaction* of T. A subtransaction T_i is also atomic sequence of methods in one object server D_i.

Each object server supports following methods to manipulate objects in the server;

1. *begin-trans*: A subtransaction starts. A log for the subtransaction is initialized. Methods issued by the subtransaction are kept in record in the log.
2. *op(o)*: A method *op* is performed on an object *o*.
3. *prepare*: The log of a subtransaction is saved in a stable memory.
4. *commit*: A database is physically updated by using the log and a subtransaction commits.
5. *abort*: A subtransaction aborts.

Suppose a pair of subtransactions T_1 and T_2 manipulate an object in an object server D_i by using methods op_1 and op_2, respectively. Here, if the result obtained by performing op_1 and op_2 depends on a computation order of op_1 and

op_2, op_1 and op_2 are referred to as *conflict* with one another on the object. For example, *read* and *write* conflict on a *file* object. A pair of methods *increment* and *decrement* do not conflict, i.e. are *compatible* on a *counter* object. On the other hand, *reset* conflicts with *increment* and *decrement* on the *counter* object. If a method from a transaction T_1 is performed before a method from another transaction T_2 and the methods conflict, every method op_1 from T_1 is required to be performed before every method op_2 from T_2 conflicting with the method op_1. This is a *serializability* property of transaction [4]. In order to realize the serializability, the locking protocol and timestamp ordering protocol [4] are used.

If a transaction manipulates objects in multiple object servers, the two-phase commitment protocol [4] is used to realize the atomic manipulation on multiple servers. After manipulating objects in the object servers by using methods, the transaction issues *prepare* messages to the servers. On receipt of *prepare*, update data of objects manipulated by the transaction is saved in the stable log of each server and then *yes* is sent back to the transaction. Unless succeeded in storing the update data in the log, the server sends *no* to the transaction and the subtransaction on the server aborts. Then, the transaction issues *commit* to the servers only if the transaction receives *yes* from all the servers. Otherwise, the transaction issues *abort* to every server which has sent *yes*. On receipt of *abort*, the log is removed and the subtransaction aborts.

Object servers may be replicated in order to make the system more reliable and available. Suppose servers D_{j1}, ..., D_{jm} ($m \geq 2$) are replicas of an object server D_j. A collection of the replicas $\{D_{j1}, ..., D_{jm}\}$ is referred to as *cluster* of D_j, denoted as $C(D_j)$.

3   Agents

3.1   Computational Model

An agent is a procedure which can be performed on one or more than one object server. An agent issues methods to an object server to manipulate objects in an object server where the agent exists. Every object server is assumed to support a platform to perform agents.

First, an agent A is initiated by an application or is autonomously initiated on an object server. The procedure and data of an agent A are first stored in the memory of an object server D_i in order to perform the agent A on D_i. If enough resource like memory to perform the agent A is allocated for the agent A on the server D_i, the agent A can be performed. Here, D_i is referred to as *current* server of A and the agent A is referred to as *land at* the server D_j. Objects in the server D_i are manipulated by the agent A through methods. In result, state of object may be changed and a part of the state may be derived. Data derived from the server D_i may be stored in the agent A. Thus, an instance A_i of the agent A on the object server D_i shows a subtransaction, i.e. a sequence of methods for manipulating objects in the server D_i. Then, the agent A finds another server D_j which has objects to be manipulated by A. Then, the agent

A moves to the server D_j. Here, the agent A may carry objects obtained from D_i as the data of A [Figure 1]. If enough resource like memory in the server D_j is allocated for the agent A, A lands at D_j.

A pair of agents A_1 and A_2 are referred to as *conflict* if A_1 and A_2 manipulate a same object through conflicting methods. For example, A_1 issues a method *reset* and A_2 issues *increment* to a *counter* object in a server D_j. Here, A_1 and A_2 conflict. The agent A is allowed to land at D_j if the following condition is satisfied:

[Landing conditions]
1. Enough resource to perform an agent A is allocated for the agent A in an object server D_j.
2. There is no agent on D_j which conflicts with A.

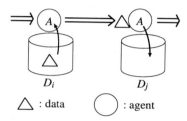

D_i D_j

$\triangle$: data $\bigcirc$: agent

Fig. 1. Agent.

3.2   Movement of Agent

Suppose an agent A is at an object server D_i and is finding an object server where the agent A can land. Suppose there are multiple possible object servers $D_{j1}, \ldots, D_{jm}$ ($m > 1$) where the agent A can land. Let $Cand_i(A)$ be a *candidate* server set, i.e. a collection of the servers $\{D_{j1}, \ldots, D_{jm}\}$ at which an agent A can land from a server D_i. For example, there are replicas $D_{j1}, \ldots, D_{jm}$ of some server D_j. $Cand_i(A)$ is a cluster $C(D_j)$ of the replicas. If an agent A only reads objects, one server, i.e. one replica D_{jk} is selected and then moves to the server D_{jk}. Here, an agent A takes another replica D_{jk} in the candidate set $Cand_i(A)$. If the agent A updates objects, all the servers in $C(D_j)$ are taken and replicas in all the servers are manipulated by A. This is similar to a famous two-phase locking (2PL) protocol [4]. On the other hand, an agent A issuing *read* takes a subset Q_r of the candidate set $Cand_i(A)$, which is a *read* quorum. The agent A issues *write* to servers in a *write* quorum Q_w. Here, $Q_r \cap Q_w \neq \phi$ and $Q_r \cup Q_w = Cand_i(A)$. This shows a quorum-based protocol [5].

In another case, the agent A is composed of multiple modules $A_1, \ldots, A_m$ ($m > 1$) which can be performed in any order and concurrently. Here, each module A_h can be performed on a server D_{jh} ($h = 1, \ldots, m$). As presented in the examples, there are two cases with respect to how many servers to be taken by an agent A at a server D_i:

1. One server in the candidate set $Cand_i(A)$ is taken.
2. Multiple servers in $Cand_i(A)$ are taken.

In the first case, we have to discuss which server in the candidate set $Cand_i(A)$ to be taken. For example, a server D_{jk} which is nearest to D_i is taken. A server which is least loaded can be also taken.

In the second case, multiple servers, possibly all the servers in the candidate set $Cand_i(A)$ are taken. In addition to discussing which servers to be taken in $Cand_i(A)$, we have to discuss how to find an optimal route to visit all the servers in the candidate set $Cand_i(A)$ [Figure 2]. For example, a route whose communication cost is the minimum is selected. This shows a serial computation. In another way, the agent A can be splitted into multiple *subagents* A_1, ..., A_m [Figure 3]. Each subagent A_k is issued to a server D_{jk} in $Cand_i(A)$. The subagents A_1, ..., A_m are concurrently performed on object servers. After manipulating objects in the servers, the subagents are merged into one agent A again. Each subagent A_k might bring data d_k obtained from an object server D_{jk}. We have to discuss where all the subagents are merged into an agent A. One idea is to take one object server D_{jk} where a subagent A_k is performed and the data d_k is the largest in all the subagents A_1, ..., A_m.

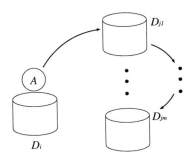

Fig. 2. Optimal routing.

3.3 Operations on Agents

As discussed here, agents are moving, splitted to multiple agents, and merged into one agent. Following operations on agents are supported by each object server:

1. $A = create\text{-}agent(\ )$: a new agent is created. An object server where an agent A is created is referred to as *home* server of A.
2. $A' = clone\text{-}agent(A)$: a clone A' of an agent A is created.
3. $split\text{-}agent(A, \{A_1, ..., A_m\})$: one agent A is splitted into multiple subagents A_1, ..., A_m $(m \geq 1)$.

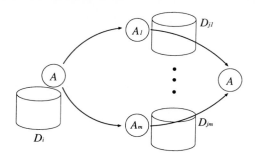

Fig. 3. Split and merge of agents.

4. *merge-agent*($\{A_1, \ldots, A_m\}$, A): multiple agents $A_1, \ldots, A_m$ are merged into an agent A.
5. *annihilate-agent*(A): an agent A is destroyed.
6. $C = cand\text{-}agent(A, D_i)$: a candidate set $C = Cand_i(A)$ is obtained.
7. $D_j = select\text{-}agent(A, D_i, C, S)$: one server D_j is selected from a candidate set C according to the strategy S. If $S = One$, an optimal server D_j is selected in C. If $S = All$, a server D_j to be visited from D_j is selected according to an optimal writing strategy. All the servers in the candidate set C are to be visited.
8. *move-agent*(A, D_i, D_j): an agent A in a server D_i is moved to another a server D_j.
9. *negotiate-agent*(A, $\{A_1, \ldots, A_m\}$): an agent A negotiates with other agents $A_1, \ldots, A_m$ to make some agreement. A decision *do, abort, block,* or *retreat* is returned.
10. *land-agent*(A, D_i): if an agent A can land at a server D_i, *true* is returned. Otherwise, *false*.
11. *conflict-agent*(A, D_i): if there is another agent which conflicts with an agent A in a server D_i, *true* is returned. Otherwise, *false*.
12. *term-agent*(A): if an agent A can finish, *yes* is returned. Otherwise, *no* is returned.
13. *commit-agent*(A): if an agent A satisfies commitment condition A *commits*. Otherwise, A *aborts*.

An agent A is performed on an object server D as follows;

1. An agent A is created at a home server D, i.e. $A = create\text{-}agent(D)$.
2. After the agent A is performed in a current server D, A moves to another server if the termination condition *term-agent*(A) is not satisfied. A candidate set C is obtained; $C = cand\text{-}agent(A, D)$.
3. If *parallel* strategy is taken, an agent A is splitted to agents $A_1, \ldots, A_n$, i.e. *split-agent*(A, $\{A_1, \ldots, A_n\}$). Then, $A_1, \ldots, A_n$ move to candidate servers in C.
4. If *serial* strategy is taken, a destination server D_i of the agent A is decided, i.e. $D_i = route\text{-}agent(A, D)$.
5. If the agent A could land at the destination server D_i, i.e. *land-agent*(A, D_i) is true, the agent A moves to a server D_i.

6. If the agent A does not conflict with agents A_1, ..., A_n in D_i, i.e. $conflict$-$agent(A, D_i)$ is $false$, A is performed on D_i. Otherwise, A negotiates with the agents A_1, ..., A_n conflicting with A, $negotiate$-$agent(A, \{A_1, \ldots, A_n\})$. If do is returned, A starts. The agent A is started, aborted, and blocks on the server D_i if do, $abort$, and $block$ are returned, respectively.

7. If the agent A is successfully performed on the server D_i, a $surrogate$ agent A_i of A is created and resides at D_i, A_i $clone$-$agent(A)$. If there is no other destination, i.e. all the object servers are manipulated, $term$-$agent(A)$ is true, the commitment procedure is started based on the consensus condition $Cons(A)$. Otherwise, go to 2.

4   Consensus among Agents

An agent A manipulates objects in multiple object servers D_1, ..., D_m ($m >$ 1). After finishing manipulating objects in all the object servers, the agent A commits if some condition C on the servers D_1, ..., D_m is satisfied. Otherwise, A aborts. For example, an atomic all-or-nothing condition is used to realize the atomicity of a transaction. That is, the agent commits only if all the object servers are successfully updated. Otherwise, the agent aborts, i.e. no update is done on the objects in any object server. The two-phase commitment (2PC) protocol [4] is used to realize the all-or-nothing principle in distributed database systems. In another example, an application would like to book one hotel. The application issues a booking request to multiple hotel objects. Here, the application can commit if at least one hotel object is obtained. Thus, if at least one of the servers is successfully manipulated, the agent A commits. There are following consensus conditions;

[**Consensus conditions**]

1. *Atomic consensus*: an agent is successfully performed on all the object servers, i.e. all-or-nothing principle. This is a condition used in the traditional two-phase commitment protocol.
2. *Majority consensus*: an agent is successfully performed on more than half of object servers.
3. *At-least-one* consensus: an agent is successfully performed on at least one object server.
4. $\binom{n}{r}$ *consensus*: an agent is successfully performed on more than r out of n servers ($r \leq n$).

The atomic, majority, and at-least-one consensus conditions are shown in forms of $\binom{n}{n}$, $\binom{n}{\lceil (n+1)/2 \rceil}$, and $\binom{n}{1}$ consensus ones, respectively. More general consensus conditions are discussed in a paper [8]. Each agent A is assumed to have a consensus condition $Cons(A)$ given by an application.

Suppose an agent A finishes manipulating objects in object servers D_1, ..., D_m. Let A_i be a *surrogate* agent of the agent A in an object server D_i [Figure 4]. Suppose another agent B might come to D_j after the agent A leaves an object server D_j. Here, the agent B negotiates with the surrogate agent A_i of the agent A if B conflicts with A. After the negotiation, the agent B might take over the surrogate A_i. Thus, when the agent A finishes visiting all the object servers, some surrogate may not exist. The agent A starts the negotiation procedure with its surrogates $A_1, \ldots, A_m$. If a consensus condition C on $A_1, \ldots, A_m$ is satisfied, the agent A commits. For example, an agent commits if all the surrogates safely exist in the atomic consensus condition. As discussed in a following section, surrogates do negotiation with agents. Then, the surrogate abort. If the surrogates exit, the computation performed by the agent is successful. Then, the surrogate agents of A are annihilated. Here, other agents conflicting with the agent A are allowed to manipulate objects.

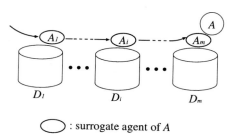

: surrogate agent of A

Fig. 4. Surrogate agents.

5   Negotiation Strategies

5.1   Protocol

Unless the landing conditions are satisfied, the agent A can not land at the server D_j. Here, the agent A can take one of the following ways:

1. The agent A waits in the current object server D_i.
2. The agent A finds another object server D_k which has objects to be possibly manipulated before D_j by A.
3. The agent A negotiates with other agents in D_j which hold resources.
4. The agent A *aborts*.

Suppose an agent lands at a current object server D_i. Here, there might be other agents $B_1, \ldots, B_k$ which are being performed on the object server D_i. Each agent B_i is an agent or surrogate agent of an agent. If the agent A conflicts with some agent B_j on an object o, A negotiates with B_j with respect to which agent A or B_j holds the object o. There are following negotiation strategies:

1. The agent A blocks until the agent B_j commits.
2. The agent A takes over B_j, i.e. B_j releases the objects and blocks. Then A starts.
3. B_j aborts and A starts.

The first way is similar to the locking protocol. An agent A blocks if some agent B holds an object o in a conflicting way with the agent A. If B waits for release of an object held by A, A and B are deadlocked. Thus, deadlock among agents may occur. When an agent A blocks in a object server D_i, a timer is started. If the timer expires, the agent A takes one of the following ways:

1. The agent A retreats to an object server D_j which A has passed over.
2. Every surrogate A_j of A initiates a deadlock detection agent $LD_j(A)$.

In the second way, an agent A takes over an agent B_j in an object server D_j if A conflicts with B_j and B_j holds an object. Here, A starts to do the negotiation with an agent B_j on D_j by using a following negotiation protocol :

[**Negotiation protocol**]

1. An agent A sends a *can-I-use* message $CIU(o, op)$ to an agent B_j on an object server D_j. This means that an agent A would like to manipulate an object o with a method op in an object server D_j.
2. On receipt of $CIU(o, op)$ from an agent A, an agent B_j sends OK to A if B_j can release the object o or B_j does not mind if A manipulates the object o. Here, there are two approaches to B_j's releasing the object o :

   a. B_j aborts if A precedes B_j.
   b. B_j rolls back to a checkpoint and then restarts if A precedes B_j. Otherwise, B_j sends No to A.
3. On receipt of OK from B_j, A starts manipulating the object o.
4. On receipt of No from B_j, there are following ways:

   a. A blocks until A receives OK/NO from B_j.
   b. A aborts.
   c. A triggers the second level negotiation protocol. □

If the agent B_j agrees with the agent A in the negotiation protocol, A can manipulate objects by taking over B_j. In the second way, the agent B_j not only releases the object but also aborts.

5.2 Resolution of Confliction

There are two types of agents;

1. Ordered agents.
2. Unordered agents.

Every pair of ordered agents manipulate objects in a well-defined way. Agents are ordered. Each agent A is assigned a *precedent* identifier $pid(A)$. An agent A_1 *precedes* another agent A_2 $(A_1 \rightarrow A_2)$ iff $pid(A_1) < pid(A_2)$. For example, a *timestamp* [4] can be used as an identifier of an agent. That is, the identifier $pid(A)$ of an agent A is time $ts(A)$ when A is initiated at the home server. An agent A_1 precedes another agent A_2 only if $ts(A_1) < ts(A_2)$. If the timestamp with identifier of home server is used as a precedent identifier of an agent, either A_1 precedes A_2 or A_2 precedes A_1 for every pair of different agents A_1 and A_2. That is, the agents are totally ordered. If a logical clock like vector clock [6] is used as precedent identifier, the agents are partially ordered. An agent A_1 is concurrent with another agent A_2 iff neither A_1 precedes A_2 nor A_2 precedes A_1.

Suppose multiple agents A_1, ..., $A_m(m{>}1)$ would like to manipulate an object o in an object server D_i and conflict with each other. The agents A_1,

..., A_m are ordered by using the precedent identifiers of the agents. Suppose $pid(A_1) < \ldots < pid(A_m)$. An agent A_s manipulates an object o before another agent A_t if $pid(A_s) < pid(A_t)$. If A_s and A_t are concurrent, A_s and A_t are allowed to be performed in any order. However, if A_s and A_t conflict on a pair of servers D_i and D_j. A_s and A_t are required to be performed in a same order at D_i and D_j. There never occurs deadlock.

Unordered agents are not ordered. Like locking protocols, an unordered agent had obtained an object if no conflicting agent obtains the object. Suppose an agent A_1 passes over an object server D_1 and is moving to another server D_2, and another agent A_2 passes over D_2 and is moving to D_1. If A_1 and A_2 conflict on each of D_1 and D_2, neither A_1 can land at D_2 nor A_2 can land at D_1. Here, deadlock occurs.

Here, an agent B_j means an "agent" or a surrogate agent in the object server D_j. An agent A would like to land at an object server D_j and conflicts with an agent B_j in D_j. First, suppose B_j is a surrogate of an agent B. The surrogate agent B_j makes a following decision depending on the commitment conditions;

1. B_j takes the at-least-one consensus principle; If B_j knows at least one surrogate exists, B_j releases the object and aborts. B_j informs the other surrogates of this abort.
2. B_j takes the majority consensus principles: If B_j knows more than half of the surrogates exist, B_j releases the object and aborts. B_j informs the other surrogates of this abort.
3. B_j takes $\binom{n}{r}$ consensus: If B_j knows more than r of the surrogate agents exist, B_j releases, the object and aborts.

As discussed here, a surrogate may be aborted in the negotiation with other agents and due to the failure of the server. There are two states of each surrogate B_j, *abortable* and *commitable*. If B_j is in *abortable* state, B_j can be aborted. For example, if another agent A conflicting with B_j takes over B_j by the negotiation between A and B_j, B_j aborts. The agent B of the surrogate B_j eventually tries to commit. B informs all the surrogates of the commitments by sending *Prepare* messages. On receipt of the *prepare* message, B_i enters *commitable* state, possibly saving update data in a log. Here, B_j does not abort in the negotiation.

6   Concluding Remarks

This paper discussed an agent model for transactions which manipulate multiple object servers. An agent first moves to an object server and then manipulates objects. The agent autonomously moves around the object servers to perform the computation. If the agent conflicts with other agents, the agent negotiates with the other agents. The negotiation is done based on the commitment conditions and types of agents, i.e. ordered and an ordered.

References

1. Aglets Software Development Kit Home, http://www.trl.ibm.com/aglets/

2. American National Standards Institute, "Database Language SQL," Document ANSI X3.135, 1986,
3. Arai, K., Tanaka, K., and Takizawa, M., "Group Protocol for Quorum-Based Replication," *Proc. of IEEE ICPADS'00*, 2000, pp.57-64.
4. Bernstein, P.A., Hadzilacos, V., and Goodman, N., "Concurrency Control and Recovery in Database Systems," *Addison Wesley*, 1987.
5. Garcia-Molina, H. and Barbara, D., "How to Assign Votes in a Distributed System," *Journal of ACM*, Vol.32, No.4, 1985, pp.841-860.
6. Mattern, F., "Virtual Time and Global States of Distributed Systems," in Parallel and Distributed Algorithms (Cosnard, M. and Quinton, P. eds.), North-Holland, Amsterdam, 1989, pp.215-226.
7. Omicini, A., Zambonelli, F., Klusch, M. and Tolksdorf, R., "Coordination of Internet Agents," *Springer-Verlag*, 2001.
8. Shimojo, I., Tachikawa, T., and Takizawa, M., "M-ary Commitment Protocol with Partially Ordered Domain," *Proc. of the 8th Int'l Conf. on Database and Expert Systems Applications (DEXA'97)*, 1997, pp.397-408.
9. Skeen, D., "Nonblocking Commitment Protocols," ACM SIGMOD, 1982, pp.133-147.
10. Tanaka, K. and Takizawa, M., "Quorum-based Locking Protocol in Nested Invocations of Methods," Proc. of *DEXA'2001*, 2001, pp.857-866.

A Data Model and Algebra for Document-Centric XML Document[1]

Sung Wan Kim [2], Pan Seop Shin, Youn Hee Kim, Jaeho Lee [3], and Hae Chull Lim

Dept. of Computer Engineering, Hong Ik University,
Sangsu-Dong 72-1, Mapo-Gu, Seoul, KOREA
{swkim, psshin, kyh, lim}@cs.hongik.ac.kr
[3] Dept. of Computer Education, Inchon National University of Education,
Gyesan-Dong San 59-12, Gyeyang-Gu, Inchon, KOREA
jhlee@mail.inue.ac.kr

Abstract. As XML is considered as a standard to represent and exchange data on the web, many researches are on going to develop the techniques on storing and searching XML documents. But the data models in these researches mainly focus data-centric XML and cannot support document-centric XML fully. A data model for document-centric XML document is proposed in this paper. Using the proposed model, it is possible to represent the semantic and structural information of XML document including extended links and some XML components which are not covered in previous XML data model fully. And some basic algebra for the proposed model is also proposed as the foundation for query language and query processing.

1 Introduction

As XML is considered as a new standard format to represent and exchange the contents and structural information of data on the web, there are many researches to develop applications based on XML. Along with these trends, many researches have studied to develop new techniques on storing, accessing, and searching XML documents [1, 2]. However, most of previous researches didn't consider the concept or the features of XML document fully. Especially, in previous researches, developing the techniques on modeling or mapping XML documents into traditional databases, such as relational or object-oriented database, are mainly focused [4]. These mapping techniques are insufficient to support the features of XML document fully since the models used in the traditional database mismatch with XML document features.

To store and search more efficiently, we need to classify XML documents according to their features, develop the corresponding techniques for modeling, storing and searching them [3]. To be compared with data-centric XML, document-centric XML is usually designed for human consumption. Hence, round-tripping is

---

[1] Research supported by Institute of Information Technology Assessment, Korea under contract IITA 2001-122-3.

[2] Working at Dept. of Computer Information, Sahmyook College, Korea also.

I. Chong (Ed.): ICOIN 2002, LNCS 2344, pp. 714–723, 2002.
© Springer-Verlag Berlin Heidelberg 2002

significant. They are characterized by less regular or irregular structure, larger grained data, and lots of mixed content. The order in which sibling elements and PCDATA occur is almost always significant. We believe, for document-centric XML, any information whether it is trivial or not should be represented in data model. The previous proposed data models, such as expanded OEM[7], XOM[8] as well as mapping to traditional database, are insufficient to support these characteristics thoroughly. Especially, extended links using XLink[12] can not supported sufficiently in previous models. In this paper, we propose a new data model which captures the features of document-centric XML fully. And we propose some basic algebra including four operations for the proposed model as the foundation for query language and query processing.

The paper is organized as follows. A related work is presented in Section 2 and the proposed data model for document-centric XML document is presented in Section 3. An modeling example is presented in Section 4 and algebra for the proposed data model is presented in Section 5. Finally, we present the conclusion in Section 6.

2   Related Work

Although they are not differ from a syntactical perspective, XML documents can be classified into data-centric XML and document-centric XML according to their goal and concept[3]. Figure 1 shows sample data-centric XML and document-centric XML respectively.

| <Flights>
   <Airline>Korean Air</Airline>
   <Origin>Seoul</Origin>
   <Destination>Cheju
   </Destination>
   <Flight>
      <Departure>09:15</Departure>
      <Arrival>10:16</Arrival>
   </Flight>
   <Flight>
      <Departure>11:15</Departure>
      <Arrival>12:16</Arrival>
   </Flight>
   <Flight>
      <Departure>13:15</Departure>
      <Arrival>14:16</Arrival>
   </Flight>
</Flights> | <Product type="software">
  <Name>&HX</Name>
  <President ID = "m1">Lim HC</President>
  <Developer advisor="m1">DB lab, HIU</Developer>
  <Description>
    <Para> The &HX is a highly <I>scalable software solution</I> that enables the creation, maintenance, and distribution of customized XML databse. It systematizes data conversion and provides an intuitive, browser-based interface for all management tasks.</Para>
    <Para>You can:</Para>
    <List>
      <Item xlink:href="item/order.html">
      Order free evaluation CD</Item>
    </List>
    <Para>The &HX costs just $100 and, if you order now, comes with some useful package as a bonus gifts</Para>
  </Description>
</Product > |

Fig. 1. Data-centric XML vs. Document-centric XML

Primitive data unit of document-centric XML is document itself but that of data-centric XML is data within XML document. In comparison with data-centric XML, document-centric XML is usually designed for human consumption. So, It is characterized by less regular or irregular structure, larger grained data, and lots of mixed content. And round-tripping is significant. Namely reconstructing the original document after modeling should result in the same document. Hence, the order in which sibling elements and PCDATA occur is significant. The differences between them are summarized in Table 1.

Table 1. Document-centric XML vs. Data-centric XML

Document-centric XML	Data-centric XML
irregular and roughly structure content	well-orderd and more finely structured content
large amount of mixed content	little or no mixed contents
significance in the sequence of elements	insignificance in the sequence of elements
human consumption	machine consumption

Most of previous researches on storing and searching XML mainly focused on data-centric XML and mapping XML into traditional DBMS. These methods model the data in XML document rather than the document itself including all semantics and information of the document. Hence, there are lost of information when modeling document-centric XML. And reconstructing the original document after modeling often results in a different document as compared with the original one. So, the well-known round-tripping problem occurs. As a result, these previously proposed data models are not fitted for document-centric XML.

On the other hand, some of working drafts, such as DOM, Infoset, XPath, defined by W3C can be used the basis of a data model. DOM(Document Object Model)[9] is not exactly data model since it provides programming interfaces to manipulate XML documents and doesn't define any operations. And XML Infoset[10] specification defines an abstract data set called the XML Information Set (Infoset). An XML document's information set consists of eleven information items which are abstract descriptions of some part of an XML document. We also think it is not perfect data model since its purpose is to provide a consistent set of information for XML documents and there is no operations to access the items. In addition, explicit reference is not supported. XPath 2.0 data model [11] based on the XML Infoset was defined with some extension to support its query language such as XSLT, XQuery, and XPath. It has no operations to access XML data items. From the view point of database, we think that these are not perfect or formal data model, however, these are good basis for defining an XML data model.

On the other hand, OEM which is well-known semi-structured data model is expanded to support XML[5,6,7]. In the expanded OEM, it is possible to model ordering and tag information and to support both tree- and graph-type modeling. However, the main considering is for elements. There is insufficiency to model document-centric XML fully since there are no support on attributes, comments, PIs, and entities and no explicit support to represent the relationship for external XML using XLink[12] and XPointer.

In XOM(eXtensible Object Model)[8], a rank concept is proposed to support some reference relationships for external XML explicitly. And it supports attributes. But there is no support on comments, PIs, entities and no considering on the extended link types for external reference using XLink[12]. Some information such as attribute type are not preserved in XOM. Even though XOM is adequate to data-centric XML for its simplicity, there is insufficiency to model document-centric XML fully also.

3 Our Data Model

The proposed data model is for the document-centric XML documents and designed to represent the semantics and structural features of XML document itself fully. Data-centric XML can be represented in our model with some modifications also. The model can be represented as a node-labeled, directed, and ordered graph. We represent XML documents as directed graph G = (N, L, V). N means nodes which are the set of main components within XML document. L means links which are the set of directed links that represent various relationships among nodes. There is a virtual root node, V, which means entire document and includes some useful information. The ordering of subnodes is left-to-right. Node is identified by unique value within a document and has its name as a label value which specifies the semantic of the node itself. Nodes are classified into five types. Links are classified into three types.

3.1 Nodes

We classified nodes into five types such as element node, attribute node, data node, entity node, and reference node. Each node has a NID(Node IDentifier) value which is identified uniquely within a document and a name which describes the semantic of node itself, namely, label. Label uses element tag name, attribute name, or entity name as its value. We think it is more appropriate that node has label than link has label. Because of labeled nodes, it is easy to define query to nodes. For easy understanding, we provide grammar-like expressions. In this grammar, ':=' means definition, '=' means description, '|' means OR, and '(', ')' pair means grouping.

N := element_node | attribute_node | data_node | entity_node | reference_node

Element node is to represent an element which has another element, attribute, and text defined by PCDATA or CDATA section as its child. Each node has a label as the tag information of element and a list which contains links to all child nodes. [ ] means an ordered list and component within the list can be appeared one and more. L, which means links, is described in next section.

element_node := (nid, e_name, [L])
e_name = element tag's name
nid = unique identifier

Data node represented as a leaf node is to represent various data within XML document. It has a data value which represents text of PCDATA or CDATA section, attribute value, and content of PI or comment. And it also has a type such as PCDATA, CDATA, ATTRIBUTE, PI, and COMMENT. And it includes data types such as integer, real, string, etc.

data_node := (nid, d_name, d_type, value)
d_name := PCDATA | CDATA | PI | COMMENT | ENTITY | ATTRIBUTE
d_type := string | integer | float | double | boolean | URI

Attribute node is to represent attribute which describes the property for an element. It has a label as the attribute name, a list which contains links to all child nodes, and a attribute type such as CDATA, ID, IDREF, IDREFS, NMTOKEN, and etc. The attribute is classified into two types normally. First is a simple attribute which has an atomic value only. Second is a complex attribute such as IDREF(S) type which has special meaning value(s), not simple data. In addition, some attributes such as NMTOKENS and IDREFS contain one more data values. In XOM, the simple attribute and element content were represented as atomic nodes. And IDREF(S)-type attribute was represented as complex node. In former case, there are lost information necessary to rebuild the same original document. Hence, we create attribute node independently to support two attribute types in a single unit and to represent full semantics in our model.

attribute_node := (nid, a_name, a_type, [L])
a_name = attribute's name
a_type := CDATA | ID | IDREF | IDREFS | ENTITY | ENTITIES | NMTOKEN | NMTOKENS

Entity node is to represent the information of entity within XML document. It includes entity name, type such as internal or external, and information for real data value to be substituted. In general, since entity is resolved before mapping XML document into a data model, there is no need to consider entity. However, in document-centric environment, information on entity usage should be maintained to rebuild the original document with the same semantics. So, we add this entity node to the child of virtual root node.

entity_node := (nid, e_name, e_type, e_value)
e_name = entity's name
e_type := SYSTEM | PUBLIC | INTERNAL
e_value = entity value

Reference node is to represent the whole or part of external XML document referenced when external reference is used (using XLink or XPointer). We define reference node independently to represent reference relationship since it is difficult to represent the real whole or part of external XML document referenced within the model directly.

reference_node := (nid, target, [L])
target = URI

The ordering among nodes, especially the ordering among sibling elements within mixed element, is significant for document-centric XML. The ordering of sibling elements within a parent element is represented as left-to-right in graph. It is possible to represent the ordering of attributes within an element like this manner.

3.2 Links

Link which represents various relationship between nodes includes link type and information of target node. We classify link into three types such as nested type link, reference type link, and linkage type link.

L := (link_type, nid)
link_type := nested | reference | linkage

Nested type link is to represent the relationship between an element and its child element, between an element and its content in a data node, and between an element and its attributes. *Reference type link* is to represent the relationship between an element and another element referenced by the former element within a document, we call it internal reference, and between an element and its real remote resource via a locator such as href attribute when extended XLink[12] is used, we call it external reference. Internal reference relationship means the case that an element with IDREF or IDREFS attribute refers another element within the same document which has the corresponding ID attribute value. External reference relationship means the case that an element with a locator attribute, such as href, refers its real remote(or external) resource whether it is XML document or not when extended XLink is used.

Finally, *linkage type link* is to represent the relationship for explicit link. According to the XLink recommendation, using XLink we can represent the relationship between resources or portions of resources. For example, XML document is a kind of resources. First, we can represent the *outbound* link which means a link from local to remote resource. This link can be represented with a simple XLink. Second, we can represent the *inbound* or *third-party* link which means a link from remote to local resource or between remote resources respectively. These can be represented with a extended XLink. These links of two types are captured with *linkage type links* in our data model directly using reference nodes.

4 Examples

Once an XML document is mapped into the proposed data model, it is possible to represent the XML document as a directed, node-labeled, ordered graph which has a virtual root node.

In the right sample document in <Figure 1>, various components of XML document including element, attribute, entity, internal reference (using IDREF), and simple external link(using XLink) are contained. <Figure 2> shows the graph representation of the sample XML document using the proposed data model. For simplicity, node's NID is omitted in this figure.

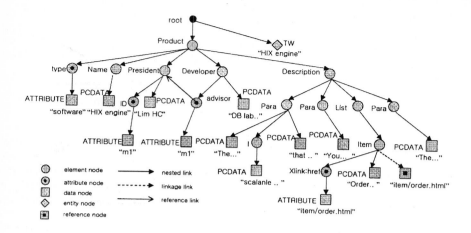

Fig. 2. A graph representation of a sample XML document

Assume, the type of "advisor" attribute in "Developer" element is IDREF. The "advisor" attribute whose value is 'm1' refers the "President" element with ID type value 'm1', which means internal reference relationship as defined in section 3.2. This internal reference is represented with a reference type link from the referring attribute to the referred element in our model. The "Item" element with locator attribute, href, means simple extended link to remote XML document (outbound type). We represent the information on the remote destination document via a reference node. This simple extended link is represented with a linkage type link from the element to reference node in our model. The ordering among nodes within a parent node is left-to-right. Especially, in case of an element like first "<Para>" element in <Figure 2> which has PCDATA text and other elements as child nodes, the ordering among child nodes is significant and should be maintained after modeling. There is no semantic and structural information loss in our data model as shown in <Figure 2> as compared with the expanded OEM or XOM . Using the expanded OEM, we cannot represent PI, entity, and extended link. Using XOM, we cannot represent PI, entity. In addition, it is impossible to know the original type for an attribute value. It is trivial problem in data-centric domain but not in document-centric domain.

```
<courseload>
    <tooltip>Course Load for swkim</tooltip>
    <person  xlink:href="stu/swkim62.xml"   xlink:label="stu62"   xlink:title="swkim" />
    <course  xlink:href="courses/cs101.xml"   xlink:label="CS-1"   xlink:title="ComSci 1" />
    <gpa xlink:label= "swkimGPA">4.2</gpa>
    <go   xlink:from="CS-1" xlink:to="stu62" />
    <goes xlink:from="stu62"  xlink:to="swkimGPA">
</courseload>
```

Fig. 3. Another XML document with extended XLink

<Figure 3> show an example which contains both external reference and extended link using XLink. As defined in section 3.2, external reference is represented via a

locator attribute, namely href, in an element. For example, in "person" element, href attribute means that the element refers a remote (external) XML reachable through URI value. The remote XML document is represented with a independent reference node. This external reference is represented with a reference type link from the element to the reference node in our model.

On the other hand, the "go" element represents an extended link between two remote XML documents using "from" and "to" attribute (third-party type). In this case, "course" element is linked to "person" element via "go" element. Note the extended linking is not related with the linked element itself but related with the remote resource externally. This third-party extended link is represented with a linkage type link between two reference nodes in our model. The "goes" element represent an extended link between a remote XML document and a local element (inbound type). This inbound extended link is represented also with a linkage type link from a reference node to a local element node. Note that XLink can be resolved with DTD or Schema.

<Figure 4> show the modeling result for the document in <Figure 3>. "title" attributes and name space (such as xlink:) for attributes are omitted. In this graph, inbound extended link from the remote reference node to the local element node "gpa" is represented directly. And third-party extended link from one remote reference node to other remote reference node is represented directly using two reference nodes also. Expanded OEM and XOM can not support these extended links.

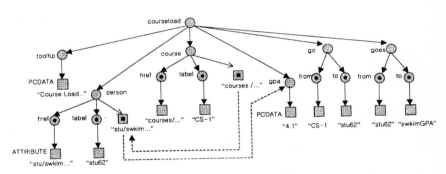

Fig. 4. A graph representation for a document in Fig. 3

5 Basic Algebra

In this section, we propose four basic operations for the proposed model as the foundation for query language, query processing, and query optimization. The proposed operations are to apply to document-centric XML document, so the ordering information among nodes is maintained within the result after applying the operations. The basic operations are Union to integrate two nodes, Select to return the node which satisfies given conditions, Project to return a node accessible through a specific path, and Join for more than one documents.

Although it is possible to define operations using various node types, we propose operations using element node only in this paper. Assume, two nodes as N1 = (nid1, name1, link1) and N2 = (nid2, name2, link2) respectively and two node lists as NList1 = [n1, n2, n3, ...] and NList2 = [n'1, n'2, n'3, ...] respectively.

5.1 Union

Union operation is to integrate two nodes into a new node. Two node are changed into child nodes with a new parent node commonly. The ordering of operands used for this operation is preserved.

N1 $\cup$ N2 = (nid, "Union", link)
link = [(E_E, nid1), (E_E, nid2)]

5.2 Select

Select operation is to choose nodes which satisfy a condition among many nodes. This operation returns a result node that has some child nodes. The child nodes satisfy a condition in the given node list. The usable conditions are various comparison operations(=, $\langle$, $\rangle$, $\leq$, $\geq$), boolean operations, and etc. Function f is to represent conditional expression and has true or false as it result value.

σ[f(NList1)] = (nid, "Result", link)
link = [(E_E, nid), (E_E, nid'), ...]

5.3 Project

Project operation is to return a result node. The result node has nodes that can be accessed through a specific path from the given node as child. It returns all accessible nodes which follow a path distinguished by the link type and the name of a target node.

Π[link_type, name, N1] = (nid, "Result", link)
link = [(E_E, nid), (E_E, nid'), ...]

5.4 Join

More than one document associable with a given condition are candidates for the join operation. After associating all nodes within each document with those within others, this operation checks the associated nodes whether to satisfy a given condition or not. And finally the associated nodes which satisfy with a given condition is related each other.

NList1 $\infty_{condition}$ NList2

6 Conclusion

In this paper, we propose a data model and algebra which are appropriate for document-centric XML. The proposed model consists of node and link. The model can be represented as a node-labeled, directed, and ordered graph which has a virtual root node.

We classify main components of XML document in detail and define them with various node types. And we define link types to represent the various relationships among nodes as well as the nested relationship for elements. Especially, extended links via XLink are captured in the proposed data model. Using the proposed model, we reduce the loss of semantic and structural information of XML document. It is possible to restore the original XML after modeling.

Some operations are proposed for the model as the foundation for query language, query processing, and query optimization. The basic operations are Union to integrate two nodes, Select to return the node which satisfies a given condition, Project to return a node accessible through a specific path, and Join for more than one documents.

Currently, we are designing and developing a native XML repository system based on main memory. The proposed model has used as a logical data model for this system.

References

1. Jennifer Widom, "Data Management for XML", IEEE Data Engineering Bulletin, Special issue on XML, Vol.22, No. 3, (1999) 44–52
2. Stefano Ceri, Piero fraternali and Stefano Paraboschi, "XML: Current Developments and Future Challenges for the Database Community", Proc. of the 7 th Int. Conf. on Extending Database Technology (EDBT), (2000) 3–17
3. Ronald Bourret, "XML and Databases"
4. JeongEun Kim, PanSeop Shin, Jaeho Lee, HaeChull Lim, "A Database Approach for Modeling and querying XML Documents", ITC-CSCC 2000 Proceedings Vol. 2 (2000) 703–706
5. Yannis Papakonstantinou, Hector Garcia-Molina and Jennifer Widom, "Object Exchange Across Heterogeneous Information Sources", Proc. of the 11th IEEE International Conference on Data Engineering, (1995)
6. J. McHugh, S. Abiteboul, R. Goldman, D. Quass, and J. Widom, "Lore: A Database Management System for Semistructured Data", ACM SIGMOD Record, Vol. 26, No. 3 (1997) 54–66
7. Roy Goldman, Jason McHugh, and Jennifer Widom, "From Semistructured Data to XML: Migrating the Lore Data Model and Query Language", Proc. of the 2nd International Workshop on the Web and Databases (WebDB '99), pp. 25–30, Philadelphia, Pennsylvania (1999)
8. Dell Zhang, Yisheng Dong, "A Data Model and Algebra for the Web", Proc. of the 10th International Workshop on Database & Expert Systems Applications (1999) 711–714
9. W3C, "Document Object Model (DOM)", http://www.w3.org
10. W3C, "XML Information Set", http://www.w3c.org/
11. W3C, "XQuery 1.0 and XPath 2.0 Data Model", http://www.w3.org
12. W3C, "XML Linking Language(XLink) Version 1.0", http://www.w3.org

Group Protocol for Delivering Requests to Replicas

Keijirou Arai, Katsuya Tanaka, and Makoto Takizawa

Dept. of Computers and Systems Engineering
Tokyo Denki University
{arai, katsu, taki}@takilab.k.dendai.ac.jp

Abstract. Distributed applications are realized by cooperation of multiple processes which manipulate data objects like databases. Objects in the systems are replicated to make the systems fault-tolerant. We discuss a system where read and write request messages are issued to replicas in a quorum-based scheme. In this paper, a quorum-based (QB) ordered (QBO) relation among request messages is defined to make the replicas consistent. We discuss a group protocol which supports a group of replicas with the QBO delivery of request messages.

1 Introduction

Distributed systems are realized in a 3-tier client server model. Users in clients initiate transactions in application servers. Transactions manipulate objects by issuing requests to data servers. Data and application servers are distributed in computers. Computers which have servers exchange request and response messages on behalf of the servers. Some computer may have both application and data servers. Thus, a collection of computers are exchanging request and response messages. Objects in data severs are replicated in order to increase performance and reliability. In this paper, we consider a system which includes replicas of simple objects like files, which supports basic *read* and *write* operations.

A transaction sends a *read* request to one replica and sends *write* to all the replicas in order to make the replicas mutually consistent in a two-phase locking protocol [3]. The two-phase locking based on read-one-write-all principle is efficient only for read dominating applications. Another way is the quorum-based scheme [3], where each of *read* and *write* requests is sent to a subset of replicas named *quorum*. The more frequently a request is issued, the smaller a quorum is.

In the group communications[4, 9, 10], a message m_1 *causally precedes* another message m_2 if the sending event of m_1 *happens before* m_2[8]. If m_1 causally precedes m_2, m_1 is required to be delivered before m_2 in every common destination of m_1 and m_2. In addition, *write* requests issued by different transactions are required to be delivered to replicas in a same order. Thus, the *totally* ordered delivery of *write* messages is also required to be supported in a group of replicas. Raynal *et al.* [1] discuss a group protocol for replicas where write requests delayed can be omitted based on the write-write semantics. The authors [5] present

I. Chong (Ed.): ICOIN 2002, LNCS 2344, pp. 724–735, 2002.
© Springer-Verlag Berlin Heidelberg 2002

a transaction-based causally ordered protocol where only messages exchanged among conflicting transactions are ordered where objects are not replicated.

Some message m transmitted in the network may be unexpectedly delayed and lost in the network. Even if messages causally/totally preceded by such a message m are received, the messages cannot be delivered until m is received. Suppose a write request w_1 and then a read request r are issued to a replica. If there exists some write request w_2 between w_1 and r, which is not destined to the replica, it is meaningless to perform r and w_1 since an obsolete data written by w_1 is read by r. Thus, it is critical to discuss what messages to be delivered to replicas in what order. Requests to be delivered are referred to as *significant*. If only significant requests are delivered in each replica, less number of requests are required to be delivered and requests stay in a queue for shorter time. We discuss a group protocol named QG (quorum-based) one where only significant message are delivered. We evaluate the QG protocol in local area network and wide area network like the Internet with respect to how many request messages can be omitted and how long each message waits in a queue.

In section 2, we present a system model. In section 3, we define a quorum-based precedent relation of messages and we discuss what messages to be ordered. In section 4, we present the QG protocol. In section 5, we discuss the evaluation of the QG protocol.

2   System Model

Computers p_1, ..., p_n are interconnected in a less reliable network where messages may be lost. Applications are realized in a 3-tier client server model. Replicas of data objects are stored in data servers and transactions in application servers issue *read* and *write* requests to data servers to [Figure 1]. Let o_t denote a replica of an object o in a computer p_t. Let $R(o)$ be a *cluster*, i.e. a set of replicas of the object o.

A pair of operations op_1 and op_2 on an object *conflict* iff op_1 or op_2 is *write*. Otherwise, op_1 and op_2 are *compatible*. On receipt of a request op from a transaction T_i, op is performed on the replica o_t in the data server of p_t if any operation conflicting with op is being neither performed nor waited. Otherwise, op is waited in the queue. This is realized by the locking protocol. Let op_i^t denote an instance of an operation op issued by T_i to a replica o_t in p_t, where op is either r(read) or w(write). After manipulating replicas, T_i issues either a *commit*(c) or *abort*(a) request message to the replicas. On receipt of c or a, every lock held by T_i is released.

A computer supports data and application servers. A computer may send requests issued by a transaction while receiving requests to the server from other computers. Thus, each computer exchanges requests with other computers. In this paper, we discuss in what order request messages received are delivered to replicas in each computer.

A transaction T_i sends *read* to N_r replicas in a read quorum Q_r and *write* to N_w replicas in a write quorum Q_w of an object o. N_r and N_w are *quorum numbers*.

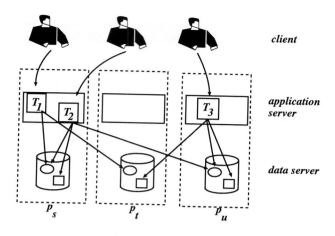

Fig. 1. System model.

$Q_r \cup Q_w = R(o)$, $N_r + N_w > q$, and $N_w + N_w > q$. Each replica o_t has a version number v_t. T_i obtains a maximum version number v_t in Q_w. v_t is incremented by one. Then, the version number of every replica in Q_w is replaced with the maximum value v_t. T_i reads the replica whose version number is maximum in Q_r.

3   Precedent Relation of Requests

3.1   Quorum-Based Precedency

A request message m from a transaction T_i is enqueued into a receipt queue RQ_t in a computer p_t. Here, let $m.op$ show an operation type op, i.e. r or w. Let $m.o$ be an object o to be manipulated by op, $m.dst$ be a set of destination computers, and $m.src$ show the source computer. A top request m in RQ_t is dequeued and then an operation $m.op$ is performed on a replica o_t of an object o ($= m.o$) in p_t.

Each computer p_u maintains a vector clock $V = \langle v_1, \ldots, v_n \rangle$ [9]. For every pair of vector clocks $A = \langle a_1, \ldots, a_n \rangle$ and $B = \langle b_1, \ldots, b_n \rangle$, $A \geq B$ if $a_t \geq b_t$ for $t = 1, \ldots, n$. If neither $A \geq B$ nor $A \leq B$, A and B are *uncomparable* ($A \parallel B$). A vector V is initially $\langle 0, \ldots, 0 \rangle$ in every computer. Each time a transaction is initiated in a computer p_u, $v_u := v_u + 1$ in p_u. When T_i is initiated, $V(T_i) := V$. A message m sent by T_i carries the vector $m.V = \langle v_1, \ldots, v_n \rangle$ ($= V(T_i)$). On receipt of m from p_u, V is manipulated in a computer p_t as $v_s := \max(v_s, m.v_s)$ for $s = 1, \ldots, n$ ($s \neq t$).

A transaction T_i initiated in p_u is given a unique identifier $tid(T_i)$. $tid(T_i)$ is a pair of the vector clock $V(T_i)$ and a computer number $no(T_i)$ of p_u. For a pair of transactions T_i and T_j, $id(T_i) < id(T_j)$ if $V(T_i) < V(T_j)$. If $V(T_i) \parallel V(T_j)$, $tid(T_i) < tid(T_j)$ if $no(T_i) < no(T_j)$. Hence, for every pair of transactions T_i and T_j, either $tid(T_i) < tid(T_j)$ or $tid(T_i) > tid(T_j)$.

Each request message m has a sequence number $m.sq$. sq is incremented by one in a computer p_t each time p_t sends a message. For each message m sent by a transaction T, $m.tid$ shows $tid(T)$.

[**Quorum-based ordering (QBO) rule**] A request m_1 *quorum-based precedes* $(Q - precedes)$ m_2 $(m_1 \prec m_2)$ if $m_1.op$ conflicts with $m_2.op$ and

1. $tid(m_1) < tid(m_2)$, or
2. $m_1.sq < m_2.sq$ and $tid(m_1) = tid(m_2)$. $\square$

$m_1 \parallel m_2$ if neither $(m_1 \prec m_2)$ nor $(m_1 \succ m_2)$. A pair of messages m_1 and m_2 received by a computer p_t are ordered $(m_1 \rightarrow_t m_2)$ in RQ_t:

- If $m_1 \prec m_2$, m_1 precedes m_2 $(m_1 \rightarrow_t m_2)$.
- Otherwise, $m_1 \rightarrow_t m_2$ if $m_1 \parallel m_2$ and m_1 is received before m_2.

"$m_1 \rightarrow_t m_2$" shows "m_1 *locally precedes* m_2 in p_t". m_1 *globally precedes* m_2 $(m_1 \rightarrow m_2)$ iff $m_1 \rightarrow_t m_2$ or $m_1 \rightarrow_t m_3 \rightarrow m_2$ in some computer p_t.

3.2   Significant Messages

Due to unexpected delay and congestions in the network, some destination computer may not receive a message m. Messages causally/totally preceding m cannot be delivered without receiving m.

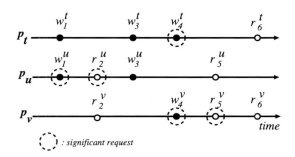

Fig. 2. Receipt sequences.

Figure 2 shows receipt queues of three computers p_t, p_u, and p_v, each of which has a replica of an object o. $N_r = N_w = 2$. For example, p_t receives write requests w_1^t, w_3^t, and w_4^t, and then a read request r_6^t, i.e. $w_1^t \rightarrow_t w_3^t \rightarrow_t w_4^t \rightarrow_t r_6^t$. $w_1^t \rightarrow r_2^u$ since $w_1^u \rightarrow_u r_2^u$. Neither $r_5^v \rightarrow_v r_6^v$ nor $r_6^v \rightarrow_v r_5^v$ since r_5^v and r_6^v are compatible.

If a read request r is performed on a replica o_t, data of o_t written by some write request w is derived by r. Here, it is significant to discuss by what write request data read by a read request is written. A read r_j^t *reads* data *written by* a write w_i^t in p_t $(w_i^t \Rightarrow_t r_j^t)$ iff $w_i^t \rightarrow_t r_j^t$ and there is no write w^t such that $w_i^t \rightarrow_t w^t \rightarrow_t r_j^t$.

A write request w_i^t is *current* for a read request r_j^t in a receipt queue RQ_t iff $w_i^t \Rightarrow_t r_j^t$ and there is no write w such that $w_i^t \rightarrow w \rightarrow r_j^t$. Here, r_j^t is also *current*. A request which is not current is *obsolete*. In addition, if a write w_2 is performed on a replica o_t after w_1 is performed, o_t is overwritten by w_2 and the data written by w_1 disappear.

- A write request w_j^t *absorbs* another write request w_i^t if $w_i^t \rightarrow_t w_j^t$ and there is no read r such that $w_j^t \rightarrow_t r \rightarrow_t w_i^t$.
- A current read request r_i^t *absorbs* another read request r_j^t iff $r_i^t \rightarrow_t r_j^t$ and there is no write w such that $r_i^t \rightarrow w \rightarrow r_j^t$.

[**Definition**] A request m is *significant* in RQ_t iff m is neither obsolete nor absorbed. □

In Figure 2, r_6^v is current but is absorbed by r_5^v. r_5^v and r_6^v are merged into one read request r_{56}^v which returns the response to the transactions T_5 and T_6. Thus, w_1^t, w_3^t, and r_6^t are insignificant in p_t. r_5^u is insignificant in p_u and r_2^v is also insignificant in p_v.

A sub sequence of significant requests obtained by removing *insignificant* requests is referred to as *significant* sequence.

4   Group Protocol

4.1   Transmission and Receipt

We present a QG (quorum-based group) protocol for a group of replicas o_1, ..., o_n of an object o in computers p_1, ..., p_n ($n \geq 1$), respectively. A quorum Q_{op} is constructed by randomly selecting N_{op} replicas in the cluster $R(o)$ each time a request op is issued. A request message m sent by a transaction T_i in p_t includes the following attributes:

$m.SSQ =$ subsequence numbers $\langle ssq_1, ..., ssq_n \rangle$.
$m.ACK =$ receipt confirmation $\langle ack_1, ..., ack_n \rangle$.
$m.V =$ vector clock, i.e. $V(T_i) = \langle v_1, ..., v_n \rangle$.
$m.C =$ write counters $\langle c_1, ..., c_n \rangle$.

There are variables $SSQ = \langle ssq_1, ..., ssq_n \rangle$, $RSQ = \langle rsq_1, ..., rsq_n \rangle$, and $RQ = \langle rq_1, ..., rq_n \rangle$ in p_t. Each time p_t sends a message m to p_u, sq and a *subsequence number* ssq_u are incremented by one. The message m carries sq and ssq_v ($v = 1, ..., n$).

The variables rq_u and rsq_s show a sequence number (sq) and a subsequence number (ssq_s) of a message which p_t expects to receive from p_u ($s = 1, ..., n$), respectively. Suppose p_t receives a message m from p_s. If $m.ssq_t = m.rsq_s$, p_t has received every message which p_s had sent to p_t before m, i.e. no message gap. Then, $rsq_s := rsq_s + 1$. $rq_s := \max(rq_s, m.sq)$. If $m.ssq_t > rsq_s$, p_t finds p_t has not received some gap message m' from p_s where $m.rsq_s \leq m'.ssq_t < m.ssq_t$. The selective retransmission is adopted.

When p_s sends a message m to p_t, $m.ack_v := rq_v$ ($v = 1, ..., n$). p_t knows p_s has accepted every message m' from p_u where $m'.sq < m.ack_u$. On receipt of m, $ACK_{su} := m.ack_{su}$ for $u=1, ..., n$. A message m from p_s is *locally ready* in a receipt queue RQ_t iff $m.ssq_t = 1$ or every message m_1 from p_s in RQ_t such that $m_1.ssq_t < m.ssq_t$ is locally ready. A message m received from p_s is locally ready in p_t if $m.ssq_t = rsq_s$. If m is locally ready in RQ_t, p_t receives every message which p_s has sent to p_t before m. m_1 *directly precedes* m_2 for p_s in RQ_t ($m_1 \rightarrow_{ts} m_2$) iff $m_1.ssq_t = m_2.ssq_t - 1$.

[Definition] Let m be a message which a computer p_t receives from p_s.

- m is *partially ready* in RQ_t iff
  1. m is locally ready or
  2. $m.op = read$ and there is a partially ready message m_1 in RQ_t such that
     - $m_1 \rightarrow_{ts} m$, and
     - $m_2.op = r$ for every message m_2 where $m_1.ssq_t < m_2.ssq_t < m.ssq_t$.

- m is *ready* in RQ_t iff
  1. m is locally ready and there is some locally ready message $m_1(\prec m)$ from every p_u ($\neq p_t$) in RQ_t, or
  2. m is partially ready, and for every p_u ($\neq p_s$), if there is no locally ready message $m_1(\succ m)$ from p_u in RQ_t, there is a partially ready message $m_2(\succ m)$ from p_u.□

Suppose p_t receives m_1 from p_s and has received no message from p_s after receiving m_1. Suppose p_t receives m_2 from another computer p_u. If $m_1.sq < m_2.ack_s$, p_t knows p_s has sent some message m_3 such that $m_1.sq < m_3.sq \leq m_2.ack_s$. However, p_t cannot know whether or not m_3 is destined to p_t.

[Definition] A message m from p_s is *uncertain* in RQ_t iff p_t does not receive m, p_t knows that some p_u ($\neq p_s$) has received m, i.e. p_t receives such a message m_1 that $m.sq < m_1.ack_s$ from p_u, and p_t does not know of $p_t \in m.dst$. □

4.2   Delivery of Requests

Suppose a computer p_t receives a message m. Let m_u denote a message sent by p_u where $m_u \prec m$ and there is no message m'_u from every computer p_u such that $m_u \prec m'_u \prec m$. Let $\max(m_1, \ldots, m_n)$ be a *maximum message* m_v such that $m_s \prec m_v$ for every m_s. Here, m_v *directly Q-precedes* m in p_t.

If m is ready in RQ_t, p_t has surely received a partially ready message m'_u from every p_u such that $m_u \prec m \prec m'_u$. The messages $m_1, \ldots, m_n$ are also partially ready. p_t can deliver m after $m_1, \ldots, m_n$. Let m'_u be a partially ready message which p_u sends to p_t such that $m_u \prec m \prec m'_u$ and there is no message m''_u from p_u such that $m \prec m''_u \prec m'_u$. If m'_u is locally ready, every message m''_u which p_u sends to p_t after sending m_u before m'_u is not destined to p_t. If m'_u is partially ready but not locally ready, m_u is uncertain. Suppose there are undestined or uncertain messages $u_1, \ldots, u_k$ such that $m_v \prec u_1 \prec \ldots \prec u_k \prec m$ as shown in Figure 3. p_t receives a message m_v ($= \max(m_1, \ldots, m_n)$) and then receives m but does not receive $u_1, \ldots, u_k$. If m is locally ready, $u_1, \ldots, u_k$ are undestined. If m is partially ready, some message u_i is uncertain. Table 1 summaries how m and m_v are insignificant.

In order to detect insignificant requests in RQ_t, p_t manipulates a vector of *write counters* $C = \langle c_1, \ldots, c_n \rangle$, where each element c_u is initially zero. Suppose

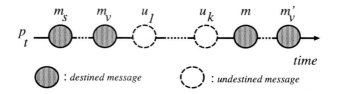

Fig. 3. Receipt sequence of messages.

Table 1. Insignificant messages.

m_v	m	$u_1, \ldots, u_k$	Insignificancy
read	read	every u_l is read.	m is insignificant (absorbed by m_v). m is merged to m_v.
		some u_l is write.	m is insignificant (obsolete).
write	read	every u_l is read.	m and m_v are significant.
		some u_l is write.	m and m_v are insignificant(obsolete).
read	write		if depends on request following m of m_v is significant.
write	write		m_v is insignificant(obsolete).

p_t sends a message m. If m is a *write* request, $c_u := c_u + 1$ for every destination p_u of m. $m.C := C$. Each message m carries write counters $m.C = \langle m.c_1, \ldots, m.c_n \rangle$. On receipt of a write request m from a computer p_s, $c_u := \max(c_u, m.c_u)(u = 1, \ldots, n)$.

[**Theorem**]Let m_1 and m_2 be messages received in a RQ_t where m_1 precedes m_2. There exists such a *write* request m_3 that $m_1 \prec m_3 \prec m_2$ if $m_1.C < m_2.C$ and $m_1.V < m_2.V$. □

A message m can be decided to be partially ready according to the following rule:

- A message m from a computer p_s is partially ready in RQ_t if

    1. $m.ssq_t = rsq_s$, i.e. m is locally ready, or
    2. $m.op = r$ and $m_1.c_t = m_2.c_t$ for a pair of requests m_1 and m_2 such that $m_1 \rightarrow_{ts} m \rightarrow_{ts} m_2$.

5   Evaluation

The QG protocol is evaluated by measuring the number of requests performed in each computer and waiting time of each message in a receipt queue through the simulation. We make the following assumptions on the simulation:

[Assumptions]

1. Each computer p_t has one replica o_t of an object o ($t = 1, \ldots, n$).
2. Transactions are initiated in each computer p_t. Each transaction issues one request, read or write request. A computer p_t sends one request issued every τ time units. τ is a random variable.
3. It takes π time units to perform one request in each computer.
4. N_r and N_w are quorum numbers for read and write, respectively. $N_r + N_w \geq n + 1$ and $n + 1 \leq 2N_w < n + 2$.
5. Each computer p_t randomly decides which replica to be included in a quorum for each request op given the quorum number N_{op}.
6. It takes δ time units to transmit a message from one computer to another. δ is summation of minimal delay time $min\delta$ and random variable ϵ.
7. It is randomly decided which type $read$ or $write$ each request is. P_r and P_w are probabilities that a request is read and write, respectively, where $P_r + P_w = 1$. □

In the QG protocol, only the significant request messages are performed on each replica. If there is at most one request in a receipt queue, all requests which arrive at the computer are performed. Thus, the more number of messages are included in the receipt queue, the more number of messages are not performed since more number of messages can be considered to be insignificant. First, we consider a group of five replicas ($n = 5$) where $N_r = N_w = 3$. We measure the ratio of significant messages (SR) to the total number of messages issued and the average waiting time (W) of each message in a receipt queue. Here, we assure $P_r = 0.8$ and $P_r = 0.2$.

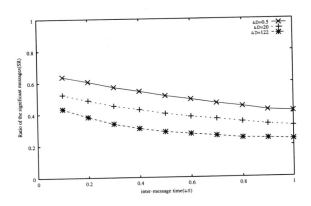

Fig. 4. Ratio of significant requests (SR).

Figures 4 and 5 show the ratio of significant messages (SR) and the average waiting time (W) for inter-transaction time τ for $n = 5$. Here, δ shows the delay time. $\delta = 0.5$[msec] means a local area network. $\delta = 20$ shows a nation-wide network, i.e. Japan, and $\delta = 120$ indicates world-wide network. For example, it take about 0.5[msec] to deliver a message from one computer to another in a local area network. It takes about 120[msec] to transmit a message from Japan to the US. In a wide area network, more number of messages are in a transmission. Hence, the larger τ is, the more number of messages arrive at each replica.

In Figure 4 the ratio of significant messages (SR) in the receipt queue is 0.6 for $\tau = 0.2$[msec], This means about 50% of the messages arriving at a computer are considered to be significant in a local area network ($\delta = 0.5$). If each computer sends a message every 0.2[msec] ($\tau = 0.2$), $SR = 0.5$ for $\delta = 0.5$ and $\tau = 0.8$. Only 50% of request messages transmitted in the network are insignificant, i.e. can be omitted in the receipt queue for $\tau = 0.6$ and $\delta = 0.5$, i.e. local area network. In the wide area network ($\delta = 122$), about 70% of request messages can be omitted in the receipt queue for $\tau = 0.6$. Thus, the more number of messages are included in the receipt queue, the more number of messages are not performed.

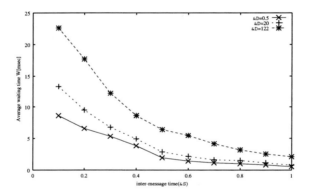

Fig. 5. Average waiting time (W).

Figure, 5 shows the average waiting time (W) of the QG protocol for τ. n/τ shows number of messages per msec which a process receives. Here, $n = 5$, the shorter τ is, the more number of messages a process receives.

Figure, 6 shows a ratio of the average waiting time of the QG protocol to the traditional group protocol.

Figure 7 shows how many requests are performed on each computer by the QG protocol where $n = 5$, $N_r = N_w = 3$, $P_r = 0.8$, $\tau = 10$ for $\pi = 0, 0.5, 1$[msec]. The vertical axis shows what percentage of requests are significant. Here, about 50% of the messages are significant. That is, half of the messages received are removed from the receipt queue. For $\pi = 1$, about 30% of the messages are significant. $\pi = 0$ shows no message stays in a receipt queue. Every request is performed. In the QG protocol, only the significant messages are delivered. This shows that fewer number of requests are performed, i.e. less computation and communication overheads in the QG protocol than the message-based protocol.

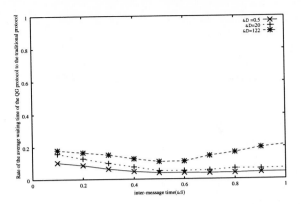

Fig. 6. Ratio of average waiting time.

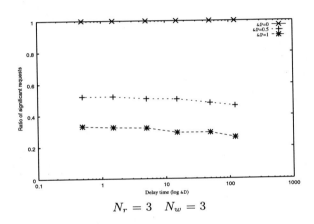

$$N_r = 3 \quad N_w = 3$$

Fig. 7. Ratio of significant requests.

Figure 8 shows average waiting time W[msec] of message in the receipt queue for number n of replicas. Here, $P_r = 0.8$, $\tau = 10$[msec], and $\pi = 0.5$[msec]. Here, $N_w = N_w = \lceil (n + 1) / 2 \rceil$. Three cases for $\delta = 0.5$, $\delta = 20$, and $\delta = 120$ of average delay time are shown. Figure 8 shows the average waiting time of each message is $O(n)$ for the number n of computers. If messages are kept in the queue according to the traditional protocols, the average waiting time is $O(n^2)$. Thus, the average waiting time can be reduced by the QG protocol. Figure 9 shows a ratio of significant messages for P_r. Here, $\pi = 0.5$[msec], $n = 5$, and $N_r = N_w = 3$. In cases $P_r = 0$ and $P_r = 1$, every request in a receipt queue is *read* and *write*, respectively. In case $P_r = 0$, a last *write* request absorbs every *write* in the queue. In case $P_r = 1$, a top *read* request absorbs every request in the queue. Here, the smallest number of requests are performed. In case "$P_r = 0.5$", the number of insignificant requests removed is the minimum.

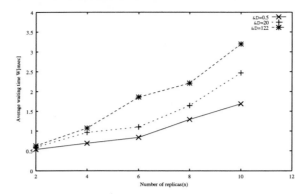

Fig. 8. Average waiting time of message.

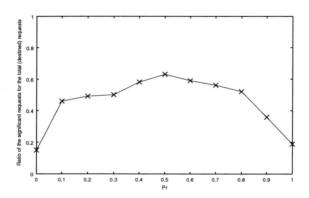

Fig. 9. Ratio of read requests(P_r).

6   Concluding Remarks

This paper discussed a group protocol for a group of computers which have replicas. The replicas are manipulated by read and write requests issued by transactions in the quorum-based scheme. We defined the quorum-based ordered delivery of messages. We defined significant messages to be ordered for a replica. We presented the QG (quorum-based group) protocol where each replica decides whether or not requests received are significant and which supports the quorum-based ordered delivery of messages. The QG protocol delivers request messages without waiting for insignificant messages. We showed how many messages to be performed and how long average waiting time of message in a receipt queue can be reduced in the QG protocol compared with the traditional group protocol.

References

1. Ahamad, M., Raynal, M., and Thia-Kime, G., "An Adaptive Protocol for Implementing Causally Consistent DistributedServices, " *Proc. of IEEE ICDCS-18*, 1998, pp.86–93.
2. Arai, K., Tanaka, K., and Takizawa, M. "Group Protocol for Quorum-Based Replication" *Proc. of IEEE ICPADS'00*, 2000, pp.57–64.
3. Bernstein, P. A., Hadzilacos, V., and Goodman, N., "Concurrency Control and Recovery in Database Systems," *Addison-Wesley*, 1987.
4. Birman, K., Schiper, A., and Stephenson, P., "Lightweight Causal and Atomic Group Multicast," *ACM Trans. Computer Systems*, Vol.9, No.3, 1991, pp.272-314.
5. Enokido, T., Tachikawa, T., and Takizawa, M., "Transaction-Based Causally Ordered Protocol for Distributed Replicated Objects," *Proc. of IEEE ICPADS'97*, 1997, pp.210–215.
6. Enokido, T., Higaki, H., and Takizawa, M., "Group Protocol for Distributed Replicated Objects," *Proc. of ICPP'98*, 1998, pp.570–577.
7. Garcia-Molina, H. and Barbara, D. "How to Assign Votes in a Distributed System," *Journal of ACM,* Vol.32, No.4, 1985, pp. 841-860.
8. Lamport, L., "Time, Clocks, and the Ordering of Events in a Distributed System," *Comm. ACM*, Vol.21, No.7, 1978, pp.558-565.
9. Mattern, F., "Virtual Time and Global States of Distributed Systems," *Parallel and Distributed Algorithms*, 1989, pp.215-226.
10. Nakamura, A. and Takizawa, M., "Causally Ordering Broadcast Protocol," *Proc. of IEEE ICDCS-14*, 1994, pp.48–55.
11. Tachikawa, T. and Takizawa, M., "Significantly Ordered Delivery of Messages in Group Communication," *Computer Communications Journal*, Vol. 20, No.9, 1997, pp. 724–731.

k-Depth Look-Ahead Task Scheduling in Network of Heterogeneous Processors

Namyoon Woo and Heon Y. Yeom

School of Computer Science and Engineering,
Seoul National University,
Seoul 151-742, KOREA
{nywoo, yeom}@dcslab.snu.ac.kr

Abstract. The objective of the task scheduling is to achieve the minimum execution time of all the tasks with their precedence requirements satisfied. Although several task scheduling heuristics for the heterogeneous environment have been presented in the literature, they overlook the various type of network and do not perform efficiently on such an environment. We present the new scheduling heuristic which is named the 'k-Depth Look-ahead.' The proposed heuristic takes the network heterogeneity into consideration and our experimental study shows that the proposed heuristic generates a better schedule especially under the condition that the network resource costs high.

1 Introduction

The objective of the task scheduling is to achieve the minimum execution time of all the tasks satisfying the task precedence requirements. One of the approaches on this research area is the task-processor mapping with weighted directed acyclic graphs (DAGs) which represent applications, and it has been known as NP-Complete [2,3,6]. Although many scheduling heuristics have been proposed for a couple of decades, most of them are only devised for a homogeneous system. Recently a heterogeneous computing system is arousing the academic and commercial interests because of the good scalability and low building cost. For the high performance of such a system an efficient task scheduling is critical but the problem is more difficult than in case of a homogeneous system. In this study, we are interested in the static task scheduling for network of heterogeneous processors. Static scheduling is done at compile-time with a directed acyclic graph (DAG) and the information of processors and a network architecture [2,3].

Several heuristics for the heterogeneous environment have been proposed. The *Dynamic-Level Scheduling* (DLS) algorithm is a greedy algorithm in that it selects a match of a task and a processor which gives the 'earliest start time' [7]. All of the ready tasks are the candidates for mapping on every iterations, so it has the relatively high time complexity. The DLS uses the earliest start time to evaluate the processor which is not recommended for a heterogeneous computing system [1]. The *Heterogeneous Earliest-Finish-Time* (HEFT) algorithm lists the

I. Chong (Ed.): ICOIN 2002, LNCS 2344, pp. 736–745, 2002.
© Springer-Verlag Berlin Heidelberg 2002

tasks according to the predefined priorities and assigns each task on the processor which allows the 'earliest finish time' [8]. Although the HEFT is designed for the heterogeneous computing system, it assumes the homogeneous network links. The *Bubble Scheduling and Allocation* (BSA) algorithm was proposed in [4]. It repeats the task-assignment and re-assignment while navigating the processors in the way of the breadth first search. The advantage of the BSA is that it does not require any routing information. However, the incremental processor-navigation may narrow a view in a processor-selection, which results in the local optimization. Usually the most heuristics assume that the LAN is used for the system and the network environment is homogeneous, but in the real world the size of the heterogeneous system becomes larger and the various network links might be used. According to our experiments, the existing heuristics do not perform efficiently under the heterogeneous network environment.

We assume that processors communicate with one another in different rates. In other words some processors may be connected by fast network links and others by slow ones. For such an environment, we propose the heuristic which is named as the "*k*-Depth Look-ahead" (*k*-DLA), which calculates the expected execution time of k successive tasks. The proposed heuristic shows remarkable performance results especially on the high-degree heterogeneous system or low-connectivity network architectures like a ring.

The remainder of this paper is organized as follows. In the next section we formally state the background of the scheduling problem and explain some definitions and parameters used. Section 3 presents the proposed scheduling algorithm for the heterogeneous environment. Section 4 contains the performance results of our algorithm compared against those of the related works. In section 5 we state the conclusions.

2 Preliminaries

Throughout the paper the following notations and parameters are used to describe the scheduling algorithms.

- A parallel application is represented by a directed acyclic graph, $\mathbf{G} = (\mathbf{T}, \mathbf{E})$ where a node set $\mathbf{T} = \{T_1, T_2, \cdots, T_n\}$ consists of n tasks and an edge set $\mathbf{E} = \{(i, j)|T_i, T_j \in \mathbf{T}\}$ contains the task-dependency constraints. We will use the term 'node' and 'task' in a same meaning.
- The weight of a node, $w(i)$ and that of an edge, $c(i, j)$ stand for the required execution cost of T_i and the communication cost of (i, j) respectively. $c(i)$ is the average of $c(i, j)$ for a fixed value i.
- The processor set, $\mathbf{P} = \{P_1, P_2, \cdots, P_p\}$ (where $|\mathbf{P}| = p$) and the network link set, $\mathbf{N} = \{(x, y)|P_x, P_y \in \mathbf{P}\}$ are also given.
- The *heterogeneity factors* h_{ix} and h'_{xy} are used to model the heterogeneous condition of systems. The actual execution cost of a task T_i on a processor P_x is $w'(i, x) = h_{ix} \cdot w(i)$. Similarly the actual cost of an edge (i, j) is $c'(i, j, x, y) = h'_{xy} \cdot c(i, j)$, if T_i executes on P_x and T_j on P_y. If P_x is same

with P_y , h'_{xy} becomes zero. In case that there is no direct link between P_x and P_y, h'_{xy} takes maximum h' among the routing links between P_x and P_y.

- The average weight of T_i is

$$\overline{w(i)} = \frac{1}{p} \cdot \sum_{x=1}^{p} w'(i,x)$$

and the average communication heterogeneity factor of P_x,

$$\overline{h'_x} = \left(\frac{1}{p} \cdot \sum_{y=1,y \neq x}^{p} \frac{1}{h'_{xy}} \right)^{-1}$$

- The *b-level* of a node is the length of the longest path from the node to the exit node and recursively defined by

$$bl(T_i) = \overline{w(i)} + \max_{T_j \in SUCC(T_i)} \{c(i,j) + bl(T_j)\}$$

, where $SUCC(T_i)$ is a set of the immediate children of T_i.
- The *critical path* (CP) is the longest path from the entry node to the exit node in a DAG. So, $bl(T_{entry})$ is the critical path length of a DAG.
- The *communication to computation ration* (CCR) is defined as the average communication cost divided by the average execution cost. For example, if the value of the CCR is 10.0, the average communication cost is ten times the average execution cost [3,8].
- The start time and the finish time of a message from T_i to T_j between P_x and P_y are denoted by $mst(i,j,x,y)$ and $mft(i,j,x,y)$ respectively. We have

$$mft(i,j,x,y) = mst(i,j,x,y) + c'(i,j,x,y)$$

- The earliest execution start time and finish time of a task T_i on a processor P_x are given by

$$est(i,x) = \max\{f(x), \max_{T_k \in PRED(T_i)} (mft(k,i,P_{T_k},x))\}$$

$$eft(i,x) = est(i,x) + w'(i,x)$$

where $PRED(T_i)$ is a set of the immediate predecessors of T_i and $f(x)$ is the available time of P_x. Scheduling heuristics purpose to minimize the $\max\{eft(i,x)\}$.

3   The Proposed Algorithm

The list scheduling heuristic mainly consists of three procedures [3,8,9]. First (1) the priority of each task is determined and tasks are listed in the order of their own priorities. After that, (2) it finds the proper processor which gives the best quality of schedule for the each task in the list. Then (3) the task is scheduled on the selected processor. In this section, we present a new metric named *Extended b-level (ebl)* and the '*k-Depth Look-ahead*' (*k*-DLA) heuristic. The proposed heuristic uses the *ebl* value to qualify the processor for a given task in step (2).

3.1   Extended B-Level

The Extended b-level (*ebl*) of a task T_i on a processor P_x within k-depth is defined as

$$ebl(i,x,k) = \begin{cases} w'(i,x) + \overline{h'_x} \cdot \overline{c(i)} \cdot \lceil \frac{|SUCC(T_i)|}{|NB(P_x)|} \rceil \\ \quad + \max_{T_j \in SUCC(T_i)} \{ebl(j,x,k-1)\} \;, \text{ if } k > 1 \text{ or } SUCC(T_i) \neq \emptyset \\ \\ w'(i,x) + \overline{h'_x} \cdot \overline{c(i)} \cdot \lceil \frac{|SUCC(T_i)|}{|NB(P_x)|} \rceil \quad\quad \text{, otherwise} \end{cases}$$

where $NB(P_x)$ is the set of neighbor processors of P_x. The *ebl* is the largest expected execution time of k successive tasks including the communication costs. It is not determined yet which processors the successors would be assigned to, so $\overline{h'_x}$, the average heterogeneity factor of network cost for P_x is used.

Figure 1 shows the notation of the *ebl* with $k = 1$ in a graphic form. T_i has five out-going messages and P_x has two links and P_y has four. The exact communication cost of each message is not known since the successor tasks are not mapped yet. We use the expected communication cost which can be obtained from the average message cost and the average heterogeneity factor of network. The number of links also may affect the messages' delivery as shown in Figure 1(b).

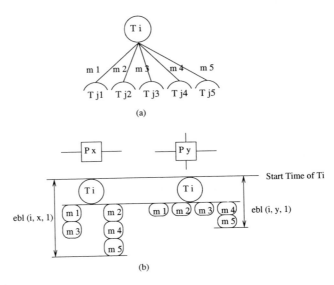

(a)

(b)

Fig. 1. (a) A task, T_i with five successors (b) The *ebl*s of T_i when assigned on P_x and P_y each

3.2   The Proposed Heuristic

The k-DLA heuristic uses the sum of the start time of T_i and the ebl to qualify the processors. The sum can be regarded as the expected start time of the k-th successor, so we name the sum the *Successor's Expected Start Time (sest)*. Algorithm 1 describes the k-DLA heuristic.

Algorithm 1 k-DLA Scheduling Heuristic

Compute the *b-level* of each task.
Sort and list the tasks in the decreasing order of *b-levels*.
while the list is not empty **do**
   Select the first task T_i and remove it from the list.
   $\forall P_x$, calculate $sest(i, x, k) = est(i, x) + ebl(i, x, k)$.
   Select P_x which gives the minimum value of the *sest*.
   Schedule T_i on P_x.
end while

If the network resource is expensive or the network contention occurs frequently, it is desirable to reduce the communication cost. Under such a condition it is apt to assign tasks to the processors which are close to each other and we can control the locality of the task assignment by adjusting the value of k. The time complexity of the k-DLA with a fixed value of k is $O(n^2 p)$ as same as that of the HEFT where n and p are the number of tasks and processors respectively [8].

4   Performance Results

4.1   Model

The randomly generated DAGs and the regular task graphs representing the various types of the parallel applications are used as input data. The number of tasks, n in random generated graphs varies from 50 to 900, and that of edges, e from $2n$ to $5n$. For each number of tasks, ten instances were generated. We used three CCRs (0.1, 1.0 and 10.0) and four heterogeneity factors H (5, 10, 20 and 40). For instance, $H = 20$ means that heterogeneity factors , h_{ix} and h'_{xy}, have a uniform distribution with a range $[1, 20]$.

Three different network topology were adopted : a ring, a mesh and a fully connected network topology with sixteen processors. The HEFT and the BSA have been tested for the comparison with three versions of k-DLA : $k = 1, 5$ and ∞.

Algorithms will be compared by the following metrics.

Normalized Schedule Length (NSL). The NSL is the schedule length divided by the summation of computation costs on the CP_{min}. The CP_{min} is the critical path of a DAG on the processor which gives the shortest critical path length. The NSL shows how close to the optimum the scheduling result is.

Running Time. The running time is the cost of the scheduling algorithm itself. As previously referred, the problem of mapping tasks on processors is NP-Complete. So the trade-off between the quality of results (NSL) and the scheduling cost should be considered.

Used Processors. Although it is not the main purpose of the scheduling heuristic to exploit the processor as less as possible, we investigate the total number of the used processors to find out the tendency or locality of task-processor mapping.

4.2   Random Graphs

Figure 2 is the NSL respect to the sizes of DAGs on three different network topology and we can find out the followings.

- As the size of problem increases, it becomes harder to find the optimal solution.
- NSLs on a ring network are larger than those on the other network topology for the high degree of the network-sharing. In other words, the messages' arrival times are delayed by the frequent network contention.

According to Figure 2(a) and (c), each k-DLA has the opposite results. In a ring, the ∞-DLA outperforms than the 1-DLA. The k-DLA heuristic assumes that the k successive tasks would be executed on the same processor. Successor nodes usually have the earliest start and finish time if they run on the processor where their predecessors run since the ring network is quite busy. So the k-DLA works better with a large k. On the other hand, in a clique each pair of processors communicates through a dedicated link, so the arrival time is not affected by the network contention. Tasks are willing to scatter all over the processors which allow their earliest finish time. In a clique the k-DLA performs worse with a large k since it ignores the tendency of task's scattering. The NSLs of the HEFT and the 1-DLA hardly differ in a clique, since the average network costs of the all processors are nearly same with each other and the decision of processor-selections in the DLA and the HEFT are similar.

Shown in Figure 3, the NSLs are affected by the heterogeneity also. As the variety of computing and network system increases, it becomes more difficult to achieve a good result. However, the k-DLA shows the best result in a ring or a mesh network. The BSA usually generates the local optimum results for it's incremental navigating among the processors. As the heterogeneity increases, the task-reassignment might not propagate. That's why the BSA performs the worst.

The CCR is related with the weight of network resources. According to Figure 4, the difference among the NSLs of the heuristics grows as the CCR value increases. The reason is that the k-DLA takes the network resources into consideration by evaluating the expected message finish time and tends to map the k successive tasks to one processor only to use the less network resources.

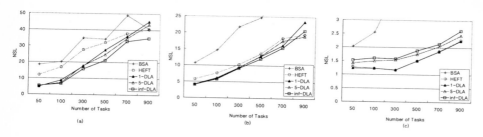

Fig. 2. Normalized schedule lengths with respect to the size of DAGs in (a) a ring (b) a mesh (c) a clique ($CCR = 1.0, H = 20$)

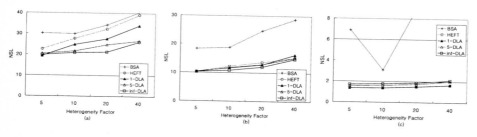

Fig. 3. Normalized schedule lengths with respect to the heterogeneity factor in (a) a ring (b) a mesh (c) a clique ($CCR = 1.0, |T| = 500$)

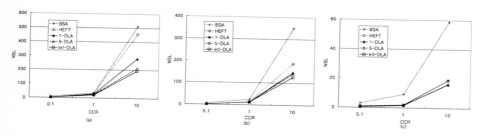

Fig. 4. Normalized schedule lengths with respect to the CCR in (a) a ring (b) a mesh (c) a clique ($H = 20, |T| = 500$)

Figure 5 represents the time required in scheduling. The BSA heuristic repeats re-assigning tasks navigating all over the processors in the breadth first search and the re-assignment of a task results in another re-assignment of successor tasks, so it takes much time. In a clique, the BSA's scheduling time gets extremely shorter because the reassigning of a task occurs much less frequently than in the other topology of network. Usually the k-DLAs with a small k and the HEFT take nearly same time in scheduling except in a ring. Contrast to our expectation, the 5-DLA takes less time than the 1-DLA and 1-DLA does than the HEFT. The anormal difference among them can be explained by the message scheduling time over the links. The 5-DLA has the small number of links to schedule the messages on than the 1-DLA or the HEFT because the 5-DLA exploits the less number of processors, which is proved in Figure 6. So the 5-DLA performs slightly better in a ring than the 1-DLA or the HEFT.

As shown in Figure 6 the k-DLA uses the less processors as k increases. In a ring and a mesh, the ∞-DLA achieves the best result in spite of using the small number of processors, which means that the tasks tend to gather close strongly to make the earliest finish time. Conversely, the HEFT exploits all the processors and suffers from the frequent network contentions. In a clique there is a dedicated network link between each pair of processors, so all the processor can be exploited without suffering from the network contention. The BSA also exploits the small number of processors. It is not caused by the network-consideration but the local navigation among the processors, so it performs the worst.

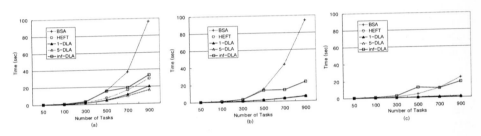

Fig. 5. Average running time of the algorithms in (a) a ring (b) a mesh (c) a clique ($CCR = 1.0, H = 20$)

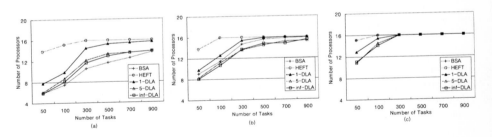

Fig. 6. The number of processors actually used in (a) a ring (b) a mesh (c) a clique ($CCR = 1.0, H = 20$)

4.3   Applications

We have experimented with the DAGs modeling parallel applications used in [5]. Target applications are *LU decomposition (32×64)*, *Laplace equation solver* and a *stencil algorithm*. The number of tasks in DAGs is 2048, 2144 and 1474, and

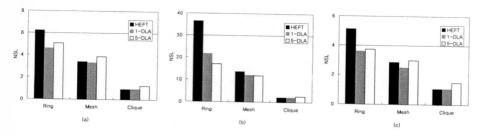

Fig. 7. Normalized schedule lengths of the DAGs modeling (a) LU decomposition (b) Stencil algorithm (c) Laplace transform ($CCR = 1.0, H = 20$)

that of edges is 4000, 4159 and 5672 respectively. The problem sizes are much bigger than those of random graphs in subsection 4.2. For each application, fifteen instances were tested. We compare only the 1-DLA, the 5-DLA and the HEFT since the ∞-DLA and the BSA are out of our concern for the high-time complexity and the poor performance respectively.

Figure 7 shows the result of three applications on each network topology with $H = 20$. A *stencil* has more edges than the other applications so it's NSL is much larger than those of the others. Outstanding issue is that the 1-DLA and 5-DLA have little difference in performance or 1-DLA performs better even in a ring. The sibling tasks of DAGs used in this subsection have the almost same weights, in other words the shapes of DAGs are not skewed. No matter how large the k is, it is not useful to find the shortest critical path within k-hop since the critical path length and the other path length are almost same. So, the 1-DLA and the 5-DLA tend to generate similar results.

5   Conclusions

We proposed the k-DLA heuristic which considers the heterogeneity of network environment. The proposed scheme outperforms particularly in a low-connectivity network as well as in the environment with the high degree of the heterogeneity or the high CCR value. The k-DLA generates the results in a comparable running time. Actually the modeling the heterogeneous network in the scheduling research area is not studied fully, and most of heuristics neglect the consideration of the network resources. Future work will include the more detail modeling and simulation of the network environment.

References

1. Almeida, V.A.F., Vasconcelos, I.M.M., Arabe, J.N.C., Menasce, D.A.: Using Random Task Graphs to Investigate the Potential Benefits of Heterogeneity in Parallel Systems. In Proceedings Supercomputing '92, pages 683–691, 1992. IEEE
2. El-Rewini, H., Lewis, T., Ali, H.: Task Scheduling in Parallel and Distributed Systems. Prentice Hall, Englewood Cliffs, New Jersey, 1994

3. Kwok, Y., Ahmad, I.: Static Scheduling Algorithms for Allocating Directed Task Graphs to Multiprocessors. ACM Computing Surveys, 31(4):406–471, 1999
4. Kwok, Y., Ahmad, I.: Link Contention-Constrained Scheduling and Mapping of Tasks and Messages to a Network of Heterogeneous Processors. Cluster Computing, 3(2):113–124, 2000
5. Radulescu, A., van Gemund, A.J.C.: Fast and Effective Task Scheduling in Heterogeneous Systems. In the 9th Heterogeneous Computing Workshop (HCW), pages 229–238, May 2000.
6. Shirazi, B., Wang, M., Pathak, G.: Analysis and Evaluation of Heuristic Methods for Static Scheduling. J. Parallel and Distributed Computing,(10):222–232, 1990
7. Sih, G.C., Lee, E.A.: A Compile-Time Scheduling Heuristic for Interconnection-Constrained Heterogeneous Processor Architectures. IEEE Transactions on Parallel and Distributed Systems, 4(2):75–87, Feb 1993
8. Topcuoglu, H., Hariri, S., Wu, M.: Task Scheduling Algorithms for Heterogeneous Processors. In the 8th IEEE Workshop on Heterogeneous Computing Systems, Apr 1999.
9. Yang, T., Gerasoulis, A.: List Scheduling with and without Communication. Parallel Computing Journal (19):1321–1344, 1993

Reflective Event Service Middleware for Distributed Component–Based Applications

Eun-young Yoon and Yong-Ik Yoon

School of Computer Science,
Sookmyung Women's University,
Seoul, Korea 140-742
{foxdiver, yiyoon }@sookmyung.ac.kr

Abstract. An Event Service is needed for providing event delivery occurring distributed component-based applications such as multimedia communication, electronic commerce, and traffic control system. It supports asynchronous communication between multiple suppliers and consumers required by the distributed applications. However, the event service specification lacks important features for user requirements reflection as follows; reflective event filtering, user quality of service (UQoS) and component management. Thus, this paper proposes a Reflective Event Service (RES) Middleware framework for distributed component-based applications. The RES middleware based on CORBA Component Model (CCM) and includes the following components; reflective event filtering component, event monitor component, and UQoS management Component. Especially, this paper concentrates on providing suitable reflective event filtering component for UQoS service.

1 Introduction

Many distributed applications exchange asynchronous requests using event-based execution models [4]. An event service is needed for providing event delivery occurring large-scale distributed applications [1], such as multimedia communication, electronic commerce, and traffic control system. It supports asynchronous communication between multiple suppliers and multiple consumers required by distributed systems. Middleware is a term that refers to a set of services that reside between the application and the operating system and aim to facilitate the development of distributed applications.

Reflective middleware is a term that describes the application of reflection to the engineering of middleware systems. reflection refers to the capability of a system to reason about and act upon itself.

In CORBA(Common Object Request Broker Architecture) standards, an Event Service interface is specified which allows asynchronous communication via event passing between Event Supplier, i.e., the objects that generate the events, and Event Consumers, i.e. the objects accept the events for processing.

This Research was supported by the Sookmyung Women's University Research Grants

I. Chong (Ed.): ICOIN 2002, LNCS 2344, pp. 746–756, 2002.
© Springer-Verlag Berlin Heidelberg 2002

An Event Channel object is defined as a mediator[3] that can dispatch events to multiple consumers on behave of multiple suppliers. However, the standard CORBA event service specification lacks important features, such as real-time management, event filtering, quality of services(QoS) control and component management, required by real-time distributed applications.

In this paper, we proposes a Reflective Event Service(RES) middleware mechanism in CORBA Component Model(CCM). Especially, this paper concentrates on providing suitable reflective event filtering component for distributed component-based applications in order to meet various user requirements under distributed system environment. It stores time constraint requirements and event filtering information input from event suppliers and consumers into UQoS repository, then processes the data through appropriate event filtering component when real-time events happen. From this mechanism, users can get the filtered event results reflected their requirements about real-time handling. It means this system provides high QoS to users. In addition, it results in decreasing network traffic as unnecessary event information is filtered from network.

2 Related Work

2.1 CORBA Event Service

In CORBA(Common Object Request Broker Architecture), the distributed object computing middleware framework being defined by a consortium of companies known as OMG(Object Management Group)[6], event service is considered one of the more important object services.

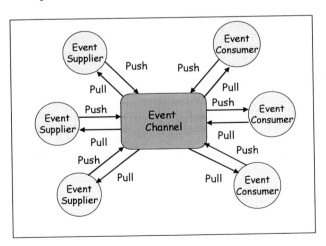

Fig. 1. CORBA Event Model

OMG has defined an Event Service interface standard for different objects in a CORBA application for use in asynchronous communication. As we can see from

Figure 1, the event service model adopted in CORBA Event Service is essentially a publisher/subscriber model, with event suppliers acting as the event publishers that generate events, and event consumers being the event subscribers that receive and process events. The event channels play the role of mediators [3] that propagate events to event consumers based on their subscription.

This CORBA Event Service provides the basic event communication interfaces in pull or push models. The CORBA event service specification provides a flexible model for asynchronous communications among objects, where objects will need to be notified when something happens to other object. Two roles for objects are defined in event service: the supplier role and the consumer role. Supplier and consumers do not need to know each other. This is the major difference between event service and traditional communication service. Real-time delivery in event service is new because event service is mostly asynchronous communication, which usually cannot guarantee the arrival of events in a given time period. Some additional event service requirements for real-time CORBA applications are discussed in.

2.2 CORBA Component Model

Component is a new basic meta-type in CORBA [7]. The component type can be specified in extended CORBA IDL and can be represented in interface repository. It encapsulates its internal representation and implementation. It includes a framework for component implementations that are hidden from clients of the component.

- **Component Levels:** There are two levels of component: basic and extended. Basic components provide a simple mechanism to "componentize" a regular CORBA object in addition to factory pattern and basic services like transaction, security supported by container. Extended components, on the other hand provide a richer set of functionality than existing CORBA object model

- **Port:** Components support a variety of surface features through which clients and other components may interact with component. These surface features are called ports. Several ports of a component model is defined as below:

  - **Facets:** Facets are distinct named interfaces provided by the component for navigation

  - **Receptacles:** Receptacles are named connection points that describe the component's ability to use a reference by some external agent.

  - **Event sources:** Event sources are named connection points that emit events of a specified type to one or more event consumers or to an event channel.

  - **Event sinks:** Event sinks are named connection points into which events of specified type may be pushed.

  - **Attributes:** Attributes are primarily intended to be used for component configuration, but might be used for other purposes also.

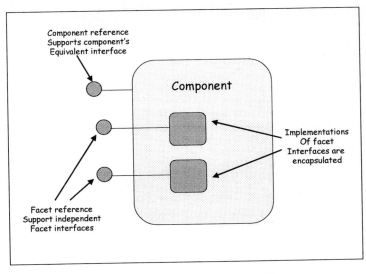

Fig. 2. CORBA Component Interfaces and Facets

The relationship between the component and its facets is characterized by the following observations:

- The implementations of the facet interfaces are encapsulated by the component, and considered to be "parts" of the component.

- Clients can navigate from any facet to the component equivalent interface, and can obtain any facet from the component equivalent interface.

- Clients can reliably determine whether any two references belong to the same component instance.

- **Client view of component:** A component provides two types of interfaces to client known as provided and supported interfaces. The component has also single distinguished interface for component reference, called component's equivalent interface that manifests component's surface feature to client. The equivalent interface allows client to navigate through facets to connect to component's port. Basic components are not allowed to offer facets, receptacles, event sources and sinks. They may offer only attributes.

- **Component homes:** The CORBA component model defines component home meta-type that manages instances of components for specified component type. Component home provides operations to manage lifecycles and optionally to manage instance of component to associate with primary key value.

The CORBA component model supports a *publish/subscribe* event model. The event model for CORBA components is designed to be compatible with the OMG Event and Notification Services. In this work, we propose that apply CORBA Component Model to our Reflective Event Service Middleware.

3   Reflective Event Service Middleware

3.1   Event Component

We define the concept of a component attending to requirements it must address. A component can be defined as:

- Binary interchangeable unit that could be installed everywhere. Interchangeably enables graceful software evolution, as another effective one without affecting others can substitute it.

- Building applications by assembling components. Components must state what they need and what do they offer. The deployment of application must be flexible enough to allow both initially distributed components and subsequent adoption to topology through application's lifetime.

- Automatic code generation based on common design and implementation patterns: code for factories, code for assemblies and code for storing and retrieving persistence object's states.

- A standard development, implementation and deployment environment covering whole application's deployment lifecycle.

The CORBA component architecture consists of several interlocking conceptual pieces that enable a complete distributed enterprise server computing architecture. These include an abstract component model, a container model, component implementation framework component implementation description language (CIDL), packaging and deployment model, integration with persistence, transactions, events and internetworking with EJB 1.0.

Applications derive specific concrete event types from this base type. When the underlying implementation of the component event mechanism provided by the container is either the CORBA Event Service or the CORBA Notification Service, event values shall be inserted into instances of the any type. The resulting any values can be used as parameters to the push parameters to the push operation on untapped event channels, or inserted into a structured event for use with the Notification Service. Our RES middleware apply from *CORBA Component Model* that have Component principles

3.2   RES Middleware Structure

Our Reflective Event Service (RES) middleware is concerned with applying techniques from the field of reflection in order to achieve flexibility and adaptability in middleware platforms. Reflective Middleware is a term that describes the application of reflection to the engineering of middleware systems. Reflection is the capability of a system to reason about and act upon itself. The RES middleware contains a representation of its own behavior, amenable to examination and change, and which is causally connected to the behavior it describes.

As shown in the figure 3, the Reflective Event Container manages a component. It creates and uses a Real-Time POA with requested features for specific component category. Client can use external interfaces of a component to interact with component and home interfaces to manage life cycle of a component

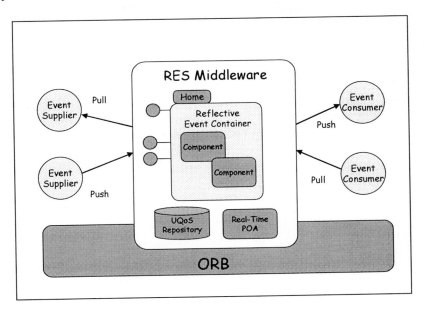

Fig. 3. RES Middleware structure

The Reflective Event Container manages component instances depending on component category. It offers all its services through simplified APIs: it becomes component's window to the outside world. The container offers series of local interfaces (internal interfaces) for establishing component's context and allows component to implement series of callback methods.

CCM containers are ideal entities to manage component QoS policies for two reasons

- **POA policy integration.** The Portable Object Adapter (POA) is a standard ORB mechanism that forwards client requests to concrete object implementations. It is also the key policy designator in the Real-Time CORBA and CORBA Messaging specifications. Each CCM container uses a dedicated POA to coordinate the policies associated with the component it manages.

- **Opportunistic component integration.** Because CCM encourages composition of unrelated objects a container provides a central repository that lets these unrelated implementation objects collaborate without explicit prior knowledge of their existence or QoS properties.

We adopt the CCM container concept and the Event Consumer and Event Supplier terms from CORBA Event Service, were an event supplier is an object that generates events, and an event consumer refers to an object that accepts event for processing. We also define an object called Reflective Event Service(RES) Middleware, which is similar with CORBA's Event Channel but provides additional reflective event container described below:

- **Event Monitoring Component**: The *even monitoring component* observes the timing behavior of the distributed application. If a real-time event is inputted to RTECP from Supplier, monitor component controls and manages flowing of events.

- **Event Filtering Component:** The *event filtering manager component* determines which consumers should receive event based on the predefined filtering and correlation rules for each event. They make the decision whether or not to propagate the consumer information towards the possible sources of events, on the statistical information about incoming events and the information about the users registered in the QoS repository.

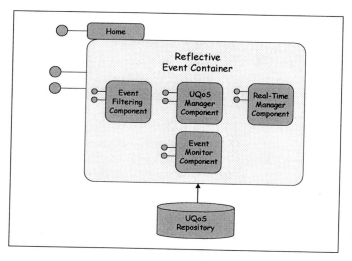

Fig. 4. Configurations of reflective Event Container

- **UQoS Manager Component:** The *UQoS manager component* manages for user requirements.

- **Real-time Manager Component:** The *real-time manager component* determines the priority of an event, puts the event to the priority queue, and then each dispatching tread dispatches an event to a linked consumer by calling the RTEvent() method of the destined consumer

- **UQoS Repository:** The *UQoS repository* manages the event consumer's information such as the object reference, real-time requirement, and the event filter information. In order to take over the information to event repository, we define the related method at event channel class.

3.3 Event Filtering

Filters are widely used in multimedia communication to perform three types of function: selective, transforming and mixing[5]. In our Event Filtering Components (EFC), we use selective and transforming filters

Configurations of event filters are crucial to the scalability of large-scale distributed application "Where to put the filters" and "how to propagate the consumer interests" are two important issues for designing event filters.

Our event filtering manager uses Filtering *Point* that classify by transmission timing to support effective filter. For the event filtering, the filtering component module determines which consumers should receive events based on the predefined filtering and correlation rules for each event. The mechanisms that perform the filtering and the correlation are referred to as event filtering component.

- **Timing Constraint Filter:** The event supplier and consumers can specify the timing constraints on event arrival or dispatch. The *Timing Constraint Filter* controls transmission using timing constraints information that receive from QoS repository when real-time events happens

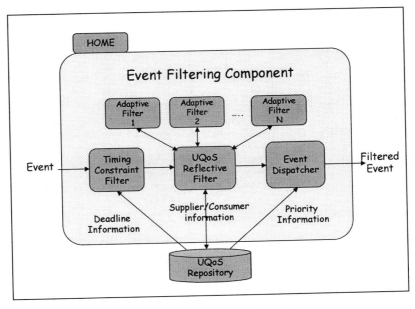

Fig. 5. Flow of Reflective Event Filtering

- **UQoS Reflective Filter:** Reflection is the capability of a system to reason about and act upon itself. The UQoS reflective filter contains a representation of its own behavior, amenable to examination and change, and which is causally connected to the behavior it describes. The UQoS Reflective Filter uses various adaptation filters to supply suitable data in event suppliers and consumers requirement

- **Event Dispatcher:** Once the event occurrences are detected, they will need to be dispatched before the deadlines, priority information to the event consumers that have registered for them

- **UQoS Repository:** All registered event's Meta information, such as timing constraints, supplier and consumer's history, are stored here.

There are many research issues associated with each components described above.

4   Reflective Event Component Implementation Framework

The Reflective Event Component Implementation Framework (RECIF) is set of classes and tools that help in implementation of components. The idea of component is binary unit is clear in CCM, which is known as executor that defines name of artifacts that can be generated by RECIF. CCM includes a declarative language (CIDL) for describing implementations of component, as well as their abstract states using Persistence State Description Language (PSDL). CIDL is superset of PSDL. The most important keyword in CIDL is "composition" that specifies how the implementation of component can be split into different executors and how the state elements are associated with each other.

```
Module Components {
  interface Reflective Event {
      EventConsumerBase
      get_consumer(in FeatureName sinkName)
          raises(InvalidName);
      Cookie subscribe(in FeatureName publisherName,
          in EventconsumerBase  subscriber)
          raises(InvalidName);
      void unsubscribe(in FeatureName publisherName,
          in Cookie ck)
          raises(InvalidName, InvalidConnection);
      void connect_consumer(in FeatureName emitterName,
          in EventConsumerBase  consumer)
          raises(InvalidName, AlreadyConnected);
              ⋮
```

Fig. 6. Reflective Event Component CIDL

A composition comprises the following elements:

- **Component home:** as specified in IDL home definition identifies the component type managed by the home as the composition's component type

- **Abstract storage binding:** identifies abstract storage type that will incarnate the component.

- **Component Executor:** may specify a number of executor segments, which are physical partitioning of component that encapsulates independent state and may be activated independently.

- **Delegation specification:** allows mapping of operations defined on home to operations on either component or abstract storage home. This feature is optional for basic component

5   Conclusion

In this paper, we discuss the requirements for Distributed Component-based Applications (DCA) in a distributed environment. Our focus is on component-based Reflective Event Service(RES) middleware.

Reflective Middleware is a term that describes the application of reflection to the engineering of middleware systems. Reflection is the capability of a system to reason about and act upon itself. The RES middleware contains a representation of its own behavior, amenable to examination and change, and which is causally connected to the behavior it describes.

The RES Middleware framework includes event filtering component, real-time manager component, event monitor component, and UQoS repository. Especially, this paper concentrates on providing suitable reflective event filtering component for real-time service in order to meet various user requirements under distributed system environment. It stores time constraint requirements and event filtering information input from event suppliers and consumers into UQoS repository, then processes the data through appropriate event filtering component when real-time events happen.

From this mechanism, users can get the filtered event results reflected their requirements about real-time handling. It means this system provides high UQoS to users. In addition, it results in decreasing network traffic as unnecessary event information is filtered from network.

References

1. D. C. Shmidt, "Scalable High Performance Event Filtering for Dynamic Multipoint Applications", 1st Int'l Workshop on High Performance Protocol Architecture, December 15, 1994, Antipolis, France.
2. G.Liu and A.K. Mok, "Real-Time Event Service Architecture for Distributed Systems", Technical Report, Real-Time System Laboratory, Department of Computer Sciences. University of Texas at Austin, 1997.
3. E. Gamma, R. Helm, R. Johnson, and J. Vlissides, Design Patterns: Elements of Reusable Object-Oriented Software. Reading MA: Addison-Wesley, 1995.
4. R. Rajkumar, M. Gagliardi, and L. Sha, "The Real-Time Publisher/Subscriber Inter Process Communication Model for Distributed Real-Time Systems: Design and Implementation," in First IEEE Real-Time Technology and Applications Symposium, May 1995.

5. D.C. Schmidt and T. Suda, "An Object-Oriented Framework for Dynamically Configuring Extensible Distributed Communication System," IEE/BCS Distributed Systems Eng. J., vol. 2, Dec. 1994, pp. 280–293

6. D.C. Schmidt et al., "Applying QoS-Enabled Distributed Object Computing Middleware to Next-Generation Distributed Applications," IEEE Communications, vol. 20, no. 10, Oct. 2000, pp. 112–123.

7. F. Kon and R.H. Campbell, "Supporting Automatic Configuration of Component-Based Distributed Systems," Proc. Fifth Conf. Object-Oriented Technologies and Systems, Usenix Assoc., Berkeley, Calif., 1999, pp. 175–187

8. J.A. Zinky, D.E. Bakken, and R. Schantz, "Architectural Support for Quality of Service for CORBA Objects," Theory and Practice of Object Systems, vol. 3, no. 1, 1997.

9. Extensible Transport Framework for Real-Time CORBA (Request for Proposal), OMG Document Orbos/2000-09-12 ed., Object Management Group, Needham, Mass., Feb. 2000.

10. F.M. Costa and G.S. Blair, "A Reflective Architecture for Middleware: Design and Implementation," Proc. 9th ECOOP Workshop for PhD Students in Object-Oriented Programming, 1999.

Automatic Specialization of Java Programs in the Distributed Environment

Jung Gyu Park and Myong-Soon Park

Internet Computing Lab.
Dept. of Computer Science and Engineering, Korea Univ., Seoul, 136-701 Korea
thejpark@yahoo.com, myongsp@ilab.korea.ac.kr

Abstract. The platform neutrality of Java programs allows them to run on heterogeneous computers. In the distributed environment, however, Java programs often cause performance problem because they are not implemented for specific clients so that they may conflict with their usage patterns at the client side. In this paper, we present a mechanism to address this problem by optimizing Java programs using program specialization technique. Unlike traditional program specialization, our specialization method does not encode the result of specialization only into run-time code. Instead, it encodes the values of multi-valued static expressions into indexed data structure that is referenced by run-time code, and single-valued static expressions into run-time code. With this approach, we can address the code explosion problem of traditional program specialization. With preliminary implementation, we achieved improvement in performance up to a factor of 9 with very low memory and space requirements and overheads.

1 Introduction

Java programming language provides platform neutrality to program code so that it can run on heterogeneous computers. In the distributed environment, Java programs may be stored in a server and migrate to a client at run-time. Because it is almost impossible to anticipate that in what circumstances the migrated Java programs run, they are usually implemented generically. At the client side, however, the Java programs often cause performance problem because they are not implemented for the specific clients so that their implementations may conflict with their usage patterns.

For example, assume that program code for mathematical dot product is being used at a client side. In general, it is implemented as a fragment of code in Fig. 1. If the client uses that code with the parameter x statically bound to {12, 5}, y to {2, 31} and scale to 2, it will be much faster for the client to use code in Fig. 2 rather than that in Fig. 1. And if the client calls that code repeatedly (e.g., in a loop) only with parameter z varying, the performance improvement gained using the latter code will be much bigger.

For a client to use specialized program code, that code must be stored in a server. But it is not possible to anticipate all the usage contexts of program code of all clients. Generate and storing specialized code for all the usage contexts is often impossible because of space and time constraints, so only some selected clients can be satisfied

I. Chong (Ed.): ICOIN 2002, LNCS 2344, pp. 757–767, 2002.
© Springer-Verlag Berlin Heidelberg 2002

with specialization. For every client to get specialized program code for its usage context, a mechanism for *specialization-on-demand* is needed.

```
float dotproduct (float[] x, float[] y, float[] z,
                                            float scale){
        if (scale != 0) {
            float result = x[0]* x[1] + y[0] * y[1] +
                                            z[0] * z[1];
            return result / scale;  }
        else return Error;
   }
```

Fig. 1. A dot product program. Those underlined are static, otherwise dynamic

```
float dotproduct (float[] z) {
        float result = 122 + z[0] * z[1];
        return result / 2;
   }
```

Fig. 2. Specialized dot product method

To provide specialization-on-demand, specialization should be performed automatically since manual specialization is error prone, tedious, and often impossible as the size of program code grows up. Program specialization (PS) enables one to obtain significant optimization by automatically specializing programs with respect to the invariants that are known at specialization-time [2, 5, 6, 7, 8]. Given a program and values of *static* (fixed) inputs, PS generates specialized version of the program that only requires *dynamic* (varying) inputs. But the size of specialized code is a problem. PS encodes the result of early computations into run-time code, so the size of run-time code may grow up exponentially as the program code evaluated with many different invariants. Assume that the dotproduct method in Fig. 1 is invoked in a loop with x and y dependent upon counting variable of the loop as shown in Fig. 3, then PS will generate 1000 instances of specialized dotproduct method. It will degrade the execution time of the specialized program because of instruction cache misses. It also takes very long time to transfer specialized program between client and server. To avoid this situation, multi-valued static expressions, expressions not depend on dynamic values but have multiple static values in its life time, (e.g., x[0]*x[1] + y[0]*y[1] in the dotproduct method) can be regarded as dynamic so that only single-valued static expressions (e.g., scale != 0 in the dotproduct method) are marked as static. With this technique, only one instance of specialized code fragment for each binding-time context is generated. In this case, however, the degree of optimization will be low.

```
for (int i = 0; i < 1000; i++){
    x[0] = x[0] + i;      x[1] = x[1] + i;
    y[0] = y[0] + i;      y[1] = y[1] + i;
    ret[i] = dotproduct (x, y, z, 2);
  }
```

Fig. 3. Method dotproduct invoked with x and y as static terms, z as a dynamic term

```
float dotproduct (float[] z, int index){
      float result = data[index] + z[0] * z[1];
      return result / 2;
  }
```

Fig. 4. Specialized code using indexed data structure for multi-valued static expressions

In this paper, we present a mechanism to address these problems by using a specialization server and our specialization technique. In this model, clients can receiver specialized program code on-demand by sending a request to a server that performs specialization. To keep the size of specialized program as small as possible without sacrificing performance, our specialization method encodes the result of early computations into not only run-time code but also indexed data structure. Our method subdivides static expressions into either single-valued or multi-valued, and encodes the results of early computations of single-valued static expressions into run-time code. However, it encodes the results of early computations of multi-valued static expressions into the indexed data structure. The values in the data structure are referenced by run-time code by accessing indexes to the data structure.

Inside a loop (e.g., the loop shown in Fig. 3), expressions whose values are dependent upon a counting variable and not upon dynamic parts of inputs are multi-valued static. For the dotproduct method in Fig. 1 is invoked in a loop with x and y dependent upon the counting variable of the loop as in Fig. 3, our method generates only one specialized method shown in Fig. 4 with 1000 data in the indexed data structure. This enables us to minimize the number of instances of specialized code fragment as small as the number of binding-time contexts without sacrificing performance. With a prototype implementation, our technique successfully performed specialization using the Java programs we tested. The size of the result of our specialization is up to 235 times smaller than that of PS in our experiments on shading procedure. We also achieved performance improvement up to a factor of 9 in the experiments.

The remainder of this paper consists of five sections. Section 2 elaborates on the motivation while Section 3 describes our algorithm in more detail. In Section 4, we analyze the costs and benefits of our approach and conclude in Section 6 after a description of related work in Section 5.

2   Overview and Motivation

Consider a fashion design visualization system on the WWW (FDVS)[1]. Its user interface and rendering module are implemented in Java and run at a client side, and 3D fashion data is stored and managed at a server side. Users load 3D fashion data stored in the server, change color and texture of clothes, properties of light, position of eye, and then render images. Since FDVS is written in Java, FDVS can be easily integrated with web page to provide more visual information about fashion design than 2D images using WWW.

The rendering is the most time-consuming process in FDVS. During the process of interactively producing computer-rendered images, users typically experiment with various image parameters such as the positions of objects, light sources, the optical characteristics of surfaces, and surface textures. Usually, users would like to see the result of rendering with such parameter changes as rapidly as possible. Some of these changes such as moving objects are likely to require significant recomputation, while others such as changing the color of objects are not. In FDVS, however, the positions of 3D objects never change, and users typically change color, texture or light source of rendering parameters. Thus, if we specialize code for rendering with respect to fixed parameters, we can obtain much faster rendering routine.

PS is a program transformation technique that optimizes programs with respect to parts of their inputs. Usually, PS processes programs in two phases: binding-time analysis (BTA) and specialization. The BTA phase takes, as its inputs, a source program and a list of binding-times of the arguments of a method in the program, and returns an annotated program in which every term in program has its binding-time. The binding-time of a term is either *static* if the value of the term can be known at the specialization-time or *dynamic* if it is unknown. In the specialization step, the annotated program is executed with the values of the static parameters, and the specialized program is returned as a result. In this sense, the fragment of code in Fig.2 can be considered as a specialized version of the code in Fig.1.

The code for rendering of FDVS is a typical candidate for PS. PS, however, has a problem when it is applied to the rendering routine: PS generates specialized code for each set of static parameter values. Because rendering routine is called repeatedly for each pixel with per-pixel invariants, PS will generate specialized rendering routines as many as the number of pixels in an image. Although the differences between each specialized rendering routine are very small, PS generates different specialized code per each pixel because PS encodes the results of early computations into specialized program code. If PS ignores per-pixel invariants, it will generate only one specialized instance of the code fragment. In this case, however, the degree of optimization will be very low. Therefore, we need to devise a method that can keep the size of the specialized program as small as possible without sacrificing performance. In the following section, we will describe a specialization method that we devised to address this problem.

---

[1] This work was supported by Korea Ministry of Information and Communication as a part of their information highway project (1996-1998).

3 Specialization

3.1 Service Model

In the distributed environment, specialization service can be provided by using a simple client-server model. A client sends a specialization server a request for specialization, which has a name of method, binding-times of input parameters, and values of static input parameters. The specialization server maintains source code for specialization. It receives requests from the client and generates a specialized program according to the message. Specialization is completed when the specialized code is compiled and its bytecode is stored. After completion of specialization, the specialization server sends a reply to the client to indicate the completion of specialization. The client then loads specialized program code in the same way it loads original program code.

3.2 Specialization Architecture

Fig.5 illustrates the overall process of our specialization.

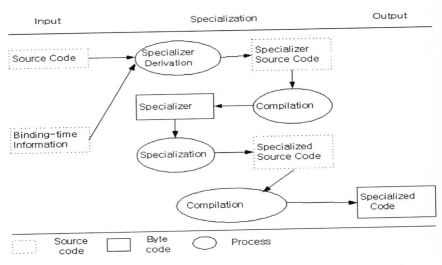

Fig. 5. Overall process of specialization

The specialization begins with binding-time analysis against an input program. Based on the results of the analysis, a specializer, which generates optimized version of the program while running, is derived. And then the specializer is executed to generate a specialized program. Thus, our specialization system can be viewed as a specializer generator.

3.3   Binding-Time Analysis (BTA)

BTA begins with parsing the source code and building a static call graph. Then, an abstract syntax tree (AST) form of each method is associated with each node in the static call graph. For the next step, it traverses the graph in top-down order performing intraprocedural BTA to the instance of the AST associated with each node of the graph. Initially static data includes object constants (e.g., string literals or statically allocated arrays) and global primitive type variables assigned with constant values. During the traversal, binding-time information propagates from one call site to another.

The intraprocedural BTA works by following the dependency model of every variable within a method along the control path. The input to the analysis is a binding-time context information of the method call. It is a tuple $<t_1,...,t_n>$, where t_i is a pair (var, bt). Here, var is one of the variables (arguments or global variables) used in the method and bt is its binding-time. The output is an AST with each term marked with its binding-time. We defined binding-times as one of the following:

static: when a term is statically bound to a single value.
multi: when a term is statically bound to multiple values.
It occurs when a method is invoked more than once with the same binding-time context or when a loop is guarded by a static predicate.
dynamic: when a term is neither static nor multi.

During specialization, the relation (method prototype, binding-time context) • AST is registered to *method table*. It is looked up at every method call site to find out whether that method is previously invoked with the same binding-time context. If no result is found, new AST and relation are generated and registered. If a result is found, the result is used to perform BTA. Every term in the code fragment is annotated with its binding-time by the following rules:

1. A term whose value or effect depends on the varying part of the input partition is marked as *dynamic*; a term that has global effect, like input or output, is marked as dynamic; a term that is a method call to an imported class is marked as dynamic if the method call has global effect; a term whose value of effect depends on dynamically loaded classes is marked as dynamic. Otherwise it is marked as static.
2. If a method is invoked more than once with the same binding-time context, variables that are static in that context with their values changed in each invocation are multi-valued static. If such a method call is found, their context is modified so that those variables are marked as *multi*. Also, the binding-time context for the body of a loop with a static predicate is constructed in such a way that variables transitively dependent upon dynamic terms in the loop are marked as dynamic and transitively dependent upon multi terms in the loop are marked as multi. After binding-time context is constructed, the body of the method or loop is labeled using that binding-time context.
3. An expression is dynamic if it has at least one dynamic operand; it is multi if it has at least one multi operand and no dynamic operands. Otherwise it is static.
4. Basically, binding-time is computed by solving data-flow equations below.

In[S]: variables that are static before statement S
Gen[S]: variables that are defined static by statement S
Kill[S]: variables that are defined dynamic by statement S
Out[S]: variables that are static after statement S
In[S] = ∩ Out[W] − C, where W ∈ control predecessor of S and C is set
        of variables in Out[W] that are control dependent on dynamic predicate.
Gen[S] = if S is a definition statement of the form 'v = expr' and expr is
        static then v else { }
Kill[S] = if S is a definition statement of the form 'v = expr' and expr is
        dynamic then v else { }
Out[S] = In[S] ∪ Gen[S] − Kill[S]

A variable v is control dependent on a predicate p if its value varies depending on the value of p. A variable that is marked as static in all branches of a conditional branch statement, if the predicate is dynamic, should be marked as dynamic outside of the statement. It is because the choice that which branch to take is determined by dynamic inputs. So we defined In[S] that it does not have variables that are control dependent on dynamic predicate.

3.4 Specializer Derivation

A specializer is derived from the AST that has been annotated by the BTA procedure. The derivation algorithm is applied to each node of a static call graph in the bottom-up order. For each node in static call graph, it produces a method that is a specializer for that node. We call the method a *specializer method*, and a method generated by specializer method a *specialized method* (a specializer is composed of specializer methods). The specializer method is composed of *execution code* and *generation code*. The execution code performs early computation and the generation code emits specialized code when the specializer method is invoked. The generation code is of the form gen (str) where str is a string that represents the generated code code (i.e., str ="code"). The derivation proceeds via a simple case analysis based on the binding-time of each term in the AST:

  static: the term is added to the generation code if it is needed for the computations of dynamic expressions (e.g., the term is an operand of a dynamic expression) or it is added to the execution code otherwise. When it is added to the generation code, string transformation is performed so that it is replaced by a constant when the specializer generates the specialized code.

  multi: the term, if it is needed for the computations of dynamic expressions, is added to both the generation code and the execution code. When it is added to the execution code, string transformation is performed to replace it by an expression that stores values into the indexed data structure. When it is added to the generation code, string transformation is performed to replace it by a reference to the indexed data structure when the specializer generates the specialized code. The transformation is restricted only to nontrivial terms so that expressions with very low execution cost are not stored in the data structure. If the term is not needed for the computations of dynamic expressions, it is added only to the execution code.

dynamic: the term is added only to the generation code.

```
float spc_dotproduct (float[] x, float[] y, float scale, int i){
        if (scale != 0) {
            gen ("float dotproduct (float[] z, int index) {");
            data[i] = x[0] * x[1] + y[0] * y[1];
            gen ("float result = data[index] + z[0] * z[1];");
            gen ("return result / " + scale + "; }");
            return Default_Float_Number;
        else return Error;
    }
```

Fig. 6. A specializer method that generates the specialized method shown in Fig. 4. The gen is implemented so that it generates code when it is first invoked and does nothing from the next invocation.

3.5 Specialization

Specializer is executed to generate specialized source code. It generates a piece of code by invoking specializer methods. For each invocation of a specializer method, a corresponding specialized method is generated. The union of specialized methods is the result of our specialization.

4 Results

In this section, we present empirical results obtained with our specialization system. The algorithm in section 3 has been implemented in a prototype system consisting of approximately 10k lines of Java and javacc, jjtree code. All measurements were conducted with the JDK 1.3 on an Intel Pentium P3-500 processor with 256 Megabytes of physical memory.

Our benchmark is a shading procedure, or shader, of FDVS. In this system, shader computes the color value for an image pixel given the pixel coordinates, various rendering information (material color, eye position, light position and illumination properties) provided by the user via GUI. It has about 100 lines of code derived from the shader code in [10].

We performed specialization against the dynamic light position, dynamic specular component of light properties, and dynamic material color. Other properties of light (e.g., ambient) and shading parameters (e.g., kd) affect similarly as specular component of light properties. We compared the results of experiments using the specialized shader with that using the original shader.

4.1 Speed Up

Fig. 7 illustrates the speedups achieved by specializing a shader. The values represent the average of multiple runs with varying dynamic parameter values. The speedups vary widely between specializations, but are always at least 1.0x.

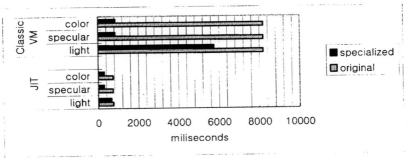

Fig. 7. Performance measurement

The complexity of computation depending on the varying parameter is different for each input so that the effect of specialization for the dynamic light position, dynamic specular component of light properties, and dynamic material color shows high variance. The specializations against dynamic value of specular and color are much faster than against light position because considerable amount of computation is related to the value of light position. When we use a classic VM, the time taken to execute specialized code with static light position is 9 times faster than that of original shader. The speedups using JIT are lower than this (up to a factor of 2.5) because of JIT optimization. The speedups only by our specialization are reflected by the results of using classic VM.

We also implemented traditional PS for comparison. In our test of dynamic light position with the resolution of image as 640 by 480, however, it generated 470 Megabytes of specialized codes that could be considered impossible for compilation. When multi-valued static expressions were marked as dynamic instead of static, only one specialized shader was generated. But in this case, the degree of specialization was very low so that it was no better than the original shader in our test.

Overheads were very low in our test. Generation of a specializer takes about 1 second because the size of shader program is small. The specialized code generation phase also takes about 1 second in our test. Specializer generation occurs less frequently than specialized code generation because specializer generation occurs when static input partition changes, but specialized code generation occurs when the values of static inputs change. In all experiments, these overheads were amortized when we use the same specialized code more than once. The size of indexed data is very small compared with the size of partially evaluated program. But the time taken to transfer the data from the server to the client is the most time-consuming process in our specialization service.

4.2 Memory Usage

The average size of indexed data for a single pixel is 8 bytes in our test. In an image resolution of 640 x 480, its space usage is 2.4 Megabytes that is well within the physical memory size of a typical PC. It is much smaller than 470 Megabytes of specialized program by PS. The size of specialized program generated by our specialization is similar to the original program.

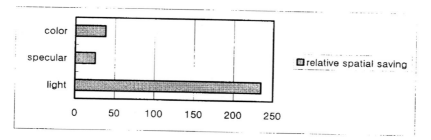

Fig. 8. Relative spatial saving compared to PS. The result of our specialization is 235 times smaller than that of PS when light position is dynamic

5   Related Works

Consel et al. have developed Tempo [2, 3], a partial evaluator for C programs. Tempo is an off-line partial evaluator. It can specialize programs not only at compile-time (source to source) but also at run-time. They demonstrated a partial evaluator for Java, named Jspec [5, 9], to specialize Java programs using Tempo and Harissa, a Java-to-C compiler.

Knoblock and Ruf suggested data specialization [7] that split a source program into cache loader and cache reader. Cache loader encodes the results of early computation into cache data structure (similar to the indexed data structure in our technique), and cache reader is run-time program that makes use of values in the cache. Their use of cache is for the rapid specialization (cache loading process) and low space overheads. Unlike our technique, it uses cache instead of constants to encode single valued static expressions. They regarded multi-valued static expressions as dynamic, so their degree of optimization is lower than ours. Furthermore, it cannot eliminate if statements with static predicate when a branch has dynamic statements because it generates run-time code before the values of static inputs are known.

Masuhara and Yonezawa proposed bytecode specialization (BCS) [8] that is a run-time specialization system for a subset of bytecode language. BCS generates programs in a bytecode code language and then translate the generated bytecode into native code by using a just-in-time (JIT) compiler. Their approach, however, has some overheads at run-time because of JIT compilation and code generation.

6 Conclusion

We have presented a mechanism for improving the performance of Java programs in the distributed environment by specializing it using specialization server. Our specialization method encodes the value of multi-valued static expressions into indexed data that is referenced by run-time code, but single-valued static expressions into run-time code directly. This approach addresses the code explosion problem of traditional PS. In our experiments on shader, the size of the result of our specialization is at most 235 times smaller than that of PS. We also gained performance improvement up to a factor of 9 with very low overheads. Although we devised this specialization mainly for distributed environment, we believe it gives a basis for a more general solution applicable to other aspects of Java applications.

References

1. Aho, A. V., Sethi, R., Ullman, J. D.: Compilers: Principals,Techniques, and Tools. Addison-Wesley, Reading, MA (1986)
2. Consel, C. and Noël, F.: A general approach to run-time specialization and its application to C. Proc. ACM SIGPLAN-SIGACT Symp. on POPL, (1996) 145–156
3. Consel, C., Hornof, L., Noyé, J., Noël, Volanschi, E.-N.: A Uniform Approach for Compile-Time and Run-Time Specialization, Proc. International Seminar on Partial Evaluation, (1996) 54–72
4. Gosling, J., Joy, B., Steele, G., Bracha, G.: The Java Language Specification, 2nd ed. Addison-Wesley, Reading, MA (2000)
5. IRISA Compose Project: Jspec Specializer. http://www.irisa.fr/compose/jspec/ (2000)
6. Jones, N. D.: An Introduction to Partial Evaluation, ACM Computing Surveys 8 (3) (1996) 480–503
7. Knoblock, T. B. and Ruf, E.: Data Specialization. Proc. ACM SIGPLAN Conf. on PLDI, (1996) 215–225
8. Masuhara, H. and Yonezawa, A.: Generating Optimized Residual Code in Run-Time Specialization. Technical Report of PE Day'99 (1999)
9. Schultz,U.P., Lawall, J., Consel, C., Muller, G.: Toward Automatic Specialization of Java Programs. Proceedings of 13th European Conference on Object-Oriented Programming (ECOOP '99) (1999) 367–390
10. Gunter, B., Knoblock, T. B., and Ruf, E.: Specializing Shader. Proc. ACM SIGGRAPH'95, (1995) 343–349.

A JMX Based Efficient Mobile Agents Platform Architecture for Network Management

Jung Hwan Lee[1] and Choong Seon Hong[2]

[1] School of Electronics and Information, Kyung Hee University,
449–701 Korea
nowsys@networking.kyunghee.ac.kr
[2] School of Electronics and Information, Kyung Hee University,
449–701 Korea
cshong@khu.ac.kr

Abstract. In order to overcome the disadvantages of existing centralized network management systems that use simple objects, dynamic object platform is proposed as alternative system. So the distributed network management systems are implemented using various distributed platforms such as CORBA and JAVA-RMI. Subsequently mobile agent-based platform is proposed. The mobile agent-based platform can additionally provide flexibility and scalability to network management system which CORBA or JAVA-RMI based platform do not support. In this paper, we address the architecture to solve the problem of the occurrence of additional traffic by using mobile agents and to save resources of network element. This paper makes a description about efficient network management architecture using mobile agents. Also we design agents using information architecture of TMN for efficient resource management of network element and improvement of operation performance.

1 Introduction

With popularization of Internet, network traffic is increasing continuously. Network can be highly jammed and can be caused to the delay of response time by effect of such increasing network traffic. It has become an essential work now that to manage network continuously and monitor, analyze and solve these problems. A lot of systems for the network management have been proposed by this necessity. ISO (International Organization for Standardization) defines CMIS/CMIP (Common Management Information Services / Common Management Information Protocol) [1]. And IETF (Internet Engineering Task Force) defines SNMP (Simple Network Management Protocol) [2]. These services and protocols are typical centralized structure using a client / server model. SNMP is used in various network management systems because the architecture that is divided as manager and agent is relatively simple and is a clear architecture. Also SNMP implementation is not difficult. So many network de

I. Chong (Ed.): ICOIN 2002, LNCS 2344, pp. 768–779, 2002.
© Springer-Verlag Berlin Heidelberg 2002

vice vendors implement these modules and the modules are loaded into network device. However, in a network such as WAN (Wide Area Network), it has a difficulty to monitor several sub-network because of bottleneck [3]. The system with decentralized architecture (Distributed Architecture) that supports MAS (Multi-Agent Systems) [4] was proposed to solve the problem of centralized management architecture.

It is a static object platform which several distributed objects are fixed in each agent using CORBA or Java-RMI. And in a contrast to this, we call it a dynamic object platform that using mobile agents. The advantages of dynamic object platform compared to static object platform are flexibility and scalability. In this paper, we make use of advantages of mobile agents which include dynamic object platform. First we studied about the weak point that mobile agents can have, namely additional traffic occurrence problem that would happen when mobile agents move. Then we studied about efficient resource management work that proposed for network management using mobile agents in chapter 2. In chapter 3, we describe about mobile agents design and stationary agents design. And in chapter 4, we describe about design and implementation of new platform architecture based on proposed item. At last, we explain about conclusion and forward subject.

2 Network Management Platform Using Mobile Agents

In this section we will describe the existing research that is proposed for network management using mobile agents platform. If we use mobile agents in network management, we can reduce unnecessary traffic over network to use network management classification factors that fault management, account management, configuration management, performance management and security management can reduce the use of unnecessary resources network element. Also if we use a java platform (JVM), it supports heterogeneous environment in network and network device. Table 1 shows merits and demerits that can have about network management using mobile agents.

Table 1. Merits and demerits of network management system using mobile agents

Item	Contents
Advantages	Efficient saving
	Support for heterogeneous environment
	Storage saving of network element by mobile agents transfer
	Extensibility
	Easy software upgrade
Disadvantages	Additional traffic occurrence when mobile agents move
	Transfer domain specification of mobile agents for mobility guarantee

2.1 Platform Architecture Applied in Network Management

General architecture is shown as figure 1. Mobile agents that are created by the network manager migrate to target node that needs the management.

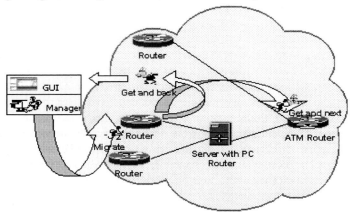

Fig. 1. General architecture that use mobile agents in network management

The general architecture which is consisted of simple operations such as GnG (Get and Go, Get and Back) [3] has a difficulty in management of mobile agents, performance management of network and efficient management of data. So the platform having new algorithms and structure proposed in order to supplement these problems.

2.2 MIAMI Project

The objectives of MIAMI (Mobile Intelligent Agents in the Management of the Information Infrastructure) [5] project are to examine the applicability of Mobile Intelligent Agents (MIAs) to network and service management. The MIAMI project, within this context, focuses on the following key objectives: [6]

■ Create a unified mobile intelligent (MIA) framework by validating, refining and enhancing the OMG MASIF standards according to the requirements for an Open European Information Infrastructure.
■ Develop mobile intelligent agent (MIA) based solutions for the management of the Open EII and for the provision of advanced communication and information services.
■ Create a reference implementation of the Unified MIA Framework and solutions in order to evaluate the service solutions in a Pan-European business environment.
■ Produce the following recommendations :
  – To infrastructure and terminal providers : when, where and how to introduce the MASIF in their future products

- To service provider : how to develop MIA based solutions for the management of the EII and for the provision of advanced communication and information service.
- Augment the results of selected ACTS projects by the integration of MIA based solution.
- Participation in the "Domain 5" cluster of ACTS agent projects for the coordination of standardization activities.
- Disseminate results by demonstrations, publications, providing input to standardization bodies, industrial forums.

In this research, the software agents which have a constrained mobility placed to solve problems of legacy mobile agents platform and to manage more effectively about network. And they proposed dynamic and efficient mobile agents platform (figure 2) that allows network's dynamic resource management and functional delimit of each management element through AVP (Active Virtual Pipe).

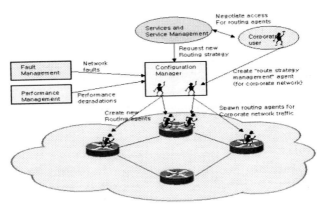

Fig. 2. The architecture of MIAMI platform

Also figure 3 shows an AVP of MIAMI. DCM (Dynamic Connectivity Manager) provides dynamic connection to each management factors (i.e., configuration management, performance management, fault management) according to types of mobile agents.

3 Design and Implementation of Agents

The agents is divided into two classifications in mobile agents platform. The one is a mobile agent and the other is a stationary agent. The mobile agents have a mobility and they can visit destination node or target node and collect necessary information. While stationary agents can't move to other area or node but it can manage dynamically specific area with mobile agents. And we basically use information architecture of TMN basically for all efficient management of agents.

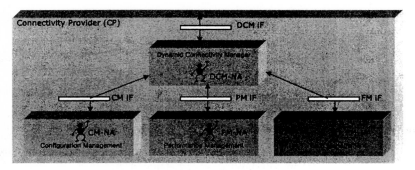

Fig. 3. MIAMI Active Virtual Pipe Domain

Also we divided by each fault management, configuration management, performance management, account management and security management using Management Information model of OSI [7]. We apply the information architecture of TMN [8] to all of agents for efficient management. Figure 4 shows Information Architecture of TMN. And we define the stationary agents class of each management using the functional management area of OSI

Managed Object

Fig. 4. Information Architecture of TMN

3.1   Design of Mobile Agents

As a mobile agents platform, we considered IKV++ company's Grasshopper platform which was strongly recommended in a result report of MIAMI project. Although there are so many platforms for mobile agent which was provided by different vendors such as Aglets [9], Grasshopper [10], Voyager [11]. But Grasshopper is a mobile agents platform that is built standard of the Object Management Group (OMG). (i.e., Mobile Agents System Interoperability Facility (MASIF)). The MASIF standard has been initiated in order to achieve interoperability between mobile agents platforms of different manufactures [12]. Grasshopper platform is composed of regions, places, agencies and different types of agents

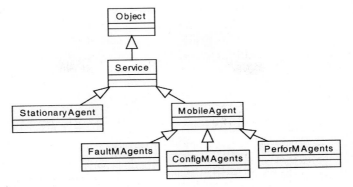

Fig. 5. Class hierarchy diagram of mobile agents.

Basically, mobile agents is object which has the operations and attributes. Therefore, mobile agents can reduce unnecessary operations and attributes so that reduce quantity of traffic which may happen additionally. So, we divided to 5 classification items that refer mobile agents over. And we subdivided by 3 classification (figure 5) that mobile agents can be applied usefully among them. Such classed mobile agents communicates with stationary agents that has associated data. (table 2) And mobile agents is created by classification according to manager's request and is sent to network element.

3.2   Design of Stationary Agents

We implemented the stationary agents using JMX (Java Management Extension) [13]. The Java Management extensions define architecture, the design patterns, the APIs, and the services for application and network management in the Java programming language. The JMX architecture is divided into three levels. (figure 6)

- Instrumentation level
- Agent level
- Distributed services level

The instrumentation level provides a specification for implementing JMX manageable resources [14]. A JMX manageable resource can be an application, an implementation of a service, a device, a user, and so forth. The instrumentation of a given resource is provided by one or more Managed Beans, or MBeans, which are either standard or dynamic. Standard MBeans are Java objects that conform to certain design patterns derived from the JavaBeans component model. Basic structure of JMX is as following. MBeans model has a similarities with Information Architecture of TMN. Information Architecture TMN uses an object oriented approach and is based on Management Information Model of OSI.

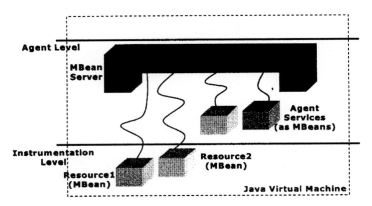

Fig. 6. Basic architecture of JMX

We designed MBeans model for efficient management of stationary agents. And we defined three MBeans. Defined MBeans are PerforMSAgents, ConfigMSAgents, FaultMSAgents. Figure 7 shows the class hierarchy diagram of stationary agents.

MIB values are defined in RFC 1213 [15]. We refractionate MIB which defined in RFC 1213 for management of MBeans by OSI functional area (FCAPS). Classifications are same as next table 2 [16]. Classified values are managed by each MBeans. And each MBeans have an operations and attributes. Also ManagedStationaryAgents contacts with PerforMSAgents, ConfigMSAgents, FaultMSAgents. SNMP API also is supported in JMX. JMX smart agents are capable of being managed through HTML browsers or by various management protocols such as SNMP and WBEM [17].

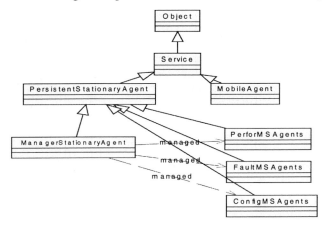

Fig. 7. Class hierarchy diagram of stationary agents

Stationary agents above in class hierarchy take the role of MBeans in JMX. Also, all stationary agents can do resources management of network elements using a persistent service of Grasshopper.

Table 2. The rearrangement of MIB

Agents Names	MIB Variables
PerforMSAgents	SnmpInPkts, snmpOutPkts, sysUpTime, ipDefaultTTL, ifInDiscards, ifOutDiscards, iFInErrors, ifOutErrors, ifInOctets, ifOutOctets, ifInUcastPkts, ifOutUcastPkts, ifInNUcastPkts, ifOutNUcastPkts, ifInUnknownProtos, ifOutQLen, ipInReceives, ipInHdrErrors, ipForwDatagrams, ipInUnknownProtos, ipInAddrErrors, ipInDiscardsm upInDelivers, ipOutDiscards, ipOutNoRoutes, ipRoutingDiscards, ipReasmReqds, ipReasmOKs, ipReasmFails, ipFragOKs, ipFragFails, ipFragCreates
FaultMSAgents	SysObjectID, sysServices, sysUptime, upInHdrErrors, ipInAddrErrors, upReasmFails, ipInReceives, ipForwDatagrams, ipInDelivers, upOutRequest, ipOutDiscards, ipOutNoRoutes, ipRoutingDiscards, ipReasmReqds, ipReasmOKs, ipReasmFails, ipFragOKs, ipFragFails, ipFragCreates
ConfigMSAgents	SysDescr, sysLocation, sysName, ifDescr, ifType, ifMtu, ifSpeed, ifAdminStatus, ipFowarding, ipAddrTable, ipRouteTable

4 Design and Implementation of Proposed Platform Architecture

As explain in chapter 3, whole platform is consisted of mobile agents and stationary agents. The whole platform is same as figure 8.

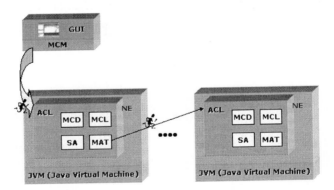

Fig. 8. Proposed platform component diagram

4.1 MCM (Mobile Code Manager)

Mobile code manager refers code repository and creates mobile agents by manager request. And Mobile code manager sends mobile agents to target node. The module composition of MCM is as following.

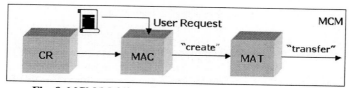

Fig. 9. MCM(Mobile code manager) component diagram

■ CR (Code Repository) : When mobile code manager creates mobile agents, code repository is referred by user request for functional network management. It stores operations and attributes for mobile agents.

■ MAC (Mobile Agents Creator) : Refer code repository and user code, it creates mobile agents.

■ MAT (Mobile Agents Transfer) : Transfer mobile agents to target node. (It depends on Grasshopper platform).

4.2 ACL (Active Class Loader)

This module exists in network elements, doing a practical network management job with mobile agents

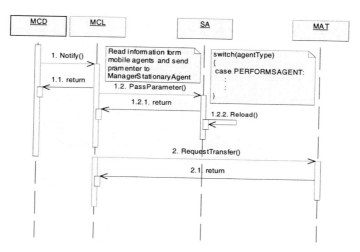

Fig. 10. Sequence diagram of Active Class Loader

■ MCD (Mobile Code Daemon) : If mobile agents arrive in destination node, alarm to node.

■ MCL (Mobile Class Loader) : Read information of mobile agents and pass data that need in stationary agents management to SA

■ MAT (Mobile Agents Transfer) : Move into other target node of mobile agents

■ SA (Stationary Agents) : Stationary agents manages practical components of network element. And stationary agents improves resources management of network element efficiency using persistent service of Grasshopper

5 Simulation Results

We compare the memory availability of stationary agents and code size of mobile agents for evaluation of proposed platform architecture. Simulation environments are next.

■ OS : Windows 2000 Professional
■ PC : Pentium III 733Mhz, RAM 512M
■ Tool : JDK 1.2.2, JDMK 4.2, Grasshopper Platform 2.2.2,

It is an optimized code to perform network management operation for mobile agents as shown in figure 5. The mobile agents code is a java class file, so we can remove unnecessary operations and attributes of mobile agents. And it can make the code more minimum size and optimum size by applying classification of functional management area.

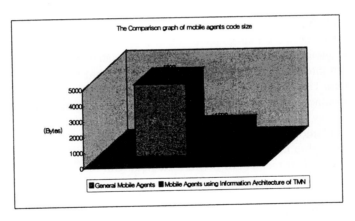

Fig. 11. Byte size comparison graph

In the above figure 11, the left bar means average code size which does not apply classification of functional management area. The right one is an average code size which applied classification of functional management area. The size of mobile agents which is applied classification of functional management area decrease about 53%. So it only needed a minimum time cost to move a destination node or manager.

Also we use a persistent service for efficient management of stationary agents. The persistence service is a part of the core functionality of Grasshopper agencies. Its purpose is to persistently store the data states of all currently hosted agents as well as runtime information about all places that exists on the agency. Persistent service deal-locates stationary agents that it is no present necessity in memory. And necessary stationary agents reallocate. Because main work of network element is packet routing, it is all-important that reduces additional resources assignment. We can see the memory measurement and number of MBeans in table 3. The API of the persistence service is divided into two parts, one provided by the agency (interface de.ikv. grasshop-per.agency.IAgentSystem) and the other provided by the persistence-supporting agent (classes Persistent MobileAgent, PersistentStationayAgent).

Table 3. The comparison of memory availability

(#N : number of stationary agents, T.M : total memory, A.M : Available memory)

	# N	T. M	A. M
SA with persistent service	3	523,744KB	231,054KB
SA without persistent service	3	523,744KB	224,529KB

6   Conclusions

The management platform applying mobile agents are flexible and extendible more than other network management platforms. But the mobile agents platform can not be applied to most network management actually up to now. Because the JVM is so heavy to integrates with node OS and platform dependency of itself. Such problems can be solved through Java Chip and so on if consider that JVM is developed by em-bedded system [18]. Mobile agents platform which proposed in this paper uses OSI management function classification based on Information Architecture of TMN. Also, we studied a way to minimize code of mobile agents and minimize network traffic that may happen additionally. The new schemes which apply persistent service to MBeans of EJB Based and to managed MIB variables by each stationary agents are more efficient in resources management of network element. Our researches till now described a general platform that use mobile agents.

For our future works, we will research algorithms about transfer path decision problem of mobile agents. And we will study about the network management scopes using mobile agents.

Acknowledgements. This work was supported by grant No. 2001-1-30300-001-2 from the Basic Research Program of Korea Science & Engineering Foundation.

References

1. IETF, "The Common Management Information Services and Protocols for the Internet (CMOT and CMIP)", RFC 1189, http://www.ietf.org
2. IETF, "A Simple Network Management Protocol", RFC 1157, http://www.ietf.org
3. D. Gavalas, M Ghanbari, M. O'Mahony, D. Greenwood, "Enabling Mobile Agents Technology for Intelligent Bulk Management Data Filtering", NOMS 2000, 623p, April 2000
4. A. Bieszczad, B. Pagurek, T. White, "Mobile Agents for Network Management", IEEE Communications Survey, Vol 1, No. 1
5. MIAMI project, "Mobile Agent Platform Assessment Report", http://www.fokus.gmd.de/research/cc/ecco/climate/documents/miami-agplatf.pdf
6. The MIAMI project at UCL, http://www.ee.ucl.uk/~dgriffin/miami/
7. Management Information Protocol of OSI, ISO DIS 10165-1: "Information Processing Systems – Open Systems Interconnection – Structure of Management Information – Part 1: Management Information Model", Geneva, 1993
8. Information Architecture of TMN, http://snmp.cs.utwente.nl/tutorials/tmn/index-15.html
9. IBM Aglets, http://www.trl.ibm.com/aglets/
10. IKV++ Grasshopper 2, http://www.grasshopper.de/index.html
11. ObjectSpace Voyager, http://www.objectspace.com/products/voyager/
12. MASIF, http://www.det.ua.pt/Projects/difference/work/D7/d7chap4.html
13. JMX, http://java.sun.com/products/JavaManagement/
14. EJB, http://java.sun.com/products/ejb/
15. IETF, "Management Information Base for Network Management of TCP/IP-based internets : MIB-II", RFC 1213, http://www.ietf.org
16. Mi-Young Kang, Sung Kim, Cheul-Young Kim, Ji-Seung Nam, "The design and implmentation of functional class for real-time network management based on java", Korea information science society, Vol. 28, No 1, April 2001
17. Web-based Enterprise Management, http://www.dmtf.org/standards/standard_wbem.php
18. Embedded Java, http://java.sun.com/products/embeddedjava/

An Efficient Coordinated Checkpointing Scheme Based on PWD Model

MaengSoon Baik[1], JinGon Shon[2], Kibom Kim[3],
JinHo Ahn[1], and ChongSun Hwang[1]

[1] Dept. of Computer Science & Engineering Korea University,
1,5-Ga, Anam-Dong, SungBuk-Gu, Seoul 136-7-1, South Korea
{msbak, jhahn, hwang}@disys.korea.ac.kr
[2] Dept. of Computer Science, Korea National Open University,
Seoul, 110-791, South Korea
jgshon@mail.knou.ac.kr
[3] Korea Advanced Integration Technology,
Seoul, 135-010, South Korea
kibom@echo.co.kr

Abstract. In this paper, we will prove that the new consistency condition for checkpoints must be needed in systems based on PWD model and propose an efficient coordinated checkpointing scheme. In our scheme, whenever process constructs a consistent global checkpoint set, it must obey the new consistency condition instead of previous consistency condition. That is, an execution of process is divided into an unique state interval occurred by non-deterministic event and checkpoint is taken in only processes happened a state interval transition. Consequently, proposed coordinated checkpointing scheme removes an unnecessary overhead emerged in previous works and guarantees the limited rollback propagation on the occurrence of a failure in systems based on PWD model.

1 Introduction

Rollback recovery protocol has been investigated as an attractive technique that offered fault tolerance to distributed systems [1,4].

Traditionally, rollback recovery protocol is classified into *checkpoint-based recovery protocol* and *log-based recovery protocol* [1,2,4]. In the checkpoint-based recovery protocol, each process saves its state information in preparation for occurrence of a failure. The saved state information is called a *checkpoint*. The method of taking checkpoint is classified into *independent checkpointing* and *coordinated checkpointing* [1,2]. Independent checkpointing minimizes the overhead at failure-free execution by independently taking checkpoint without synchronizing with other processes. On the other hand, coordinated checkpointing forces the synchronization with other process whenever takes checkpoint[1,2,8,12,13, 14].

Log-based recovery protocol is usually classified into *optimistic message logging* and *pessimistic message logging*. In the optimistic message logging, the

I. Chong (Ed.): ICOIN 2002, LNCS 2344, pp. 780–790, 2002.
© Springer-Verlag Berlin Heidelberg 2002

determinants of received messages are logged in volatile memory and logged in stable storage at adequate time of process [3,6,7,10]. On the other hand, pessimistic message logging enables to recover until the state before occurrence of a failure by logging determinant of all nondeterministic events on stable storage before processing [1,3].

Because the works that must be replayed on occurrence of a failure come to be increased, log-based recovery protocol is combined with checkpoint-based recovery protocol. There has been works that independent checkpointing is better suitable for log-based recovery protocol than coordinated checkpointing. In the recent works, however, the combination log-based recovery protocol with coordinated checkpointing yields the better effect as a respect of simplicity and performance than other combination [2,8]. However, because related works [2], [8] followed the classical coordinated checkpointing investigated without consideration of PWD model, they must take the unnecessary checkpoint. Although related works [15], [16] change the log-based protocol according to the PWD model, they do not remove the overhead that takes unnecessary checkpoint at the construction of a consistent global checkpoint set. Consequently, related works [2], [8], [15], [16] do not yield better effect than incomplete combination message logging with coordinated checkpointing proposed in systems which is not based on PWD model.

In our papers, we will show that new consistency condition for checkpoint must be needed in systems based on PWD model and propose an efficient coordinated checkpointing suitable for PWD model. Proposed coordinated checkpointing reduces the number of checkpoint must be taken by process and guarantees the limited rollback propagation on the occurrence of a failure. Also, the overhead for interaction with outside world and stable storage comes to be reduced.

2   System Model

System consists of processes which execute an application program and communication channel between processes [1]. Communication channel is assumed that is reliable and guarantees the FIFO message delivery order [11]. An execution of each process is based on PWD model, which postulates that all nondeterministic events that a process executes can be identified and that the information necessary to replay each event during recovery can be logged in the event's dterminant [1] . Also, all nondeterministic event occurred in process is restricted within a delivery of messages. In systems based on PWD model, an execution of each process is determined by the initial state of process, the delivery order and contents of received messages. Thus, if the delivery order and contents of received messages is maintained, then the execution occurred by received messages is able to be replayed [3,5,6,7,9].

In figure 1, the whole execution of process p_1 is determined by the initial state, delivery orders and contents of received messages $m1$, $m4$, $m5$ and $m7$. The formal representation of process's execution in systems based on PWD model is as follows.

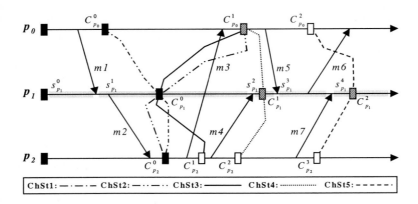

Fig. 1. The execution of process and checkpoint set in systems composed of three processes

$$Ex(p_i) = Init(p_i) + \biguplus_{j \in I} \overrightarrow{s_{p_i}^j(m_k)} \tag{1}$$

- $Ex(p_i)$: the whole execution state of process p_i,
- $Init(p_i)$: the initial state of process p_i,
- $+$: sequential work flow of a process,
- $\biguplus_{j \in I}^{\rightarrow}$: the set of sequential work flows of a process having ascending order j (I : state interval number),
- $s_{p_i}^j(m_k)$: the j-*th* state interval of process p_i occurred by message m_k.

Also, process halts the its execution on the occurrence of a failure [1,4,10]. System automatically perceives of the failure of process and recovers the process occurred failure by means of using the information saved on the stable storage.

3   Motivation

The most important issues of previous coordinated checkpointing is how each process to take the checkpoint without creation of an *orphan message*. Orphan message means that the delivering event of message is recorded in receiver process, but the sending event of message not recorded in sender process [5,11,12]. The definition of orphan message is as follows

Definition 1. *Message* $m_{j \to i}$ *is an orphan message* ***if and only if*** *receive*$(m_{j \to i}) \in C_{p_i}$ *and send*$(m_{j \to i}) \notin C_{p_j}$ *(i $\neq$ j:process id)*

In the case that message m is transferred from process p_j to process p_i, message m is indicated by $m_{j \to i}$. And the checkpoint taken in process p_i is indicated by C_{p_i}. If the delivery event of message $m_{j \to i}$ is written at the checkpoint C_{p_i} of receiver process p_i and the sending event of message $m_{j \to i}$ not written at the checkpoint C_{p_j} of sender process p_j, message $m_{j \to i}$ comes to be an orphan message.

Also, if the set of checkpoints maintained by each process doesn't make an orphan message, it is called by a *consistent global checkpoint set*. The definition of a consistent global checkpoint set in the systems consisted of n-*th* processes is as follows.

Definition 2. $\bigcup_{k=0}^{n-1} C_{p_k}$ *is a consistent global checkpoint set **if and only if*** $\forall i, j \ \neg \exists m_{j \to i} \ receive(m_{j \to i}) \in C_{p_i} \ and \ send(m_{j \to i}) \notin C_{p_j} \ (i \neq j : process \ id)$

In figure 1, processes p_0, p_1 and p_2 in checkpoint set **ChSt1** maintain consistent global checkpoints $C_{p_0}^0$, $C_{p_1}^0$ and $C_{p_2}^0$. The $k-th$ checkpoint of process p_i is indicated by $C_{p_i}^k$. When process p_0 takes the new checkpoint $C_{p_0}^1$, the checkpoint set is advanced from **ChSt1** to **ChSt2**. In the checkpoint set **ChSt2**, the delivery event of message $m3$ sent by process p_2 is recorded on checkpoint $C_{p_0}^1$ newly taken by process p_1, but the sending event of message $m3$ is not recorded on checkpoint $C_{p_2}^0$ of process p_2. That is, the formal expression for message $m3$ is as follows.

$$deliver(m3_{p_2 \to p_0}) \in C_{p_0}^1 \quad and \quad send(m3_{p_2 \to p_0}) \notin C_{p_0}^0 \tag{2}$$

After all, message $m3$ comes to be an orphan message by **Definition 1** and the checkpoint set **ChSt2** does not construct a consistent global checkpoint set by **Definition 2**. While checkpoint set **ChSt1** is advanced to checkpoint set **ChSt5**, processes p_0, p_1 and p_2 must take the checkpoint $C_{p_2}^1$ in checkpoint set **ChSt3** , checkpoint $C_{p_2}^2$ in checkpoint set **ChSt4** and checkpoints $C_{p_0}^2$ and $C_{p_2}^3$ in checkpoint set **ChSt5** in order to constitute a consistent global checkpoint set.

However, the creation condition of orphan message and the consistency condition between checkpoints are changed in PWD model. In the systems based on PWD model, the sending event in state interval occurred by non-deterministic event is able to be replayed determinedly.

4   New Consistency Condition

In the execution of processes in systems not based on PWD model, since the execution after checkpoint is progressed non-deterministicedly, the information of event such as sending of messages and internal state transition must be included in checkpoints in order to satisfy the consistency condition of checkpoints [11, 12].

In figure 2, sending event of message $m1$ by process p_0 and delivery event of message $m1$ by process p_1 are included in checkpoints $C_{p_0}^1$ and $C_{p_1}^1$. Also, sending event of message $m2$ by process p_1 and delivery event of message $m3$ by process p_2 are included in checkpoints $C_{p_1}^1$ and $C_{p_2}^1$. Thus, checkpoint set **ChSet1** constitutes a consistent global checkpoint set.

Even if process p_2 includes the information of sending event of message $m3$ in its checkpoint $C_{p_2}^1$ and process p_2 doesn't include the information of delivery event of message $m3$ in its checkpoint $C_{p_0}^1$, the consistency condition of those checkpoints are not violated. The reason of this result is that process p_2 guarantees the deterministic resending of message $m3$ because it contains the information of sending message $m3$. This message $m3$ is called the *in-transit message* [1].

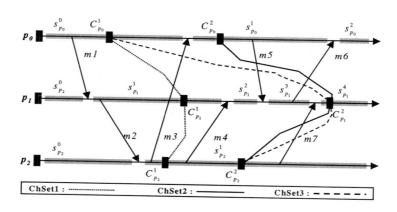

Fig. 2. The convertible message and orphan message in systems based on PWD model

However, checkpoint set **ChSet2** does not constitute a consistent global checkpoint set. The reason of this result is that the delivery event of messages $m5$ and $m7$ is recorded in checkpoint $C_{p_1}^2$, on the other hand, the sending of messages $m5$ and $m7$ not recorded in checkpoints $C_{p_0}^2$ and $C_{p_2}^2$. That is, processes p_0 and p_2 doesn't guarantee the deterministic resending of messages $m5$ and $m7$ and these messages come to be an orphan messages. To **ChSet2** be a consistent global checkpoint set, processes p_0 and p_2 must record the sending event of messages $m5$ and $m7$ in checkpoints $C_{p_0}^2$ and $C_{p_2}^2$. The formal description of this consistency condition is as follows.

Proposition 1. $\forall i, j \ deliver(m_{j \to i}) \in C_{p_i} \Rightarrow send(m_{j \to i}) \in C_{p_j}$ **if and only if** C_{p_i} *is consistent with* C_{p_j}

That is, if checkpoint C_{p_i} of process p_i contains the delivery event of message $m_{j \to i}$, then checkpoint C_{p_j} must contain the sending event of message $m_{j \to i}$. Assume that a checkpoint set satisfy the **Proposition 1**, then a consistent global checkpoint set $\sum C_{p_k}$ is represented as follows.

Lemma 1. *A consistent global checkpoint set* $\sum C_{p_k}$ *is* $\bigcup_{k \in Id} C_{p_k}$ *such that* $\forall i, j \in ID$, C_{p_i} *and* C_{p_j} *satisfy* **Proposition 1**, *Id: process Id*

However, if the execution of process constituting systems is based on PWD model, then the consistency condition between checkpoints proposed in **Proposition 1** and a consistent global checkpoint set in **Lemma 1** must be changed.

In figure 2, checkpoint set **ChSet2** is not able to constitute a consistent global checkpoint set in systems not based on PWD model because of violating the **Proposition 1** by messages $m5$ and $m7$ being an orphan message.

But, checkpoint set **ChSet2** can constitute a consistent global checkpoint set in systems based on PWD model. The reason of this phenomenon is that the sending event and internal state transition occurred by non-deterministic event is able to be determinedly replayed in PWD model. That is, process p_0 can determinedly replay the state interval $s_{p_0}^1$ containing the sending event of message $m5$ because it record the determinant of message $m3$ which generates the state interval sending the message $m5$. This reason is applied equally process p_2. In systems based on PWD model, the formal representation of changed consistency condition between checkpoints is as follows.

Proposition 2. $\forall i,j \ deliver(m_{j\rightarrow i}) \in C_{p_i} \Rightarrow Det(ss_{p_j}(m_{j\rightarrow i})) \in C_{p_j}$ **if and only if** C_{p_i} *is consistent with* C_{p_i}

In **Proposition 2**, $ss_{p_j}(m_{j\rightarrow i})$ is the state interval of process p_j sending the message $m_{j\rightarrow i}$ and $Det(ss_{p_j}(m_{j\rightarrow i}))$ is the determinant of message generating the state interval $ss_{p_j}(m_{j\rightarrow i})$. That is, if process p_i contains the delivery event of message $m_{j\rightarrow i}$ sent by process p_j in its checkpoint and the determinant of state interval $ss_{p_j}(m_{j\rightarrow i})$ of process p_j sending message $m_{j\rightarrow i}$ is contained in checkpoint of process p_j, then two checkpoints maintain the consistent relation.

Like this, the definition of previous orphan message must come to be changed in systems based on PWD model. In figure 2, messages $m5$ and $m7$ in checkpoint set **ChSet3** are orphan messages, but only message $m5$ comes to be an orphan message in systems based on PWD model. The reason of message $m7$ not to be an orphan message in systems based on PWD model is that process p_2 contains the determinant of message $m2$ in its checkpoint. We call the message $m7$ by a *Convertible message*. It has been classified into an orphan message in previous systems, but not classified in systems based on PWD model. The definition of a Convertible message is as follows.

Definition 3. *Message* $m_{j\rightarrow i}$ *is a Convertible message* **if and only if** $\exists \ i,j$ $delivery(m_{j\rightarrow i}) \in C_{p_i}$ *and* $Det(ss(m_{j\rightarrow i})) \in C_{p_j}$ $(i \neq j: process \ id)$

The representation of changed definition of orphan message in systems based on PWD is as follows.

Definition 4. *Message* $m_{j\rightarrow i}$ *is a orphan message* **if and only if** $\exists \ i,j$ $delivery(m_{j\rightarrow i}) \in C_{p_i}$ *and* $Det(ss(m_{j\rightarrow i})) \notin C_{p_j}$ $(i \neq j: process \ id)$

After all, the consistency condition of checkpoints in systems based on PWD model comes to be changed in aspect of the construction of orphan message. In previous systems, convertible messages are contained in orphan messages, but they are not contained in orphan messages in systems based on PWD model. Thus, a consistent global checkpoint set in systems based on PWD model is as follows.

Lemma 2. *A consistent global checkpoint set based on PWD model* $\sum_{PWD} C_{p_k}$ *is* $\bigcup_{k\in Id} C_{p_k}$ *such that* $\forall \ i,j \in ID \ C_{p_i}$ *and* C_{p_j} *satisfy* **Proposition2**, *Id: process Id*

5   Coordinated Checkpointing Algorithm

5.1   Data Structure

Data structures held by process p_i to recover in a failure-free execution are as follows three.

– $DeV_{p_i}^k$: process p_i maintains the transitive dependency vector $DeV_{p_i}^k$ in order to distinguish the state interval of other process having the dependency with its k-*th* state interval[13].
– ϵ_{p_i}: process p_i maintains the variable ϵ_{p_i} that indicates the state interval containing the current checkpoint. Whenever process p_i takes the new checkpoint, it updates the variable ϵ_{p_i} to state interval containing current checkpoint.
– CT_{p_i}: each process p_i maintains the variables CT_{p_i} that is increased by coordinator process on termination of algorithm in order to determine the number of a consistent global checkpoint set containing its checkpoint in systems.

5.2   Algorithm

The coordinated checkpointing algorithm proposed in our papers takes the checkpoint satisfying the **Proposition 2** instead of **Proposition 1**. This reduces the number of checkpoints forced to be taken in systems.

Coordinator Process p_i: Process p_i wants to take new checkpoint comes to be a coordinator process and takes tentative checkpoint through recording from previous checkpoint to current state interval. And it determines process p_j that construct new dependency relation with it through comparing its variables ϵ_{p_i} with the dependency vector of state interval $s_{p_i}^\delta$ taken tentative checkpoint. Selection of process p_j is as follows.

$$forced\ checkpoint(p_j) = \begin{cases} p_j & if(DeV_{p_i}^{\epsilon_{p_i}})_j < (DeV_{p_i}^\delta)_j \\ \text{``No''} & otherwise \end{cases} \qquad (3)$$

Coordinator process p_i send the *request-message* attached its variables CT_{p_i} added by one to process p_j constructing the new dependency after the most recent checkpoint. Coordinator process p_i must delay while it receives *ack-message* from all process sent request-message. And if it receives the all ack-message from sent process sent request-message, it convert its tentative checkpoint to permanent checkpoint and updates the variables ϵ_{p_i} to state interval taken current permanent checkpoint. Also, variables CT_{p_i} is increased by one. Finally coordinator process p_i broadcast the *decision-message* attached increased variables CT_{p_i} to all processes. If coordinator process has not received from all processes, it discard its tentative checkpoint and decrease variables CT_{p_i} by one. And it sends the *discard-message* attached changed value of variables CT_{p_i} to only processes sent request-message.

Non-coordinator Process p_j: Each process received request-message decides its execution by comparing its state interval ϵ_{p_j} containing current checkpoint with current executing state interval.

- $\epsilon_{p_j} < s_{p_j}^{\delta}$: process p_j takes tentative checkpoint in current state interval and updates the value of variables CT_{p_j} to value of new variables CT_{p_i} attached in request-message. And process p_j propagates the request-message to only processes constructing new dependency after the checkpoint indicated by ϵ_{p_j}. Only if process p_j receive the process existing the dependency relation with it, it sends the ack-message to coordinator process.
- $\epsilon_{p_j} = s_{p_j}^{\delta}$: process p_j updates value of CT_{p_j} to value of CT_{p_i}, immediately sends ack-message to coordinator process.

If process p_j receives the request-message sent from process p_k participating same algorithm after taking permanent checkpoint, it immediately sends ack-message to coordinator process without taking additional checkpoint because the values of variable CT_{p_i} attached in two request-message are equal

If process p_j receives the decision-message, it executes as follows.

- $\epsilon_{p_j} < s_{p_j}^{\delta}$: process p_j received decision-message from coordinator process p_i convert current tentative checkpoint to permanent checkpoint. And it updates the value of its variable CT_{p_j} to the value of CT_{p_i} attached in decision-message.
- $\epsilon_{p_j} = s_{p_j}^{\delta}$: process p_j only updates the value of its variable CT_{p_j} to the value of CT_{p_i} attached in decision-message.

If process p_j receives the discard-message, it executes as follows.

- $\epsilon_{p_j} < s_{p_j}^{\delta}$: process p_j discards its tentative checkpoint and updates the value of its variable CT_{p_j} to the value of attached variable CT_{p_i}. And it sends the discard-message to processes propagated the request-message.
- $\epsilon_{p_j} = s_{p_j}^{\delta}$: process p_j updates the value of its variable CT_{p_i} to CT_{p_i}.

Non-Participant Process p_k: If process p_x not participated in algorithm received the decision-message from coordinator process, it means that algorithm is regularly terminated. Because it means that a consistent global checkpoint set is advanced, process p_x updates the value of its variable CT_{p_x} to the value of CT_{p_i} attached in decision-message.

5.3   Proof of Correctness

The correctness of proposed algorithm is proved in theorem 1 through three lemmas.

Lemma 3. *If process p_i take new checkpoint $C_{p_i}^{new}$ in proposed algorithm, then newly taken checkpoint $C_{p_i}^{new}$ of process p_i is a consistent checkpoint.*

Lemma 4. *If process p_i maintain previous checkpoint $C_{p_i}^{unchanged}$ in proposed algorithm, then current checkpoint $C_{p_i}^{unchanged}$ of process p_i is a consistent checkpoint.*

Lemma 5. *If process p_i does not participate in proposed algorithm and maintain previous checkpoint $C_{p_i}^{non-part}$, then current checkpoint $C_{p_i}^{non-part}$ of process p_i is a consistent checkpoint.*

Theorem 1. *If process p_i maintain checkpoint C_{p_i}, then the checkpoint C_{p_i} is always the consistent checkpoint.*

6   Simulation

We simulates the number of checkpoints in proposed coordinated checkpointing and compares this result with two previous coordinated checkpointing schemes [1], [12] through Parsec Simulation Language.

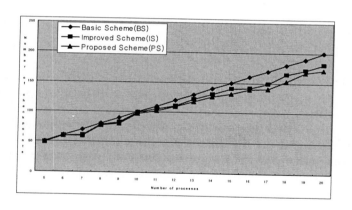

Fig. 3. The results of simulation in three coordinated checkpointing BS, IS and PS

Figure 3 illustrates this results. In figure 3, Basic Scheme(BS) takes the checkpoint without consideration of communication and state interval transition [1]. And Improved Scheme(IS) takes the checkpoint only process directly or transitively sending messages to coordinator processes after its recent checkpoint [12]. The result of simulation is summarized in **Theorem2**.

Theorem 2. *Proposed Scheme PS constructs checkpoints less than IS and BS*

Proof. Assume that variables ρ_{p_i} is the probability variables indicated probability of sending messages to coordinator processes after recent checkpoint by process p_i. However, this variables is suitable for previous scheme [12]. Thus, we classify variables ρ_{p_i} into $\xi_{p_i}^{only-send}$ and $\xi_{p_i}^{after-receive}$. The variables $\xi_{p_i}^{only-send}$ indicates the probability that process p_i directly or transitively sends messages to coordinator process without receiving messages and $\xi_{p_i}^{after-receive}$ indicates the probability that process p_i directly or transitively sends messages to coordinator process after receiving messages. That is, the probability variables ρ_{p_i} is as follows.

$$\rho_{p_i} = \xi_{p_i}^{only-send} + \xi_{p_i}^{after-receive} \tag{4}$$

In previous scheme [12], the number of checkpoints in systems composed of N-th processes is $N \times \sum_{i=0}^{N-1} \rho_{p_i}$. However, our scheme PS takes the number of checkpoints $N \times \sum_{i=0}^{N-1} \xi_{p_i}^{after-receive}$. After all, equation(5) is drew from this result.

$$
\begin{aligned}
&\rho_{p_i} \leq 1 \\
&N \times \sum_{i=0}^{N-1} p_i \geq N \times \sum_{i=0}^{N-1} \rho_{p_i} \\
&\xi_{p_i}^{only-send} \leq 1 \\
&\xi_{p_i}^{after-receive} \leq 1 \\
&\rho_{p_i} = \xi_{p_i}^{only-send} + \xi_{p_i}^{after-receive} \\
&\rho_{p_i} \geq \xi_{p_i}^{after-receive} \\
&N \times \sum_{i=0}^{N-1} p_i \geq N \times \sum_{i=0}^{N-1} \rho_{p_i} \geq N \times \sum_{i=0}^{N-1} \xi_{p_i}^{after-receive}
\end{aligned}
\tag{5}
$$

7   Conclusion

In our papers, we prove that the new consistency condition of checkpoints must be needed in systems based on PWD model and the construction condition of an orphan message must be changed. We classified an orphan message into a convertible message and an orphan message. In systems based on PWD model, a convertible message is not classified into an orphan message. Also, the checkpoint set having a convertible messages is able to construct a consistent global checkpoint set. And, we propose new coordinated checkpointing scheme that only process happened state interval transition must take the new checkpoint, it reduces an unnecessary checkpointing overhead.

References

1. E. N. Elnozahy, D. B. Johnson and Y. M. Wang, "A Survey of Rollbalck-Recovery Protocols in Message Passing Systems," *CMU Technical Report CMU-CS-99-148*, June. 1999.
2. E. N. Elnozahy and W. Zwaenepoel, "On the use and implementation of message logging," *In Proceedings of the 24-th International Symposium on Fault-Tolerant Computing*, pp. 298–307, Jun. 1994.
3. R. E. Strom and S. Yemini, "Optimistic recovery in distributed systems," *ACM Trans. on Computer Systemss*, vol. 3, No. 3, pp. 204226, Aug. 1985.
4. L. Alvisi, "Understanding the message logging paradigm ofr mssking process crashes," *Ph. D. Thesis*, Department of Computer Science, Comell University, Jan. 1996.

5. L. Alvisi and K. Marzullo, "Message Logging: Pessimistic, Optimistic, Causal and Optimal," *IEEE Trans. on Software Engineering*, vol. 24, pp. 149–159, Feb. 1998.
6. S. Rao, L. Alvisi and H. M. Vin, "The cost of recovery in message logging protocols," *In Proceedings of the 17-th IEE-E Symposium on Reliable Distributed Systems(SRDS)*, pp. 10-18, 1998.
7. D. B. Johnson and W. Zwaenepoel, "Recovery in distributed systems using optimistic message logging and checkpointing," *Journal of Algorithms*, vol. 11, pp. 462–491, Sept. 1990.
8. E. N. Elnozahy, D. B. Johnson and W. Zwaenepoel, "The Performance of Consistent Checkpointing," *Proc. 11th Symp. on Reliable Distributed Systems*, pp. 39–47, Oct. 1992.
9. D. B. Johnson and W. Zwaenepoel, "Sender-based message logging," *In Proceedings of the 17th International Symposium on Fault=Tolerant Computing*, pp. 14–19, June. 1987.
10. D. B. Johnson, "Distributed system fault tolerance using message logging and checkpointing," *Ph. D. Thesis*, Rice University, Dec. 1989.
11. K. M. Chandy and L. Lamport, "Distributed snapshots: Determining global states of distributed systems," *ACM Trans. on Computer Systems*, vol. 3, No. 1, pp. 63–75, Feb. 1985.
12. R. Koo and S. Toueg, "Checkpointing and rollback-recovery for distributed systems," *IEEE Trans. on Software Engineering*, vol. SE-13, No. 1, pp. 23–31, Jan. 1987.
13. G. Cao and M. Singhal, "On Coordinated Checkpointing in Distributed Systems," *IEEE Trans. on Parallel and Distributed Systems*, vol. 9, No. 12, pp. 1213–1225, Dec. 1998.
14. H. J. Michel, R. H. Netzer and M. Raynal, "Consistency Issues in Distributed Checkpoints," *IEEE Trans. on Software Engineering*, vol. 25, No. 2, pp. 274–281, March/April. 1999.
15. Nitin H. Vaidya, "Staggered Consistent Checkpointing," *IEEE Trans. on Parallel and Distributed Systems*, vol. 10, No. 7, pp. 694–702, July. 1999.
16. J. Fowler and W. Zwaenepoel, "Causal Distributed Breakpoints," *Proc. Int'l Conf. Distributed Computing Systems*, pp. 134–141, May. 1990.

Extensions to DNS for Supporting Internationalized Domain Names

Dongman Lee[1], Hyewon Shin[1] Soon J. Hyun[1], Younghee Lee[1], Myoungchurl Kim[1], and Hee Yong Youn[2]

[1] Information and Communications University
Taejon, Korea
{dlee, hwshin, shyun, yhlee, mckim}@icu.ac.kr
[2] School of Electrical and Computer Engineering
Sungkyunkwan University, Suwon 440-746, Korea
youn@ece.skku.ac.kr

Abstract. As the Internet become popular around the globe, there have been several approaches to support internationalized domain names, that is, names consisting of non-ASCII characters. However, they require modification of existing applications or significant changes in the existing DNS protocol. In this paper, we propose an extension to the existing DNS protocol for supporting internationalized domain names while maintaining interoperability and compatibility with the DNS. For this, we exploit the reserved field of Opcode in the DNS query to support Internationalized Domain Names. The proposed scheme uses a UTF-8 encoding scheme as a wire format since it allows ASCII names to remain intact. We implement the proposed scheme using BIND 8.2.1 on Solaris, Sun Unix OS. The performance results show that the proposed scheme does not incur any significant overhead while it maintains the interoperability and compatibility with the existing DNS.

1 Introduction

As the Internet becomes popular around the globe, there have been several efforts to allow users to use their mother tongue instead of English in accessing the Internet. Internet protocols such as HTML, XML, IMAP, FTP, and many other text-based protocols were at least partially internationalized [3]. However, the domain name system (DNS) is yet to follow this paradigm shift. It only allows names in ASCII characters, so called alphanumeric characters, which are difficult to remember or use for the people who are not familiar with English.

There have been several approaches to support internationalized domain names, that is, names consisting of non-ASCII characters. They can be classified into two approaches: application extension and DNS protocol extension. The former supports internationalized domain names without modifying or changing the current DNS protocol. IDNA [4] proposes a scheme in which the application encodes non-English domain names (hereafter, we call them internationalized domain names, IDNs) into names compatible with the existing DNS, that is ASCII characters. This allows users

I. Chong (Ed.): ICOIN 2002, LNCS 2344, pp. 791–801, 2002.
© Springer-Verlag Berlin Heidelberg 2002

to use IDNs at their application's user interface and yet these names to be handled by the existing DNS servers. However, this approach requires the modification of existing applications using the DNS protocol. UDNS [15] also requires the application to encode an IDN into ASCII-compatible names like IDNA when the underlying DNS does not support non-ASCII characters. Some approaches propose using a UTF-5 based encoding scheme [16] that converts non-ASCII characters into ASCII-compatible ones. However, this scheme requires even ASCII character based names to be encoded as well. Stuart Kwan suggests using a UTF-8 encoding scheme in which ASCII-based names remain preserved [12]. However, it does not provide interoperability with the systems supporting only ASCII-based names [6][20]. J. Klensin suggests a new class for IDNs [9][10]. Unlike the approaches described earlier, this allows non-ASCII names to be handled separately from ASCII names. However, with this scheme, the current DNS query format should be extended, which in turn the whole DNS components should be modified as well as other related protocol [18].

We propose an extension to the existing DNS for supporting IDNs. We call the extension EDIDN *. As discussed above, we should consider interoperability and compatibility with the current DNS. For this, we exploit the reserved field of Opcode in the DNS query to support IDNs. The proposed scheme uses UTF-8 encoding as a wire format because UTF-8 encoding does not impose any changes into ASCII-based names. However, to support interoperability with conventional DNS servers or resolvers, it only uses a new Opcode when non-ASCII names are asked to resolve. If a query containing a non-ASCII name is sent to a server with the current DNS, that is, supporting only ASCII names, the resolver in the proposed scheme resends the query to servers supporting EDIDN. To distinguish errors from the servers supporting EDIDN, we use the reserved filed of RCODE in a DNS response message. We implement the proposed scheme using BIND 8.2.1 [8] on Solaris, Sun Unix OS. The performance results show that the proposed scheme does not incur any significant overhead to the existing DNS.

The rest of this paper is organized as follows. We discuss the existing approaches for supporting internationalized domain names in Section 2. Section 3 describes the proposed scheme in detail. Section 4 explains the implementation and performance analysis. Conclusion follows in Section 5.

2   Related Work

The current DNS infrastructure does not support internationalized domain names. Approaches for supporting internationalized domain names in DNS can be divided into application extension and protocol extension. In this section, we describe both approaches including issues in them.

---

*Extensions to DNS for supporting Internationalized Domain Names.

2.1 IDNA (Internationalized Domain Name Application)

IDNA [4] enables internationalized domain names in DNS name labels. IDNA works by allowing applications to use certain ASCII name labels to represent non-ASCII name labels. In IDNA, applications perform the processing of internationalized domain names from users, and display internationalized host names to users. The IDNA protocol does not affect the existing DNS protocol. However all applications must have libraries for performing conversion from IDNs to ACE (ASCII-compatible encoding) names and vice versa.

2.2 IDNE (Internationalized Domain Name Using EDNS)

IDNE [2] describes an extended mechanism based on EDNS that enables to the use of IDN without causing harm to the current DNS. IDNE enables IDN host names with as many characters as current ASCII-only host names. It fully supports UTF-8 character encoding set and conforms to the IDN requirements. The major drawback of IDNE is that all protocols, applications and DNS servers will have to be upgraded to support this proposal.

2.3 Internationalizing the DNS, New Class

J. Klensin proposed a new Class and a new DNS lookup schemes [9][10][11]. The proposal considered future extension through the use of defining new Classes. All current usage in the Internet environment uses the "IN" class [10]. A new extended mechanism via a new Class scheme may have several reasons that extended and conventional DNS name labels and other fields are defined as an internationalization approach, not only ASCII. Extend DNS with new RR types and new class requires changes to name server, resolver, applications and all other related components. Potential future DNS extension does not allow conventional DNS to handle the new class and other related specific extension part. It requires changes in the current DNS protocols and related Internet protocols.

3 Design

In this section, we describe the design considerations and the details of EDIDN protocol and how the existing DNS protocol is integrated into it. We present the proposed protocol including the resolver and name server specification.

3.1 Design Consideration

3.1.1 Internationalization

The encoding of domain names in the Internet is restricted to a subset of 7 bits ASCII [13]. A domain name must start with a letter, end with a letter or digit, and have as interior characters only letters, digits, or hyphens. There are also some restrictions on the length. Labels must be 63 characters or less [13]. They are restricted to DNS itself and many other protocols on Internet. In accordance with the use of internationalized domain name, the EDIDN protocol will require some changes in comparison with the current DNS. All internationalized domain names must be represented in one standard format. We consider the existing host name rule that is compatible with the current DNS encoding, complexity and range of characteristics of internationalized script. So the extended DNS protocol should process the Universal Character Set as ISO 10646 [17]. We select UTF-8 encoding as CES (Coded Encoding Set). It is compatible with the current ASCII-based protocols and it does not process ambiguous characters. The internationalized domain names may use several local character sets, but they must be encoded into Unicode and UTF-8 before being sent over a wire.

3.1.2 Compatibility and Interoperability

To support internationalized domain names, DNS should also allow any UTF-8 octet characters to be processed and passed over the network. Therefore, the extended DNS protocol for internationalization must provide backward compatibility and interoperability with the current DNS [3][19]. In other words, the new protocol must be used with the current protocol with no or minimum changes. For this, we exploit the reserved field of the Opcode and RCODE in a DNS query since it requires no extension to the DNS protocol itself. The extension should also allow Internet protocols to handle names without conversion.

3.1.3 Canonicalization

DNS indexes and matches domain names to look up a domain name from zone data [5]. In the conventional DNS, canonicalization is subject to US-ASCII only. However, every internationalized character must be canonicalized in its own rules for a DNS standardized matching policy, e.g. case-insensitive matching rule.

3.1.4 Operational Issues

EDIDN needs an operation for interoperability with the current DNS. Therefore, it is needed to specify the operational guidelines for EDIDN. There could be three different cases when the extended protocol and the existing DNS protocol are used together: the existing resolver and the extended server, the extended resolver and the existing server, and the extended resolver and the extended server. The first and third cases should work with no modification or changes since the extended server can support ASCII-based domain name as well as IDNs. The second case requires an extra consideration since the existing server does not handle non-ASCII based names.

3.2 Proposed Extension to DNS for Supporting Internationalized DNS

3.2.1 Protocol Extension

EDIDN supports internationalized domain names while maintaining interoperability with the existing DNS and other protocols. For this, we use the existing DNS header format without modification and exploit the unused values of Opcode and RCODE in a DNS query / response message.

Table 1 shows the Opcode field of a DNS query header. The Opcode section is composed of a four bit field that specifies the type of a DNS query. The field is set by the originator of a query and copied into the request [14]. To indicate an EDIDN query (EQUERY), we use an unused value which is 3. That is, if the value is 3, a given query contains an international domain name.

Table 1. Opcode Field of EDIDN Query

Bit Value	Specifies kind of query
0	A standard query question (QUERY)
1	An inverse query (IQUERY)
2	A server status request (STATUS)
3	An extended Query (EQUERY)
4 - 15	Reserved for future use

A conventional RCODE is composed of six bits. The values, 0 to 5, are assigned for a conventional DNS response query [14]. There can be two error cases when an EDIDN sends a query to a conventional DNS server, that is, a server that cannot handle an international domain name: the DNS server only processes the opcode value of 0, 1, or 2 and the DNS server does handle the opcode value greater than 2 as well. In the former case, the server will return the RCODE with 4, "not implemented," while the server will return the RCODE with 3, "name error" in the latter case. To allow a query with international domain names to be processed, we do not treat these RCODE values as an error. Instead, the EDIDN resolver resends the query to the next authoritative server until the name is resolved. To distinguish the same RCODE return values from an EDIDN server, which should be treated as an error, we use the unused RCODE values, 6 and 7 for this purpose. Table 2 shows the values of the RCODE field in an EDIDN query and their interpretation.

Table 2. Field of Extended RCODE Type

Current NS	EDIDN NS	Response Message	Response Message Interpretation
0	0	No error condition	No error condition
1		Format error	The name server was unable to interpret the query.
2		Server failure error	The name server was unable to process this query due to a problem with the name server.
3	6	Name error	Meaningful only for responses from an authoritative name server, this code signifies that the domain name referenced in the query does not exist.
4	7	Not Implemented	The name server does not support the requested kind of query.
5		Refused	The name server refuses to perform the specified operation for policy reasons. For example, a name server may not wish to provide the information to the particular requester, or a name server may not wish to perform a particular operation (e.g., zone transfer) for particular data

3.2.2 Resolver Extension

If a name server is an EDIDN name server, it can process and save an extended query and reply an answer to the original EDIDN resolver. If a name server is a conventional name server, it cannot process and save an EDIDN query, and then sends an error status to the original EDIDN resolver. The EDIDN resolver sends a recursive query to other name servers and the forwarder to get a right answer as described above.

3.3 Operational Model

We identify the following important design goals for EDIDN that enables internationalized domain name and conventional domain name at the same time. We should consider three cases of the usage of between EDIDN and conventional DNS: (i) it is interoperate with extended resolvers and conventional name servers, allowing users to seamlessly handle the internationalized domain name; (ii) it is interoperate with extended resolver and extended name servers. The extended resolver must continue to allow any queries and responses that result of EDIDN protocol and conventional DNS protocol. (iii) it is interoperate with conventional resolver and extended name servers. Fig 1, Fig.2 show EDIDN and DNS processing. [14].

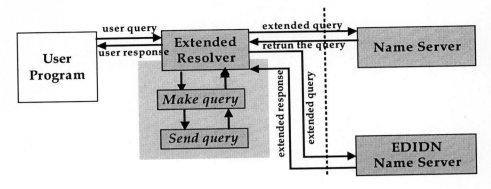

Fig.1. Extended Resolver with Conventional name server and extended Nam server

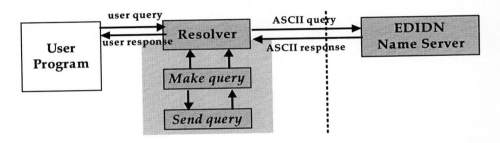

Fig.2. Conventional resolver with Extended Name Server

4 Implementation

We implement EDIDN on Solaris, written in C. We use BIND 8.2.1 [8] which is an implementation of the Domain Name System protocols and provides an openly redistributable reference of implementation of the major components of the Domain Name System. In this section, we present the details of two main components of EDIDNS: resolver and name server. Fig 3 shows how these two components interact with each other to resolve domain names.

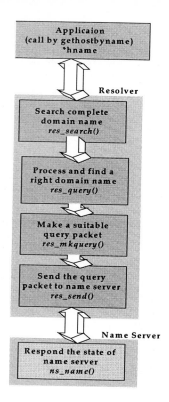

Fig. 3. Resolver and Name Server Library Routines

4.1 Resolver

EDIDN uses a resolver library to communicate with a name server. The resolve issues to its name server a query that is used to resolve an internationalized domain name. It converts the local character set to UCS and then to UTF-8 using *ucs4_to_utf8()*. For reverse mapping, *utf8_to_ucs4()* is provided. *res_search()* is extended to search an internationalized domain name and made up full domain name and to call the res_query(). *res_query()* builds a query, sends and awaits a response from a server. This routine is extended to handle EDIDN queries. If a query contains an international domain name, the query's opcode is set to EQUERY by *res_enmkquery()*, which is added for EDIDN. Then, *res_query()* waits for a response from a server. If the RCODE value of the response is NXDOMAIN or NOTIMP, *res_query* is extended to try again for recursive query.

4.2 Name Server

We modify the name server library routine because we distinguish between conventional responses and EDIDN ones. We modify the opcode interpretation and RCODE assigning parts for handling EDIDN queries. *ns_main()* is extended to handle domain names encoded in UTF-8. We also extend the zone file editor which allows international domain names encoded in UTF-8 to be added into a zone file.

4.3 Experimental Results and Analysis

We performed a set of experiments to measure the performance of the proposed extension, EDIDN System. The important metric for the experiments is execution time for resolving a query, compared with the current DNS. The experiments were run on PC as a client and on Solaris as a name server connected by 10 Mbps Ethernet. We measured the execution time in the three cases as follows: between current DNS resolver and DNS name server, between EDIDN resolver and EDIDN name server and between EDIDN resolver and DNS name server. We run the experiments 50 times for each case and the results were averaged out of them.

Table 3 shows the execution time in each case. The execution time consists of query preparation and display at resolver, query transmission, and query resolution at server.

Table 3. Execution Time of Each Case

	Resolver (ms)	Transmission (ms)	Server (ms)	Total Execution Time (ms)
DNS Resolver and DNS Server	1.289	1.324	1.022	3.636
EDIDN resolver and EDIDN server	1.384	1.374	1.023	3.783

As Table 3 shows, the total extension time for query resolution by the EDIDN resolver and server is almost the same (less than 4 % overhead) as that by the current DNS resolver and server. It is because we use the reserved bits values in the current DNS message format. The EDIDN resolver takes a little more time than the existing resolver since conversion from a local code to UCS to UTF-8 is required in the EDIDN resolver. Therefore, EDIDN does not incur any significant overhead into the current DNS system.

5 Conclusion

In this paper, we propose an extension to DNS for supporting an international domain name while supporting interoperability and compatibility with the existing DNS. For internationalized domain names, we use Unicode as character coding set and UTF-8 as an encoding scheme. For interoperability and compatibility, we consider three cases: the existing resolver and the extended server, the extended resolver and the existing server, and the extended resolver and the extended server. For this, we exploit the unused values of Opcode and RCODE in a query and a response, respectively, which requires no modification to the existing DNS. To indicate an EDIDN query (EQUERY), we use a separate Opcode value, 3. To distinguish the same RCODE return values from an EDIDN server, which should be treated as an error, we use the unused RCODE values, 6 and 7 when an EDIDN sends a query to a conventional DNS server, that is, a server that cannot handle an international domain name. In this case, the EDIDN resolver sends a query iteratively until a given internationalized domain name gets resolved. We implement EDIDN on Solaris, written in C using BIND 8.2.1. The performance results show that the proposed scheme does not incur any significant overhead while it maintains interoperability and compatibility with the existing DNS.

References

1. Albits, P., Liu, C.: DNS and Bind.3rd ed., O'REILLY (1998)
2. Blanchet, M., Hoffman, P.: internationalized domain names using EDNS using EDNS (IDNE), Internet drafts, draft-ietf-idn-idne-02.txt, March 2001
3. Braden, R.: Requirements for Internet Hosts -- Application and Support. Internet Engineering Task Force, RFC 1123 (1989)
4. Faltstrom, P., Hoffman, P.: Internationalizing Host Names in Applications. Internet drafts, draft-ietf-idn-idna-05.txt (2001)
5. Gustafsson, A.: DNS Zone Transfer Protocol Clarifications. Internet draft, draft-ietf-dnsext-axfr-clarify-03.txt (July 2001)
6. Hoffman, P.: Comparison of Internationalized Domain Name Proposals. Internet drafts, draft-ietf-idn-compare-01.txt (2000)
7. Hoffman, P.: Internationalized Host Names Using Resolvers and Applications (IDNRA). Internet drafts, draft-ietf-idn-idnra-00.txt (2000)
8. ISC Web Site: http://www.isc.org/products/BIND/, BIND Version 8.2.1.
9. Klensin, J.C.: A Search-based access model for the DNS. Internet draft, draft-klensin-dns-search-02.txt (2001)
10. Klensin, J.C.: Internationalizing the DNS -- A New Class. Internet draft, draft-klensin-i18n-newclass-01.txt, (2001)
11. Klensin, J.C.: Role of the Domain Name System. Internet draft, draft-klensin-dns-role-01.txt (2001)
12. Kwan, S., Gilroy, J.: Using the UTF-8 Character Set in the Domain Name System. Internet drafts, draft-skwan-utf8-dns-04.txt (2000)
13. Mockapetris, P.: Domain Names - Concepts and Facilities. Internet Engineering Task Force, RFC 1034 (1987)
14. Mockapetris, P.: Domain Names - Implementation and Specification. Internet Engineering Task Force, RFC 1035 (1987)

15. Oscarsson, D.: Using the Universal Character Set in the Domain Name System. Internet drafts, draft-ietf-idn-udns-03.txt (2001)
16. Seng, J., Duerst, M., Tan,T.W.: UTF-5, a transformation format of Unicode and ISO 10646. Internet drafts, draft-jseng-utf5-01.txt (2001)
17. The   Unicode   Code   Consortium: The   Unicode   Standard, http://www.unicode.org/unicode/standard/standard.html
18. Weider, C., Preston, C., Simonsen, S.: The Report of the IAB Character Set Workshop. Internet Engineering Task Force, RFC 2130 (1997)
19. Wenzel, Z. Seng, J.: Requirements of Internationalized Domain Names. Internet drafts, draft-ietf-idn-requirements-08 (2001)
20. Yergeau, F.: UTF-8, a transformation format of ISO 10646. Internet Engineering Task Force, RFC 2279 (1998)

An Architecture of Agent Repository for Adaptive Multiagent System

Takahiro Uchiya[1], Takashi Katoh[1], Takuo Suganuma[1], Tetsuo Kinoshita[2], and Norio Shiratori[1]

[1] Research Institute of Electrical Communication / Graduate School of Information Science, Tohoku University, 2-1-1 Katahira, Aoba-ku, Sendai, 980-8577 JAPAN
{takahiro, p-katoh, suganuma, norio}@shiratori.riec.tohoku.ac.jp
[2] Information Synergy Center, Tohoku University, 2-1-1 Katahira, Aoba-ku, Sendai, 980-8577 JAPAN
kino@shiratori.riec.tohoku.ac.jp

Abstract. It is important to use agents' behavioral history in a multiagent system effectively as a way to realize advanced service organization / adjustment in order to make the multiagent system adaptable to users and environments. However, it is difficult for multiagent system to keep and manage its behavioral history and to reflect the history to the entire system's behavior effectively after some task executions in agent workplace. To cope with this difficulty, in this paper, we propose a new multiagent framework to improve performance of the agents' cooperative works by using the behavioral history of each agent. A new architecture and some mechanisms are introduced to the agent repository in the framework. We designed and implemented a new agent repository based on the proposal, and some results of experiments using the prototype system show the effectiveness of our approach.

1 Introduction

In global dispersal networked environment such as the Internet, generally, user requirements on heterogeneous network / platform environments are highly diverse. In spite of growth of network bandwidth and computer performance, these diversities would be gradually increased. These situations may cause difficulties of resource allocation of network / platform and suitable / smooth service provision to users [1].

In this context, the multiagent framework is proposed as a new approach corresponds to complex user requirements and heterogeneity of environments [2,3,4,5]. In the multiagent system, agents cooperate with each other and perform organic actions such as configuration of their organization, reconfiguration of the organization and negotiation. According to these properties, multiagent system is regarded as the system that provides dynamical service configuration and tuning function for users.

I. Chong (Ed.): ICOIN 2002, LNCS 2344, pp. 802–813, 2002.
© Springer-Verlag Berlin Heidelberg 2002

In service configuration and tuning process for the specific users and environments, it is significant to utilize agents' behavioral history effectively in order to improve performance of the agents' cooperation. For example, the agents that give the video-conference services [4, 5] adapt to users and environments by cooperatively changing the parameters of the quality of services (video resolution, video frame rate, audio quality, etc). If the agents' behavioral history can be utilized after their execution in the network environment, in case that a similar videoconference service requirement occurred on the similar condition of network and platform, it would be possible to omit the excess message exchanging and knowledge process, in the phase of suitable agent selection and initial parameter setting for video-conferencing. Consequently, the effective cooperative behavior is attained.

Nevertheless, it is difficult for multiagent system to keep and manage its behavioral history and to reflect the history to the entire system's behavior effectively after some task executions in agent working environment. Since the agents exist dispersedly on the network environment, the means of collecting and reusing the individual agent's behavioral history is highly difficult to accomplish.

To cope with this difficulty, we focus on a repository-based multiagent framework [1, 2] that carries out the organization and reorganization of agents using the agent repository. The repository-based multiagent framework is the system framework to provide effective use/reuse of software components and expertise on the components. The software components and related knowledge are stored in the repository as agents. By configuring and adjusting the agent organization from the repository, suitable services are provided dynamically according to the situation of user requirements and network / platform environments. We focus on the function-centralization property of the agent repository, and introduce the agents' behavioral history utilization mechanism in the repository of the above framework.

In this paper, we propose a new multiagent framework to improve performance of the agents' cooperative works by using the behavioral history of each agent. A new architecture and some mechanisms are introduced to the agent repository in the framework. We embed the following mechanisms in the agent repository.

(M1) Agent Lifecycle Management Mechanism

The mechanism that manages the agents which are once instantiated and perform the task in the working environment. The mechanism manages the agents' whole life-cycle.

(M2) History Utilization Mechanism

The mechanism that utilizes agents' behavioral history and adjusts behavior of agents in the configuration phase of the agent organization to improve efficiency of the service provision.

(M3) Repository Management Agent

The agent that plays a role of maintenance of the whole repository system.

We call the new repository we propose in this paper as Active Agent Repository (AAR). The AAR extracts the agents' behavioral history after their executions, acquires the operational knowledge from the history, and reuses them, by close cooperation among the three mechanisms described above.

This paper is organized as follows. Firstly, in section 2, we propose the concept of AAR and the AAR-based multiagent framework. Subsequently, in section 3, we describe the design of AAR. In section 4, we illustrate the implemented prototype system based on the proposed architecture. We also demonstrate the experimental results to show the effectiveness of the proposed architecture in section 5. Finally we conclude this paper.

2   Concept of Active Agent Repository

2.1   Repository-Based Multiagent Framework

We briefly explain the concept of the ADIPS framework [2,3] as a typical instance of the repository-based multiagent framework. ADIPS (Agent-based Architecture of Distributed Information Processing Systems) is the framework of constructing the multiagent system on the dispersal environment.

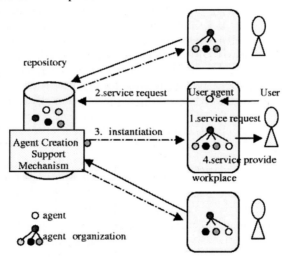

Fig. 1. Service Configuration Function of ADIPS framework

In the ADIPS framework, dynamical service provision is realized by the Service Configuration Function of the framework. Figure 1 shows the model of the Service Configuration Function of the ADIPS framework. In this model, a user requests the service to the agent called 'user agent', which is responsible for acquisition of the user's requirements. The user agent transfers the request to the agent server called 'repository'.

In the repository, agent group that adapts to the user request and environment is selected and the organization is configured by cooperative works among the agents. These agents are instantiated onto the agents' working environment called 'workplace'. The Service Configuration Function is achieved by the Agent Creation Support

Mechanism in the repository and negotiation capability to handle the contract-net [6] based protocol in each agent.

A remarkable feature of the repository-based multiagent framework, such as ADIPS framework, is the centralized management of service function of the agents. That is, if we adjust or tune the agent specification in the repository, then it is possible to change the behavior of the whole multiagent system. By effective use of this property, we can make the behavior and performance of whole multiagent system efficient. So far, this type of behavior tuning is actually performed by human administrator or designer of the repository based on the results of careful observation of the agents' behavior in the workplaces. Although this behavior tuning is very effective for adequate operation of the repository, it markedly increases administrative tasks of the repository. The lack of support functions for this type of tuning has been pointed out.

2.2 AAR-Based Multiagent Framework

In order to address the problem described in the previous subsection, we consider the automatic behavior tuning of agents in the repository to attain the improvement of performance of the agents' cooperative works. From analysis of an administrator's activities concerning maintenance of the repository, we decided to extend the functions of existing agent repository in order to perform the maintenance tasks automatically. We call the extended repository as the Active Agent Repository (AAR), and designed a new multiagent framework based on the AAR.

Figure 2 shows the model of the AAR-based multiagent framework.

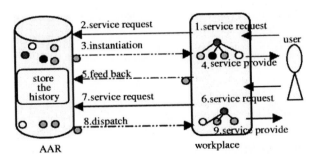

Fig. 2. AAR-based multiagent framework

In this framework, the agent organization that is instantiated in the workplace provides service to user same as the traditional framework. After this basic action, they feed back onto AAR keeping their behavioral history by themselves. In AAR, the agents' behavioral history is stored. If necessary the history is extracted and processed, and it is reflected to agents in form of agent knowledge. Therefore, in case that a user requirement arrives to the repository on the same condition as the service provided previously, complex cooperative works among the agents in AAR is simplified. This would make the selecting and dispatching the agent group smooth.

3 Design of Active Agent Repository

Figure 3 shows the architecture of the AAR. AAR consists of four mechanisms, i.e., the Agent Creation Support Mechanism (ACSM), the Agent Lifecycle Management Mechanism (ALMM), the History Utilization Mechanism (HUM) and the Repository Management Agent (RMA). ACSM is the mechanism which is responsible for the configuration / reconfiguration of the agent organization and the instantiation of them. It plays a same role of ACSM in the traditional repository. Thus, in this section we describe the design detail of newly added three mechanism of the AAR, i.e., ALMM, HUM and RMA.

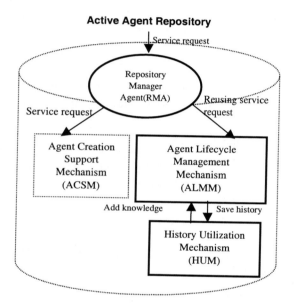

Fig. 3. Architecture of AAR

3.1 History Utilization Mechanism

The History Utilization Mechanism (HUM) is the mechanism in order to reuse the agents' behavioral history by close cooperation with the Agent Lifecycle Management Mechanism (ALMM). It extracts and processes the history, and appends it to the agents as their knowledge for the sake of effective cooperation among agents. Figure 4 shows the internal structure of the HUM. HUM consists of a reasoning engine for utilization of the behavioral history and a database of the behavioral history.

When an agent feeds back to the repository from the workplace, ALMM sends the agent's behavioral history data to HUM. The HUM stores the data into the database of the behavioral history. The history data is also put to the working memory in the form of the 'facts'. According to this change operation on the facts in the working memory, the reasoning engine starts its inference. The reasoning engine extracts information that is useful in tuning of configuration process of the agent organization from the history data. The engine also makes some operations on knowledge of agents in AAR. In this reasoning engine, the inference mechanism is implemented as a production system.

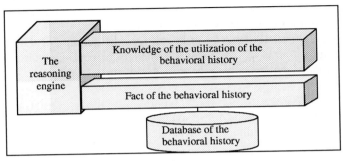

Fig. 4. Internal structure of HUM

3.1.1 Database of the Behavioral History

The HUM stores and manages the data of the behavioral history using a database scheme. In the database, a data of each agent's behavioral history is represented as a record to realize effective search and retrieve of amount of data. Figure 5 shows the model of the history data (HD).

```
HD = <RD,HDB>
RD = repository description
HDB = {hd | hd = <hddesc, aname,placename,uname,OCDesc,WM,WMH,MH>}
hddesc = meta data description
aname = {an | an = <rname,wname>}
rname = agent name in repository
wname = agent name in workplace
placename = workplace name
uname = user name
OCDesc = organize construction information
WM = workingmemory description
WMH = workingmemory history
MH = message history
```

Fig. 5. Model of history data

3.1.2 Reasoning Engine

Here we explain the reasoning engine for the utilization of the behavioral history that adds the knowledge dynamically to the agent. The reasoning engine is constructed as a rule-based system that executes the action part in a 'rule' whose condition part matches to some 'facts' in the working memory. In this reasoning engine, on the one hand, the history data is set in the working memory as the facts. The knowledge to use the history data, on the other hand, is written by the agent developer / designer in the form of rule into the rule memory. Figure 6 shows an example of the knowledge of the utilization of the behavioral history.

To realize the function of HUM, we added the following two action commands used in the action part in the rule description.

a)makelink

(makelink performative ManagerAg Ag ...)

An action to adjust the behavior of agents for configuration of the agent organization. When the action is invoked, new knowledge for configuration of the agent organization is added dynamically to the manager agent.

b)makerule

(makerule Ag addRule)

An action to adjust the behavior of agents. This action is used for general purpose knowledge operation by adding rules dynamically to the agents.

```
(historyknowledge data01
(knowledge
 (rule rule1
  (data :RNAME video :performance max ) = ?data1
  (data :RNAME board :performance max) = ?data2
  (data :RNAME fvcs) = ?ma1
 -->
 (makelink best-performance-videogroup ?ma1:wname
                 (?data1:WNAME ?data2:WNAME))))
```

Fig. 6. An example of knowledge of the utilization of the behavioral history

3.2 Agent Lifecycle Management Mechanism

The Agent Lifecycle Management Mechanism (ALMM) holds the agents which feed back from the workplace after providing the services. It also manages the agents' lifecycle after their feed back. ALMM consists of the following three functions,

(1) Agent transport function

The function to receive the agents that feed back from the workplace and to be resided the agents in the agent resting place. Moreover, this function dispatches the agents from the agent resting place to the workplace again.

(2) Agent resting place

To hold the agents which feed back from the workplace, the workplace-like agent working environment is needed in the repository to let the fed back agents running. The agent resting place provides the environment to be able to handle the behavioral history with HUM and to act cooperatively for adjusting the knowledge.

(3) Lifecycle management function

The function to manage the agents' lifecycle in the agent resting place. Concretely, this function puts unused agents in order, or unifies the agents which have same knowledge. This function is very important for effective operation of the repository, but we do not discuss this in this paper.

3.3 Repository Manager Agent

This agent monitors and manages the status of the whole repository. In particular, RMA plays the part of handling and passing the request message to adequate mechanisms such as ACSM and ALMM, when a message arrives to the repository. The followings are the functions of the RMA.

(1) Agent Search Function

When a agent search request arrives from the workplace with the performative (the specifier of agent communication primitive) <search-agent>, RMA returns the value of 'true' or 'false' after checking the existence of the agent by using the Agent Management Table.

(2) Agent List Function

When a request to list up the available agent names arrives from the workplace with the performative <request-agentnames> or <request-servicenames>, RMA returns the list of the available agent names.

(3) Agent Instantiation Invocation Function

When a request of use of the agent arrives from the workplace with the performative <request-agent> or <request-service>, RMA decides to pass the request message to either ACSM or ALMM. If the agent who is used previously will be available, RMA passes the request to ALMM, otherwise RMA transfers the request to ACSM.

The RMA holds the four tables representing the status of repository to deal with many types of requests sent from the workplace, i.e., the Agent Management Table (AMT), the Service Name Table (SNT), the Service Management Table (SMT), and the Performative Management Table (PMT). Figure 7 shows the outline of each table. Figure 8 shows the message handling algorithm of RMA based on the tables. RMA handles the messages from workplace by checking the performatives of the request message and it can distinguish whether the service has been used or not by checking the SMT.

```
AMT = { a_desc la_desc = <aname, astate>}
   aname = agent name
   astate = agent state
SNT = {s_desc ls_desc = < sname>}
   name = service name
SMT = {ss_desc lss_desc = <sname,uname,wpname,oinfo>}
   wpname = workplace name
   uname = user name
   oinfo = organization information
PMT = {a_perf}
   a_perf = available performative
```

Fig. 7. Tables of the RMA

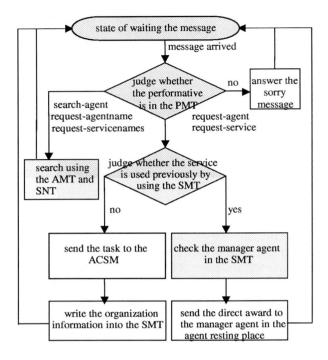

Fig. 8. Message handling algorithm of the RMA

4 Implementation

We implemented a prototype of the Active Agent Repository based on the design described in section 3. The implementation environment of the AAR is configured on a Sun Ultra SPARCStation with the Java2 programming language. For the basis of the agent framework, we used the TAF [7] agent platform. The TAF is one of the derived versions of the ADIPS framework and is developed aiming at support of studying and teaching of agent system development. This agent platform is originally designed for the use in local environment to reduce the excess complexities for novice users of agent systems. Thus, we extended the communication capability between the repository and the workplace by using the JavaRMI. This extension allows us to use TAF over the network. Subsequently, we extended the repository of TAF to AAR-based on the design in section 3. The reasoning engine of the HUM was implemented by using the rule-base system provided by TAF. Because of the compatibility to the TAF rule-base system, the knowledge description for HUM is basically in conformity with the rule description of the TAF, and we just extended the part of the actions as described in 3.1.2. The total number of Java classes for AAR is 50 classes.

5 Experiments and Evaluation

5.1 Experiments

As the application of the experiment of our proposed system, we selected the Flexible Videoconference System [8,9] (FVCS). We performed two types of experiments using FVCS, i.e., feasibility evaluation experiment (Ex.1) and the efficiency evaluation experiment (Ex.2). In the Ex.1, we examine that the prototype system is built properly according to the proposed design. In the experiment, we send some agent use request to AAR so as to be dispatched the fed back agents. In the Ex.2, we examine whether the configuration time of agent organization will be abbreviated or not, by using the information obtained from the history data, in the case that the service request arrives from the same user.

5.1.1 Experiment environment. The hardware environment of the FVCS consists of three Ultra SPARCStations connected with a switching hub of 100Base-T. On this hardware environment, we installed the AAR-based multiagent framework implemented by Java programming language. One of the SPARCStations is used for the AAR, and other two SPARCStations are used for workplace-A and workplace-B.

```
(historyknowledge FVCSKnowledge
(knowledge
 (rule speedup-organization
 (data :RNAME VideoCoferenceManager
       :PLACENAME ?place :OSDsec (Members :manager "(interface)" :member1 ?m1
       :member2 ?m2 :member3 ?m3 :member4 ?m4)) = ?ag1
 (data :RNAME ?m1) = ?ag2
 (data :RNAME ?m2) = ?ag3
 ...
 -->
 (makelink direct-award ?ag1:WNAME (?data2:WNAME ?data3:WNAME ...)))))
```

Fig. 9. The part of knowledge used in the experiment

5.1.2 Condition of the experiment. Figure 9 shows the example of the knowledge of the utilization of the behavioral history used in AAR. This description pattern is in conformity with the rule description of TAF [3]. In this example, the part of the knowledge is shown to speedup the configuration process of the agent organization by using the direct-award protocol instead of using the contract net protocol, based on the behavioral history data of the FVCS.

5.1.3 Experiment processes.
The experiment executed as the following processes.

(1) FVCS starts between the workplace-A and the workplace-B. Firstly, the user starts the workplace by sending the videoconference service request to the interface. The service request is sent to the repository through the network. In the repository the configuration of the organization of the FVCS is executed and the agent group is instantiated to the workplace to start providing the service.

(2) When the videoconference finished, the agents feed back automatically to the repository from the workplace.

(3) After the feedback of agents is completed, user issues the second service request of the FVCS via the interface.

5.2 Evaluation

5.2.1 Feasibility evaluation.
After finishing the FVCS, the agents of FVCS terminated the action and fed back to the AAR. The behavioral history of the FVCS agents was put in the working memory, as well as it was stored in the database of the behavioral history data. Next, when the FVCS started again, the second request of the configuration of the organization was sent with the performative of the direct-award toward the agent group that acts previously. Each agent instantiated again to the workplace and provided the videoconference service with the reasonable quality. From this behavior observation, we confirmed the validity of the design and the implementation.

5.2.2 Efficiency Evaluation.
After saving the agent's behavioral history into the database, the history data was added to the working memory of the reasoning engine. By using the behavioral history, a rule for speedup-organization was fired and *make-link* action was executed. The new knowledge of configuration of the organization with the direct-award performative was added to the manager agent of the FVCS. When a user issued the second service request, the RMA recognized that the request had the same conditions as the previous request by referring the SMT. The message was passed to the ALMM and the manager agent in ALMM executed the configuration of the organization. In this phase, the time from the request issued till the configuration was completed, was 13.6 seconds. The first configuration was performed in 17.8 seconds, so we confirmed the speed-up by using the behavioral history for the configuration of the agent organization.

The above results show that the AAR-based multiagent framework with the Active Agent Repository proposed in this paper is possible to configure the agent organization efficiently compare to the traditional framework.

6 Conclusion

In this paper, we proposed a new multiagent framework to improve performance of the agents' cooperative works by using the behavioral history of each agent. A new architecture and some mechanisms were introduced to the agent repository in the framework. We designed and implemented the new agent repository based on the proposal. Through the evaluation experiments using FVCS, we showed the capability of the efficient configuration of the agent organization by introducing the Active Agent Repository.

We have following future works,

– introducing the template for the utilization of the behavioral history

– introducing the mechanism of automatic acquisition of knowledge of the utilization of the agents' behavioral history.

References

1. Kurt Geihs, "Middleware Challenges Ahead", IEEE Computer, Vol.34, No.6, pp. 24–31, 2001.
2. S. Fujita, K. Sugawara, T. Kinoshita, and N. Shiratori,, "Agent-based Architecture of Distributed Information Processing Systems", Transactions of IPSJapan, Vol.37, No.5, pp.840–852, 1996. (In Japanese)
3. S. Fujita, H. Hara, K. Sugawara, T. Kinoshita, and N. Shiratori, "Agent-based Design Model of Adaptive Distributed Systems, The International Journal of Artificial Intelligence Neural Networks and Complex Problem-Solving Technologies", vol.9, No.1, pp.57–70, July/August 1998.
4. Takuo Suganuma, Tetsuo Kinoshita and Norio Shiratori, "Flexible Network Layer in Dynamic Networking Architecture", Proc. of International Workshop on Flexible Network and Cooperative Distributed Agents (FNCDA2000), pp.473–478, 2000.
5. Martin L.Griss, Glida Pour, "Accelerating Development with Agent Components", IEEE Computer, Vol.34, No.5, pp.37–43, 2001
6. R. G. Smith: "The contract net protocol: High-level communication and control in a distributed problem solver", IEEE Trans. Computer, vol.29, No.12, pp.1104–1113, 1980.
7. H. Hara, S. Konno, K. Sugawara, and T. Kinoshita, "Design and Implementation of Training–system for Agent Framework TAF", Proc. of Workshop on Software Agent and its Applications (SAA2000), pp.183–190, 2001. (In Japanese)
8. T. Suganuma, S. Fujita, K. Sugawara, T. Kinoshita, and N. Shiratori, "Flexible Videoconference System Based on Multiagent-based Architecture", Transactions of IPSJapan, Vol.38, No.6, pp.1214–1224, 1997. (In Japanese)
9. T. Suganuma, S. Lee, T. Kinoshita, and N. Shiratori, "An Agent Architecture for Strategy-centric Adaptive QoS Control in Flexible Videoconference System", Next Generation Computing, Vol.19, No.2, pp.173–191, 2001.

Round-Robin Scheduling Algorithm with Multiple Distributed Windows

Jin Seek Choi[1], Bum Sik Bae[2], Hyeong Ho Lee[3], and Hyeung Sub Lee[3]

[1] School of Engineering
Information and Communications University,
P.O.Box 77, Yusong, Daejeon 305-600, KOREA
(Tel : +82-42-866-6144, Fax : +82-42-866-6110, jin@icu.ac.kr)[†]
[2] Information and Communication Business,
Samsung Electronics Co. LTD
(Tel : +82-31-770-6726, bsbae@samsung.com)
[3] Router Technology Department
Electronics and Telecommunications Research Institute (ETRI), Daejeon, Korea
(Tel:+82-42-860-6130, Fax: +82-42-860-5410, {holee,leehs}@etri.re.kr)

Abstract. This paper proposes a new efficient cell scheduling scheme based on the round-robin scheme with multiple distributed windows. The proposed scheduler consists of a set of windows. Each input queue may be linked with windows in proportional to the reserved bandwidth. The scheduler rotates along the windows in a round-robin manner and each window generates a token. Token enables a cell to get service in a slot time. The proposed scheme can be implemented with simple hardware logic of O(1) round-robin technique and can reduce the maximum delay. The proposed scheme also provides good fairness of weighted fair queueing.

Keywords. Integrated services network, Bandwidth guarantee, Scheduling, Round-robin, Quality-of-service

1   Introduction

Integrated services consist of heterogeneous traffics with a wide range of quality-of-service (QoS) requirements such as delay, delay jitter, loss and so on. These requirements have to be preserved while traffics are going through the network. QoS is mainly dependent on the service order and service rate. For supporting integrated services, therefore the network must have an efficient scheduling algorithm that can guarantee the bandwidth-on-demand, good fairness properties and simplicity of implementation.

Until now, some scheduling algorithms have been proposed, such as generalized processor sharing (GPS), weighted round-robin (WRR) and so on [1,2].

[†] This work was supported in part by ETRI and OIRC-ERC program funded by KOSEF.

GPS is an ideal method since it guarantees bandwidth of all connections according to their reservation. However, it cannot be applied to the physical network due to the implementation problem of the fluid flow operation. Recently, some approximation algorithms such as weighted fair queueing (WFQ), self-clocked fair queueing (SCFQ) and start-time fair queueing have been proposed [3,5,4,8]. They approximate the fluid flow operation as a packet-by-packet operation.

Nevertheless, they still have the implementation complexity of sorting capability for fair queueing. The insertion for a sorted priority queue has an intrinsic computation complexity of $O(logN)$, where N is the number of connections in the system [2]. So, they are not suitable for implementing in high-speed network. To reduce the computation complexity, WRR and its modifications have been proposed for high-speed network such as Asynchronous Transfer Mode (ATM) networks. Since the round-robin and its derivative techniques do not require any arrangement of packet order at every slot time, the computation complexity will be $O(1)$ per connection. They are suitable for implementing in high-speed network. However, they may transmit the burst of cells in proportion to the reservation. They can not overcome the problem of fairness. Moreover, the maximum delay will not be bounded as much as that of GPS. Even though the WRR with save and borrow and deficit round-robin algorithms improve delay characteristics a little, it still has the problem of the maximum delay bound that may grow linearly with the number of connection s sharing the link [6,7].

In this paper, we propose a new cell scheduling scheme based on the round-robin technique for high-speed networks. The proposed scheme is a derivative of the weighted round-robin technique. The round-robin technique is computation efficient. By distributing the weight into the multiple distributed windows, the proposed protocol can reduce the maximum delay and provide good fairness as much as that of WFQ.

2   Proposed Algorithm

Fig. 1 shows the structure of the proposed scheduler. It consists of input queues, a set of windows and a scheduler. Input queues store incoming cells and transmit outgoing cells. The input queues are linked with windows. Multiple windows can be dedicated to one input queue in proportional to the reserved bandwidth. Comparing to the WRR scheme, the proposed algorithm allocates the windows in distributed fashion (see Fig. 2). The scheduler rotates along the windows in a round-robin technique and each window generates a token. Here, a token means that one cell is permitted to get service in a slot time.

When a new connection is established, an input queue is assigned to the connection and a number of windows are allocated to the input queue according to the reserved bandwidth. The allocated windows are linked with the queue. If there are not enough windows (or bandwidth), the connection is refused. When the connection is released, the system removes the queue and releases the links between the input queue and the windows.

The input queues will be logically created. The maximum number of the input queues will be bounded by the number of connections provided by the system. On the other hand, the number of windows is fixed. The number is determined by the total link capacity divided by the unit of reserved bandwidth. Each window means the unit bandwidth. For example, under the link capacity of $10Gigabits/s$, the number of windows becomes 100 when the unit of reserved bandwidth is $100Mbits/s$. In this case, if a connection is established with the reserved bandwidth of $500Mbits/s$, 5 windows are assigned for the connection and they are evenly distributed. The window may restrict the minimum of reserved bandwidth as $100Mbits/s$.

The window is represented by a counter. All windows will have the initial value of 0. When a packet arrives in the input queue, one is added to the value of the windows linked with the input queue. On the other hand, when a packet is transmitted from the input queue, one is decreased from the value of the windows linked with the input queue. The windows linked with the same queue have the same value. Only when the value of the counter is greater than zero, the window is activated. The scheduler rotates along active windows in a round-robin manner and an active window generates a token. The windows which are not active should be skipped in the rotation. If there is no active window, which means that there is no waiting cells in the system, then the scheduler stops at the current window position, and waits for a window to be active.

The cycle time of the scheduler is defined as the shortest time interval that the scheduler revisits the same window. This cycle time is bounded by the transmission time of all windows. If some windows are not active, the cycle time may be reduced. The cycle time will be varied from zero to the maximum boundary. For each connection, the scheduler can guarantee at least the number of cell transmission as much as the number of reserved windows in a cycle time. The maximum delay for the queue is also bounded by the time interval that the scheduler visits the adjacent reserved windows for the queue.

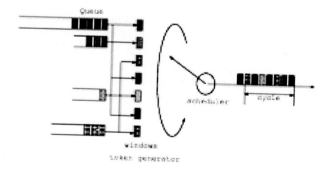

Fig. 1. The system structure of the proposed scheme.

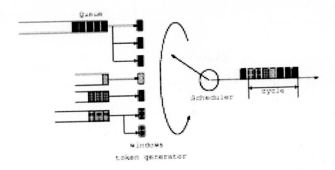

Fig. 2. The system structure of the weighted round-robin scheme.

3   Performance Analysis

3.1   Implementation Complexity

The proposed scheme can be implemented easily to operate at high speed. WFQ needs computation complexity and has to use the mechanism of a sorted priority queue. On the other hand, the proposed algorithm does not need sorting and computational operations of virtual times [3]. So it has a complexity of $O(1)$ the same as the WRR. Moreover, each window is represented by a counter. The counter is increased and decreased when a cell is incoming and sending, respectively. While the value of the window is greater than 0, it normally generates a token in the round-robin mechanism. If it becomes 0, the window is de-activated and it is skipped from the rotation of the scheduler. It preserves work-conserving.

3.2   Fairness

The fairness of a scheduling algorithm means the difference of the normalized service received by any two backlogged connections during any time interval. Let T be the total number of windows in the system and T_i be the number of windows for connection i. We assume that every cell has constant size of L. The channel capacity is denoted by C. Let $W_i(t_1, t_2)$ be the amount of connection i traffic served in the interval $[t_1, t_2]$, then it can be given by

$$W_i(t_1, t_2) \geq \lfloor \frac{(t_2 - t_1)/(L/C)}{T/T_i} \rfloor L = (t_2' - t_1)r_i, \tag{1}$$

where r_i is the reserved bandwidth for connection i, i.e. $r_i = T_i C/T$, and t_2' is the closest value to t_2 while satisfying $t_2' = t_1 + n/(L/r_i)$ for any integer n. $\lfloor z \rfloor$ indicates the largest integer not greater than z. Let $w_i(t_1, t_2)$ be the amount of connection i traffic normalized by its reserved bandwidth, then it can be given by

$$w_i(t_1, t_2) \geq (t_2' - t_1). \tag{2}$$

Let $w(t_1, t_2)$ be total amount of traffic served by the system in the interval $[t_1, t_2]$, then it can be given by

$$w(t_1, t_2) = \lfloor \frac{t_2 - t_1}{L/C} \rfloor \frac{L}{C}. \tag{3}$$

The service lag, $\delta_i(t_1, t_2)$, of connection i in the system is bounded as follows:

$$\delta_i(t_1, t_2) \equiv |w(t_1, t_2) - w_i(t_1, t_2)| \\ = \lfloor \frac{t_2 - t_2'}{L/C} \rfloor \leq \frac{T}{T_i} \frac{L}{C} = \frac{L}{r_i}. \tag{4}$$

Given any time interval (t_1, t_2) and any pair of backlogged connections i and j, the following bound holds

$$|w_i(t_1, t_2) - w_j(t_1, t_2)| \leq \frac{L}{r_i} + \frac{L}{r_j}. \tag{5}$$

Thus, the difference of normalized amount of services received by any two back-logged connections is limited and WFQ shows the same result in [3].

3.3   Maximum Delay

In the case of real-time services, maximum delay is a more important factor than average delay. To obtain limited maximum delay, a minimum bandwidth property parameter must be calculated. It is defined in [9], and the proposed algorithm has minimum bandwidth property with a parameter μ, which is given by

$$\mu = t + \lceil \rho^{-1} \rceil - (t + \rho^{-1}) = \lceil \rho^{-1} \rceil - \rho^{-1} < 1 \tag{6}$$

where ρ^{-1} is the reserved bandwidth. $\lceil x \rceil$ is the smallest integer greater than or equal to the real number x. We assume that an input traffic is tightly (σ, ρ)-constrained [9]. In using the theorem in [9], the maximum delay d is bounded by

$$d \leq (1 + \sigma)\rho^{-1} + (n - 1)\rho^{-1} + \sum_{j=1}^{n} \mu_j + \Pi_n, \tag{7}$$

where n is the number of intermediate switches and μ_j is the minimum bandwidth property parameter at switch j, and Π_n is the propagation delay from the source to switch n.

Table 1 shows numerical results for four algorithms. We assume that the input traffic is $(50cells, 2Mbps)$-constrained and there are five intermediate switches. There are 50 different connections at each switch of which the transmission rate is $155Mbps$. In WRR, the frame size is $200cells$ and they are equally assigned to all connections. Maximum delays in WFQ, SCFQ and WRR are obtained by using the results in [2,4,9]. We can see that the maximum delay in the proposed scheme is slightly greater than that in WFQ. However, the proposed algorithm shows much better performance than SCFQ and WRR.

4   Conclusions

In this paper, we have proposed a new efficient scheduling scheme, which adopts a round-robin scheduling algorithm with multiple windows. We have shown that the proposed scheme has the performance of fairness as much as WFQ, and yields better delay bound than WRR. Moreover, the proposed scheme has no computational overheads. So, it can be easily implemented in high-speed networks.

Table 1. Maximum delays in proposed, WFQ, SCFQ, and WRR.

	Proposed	WFQ	SCFQ	WRR
maximum delay	11.690 msec	11.676 msec	12.333 msec	13.673 msec
delay jitter	11.669 msec	11.655 msec	12.312 msec	13.625 msec

References

1. M. W. Garrett, "A service architecture for ATM: From Applications to scheduling", IEEE Netw. Mag., May/June, 1996.
2. H. Zhang, "Service disciplines for guaranteed performance service in packet- switching networks", Proc. of the IEEE, Vol. 83, No. 10, Oct. 1995.
3. A. K. Parekh and R. G. Gallager, "A generalized processor sharing approach to flow control in integrated services networks: the multiple node case", IEEE/ATM Transactions on Networking, Vol. 2, No. 2, pp. 137–150, 1994.
4. S. J. Golestani, "A self-clocked fair queueing scheme for broadband application", Proc. of IEEE INFOCOM'94, pp. 636 646, 1994.
5. P. Goyal, H. M. Vin and H. Cheng, "Start-time fair queuing: a scheduling algorithm for integrated services packet switching networks", IEEE/ATM Transactions on Networking, Vol. 5, No. 5, pp. 690–704, 1997.
6. Hideyuki Shimonishi, Makiko Yoshida, Ruixue Fan, and Hiroshi Suzuki, "An improvement of weighted round robin cell scheduling in ATM networks", Proced. of IEEE GLOBECOM'97, pp. 1119–1123, 1997.
7. M. Shreedhar and G. Barghese, "Efficient fair queueing using deficit round-robin", IEEE/ACM Trans. Networking, Vol. 4, No. 3, pp. 375–385, June 1996.
8. J.C.R. Bennett and H. Zhang, "WF^{2}Q: Worst-case fair weighted fair queueing", Proc. of IEEE INFOCOM'96, pp. 120–128, 1996.
9. George Kesidis, ATM Network Performance, Kluwer Academic Publishers, 1996.

Author Index